GEOMETRIC APPLICATIONS

PHYSICAL SCIENCE

Elementary Algebra

Concepts and Applications

FOURTH EDITION

Elementary Algebra

Concepts and Applications

FOURTH EDITION

MARVIN L. BITTINGER
Indiana University–Purdue University at Indianapolis

MERVIN L. KEEDY
Purdue University

DAVID ELLENBOGEN
St. Michael's College

ADDISON-WESLEY PUBLISHING COMPANY
Reading, Massachusetts • Menlo Park, California • New York
Don Mills, Ontario • Wokingham, England • Amsterdam
Bonn • Sydney • Singapore • Tokyo • Madrid
San Juan • Milan • Paris

Sponsoring Editor	Melissa Acuña
Managing Editor	Karen Guardino
Production Supervisor	Jack Casteel
Design, Editorial, and Production Services	Quadrata, Inc.
Illustrator	Scientific Illustrators and Leo Harrington
Manufacturing Supervisor	Roy Logan
Cover Designer	Geri Davis, Quadrata, Inc., and Hannus Design Associates
Cover Photograph	Jurgen Vogt, The Image Bank

Photo Credits

p. 1, ©Barrie Rokeach, The Image Bank **p. 61,** Dick Morton **p. 113,** David F. Wisse **p. 147,** Leo de Wys, Inc./Steve Brown **p. 209,** ©Grant Faint, The Image Bank **p. 257,** ©Addison Geary 1991, Stock/Boston **p. 297,** ©Todd Phillips, Third Coast Stock Source, Inc. **p. 317,** ©David Ball, The Stock Market **p. 318,** AP/Wide World Photos **p. 340,** AP/Wide World Photos **p. 351,** ©92 Mark Ferri, The Stock Market **p. 385,** Leo de Wys, Inc./Paul Gerda **p. 389,** ©89 Henley & Savage, The Stock Market **p. 410,** Comstock **p. 425,** ©Bill Ross, All Stock **p. 440,** UPI/Bettmann Newsphotos **p. 451,** NASA **p. 453,** ©86 Jeffry W. Myers, The Stock Market

Library of Congress Cataloging-in-Publication Data

Bittinger, Marvin L.
Elementary algebra : concepts and applications / Marvin L. Bittinger, Mervin L. Keedy, David Ellenbogen.—4th ed.
p. cm
Includes index.
ISBN 0-201-53782-6
1. Algebra. I. Keedy, Mervin Laverne. II. Ellenbogen, David. III. Title.
QA152.2.B5795 1994
512.9—dc20
93-5667
CIP

Reprinted with corrections, September 1995.

6 7 8 9 10—DOW—9695

For Peggy

Who has shown that it is possible to be a successful student, nurturing mother, and loving wife all at once.

D.J.E.

PREFACE

Appropriate for a one-term course in elementary algebra, this text is intended for students who have a firm background in arithmetic. It is the first of two texts in an algebra series that also includes *Intermediate Algebra: Concepts and Applications,* Fourth Edition, by Bittinger/Keedy/Ellenbogen. *Elementary Algebra: Concepts and Applications,* Fourth Edition, is a significant revision of the Third Edition with respect to design, contents, pedagogy, and an expanded supplements package. This series is designed to prepare students for any mathematics course at the college algebra level.

APPROACH

Our approach, which has been developed over many years, is designed to help today's students both learn and retain mathematical concepts. Our goal in preparing this revision was to address the major challenges for teachers of developmental mathematics courses that we have seen emerging during the early 1990s. The first challenge is to prepare students of developmental mathematics to make the transition from ''skills-oriented'' elementary and intermediate algebra courses to the more ''concept-oriented'' presentation of college algebra or other college-level mathematics courses. The second is to teach these same students critical-thinking skills: to reason mathematically, to communicate mathematically, and to solve mathematical problems. The third challenge is to reduce the amount of content overlap between elementary algebra and intermediate algebra texts.

Following are some aspects of the approach that we have used in this revision to help meet the challenges we all face teaching developmental mathematics.

PROBLEM SOLVING

One distinguishing feature of our approach is our treatment of and emphasis on problem solving. We use problem solving and applications to motivate the material wherever possible, and we include real-life applications and problem-solving techniques throughout the text. We feel that problem solving encourages students to think about how mathematics can be used. It also challenges students and helps to prepare them for more difficult material in later courses.

- In Chapter 2, we introduce the five-step process for solving problems: (1) Familiarize, (2) Translate, (3) Carry out, (4) Check, and (5) State the answer. These steps are used throughout the text whenever we encounter a problem-solving situation. Repeated use of this algorithm gives students a sense that they have a starting point for any type of problem they encounter, and frees them to focus on the mathematics necessary to successfully translate the problem situation. (See pages 84–86, 302, 303, 341, and 342.)

CONTENT

Chapter 1 includes a brief review of arithmetic topics and then moves quickly into elementary algebra topics. This allows instructors sufficient time to cover the topics necessary to prepare students for intermediate algebra.

- Chapter 3 contains an intuitive introduction to graphing, a topic that is integrated throughout the text. This helps students to visualize the mathematics of many concepts while at the same time allowing them to develop facility with graphing throughout the course.

PEDAGOGY

Skill Maintenance Exercises and *Cumulative Reviews.* Retention of skills is critical to the future success of our students. In nearly all exercise sets, we include carefully chosen exercises that review skills and concepts from preceding chapters of the text. Each chapter test includes Skill Maintenance Exercises selected from the three or four text sections that are identified at the beginning of each chapter. After Chapters 3, 6, and 10, we have also included a Cumulative Review, which reviews skills and concepts from all preceding chapters of the text. (See pages 114, 227, 315, 316, and 363.)

Synthesis Exercises. Each exercise set ends with a set of synthesis exercises. These problems can offer opportunities for students to synthesize skills and concepts from earlier sections with the present material, or can provide students with deeper insights into the current topic. Synthesis exercises are generally more challenging than those in the main body of the exercise set. (See pages 21, 79, 141, and 233.)

Verbalization Skills. Wherever appropriate throughout the text, we have discussed how mathematical terms are used in language. The Summary and Review sections emphasize key terms and important properties and formulas. In addition, thinking and writing exercises are included in the Synthesis Exercises. These encourage students to verbalize mathematical concepts, leading to better understanding. (See pages 45 and 253.)

WHAT'S NEW IN THE FOURTH EDITION?

We have rewritten many key topics in response to user and reviewer feedback and have made significant improvements in design and pedagogy. Detailed information about the content changes is available in the form of a Conversion Guide. Please ask your local Addison-Wesley sales representative for more information. Following is a list of the major changes in this revision.

New Design

- The new design is more open and readable. Pedagogical use of color makes it easier to see where exercises, explanations, and examples begin and end.
- The entire art program is new for this edition. We have ensured the accuracy of the graphical art through the use of computer-generated graphs. Color in the graphical art is used pedagogically and precisely to help the student visualize the mathematics. (See pages 114 and 338.)

Technology Connections

- These features integrate technology, increase the understanding of concepts through visualization, encourage exploration, and motivate discovery learning. Optional Technology Connection exercises occur in many exercise sets. (See page 330.)

Writing Exercises

- Nearly every set of Synthesis Exercises begins with two writing exercises. These exercises are usually not as difficult as other synthesis exercises, but require written answers that aid in student comprehension, critical thinking, and conceptualization. Because some instructors may collect answers to writing exercises, and because more than one answer may be correct, answers to writing exercises are provided only for the chapter reviews. (See pages 30 and 41.)

Content Changes. A variety of content changes have been made. Some of the more significant changes are listed below.

- Graphing is now introduced in Chapter 3 and used as a tool for solving problems there and in later chapters.
- Negative exponents are now covered at the *end* of Chapter 4. This change simplifies the development of the properties of exponents and enables instructors to cover negative exponents at their option.
- Although our fear of students performing ''illegal'' cancellations is as acute as ever, we now use canceling as a way to simplify rational expressions. We do so in recognition of the fact that we use canceling when working on our own. Whenever canceling is used, we point out that we are effectively ''removing'' a factor of 1. (See pages 260–261.)
- Differences of squares are now factored *after* polynomials of the type $x^2 + bx + c$ and $ax^2 + bx + c$. This change should better enable students to view differences of squares as a special type of quadratic polynomial.
- Because so many students remain convinced that they ''cannot do word problems,'' we have made increased use of guessing as a means of familiarizing oneself with a problem-solving situation. By checking to see if a guess is correct, students can more easily discover an algebraic translation of the problem. (See pages 88–91.)
- The use of geometry as a way of visualizing concepts has been expanded (see

Sections 4.4 and 4.5). Similar triangles are now covered in Section 6.8, ''Problem Solving: Rational Equations and Proportions.''

- Throughout the text, we have included a variety of new applications that appeal to a large cross section of the student population. By emphasizing applications that students and faculty find interesting, we hope that we have made the text enjoyable to use. (See pages 114, 116, and 297).

SUPPLEMENTS FOR THE INSTRUCTOR

INSTRUCTOR'S SOLUTIONS MANUAL
by Judith A. Penna

This supplement contains worked-out solutions to all exercises in the text.

INSTRUCTOR'S RESOURCE GUIDE
by Donna DeSpain

This supplement contains the following:

- Extra practice problems for challenging topics in the text
- Black-line masters of grids and number lines for transparency masters or test preparation
- Videotape index and section cross references to the tutorial software packages available with this text
- Conversion guide from the Third Edition to the Fourth Edition

PRINTED TEST BANK
by Donna DeSpain

This supplement contains the following:

- Six alternative test forms for each chapter and six final examinations
- Two multiple-choice versions of each chapter test

All test forms have been completely rewritten.

COMPUTERIZED TESTING

Omnitest II (for IBM and Macintosh). This computerized test bank allows you to create up to 99 versions of a customized test with just a few keystrokes, and allows the option of choosing items by chapter, section, or objective. It contains over 400 multiple-choice and open-ended algorithms. You may enter your own test items, edit existing items, and determine the level of difficulty of problems.

SUPPLEMENTS FOR THE STUDENT

STUDENT'S SOLUTIONS MANUAL
by Judith A. Penna

This manual contains completely worked-out solutions with step-by-step annotations for all the odd-numbered exercises in the text, and answers for all even-numbered exercises in the text.

VIDEOTAPES

Developed especially for the Bittinger/Keedy/Ellenbogen texts, these videotapes feature an engaging team of lecturers presenting material from each section of the text in an interactive format that includes a group of students. The lecturers' presentations often incorporate slides, sophisticated computer-generated graphics, and a white board to support an approach that emphasizes visualization and problem solving.

TUTORIAL SOFTWARE

THE MATHLAB+ **(IBM and Macintosh).** This software combines a unique combination of drill and practice modules with an interactive and easy-to-use graphing tool. The drill and practice segments feature feedback for wrong answers and detailed record keeping. The graphing tool allows students to graph and explore a wide variety of two-dimensional functions.

Algebra Problem Solver (IBM). After selecting a topic and an exercise type, students can enter their own exercises or request an exercise from the computer. In each case, the student is given detailed, annotated, step-by-step solutions.

ACKNOWLEDGMENTS

We wish to express our appreciation to the many people who helped with the development of this book. Barbara Johnson and Laurie A. Hurley deserve special thanks for their many fine suggestions. Their proofreadings of the text, in spite of almost endless time pressure, contributed immeasurably to the accuracy and readability of the text. Judy Penna also merits special thanks for her preparation of the *Student's Solution Manual,* the *Instructor's Solution Manual,* and the indexes. Judy's work is always performed with a thoroughness that amounts to another proofreading of the book and for that we are grateful. We are also indebted to Stuart Ball for his expert guidance in preparing the Technology Connections and the associated artwork.

This book's sponsoring editor, Melissa Acuña, performed admirably in coordinating the many intricacies of this project; Lenore Parens, as developmental editor; contributed many useful ideas and suggestions; George and Brian Morris of Scientific Illustrators generated a remarkable set of graphs and illustrations that are both precise and easily understood; and Leo Harrington drew the many fine sketches that enhance our exercises and examples. Geri Davis and Martha Morong of Quadrata, Inc., provided design, editorial, and production services second to none, ensuring that every last detail has been taken care of. To all of these people, we offer our deepest thanks.

In addition, we thank the following professors for their thoughtful reviews and insightful comments.

Dick J. Clark, *Portland Community College*
John Coburn, *St. Louis Community College*
Linda Crabtree, *Longview Community College*
Susan Dimick, *Spokane Community College*
Arthur Dull, *Diablo Valley College*

Margaret Finster, *Erie Community College South*
Kenneth Grace, *Anoka–Ramsey Community College*
Jay Graening, *University of Arkansas*
Steve Green, *Tyler Junior College*
Susan Haller, *St. Cloud University*
Lonnie Hass, *Northern Dakota State University*
Elaine Hubbard, *Kennesaw State College*
Carolyn Krause, *Delaware State College*
Pam Littleton, *Tarleton State University*
William Livingston, *Missouri Southern State College*
Robert Malena, *Community College of Allegheny—South*
Mauricio Marroquin, *Los Angeles Valley College*
Myrna Mitchell, *Pima County Community College*
Carla Moldavan, *Kennesaw State College*
Jane Pinnow, *University of Wisconsin—Parkside*
William Rundburg, *College of San Mateo*
Lawrence Runyan, *Shoreline Community College*
Debra Singleton, *Lexington Community College*
Barbara Jane Sparks, *Camden Community College*
Beverly Weatherwax, *Southwest Missouri State University*

Finally, a special thank you to all those who so generously agreed to discuss their professional uses for mathematics in our chapter openers. These dedicated people, none of whom we knew prior to writing this text, all share a desire to make math more meaningful to students. We cannot imagine a finer set of role models.

M.L.B.
M.L.K.
D.J.E.

CONTENTS

1 Introduction to Algebra and Algebraic Expressions

2 Equations, Inequalities, and Problem Solving

3 Introduction to Graphing

4 Polynomials

5 Polynomials and Factoring

6 Rational Expressions and Equations

10 Quadratic Equations

Elementary Algebra
Concepts and Applications

FOURTH EDITION

CHAPTER 1

Introduction to Algebra and Algebraic Expressions

AN APPLICATION

In the course of one four-month period, the water level of Lake Champlain went down 2 ft, up 1 ft, down 5 ft, and up 3 ft. How much had the lake level changed at the end of the four months?

This problem appears as Example 7 in Section 1.5.

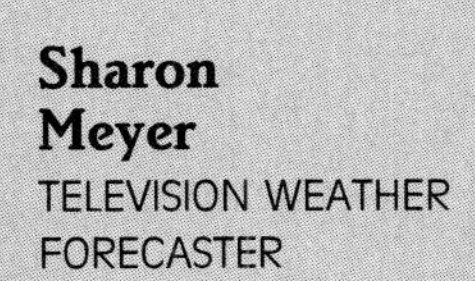

Sharon Meyer
TELEVISION WEATHER FORECASTER

"Whether it's in calculating the number of heating degree days, the probability of precipitation, or changes in lake level, a forecaster's career relies heavily on a solid understanding of mathematics. I use math every single day as part of my job."

Problem solving is the focus of this text. In this chapter we discuss some preliminaries that are needed for the problem-solving approach that we develop and begin to use in Chapter 2. We also review some arithmetic, discuss real numbers and their properties, and examine how real numbers are added, subtracted, multiplied, divided, and raised to powers.

1.1 Introduction to Algebra

Algebraic Expressions • Translating to Algebraic Expressions • Translating to Equations

This section introduces some basic concepts of algebra. Since equation solving is central to the study of algebra, we concentrate on the expressions that appear in equations and some important words for translating English to mathematics.

Algebraic Expressions

Probably the greatest difference between arithmetic and algebra is the extensive use of *variables* in algebra. When a letter is used to stand for any number chosen from a variety of numbers, we call the number a **variable.** For example, if n represents the number of students registered for a college's 8 AM section of Elementary Algebra, the number n will vary from semester to semester, if not from day to day. If each student in the 8:00 AM Elementary Algebra section paid $500 to take the class, the college would collect a total of $500 \cdot n$ dollars. Since the cost per student, $500, is the same regardless of how many students are registered, the number 500 is called a **constant.** The number of registered students n can vary, so n is a variable.

Cost Per Student (in Dollars)	Number of Students Registered	Total Collected (in Dollars)
500	n	$500 \cdot n$

The expression $500 \cdot n$ is called a **variable expression** because its value varies with the choice of n. Of course, the total amount collected, $500 \cdot n$, will grow as the number of students registered grows. We now replace n with a variety of values and compute the total amount collected. In doing so, we say that we are **evaluating** the expression $500 \cdot n$.

Cost Per Student (in Dollars) 500	Number of Students Registered n	Total Collected (in Dollars) $500 \cdot n$
500	20	10,000
500	25	12,500
500	30	15,000

Variable expressions are examples of *algebraic expressions*. An **algebraic expression** consists of variables, numerals, operation signs, and/or grouping symbols. Examples of algebraic expressions are

$$x + 35, \quad 7 \cdot t, \quad 13a - b, \quad 15 \div z, \quad \tfrac{9}{7}, \text{ and } 3x(a + b).$$

Note that a fraction bar is a division symbol: $\frac{9}{7}$ means $9 \div 7$. Similarly, multiplication can be written in several ways. For example, "7 times t" can be written as $7 \cdot t$, $7 \times t$, $7(t)$, or simply $7t$.

To evaluate an algebraic expression, we **substitute** a number for each variable in the expression.

EXAMPLE 1

Evaluate the expression for the given values: **(a)** $x + y$ when $x = 37$ and $y = 28$; **(b)** $5ab$ when $a = 2$ and $b = 3$.

Solution

a) We substitute 37 for x and 28 for y and carry out the addition:

$$x + y = 37 + 28 = 65.$$

The number 65 is called the **value** of the expression.

b) We substitute 2 for a and 3 for b and multiply:

$$5ab = 5 \cdot 2 \cdot 3 = 10 \cdot 3 = 30.$$

❑

EXAMPLE 2

The area A of a rectangle of length l and width w is given by the formula $A = lw$. Find the area when l is 24.5 in. and w is 10 in.

Solution We evaluate, using 24.5 in. for l and 10 in. for w and carry out the multiplication:

$$\begin{aligned} A = lw &= (24.5 \text{ in.})(10 \text{ in.}) \\ &= (24.5)(10)(\text{in.})(\text{in.}) \\ &= 245 \text{ in}^2, \quad \text{or 245 square inches.} \end{aligned}$$

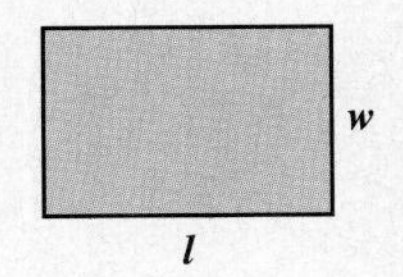

Note that (in.)(in.) = in^2. Exponents are discussed in detail in Section 1.8. ❑

EXAMPLE 3

The area of a triangle with a base of length b and a height of length h is given by the formula $A = \frac{1}{2}bh$. Find the area when b is 8 m and h is 6.4 m.

Solution We substitute 8 m for b and 6.4 m for h and carry out the multiplication:

$$\begin{aligned} A = \tfrac{1}{2}bh &= \tfrac{1}{2}(8 \text{ m})(6.4 \text{ m}) \\ &= \tfrac{1}{2}(8)(6.4)(\text{m})(\text{m}) \\ &= 4(6.4) \text{ m}^2 \\ &= 25.6 \text{ m}^2, \quad \text{or 25.6 square meters.} \end{aligned}$$

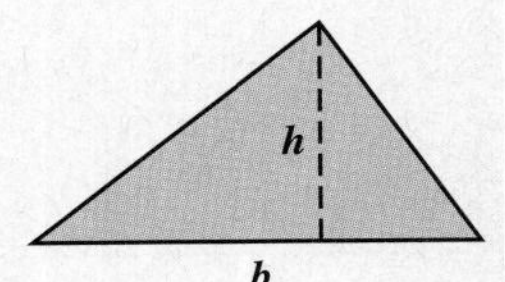

❑

Translating to Algebraic Expressions

Before attempting to translate problems to equations, we need to be able to translate certain phrases to algebraic expressions.

KEY WORDS

Addition (+)	Subtraction (−)	Multiplication (·)	Division (÷)
add sum plus more than increased by greater than	subtract difference minus less than decreased by take from	multiply product times twice of	divide quotient divided by ratio of

EXAMPLE 4

Translate each phrase to an algebraic expression.

a) Twice (or two times) some number
b) Seven less than some number
c) Eighteen more than a number
d) A number divided by five

Solution

a) Think of some number, say 8. What number is twice 8? It is 16. To get 16, you multiplied by 2. Now consider a variable. Use y to represent "some number" and multiply by 2. The expression

$$y \times 2, \qquad 2 \times y, \qquad 2 \cdot y, \quad \text{or} \quad 2y$$

is the translation of "Twice (or two times) some number."

b) We let x represent "some number." Now if the number were 23, then the translation would be $23 - 7$. If the number were 345, the translation would be $345 - 7$. If the number is x, the translation of "Seven less than some number" is

$$x - 7.$$

c) If we knew the number to be 10, the translation would be $10 + 18$, or $18 + 10$. We let t represent "a number," so the translation of "Eighteen more than a number" is

$$t + 18, \quad \text{or} \quad 18 + t.$$

d) We let m represent "a number." If the number were 9, the translation would be $9 \div 5$, or $\frac{9}{5}$. Thus our translation of "A number divided by five" is

$$m \div 5, \quad \text{or} \quad \frac{m}{5}.$$

❑

CAUTION! Because the order in which we subtract and divide affects the answer, answering $7 - x$ or $5 \div m$ in Examples 4(b) and 4(d) is incorrect.

EXAMPLE 5

Translate each of the following.

a) Some number, increased by 5
b) Half of a number
c) Five more than three times some number
d) Six less than the product of two numbers
e) Seventy-six percent of some number

Solution

	Phrase	*Algebraic Expression*
a)	Some number, increased by 5	$n + 5$, or $5 + n$
b)	Half of a number	$\frac{1}{2}t$, or $\frac{t}{2}$
c)	Five more than three times some number	$3p + 5$, or $5 + 3p$
d)	Six less than the product of two numbers	$mn - 6$
e)	Seventy-six percent of some number	$76\%z$, or $0.76z$

❑

Translating to Equations

The symbol = (''equals'') is used to indicate that the algebraic expressions on either side of the equals sign represent, or name, the same number. An **equation** is a number sentence with the verb = . Equations may be true, false, or neither true nor false.

EXAMPLE 6

Determine whether the equation is true, false, or neither: **(a)** $3 + 2 = 5$; **(b)** $7 - 2 = 4$; **(c)** $x + 6 = 13$.

Solution

a) $3 + 2 = 5$ The equation is *true*.

b) $7 - 2 = 4$ The equation is *false*.

c) $x + 6 = 13$ The equation is *neither* true nor false, because we do not know what number x represents. ❑

Solution

A replacement or substitution that makes an equation true is called a *solution*. Some equations may have more than one solution, and some may have no solution. When we have found all the solutions, we say that we have *solved* the equation.

One way to determine whether a number is a solution of an equation is to use that number to evaluate the expressions on each side of the equation by substitution. If the values are the same, then the number is a solution.

EXAMPLE 7

Determine whether 7 is a solution of $x + 6 = 13$.

Solution

$x + 6 = 13$	Writing the equation
$7 + 6 \ ? \ 13$	Substituting 7 for x
$13 \mid 13$	13 = 13 is TRUE.

Since the left-hand and the right-hand sides are the same, we have a solution. No other number makes the equation true, so the only solution is the number 7. ❑

Although we do not study solving equations until Chapter 2, we can translate certain problem situations to equations now.

EXAMPLE 8

Translate the following problem to an equation.

What number plus 478 is 1019?

Solution We let y represent the unknown number. In this example, the translation comes almost directly from the English sentence.

What number plus 478 is 1019?

$$y + 478 = 1019$$

Note that "is" translates to "=" and "plus" translates to "+." ❑

Sometimes it helps to reword a problem before translating.

EXAMPLE 9

Translate the following problem to an equation.

The elevation of Denver, 5280 ft, is 88 times the elevation of Houston. What is the elevation of Houston?

Solution We let h represent the elevation of Houston. The rewording and translation follow:

Rewording: 88 times the elevation of Houston is 5280.

Translating: $88 \cdot h = 5280$

Note that "times" translates to "·." ❑

EXERCISE SET 1.1

Evaluate.

1. $6x$, when $x = 7$

2. $7y$, when $y = 7$

3. $9 + a$, when $a = 7$

4. $15 - a$, when $a = 11$

5. $\frac{3p}{q}$, when $p = 2$ and $q = 6$

6. $\frac{5y}{z}$, when $y = 15$ and $z = 25$

7. $\frac{x + y}{5}$, when $x = 10$ and $y = 20$

8. $\frac{p + q}{2}$, when $p = 2$ and $q = 16$

9. $\frac{x - y}{8}$, when $x = 20$ and $y = 4$

10. $\frac{m - n}{5}$, when $m = 16$ and $n = 6$

11. $\frac{x}{y}$, when $x = 3$ and $y = 6$

12. $\frac{p}{q}$, when $p = 4$ and $q = 16$

13. $\frac{5z}{y}$, when $z = 8$ and $y = 2$

14. $\frac{9m}{q}$, when $m = 4$ and $q = 18$

Substitute to find the value of each expression.

15. A driver who drives at a speed of r miles per hour for t hours will travel a distance of rt miles. How far will a driver travel at a speed of 55 mph for 4 hr?

16. A baseball player's batting average is h/a, where h is the number of hits and a is the number of "at bats." Find the batting average of a batter who had 13 hits in 25 at bats.

17. The area of a parallelogram with base b and height h is bh. Find the area of the parallelogram when the height is 15.4 cm (centimeters) and the base is 6.5 cm.

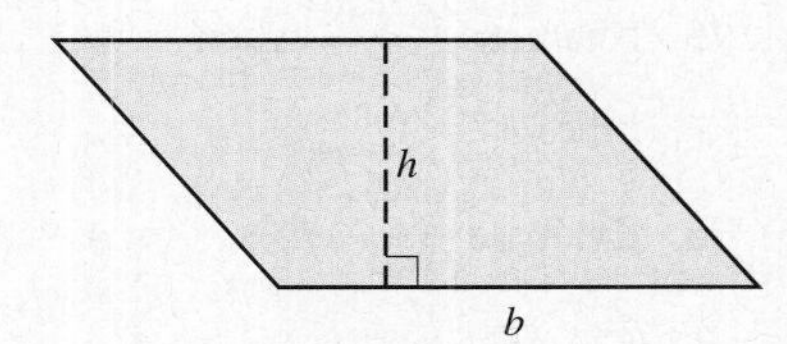

18. A satellite orbiting 300 mi above the earth's surface travels about 27,000 mi in one orbit. The time, in hours, that it takes to orbit the earth one time is given by

$$\frac{27{,}000}{v},$$

where v is the velocity of the satellite in miles per hour. Find the orbiting time of the satellite when the velocity v is 10,000 mph.

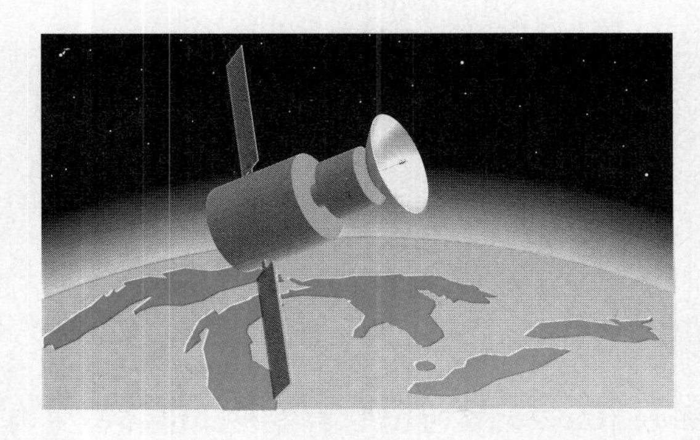

19. Enrico takes five times as long to do a job as Rosa does. Suppose t represents the time it takes Rosa to do the job. Then $5t$ represents the time it takes Enrico. How long does it take Enrico if Rosa takes 30 sec? 90 sec? 2 min?

20. Theresa is six years younger than her husband Frank. Suppose the variable x represents Frank's age. Then $x - 6$ represents Theresa's age. How old is Theresa when Frank is 29? 34? 47?

Translate to an algebraic expression.

21. 6 more than b

22. 8 more than t

23. 9 less than c

24. 4 less than d

25. 6 increased by q

26. 11 increased by z

27. b more than a

28. c more than d

29. x less than y

30. c less than h

31. x divided by w

32. s divided by t

33. m subtracted from n

34. p subtracted from q

35. The sum of r and s

36. The sum of d and f

37. Twice x

38. Three times p

39. One third of t

40. One quarter of d

41. 97% of some number

42. 43% of some number

43. Mandy had d dollars before going to the bookstore. She bought a book for $29.95. How much did Mandy have after the purchase?

44. Dan drove at a speed of 65 mph for t hours. How far did Dan travel?

Determine whether the given number is a solution of the given equation.

45. 15; $x + 17 = 32$

46. 75; $y + 28 = 93$

47. 21; $x - 7 = 12$

48. 27; $y - 8 = 19$

49. 7; $6x = 54$

50. 9; $8y = 72$

51. 30; $\frac{x}{6} = 5$

52. 49; $\frac{y}{8} = 6$

53. 19; $5x + 7 = 107$

54. 9; $9x + 5 = 86$

Translate each problem to an equation. Do not solve.

55. What number added to 60 is 112?

56. Seven times what number is 2233?

57. When 42 is multiplied by a number, the result is 2352. Find the number.

58. When 345 is added to a number, the result is 987. Find the number.

59. A game board has 64 squares. If you win 35 squares and your opponent wins the rest, how many does your opponent get?

60. A consultant charges $80 an hour. How many hours did the consultant work in order to make $53,400?

61. In a recent year, the cost of four 12-oz boxes of Post® Raisin Bran was $7.96. How much did one box cost?

62. The total amount spent on women's blouses in a recent year was $6.5 billion. This was $0.2 billion more than was spent on women's dresses. How much was spent on women's dresses?

Synthesis

To the student and the instructor: Synthesis exercises are designed to challenge students to extend the concepts or skills studied in that section. Some synthesis exercises will require the assimilation of skills and concepts from several sections.

Writing exercises, denoted by ◈, should be answered using one or more complete English sentences. In virtually every section, two writing exercises appear as the first synthesis exercises. These two exercises are not as challenging as those exercises appearing later in the exercise set and can be assigned to students who might not otherwise attempt synthesis exercises. Because many writing exercises are open-ended, their solutions are not listed in the answer section.

63. ◈ How does a variable expression differ from a variable, and how does it differ from an equation?

64. ◈ Write a problem, in words, that translates to the equation $35x = 840$.

Translate to an algebraic expression.

65. A number y plus two times x

66. A number a plus 2 plus b

67. A number that is 3 less than twice x

68. Your age in 5 years, if you are a years old now

69. Your age two years ago, if you are b years old now

70. Some number x increased by itself

71. The perimeter of a square with side s (perimeter means distance around)

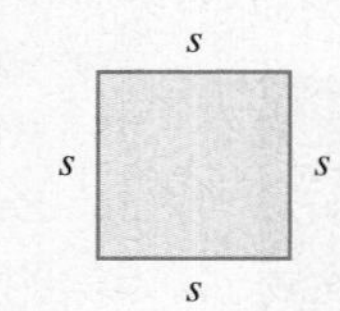

72. The perimeter of a rectangle with length l and width w

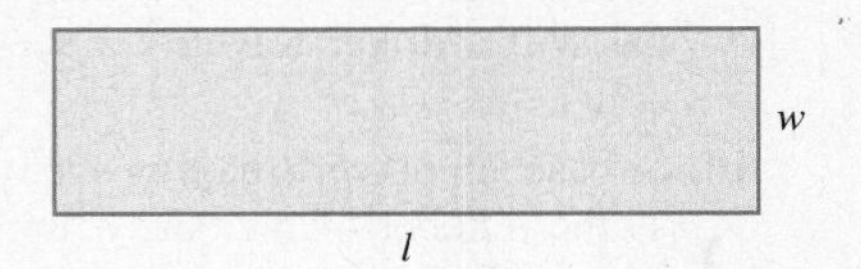

73. Evaluate $\frac{x + y}{4}$ when $y = 8$ and x is twice y.

74. Evaluate $\frac{x - y}{7}$ when $y = 35$ and x is twice y.

75. Evaluate $\frac{y - x}{3}$ when $x = 9$ and y is three times x.

76. Evaluate $\frac{y + x}{2} + \frac{3 \cdot y}{x}$ for $x = 2$ and $y = 4$.

Answer each question with an algebraic expression.

77. If $w + 3$ is a whole number, what is the next whole number after it?

78. If $d + 2$ is an odd number, what is the preceding odd number?

79. The difference between two numbers is 3. One number is t. What are two possible values for the other number?

80. You invest n dollars at 10% interest per year. Write an expression for the number of dollars in the bank a year from now.

1.2 The Commutative, Associative, and Distributive Laws

Equivalent Expressions • The Commutative Laws • The Associative Laws • Using the Laws Together • The Distributive Law • The Distributive Law and Factoring

In order to solve equations, it is important to be able to manipulate algebraic expressions. The commutative, associative, and distributive laws discussed in this section enable us to write *equivalent expressions* that can streamline our work.

Equivalent Expressions

In arithmetic we learned that expressions like $4 + 4 + 4$, $3 \cdot 4$, and $4 \cdot 3$ all represent the same number, 12. Expressions that represent the same number are said to be **equivalent.** The expressions $t + 18$ and $18 + t$ were used in Section 1.1 when we translated "eighteen more than some number." To illustrate that these expressions are equivalent, we can make some choices for t:

When $t = 3$, $t + 18 = 3 + 18 = 21$ and $18 + t = 18 + 3 = 21$.
When $t = 40$, $t + 18 = 40 + 18 = 58$ and $18 + t = 18 + 40 = 58$.

The Commutative Laws

We have seen that changing the order in addition or multiplication does not change the result. Equations like $3 + 18 = 18 + 3$ and $3 \cdot 4 = 4 \cdot 3$ illustrate this idea, and show that addition and multiplication are **commutative.**

The Commutative Laws

For Addition. For any numbers a and b,

$$a + b = b + a.$$

(We can change the order when adding without affecting the answer.)

For Multiplication. For any numbers a and b,

$$ab = ba.$$

(We can change the order when multiplying without affecting the answer.)

EXAMPLE 1

Use the commutative laws to write an expression equivalent to each of the following: **(a)** $y + 5$; **(b)** $9x$; **(c)** $7 + ab$.

Solution

a) An expression equivalent to $y + 5$ is $5 + y$ by the commutative law of addition.
b) An expression equivalent to $9x$ is $x9$ by the commutative law of multiplication.
c) An expression equivalent to $7 + ab$ is $ab + 7$ by the commutative law of *addition*.

Another expression equivalent to $7 + ab$ is $7 + ba$ by the commutative law of *multiplication*.

Also equivalent to $7 + ab$ is $ba + 7$ by both commutative laws. ❑

The Associative Laws

Parentheses are used to indicate groupings. We normally simplify within the parentheses first. For example,

$$3 + (8 + 4) = 3 + 12 = 15$$

and

$$(3 + 8) + 4 = 11 + 4 = 15.$$

Similarly,

$$4(2 \cdot 3) = 4(6) = 24$$

and

$$(4 \cdot 2)3 = (8)3 = 24.$$

Note that, so long as only addition or only multiplication appears in an expression, changing the grouping does not change the result. Equations such as $3 + (8 + 4) = (3 + 8) + 4$ and $4(2 \cdot 3) = (4 \cdot 2)3$ illustrate that addition and multiplication are **associative.**

The Associative Laws

For Addition. For any numbers a, b, and c,

$$a + (b + c) = (a + b) + c.$$

(Numbers can be grouped in any manner for addition.)

For Multiplication. For any numbers a, b, and c,

$$a \cdot (b \cdot c) = (a \cdot b) \cdot c.$$

(Numbers can be grouped in any manner for multiplication.)

EXAMPLE 2

Use an associative law to write an expression equivalent to each of the following: **(a)** $y + (z + 3)$; **(b)** $(8x)y$.

Solution

a) An expression equivalent to $y + (z + 3)$ is $(y + z) + 3$ by the associative law of addition.

b) An expression equivalent to $(8x)y$ is $8(xy)$ by the associative law of multiplication. ❑

Using the Laws Together

When only additions or only multiplications are involved, parentheses can be placed any way we please. For that reason, we often omit them. For example,

$x + (y + 7)$ means $x + y + 7$, and $l(wh)$ means lwh.

A sum such as $(5 + 1) + (3 + 5) + 9$ can be simplified by looking for pairs of numbers that add to 10:

$$\begin{aligned}(5 + 1) + (3 + 5) + 9 &= 5 + 5 + 9 + 1 + 3 \\ &= 10 + 10 + 3 = 23.\end{aligned}$$

EXAMPLE 3

Use the commutative and/or associative laws of addition to write at least two expressions equivalent to $(x + 5) + y$.

Solution

a) $(x + 5) + y = x + (5 + y)$ Using the associative law

$\qquad = x + (y + 5)$ Using the commutative law

b) $(x + 5) + y = y + (x + 5)$ Using the commutative law

$= y + (5 + x)$ Using the commutative law again ❑

EXAMPLE 4

Use the commutative and/or associative laws of multiplication to rewrite $2(x3)$ as $6x$. Show and give reasons for each step.

Solution

$2(x3) = 2(3x)$ Using the commutative law

$= (2 \cdot 3)x$ Using the associative law

$= 6x$ Simplifying ❑

The Distributive Law

The *distributive law* is probably the single most important law for manipulating algebraic expressions. Unlike the commutative and associative laws, the distributive law uses multiplication together with addition.

You have already used the distributive law although you probably didn't realize it. To see this, try to multiply $3 \cdot 21$ mentally. Most people find the product, 63, by thinking of 21 as $20 + 1$ and then multiplying 20 by 3 and 1 by 3. The sum of the two products, $60 + 3$, is 63. Note that if the 3 is not used as a multiplier twice, the result will not be correct.

EXAMPLE 5

Compute in two ways: $4(3 + 2)$.

Solution

a) As in the multiplication of $3(20 + 1)$ above, we can multiply by 4 twice and add the results:

$4(3 + 2) = 4 \cdot 3 + 4 \cdot 2$ Multiplying both 3 and 2 by 4

$= 12 + 8 = 20.$ Adding

b) By first adding inside the parentheses, we get the same result in a different way:

$4(3 + 2) = 4(5)$ Adding; $3 + 2 = 5$

$= 20.$ Multiplying ❑

The Distributive Law

For any numbers a, b, and c,

$$a(b + c) = ab + ac.$$

EXAMPLE 6

Multiply: $3(x + 2)$.

Solution Since $x + 2$ cannot be simplified unless a value for x is given, we use the distributive law:

$3(x + 2) = 3x + 3 \cdot 2$ Using the distributive law

$= 3x + 6.$ ❑

In the expression $x + 2$, the parts separated by the plus sign are called **terms.*** The distributive law can also be used when more than two terms are being multiplied.

EXAMPLE 7

Multiply: $6(s + 2 + 5w)$.

Solution

$$6(s + 2 + 5w) = 6s + 6 \cdot 2 + 6 \cdot 5w \quad \text{Using the distributive law}$$
$$= 6s + 12 + (6 \cdot 5)w \quad \text{Using the associative law for multiplication}$$
$$= 6s + 12 + 30w$$

❑

Because of the commutative law of multiplication, the distributive law can be used on the ''right'': $(b + c)a = ba + ca$.

EXAMPLE 8

Multiply: $(c + 4)5$.

Solution

$$(c + 4)5 = c \cdot 5 + 4 \cdot 5 \quad \text{Using the distributive law on the right}$$
$$= 5c + 20$$

❑

CAUTION! The distributive law provides a useful way of removing parentheses. However, do not forget to multiply each number inside the parentheses by the number outside:

$$a(b + c) \neq ab + c.$$

The Distributive Law and Factoring

If we reverse the statement of the distributive law, we have the basis of a process called **factoring:** $ab + ac = a(b + c)$. To **factor** an expression means to write an equivalent expression that is a product. The parts of the product are then called **factors.**

EXAMPLE 9

Use the distributive law to factor: **(a)** $3x + 3y$; **(b)** $7x + 21y + 7$.

Solution

a) By the distributive law,

$$3x + 3y = 3(x + y). \quad \text{The } \textit{common factor} \text{ is 3.}$$

b) $7x + 21y + 7 = 7 \cdot x + 7 \cdot 3y + 7 \cdot 1$ The common factor is 7.

$= 7(x + 3y + 1)$ Using the distributive law

Be sure not to omit the 1 or the common factor, 7.

❑

*Terms are discussed in greater detail in Sections 1.5–1.8.

To check our factoring, we can multiply and see if the original expression is obtained. For example, since

$$7(x + 3y + 1) = 7x + 7 \cdot 3y + 7 \cdot 1$$
$$= 7x + 21y + 7,$$

the factoring of Example 9(b) is correct.

EXERCISE SET 1.2

Use the commutative law of addition to write an equivalent expression.

1. $y + 5$
2. $x + 6$
3. $5 + ab$
4. $x + 3y$
5. $9x + 3y$
6. $3a + 7b$
7. $2(a + 3)$
8. $9(x + 5)$

Use the commutative law of multiplication to write an equivalent expression.

9. rt
10. mn
11. $5a$
12. $7b$
13. $5 + ab$
14. $x + 3y$
15. $2(a + 3)$
16. $9(x + 5)$

Use the associative law of addition to write an equivalent expression.

17. $x + (y + 2)$
18. $(a + 3) + b$
19. $(9 + m) + 2$
20. $x + (2 + y)$
21. $(ab + c) + d$
22. $(m + np) + r$

Use the associative law of multiplication to write an equivalent expression.

23. $5(ab)$
24. $(7x)y$
25. $(6m)n$
26. $9(rp)$
27. $3[2(a + b)]$
28. $5[x(2 + y)]$

Use the commutative and/or associative laws to write two equivalent expressions.

29. $(a + b) + 2$
30. $5 + (v + w)$
31. $7(ab)$
32. $(xy)3$

Use the commutative and/or associative laws to rewrite each of the following. Label each step with a reason, as in Example 4.

33. $(3a)4$ as $12a$
34. $(2 + m) + 3$ as $m + 5$
35. $5 + (2 + x)$ as $x + 7$
36. $(a3)5$ as $15a$

Multiply.

37. $2(b + 5)$
38. $4(x + 3)$
39. $7(1 + t)$
40. $6(v + 4)$
41. $3(x + 1)$
42. $7(x + 8)$
43. $4(1 + y)$
44. $9(s + 1)$
45. $6(5x + 2)$
46. $9(6m + 7)$
47. $7(x + 4 + 6y)$
48. $4(5x + 8 + 3p)$
49. $(a + b)2$
50. $(x + 2)7$
51. $(x + y + 2)5$
52. $(2 + a + b)6$

Use the distributive law to factor each of the following. Check by multiplying.

53. $2x + 2y$
54. $5y + 5z$
55. $5 + 5y$
56. $13 + 13x$
57. $3x + 12y$
58. $5x + 20y$
59. $5x + 10 + 15y$
60. $3 + 27b + 6c$
61. $9x + 9$
62. $6x + 6$
63. $9x + 3y$
64. $15x + 5y$
65. $2a + 16b + 64$
66. $5 + 20x + 35y$
67. $11x + 44y + 121$
68. $7 + 14b + 56w$

Skill Maintenance

To the student and the instructor: ***Skill maintenance exercises*** review skills studied in earlier sections of the text. These exercises appear in almost every exercise set.

Translate to an algebraic expression.

69. 9 less than t
70. Half of m

Synthesis

71. ◈ Are subtraction and division commutative? Why or why not?

72. ◈ Are subtraction and division associative? Why or why not?

Tell whether the following expressions are equivalent. Also, explain why.

73. $5m + 6$ and $6 + 5m$
74. $3(2 + x + y)$ and $6 + 3(x + y)$
75. $axy + ax$ and $xa(1 + y)$
76. $3a(b + c)$ and $(ca + ba)3$
77. ◈ Factor $17x + 34$. Then evaluate $17x + 34$ and its factorization when $x = 10$. Do your results indicate that $17x + 34$ and its factorization are equivalent? Why or why not?
78. When you put money in the bank and draw simple interest, the amount in your account later on is given by the expression $P + Prt$, where P is the principal, r is the rate of interest, and t is the time. Factor the expression.
79. ◈ Evaluate the expressions $3(2 + x)$ and $6 + x$ when $x = 0$. Do your results indicate that $3(2 + x)$ and $6 + x$ are equivalent? Why or why not?

1.3 Fractional Notation

Factors and Prime Factorizations • Fractional Notation • Multiplication and Simplification • Canceling • Addition, Subtraction, and Division

This section reviews multiplication, addition, subtraction, and division with fractional notation. Although much of this may be review, note that fractional expressions that contain variables are also introduced.

Factors and Prime Factorizations

In order to be able to study addition and subtraction using fractional notation, we first review how *natural numbers* are factored. **Natural numbers** can be thought of as the counting numbers:

$$1, 2, 3, 4, 5, \ldots .$$

The dots indicate that the pattern of the preceding numbers continues without ending.*

In Section 1.2, we factored expressions by writing an equivalent product. For example, $3x + 3y$ was factored as $3(x + y)$. Natural numbers can also be factored by writing an equivalent product. For instance, 30 can be factored as $3 \cdot 10$. To **factor** a number means to write it as a product; each number appearing in a product is called **a factor.** Note that the word factor can be used as both a verb and a noun.

EXAMPLE 1

Factor the number 8. List the factors.

Solution The number 8 can be factored in several ways:

$$2 \cdot 4, \qquad 1 \cdot 8, \qquad 2 \cdot 2 \cdot 2.$$

The factors of 8 are 1, 2, 4, and 8. ❑

*A less frequently used set of numbers, the **whole numbers,** includes 0: 0, 1, 2, 3, ...

A collection of symbols expressing a number as a product is called a **factorization** of the number.

EXAMPLE 2

Write several factorizations of the number 12.

Solution

$1 \cdot 12, \quad 2 \cdot 6, \quad 3 \cdot 4, \quad 2 \cdot 2 \cdot 3$ ☐

Some numbers have only two factors, the number itself and 1. Such numbers are called **prime.**

Prime Number

A *prime number* is a natural number that has exactly two different factors.

EXAMPLE 3

Which of these numbers are prime? 7, 4, 1

Solution

7 is prime. It has exactly two different factors, 7 and 1.

4 is not prime. It has three different factors, 1, 2, and 4.

1 is not prime. It does not have two *different* factors. ☐

If a natural number, other than 1, is not prime, we call it **composite.** Every composite number can be factored into a product of prime numbers. Such a factorization is called a **prime factorization.**

EXAMPLE 4

Find the prime factorization of 36.

Solution We begin by factoring 36 in any way that we can. One way is like this:

$$36 = 4 \cdot 9.$$

The factors 4 and 9 are not prime, so we factor them:

$$36 = 4 \cdot 9 = 2 \cdot 2 \cdot 3 \cdot 3. \quad \text{2 and 3 are both prime.}$$

The prime factorization of 36 is $2 \cdot 2 \cdot 3 \cdot 3$. ☐

Fractional Notation

An example of **fractional notation** for a number is:

Numerator

$$\frac{2}{3}.$$

Denominator

The top number is called the **numerator,** and the bottom number is called the **denominator.** When the numerator and the denominator are the same nonzero number, we have fractional notation for the number 1.

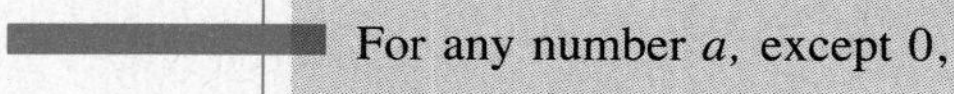

For any number a, except 0,

$$\frac{a}{a} = 1.$$

(Any nonzero number divided by itself is 1.)

Multiplication and Simplification

Recall from arithmetic that fractions are multiplied according to the following rule.

Multiplication of Fractions

For any two fractions a/b and c/d,

$$\frac{a}{b} \cdot \frac{c}{d} = \frac{ac}{bd}.$$

(The numerator of the product is the product of the individual numerators. The denominator of the product is the product of the individual denominators.)

EXAMPLE 5

Multiply: **(a)** $\frac{2}{3} \cdot \frac{7}{5}$; **(b)** $\frac{4}{x} \cdot \frac{8}{y}$.

Solution

a) We multiply numerators as well as denominators:

$$\frac{2}{3} \cdot \frac{7}{5} = \frac{2 \cdot 7}{3 \cdot 5} = \frac{14}{15}.$$

b) $\frac{4}{x} \cdot \frac{8}{y} = \frac{4 \cdot 8}{x \cdot y} = \frac{32}{xy}$ ❑

When one of the fractions being multiplied is 1, multiplying yields an equivalent expression because of the *identity property of* 1.

The Identity Property of 1

For any number a,

$$a \cdot 1 = a.$$

(Multiplying a number by 1 gives that same number.)

EXAMPLE 6

Multiply: $\frac{4}{5} \cdot \frac{6}{6}$.

Solution Since $\frac{6}{6} = 1$, the expression $\frac{4}{5} \cdot \frac{6}{6}$ is equivalent to $\frac{4}{5} \cdot 1$, or simply $\frac{4}{5}$. We have

$$\frac{4}{5}\cdot\frac{6}{6}=\frac{4\cdot 6}{5\cdot 6}=\frac{24}{30}.$$

Note that $\frac{24}{30}$ is equivalent to $\frac{4}{5}$. ❑

The steps of Example 6 can be reversed by "removing a factor of 1"—in this case, $\frac{6}{6}$. By removing a factor of 1, we can *simplify* an expression like $\frac{24}{30}$ to an equivalent expression like $\frac{4}{5}$.

To simplify, we factor the numerator and the denominator, looking for the largest factor common to both. This is sometimes made easier by writing the prime factorizations. Once common factors have been identified, the fraction can be expressed as a product of two fractions, one of which is in the form a/a.

EXAMPLE 7

Simplify: **(a)** $\frac{15}{40}$; **(b)** $\frac{36}{24}$.

Solution

a) Observe that 5 is a factor of both 15 and 40:

$$\frac{15}{40}=\frac{3\cdot 5}{8\cdot 5}$$ Factoring the numerator and the denominator, using the common factor, 5

$$=\frac{3}{8}\cdot\frac{5}{5}$$ Rewriting as a product of two fractions

$$=\frac{3}{8}\cdot 1$$ $\frac{5}{5}=1$

$$=\frac{3}{8}.$$ Using the identity property of 1 (removing a factor of 1)

b) $$\frac{36}{24}=\frac{2\cdot 2\cdot 3\cdot 3}{2\cdot 2\cdot 2\cdot 3}$$ Writing the prime factorizations and identifying common factors

$$=\frac{3}{2}\cdot\frac{2\cdot 2\cdot 3}{2\cdot 2\cdot 3}$$ Rewriting as a product of two fractions

$$=\frac{3}{2}\cdot 1$$ $\frac{2\cdot 2\cdot 3}{2\cdot 2\cdot 3}=1$

$$=\frac{3}{2}$$ Using the identity property of 1 ❑

It is always wise to check your result to see if any common factors of the numerator and the denominator remain. (This will never happen if prime factorizations are used correctly.) If common factors remain, repeat the process and simplify again.

Canceling

Canceling is a shortcut that you may have used for removing a factor of 1 when working with fractional notation. With *great* concern, we mention it as a possible way to speed up your work. You should use canceling only when removing common factors in numerators and denominators. Canceling *may not* be done in sums or when adding expressions together. Our concern is that "canceling" be done with care and understanding. Example 7(b) might have been done faster as follows:

$$\frac{36}{24} = \frac{\not{2}\cdot\not{2}\cdot 3\cdot\not{3}}{\not{2}\cdot\not{2}\cdot 2\cdot\not{3}} = \frac{3}{2}, \quad \text{or} \quad \frac{36}{24} = \frac{3\cdot\not{12}}{2\cdot\not{12}} = \frac{3}{2}, \quad \text{or} \quad \frac{\overset{\overset{3}{\not{18}}}{\not{36}}}{\underset{\underset{2}{\not{12}}}{\not{24}}} = \frac{3}{2}$$

CAUTION! The difficulty with canceling is that it is often applied incorrectly:

$$\underbrace{\frac{\not{2}+3}{\not{2}} = 3,}_{\text{Wrong!}} \quad \underbrace{\frac{\not{4}+1}{\not{4}+2} = \frac{1}{2},}_{\text{Wrong!}} \quad \underbrace{\frac{1\not{5}}{\not{5}4} = \frac{1}{4}.}_{\text{Wrong!}}$$

$$\frac{2+3}{2} = \frac{5}{2} \qquad \frac{4+1}{4+2} = \frac{5}{6} \qquad \frac{15}{54} = \frac{5\cdot 3}{18\cdot 3} = \frac{5}{18}$$

In each of these situations, the expressions canceled out were *not* factors. Factors are parts of products. For example, in $2 \cdot 3$, 2 and 3 are factors, but in $2 + 3$, 2 and 3 are *not* factors. **If you can't factor, you can't cancel! If in doubt, don't cancel!**

The number of factors in the numerator and the denominator may not always be the same. If not, the identity property of 1 allows us to insert the number 1 as a factor.

EXAMPLE 8

Simplify: $\frac{9}{72}$.

Solution

$$\frac{9}{72} = \frac{1\cdot 9}{8\cdot 9} \qquad \text{Factoring and using the identity property of 1 to write 9 as } 1\cdot 9$$

$$= \frac{1\cdot\not{9}}{8\cdot\not{9}} \qquad \text{Removing a factor of 1: } \frac{9}{9} = 1$$

$$= \frac{1}{8} \qquad \text{Simplifying}$$

❑

Addition, Subtraction, and Division

When denominators are the same, fractions are added or subtracted by adding or subtracting numerators and keeping the same denominator.

EXAMPLE 9

Add and simplify: $\frac{4}{8} + \frac{5}{8}$.

Solution The common denominator is 8. We add the numerators and keep the common denominator:

$$\frac{4}{8} + \frac{5}{8} = \frac{4+5}{8} = \frac{9}{8}.$$

❑

In arithmetic, you usually write $1\frac{1}{8}$ rather than the *improper* fraction $\frac{9}{8}$. In algebra, symbols such as $\frac{9}{8}$ are more useful and are quite "proper" for our purposes.

When denominators are different, we use the property of 1 and multiply to find a common denominator.

EXAMPLE 10

Add or subtract as indicated: **(a)** $\frac{7}{8} + \frac{5}{12}$; **(b)** $\frac{9}{8} - \frac{4}{5}$.

Solution

a) The number 24 is divisible by both 8 and 12. We multiply both $\frac{7}{8}$ and $\frac{5}{12}$ by suitable factors of 1 to obtain two fractions with denominators of 24:

$$\frac{7}{8} + \frac{5}{12} = \frac{7}{8}\cdot\frac{3}{3} + \frac{5}{12}\cdot\frac{2}{2}$$

Multiplying by 1. Since $3 \cdot 8 = 24$, we multiply the first number by $\frac{3}{3}$. Since $2 \cdot 12 = 24$, we multiply the second number by $\frac{2}{2}$.

$$= \frac{21}{24} + \frac{10}{24}$$

Performing the multiplications

$$= \frac{31}{24}.$$

b) $$\frac{9}{8} - \frac{4}{5} = \frac{9}{8}\cdot\frac{5}{5} - \frac{4}{5}\cdot\frac{8}{8}$$

Using 40 as a common denominator

$$= \frac{45}{40} - \frac{32}{40}$$

$$= \frac{13}{40}$$

❑

Two numbers whose product is 1 are called **reciprocals,** or **multiplicative inverses,** of each other. All numbers, except zero, have reciprocals. For example,

the reciprocal of $\frac{2}{3}$ is $\frac{3}{2}$ because $\frac{2}{3}\cdot\frac{3}{2} = \frac{6}{6} = 1$,
the reciprocal of 9 is $\frac{1}{9}$ because $9\cdot\frac{1}{9} = \frac{9}{9} = 1$, and
the reciprocal of $\frac{1}{4}$ is 4 because $\frac{1}{4}\cdot 4 = 1$.

Any division problem can be rewritten as multiplication.

Division of Fractions

To divide, multiply by the reciprocal of the divisor:

$$\frac{a}{b} \div \frac{c}{d} = \frac{a}{b}\cdot\frac{d}{c}.$$

EXAMPLE 11

Divide: $\frac{1}{2} \div \frac{3}{5}$.

Solution

$$\frac{1}{2} \div \frac{3}{5} = \frac{1}{2}\cdot\frac{5}{3}$$

$\frac{5}{3}$ is the reciprocal of $\frac{3}{5}$

$$= \frac{5}{6}$$

❑

After we have performed an operation of multiplication, addition, subtraction, or division, the answer may need to be simplified.

EXAMPLE 12

Perform the indicated operation and simplify: **(a)** $\frac{7}{10} - \frac{1}{5}$; **(b)** $\frac{5}{6} \cdot \frac{9}{25}$; **(c)** $\frac{2}{3} \div \frac{4}{9}$.

Solution

a) $\frac{7}{10} - \frac{1}{5} = \frac{7}{10} - \frac{1}{5} \cdot \frac{2}{2}$ Using 10 as the common denominator

$= \frac{7}{10} - \frac{2}{10}$

$= \frac{5}{10} = \frac{1 \cdot \not{5}}{2 \cdot \not{5}} = \frac{1}{2}$ Removing a factor of 1: $\frac{5}{5} = 1$

b) $\frac{5}{6} \cdot \frac{9}{25} = \frac{5 \cdot 9}{6 \cdot 25}$ Multiplying numerators and denominators

$= \frac{5 \cdot 3 \cdot 3}{2 \cdot 3 \cdot 5 \cdot 5}$ Factoring

$= \frac{3 \cdot \not{3} \cdot \not{5}}{2 \cdot 5 \cdot \not{3} \cdot \not{5}}$ Removing a factor of 1: $\frac{3 \cdot 5}{3 \cdot 5} = 1$

$= \frac{3}{10}$ Simplifying

c) $\frac{2}{3} \div \frac{4}{9} = \frac{2}{3} \cdot \frac{9}{4}$ Multiplying by the reciprocal of the divisor

$= \frac{\not{2} \cdot \not{3} \cdot 3}{\not{3} \cdot \not{2} \cdot 2}$ Factoring and removing a factor of 1: $\frac{2 \cdot 3}{2 \cdot 3} = 1$

$= \frac{3}{2}$ ❑

EXERCISE SET 1.3

Write at least one factorization of each number. There can be more than one correct answer.

1. 56 **2.** 102
3. 93 **4.** 144

Find the prime factorization of each number. If the number is prime, state so.

5. 14 **6.** 15 **7.** 33
8. 55 **9.** 9 **10.** 25
11. 49 **12.** 121 **13.** 18
14. 24 **15.** 40 **16.** 56
17. 90 **18.** 120 **19.** 210
20. 330 **21.** 79 **22.** 143
23. 119 **24.** 221

Simplify.

25. $\frac{18}{45}$ **26.** $\frac{16}{56}$ **27.** $\frac{49}{14}$
28. $\frac{72}{27}$ **29.** $\frac{6}{42}$ **30.** $\frac{13}{104}$
31. $\frac{56}{7}$ **32.** $\frac{132}{11}$ **33.** $\frac{19}{76}$
34. $\frac{17}{51}$ **35.** $\frac{100}{20}$ **36.** $\frac{150}{25}$

37. $\dfrac{425}{525}$ **38.** $\dfrac{625}{325}$ **39.** $\dfrac{2600}{1400}$

40. $\dfrac{4800}{1600}$ **41.** $\dfrac{8 \cdot x}{6 \cdot x}$ **42.** $\dfrac{13 \cdot v}{39 \cdot v}$

Perform the indicated operation and simplify.

43. $\dfrac{1}{4} \cdot \dfrac{1}{2}$ **44.** $\dfrac{11}{10} \cdot \dfrac{8}{5}$ **45.** $\dfrac{17}{2} \cdot \dfrac{3}{4}$

46. $\dfrac{11}{12} \cdot \dfrac{12}{11}$ **47.** $\dfrac{1}{2} + \dfrac{1}{2}$ **48.** $\dfrac{1}{2} + \dfrac{1}{4}$

49. $\dfrac{4}{9} + \dfrac{13}{18}$ **50.** $\dfrac{4}{5} + \dfrac{8}{15}$ **51.** $\dfrac{3}{a} \cdot \dfrac{b}{7}$

52. $\dfrac{x}{5} \cdot \dfrac{y}{z}$ **53.** $\dfrac{3}{x} + \dfrac{2}{x}$ **54.** $\dfrac{7}{a} - \dfrac{5}{a}$

55. $\dfrac{3}{10} + \dfrac{8}{15}$ **56.** $\dfrac{9}{8} + \dfrac{7}{12}$ **57.** $\dfrac{5}{4} - \dfrac{3}{4}$

58. $\dfrac{12}{5} - \dfrac{2}{5}$ **59.** $\dfrac{13}{18} - \dfrac{4}{9}$ **60.** $\dfrac{13}{15} - \dfrac{8}{45}$

61. $\dfrac{11}{12} - \dfrac{2}{5}$ **62.** $\dfrac{15}{16} - \dfrac{2}{3}$ **63.** $\dfrac{7}{6} \div \dfrac{3}{5}$

64. $\dfrac{7}{5} \div \dfrac{3}{4}$ **65.** $\dfrac{8}{9} \div \dfrac{4}{15}$ **66.** $\dfrac{3}{4} \div \dfrac{3}{7}$

67. $\dfrac{1}{4} \div \dfrac{1}{2}$ **68.** $\dfrac{1}{10} \div \dfrac{1}{5}$ **69.** $\dfrac{\frac{13}{12}}{\frac{39}{5}}$

70. $\dfrac{\frac{17}{6}}{\frac{3}{8}}$ **71.** $100 \div \dfrac{1}{5}$ **72.** $78 \div \dfrac{1}{6}$

73. $\dfrac{3}{4} \div 10$ **74.** $\dfrac{5}{6} \div 15$ **75.** $\dfrac{5}{3} \div \dfrac{a}{b}$

76. $\dfrac{x}{7} \div \dfrac{4}{y}$ **77.** $\dfrac{x}{6} - \dfrac{1}{3}$ **78.** $\dfrac{9}{10} + \dfrac{x}{2}$

Skill Maintenance

Use a commutative law to write an equivalent expression. There can be more than one correct answer.

79. $5(x + 3)$ **80.** $7 + (a + b)$

Synthesis

81. ◈ Is multiplication of fractions commutative? Why or why not?

82. ◈ Use the word factor in two sentences—once as a noun and once as a verb.

Simplify.

83. $\dfrac{128}{192}$ **84.** $\dfrac{pqrs}{qrst}$

85. $\dfrac{33sba}{2(11a)}$ **86.** $\dfrac{4 \cdot 9 \cdot 16}{2 \cdot 8 \cdot 15}$

87. $\dfrac{36 \cdot (2rh)}{8 \cdot (9hg)}$ **88.** $\dfrac{3 \cdot (4xy) \cdot (5)}{2 \cdot (3x) \cdot (4y)}$

89. A candy company uses two sizes of boxes, 6 in. and 8 in. long. These are packed in bigger cartons to be shipped. What is the shortest-length carton that will accommodate boxes of either size without any room left over? (Each carton can contain only boxes of one size; no mixing is allowed.)

90. In the following table, the top number has been factored in such a way that the sum of the factors is the bottom number. For example, in the first column, 56 has been factored as $7 \cdot 8$, and $7 + 8 = 15$, the bottom number. Find the missing numbers in the table.

Product	56	63	36	72	140	96
Factor	7					
Factor	8					
Sum	15	16	20	38	24	20

Product		168	110			
Factor				9	24	3
Factor	8	8		10	18	
Sum	14		21			24

Find the area of each figure.

91.

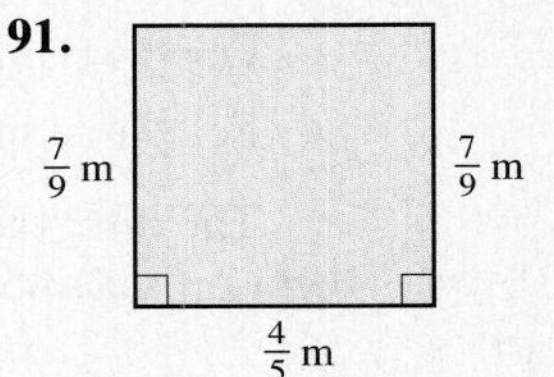

92.

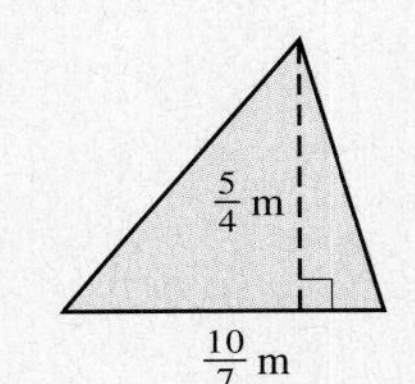

93. Find the perimeter of a square with sides of length $\frac{5}{9}$ m.

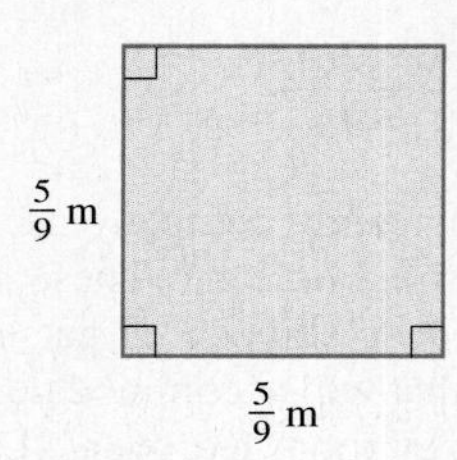

94. Find the perimeter of the rectangle in Exercise 91.

95. ◈ Make use of the properties and laws discussed in Sections 1.2 and 1.3 to explain why $x + y$ is equivalent to $(2y + 2x)/2$.

1.4 Positive and Negative Real Numbers

The Integers • Integers and the Real World • The Rational Numbers • The Real Numbers and Order • Translating to Inequalities • Absolute Value

A **set** is a collection of objects. The set containing the numbers 1, 3, and 7 is generally written $\{1, 3, 7\}$ and is said to be a **subset** if it is part of some other set. In this section, we examine the set of *real numbers* and some of its important subsets. More on sets can be found in Appendix A.

The Integers

Two sets of numbers have already been discussed. We represent these sets using dots on a number line.

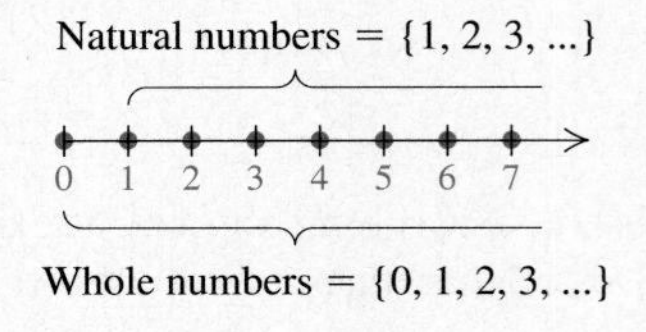

To create a new set, called the *integers,* we start with the whole numbers, 0, 1, 2, 3, and so on. For each natural number 1, 2, 3, and so on, we include in the set a new number the same distance to the left of 0 on the number line:

For the number 1, include the *opposite* number -1 (negative 1).

For the number 2, include the *opposite* number -2 (negative 2).

For the number 3, include the *opposite* number -3 (negative 3), and so on.

The **integers** consist of the whole numbers and these new numbers. We picture them on a number line as follows.

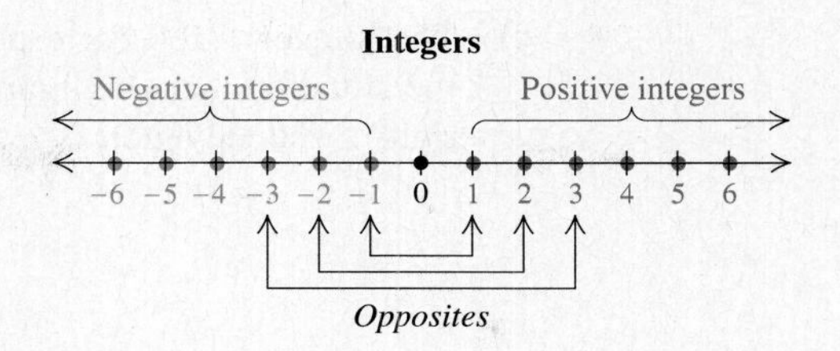

The new numbers to the left of 0 are called *negative integers*. The natural numbers are called *positive integers*. Zero is neither positive nor negative, so the whole numbers are sometimes referred to as *nonnegative integers*. Numbers like -1 and 1 or 3 and -3 are said to be *opposites* of each other, with 0 acting as its own opposite.

Set of Integers

The set of integers = {. . . , −5, −4, −3, −2, −1, 0, 1, 2, 3, 4, 5, . . .}.

Integers and the Real World

Integers are associated with many real-world problems and situations.

EXAMPLE 1

State the integer that corresponds to each situation: **(a)** The temperature is 3 degrees below zero; **(b)** Losing 21 points in a card game; **(c)** Death Valley is 280 ft below sea level; **(d)** A business made \$145 on Monday, but lost \$68 on Tuesday.

Solution

a) Since 3° below zero is −3°, the corresponding integer is −3.

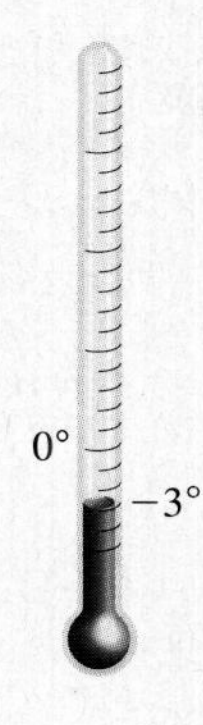

b) Losing 21 points in a card game gives you −21 points.

c) The integer -280 corresponds to the situation (see the figure at right). The elevation is -280 ft.

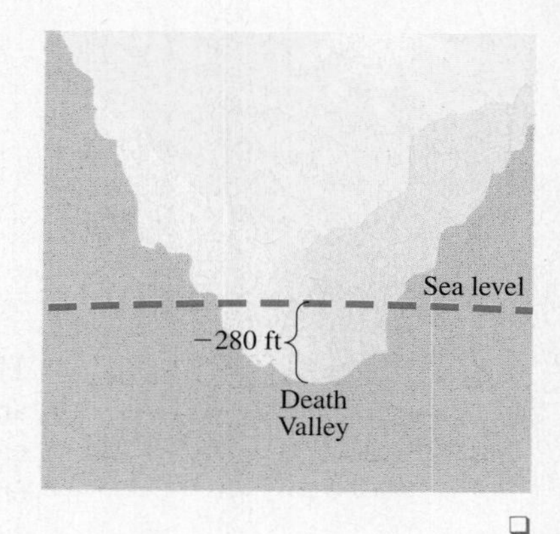

d) The integer 145 corresponds to the profit on Monday and -68 corresponds to the loss on Tuesday. ❑

The Rational Numbers

We now examine a set of numbers that contains the set of integers as a subset. This set, the **rational numbers,** contains fractions and decimals in addition to the integers. The following are rational numbers:

$$\frac{2}{3},\quad -\frac{2}{3},\quad \frac{7}{1},\quad 4,\quad -3,\quad 0,\quad \frac{23}{-8},\quad 2.4,\quad -0.17.$$

The number $-\frac{2}{3}$ (read "negative two-thirds") can also be named $\frac{2}{-3}$ or $\frac{-2}{3}$. The number 2.4 can be named $\frac{24}{10}$ or $\frac{12}{5}$, and -0.17 can be named $-\frac{17}{100}$.

Note that the set of rational numbers contains the whole numbers, the integers, and all fractions and decimals commonly seen in arithmetic. We cannot list all rational numbers, but we can describe the set of rational numbers as follows.

Set of Rational Numbers

The set of rational numbers $= \left\{\frac{a}{b} \,\middle|\, a \text{ and } b \text{ are integers and } b \neq 0\right\}$.

This is read "the set of all numbers $\frac{a}{b}$, where a and b are integers and $b \neq 0$."

Every rational number can be **graphed** by marking its location on a number line.

EXAMPLE 2

Graph each of the following rational numbers: **(a)** $\frac{5}{2}$; **(b)** -3.2; **(c)** $\frac{11}{8}$.

Solution

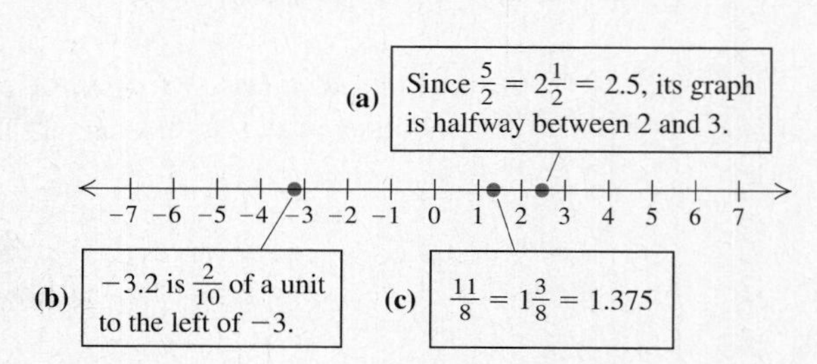

❑

Every rational number can be written using fractional or decimal notation.

EXAMPLE 3

Convert to decimal notation: $-\frac{5}{8}$.

Solution We first find decimal notation for $\frac{5}{8}$. Since $\frac{5}{8}$ means $5 \div 8$, we divide.

```
   0.6 2 5
8)5.0 0 0
  4 8
    2 0
    1 6
      4 0
      4 0
        0
```

Thus, $\frac{5}{8} = 0.625$, so $-\frac{5}{8} = -0.625$. ❑

Decimal notation for $-\frac{5}{8}$ is -0.625. We consider -0.625 to be a **terminating decimal** because we reached a remainder of 0. Decimal notation for some numbers, however, repeats.

EXAMPLE 4

Convert to decimal notation: $\frac{7}{11}$.

Solution We divide:

```
      0.6 3 6 3 . . .
1 1)7.0 0 0 0
    6 6
      4 0
      3 3
        7 0
        6 6
          4 0
          3 3
            7
```

We abbreviate repeating decimals by writing a bar over the repeating part, in this case, $0.\overline{63}$. ❑

The Real Numbers and Order

Every rational number has a point on the number line. However, not every point on the number line corresponds to a rational number. Some points correspond to what are called **irrational numbers.**

What kinds of numbers are irrational numbers? One example is the number π, which is used to find the area and circumference of a circle: $A = \pi r^2$ and $C = 2\pi r$.

Another irrational number, $\sqrt{2}$ (read "the square root of 2"), is the length of the diagonal of a square with sides of length 1. It is also the number that, when multiplied by itself, gives 2. No rational number can be multiplied by itself to get 2, although the following

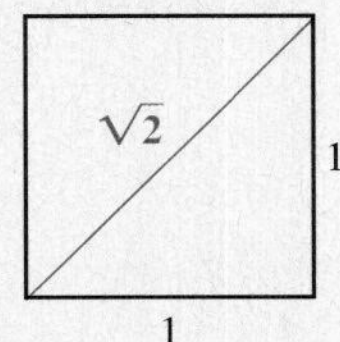

approximations come close:

1.4 is an *approximation* of $\sqrt{2}$ because $(1.4)^2 = 1.96$;

1.41 is a better approximation because $(1.41)^2 = 1.9881$;

1.4142 is an even better approximation because $(1.4142)^2 = 1.99996164$.

To approximate $\sqrt{2}$ on most calculators, simply press [2] and then the [√] key. On other calculators, press [√], [2], and [ENTER], or consult an owner's manual.

Decimal notation for rational numbers *either* terminates *or* repeats. Decimal notation for irrational numbers *neither* terminates *nor* repeats. Examples of irrational numbers are π, $\sqrt{3}$, $-\sqrt{8}$, $\sqrt{11}$, and 0.121221222122221 . . . .

The rational numbers and the irrational numbers together correspond to all the points on a number line and make up what is called the **real-number system.**

Set of Real Numbers

The set of real numbers = The set of all numbers corresponding to points on the number line.

The following figure shows the relationships among various kinds of numbers.

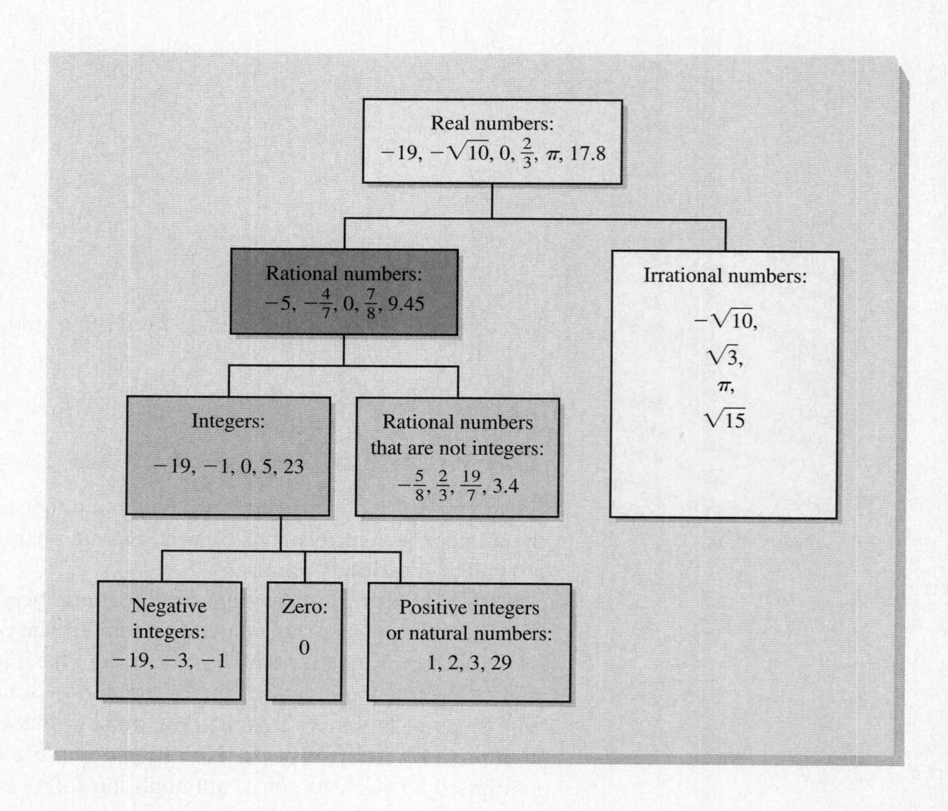

EXAMPLE 5

Graph the real number $\sqrt{3}$ on a number line.

Solution We use a calculator or Table 1 at the back of the book and approximate: $\sqrt{3} \approx 1.732$ ("$\approx$" means "approximately equals"). Then we locate this number on a number line.

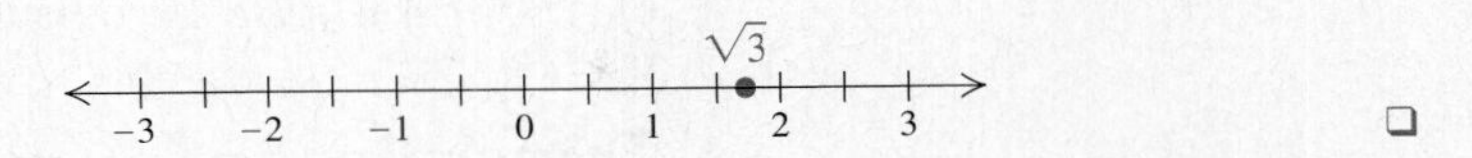

❑

Real numbers are named in order on the number line, with larger numbers further to the right. For any two numbers on the line, the one to the left is less than the one to the right. We use the symbol $<$ to mean "**is less than.**" The sentence $-8 < 6$ means "-8 is less than 6." The symbol $>$ means "**is greater than.**" The sentence $-3 > -7$ means "-3 is greater than -7."

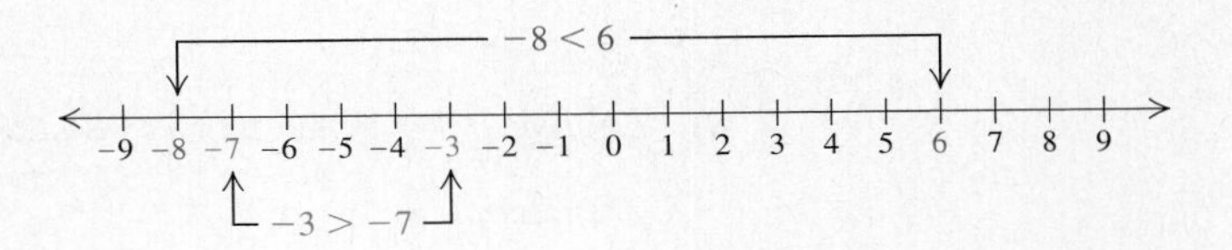

EXAMPLE 6

Use either $<$ or $>$ for ■ to write a true sentence: **(a)** 2 ■ 9; **(b)** -3.45 ■ 1.32; **(c)** 6 ■ -12; **(d)** -18 ■ -5; **(e)** $\frac{7}{11}$ ■ $\frac{5}{8}$.

Solution

a) Since 2 is to the left of 9, 2 is less than 9, so $2 < 9$.
b) Since -3.45 is to the left of 1.32, we have $-3.45 < 1.32$.
c) Since 6 is to the right of -12, we have $6 > -12$.
d) Since -18 is to the left of -5, we have $-18 < -5$.
e) We convert to decimal notation: $\frac{7}{11} = 0.\overline{63}\ldots$ and $\frac{5}{8} = 0.625$. Thus, $\frac{7}{11} > \frac{5}{8}$. We also could have used a common denominator: $\frac{7}{11} = \frac{56}{88} > \frac{55}{88} = \frac{5}{8}$. ❑

Sentences like "$a < -5$" and "$-3 > -8$" are called **inequalities.** It is useful to remember that every inequality can be written two ways. For instance,

$$-3 > -8 \quad \text{has the same meaning as} \quad -8 < -3.$$

It may be helpful to think of an inequality sign as an "arrow" with the smaller side pointing to the smaller number.

Note that all positive real numbers are greater than zero and all negative real numbers are less than zero.

> If x is a positive real number, then $x > 0$.
> If x is a negative real number, then $x < 0$.

Expressions like $a \leqslant b$ and $b \geqslant a$ are also inequalities. We read $a \leqslant b$ as "*a* **is less than or equal to** *b*." We read $a \geqslant b$ as "*a* **is greater than or equal to** *b*."

EXAMPLE 7

Write true or false for each inequality: **(a)** $-3 \leq 5$; **(b)** $-3 \leq -3$; **(c)** $-5 \geq 4$.

Solution

a) $-3 \leq 5$ is *true* because $-3 < 5$ is true.

b) $-3 \leq -3$ is *true* because $-3 = -3$ is true.

c) $-5 \geq 4$ is *false* since neither $-5 > 4$ nor $-5 = 4$ is true. ❑

Translating to Inequalities

In the following example, we see some ways in which English sentences can be translated to inequalities.

EXAMPLE 8

Translate each sentence to mathematical language: **(a)** A number is less than -3; **(b)** The temperature is at least 75°; **(c)** A debt of \$150 is worse than a debt of \$100.

Solution

a) *English:* A number is less than -3.

Translation: $x < -3$.

b) *English:* The temperature is at least 75°.

Translation: $t \geq 75$.

c) *English:* A debt of \$150 is worse than a debt of \$100.

Translation: $-150 < -100$. ❑

Absolute Value

In Section 1.5, we will need terminology for a number's distance from zero. We call a number's distance from zero on a number line its **absolute value.** Thus the absolute value of 4 is 4 and the absolute value of -4 is also 4.

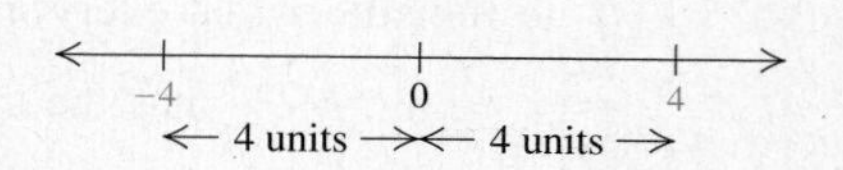

Absolute Value

We write $|a|$, read "the absolute value of a," to represent the number of units that a is from zero.

EXAMPLE 9

Find the absolute value: **(a)** $|-3|$; **(b)** $|7.2|$; **(c)** $|0|$.

Solution

a) $|-3| = 3$ since -3 is 3 units from 0.

b) $|7.2| = 7.2$ since 7.2 is 7.2 units from 0.

c) $|0| = 0$ since 0 is 0 units from itself. ❑

Distance is never negative, so numbers that are opposites have the same absolute value. If a number is nonnegative, its absolute value is the number itself. If a number is negative, its absolute value is its opposite.

EXERCISE SET 1.4

Tell which real numbers correspond to each situation.

1. In a game, Mo won 5 points. In the next game, he lost 12 points.

2. The temperature on Wednesday was 18° above zero. On Thursday, it was 2° below zero.

3. A family owes \$170. The same family has \$950 in its bank account.

4. A printer earned \$1200 one week and lost \$560 the next.

5. The Dead Sea is 1286 feet below sea level, whereas Mt. Everest is 29,028 feet above sea level.

6. In bowling, the Jets are 34 pins behind the Strikers after the first game. Describe the situation in two ways.

7. Janice deposited \$750 in a savings account. Two weeks later, she withdrew \$125.

8. During a certain time period, the United States had a deficit of \$3 million in foreign trade.

9. During a video game, Cindy intercepted a missile worth 20 points, lost a starship worth 150 points, and captured a base worth 300 points.

10. Ignition occurs 10 seconds before liftoff. A spent fuel tank is detached 235 seconds after liftoff.

Graph the rational number on a number line.

11. $\frac{10}{3}$ **12.** $-\frac{17}{5}$ **13.** -4.3

14. 3.87 **15.** -2 **16.** 5

Find decimal notation.

17. $-\frac{3}{8}$ **18.** $-\frac{1}{8}$ **19.** $\frac{5}{3}$

20. $\frac{5}{6}$ **21.** $\frac{7}{6}$ **22.** $\frac{5}{12}$

23. $\frac{2}{3}$ **24.** $\frac{1}{4}$ **25.** $-\frac{1}{2}$

26. $\frac{5}{8}$ **27.** $\frac{1}{10}$ **28.** $-\frac{7}{20}$

Write a true sentence using either $<$ or $>$.

29. 5 ■ 0 **30.** 9 ■ 0 **31.** -9 ■ 5

32. 8 ■ -8 **33.** -6 ■ 6 **34.** 0 ■ -7

35. -8 ■ -5 **36.** -4 ■ -3

37. -5 ■ -11 **38.** -3 ■ -4

39. -12.5 ■ -9.4 **40.** -10.3 ■ -14.5

41. 2.14 ■ 1.24 **42.** -3.3 ■ -2.2

43. $\frac{5}{12}$ ■ $\frac{11}{25}$ **44.** $-\frac{14}{17}$ ■ $-\frac{27}{35}$

Write an inequality with the same meaning as each of the following.

45. $-6 > x$ **46.** $x < 8$

47. $-10 \leq y$ **48.** $12 \geq t$

Write true or false.

49. $-3 \geq -11$ **50.** $5 \leq -5$

51. $0 \geq 8$ **52.** $-5 \leq 7$

53. $-8 \leq -8$ **54.** $8 \geq 8$

Translate to mathematical language.

55. -5 is greater than some number.

56. Some number is less than -1.

57. In cards, a score of 120 is better than one of -20.

58. A deposit of \$20 in a savings account is better than a withdrawal of \$25.

59. In trade, a deficit of \$500,000 is worse than an excess of \$1,000,000.

60. In bowling, it is better to be 60 pins ahead than to be 20 pins ahead.

61. Alice's test score was at most 95.

62. Some number never exceeds -9.

63. Fran's Franks considers profits to be poor if they don't exceed \$15,000.

64. A number is at most zero.

Find the absolute value.

65. $|-3|$ **66.** $|-7|$ **67.** $|10|$

68. $|11|$ **69.** $|0|$ **70.** $|-4|$

71. $|-24|$ **72.** $|325|$ **73.** $\left|-\frac{2}{3}\right|$

74. $\left|-\frac{10}{7}\right|$ **75.** $|43.9|$ **76.** $|14.8|$

77. $|x|$ when $x = 5$ **78.** $|b|$ when $b = -\frac{7}{8}$

79. List ten examples of rational numbers.

80. List ten examples of rational numbers that are *not* integers.

81. List three examples of irrational numbers.

82. List three examples of negative integers.

Skill Maintenance

83. Multiply and simplify: $\frac{21}{5} \cdot \frac{1}{7}$.

84. Evaluate $3xy$ when $x = 2$ and $y = 7$.

85. Use a commutative law to write an expression equivalent to $ab + 5$.

86. Factor: $3x + 9 + 12y$.

Synthesis

87. ◈ Is every nonnegative integer a whole number? Why or why not?

88. ◈ Why is it impossible for the absolute value of a number to be negative?

List in order from least to greatest.

89. 13, -12, 5, -17 **90.** -23, 4, 0, -17

91. $\frac{4}{5}, \frac{4}{3}, \frac{4}{8}, \frac{4}{6}, \frac{4}{9}, \frac{4}{2}, -\frac{4}{3}$

92. $-\frac{2}{3}, \frac{1}{2}, -\frac{3}{4}, -\frac{5}{6}, \frac{3}{8}, \frac{1}{6}$

Write a true sentence using either $<$, $>$, or $=$.

93. $|-5| \quad |-2|$ **94.** $|4| \quad |-7|$

95. $|-8| \quad |8|$ **96.** $|23| \quad |-23|$

97. $|-3| \quad |5|$ **98.** $|-19| \quad |-27|$

Solve.

99. $|x| = 7$

100. $|x| < 2$ (Consider only integer replacements.)

101. We know that 0.3333. . . is $\frac{1}{3}$ and 0.6666. . . is $\frac{2}{3}$. What rational number is named by each of the following?

a) 0.5555. . . **b)** 0.1111. . .

c) 0.2222. . . **d)** 0.9999. . .

1.5

Addition of Real Numbers

Adding with a Number Line • Adding without a Number Line • Problem Solving • Collecting Like Terms

We now consider addition of real numbers. To gain understanding, we will use a number line first. After we observe the principles involved, we will develop rules that will enable us to work more quickly.

Adding with a Number Line

To perform the addition $a + b$ on a number line, we start at a and then move according to b.

a) If b is positive, we move to the right (the positive direction).
b) If b is negative, we move to the left (the negative direction).
c) If b is 0, we stay at a.

EXAMPLE 1

Add: $-4 + 9$.

Solution To add on a number line, we locate the first number, -4, and then move 9 units to the right. Note that it requires 4 units to reach 0. The difference between 9 and 4 is where we finish.

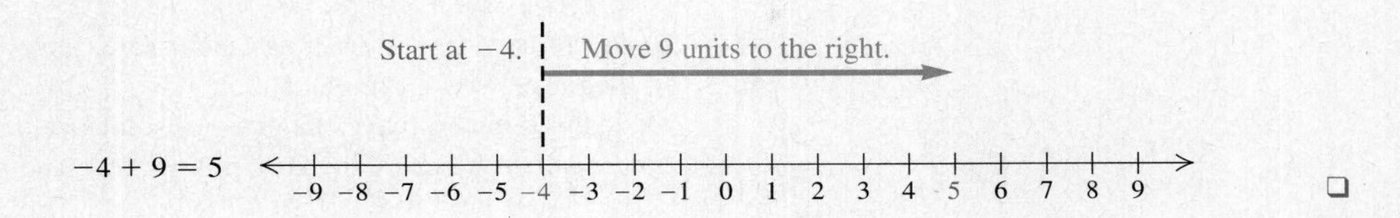

EXAMPLE 2

Add: $3 + (-5)$.

Solution We locate the first number, 3, and then move 5 units to the left. Note that it requires 3 units to reach 0. The difference between 5 and 3 is 2, so we finish 2 units to the left of 0.

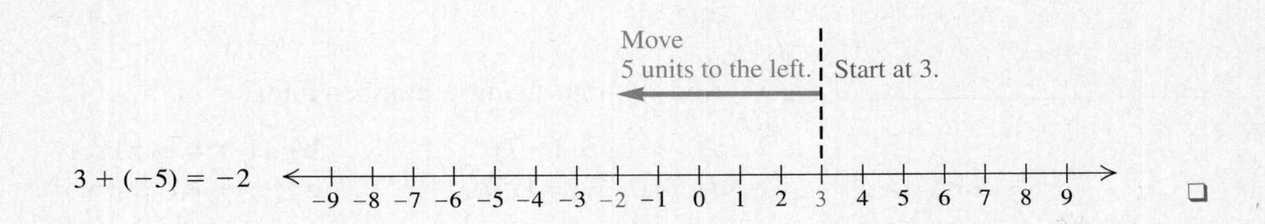

EXAMPLE 3

Add: $-4 + (-3)$.

Solution After locating -4, we move 3 units to the left. We finish a total of 7 units to the left of 0.

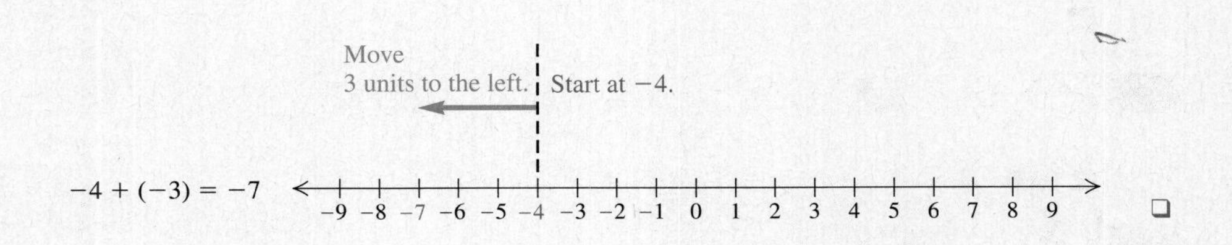

EXAMPLE 4

Add: $-5.2 + 0$.

Solution We locate -5.2 and move 0 units. Thus, we finish where we started.

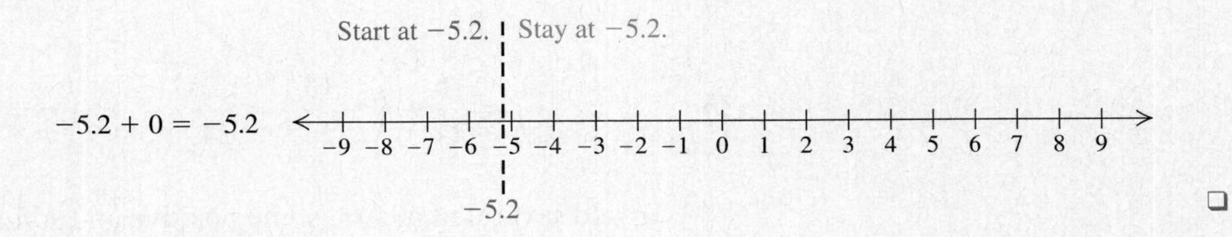

From Examples 1–4, we develop the following rules.

Rules for Addition of Real Numbers

1. *Positive numbers:* Add as usual. The answer is positive.
2. *Negative numbers:* Add absolute values and make the answer negative (see Example 3).
3. *A positive and a negative number:* Subtract absolute values. Then:
   a) If the positive number has the greater absolute value, the answer is positive (see Example 1).
   b) If the negative number has the greater absolute value, the answer is negative (see Example 2).
   c) If the numbers have the same absolute value, the answer is 0.
4. *One number is zero:* The sum is the other number (see Example 4).

Rule 4 is known as the **Identity Property of 0.** It says that for any real number a, $a + 0 = a$.

Adding without a Number Line

The rules for addition that are listed above can be used without drawing a number line.

EXAMPLE 5

Add without using a number line.

a) $-12 + (-7)$ **b)** $-1.4 + 8.5$ **c)** $-36 + 21$
d) $1.5 + (-1.5)$ **e)** $-\frac{7}{8} + 0$ **f)** $\frac{2}{3} + \left(-\frac{5}{8}\right)$

Solution

a) $-12 + (-7) = -19$

Two negatives. *Think:* Add the absolute values, 12 and 7, to get 19. Make the answer *negative,* -19.

b) $-1.4 + 8.5 = 7.1$

A negative and a positive. *Think:* The difference of absolute values is $8.5 - 1.4$, or 7.1. The positive number has the larger absolute value, so the answer is *positive,* 7.1.

c) $-36 + 21 = -15$

A negative and a positive. *Think:* The difference of absolute values is $36 - 21$, or 15. The negative number has the larger absolute value, so the answer is *negative,* -15.

d) $1.5 + (-1.5) = 0$

A negative and a positive. *Think:* Since the numbers are opposites, they have the same absolute value and the answer is 0.

e) $-\frac{7}{8} + 0 = -\frac{7}{8}$

One number is zero. The sum is $-\frac{7}{8}$.

f) $\frac{2}{3} + \left(-\frac{5}{8}\right) = \frac{16}{24} + \left(-\frac{15}{24}\right) = \frac{1}{24}$

A negative and a positive ❑

To add several numbers, some positive and some negative, the commutative and associative laws can be used. We add all the positives, then all the negatives, and then add the results. Of course, we can also add from left to right, if we prefer.

EXAMPLE 6

Add: $15 + (-2) + 7 + 14 + (-5) + (-12)$.

Solution

$15 + (-2) + 7 + 14 + (-5) + (-12)$

$= 15 + 7 + 14 + (-2) + (-5) + (-12)$ Using the commutative law of addition

$= (15 + 7 + 14) + [(-2) + (-5) + (-12)]$ Using the associative law of addition

$= 36 + (-19)$ Adding the positives; adding the negatives

$= 17$ Adding a positive and a negative ❑

Problem Solving

The addition of real numbers is used in many real-world applications.

EXAMPLE 7

In the course of one four-month period, the water level of Lake Champlain went down 2 ft, up 1 ft, down 5 ft, and up 3 ft. How much had the lake level changed at the end of the four months?

Solution The problem translates to a sum:

Rewording:	The 1st change	plus	the 2nd change	plus	the 3rd change	plus	the 4th change	is	the total change.
	↓	↓	↓	↓	↓	↓	↓	↓	↓
Translating:	-2	$+$	1	$+$	(-5)	$+$	3	$=$	Total change.

Since

$$\begin{aligned} -2 + 1 + (-5) + 3 &= -1 + (-5) + 3 \\ &= -6 + 3 \\ &= -3, \end{aligned}$$

the lake level has dropped 3 ft at the end of the four months. ❑

Collecting Like Terms

The rules for addition apply to variable expressions as well as to numbers. When two terms have variable factors that are exactly the same, like $5ab$ and $7ab$, the terms are called **like, or similar, terms.** The distributive law enables us to **collect, or combine, like terms.**

EXAMPLE 8

Collect like terms.

a) $-7x + 9x$ **b)** $2a + (-3b) + (-5a) + 9b$
c) $7 + y + (-3.5y) + 2$

Solution

a) $-7x + 9x = (-7 + 9)x$ Using the distributive law

$= 2x$ Adding

b) $2a + (-3b) + (-5a) + 9b = 2a + (-5a) + (-3b) + 9b$ Using the commutative law of addition

$= (2 + (-5))a + (-3 + 9)b$ Using the distributive law

$= -3a + 6b$ Adding

c) $7 + y + (-3.5y) + 2 = y + (-3.5y) + 7 + 2$ Using the commutative law of addition

$= (1 + (-3.5))y + 7 + 2$ Using the distributive law

$= -2.5y + 9$ Adding ❑

With practice we can leave out some steps, collecting like terms mentally. Numbers like 7 and 2 in the expression $7 + y + (-3.5y) + 2$ are constants and are also considered to be like terms.

EXERCISE SET 1.5

Add using a number line.

1. $-9 + 2$

2. $2 + (-5)$

3. $-10 + 6$

4. $8 + (-3)$

5. $-8 + 8$

6. $6 + (-6)$

7. $-3 + (-5)$

8. $-4 + (-6)$

Add. Do not use a number line except as a check.

9. $-7 + 0$

10. $-13 + 0$

11. $0 + (-27)$

12. $0 + (-35)$

13. $17 + (-17)$

14. $-15 + 15$

15. $-17 + (-25)$

16. $-24 + (-17)$

17. $-18 + 18$

18. $11 + (-11)$

19. $8 + (-5)$

20. $-7 + 8$

21. $-4 + (-5)$

22. $10 + (-12)$

23. $13 + (-6)$

24. $-3 + 14$

25. $11 + (-9)$

26. $-14 + (-19)$

27. $-20 + (-6)$

28. $19 + (-19)$

29. $-15 + (-7)$

30. $23 + (-5)$

31. $40 + (-8)$

32. $-23 + (-9)$

33. $-25 + 25$

34. $40 + (-40)$

35. $63 + (-18)$

36. $85 + (-65)$

37. $-6.5 + 4.7$

38. $-3.6 + 1.9$

39. $-2.8 + (-5.3)$

40. $-7.9 + (-6.5)$

41. $-\frac{3}{5} + \frac{2}{5}$

42. $-\frac{4}{3} + \frac{2}{3}$

43. $-\frac{3}{7} + \left(-\frac{5}{7}\right)$

44. $-\frac{4}{9} + \left(-\frac{6}{9}\right)$

45. $-\frac{5}{8} + \frac{1}{4}$

46. $-\frac{5}{6} + \frac{2}{3}$

47. $-\frac{3}{7} + \left(-\frac{2}{5}\right)$

48. $-\frac{5}{8} + \left(-\frac{1}{3}\right)$

49. $75 + (-14) + (-17) + (-5)$

50. $28 + (-44) + 17 + 31 + (-94)$

51. $-44 + \left(-\frac{3}{8}\right) + 95 + \left(-\frac{5}{8}\right)$

52. $24 + 3.1 + (-44) + (-8.2) + 63$

53. $98 + (-54) + 113 + (-998) + 44 + (-612) + (-18) + 334$

54. $-455 + (-123) + 1026 + (-919) + 213 + 111 + (-874)$

Problem Solving

55. In a college football game, the quarterback attempted passes with the following results.

First try	13-yd gain
Second try	incomplete
Third try	12-yd loss
Fourth try	21-yd gain
Fifth try	14-yd loss

Find the total gain (or loss).

56. The following table shows the profits and losses of a small business over a five-year period. Find the profit or loss after this period of time.

Year	*Profit or loss*
1989	+\$32,056
1990	−\$2,925
1991	+\$81,429
1992	−\$19,365
1993	−\$13,875

57. The barometric pressure at Omaha dropped 6 millibars (mb); then it rose 3 mb. After that, it dropped 14 mb and then rose 4 mb. What was the total change in pressure?

58. Monday the value of a share of IBM stock dropped $\$\frac{1}{4}$. Tuesday it rose in value $\$\frac{5}{8}$ and on Wednesday it lost $\$\frac{3}{8}$. How much did the stock's value rise or fall at the end of the three-day period?

59. Kyle's credit card bill is \$470. She sends a check to the credit card company for \$45, charges another \$160 in merchandise, and then pays off another \$500 of her bill. How much does either Kyle owe the company or the company owe Kyle?

60. Tony has \$460 in a checking account. He writes a check for \$530, makes a deposit of \$75, and then writes a check for \$90. What is the balance in the account?

Collect like terms.

61. $3a + 8a$

62. $4x + 8x$

63. $-2x + 15x$

64. $2m + (-7m)$

65. $4x + 7x$

66. $5a + 9a$

67. $7m + (-9m)$

68. $-4x + 9x$

69. $-6a + 10a$

70. $10n + (-17n)$

71. $-3 + 8x + 4 + (-10x)$

72. $8a + 5 + (-a) + (-3)$

Find the perimeter of the figure.

73.

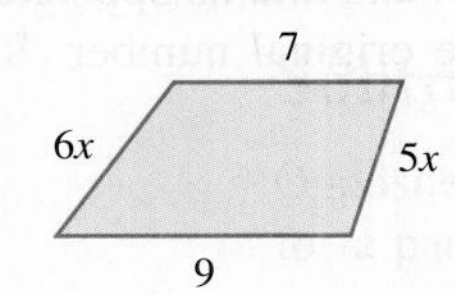

74.

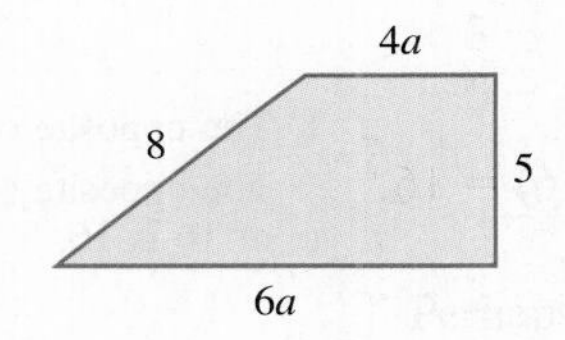

75.

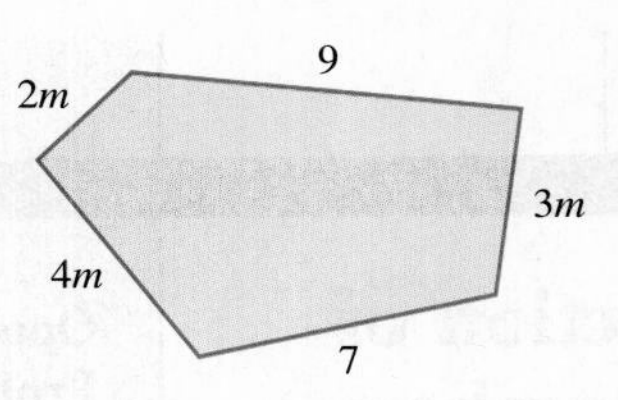

76.

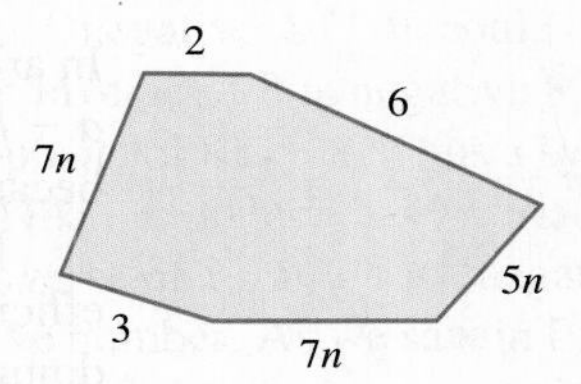

Skill Maintenance

77. Multiply: $7(3z + y + 2)$.

78. Divide and simplify: $\frac{7}{2} \div \frac{3}{8}$.

Synthesis

79. ◈ Without performing the actual addition, explain why the sum of all integers from −50 to 50 is 0.

80. ◈ Write a problem for a classmate to solve. Devise the problem so that it translates to a sum of negative and positive integers.

81. A stock's value rose $\$2\frac{3}{8}$ and then dropped $\$3\frac{1}{4}$ before finishing at $\$64\frac{3}{8}$. What was the stock's original value?

82. A sports card's value dropped \$12 and then rose \$3.70 before settling at \$32.50. What was its original value?

Find the missing term.

83. $7x + \_\_\_\_ + (-9x) + (-2y) = -2x - 7y$

84. $-3a + 9b + \_\_\_\_ + 5a = 2a - 2b$

85. $3m + 2n + \_\_\_\_ + (-2m) = 2n + (-6m)$

86. $\_\_\_\_ + 9x + (-4y) + x = 10x - 7y$

87. The perimeter of a rectangle is $6x + 10$. If the rectangle's length is 5, determine its width.

88. After five rounds of golf, a golf pro was 3 under par twice, 2 over par once, 2 under par once, and 1 over par once. On average, how far above or below par was the golfer?

For any real numbers a and b,

$$a - b = a + (-b).$$

(To subtract, add the opposite, or additive inverse, of the number being subtracted.)

EXAMPLE 5

Subtract the following and then check by addition: **(a)** $2 - 6$; **(b)** $4 - (-9)$; **(c)** $-4.2 - (-3.6)$.

Solution

a) $2 - 6 = 2 + (-6) = -4$ — The opposite of 6 is -6. We change the subtraction to addition and add the opposite. *Check:* $-4 + 6 = 2$.

b) $4 - (-9) = 4 + 9 = 13$ — The opposite of -9 is 9. We change the subtraction to addition and add the opposite. *Check:* $13 + (-9) = 4$.

c) $-4.2 - (-3.6) = -4.2 + 3.6 = -0.6$ — Adding the opposite. *Check:* $-0.6 + (-3.6) = -4.2$. ❑

EXAMPLE 6

Subtract $-\frac{3}{5}$ from $-\frac{2}{5}$.

Solution A common denominator already exists so we subtract as follows:

$$-\frac{2}{5} - \left(-\frac{3}{5}\right) = -\frac{2}{5} + \frac{3}{5} \quad \text{Adding the opposite}$$

$$= \frac{-2+3}{5} = \frac{1}{5}.$$

Check: $\frac{1}{5} + \left(-\frac{3}{5}\right) = \frac{1+(-3)}{5} = \frac{-2}{5}.$ ❑

The symbol "$-$" is read differently depending on where it appears. For instance, the expression $-5 - (-x)$ is read "negative five minus the opposite of x."

EXAMPLE 7

Read each of the following and then subtract: **(a)** $3 - 5$; **(b)** $-4.6 - (-9.8)$; **(c)** $-\frac{3}{4} - \frac{7}{5}$.

Solution

a) $3 - 5$; Read "three minus five"

$3 - 5 = 3 + (-5) = -2$ — Adding the opposite

b) $-4.6 - (-9.8)$; Read "negative four point six minus negative nine point eight"

$-4.6 - (-9.8) = -4.6 + 9.8 = 5.2$ — Adding the opposite

c) $-\frac{3}{4} - \frac{7}{5}$; Read "negative three-fourths minus seven-fifths"

$-\frac{3}{4} - \frac{7}{5} = -\frac{15}{20} + \left(-\frac{28}{20}\right) = -\frac{43}{20}$ — Finding a common denominator and adding the opposite ❑

When several additions and subtractions occur together, we can make them all additions.

EXAMPLE 8

Simplify: $8 - (-4) - 2 - (-5) + 3$.

Solution

$$8 - (-4) - 2 - (-5) + 3 = 8 + 4 + (-2) + 5 + 3 \quad \text{To subtract, we add the opposite.}$$
$$= 18$$

The **terms** of an algebraic expression are separated by plus signs. For instance, the terms of the expression $5x - 7y - 9$ are $5x$, $-7y$, and -9 since $5x - 7y - 9 = 5x + (-7y) + (-9)$.

EXAMPLE 9

Identify the terms of the expression $4 - 2ab + 7a - 9$.

Solution We have

$$4 - 2ab + 7a - 9 = 4 + (-2ab) + 7a + (-9), \quad \text{Rewriting as addition}$$

so the terms are 4, $-2ab$, $7a$, and -9.

EXAMPLE 10

Collect like terms.

a) $1 + 3x - 7x$
b) $-5a - 7b - 4a + 10b$
c) $9 - 3m - 14 + 7m$

Solution

a) $1 + 3x - 7x = 1 + 3x + (-7x)$ Adding the opposite
$= 1 + (3 + (-7))x$ Using the distributive law
$= 1 + (-4x)$
$= 1 - 4x$ Rewriting as subtraction to be more concise

b) $-5a - 7b - 4a + 10b = -5a + (-7b) + (-4a) + 10b$ Rewriting as addition
$= -5a + (-4a) + (-7b) + 10b$ Using the commutative law of addition
$= -9a + 3b$ Adding like terms mentally

c) $9 - 3m - 14 + 7m = 9 + (-3m) + (-14) + 7m$ Rewriting
$= 9 + (-14) + (-3m) + 7m$ Using a commutative law
$= -5 + 4m$

Problem Solving

Subtraction is used to solve problems involving differences.

EXAMPLE 11

The lowest point in Asia is the Dead Sea, which is 400 m below sea level. The lowest point in the United States is Death Valley, which is 86 m below sea level. What is the difference in elevation between the Dead Sea and Death Valley?

Solution It is helpful to draw a picture of the situation.

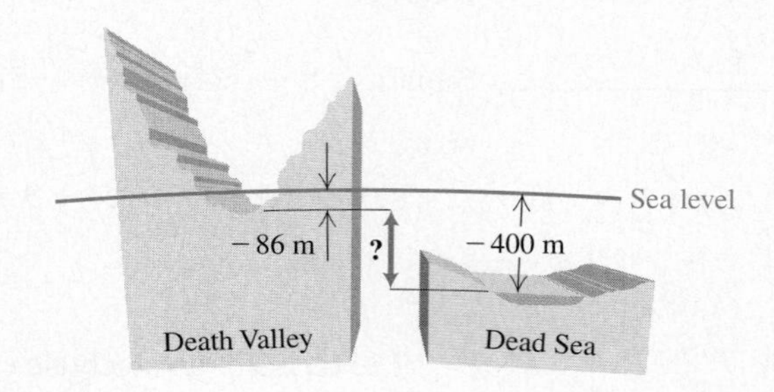

To find the difference in elevation, we always subtract the lower elevation, -400 m, from the higher elevation, -86 m:

$$-86 - (-400) = -86 + 400 = 314.$$

Death Valley is 314 m higher than the Dead Sea. ❑

EXERCISE SET 1.6

Find the opposite, or additive inverse.

1. 24 **2.** -64 **3.** -9
4. $\frac{7}{2}$ **5.** -26.9 **6.** 48.2

Find $-x$ when x is each of the following.

7. 9 **8.** -26 **9.** $-\frac{14}{3}$
10. $\frac{1}{328}$ **11.** 0.101 **12.** 0

Find $-(-x)$ when x is each of the following.

13. -65 **14.** 29
15. $\frac{5}{3}$ **16.** -9.1

Change the sign. (Find the opposite.)

17. -1 **18.** -7
19. 7 **20.** 10

Subtract.

21. $3 - 7$ **22.** $4 - 9$
23. $0 - 7$ **24.** $0 - 10$
25. $-8 - (-2)$ **26.** $-6 - (-8)$
27. $-10 - (-10)$ **28.** $-8 - (-8)$
29. $12 - 16$ **30.** $14 - 19$
31. $20 - 27$ **32.** $30 - 4$
33. $-9 - (-3)$ **34.** $-7 - (-9)$
35. $-40 - (-40)$ **36.** $-9 - (-9)$
37. $7 - 7$ **38.** $9 - 9$
39. $7 - (-7)$ **40.** $4 - (-4)$
41. $8 - (-3)$ **42.** $-7 - 4$
43. $-6 - 8$ **44.** $6 - (-10)$
45. $-4 - (-9)$ **46.** $-14 - 2$
47. $-6 - (-5)$ **48.** $-4 - (-3)$
49. $8 - (-10)$ **50.** $5 - (-6)$
51. $0 - 5$ **52.** $0 - 6$
53. $-5 - (-2)$ **54.** $-3 - (-1)$
55. $-7 - 14$ **56.** $-9 - 16$
57. $0 - (-5)$ **58.** $0 - (-1)$
59. $-8 - 0$ **60.** $-9 - 0$
61. $7 - (-5)$ **62.** $20 - (-15)$
63. $2 - 25$ **64.** $18 - 63$
65. $-42 - 26$ **66.** $-18 - 63$
67. $-71 - 2$ **68.** $-49 - 3$
69. $24 - (-92)$ **70.** $48 - (-73)$
71. $-50 - (-50)$ **72.** $-70 - (-70)$
73. $\frac{3}{8} - \frac{5}{8}$ **74.** $\frac{3}{9} - \frac{9}{9}$
75. $\frac{3}{4} - \frac{2}{3}$ **76.** $\frac{5}{8} - \frac{3}{4}$

77. $-\frac{3}{4}-\frac{2}{3}$ **78.** $-\frac{5}{8}-\frac{3}{4}$

79. $-2.8-0$ **80.** $6.04-1.1$

81. $0.99-1$ **82.** $0.87-1$

83. $\frac{1}{6}-\frac{2}{3}$ **84.** $-\frac{3}{8}-\left(-\frac{1}{2}\right)$

85. $-\frac{4}{7}-\left(-\frac{10}{7}\right)$ **86.** $\frac{12}{5}-\frac{12}{5}$

Translate the phrase to mathematical language and simplify. See Example 11.

87. The difference between 1.5 and -3.5

88. The difference between -2.1 and -5.9

89. The difference between -79 and 114

90. The difference between 23 and -17

91. Subtract 41 from -13.

92. Subtract 19 from -7.

93. Subtract -25 from 9.

94. Subtract -31 from -5.

Write words for each of the following and then perform the subtraction.

95. $-3.2-5.8$ **96.** $-2.7-5.9$

97. $-230-(-500)$ **98.** $-350-(-1000)$

Simplify.

99. $18-(-15)-3-(-5)+2$

100. $22-(-18)+7+(-42)-27$

101. $-31+(-28)-(-14)-17$

102. $-43-(-19)-(-21)+25$

103. $-34-28+(-33)-44$

104. $39+(-88)-29-(-83)$

105. $-93-(-84)-41-(-56)$

106. $84+(-99)+44-(-18)-43$

Identify the terms in each expression.

107. $3x-2y$ **108.** $7a-9b$

109. $-5+3m-6mn$ **110.** $-9-4t+10rt$

111. $5-a-6b+2$ **112.** $-2+3x-y-8$

Collect like terms.

113. $7a-12a$ **114.** $3x-15x$

115. $-3m-5+m$ **116.** $-7+9n-8$

117. $3x+5-9x$ **118.** $2+3a-7$

119. $2-6t-9-2t$ **120.** $-5+3b-7-5b$

121. $-5-(-3x)+3x+4x-(-12)$

122. $14-(-5x)+2x-(-32)$

123. $13x-(-2x)+45-(-21)$

124. $8x-(-2x)-14-(-5x)+53$

Problem Solving

125. Your total assets are \$619.46. You borrow \$950 for the purchase of a stereo system. What are your total assets now?

126. You owe a friend \$420. The friend decides to cancel \$156 of the debt. How much do you owe now?

127. In Churchill, Manitoba, Canada, the average daily low temperature in January is $-31°C$. The average daily low temperature in Key West, Florida, is 19°C. What is the difference in the average daily low temperature of the two cities?

128. On a winter night, the temperature dropped from 5°C to $-12°C$. How many degrees did it drop?

129. The lowest point in Africa is Lake Assal, which is 156 m below sea level. The lowest point in South America is the Valdes Peninsula, which is 40 m below sea level. How much lower is Lake Assal than the Valdes Peninsula?

130. The deepest point in the Pacific Ocean is the Marianas Trench, with a depth of 10,415 m. The deepest point in the Atlantic Ocean is the Puerto Rico Trench, with a depth of 8648 m. What is the difference in elevation of the two trenches?

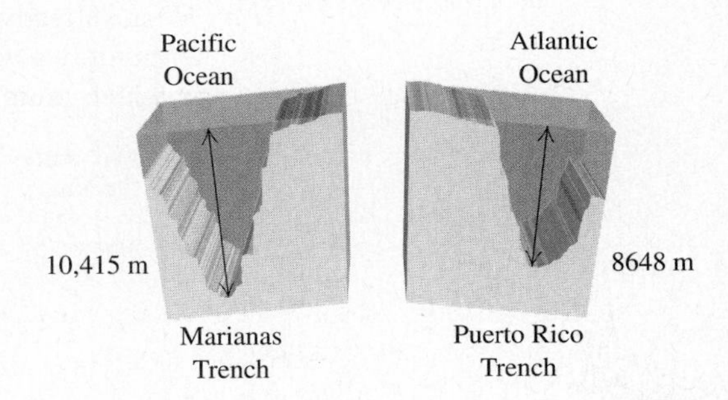

Skill Maintenance

131. Find the area of a rectangle when the length is 36 ft and the width is 12 ft.

132. Find the prime factorization of 864.

Synthesis

133. ◈ Explain why $-a+b$ is the opposite of $a+(-b)$.

134. ◈ A student claims to be able to add real numbers but unable to subtract them. What advice would you offer this student?

Tell whether each statement is true or false for all real numbers m and n. Use various replacements for m and n to support your answer.

135. If $m > n$, then $m - n > 0$.

136. If $m > n$, then $m + n > 0$.

137. If m and n are opposites, then $m - n = 0$.

138. If $m = -n$, then $m + n = 0$.

139. ◈ A gambler loses a wager and then loses "double or nothing" (meaning the gambler owes twice as much) twice more. After the three losses, the gambler's assets are $-\$20$. Explain how much the gambler originally bet and how the \$20 debt occurred.

140. ◈ If n is positive and m is negative, what is the sign of $n + (-m)$? Why?

1.7 Multiplication and Division of Real Numbers

Multiplication • **Division**

We now develop rules for multiplication and division of real numbers. Since multiplication and division are closely related, it should come as no surprise that the rules are quite similar.

Multiplication

We are already familiar with how to multiply two nonnegative numbers. To see how to multiply a positive number and a negative number, consider the following pattern in which multiplication is regarded as repeated addition:

This number decreases by 1 each time. →

$$\begin{aligned} 4(-5) &= (-5) + (-5) + (-5) + (-5) = -20 \\ 3(-5) &= (-5) + (-5) + (-5) = -15 \\ 2(-5) &= (-5) + (-5) = -10 \\ 1(-5) &= (-5) = -5 \\ 0(-5) &= 0 = 0 \end{aligned}$$

← This number increases by 5 each time.

This pattern illustrates that the product of a negative number and a positive number is negative.

> To multiply a positive number and a negative number, multiply their absolute values. The answer is negative.

EXAMPLE 1

Multiply: **(a)** $8(-5)$; **(b)** $-\frac{1}{3} \cdot \frac{5}{7}$.

Solution

a) $8(-5) = -40$

b) $-\frac{1}{3} \cdot \frac{5}{7} = -\frac{5}{21}$ ❑

The pattern developed above includes not just products of positive and negative numbers, but a product involving zero as well.

The Multiplicative Property of Zero

For any real number a,

$$0 \cdot a = a \cdot 0 = 0.$$

(The product of 0 and any real number is 0).

EXAMPLE 2

Multiply: $173(-452)0$.

Solution

$$\begin{aligned} 173(-452)0 &= 173[(-452)0] && \text{Using the associative law of multiplication} \\ &= 173[0] && \text{Using the multiplicative property of zero} \\ &= 0 && \text{Using the multiplicative property of zero again} \end{aligned}$$

Note that whenever 0 appears as a factor, the product will be 0. ❑

We can extend the above pattern still further to examine the product of two negative numbers.

This number decreases by 1 each time. →

$$\begin{aligned} 2(-5) &= (-5) + (-5) = -10 \\ 1(-5) &= (-5) = -5 \\ 0(-5) &= 0 = 0 \\ -1(-5) &= -(-5) = 5 \\ -2(-5) &= -(-5) - (-5) = 10 \end{aligned}$$

← This number increases by 5 each time.

According to the pattern, the product of two negative numbers is positive.

To multiply two negative numbers, multiply their absolute values. The answer is positive.

EXAMPLE 3

Multiply: **(a)** $(-5)(-7)$; **(b)** $(-1.2)(-3)$.

Solution

a) The absolute value of -5 is 5 and the absolute value of -7 is 7. Thus,

$$\begin{aligned} (-5)(-7) &= 5 \cdot 7 && \text{Multiplying absolute values} \\ &= 35. \end{aligned}$$

b) $$\begin{aligned} (-1.2)(-3) &= (1.2)(3) && \text{Multiplying absolute values} \\ &= 3.6 && \text{Try to go directly to this step.} \end{aligned}$$ ❑

When three or more numbers are multiplied, we can order and group the numbers as we please, because of the commutative and associative laws.

EXAMPLE 4

Multiply: **(a)** $-3(-2)(-5)$; **(b)** $-4(-6)(-1)(-2)$.

Solution

a) $-3(-2)(-5) = 6(-5)$ Multiplying the first two numbers. The product of two negatives is positive.

$= -30$ The product of a positive and a negative is negative.

b) $-4(-6)(-1)(-2) = 24 \cdot 2$ Multiplying the first two numbers and the last two numbers

$= 48$ ❑

We can see the following pattern in the results of Example 4.

The product of an even number of negative numbers is positive.
The product of an odd number of negative numbers is negative.

Division

Because the definition of division makes use of multiplication, the rules for multiplication are used to develop rules for division.

The quotient $\frac{a}{b}$ (or $a \div b$) is the number, if there is one, that when multiplied by b gives a. ($a \div b = c$ if $c \cdot b = a$.)

EXAMPLE 5

Divide, if possible, and check your answer: **(a)** $14 \div (-7)$; **(b)** $\frac{-32}{-4}$; **(c)** $\frac{-10}{7}$; **(d)** $\frac{-17}{0}$.

Solution

a) $14 \div (-7) = -2$ We look for a number that when multiplied by -7 gives 14. That number is -2. *Check:* $(-2)(-7) = 14$.

b) $\frac{-32}{-4} = 8$ We look for a number that when multiplied by -4 gives -32. That number is 8. *Check:* $8(-4) = -32$.

c) $\frac{-10}{7} = -\frac{10}{7}$ We look for a number that when multiplied by 7 gives -10. That number is $-\frac{10}{7}$. *Check:* $-\frac{10}{7} \cdot 7 = -10$.

d) $\frac{-17}{0}$ is **undefined.** We look for a number that when multiplied by 0 gives -17. There is no such number because the product of 0 and *any* number is 0, not -17. ❑

The rules for division are the same as those for multiplication. We state them together.

Rules for Multiplication and Division

To multiply or divide two real numbers:

1. Using the absolute values, multiply or divide, as indicated.
2. If the signs are the same, the answer is positive.
3. If the signs are different, the answer is negative.

Had Example 5(a) been written as $-14 \div 7$ instead of $14 \div (-7)$, the result would have still been -2. Similarly, had Example 5(c) been written as $10/-7$ instead of $-10/7$, the result, $-\frac{10}{7}$, would not have changed. In short, our rules for division give us the following:

$$\frac{-a}{b} = \frac{a}{-b} = -\frac{a}{b} \quad \textbf{and} \quad \frac{-a}{-b} = \frac{a}{b}.$$

EXAMPLE 6

Rewrite each of the following in two equivalent forms: **(a)** $\frac{5}{-2}$; **(b)** $-\frac{3}{10}$.

Solution We use the property listed above.

a) $\dfrac{5}{-2} = \dfrac{-5}{2}$ and $\dfrac{5}{-2} = -\dfrac{5}{2}$

b) $-\dfrac{3}{10} = \dfrac{-3}{10}$ and $-\dfrac{3}{10} = \dfrac{3}{-10}$ ❑

In some situations, it may help to rewrite a fraction that has a negative sign in an equivalent form.

EXAMPLE 7

Perform the indicated operation: **(a)** $(-\frac{4}{5})(\frac{-7}{3})$; **(b)** $-\frac{2}{7} + \frac{9}{-7}$.

Solution

a) $\left(-\dfrac{4}{5}\right)\left(\dfrac{-7}{3}\right) = \left(\dfrac{-4}{5}\right)\left(\dfrac{-7}{3}\right)$ Rewriting $-\dfrac{4}{5}$ as $\dfrac{-4}{5}$

$= \dfrac{28}{15}$ Try to go directly to this step.

b) Given a choice, we generally choose a positive denominator, although this is not a "must":

$-\dfrac{2}{7} + \dfrac{9}{-7} = \dfrac{-2}{7} + \dfrac{-9}{7}$ Rewriting both fractions with a common denominator of 7

$= \dfrac{-11}{7}$, or $-\dfrac{11}{7}$. ❑

EXAMPLE 8

Find the reciprocal: **(a)** -27; **(b)** $\frac{-3}{4}$; **(c)** $-\frac{1}{5}$.

Solution

a) The reciprocal of -27 is $\frac{1}{-27}$. More often, this number is written as $-\frac{1}{27}$.
b) The reciprocal of $\frac{-3}{4}$ is $\frac{4}{-3}$, or, equivalently, $-\frac{4}{3}$.
c) The reciprocal of $-\frac{1}{5}$ is -5. ❑

Keep in mind that the opposite, or additive inverse, of a number is what we add to the number to get 0, whereas a reciprocal is what we multiply the number by to get 1. Compare the following.

Number	**Opposite (Change the sign.)**	**Reciprocal (Invert but do not change the sign.)**
$-\frac{3}{8}$	$\frac{3}{8}$	$-\frac{8}{3}$
19	-19	$\frac{1}{19}$
0	0	Undefined

$\left(-\frac{3}{8}\right)\left(-\frac{8}{3}\right) = 1$

$-\frac{3}{8} + \frac{3}{8} = 0$

When dividing with fractional notation, it is usually easier to multiply by a reciprocal. With decimal notation, it is usually easier to carry out long division.

EXAMPLE 9

Divide: **(a)** $-\frac{2}{3} \div \left(-\frac{5}{4}\right)$; **(b)** $-\frac{3}{4} \div \frac{3}{10}$; **(c)** $27.9 \div (-3)$.

Solution

a) $-\frac{2}{3} \div \left(-\frac{5}{4}\right) = -\frac{2}{3} \cdot \left(-\frac{4}{5}\right) = \frac{8}{15}$ Multiplying by the reciprocal

Be careful not to change the sign when taking a reciprocal!

b) $-\frac{3}{4} \div \frac{3}{10} = -\frac{3}{4} \cdot \left(\frac{10}{3}\right) = -\frac{30}{12} = -\frac{5}{2} \cdot \frac{6}{6} = -\frac{5}{2}$ Removing a factor of 1: $\frac{6}{6} = 1$

c) $27.9 \div (-3) = \frac{27.9}{-3} = -9.3$ Do the long division $3\overline{)27.9}$ (quotient 9.3). The answer is negative. ❑

In Example 5(d), we explained why we cannot divide -17 by 0. This also explains why *no* nonzero number b can be divided by 0: Consider $b \div 0$. Is there a number that when multiplied by 0 gives b? No, because the product of 0 and any number is 0, not b. We say that $b \div 0$ is **undefined** for $b \neq 0$.

On the other hand, if we divide 0 by 0, we look for a number r such that $0 \div 0 = r$ and $r \cdot 0 = 0$. But, $r \cdot 0 = 0$ for *any* number r. Thus it appears that $0 \div 0$ could be any number we choose. Getting any answer we want when we divide 0 by 0 would lead to contradictions. Thus we say that $0 \div 0$ is **indeterminate.**

Finally, note that $0 \div 7 = 0$ since $0 \cdot 7 = 0$. This can be written $0/7 = 0$.

EXAMPLE 10

Divide, if possible: **(a)** $\frac{0}{-2}$; **(b)** $\frac{5}{0}$; **(c)** $\frac{0}{0}$.

Solution

a) $\frac{0}{-2} = 0$ *Check:* $0(-2) = 0$.

b) $\frac{5}{0}$ is undefined.

c) $\frac{0}{0}$ is indeterminate. ❑

Division Involving Zero

For any nonzero real number a,

$$\frac{0}{a} = 0 \quad \text{and} \quad \frac{a}{0} \text{ is undefined.}$$

The expression $\frac{0}{0}$ is indeterminate.

EXERCISE SET 1.7

Multiply.

1. $-8 \cdot 2$
2. $-2 \cdot 5$
3. $-7 \cdot 6$
4. $-9 \cdot 2$
5. $8 \cdot (-3)$
6. $9 \cdot (-5)$
7. $-9 \cdot 8$
8. $-10 \cdot 3$
9. $-8 \cdot (-2)$
10. $-2 \cdot (-5)$
11. $-7 \cdot (-6)$
12. $-9 \cdot (-2)$
13. $15 \cdot (-8)$
14. $-12 \cdot (-10)$
15. $-14 \cdot 17$
16. $-13 \cdot (-15)$
17. $-25 \cdot (-48)$
18. $39 \cdot (-43)$
19. $-3.5 \cdot (-28)$
20. $97 \cdot (-2.1)$
21. $9 \cdot (-8)$
22. $7 \cdot (-9)$
23. $-7 \cdot (-3.1)$
24. $-4 \cdot (-3.2)$
25. $\frac{2}{3} \cdot \left(-\frac{3}{5}\right)$
26. $\frac{5}{7} \cdot \left(-\frac{2}{3}\right)$
27. $-\frac{3}{8} \cdot \left(-\frac{2}{9}\right)$
28. $-\frac{5}{8} \cdot \left(-\frac{2}{5}\right)$
29. -6.3×2.7
30. -4.1×9.5
31. $-\frac{5}{9} \cdot \frac{3}{4}$
32. $-\frac{8}{3} \cdot \frac{9}{4}$
33. $7 \cdot (-4) \cdot (-3) \cdot 5$
34. $9 \cdot (-2) \cdot (-6) \cdot 7$
35. $-\frac{2}{3} \cdot \frac{1}{2} \cdot \left(-\frac{6}{7}\right)$
36. $-\frac{1}{8} \cdot \left(-\frac{1}{4}\right) \cdot \left(-\frac{3}{5}\right)$
37. $-3 \cdot (-4) \cdot (-5)$
38. $-2 \cdot (-5) \cdot (-7)$
39. $-2 \cdot (-5) \cdot (-3) \cdot (-5)$
40. $-3 \cdot (-5) \cdot (-2) \cdot (-1)$
41. $(-14) \cdot (-27) \cdot 0$
42. $7 \cdot (-6) \cdot 5 \cdot (-4) \cdot 3 \cdot (-2) \cdot 1 \cdot 0$
43. $(-8)(-9)(-10)$
44. $(-7)(-8)(-9)(-10)$
45. $(-6)(-7)(-8)(-9)(-10)$
46. $(-5)(-6)(-7)(-8)(-9)(-10)$

Divide, if possible, and check. If a quotient is undefined or indeterminate, state so.

47. $36 \div (-6)$
48. $\dfrac{28}{-7}$
49. $\dfrac{26}{-2}$
50. $26 \div (-13)$
51. $\dfrac{-16}{8}$
52. $-22 \div (-2)$
53. $\dfrac{-48}{-12}$
54. $-63 \div (-9)$
55. $\dfrac{-72}{9}$
56. $\dfrac{-50}{25}$
57. $-100 \div (-50)$
58. $\dfrac{-200}{8}$
59. $-108 \div 9$
60. $\dfrac{-64}{-7}$
61. $\dfrac{200}{-25}$
62. $-300 \div (-13)$
63. $\dfrac{75}{0}$
64. $\dfrac{0}{-5}$
65. $\dfrac{88}{-9}$
66. $\dfrac{0}{0}$
67. $\dfrac{0}{-9}$
68. $\dfrac{-35}{0}$
69. $0 \div 0$
70. $0 \div (-47)$

Write the number in two equivalent forms, as in Example 6.

71. $\dfrac{9}{-5}$

72. $\dfrac{-12}{7}$

73. $\dfrac{-36}{11}$

74. $\dfrac{9}{-14}$

75. $-\dfrac{7}{3}$

76. $-\dfrac{4}{15}$

77. $\dfrac{-x}{2}$

78. $\dfrac{9}{-a}$

Find the reciprocal.

79. $\dfrac{-3}{7}$

80. $\dfrac{2}{-9}$

81. $-\dfrac{47}{13}$

82. $-\dfrac{31}{12}$

83. -10

84. 13

85. 4.3

86. -8.5

87. $\dfrac{5}{-3}$

88. $\dfrac{-6}{11}$

89. -1

90. $\dfrac{1}{1/2}$

Perform the indicated operation and simplify, if possible. If a quotient is undefined or indeterminate, state so.

91. $\left(-\frac{3}{7}\right)\left(\frac{2}{-5}\right)$

92. $\left(\frac{-4}{9}\right)\left(-\frac{2}{3}\right)$

93. $\left(\frac{7}{-2}\right)\left(\frac{-5}{6}\right)$

94. $\left(\frac{-6}{5}\right)\left(\frac{2}{-11}\right)$

95. $\frac{-4}{5} + \frac{7}{-5}$

96. $\frac{3}{-8} + \frac{-5}{8}$

97. $\left(-\frac{2}{7}\right)\left(\frac{5}{-8}\right)$

98. $\left(\frac{-9}{5}\right)\left(-\frac{10}{7}\right)$

99. $\frac{-9}{7} + \left(-\frac{4}{7}\right)$

100. $\left(-\frac{3}{11}\right) + \frac{5}{-11}$

101. $\frac{3}{4} \div \left(-\frac{2}{3}\right)$

102. $\frac{7}{8} \div \left(-\frac{1}{2}\right)$

103. $\frac{-5}{12} \cdot \frac{7}{15}$

104. $\frac{9}{5} \cdot \frac{-20}{3}$

105. $\left(-\frac{12}{5}\right) + \left(-\frac{3}{5}\right)$

106. $\left(-\frac{18}{7}\right) + \left(-\frac{3}{7}\right)$

107. $-\frac{5}{4} \div \left(-\frac{3}{4}\right)$

108. $-\frac{5}{9} \div \left(-\frac{5}{6}\right)$

109. $-6.6 \div 3.3$

110. $-44.1 \div (-6.3)$

111. $\frac{-3}{7} - \frac{2}{7}$

112. $\frac{-5}{9} - \frac{2}{9}$

113. $\frac{-5}{9} + \frac{2}{-3}$

114. $\frac{-3}{10} + \frac{2}{-5}$

115. $\left(\frac{-3}{5}\right) \div \frac{6}{15}$

116. $\frac{7}{10} \div \left(\frac{-3}{5}\right)$

117. $\frac{4}{9} - \frac{1}{-9}$

118. $\frac{5}{7} - \frac{1}{-7}$

119. $\frac{3}{-10} + \frac{-1}{5}$

120. $\frac{-4}{15} + \frac{2}{-3}$

121. $\frac{-2}{3} - \frac{1}{-6}$

122. $\frac{-7}{10} - \frac{1}{-5}$

Skill Maintenance

123. Simplify: $\frac{264}{468}$.

124. Collect like terms: $x + 12y + 11x - 14y - 9$.

Synthesis

125. ◈ Most calculators have a key, often appearing as [1/x], for finding reciprocals. To use this key, enter a number and then press [1/x] to find its reciprocal. What should happen if you enter a number on a calculator and press the reciprocal key twice? Why?

126. ◈ What advice would you offer a student who claims to be able to multiply, but not divide, any two real numbers?

127. Determine those real numbers a for which the opposite of a is the same as the reciprocal of a.

128. Determine those real numbers that are their own reciprocals.

Tell whether the expression represents a positive number or a negative number when m and n are negative.

129. $\dfrac{-n}{m}$

130. $\dfrac{-n}{-m}$

131. $-\left(\dfrac{-n}{m}\right)$

132. $-\left(\dfrac{n}{-m}\right)$

133. $-\left(\dfrac{-n}{-m}\right)$

134. What must be true of m and n if $-mn$ is to be **(a)** positive? **(b)** zero? **(c)** negative?

135. The following is a proof that a positive number times a negative number is negative. Explain the reason for each step. Assume that $a > 0$ and $b > 0$.

$$\begin{aligned} a(-b) + ab &= a[-b + b] \\ &= a(0) \\ &= 0 \end{aligned}$$

Therefore, $a(-b) = -ab$.

136. ◈ Is it true that for any numbers a and b, if a is larger than b, then the reciprocal of a is smaller than the reciprocal of b? Why or why not?

1.8 Exponential Notation and Order of Operations

Exponential Notation • Order of Operations • Simplifying and the Distributive Law • The Opposite of a Sum

Algebraic expressions often contain *exponential notation*. In this section, we learn how to use exponential notation as well as rules for the *order of operations* in performing certain algebraic manipulations.

Exponential Notation

A product like $3 \cdot 3 \cdot 3 \cdot 3$, in which the factors are the same, is called a **power.** Powers occur often enough that a simpler notation called **exponential notation** is often used. For

$$\underbrace{3 \cdot 3 \cdot 3 \cdot 3}_{4 \text{ factors}}, \quad \text{we write} \quad 3^4.$$

This is read "three to the fourth power," or, simply, "three to the fourth." The number 4 is called an **exponent** and the number 3 a **base.**

Expressions like s^2 and s^3 are usually read "s squared" and "s cubed," respectively. This comes from the fact that a square of side s has an area A given by $A = s^2$ and a cube of side s has a volume V given by $V = s^3$.

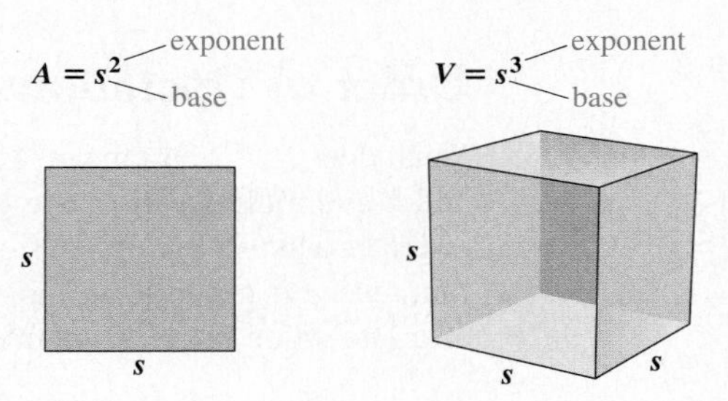

EXAMPLE 1

Write exponential notation for $10 \cdot 10 \cdot 10 \cdot 10 \cdot 10$.

Solution

Exponential notation is 10^5. 5 is the *exponent*. 10 is the *base*. ❑

EXAMPLE 2

Evaluate: **(a)** 5^2; **(b)** $(-5)^3$; **(c)** $(2n)^3$.

Solution

a) $5^2 = 5 \cdot 5 = 25$ The second power indicates two factors of 5.

b) $(-5)^3 = (-5)(-5)(-5)$

$= 25(-5)$ Using the associative law of multiplication

$= -125$

c) $(2n)^3 = (2n)(2n)(2n)$

$= 2 \cdot 2 \cdot 2 \cdot n \cdot n \cdot n$ Using the associative and commutative laws of multiplication

$= 8n^3$ ❑

Look for a pattern in the following:

$8 \cdot 8 \cdot 8 \cdot 8 = 8^4$

$8 \cdot 8 \cdot 8 = 8^3$ We divide by 8 each time.

$8 \cdot 8 = 8^2$

$8 = 8^?$.

The exponents decrease by 1 each time. To continue the pattern, we say that

$8 = 8^1$.

Exponential Notation

$b^1 = b$, for any number b.

For any natural number n greater than or equal to 2,

$$b^n \text{ means } \overbrace{b \cdot b \cdot b \cdot b \cdots b}^{n \text{ factors}}.$$

Order of Operations

What does $5 \times 2 + 4$ mean? If we multiply 5 by 2 and add 4, we get 14. If we add 2 and 4 and multiply by 5, we get 30. Since our results are different, we see that the order in which we carry out operations is important. To tell which operation to do first, we use grouping symbols such as parentheses ( ), brackets [ ], braces { }, absolute value bars | |, or fraction bars. For example,

$(3 \times 5) + 6 = 15 + 6 = 21$,

but

$3 \times (5 + 6) = 3 \times 11 = 33$.

Besides grouping symbols, there are rules for the order in which operations should be done.

Rules for Order of Operations

1. Perform all calculations within grouping symbols.
2. Evaluate all exponential expressions.
3. Perform all multiplications and divisions in order from left to right.
4. Perform all additions and subtractions in order from left to right.

EXAMPLE 3

Simplify: $15 - 2 \times 5 + 3$.

Solution When no groupings or exponents appear, we always multiply or divide before adding or subtracting:

$$15 - 2 \times 5 + 3 = 15 - 10 + 3 \quad \text{Multiplying}$$
$$\left.\begin{aligned} &= 5 + 3 \\ &= 8. \end{aligned}\right\} \quad \text{Subtracting and adding from left to right}$$

❑

Always calculate within parentheses first. When there are exponents and no parentheses, simplify powers before multiplying or dividing.

EXAMPLE 4

Simplify: **(a)** $(3 \times 4)^2$; **(b)** 3×4^2.

Solution

a) $(3 \times 4)^2 = (12)^2$ Working within parentheses first
$= 144$

b) $3 \times 4^2 = 3 \times 16$ Simplifying the power
$= 48$ Multiplying

Note that $(3 \times 4)^2 \neq 3 \times 4^2$. ❑

CAUTION! Example 4 illustrates that, in general, $(ab)^2 \neq ab^2$.

EXAMPLE 5

Evaluate for $x = 5$: **(a)** $(-x)^2$; **(b)** $-x^2$.

Solution

a) $(-x)^2 = (-5)^2 = (-5)(-5) = 25$ Substitute 5 for x, take the opposite, and then evaluate the power.

b) $-x^2 = -(5)^2 = -25$ Substitute 5 for x. Evaluate the power. Then find the opposite. ❑

CAUTION! Example 5 illustrates that, in general, $(-x)^2 \neq -x^2$.

EXAMPLE 6

Evaluate $-15 \div 3(6 - a)^3$ for $a = 4$.

Solution

$$-15 \div 3(6 - a)^3 = -15 \div 3(6 - 4)^3 \quad \text{Substituting 4 for } a$$
$$= -15 \div 3(2)^3 \quad \text{Working within parentheses first}$$
$$= -15 \div 3 \cdot 8 \quad \text{Simplifying the exponential expression}$$
$$\left.\begin{aligned} &= -5 \cdot 8 \\ &= -40 \end{aligned}\right\} \quad \text{Dividing and multiplying from left to right}$$

❑

When combinations of grouping symbols are used, the rules still apply. We begin with the innermost grouping symbols and work to the outside.

EXAMPLE 7

Simplify: $16 \div (-2) + 3[10 + 2(3 - 5)^3]$.

Solution

$16 \div (-2) + 3[10 + 2(3 - 5)^3]$	
$= 16 \div (-2) + 3[10 + 2(-2)^3]$	Doing the calculations in the innermost parentheses first
$= 16 \div (-2) + 3[10 + 2(-8)]$	$(-2)^3 = (-2)(-2)(-2) = -8$
$= 16 \div (-2) + 3[10 + (-16)]$	
$= 16 \div (-2) + 3[-6]$	Completing the calculations within the brackets
$= -8 + (-18)$	Multiplying and dividing from left to right
$= -26$	❑

EXAMPLE 8

Calculate: $\dfrac{12(9 - 7) + 4 \cdot 5}{3^4 + 2^3}$.

Solution An equivalent expression with brackets as grouping symbols is

$$[12(9 - 7) + 4 \cdot 5] \div [3^4 + 2^3].$$

What this shows, in effect, is that we do the calculations in the numerator and then in the denominator, and divide the results:

$$\frac{12(9 - 7) + 4 \cdot 5}{3^4 + 2^3} = \frac{12(2) + 4 \cdot 5}{81 + 8} = \frac{24 + 20}{89} = \frac{44}{89}.$$

❑

Simplifying and the Distributive Law

Sometimes we cannot simplify within parentheses. When a sum or difference is within the parentheses, we can use the distributive law to help simplify.

EXAMPLE 9

Simplify: $5x - 9 + 2(4x + 5)$.

Solution

$5x - 9 + 2(4x + 5) = 5x - 9 + 8x + 10$	Using the distributive law
$= 13x + 1$	Collecting like terms ❑

Now that exponents have been introduced, we can make our definition of *like* or *similar terms* more precise. **Like, or similar, terms** are either constant terms or terms containing the same variable(s) raised to the same power(s). Thus, 5 and -7, $19xy$ and $-xy$, and $4a^3b$ and $7a^3b$ are all pairs of like terms.

EXAMPLE 10

Simplify: $7x^2 + 3(x^2 + 2x) - 5x$.

Solution

$$7x^2 + 3(x^2 + 2x) - 5x = 7x^2 + 3x^2 + 6x - 5x \quad \text{Using the distributive law}$$
$$= 10x^2 + x \quad \text{Collecting like terms}$$ ❑

The Opposite of a Sum

Multiplication by -1 changes a number's sign. For example, $-1(7) = -7$ and $-1(-5) = 5$. Thus, when a number is multiplied by -1, we get the opposite of that number.

The Property of -1

For any real number a,

$$-1 \cdot a = -a.$$

(Negative one times a is the opposite of a.)

When grouping symbols are preceded by a "$-$" symbol, we can multiply the grouping by -1 and use the distributive law. In this manner, we can find the *opposite*, or *additive inverse, of a sum*.

EXAMPLE 11

Write an expression equivalent to $-(3x + 2y + 4)$ without using parentheses.

Solution

$$-(3x + 2y + 4) = -1(3x + 2y + 4) \quad \text{Using the property of } -1$$
$$= -1(3x) + (-1)(2y) + (-1)4 \quad \text{Using the distributive law}$$
$$= -3x - 2y - 4 \quad \text{Using the property of } -1$$ ❑

Example 11 illustrates an important property of real numbers.

The Opposite of a Sum

For any real numbers a and b,

$$-(a + b) = -a + (-b).$$

(The opposite of a sum is the sum of the opposites.)

To remove parentheses from an expression like $-(x - 7y + 5)$, we can first rewrite the subtraction as addition:

$$-(x - 7y + 5) = -(x + (-7y) + 5) \quad \text{Rewriting as addition}$$
$$= -x + 7y - 5. \quad \text{Taking the opposite of a sum}$$

This procedure is normally streamlined to one step in which we find the opposite by "removing parentheses and changing the sign of every term":

$$-(x - 7y + 5) = -x + 7y - 5.$$

EXAMPLE 12

Simplify: $3x - (4x + 2)$.

Solution

$3x - (4x + 2) = 3x + [-(4x + 2)]$	Adding the opposite of $(4x + 2)$
$= 3x + [-4x - 2]$	Changing the sign of each term inside the parentheses
$= 3x - 4x - 2$	
$= -x - 2$	Collecting like terms

❑

In practice, the first two steps of Example 12 are often skipped.

EXAMPLE 13

Simplify: **(a)** $5y - (3y + 4)$; **(b)** $3y - 2 - (2y - 4)$.

Solution

a)

$5y - (3y + 4) = 5y - 3y - 4$	Removing parentheses by changing the sign of every term inside the parentheses
$= 2y - 4$	Collecting like terms

b) $3y - 2 - (2y - 4) = 3y - 2 - 2y + 4$
$= y + 2$

❑

Expressions such as $7 - 3(x + 2)$ can be simplified as follows:

$7 - 3(x + 2) = 7 + [-3(x + 2)]$	Adding the opposite of $3(x + 2)$
$= 7 + [-3x - 6]$	Multiplying $x + 2$ by -3
$= 7 - 3x - 6$	Try to go directly to this step.
$= 1 - 3x.$	Collecting like terms

EXAMPLE 14

Simplify: **(a)** $3y - 2(4y - 5)$; **(b)** $2a + 3b - 7 - 4(-5a - 6b + 12)$.

Solution

a)

$3y - 2(4y - 5) = 3y - 8y + 10$	Multiplying each term in the parentheses by -2
$= -5y + 10$	

b) $2a + 3b - 7 - 4(-5a - 6b + 12) = 2a + 3b - 7 + 20a + 24b - 48$
$= 22a + 27b - 55$

❑

EXERCISE SET 1.8

Write exponential notation.

1. $10 \times 10 \times 10$

2. $6 \times 6 \times 6 \times 6$

3. $x \cdot x \cdot x \cdot x \cdot x \cdot x \cdot x$

4. $y \cdot y \cdot y \cdot y \cdot y \cdot y$

5. $3y \cdot 3y \cdot 3y \cdot 3y$

6. $5m \cdot 5m \cdot 5m \cdot 5m \cdot 5m$

Simplify.

7. 2^4 **8.** 5^3 **9.** $(-3)^2$

10. $(-7)^2$ **11.** 1^5 **12.** $(-1)^5$

13. 4^3
14. 9^1
15. $(-4)^3$
16. 5^4
17. 7^1
18. 1^7
19. $(4a)^2$
20. $(3x)^2$
21. $(-7x)^3$
22. $(-5x)^4$
23. $7 + 2 \times 6$
24. $11 + 4 \times 4$
25. $8 \times 7 + 6 \times 5$
26. $10 \times 5 + 1 \times 1$
27. $19 - 5 \times 3 + 3$
28. $14 - 2 \times 6 + 7$
29. $9 \div 3 + 16 \div 8$
30. $32 - 8 \div 4 - 2$
31. $7 + 10 - 10 \div 2$
32. $(2 - 5)^2$
33. $(3 - 5)^3$
34. $3 \cdot 2^3$
35. $8 - 2 \cdot 3 - 9$
36. $8 - (2 \cdot 3 - 9)$
37. $(8 - 2 \cdot 3) - 9$
38. $(8 - 2)(3 - 9)$
39. $(-24) \div (-3) \cdot \left(-\frac{1}{2}\right)$
40. $32 \div (-2) \cdot (-2)$
41. $16 \cdot (-24) + 50$
42. $10 \cdot 20 - 15 \cdot 24$
43. $2^4 + 2^3 - 10$
44. $40 - 3^2 - 2^3$
45. $5^3 + 26 \cdot 71 - (16 + 25 \cdot 3)$
46. $4^3 + 10 \cdot 20 + 8^2 - 23$
47. $[2 \cdot (5 - 3)]^2$
48. $5^3 - 7^2$
49. $\dfrac{7 + 2}{5^2 - 4^2}$
50. $\dfrac{5^2 - 3^2}{2 \cdot 6 - 4}$
51. $8(-7) + |6(-5)|$
52. $|10(-5)| + 1(-1)$
53. $19 - 5(-3) + 3$
54. $14 - 2(-6) + 7$
55. $9 \div (-3) \cdot 16 \div 8$
56. $-32 - 8 \div 4 \cdot (-2)$
57. $20 + 4^3 \div (-8) \cdot 2$
58. $2 \times 10^3 - 5000$
59. $8|(6 - 13) - 11|$
60. $6|9 - (3 - 4)|$
61. $256 \div (-32) \div (-4)$
62. $-1000 \div (-100) \div 10$
63. $\dfrac{5^2 - 4^3 - 3}{9^2 - 2^2 - 1^5}$
64. $\dfrac{3(6 - 7) - 5 \cdot 4}{6 \cdot 7 - 8(4 - 1)}$
65. $\dfrac{20(8 - 3) - 4(10 - 3)}{10(2 - 6) - 2(5 + 2)}$
66. $\dfrac{2^3 - 3^2 + 12 \cdot 5}{-32 \div (-16) \div (-4)}$

Evaluate.

67. $7 - 3x$, when $x = 5$
68. $9 - x^2$, when $x = 4$
69. $a \div 6 \cdot 2$, when $a = 12$
70. $10 \div a \cdot 5$, when $a = 2$
71. $-20 \div t^2 - 3(t - 1)$, when $t = -4$
72. $-30 \div t(t + 4)^2$, when $t = -6$
73. $-x^2 - 5x$, when $x = -3$
74. $(-x)^2 - 5x$, when $x = -3$

Rename the expression without using parentheses.

75. $-(2x + 7)$
76. $-(3x + 5)$
77. $-(5x - 8)$
78. $-(6x - 7)$
79. $-(4a - 3b + 7c)$
80. $-(5x - 2y - 3z)$
81. $-(3x^2 + 5x - 1)$
82. $-(8x^3 - 6x + 5)$

Remove parentheses and simplify.

83. $9x - (4x + 3)$
84. $7y - (2y + 9)$
85. $2a - (5a - 9)$
86. $11n - (3n - 7)$
87. $2x + 7x - (4x + 6)$
88. $3a + 2a - (4a + 7)$
89. $2x - 4y - 3(7x - 2y)$
90. $3a - 7b - 1(4a - 3b)$
91. $15x - y - 5(3x - 2y + 5z)$
92. $4a - b - 4(5a - 7b + 8c)$
93. $3x^2 + 7 - (2x^2 + 5)$
94. $7x^4 + 9x - (5x^4 + 3x)$
95. $9x^3 + x - 2(x^3 + 3x)$
96. $-7x^2 + 5x - 3(x^2 - 4x)$
97. $12a^2 - 3ab + 5b^2 - 5(-5a^2 + 4ab - 6b^2)$
98. $-8a^2 + 5ab - 12b^2 - 6(2a^2 - 4ab - 10b^2)$
99. $-7t^3 - t^2 - 3(5t^3 - 3t)$
100. $9t^4 + 7t - 5(9t^3 - 2t)$
101. $[10(x + 3) - 4] + [2(x - 1) + 6]$
102. $[9(x + 5) - 7] + [4(x - 12) + 9]$
103. $[7(x^2 + 5) - 19] - [4(x^2 - 6) + 10]$
104. $[6(x^3 + 4) - 12] - [5(x^3 - 8) + 11]$
105. $3\{[7(x - 2) + 4] - [2(2x - 5) + 6]\}$
106. $4\{[8(x - 3) + 9] - [4(3x - 7) + 2]\}$
107. $4\{[5(x^3 - 3) + 2] - 3[2(x^3 + 5) - 9]\}$
108. $3\{[6(x^2 - 4) + 5] - 2[5(x^2 + 8) - 10]\}$

Skill Maintenance

Translate to an algebraic expression.

109. Nine more than twice a number
110. Half of the sum of two numbers

Synthesis

111. ◈ Write the sentence $(-x)^2 \neq -x^2$ in words. Explain why $(-x)^2$ and $-x^2$ are not equivalent.

112. ◈ Write the sentence $-|x| \neq -x$ in words. Explain why $-|x|$ and $-x$ are not equivalent.

Simplify.

113. $z - \{2z - [3z - (4z - 5z) - 6z] - 7z\} - 8z$

114. $\{x - [f - (f - x)] + [x - f]\} - 3x$

115. $x - \{x - 1 - [x - 2 - (x - 3 - \{x - 4 - [x - 5 - (x - 6)]\})]\}$

116. ◈ Determine whether it is true that, for any real numbers a and b, $ab = (-a)(-b)$. Explain why or why not.

117. ◈ Determine whether it is true that, for any real numbers a and b, $-(ab) = (-a)b = a(-b)$. Explain why or why not.

If $n > 0$, $m > 0$, and $n \neq m$, determine whether each of the following is true.

118. $-n + m = n - m$

119. $-n + m = -(n + m)$

120. $-n - m = -(n + m)$

121. $-n - m = -(n - m)$

122. $n(-n - m) = -n^2 + nm$

123. $-m(n - m) = -(mn + m^2)$

124. $-m(-n + m) = m(n - m)$

125. $-n(-n - m) = n(n + m)$

SUMMARY AND REVIEW 1

KEY TERMS

Variable, p. 2
Constant, p. 2
Variable expression, p. 2
Evaluate an expression, p. 2
Algebraic expression, p. 3
Substitute, p. 3
Value of an expression, p. 3
Equation, p. 5
Equivalent expressions, p. 9
Term, p. 12
Factors, p. 12
Natural number, p. 14
Whole number, p. 14
Factorization, p. 15
Prime number, p. 15
Composite number, p. 15
Prime factorization, p. 15
Fractional notation, p. 15
Numerator, p. 15
Denominator, p. 15
Simplifying, p. 17
Reciprocal, p. 19
Set, p. 22
Subset, p. 22
Integer, p. 22
Negative integer, p. 23
Positive integer, p. 23
Rational number, p. 24
Terminating decimal, p. 25
Irrational number, p. 25
Real-number system, p. 26
Less than, p. 27
Greater than, p. 27
Inequality, p. 27
Absolute value, p. 28
Collect like terms, p. 33
Opposite, p. 36
Undefined, p. 46
Indeterminate, p. 46
Power, p. 49
Exponential notation, p. 49
Exponent, p. 49
Base, p. 49
Like terms, p. 52

IMPORTANT PROPERTIES AND FORMULAS

Area of a rectangle:	$A = lw$
Area of a triangle:	$A = \frac{1}{2}bh$
Area of a parallelogram:	$A = bh$
Commutative laws:	$a + b = b + a, \quad ab = ba$

Associative laws:	$a + (b + c) = (a + b) + c,$	$a(bc) = (ab)c$
Distributive law:	$a(b + c) = ab + ac$	
Identity property of 1:	$1 \cdot a = a \cdot 1 = a$	
Law of opposites:	$a + (-a) = 0$	
Multiplicative property of 0:	$0 \cdot a = a \cdot 0 = 0$	

$$\frac{-a}{b} = \frac{a}{-b} = -\frac{a}{b}, \quad \frac{-a}{-b} = \frac{a}{b}$$

Property of -1:	$-1 \cdot a = -a$
Opposite of a sum:	$-(a + b) = -a + (-b)$

Rules for Order of Operations

1. Perform all calculations within grouping symbols.
2. Evaluate all exponential expressions.
3. Perform all multiplications and divisions in order from left to right.
4. Perform all additions and subtractions in order from left to right.

REVIEW EXERCISES

Evaluate.

1. $3a$, when $a = 5$
2. $\frac{x}{y}$, when $x = 12$ and $y = 2$
3. $\frac{2 \cdot p}{q}$, when $p = 20$ and $q = 8$
4. $\frac{x - y}{3}$, when $x = 17$ and $y = 5$
5. $10 - y^2$, when $y = 5$
6. $-10 + a^2 \div (b + 1)$, when $a = 5$ and $b = 4$

Translate to an algebraic expression.

7. 8 less than z
8. Three times x
9. One-third of y

10. Determine whether 35 is a solution of $x/5 = 8$.
11. Translate to an equation: Six times what number is 6768?
12. Use the commutative law of addition to write an expression equivalent to $2x + y$.
13. Use the commutative law of multiplication to write an expression equivalent to $2x + y$.
14. Use the associative law of addition to write an expression equivalent to $(2x + y) + z$.
15. Use the commutative and associative laws to write three expressions equivalent to $4(xy)$.

Multiply.

16. $6(3x + 5y)$
17. $8(5x + 3y + 2)$

Factor.

18. $21x + 7y$
19. $35x + 14 + 7y$

Simplify.

20. $\frac{20}{48}$
21. $\frac{180}{18}$

Perform the indicated operation and simplify.

22. $\frac{4}{9} + \frac{5}{12}$
23. $\frac{3}{4} \div 3$
24. $\frac{2}{3} - \frac{1}{15}$
25. $\frac{9}{10} \cdot \frac{16}{5}$

26. Tell which integers correspond to this situation: Renir has a debt of $45 and Raoul has $72 in his savings account.
27. Translate to mathematical language: A bowling score is at most 300.
28. Graph on a number line: $\frac{-1}{3}$.
29. Write an inequality with the same meaning as $-3 < x$.
30. Write true or false: $0 \leq -1$.
31. Find decimal notation: $-\frac{7}{8}$.

32. Find the absolute value: $|-1|$.

33. Find $-(-x)$ when x is -5.

Simplify.

34. $4 + (-7)$

35. $-\frac{2}{3} + \frac{1}{12}$

36. $6 + (-9) + (-8) + 7$

37. $-3.8 + 5.1 + (-12) + (-4.3) + 10$

38. $-3 - (-7)$

39. $-\frac{9}{10} - \frac{1}{2}$

40. $-3.8 - 4.1$

41. $-9 \cdot (-6)$

42. $-2.7(3.4)$

43. $\frac{2}{3} \cdot \left(-\frac{3}{7}\right)$

44. $3 \cdot (-7) \cdot (-2) \cdot (-5)$

45. $35 \div (-5)$

46. $-5.1 \div 1.7$

47. $-\frac{3}{5} \div \left(-\frac{4}{5}\right)$

48. $|-3 \cdot 4 - 12 \cdot 2| - 8(-7)$

49. $|-12(-3) - 2^3 - (-9)(-10)|$

50. $120 - 6^2 \div 4 \cdot 8$

51. $(120 - 6^2) \div 4 \cdot 8$

52. $(120 - 6^2) \div (4 \cdot 8)$

53. $\dfrac{4(18 - 8) + 7 \cdot 9}{9^2 - 8^2}$

Collect like terms.

54. $11a + 2b + (-4a) + (-5b)$

55. $7x - 3y - 9x + 8y$

56. Find the opposite of -7.

57. Find the reciprocal of -7.

58. Write exponential notation for $2x \cdot 2x \cdot 2x \cdot 2x$.

59. Simplify: $(-3y)^3$.

Remove parentheses and simplify.

60. $2a - (5a - 9)$

61. $3(b + 7) - 5b$

62. $3[11x - 3(4x - 1)]$

63. $2[6(y - 4) + 7]$

64. $[8(x + 4) - 10] - [3(x - 2) + 4]$

65. $5\{[6(x - 1) + 7] - [3(3x - 4) + 8]\}$

Synthesis

66. Explain at least three uses of the distributive law considered in this chapter.

67. Devise a rule for determining the sign of a negative quantity raised to a power.

68. Evaluate $a^{50} - 20a^{25}b^4 + 100b^8$ when $a = 1$ and $b = 2$.

69. If $0.090909\ldots = \frac{1}{11}$ and $0.181818\ldots = \frac{2}{11}$, what rational number is named by each of the following?

a) $0.272727\ldots$ **b)** $0.909090\ldots$

Simplify.

70. $-\left|\frac{7}{8} - \left(-\frac{1}{2}\right) - \frac{3}{4}\right|$

71. $(|2.7 - 3| + 3^2 - |-3|) \div (-3)$

CHAPTER TEST 1

1. Evaluate $\dfrac{3x}{y}$ when $x = 10$ and $y = 5$.

2. Write an algebraic expression: Nine less than some number.

3. Find the area of a triangle when the height h is 30 ft and the base b is 16 ft.

4. Use the commutative law of addition to write an expression equivalent to $3p + q$.

5. Use the associative law of multiplication to write an expression equivalent to $x \cdot (4 \cdot y)$.

Multiply.

6. $3(6 - x)$

7. $-5(y - 1)$

Factor.

8. $11 - 44x$

9. $7x + 21 + 14y$

10. Find the prime factorization of 300.

Write a true sentence using either $<$ or $>$.

11. $-4 \blacksquare 0$

12. $-3 \blacksquare -8$

13. $-0.78 \blacksquare -0.87$

14. $-\frac{1}{8} \blacksquare \frac{1}{2}$

Find the absolute value.

15. $|-7|$

16. $\left|\frac{9}{4}\right|$

17. $|-2.7|$

18. Find the opposite of $\frac{2}{3}$.

19. Find the reciprocal of $-\frac{4}{7}$.

20. Find $-x$ when x is -8.

21. Write an inequality with the same meaning as $x \leqslant -2$.

Compute and simplify.

22. $3.1 - (-4.7)$

23. $-8 + 4 + (-7) + 3$

24. $-\frac{1}{5} + \frac{3}{8}$

25. $2 - (-8)$

26. $3.2 - 5.7$

27. $\frac{1}{8} - \left(-\frac{3}{4}\right)$

28. $4 \cdot (-12)$

29. $-\frac{1}{2} \cdot \left(-\frac{3}{8}\right)$

30. $-45 \div 5$

31. $-\frac{3}{5} \div \left(-\frac{4}{5}\right)$

32. $4.864 \div (-0.5)$

33. $-2(16) - |2(-8) - 5^3|$

34. $6 + 7 - 4 - (-3)$

35. $256 \div (-16) \div 4$

36. $2^3 - 10[4 - (-2 + 18)3]$

37. Collect like terms: $18y + 30a - 9a + 4y$.

38. Simplify: $(-2x)^4$.

Remove parentheses and simplify.

39. $5x - (3x - 7)$

40. $4(2a - 3b) + a - 7$

41. $4\{3[5(y - 3) + 9] + 2(y + 8)\}$

Synthesis

42. Evaluate $\dfrac{5y - x}{4}$ when $x = 20$ and y is 4 less than x.

Simplify.

43. $\dfrac{13{,}800}{42{,}000}$

44. $|-27 - 3(4)| - |-36| + |-12|$

45. $a - \{3a - [4a - (2a - 4a)]\}$

CHAPTER 2

Equations, Inequalities, and Problem Solving

AN APPLICATION

A rectangular community vegetable garden is to be enclosed with 92 m of wooden fencing. In order to allow for compost storage, the garden must be 4 m longer than it is wide. Determine the dimensions of the garden.

This problem appears as Example 5 in Section 2.5.

Robert E. Romero
ZONING ENFORCEMENT MANAGER

"My primary responsibility is to enforce city ordinances relating to the development and use of private property. I use mathematics every day—for example, when determining lot size for irregularly shaped lots, solar access angles for tall buildings, parking layouts, and area available for signage."

Solving equations and inequalities is a recurring theme in much of mathematics. In this chapter, we will study some of the principles used to solve equations and inequalities. Then we will use equations and inequalities to solve applied problems.

In addition to material from this chapter, the review and test for Chapter 2 include material from Sections 1.1, 1.2, 1.7, and 1.8.

2.1 Solving Equations

Equations and Solutions • The Addition Principle • The Multiplication Principle

Solving equations is essential for problem solving in algebra. In this section, we study two of the most important principles used to solve equations.

Equations and Solutions

We have already seen that an equation is a number sentence stating that the expressions on either side of the equals sign represent the same number. Some equations, like $3 + 2 = 5$ or $2x + 6 = 2(x + 3)$, are *always* true and some, like $3 + 2 = 6$ or $x + 2 = x + 3$, are *never* true. In this text, we will concentrate on equations like $x + 6 = 13$ or $7x = 141$ that are either true or false, depending on the replacement value.

Solution of an Equation

Any replacement for the variable that makes an equation true is called a *solution* of the equation. To solve an equation means to find *all* of its solutions.

One way to determine whether a number is a solution of an equation is to evaluate the algebraic expression on each side of the equation by substitution. If the values are the same, then the number is a solution.

EXAMPLE 1

Determine whether 7 is a solution of $x + 6 = 13$.

Solution We have

$x + 6 = 13$		Writing the equation
$7 + 6 \ ?\ 13$		Substituting 7 for x
13	13 TRUE	

Since the left-hand and the right-hand sides are the same, 7 is a solution. ❑

EXAMPLE 2

Determine whether 19 is a solution of $7x = 141$.

Solution We have

$7x = 141$	Writing the equation
$7(19) \ ? \ 141$	Substituting 19 for x
$133 \mid 141$ FALSE	

Since the left-hand and the right-hand sides are not the same, 19 is not a solution. ❑

The Addition Principle

Consider the equation

$$x = 7.$$

We can easily see that the solution of this equation is 7. If we replace x by 7, we get

$$7 = 7, \quad \text{which is true.}$$

Now consider the equation of Example 1:

$$x + 6 = 13.$$

There we discovered that the solution of $x + 6 = 13$ is also 7, but the solution of $x = 7$ seems more obvious. We now begin to consider principles that allow us to start with one equation and end up with an equation like $x = 7$, in which the variable is alone on one side and for which the solution is easy to see. The equations $x + 6 = 13$ and $x = 7$ are **equivalent.**

Equivalent Equations

Equations with the same solutions are called *equivalent equations.*

One of the principles that we use in solving equations concerns adding. An equation $a = b$ says that a and b stand for the same number. Suppose this is true, and we add a number c to the number a. We get the same answer if we add c to b, because a and b are the same number.

The Addition Principle

For any real numbers a, b, and c,

$$\text{if } a = b, \quad \text{then} \quad a + c = b + c.$$

To visualize the addition principle, consider a balance similar to one a jeweler might use. When the two sides of the balance hold

quantities of equal weight, the balance is level. If weight is added or removed, equally, on both sides, the balance will remain level.

When we use the addition principle, we sometimes say that we "add the same number on both sides of an equation." This is also true for subtraction, since every subtraction can be regarded as the addition of an opposite.

EXAMPLE 3

Solve: $x + 5 = -7$.

Solution

$$x + 5 = -7$$

$$x + 5 - 5 = -7 - 5$$ Using the addition principle: adding -5 on both sides or subtracting 5 on both sides

$$x + 0 = -12$$ Simplifying

$$x = -12$$ Identity property of 0

It is obvious that the solution of $x = -12$ is the number -12. To check the answer in the original equation, we substitute.

Check:

$$\begin{array}{r|l} x + 5 & = -7 \\ \hline -12 + 5 & ?\ -7 \\ -7 & -7 \end{array}$$ TRUE

The solution of the original equation is -12. ❑

In Example 3, to get x alone, we used the addition principle and subtracted 5 on both sides. This eliminated the 5 on the left and produced a simpler, but equivalent, equation, $x = -12$.

Next we use the addition principle to solve a subtraction problem.

EXAMPLE 4

Solve: $-6.5 = y - 8.4$.

Solution

$$-6.5 = y - 8.4$$

$$-6.5 + 8.4 = y - 8.4 + 8.4$$ Using the addition principle: adding 8.4 on both sides eliminates -8.4 on the right

$$1.9 = y$$

Check:

$$\begin{array}{r|l} -6.5 & = y - 8.4 \\ \hline -6.5 & ?\ 1.9 - 8.4 \\ -6.5 & -6.5 \end{array}$$ TRUE

The solution is 1.9. ❑

Note that the equations $a = b$ and $b = a$ have the same meaning. Thus to solve $-6.5 = y - 8.4$, we can reverse it and solve $y - 8.4 = -6.5$ if we wish.

EXAMPLE 5

Solve: $-\frac{2}{3} + x = \frac{5}{2}$.

Solution

$$-\tfrac{2}{3} + x = \tfrac{5}{2}$$

$$-\tfrac{2}{3} + x + \tfrac{2}{3} = \tfrac{5}{2} + \tfrac{2}{3} \qquad \text{Adding } \tfrac{2}{3}$$

$$x = \tfrac{5}{2} + \tfrac{2}{3}$$

$$x = \tfrac{5}{2} \cdot \tfrac{3}{3} + \tfrac{2}{3} \cdot \tfrac{2}{2} \qquad \text{Multiplying by 1 to obtain a common denominator}$$

$$x = \tfrac{15}{6} + \tfrac{4}{6}$$

$$x = \tfrac{19}{6}$$

The check is left to the student. The solution is $\frac{19}{6}$. ❑

The Multiplication Principle

An equation like $\frac{5}{4}x = \frac{3}{8}$ says that $\frac{5}{4}x$ and $\frac{3}{8}$ represent the same number. Because of this, if $\frac{5}{4}x$ and $\frac{3}{8}$ are both multiplied by some number c, the products $c \cdot \frac{5}{4}x$ and $c \cdot \frac{3}{8}$ will represent the same number.

The Multiplication Principle

For any real numbers a, b, and c,

if $a = b$, then $c \cdot a = c \cdot b$.

EXAMPLE 6

Solve: $\frac{5}{4}x = \frac{3}{8}$.

Solution

$$\tfrac{5}{4}x = \tfrac{3}{8} \qquad \text{Note that the reciprocal of } \tfrac{5}{4} \text{ is } \tfrac{4}{5} \text{ and that } \tfrac{4}{5} \cdot \tfrac{5}{4} = 1.$$

$$\tfrac{4}{5} \cdot \tfrac{5}{4}x = \tfrac{4}{5} \cdot \tfrac{3}{8} \qquad \text{Using the multiplication principle: multiplying on both sides by } \tfrac{4}{5} \text{ eliminates } \tfrac{5}{4} \text{ on the left}$$

$$1 \cdot x = \tfrac{3}{10} \qquad \text{Simplifying}$$

$$x = \tfrac{3}{10} \qquad \text{Using the identity property of 1: } 1 \cdot x = x$$

Check:

$\frac{5}{4}x = \frac{3}{8}$	
$\frac{5}{4}\left(\frac{3}{10}\right)$?	$\frac{3}{8}$
$\frac{3}{8}$	$\frac{3}{8}$ TRUE

The solution is $\frac{3}{10}$. ❑

In Example 6, to get x alone, we multiplied by the *multiplicative inverse,* or *reciprocal,* of $\frac{5}{4}$. When we multiplied, we got the *multiplicative identity,* 1, times x, which simplified to x. This enabled us to eliminate the $\frac{5}{4}$ on the left.

Because division is the same as multiplying by a reciprocal, the multiplication principle also tells us that we can "divide on both sides by the same nonzero number." That is,

$$\text{if } a = b, \text{ then } \frac{1}{c} \cdot a = \frac{1}{c} \cdot b \quad \text{and} \quad \frac{a}{c} = \frac{b}{c} \quad (\text{provided } c \neq 0).$$

In an expression like $3x$, the number 3 is called the **coefficient.** In practice, it is usually more convenient to divide on both sides of an equation if the coefficient of the variable is in decimal notation or is an integer. When the coefficient is in fractional notation, it is more convenient to multiply by a reciprocal.

EXAMPLE 7

Solve: **(a)** $-4x = 92$; **(b)** $12.6 = 3x$; **(c)** $-x = 9$.

Solution

a) $-4x = 92$

$\frac{-4x}{-4} = \frac{92}{-4}$ Using the multiplication principle. Dividing on both sides by -4 is the same as multiplying by $-\frac{1}{4}$.

$1 \cdot x = -23$ Simplifying

$x = -23$ Using the identity property of 1

Check:

$$\begin{array}{r|l} -4x & 92 \\ \hline -4(-23) & ?\ 92 \\ 92 & 92 \end{array}$$ TRUE

The solution is -23.

b) $12.6 = 3x$

$\frac{12.6}{3} = \frac{3x}{3}$ Dividing on both sides by 3 or multiplying by $\frac{1}{3}$

$4.2 = 1x$

$4.2 = x$ Simplifying

Check:

$$\begin{array}{r|l} 12.6 & 3x \\ \hline 12.6 & ?\ 3 \cdot 4.2 \\ 12.6 & 12.6 \end{array}$$ TRUE

The solution is 4.2.

c) To solve an equation like $-x = 9$, remember that when an expression is multiplied or divided by -1, its sign is changed. Here we multiply on both sides by -1:

$-x = 9$

$(-1)(-x) = (-1) \cdot 9$ Multiplying by -1 on both sides

$x = -9.$ Note that $(-1)(-x)$ is the same as $(-1)(-1)x$.

Check:

$$\begin{array}{r|l} -x & 9 \\ \hline -(-9) & ?\ 9 \\ 9 & 9 \end{array}$$ TRUE

The solution is -9. ❑

Consider an equation like $y/(-9) = 14$. The left side can be rewritten as $(1/-9) \cdot y$. Using the multiplication principle, we can multiply on both sides by -9 to solve, as in the following example.

EXAMPLE 8

Solve: $\frac{y}{-9} = 14$.

Solution

$$\frac{y}{-9} = 14$$

$$\frac{1}{-9} \cdot y = 14 \qquad \text{Rewriting division as multiplication}$$

$$-9\left(\frac{1}{-9}\right)y = -9 \cdot 14 \qquad \text{Multiplying by } -9 \text{ on both sides}$$

$$y = -126 \qquad \text{Simplifying}$$

Check:

$\frac{y}{-9} = 14$	
$\frac{-126}{-9}$	? 14
14	14 TRUE

The solution is -126. ❑

EXERCISE SET 2.1

Solve using the addition principle. Don't forget to check!

1. $x + 2 = 6$
2. $x + 5 = 8$
3. $x + 15 = -5$
4. $y + 9 = 43$
5. $x + 6 = -8$
6. $t + 9 = -12$
7. $-2 = x + 16$
8. $-6 = y + 25$
9. $x - 9 = 6$
10. $x - 8 = 5$
11. $x - 7 = -21$
12. $x - 3 = -14$
13. $5 + t = 7$
14. $8 + y = 12$
15. $13 = -7 + y$
16. $15 = -9 + z$
17. $-3 + t = -9$
18. $-6 + y = -21$
19. $r + \frac{1}{3} = \frac{8}{3}$
20. $t + \frac{3}{8} = \frac{5}{8}$
21. $m + \frac{5}{6} = -\frac{11}{12}$
22. $x + \frac{2}{3} = -\frac{5}{6}$
23. $x - \frac{5}{6} = \frac{7}{8}$
24. $y - \frac{3}{4} = \frac{5}{6}$
25. $-\frac{1}{5} + z = -\frac{1}{4}$
26. $-\frac{1}{8} + y = -\frac{3}{4}$
27. $x + 2.3 = 7.4$
28. $y + 4.6 = 9.3$
29. $-9.7 = -4.7 + y$
30. $-7.8 = 2.8 + x$

Solve using the multiplication principle. Don't forget to check!

31. $6x = 36$
32. $3x = 39$
33. $5x = 45$
34. $9x = 72$
35. $84 = 7x$
36. $56 = 8x$
37. $-x = 40$
38. $100 = -x$
39. $-x = -1$
40. $-68 = -r$
41. $7x = -49$
42. $9x = -36$
43. $-12x = 72$
44. $-15x = 105$
45. $-21x = -126$
46. $-13x = -104$
47. $\frac{t}{7} = -9$
48. $\frac{y}{-8} = 11$
49. $\frac{3}{4}x = 27$
50. $\frac{4}{5}x = 16$
51. $\frac{-t}{3} = 7$
52. $\frac{-x}{6} = 9$
53. $\frac{1}{5} = -\frac{m}{3}$
54. $\frac{1}{9} = -\frac{z}{7}$

55. $-\frac{3}{5}r = -\frac{9}{10}$

56. $-\frac{2}{5}y = -\frac{4}{15}$

57. $\frac{-3r}{2} = -\frac{27}{4}$

58. $\frac{5x}{7} = -\frac{10}{14}$

59. $6.3x = 44.1$

60. $2.7y = 54$

Solve.

61. $3.7 + t = 8.2$

62. $\frac{3}{4}x = 18$

63. $18 = -\frac{2}{3}x$

64. $t - 7.4 = -12.9$

65. $17 = y + 29$

66. $96 = -\frac{3}{4}t$

67. $y - \frac{2}{3} = -\frac{1}{6}$

68. $-\frac{x}{7} = \frac{2}{9}$

69. $-24 = \frac{8x}{5}$

70. $\frac{1}{5} + y = -\frac{3}{10}$

71. $-4.1t = 10.25$

72. $\frac{19}{23} = -x$

Skill Maintenance

Collect like terms.

73. $3x + 4x$

74. $6x + 5 - 7x$

Remove parentheses and simplify.

75. $3x - (4 + 2x)$

76. $2 - 5(x + 5)$

Synthesis

77. ◈ Why are equivalent equations useful when solving equations?

78. ◈ Explain why it is not necessary to prove a subtraction principle: If $a = b$, then $a - c = b - c$.

Solve. The icon ▦ indicates an exercise designed to give practice in the use of a calculator.

79. ▦ $-356.788 = -699.034 + t$

80. ▦ $-0.2344m = 2028.732$

81. $0 \cdot x = 9$

82. $x + 3 = 3 + x$

83. $4|x| = 48$

84. $2|x| = -12$

85. $0 \cdot x = 0$

86. $x + x = x$

87. $x + 4 = 5 + x$

88. $|3x| = 6$

Solve for x.

89. $ax = 5a$

90. $x - 4 = a$

91. $3x = \frac{b}{a}$

92. $cx = a^2 + 1$

93. $1 - c = a + x$

94. $|x| + 6 = 19$

95. If $x - 4720 = 1634$, find $x + 4720$.

96. ▦ A student makes a calculation and gets an answer of 22.5. On the last step, the student multiplies by 0.3 when a division by 0.3 should have been done. What should the correct answer be?

97. ◈ Are the equations $x = 5$ and $x^2 = 25$ equivalent? Why or why not?

2.2 Using the Principles Together

Applying Both Principles • Collecting Like Terms • Clearing Fractions and Decimals • Equations Containing Parentheses

The equations in Section 2.1 required use of *either* the addition principle *or* the multiplication principle. Now we will consider equations in which *both* principles are used.

Applying Both Principles

Consider the equation $3x + 5 = 17$. To solve such an equation, we first isolate the variable term, $3x$, using the addition principle. We then use the multiplication principle to get the variable by itself.

EXAMPLE 1

Solve: $3x + 5 = 17$.

Solution

$$3x + 5 = 17$$

$$3x + 5 - 5 = 17 - 5$$ Using the addition principle: subtracting 5 on both sides (adding -5)

First isolate the x-term.

$$3x = 12$$ Simplifying

$$\frac{3x}{3} = \frac{12}{3}$$ Using the multiplication principle: dividing by 3 on both sides (multiplying by $\frac{1}{3}$)

Then isolate x.

$$x = 4$$ Simplifying

Check:

$3x + 5 = 17$	
$3 \cdot 4 + 5$?	17
$12 + 5$	
17	17 TRUE

We use the rules for order of operations: Find the product, $3 \cdot 4$, and then add.

The solution is 4. ❑

EXAMPLE 2

Solve: $-5x - 6 = 16$.

Solution

$$-5x - 6 = 16$$

$$-5x - 6 + 6 = 16 + 6$$ Adding 6 on both sides

$$-5x = 22$$

$$\frac{-5x}{-5} = \frac{22}{-5}$$ Dividing on both sides by -5

$$x = -\frac{22}{5}, \quad \text{or } -4\frac{2}{5}$$ Simplifying

Check:

$-5x - 6 = 16$	
$-5(-\frac{22}{5}) - 6$?	16
$22 - 6$	
16	16 TRUE

The solution is $-\frac{22}{5}$. ❑

EXAMPLE 3

Solve: $45 - t = 13$.

Solution

$$45 - t = 13$$

$$45 - t - 45 = 13 - 45$$ Subtracting 45 on both sides

$$-t = -32$$

$$(-1)(-t) = (-1)(-32)$$ Multiplying on both sides by -1 (Dividing on both sides by -1 would also change the sign on both sides.)

$$t = 32$$

We leave the check to the student. The solution is 32. ❑

As we improve our equation-solving skills, we begin to shorten some of our writing. Thus we may not always write a number being added, subtracted, multiplied, or divided on both sides. We simply write it on the opposite side, as in the following example.

EXAMPLE 4

Solve: $16.3 - 7.2y = -8.18$.

Solution

$$16.3 - 7.2y = -8.18$$

$$-7.2y = -8.18 - 16.3$$ Subtracting 16.3 on both sides. We write the subtraction of 16.3 on the right side and remove 16.3 on the left side.

$$-7.2y = -24.48$$

$$y = \frac{-24.48}{-7.2}$$ Dividing by -7.2 on both sides. We write the division by -7.2 on the right side and remove the -7.2 on the left side.

$$y = 3.4$$

Check:

$16.3 - 7.2y$	$= -8.18$
$16.3 - 7.2(3.4)$	? -8.18
$16.3 - 24.48$	
-8.18	-8.18 TRUE

The solution is 3.4. ❑

Collecting Like Terms

If like terms appear in an equation, we first collect them and then solve. When the like terms are not on the same side of an equation, we can use the addition principle to write all variable terms on one side.

EXAMPLE 5

Solve.

a) $3x + 4x = -14$ **b)** $2x - 4 = -3x + 1$
c) $6x + 5 - 7x = 10 - 4x + 3$

Solution

a) $$3x + 4x = -14$$

$$7x = -14$$ Collecting like terms

$$x = \frac{-14}{7}$$ Dividing by 7 on both sides

$$x = -2$$

The check is left to the student. The solution is -2.

b)

Isolate variable terms on one side and constant terms on the other side.

$$2x - 4 = -3x + 1$$

$$2x - 4 + 4 = -3x + 1 + 4$$ Adding 4

$$2x = -3x + 5$$ Simplifying

$$2x + 3x = -3x + 3x + 5$$ Adding $3x$

$$5x = 5$$ Collecting like terms and simplifying

and

$$\frac{5x}{5} = \frac{5}{5} \qquad \text{Dividing by 5}$$

$$x = 1 \qquad \text{Simplifying}$$

Check:

$$\begin{array}{r|l} 2x - 4 & = -3x + 1 \\ \hline 2 \cdot 1 - 4 & ?\ -3 \cdot 1 + 1 \\ 2 - 4 & -3 + 1 \\ -2 & -2 \end{array} \qquad \text{TRUE}$$

The solution is 1.

c)

$$\begin{aligned} 6x + 5 - 7x &= 10 - 4x + 3 & & \\ -x + 5 &= 13 - 4x & & \text{Collecting like terms} \\ -x + 5 + 4x &= 13 - 4x + 4x & & \text{Adding } 4x \\ 3x + 5 &= 13 & & \text{Simplifying} \\ 3x + 5 - 5 &= 13 - 5 & & \text{Subtracting 5} \\ 3x &= 8 & & \text{Simplifying} \\ \frac{3x}{3} &= \frac{8}{3} & & \text{Dividing by 3} \\ x &= \frac{8}{3} & & \text{Simplifying} \end{aligned}$$

The student can confirm that $\frac{8}{3}$ checks and is the solution. ❑

Clearing Fractions and Decimals

Equations are usually easier to solve when they do not contain fractions or decimals. Consider, for example,

$$\tfrac{1}{2}x + 5 = \tfrac{3}{4} \quad \text{and} \quad 2.3x + 7 = 5.4.$$

If we multiply on both sides of the first equation by 4 and on both sides of the second equation by 10, we have

$$4\left(\tfrac{1}{2}x + 5\right) = 4 \cdot \tfrac{3}{4} \quad \text{and} \quad 10(2.3x + 7) = 10 \cdot 5.4,$$

or

$$2x + 20 = 3 \qquad \text{and} \qquad 23x + 70 = 54.$$

The first equation has been "cleared of fractions" and the second equation has been "cleared of decimals."

The easiest way to clear an equation of fractions is to multiply *every term on both sides* of the equation by the smallest, or *least,* common denominator.

EXAMPLE 6

Solve: $\frac{2}{3}x - \frac{1}{6} = 2x$.

Solution The number 6 is the least common denominator, so we multiply by 6 on both sides.

$$6\left(\frac{2}{3}x - \frac{1}{6}\right) = 6 \cdot 2x \qquad \text{Multiplying by 6 on both sides}$$

and

$$6 \cdot \frac{2}{3}x - 6 \cdot \frac{1}{6} = 6 \cdot 2x$$ Using the distributive law. (*Caution*! Be sure to multiply *all* the terms by 6.)

$$4x - 1 = 12x$$ Simplifying. Note that the fractions are cleared.

$$-1 = 8x$$ Subtracting $4x$ on both sides

$$-\frac{1}{8} = x$$ Dividing by 8

The number $-\frac{1}{8}$ checks and is the solution. ❑

To clear an equation of decimals, we count the greatest number of decimal places in any one number. If the greatest number of decimal places is 1, we multiply both sides by 10^1, or 10; if it is 2, we multiply by 10^2, or 100; and so on.

EXAMPLE 7

Solve: $16.3 - 7.2y = -8.18$.

Solution The greatest number of decimal places in any one number is *two*. Multiplying by 100, which has *two* 0's, will clear *all* decimals.

$$100(16.3 - 7.2y) = 100(-8.18)$$ Multiplying by 100 on both sides

$$100(16.3) - 100(7.2y) = 100(-8.18)$$ Using the distributive law

$$1630 - 720y = -818$$ Simplifying. Note that the decimals are cleared.

$$-720y = -818 - 1630$$ Subtracting 1630 on both sides

$$-720y = -2448$$ Collecting like terms

$$y = \frac{-2448}{-720}$$ Dividing by -720 on both sides

$$y = 3.4$$

In Example 4, the same solution was found without clearing decimals. Finding the same answer in two different ways is a good check. The solution is 3.4. ❑

Equations Containing Parentheses

To solve equations that contain parentheses, we can use the distributive law to first remove the parentheses. Then we proceed as before.

EXAMPLE 8

Solve: **(a)** $3x = 2(5x - 7)$; **(b)** $2 - 5(x + 5) = 3(x - 2) - 1$.

Solution

a)

$$3x = 2(5x - 7)$$

$$3x = 10x - 14$$ Using the distributive law

$$3x - 10x = -14$$ Subtracting $10x$ to get all x-terms on one side

$$-7x = -14$$ Collecting like terms

$$\frac{-7x}{-7} = \frac{-14}{-7}$$ Dividing by -7

$$x = 2$$

Check:

$$\begin{array}{r|l} 3x & = 2(5x - 7) \\ \hline 3\cdot 2 \;?& 2(5\cdot 2 - 7) \\ 6 & 2(10 - 7) \\ & 2\cdot 3 \\ 6 & 6 \end{array} \quad \text{TRUE}$$

The solution is 2.

b) $2 - 5(x + 5) = 3(x - 2) - 1$

$2 - 5x - 25 = 3x - 6 - 1$ — Using the distributive law to multiply and remove parentheses

$-5x - 23 = 3x - 7$ — Simplifying

$-23 + 7 = 3x + 5x$ — Adding $5x$ and 7 to get all x-terms on one side and all constant terms on the other side

$-16 = 8x$ — Simplifying

$-2 = x$ — Dividing by 8

Check:

$$\begin{array}{r|l} 2 - 5(x + 5) & = 3(x - 2) - 1 \\ \hline 2 - 5(-2 + 5) \;?& 3(-2 - 2) - 1 \\ 2 - 5(3) & 3(-4) - 1 \\ 2 - 15 & -12 - 1 \\ -13 & -13 \end{array} \quad \text{TRUE}$$

The solution is -2. ❑

Here is a procedure for solving the types of equations discussed in this section.

An Equation-Solving Procedure

1. If necessary, use the distributive law to remove parentheses. Then collect like terms on each side.
2. Multiply on both sides to clear fractions or decimals. (This is optional, but it can ease computations.)
3. Get all terms with variables on one side and all constant terms on the other side, using the addition principle.
4. Collect like terms again, if necessary.
5. Multiply or divide to solve for the variable, using the multiplication principle.
6. Check all possible solutions in the original equation.

EXERCISE SET 2.2

Solve and check.

1. $5x + 6 = 31$

2. $3x + 6 = 30$

3. $8x + 4 = 68$

4. $7z + 9 = 72$

5. $4x - 6 = 34$

6. $6x - 3 = 15$

7. $3x - 9 = 33$

8. $5x - 7 = 48$

9. $7x + 2 = -54$

10. $5x + 4 = -41$

11. $-45 = 3 + 6y$
12. $-91 = 9t + 8$
13. $-4x + 7 = 35$
14. $-5x - 7 = 108$
15. $-7x - 24 = -129$
16. $-6z - 18 = -132$
17. $5x + 7x = 72$
18. $4x + 5x = 45$
19. $8x + 7x = 60$
20. $3x + 9x = 96$
21. $4x + 3x = 42$
22. $6x + 19x = 100$
23. $-6y - 3y = 27$
24. $-4y - 8y = 48$
25. $-7y - 8y = -15$
26. $-10y - 3y = -39$
27. $10.2y - 7.3y = -58$
28. $6.8y - 2.4y = -88$
29. $x + \frac{1}{3}x = 8$
30. $x + \frac{1}{4}x = 10$
31. $8y - 35 = 3y$
32. $4x - 6 = 6x$
33. $8x - 1 = 23 - 4x$
34. $5y - 2 = 28 - y$
35. $2x - 1 = 4 + x$
36. $5x - 2 = 6 + x$
37. $6x + 3 = 2x + 11$
38. $5y + 3 = 2y + 15$
39. $5 - 2x = 3x - 7x + 25$
40. $10 - 3x = 2x - 8x + 40$
41. $4 + 3x - 6 = 3x + 2 - x$
42. $5 + 4x - 7 = 4x - 2 - x$
43. $4y - 4 + y + 24 = 6y + 20 - 4y$
44. $5y - 7 + y = 7y + 21 - 5y$

Solve and check. Clear fractions or decimals first.

45. $\frac{7}{2}x + \frac{1}{2}x = 3x + \frac{3}{2} + \frac{5}{2}x$
46. $\frac{7}{8}x - \frac{1}{4} + \frac{3}{4}x = \frac{1}{16} + x$
47. $\frac{2}{3} + \frac{1}{4}t = 6$
48. $-\frac{3}{2} + x = -\frac{5}{6} - \frac{4}{3}$
49. $\frac{2}{3} + 3y = 5y - \frac{2}{15}$
50. $\frac{1}{2} + 4m = 3m - \frac{5}{2}$
51. $\frac{5}{3} + \frac{2}{3}x = \frac{25}{12} + \frac{5}{4}x + \frac{3}{4}$
52. $1 - \frac{2}{3}y = \frac{9}{5} - \frac{1}{5}y + \frac{3}{5}$
53. $2.1x + 45.2 = 3.2 - 8.4x$
54. $0.96y - 0.79 = 0.21y + 0.46$
55. $1.03 - 0.6x = 0.71 - 0.2x$
56. $1.7t + 8 - 1.62t = 0.4t - 0.32 + 8$
57. $\frac{2}{7}x - \frac{1}{2}x = \frac{3}{4}x + 1$
58. $\frac{5}{16}y + \frac{3}{8}y = 2 + \frac{1}{4}y$

Solve and check.

59. $3(2y - 3) = 27$
60. $4(2y - 3) = 28$
61. $40 = 5(3x + 2)$
62. $9 = 3(5x - 2)$
63. $2(3 + 4m) - 9 = 45$
64. $3(5 + 3m) - 8 = 88$
65. $5r - (2r + 8) = 16$
66. $6b - (3b + 8) = 16$
67. $6 - 2(3x - 1) = 2$
68. $10 - 3(2x - 1) = 1$
69. $5(d + 4) = 7(d - 2)$
70. $3(t - 2) = 9(t + 2)$
71. $8(2t + 1) = 4(7t + 7)$
72. $7(5x - 2) = 6(6x - 1)$
73. $3(r - 6) + 2 = 4(r + 2) - 21$
74. $5(t + 3) + 9 = 3(t - 2) + 6$
75. $19 - (2x + 3) = 2(x + 3) + x$
76. $13 - (2c + 2) = 2(c + 2) + 3c$
77. $\frac{1}{3}(6x + 24) - 20 = -\frac{1}{4}(12x - 72)$
78. $\frac{1}{4}(8y + 4) - 17 = -\frac{1}{2}(4y - 8)$
79. $2[4 - 2(3 - x)] - 1 = 4[2(4x - 3) + 7] - 25$
80. $5[3(7 - t) - 4(8 + 2t)] - 20 = -6[2(6 + 3t) - 4]$
81. $\frac{2}{3}(2x - 1) = 10$
82. $\frac{4}{5}(3x + 4) = 20$
83. $\frac{3}{4}\left(3x - \frac{1}{2}\right) - \frac{2}{3} = \frac{1}{3}$
84. $\frac{2}{3}\left(\frac{7}{8} - 4x\right) - \frac{5}{8} = \frac{3}{8}$
85. $0.7(3x + 6) = 1.1 - (x + 2)$
86. $0.9(2x + 8) = 20 - (x + 5)$
87. $a + (a - 3) = (a + 2) - (a + 1)$
88. $0.8 - 4(b - 1) = 0.2 + 3(4 - b)$

Skill Maintenance

89. Divide: $-22.1 \div 3.4$.
90. Factor: $7x - 21 - 14y$.
91. Use $<$ or $>$ for ■ to write a true sentence: -15 ■ -13.
92. Find $-(-x)$ when $x = -14$.

Synthesis

93. What procedure would you follow for solving an equation like $0.23x + \frac{17}{3} = -0.8 + \frac{3}{4}x$? Could your procedure be streamlined? If so, how?

94. Dave is determined to solve the equation $3x + 4 = -11$ by first using the multiplication principle to "eliminate" the 3. How should Dave proceed and why?

Solve.

95. $0.008 + 9.62x - 42.8 = 0.944x + 0.0083 - x$
96. $-2[3(x - 2) + 4] = 4(1 - x) + 8$
97. $0 = y - (-14) - (-3y)$
98. $3(x + 4) = 3(4 + x)$
99. $475(54x + 7856) + 9762 = 402(83x + 975)$

100. $30{,}000 + 20{,}000x = 55{,}000(1 + 12{,}500x)$
101. $x(x - 4) = 3x(x + 1) - 2(x^2 + x - 5)$
102. $0.05y - 1.82 = 0.708y - 0.504$
103. $-2y + 5y = 6y$
104. $3x = 4x$
105. $\frac{5 + 2y}{3} = \frac{25}{12} + \frac{5y + 3}{4}$
106. $\frac{4 - 3x}{7} = \frac{2 + 5x}{49} - \frac{x}{14}$

2.3 Formulas

Evaluating Formulas • Solving for a Letter

Many applications of mathematics involve relationships between two or more quantities. An equation that represents such a relationship will use two or more letters and is known as a **formula.**

EXAMPLE 1

The formula $M = \frac{1}{5}n$ can be used to determine how far you are from a thunderstorm. Here n is the number of seconds that it takes the sound of thunder to reach you once lightning appears and M is the distance, in miles, that you are from the lightning. If it takes 10 sec for the sound of thunder to reach you after you have seen lightning, how far away is the storm?

Solution We substitute 10 for n and calculate M: $M = \frac{1}{5}n = \frac{1}{5}(10) = 2$. The storm is 2 mi away. ❑

Suppose that we are told how far away a storm is and we want to predict how long it will take the sound of thunder to reach us. We could substitute the distance, say 2, for M, and then solve for n:

$2 = \frac{1}{5}n$ Replacing M with 2

$10 = n.$ Multiplying by 5

Were we to do this for a variety of distances, it might be easier to first solve for n and then substitute values for M.

EXAMPLE 2

Solve for n: $M = \frac{1}{5}n$.

Solution We have

$$M = \frac{1}{5}n \qquad \text{We want this letter alone.}$$
$$5 \cdot M = 5 \cdot \frac{1}{5}n \qquad \text{Multiplying on both sides by 5}$$
$$5M = n.$$

The equation $5M = n$ gives a quick, easy way to find the number of seconds it takes thunder to reach us when a storm is M miles away. ❑

To see how the addition and multiplication principles apply to formulas, compare the following. In (A), we solve as we did before; in (B), we do not carry out the calculations; and in (C), we cannot carry out calculations since the numbers are unknown.

A.
$$5x + 2 = 12$$
$$5x = 12 - 2$$
$$5x = 10$$
$$x = \frac{10}{5} = 2$$

B.
$$5x + 2 = 12$$
$$5x = 12 - 2$$
$$x = \frac{12 - 2}{5}$$

C.
$$ax + b = c$$
$$ax = c - b$$
$$x = \frac{c - b}{a}$$

EXAMPLE 3

Solve for r: $C = 2\pi r$.

Solution Although π is a constant (approximately 3.14), to keep our solution precise we do not replace it with an approximation.

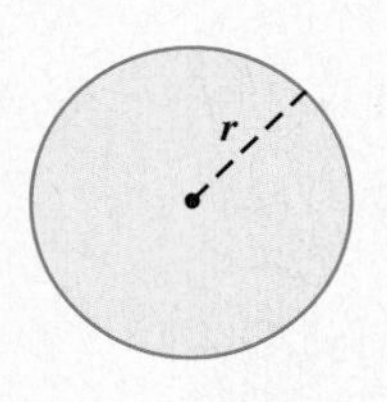

Given a radius, r, we can use this equation to find a circle's circumference, C.

$$C = 2\pi r \qquad \text{We want this letter alone.}$$
$$\frac{C}{2\pi} = \frac{2\pi r}{2\pi} \qquad \text{Dividing by } 2\pi$$
$$\frac{C}{2\pi} = r$$

Given a circle's circumference, C, we can use this equation to find the radius, r. ❑

EXAMPLE 4

Solve for a: $T = \dfrac{a + b + c}{3}$.

Solution This is a formula for the average, T, of three numbers a, b, and c.

$$T = \frac{a + b + c}{3} \qquad \text{We want the letter } a \text{ alone.}$$
$$3T = a + b + c \qquad \text{Multiplying by 3 to clear the fraction}$$
$$3T - b - c = a \qquad \text{Using the addition principle to get } a \text{ by itself (subtracting } b \text{ and subtracting } c \text{ on both sides)}$$

This formula can be used when two test grades are known and we wish to know what grade we need on a third test in order to have an average of T after three tests. ❑

EXAMPLE 5

Solve for C: $Q = \dfrac{100M}{C}$.

Solution This is a formula used in psychology for intelligence quotient Q, where M is mental age and C is chronological, or actual, age.

$$Q = \frac{100M}{C} \qquad \text{We want the letter } C \text{ alone.}$$

$$CQ = 100M \qquad \text{Multiplying by } C \text{ to clear the fraction}$$

$$C = \frac{100M}{Q} \qquad \text{Dividing by } Q$$

This formula can be used to determine a person's chronological, or actual, age once the person's mental age and intelligence quotient are known. ❑

With the formulas in this section, we can use a procedure like that described in Section 2.2 to solve for a given letter.

A Formula-Solving Procedure

To solve a formula for a given letter:

1. Identify the letter being solved for and multiply on both sides to clear fractions or decimals, if necessary.
2. Collect like terms on each side, if necessary. This may require factoring.
3. Get all terms with the letter to be solved for on one side of the equation and all other terms on the other side.
4. Collect like terms again, if necessary. This may require factoring.
5. Multiply or divide to solve for the variable in question.

EXAMPLE 6

Solve for x: $y = ax + bx - 4$.

Solution We solve as follows:

$$y = ax + bx - 4 \qquad \text{We want this letter alone.}$$

$$y + 4 = ax + bx \qquad \text{Adding 4}$$

$$y + 4 = (a + b)x \qquad \text{Collecting like terms by factoring out } x$$

$$\frac{y + 4}{a + b} = x. \qquad \text{Multiplying by } \frac{1}{a + b}$$

We can also write this as

$$x = \frac{y + 4}{a + b}.$$ ❑

CAUTION! Had we performed the following steps in Example 6, we would *not* have solved for x:

$$y = ax + bx - 4$$

$y - ax + 4 = bx$ Subtracting ax and adding 4

Two occurrences of x

$\dfrac{y - ax + 4}{b} = x$ Dividing by b

The mathematics of each step is correct, but note that x occurs on both sides of the formula. Thus *we have not solved the formula for x*. Remember that the variable being solved for should be alone on one side of the equation, with *no* occurrence of that variable on the other side!

EXERCISE SET 2.3

Solve the formula for the indicated letter.

1. $A = bh$, for b
(Area of a parallelogram with base b and height h)

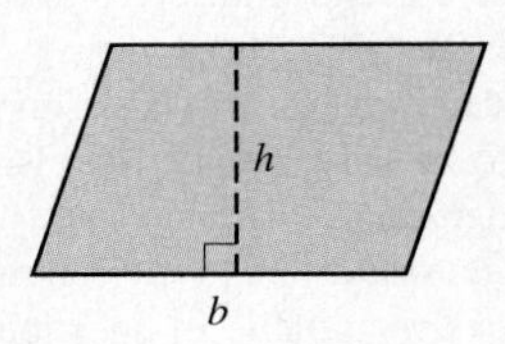

2. $A = bh$, for h

3. $d = rt$, for r
(A distance formula, where d is distance, r is speed, and t is time)

4. $d = rt$, for t

5. $I = Prt$, for P
(Simple-interest formula, where I is interest, P is principal, r is interest rate, and t is time)

6. $I = Prt$, for t

7. $F = ma$, for a
(A physics formula, where F is force, m is mass, and a is acceleration)

8. $F = ma$, for m

9. $P = 2l + 2w$, for w
(Perimeter of a rectangle of length l and width w)

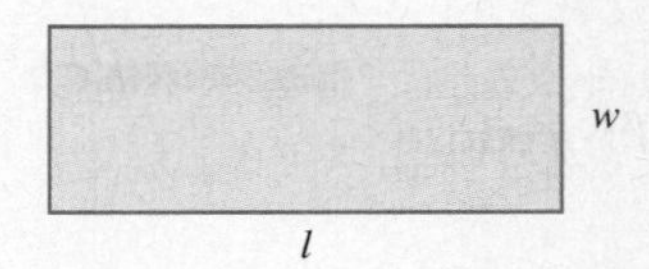

10. $P = 2l + 2w$, for l

11. $A = \pi r^2$, for r^2
(Area of a circle with radius r)

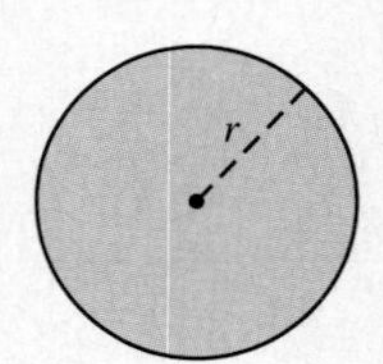

12. $A = \pi r^2$, for π

13. $A = \frac{1}{2}bh$, for b
(Area of a triangle with base b and height h)

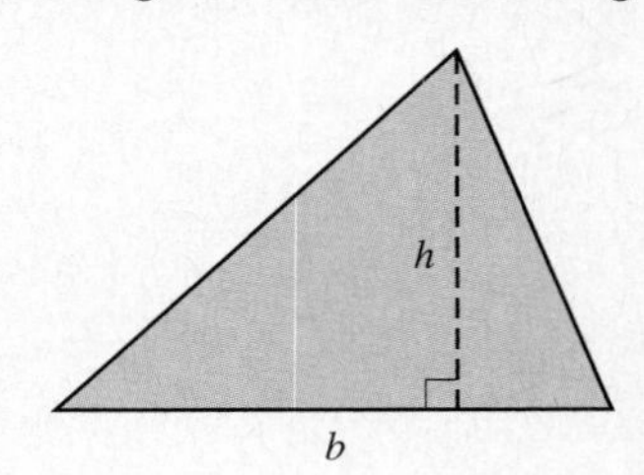

14. $A = \frac{1}{2}bh$, for h

15. $E = mc^2$, for m
(A relativity formula)

16. $E = mc^2$, for c^2

17. $Q = \dfrac{c + d}{2}$, for d

18. $Q = \dfrac{p - q}{2}$, for p

19. $A = \dfrac{a + b + c}{3}$, for b

20. $A = \dfrac{a + b + c}{3}$, for c

21. $v = \dfrac{3k}{t}$, for t

22. $P = \dfrac{ab}{c}$, for c

23. $Ax + By = C$, for y

24. $Ax + By = C$, for x

25. $A = \frac{1}{2}ah + \frac{1}{2}bh$, for b

26. $A = \frac{1}{2}ah - \frac{1}{2}bh$, for a; for h

27. $Q = 3a + 5ca$, for a

28. $P = 4m + 7mn$, for m

29. The formula

$$A = P + Prt$$

is used to find the amount A in an account when simple interest is added to an investment of P dollars (see Exercise 5). Solve for P.

30. The formula

$$S = P - P(0.01r)$$

can be used to find the sale price S of an item when the regular price P is reduced r percent. Solve for P.

31. The area of a sector of a circle is given by

$$A = \frac{\pi r^2 S}{360},$$

where r is the radius and S is the angle measure, in degrees, of the sector. Solve for S.

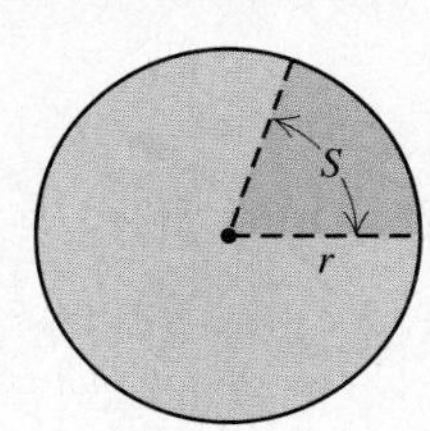

32. Solve for r^2: $A = \dfrac{\pi r^2 S}{360}$.

33. The formula $R = -0.0075t + 3.85$ can be used to estimate the world record in the 1500-m run t years after 1930. Solve for t.

34. The formula $F = \frac{9}{5}C + 32$ can be used to convert from Celsius, or Centigrade, temperature C to Fahrenheit temperature F. Solve for C.

Skill Maintenance

Multiply.

35. $7(-3)2$

36. $-\frac{2}{3} \cdot \frac{9}{10}$

Simplify.

37. $10 \div (-2) \cdot 5 - 4$

38. $3|7 - (2 - 5)|$

Synthesis

39. ◈ Devise an application in which it would be useful to solve the equation $d = rt$ for r (see Exercise 3).

40. ◈ A meteorologist claims to be able to convert Celsius temperatures to Fahrenheit temperatures but not Fahrenheit to Celsius. What advice would you offer the meteorologist?

41. Solve for y:

$$\frac{\left(\frac{y}{z}\right)}{\left(\frac{z}{t}\right)} = 1.$$

42. Solve for F: $\dfrac{1}{E + F} = G$.

43. Solve for t: $q = r(s + t)$.

44. Solve for c: $ac = bc + d$.

45. Solve for x: $a = c(x + y) + bx$.

46. Solve for a: $3a = c - a(b + d)$.

47. ◈ In the formula $A = l \cdot w$, suppose that l and w are both doubled. What is the effect on A? Why?

48. ◈ The equations

$$P = 2l + 2w \quad \text{and} \quad w = \frac{P}{2} - l$$

are equivalent formulas involving the perimeter P, length l, and width w of a rectangle. Devise a problem for which the second of the two formulas would be more useful.

2.4 Applications with Percent

Converting Between Percent Notation and Decimal Notation • Solving Percent Problems

Formulas like those in Section 2.3 appear in a wide variety of applications. In this section, we examine applications involving percent.

Suppose that Village Stationers installs a new cash register and the sales clerks inadvertently print out "totals" on each receipt without separating each transaction into "merchandise" and the five-percent "sales tax." In order to process any refunds, the shop needs a formula for separating each total into the amount spent on merchandise and the amount spent on tax.

Before such a formula can be developed, we must review the basics of percent problems.

Converting Between Percent Notation and Decimal Notation

The average family spends 26% of its income for food. What does this mean? It means that out of every \$100 earned, \$26 is spent for food. Thus 26% is a ratio of 26 to 100.

The percent symbol % means "per hundred." We can regard the percent symbol as part of a name for a number. For example,

$$26\% \quad \text{is defined to mean} \quad 26 \times 0.01, \quad \text{or} \quad 26 \times \frac{1}{100}, \quad \text{or} \quad \frac{26}{100}.$$

In general,

Percent Notation

$$n\% \quad \text{means} \quad n \times 0.01, \quad \text{or} \quad n \times \frac{1}{100}, \quad \text{or} \quad \frac{n}{100}.$$

EXAMPLE 1 Convert to decimal notation: **(a)** 78%; **(b)** 1.3%.

Solution

a) $78\% = 78 \times 0.01$ Replacing % by $\times$ 0.01

$= 0.78$

b) $1.3\% = 1.3 \times 0.01$ Replacing % by $\times\, 0.01$

$= 0.013$ ❑

> To convert from percent notation to decimal notation, move the decimal point two places to the left and drop the percent symbol.

EXAMPLE 2

Convert 43.67% to decimal notation.

Solution

43.67% 0.43.67 $43.67\% = 0.4367$

Move the decimal point two places to the left. ❑

The procedure used in Example 2 can be reversed. Consider 0.38:

$0.38 = \frac{38}{100}$ Converting to fractional notation

$= 38\%.$ $\frac{n}{100}$ means $n\%$.

> To convert from decimal notation to percent notation, move the decimal point two places to the right and write a percent symbol.

EXAMPLE 3

Convert to percent notation: **(a)** 1.27; **(b)** $\frac{1}{4}$; **(c)** 0.3.

Solution

a) We first move the decimal point two places to the right: 1.27.

and then write a % symbol: 127%

b) Note that $\frac{1}{4} = 0.25$. We move the decimal point two places to the right: 0.25.

and then write a % symbol: 25%

c) We first move the decimal point two places to the right (recall that $0.3 = 0.30$): 0.30.

and then write a % symbol: 30% ❑

There is a table of fractional, decimal, and percent equivalents at the back of the text. If you do not already know these facts, it might be helpful to memorize some or all of them.

Solving Percent Problems

To solve problems involving percents, we translate to mathematical language and then solve an equation.

EXAMPLE 4

What is 11% of 49?

Solution

Translate: What is 11% of 49?

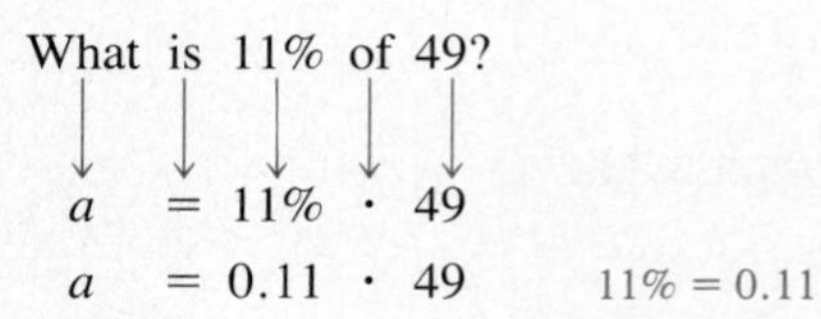

$$a = 11\% \cdot 49$$

$$a = 0.11 \cdot 49 \qquad 11\% = 0.11$$

Since a is already by itself, we just multiply:

$$\begin{array}{r} 49 \\ \times\, 0.11 \\ \hline 49 \\ 490 \\ \hline a = 5.39 \end{array}$$

> A way of checking answers is by estimating as follows:
>
> $$11\% \times 49 \approx 10\% \times 50$$
> $$= 0.10 \times 50 = 5.$$
>
> Since 5 is close to 5.39, our answer is reasonable.

Thus, 5.39 is 11% of 49. The answer is 5.39. ❑

EXAMPLE 5

3 is 16 percent of what?

Solution Translate: 3 is 16 percent of what?

$$3 = 16\% \cdot y$$

We solve the equation:

$$3 = 0.16y$$

$$\frac{3}{0.16} = y \qquad \text{Dividing on both sides by 0.16}$$

$$18.75 = y.$$

Thus, 3 is 16 percent of 18.75. The answer is 18.75. ❑

EXAMPLE 6

What percent of \$50 is \$16?

Solution Translate: What percent of \$50 is \$16?

$$n \cdot 50 = 16$$

We solve the equation and then convert to percent notation:

$$n \cdot 50 = 16$$

$$n = \frac{16}{50} \qquad \text{Dividing on both sides by 50}$$

$$n = 0.32 = 32\%.$$

Thus, \$16 is 32% of \$50. The answer is 32%. ❑

Examples 4–6 represent the three basic types of percent problems.

EXAMPLE 7

A receipt from Village Stationers indicates the total paid (including tax), but not the price of the merchandise. If the sales tax is 5%, find the following:

a) The cost of the merchandise when the total is \$31.50.
b) A formula for the cost of the merchandise c when the total is T dollars.

Solution

a) When tax is added to the cost of an item, the customer actually pays more than 100% of the item's price. When sales tax is 5%, the total paid is 105% of the price of the merchandise. Thus if $c =$ the cost of the merchandise, we have

$$\underbrace{\$31.50}_{\downarrow} \text{ is } 105\% \text{ of } c$$

$$31.50 = 1.05 \cdot c$$

$\frac{31.50}{1.05} = c$ Dividing by 1.05

$30 = c.$ Simplifying

The merchandise cost \$30 before tax.

b) When the total is T dollars, we modify the approach used in part (a):

$$T = 1.05c$$

$\dfrac{T}{1.05} = c.$ Dividing by 1.05

To check this result, note that the answer in part (a) shows that when T is replaced with \$31.50, $c =$ \$30. The sales tax of 5% would add $0.05 \cdot \$30$, or \$1.50, to the bill, for a total of \$30 + \$1.50 = \$31.50. Thus our answer in part (a) checks and the formula $T/1.05 = c$ appears to yield correct values for c.

The formula $c = T/1.05$ can be used to find the cost of the merchandise when the total T is known and the sales tax is 5%. ❑

EXERCISE SET 2.4

Find decimal notation.

1. 76% **2.** 54% **3.** 54.7%

4. 96.2% **5.** 100% **6.** 1%

7. 0.61% **8.** 125% **9.** 240%

10. 0.73% **11.** 3.25% **12.** 2.3%

Find percent notation.

13. 4.54 **14.** 1 **15.** 0.998

16. 0.73 **17.** 2 **18.** 0.0057

19. 0.072 **20.** 1.34 **21.** 9.2

22. 0.013 **23.** 0.0068 **24.** 0.675

25. $\frac{1}{8}$ **26.** $\frac{1}{3}$ **27.** $\frac{17}{25}$

28. $\frac{11}{20}$ **29.** $\frac{3}{4}$ **30.** $\frac{2}{5}$

31. $\frac{7}{10}$ **32.** $\frac{8}{10}$ **33.** $\frac{3}{5}$

34. $\frac{17}{50}$ **35.** $\frac{2}{3}$ **36.** $\frac{3}{8}$

Solve.

37. What percent of 68 is 17?

38. What percent of 75 is 36?

39. What percent of 125 is 30?

40. What percent of 300 is 57?

41. 45 is 30% of what number?

42. 20.4 is 24% of what number?

43. 0.3 is 12% of what number?

44. 7 is 175% of what number?

45. What number is 65% of 840?

46. What number is 1% of a million?

47. What percent of 80 is 100?

48. What percent of 10 is 205?

49. What is 2% of 40?

50. What is 40% of 2?

51. 2 is what percent of 40?

52. 40 is 2% of what number?

53. The FBI annually receives 16,000 applications for agents. It accepts 600 of these applicants. What percent does it accept?

54. The U.S. Postal Service reports that we open and read 78% of the junk mail that we receive. A business sends out 9500 advertising brochures. How many of them can it expect to be opened and read?

55. It has been determined by sociologists that 17% of the population is left-handed. Each week 160 bowlers enter a tournament conducted by the Professional Bowlers Association. How many would you expect to be left-handed? Round to the nearest one.

56. In a medical study, it was determined that if 800 people kiss someone else who has a cold, only 56 will actually catch the cold. What percent is this?

57. On a test of 88 items, a student got 76 correct. What percent were correct?

58. A baseball player had 13 hits in 25 times at bat. What percent were hits?

59. A bill at Officeland totaled $37.80. How much did the merchandise cost if the sales tax is 5%?

60. Doreen's checkbook shows that she wrote a check for $987 for building materials. What was the price of the materials if the sales tax was 5%?

Skill Maintenance

61. Convert to decimal notation: $\frac{23}{25}$.

62. Add: $-23 + (-67)$.

63. Subtract: $-45.8 - (-32.6)$.

64. Remove parentheses and simplify:

$$4a - 8b - 5(5a - 4b).$$

Synthesis

65. ◈ Mary Alice bought a dress for $120 after it had been reduced 20%. When Mary Alice returned the dress to the shop, the salesclerk incorrectly said that the original price must have been $144. What mistake did the salesclerk make?

66. ◈ Would it be better to receive a 5% raise and then an 8% raise or the other way around? Why?

67. Rollie's Music charges $11.99 for a compact disc. Sound Warp charges $13.99, but you have a coupon for $2 off. In both cases, a 7% sales tax is charged on the *regular* price. How much does the disc cost at each store?

68. The weather report is "a 60% chance of showers during the day, 30% tonight, and 5% tomorrow morning." What are the chances that it won't rain during the day? tonight? tomorrow morning?

69. If x is 160% of y, y is what percent of x?

70. The new price of a car is 25% higher than the old price of $10,400. The old price is what percent lower than the new price?

71. Generalize the result of Example 7. That is, find a formula for the cost of the merchandise c when the total is T dollars and the tax rate is r.

2.5 Problem Solving

Five Steps for Problem Solving • Applying the Five Steps

One of the most important uses of algebra is as a tool for problem solving. In this section, we develop a problem-solving approach that will be used throughout the remainder of the text.

Five Steps for Problem Solving

In Section 2.4, we solved a problem in which Village Stationers needed a formula. Our solution of the problem required us to *familiarize* ourselves with percent notation so that we could then *translate* a percent problem into an equation. We then *solved* the equation, *checked* the solution, and *stated* the answer to the problem at the end of the section.

Five Steps for Problem Solving in Algebra

1. *Familiarize* yourself with the problem situation.
2. *Translate* to mathematical language. (This often means write an equation.)
3. *Carry out* some mathematical manipulation. (This often means *solve* an equation.)
4. *Check* your possible answer in the original problem.
5. *State* the answer clearly.

Of the five steps, the most important is probably the first one: becoming familiar with the problem situation. Here are some hints for familiarization.

To familiarize yourself with the problem situation:

1. Read the problem carefully.
2. Reread the problem, perhaps aloud. Try to visualize the problem.
3. List the information given and the questions to be answered. Choose a variable (or variables) to represent the unknown and clearly state what the variable represents. For example, let L = length in centimeters, d = distance in miles, and so on.
4. Gather further information. Look up a formula in this book or in a reference book. Talk to a reference librarian or an expert in the field.
5. Make a table of the given information and the information you have collected. Look for patterns that may help in the translation to an equation.
6. Make a drawing and label it with known and unknown information, using specific units if given.
7. Guess the answer and check the guess. Observe the manner in which the guess is checked.

EXAMPLE 1

A 72-in. board is cut into two pieces. One piece is twice as long as the other. How long are the pieces?

Solution

1. FAMILIARIZE. We first draw a picture. We let

x = the length of the shorter piece.

Then

$2x$ = the length of the longer piece.

(We can also let y = the length of the longer piece. Then $\frac{1}{2}y$ = the length of the shorter piece. This, however, introduces fractions and will make the solution somewhat more difficult.)

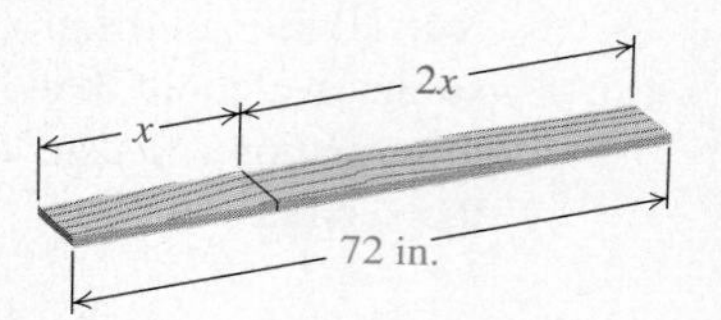

We can further familiarize ourselves with the problem by making a guess. Suppose $x = 31$ in. Then $2x = 62$ in. and $x + 2x = 93$ in. This is not correct but making the guess does help us become familiar with the problem.

2. **TRANSLATE.** From the figure, we can see that the lengths of the two pieces must add up to 72 in. This gives us our translation.

Length of one piece plus length of other is 72

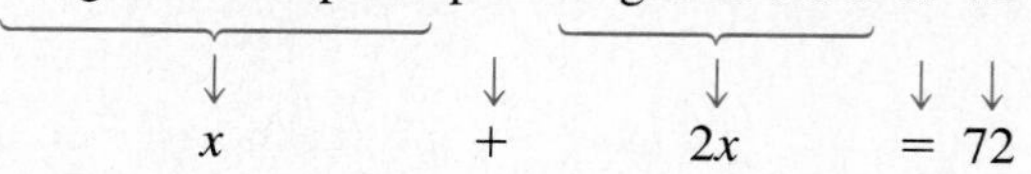

$$x \quad + \quad 2x \quad = 72$$

3. **CARRY OUT.** We solve the equation:

$$x + 2x = 72$$
$$3x = 72 \qquad \text{Collecting like terms}$$
$$x = 24. \qquad \text{Dividing by 3}$$

4. **CHECK.** If one piece is 24 in. long, the other, to be twice as long, must be 48 in. long. The lengths of the pieces add up to 72 in. This checks.

5. **STATE.** One piece is 24 in. long, and the other is 48 in. long. ❑

EXAMPLE 2

Five more than twice a number is nineteen. What is the number?

Solution

1. **FAMILIARIZE.** Let $x =$ the number. Then "twice a number" translates to $2x$ and "five more than twice a number" translates to $2x + 5$.

2. **TRANSLATE.** The familiarization leads us to the following translation:

Five more than twice a number is nineteen

$$2x + 5 \quad = \quad 19.$$

3. **CARRY OUT.** We solve the equation:

$$2x + 5 = 19$$
$$2x = 14 \qquad \text{Subtracting 5}$$
$$x = 7. \qquad \text{Dividing by 2}$$

4. **CHECK.** Twice, or two times, 7 is 14. Adding 5 to 14, we get 19. This checks.

5. **STATE.** The number is 7. ❑

The following are examples of **consecutive integers:** 16, 17, 18, 19; and

-31, -30, -29. Unknown consecutive integers can be represented in the form x, $x + 1$, $x + 2$, and so on.

The following are examples of **consecutive even integers:** 16, 18, 20, 22; and -52, -50, -48. Unknown consecutive even integers can be represented in the form x, $x + 2$, $x + 4$, and so on.

The following are examples of **consecutive odd integers:** 21, 23, 25, 27; and -71, -69, -67. Unknown consecutive odd integers can be represented in the form x, $x + 2$, $x + 4$, and so on.

EXAMPLE 3

A book is opened. The sum of the page numbers on the facing pages is 233. Find the page numbers.

Solution

1. FAMILIARIZE. Page numbers on facing pages are consecutive integers. Thus if we let $x =$ the smaller number, then $x + 1 =$ the larger number. (Another way such numbers could be represented is to let $y =$ the larger number and $y - 1 =$ the smaller.) Had this problem appeared and you did not understand the word "sum," you would have needed to learn its meaning. This is a general problem-solving tip: You may need to look up a definition or a formula or some other piece of information in order to solve a problem.

   To become more familiar with the problem, we can make a table. How do we get the entries in the table? First, we just guess a value for x. Then we find $x + 1$. Finally, we add the two numbers to find their sum.

x	$x+1$	Sum of x and $x+1$	
14	15	29	⟵ This guess was much too small.
102	103	205	⟵ This guess was a bit too small.

   Our second guess leads us to suspect that the value of x is not much greater than 102. The problem could actually be solved with further guessing, but we need to practice using algebra.

2. TRANSLATE. We reword the problem and translate as follows.

   Rewording: First integer + second integer = 233

   $$\begin{array}{lccccc} \textit{Translating:} & x & + & (x+1) & = & 233 \end{array}$$

3. CARRY OUT. We solve the equation:

$$x + (x + 1) = 233$$

$$2x + 1 = 233 \qquad \text{Using an associative law and collecting like terms}$$

$$2x = 232 \qquad \text{Subtracting 1}$$

$$x = 116. \qquad \text{Dividing by 2}$$

   If x is 116, then $x + 1$ is 117.

4. CHECK. Our possible answers are 116 and 117. These are consecutive integers and their sum is 233, so the answers check in the original problem.

5. STATE. The page numbers are 116 and 117. ❑

EXAMPLE 4

Truck-Rite Rentals rents trucks at a daily rate of \$49.95 plus 39¢ per mile. Concert Productions has budgeted \$100 for renting a truck to haul equipment to an upcoming concert. How many miles can a rental truck be driven on a \$100 budget?

Solution

1. **FAMILIARIZE.** Suppose Concert Productions drives 75 mi. Then the cost is

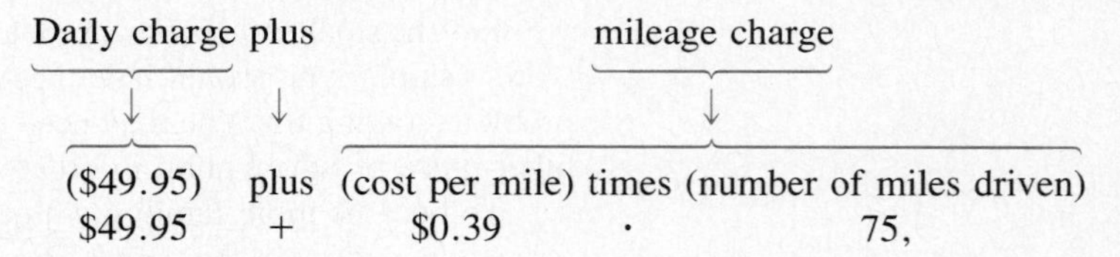

which is \$49.95 + \$29.25, or \$79.20. This familiarizes us with the way in which a calculation is made. Note that we convert 39 cents to \$0.39 so that we are using the same units, dollars, throughout the problem.

Let m = the number of miles that can be driven for \$100.

2. **TRANSLATE.** We reword the problem and translate as follows.

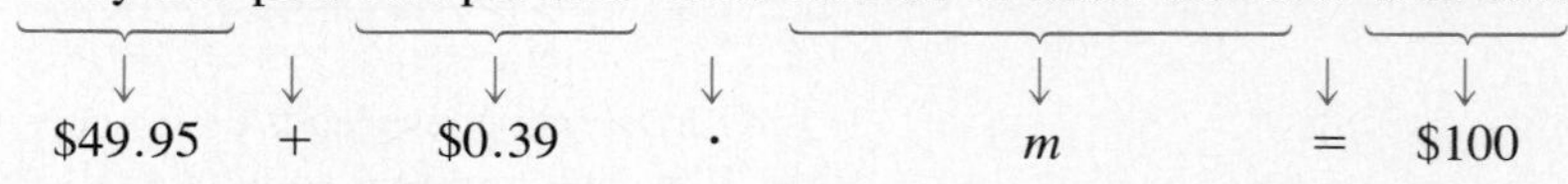

3. **CARRY OUT.** We solve the equation:

$$49.95 + 0.39m = 100$$

$$100(49.95 + 0.39m) = 100(100) \quad \text{Multiplying by 100 on both sides to clear decimals}$$

$$100(49.95) + 100(0.39m) = 10{,}000 \quad \text{Using the distributive law}$$

$$4995 + 39m = 10{,}000$$

$$39m = 5005 \quad \text{Subtracting 4995}$$

$$m = \frac{5005}{39} \quad \text{Dividing by 39}$$

$$m \approx 128.3 \quad \text{Rounding to the nearest tenth}$$

4. **CHECK.** We check in the original problem. We multiply 128.3 by \$0.39, getting \$50.037. Then we add \$50.037 to \$49.95 and get \$99.987, which is just about the \$100 allotted.

5. **STATE.** The truck can be driven about 128.3 mi on the rental allotment of \$100. ❑

EXAMPLE 5

A rectangular community garden is to be enclosed with 92 m of fencing. In order to allow for compost storage, the garden must be 4 m longer than it is wide. Determine the dimensions of the garden.

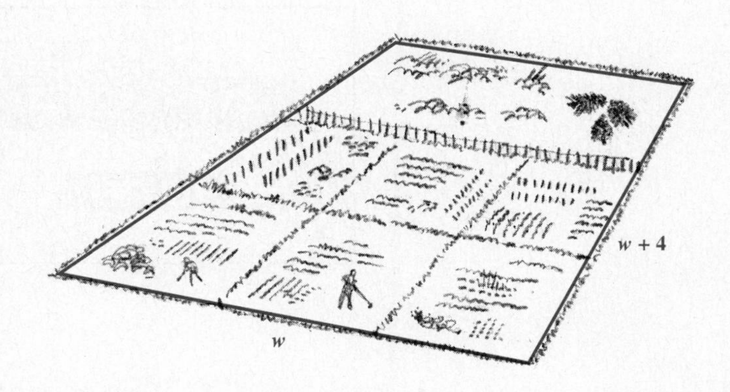

Solution

1. FAMILIARIZE. Suppose the garden were 30 m wide. The length would then be $30 + 4$ m, or 34 m, and the perimeter would be $2 \cdot 30 + 2 \cdot 34$ m, or 128 m. Our guess was too big, but we have at least gained familiarity with the problem.

We let $w =$ the width of the garden. Since the garden is "4 m longer than it is wide," we have $w + 4 =$ the length. Recall that the perimeter P of a rectangle is the distance around it and is given by the formula $2l + 2w = P$, where $l =$ the length and $w =$ the width.

2. TRANSLATE. To translate the problem, we substitute $w + 4$ for l and 92 for P, as follows:

$$2l + 2w = P$$
$$2(w + 4) + 2w = 92.$$

3. CARRY OUT. We solve the equation:

$$2(w + 4) + 2w = 92$$
$$2w + 8 + 2w = 92 \quad \text{Using the distributive law}$$
$$4w + 8 = 92 \quad \text{Collecting like terms}$$
$$4w = 84$$
$$w = 21.$$

The dimensions appear to be $w = 21$ m and l, or $w + 4$, $= 25$ m.

4. CHECK. If the width is 21 m and the length 25 m, the perimeter is 2(25 m) + 2(21 m), or 92 m. Since 92 m of fencing is available, the dimensions 21 m and 25 m check.

5. STATE. The garden should be 21 m wide and 25 m long. ❑

EXAMPLE 6

The price of a car was reduced to a sale price of \$13,559. This was a 9% reduction. What was the original price?

Solution

1. FAMILIARIZE. Suppose the original price were \$16,000. A 9% reduction can be found by taking 9% of \$16,000, that is,

$$9\% \text{ of } \$16{,}000 = 0.09(\$16{,}000) = \$1440.$$

The sale price is then found by subtracting the amount of reduction:

$$\$16{,}000 - \$1440 = \$14{,}560.$$

In becoming familiar with the problem, we find that our guess of \$16,000 was too high. We let $x =$ the original price of the car. Because it is reduced by 9%, the sale price $= x - 9\%x$.

2. TRANSLATE. We reword and then translate.

Rewording: Original price − reduction is sale price

Translating: $x \quad - \quad 9\%x \quad = \quad \$13{,}559$

3. CARRY OUT. We solve the equation:

$$x - 9\%x = 13{,}559$$

$$x - 0.09x = 13{,}559 \qquad \text{Converting to decimal notation}$$

$$1x - 0.09x = 13{,}559$$

$$(1 - 0.09)x = 13{,}559 \qquad \text{Factoring out the } x$$

$$0.91x = 13{,}559$$

Collecting like terms. Had we noted that the sale price is 91% of the original price, we could have begun with this equation.

$$x = \frac{13{,}559}{0.91} \qquad \text{Dividing by 0.91}$$

$$x = 14{,}900.$$

4. CHECK. To check, we find 9% of \$14,900 and subtract:

$$9\% \times \$14{,}900 = 0.09 \times \$14{,}900 = \$1341$$

$$\$14{,}900 - \$1341 = \$13{,}559.$$

Since we get the sale price, \$13,559, the \$14,900 checks.

5. STATE. The original price was \$14,900. ❑

CAUTION! A common error in a problem like this is to take 9% of the sale price and subtract or add. Note that 9% of the original price is not equal to 9% of the sale price!

EXAMPLE 7

The second angle of a triangle is 20° greater than the first. The third angle is twice as large as the first. How large are the angles?

Solution

1. FAMILIARIZE. We draw a picture. Here the measure of the first angle $= x$, the measure of the second angle $= x + 20$, and the measure of the third angle $= 2x$.

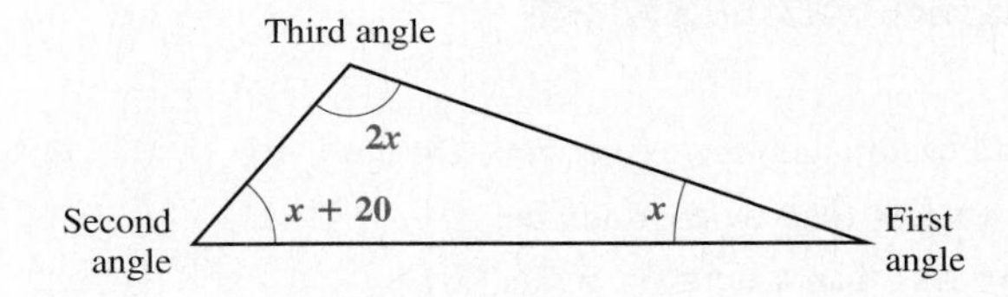

2. TRANSLATE. To translate, we need to recall a geometric fact (you might, as part of step 1, look it up in a geometry book or in the list of formulas at the back of this book). The measures of the angles of any triangle add up to 180°.

$$\underbrace{\text{Measure of first angle}}_{x} + \underbrace{\text{measure of second angle}}_{(x+20)} + \underbrace{\text{measure of third angle}}_{2x} = 180^\circ$$

$$x + (x + 20) + 2x = 180.$$

3. CARRY OUT. We solve:

$$\begin{aligned} x + (x + 20) + 2x &= 180 \\ 4x + 20 &= 180 \\ 4x &= 160 \\ x &= 40. \end{aligned}$$

The measures for the angles appear to be:

First angle: $x = 40^\circ$,
Second angle: $x + 20 = 40 + 20 = 60^\circ$,
Third angle: $2x = 2(40) = 80^\circ$.

4. CHECK. Consider 40°, 60°, and 80°. The second is 20° greater than the first, the third is twice the first, and the sum is 180°. These numbers check.

5. STATE. The measures of the angles are 40°, 60°, and 80°. ❑

We close this section with some tips to aid you in problem solving.

Problem-Solving Tips

1. The more problems you solve, the more your skills will improve.
2. Look for patterns when solving problems. Each time you study an example in a text, you may observe a pattern for problems that you will encounter later in the exercise sets or some other practical situation.
3. When translating to mathematics, consider the dimensions of the variables and constants in the equation. The variables that represent length should all be in the same unit, those that represent money should all be in dollars or all in cents, and so on.
4. Make sure that units appear in the answer whenever appropriate.

EXERCISE SET 2.5

Translate to an algebraic expression. Do not solve.

1. Three less than twice a number
2. Five less than a number divided by 8
3. One half of the product of a number and 7
4. Two fewer than ten times a number
5. Five times the sum of 3 and some number
6. The sum of two numbers times 6
7. An 8-ft board is cut into two pieces. One piece is 2 ft longer than the other. Let L = the length of the shorter piece. Write an expression for the longer piece.

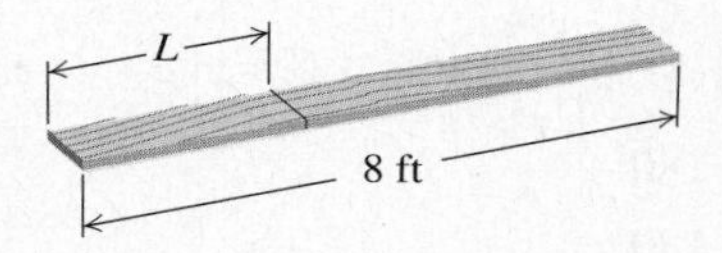

8. A 240-in. pipe is cut into two pieces. One piece is three times the length of the other. Let x = the length of the longer piece. Write an expression for the shorter piece.

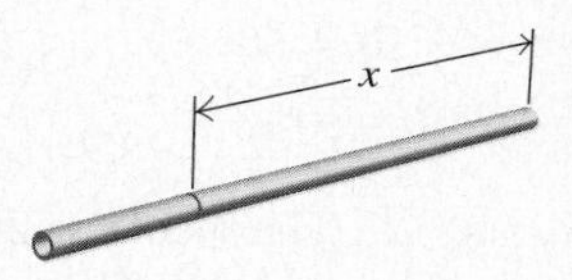

9. The price of a book is decreased by 30% during a sale. Let b = the price of the book before the reduction. Write an expression for the sale price.
10. On sale, the price of a compact disc player was reduced by 20%. Let p = the price before the reduction. Write an expression for the sale price.
11. Write an expression for the sum of three consecutive even integers.
12. Write an expression for the sum of three consecutive integers.
13. Jetz rents compact cars at a rate of \$34.95 plus 27¢ per mile. Let m = the total number of miles driven. Write an expression for the cost of driving m miles.
14. Stacy scored two more points than half the number scored by the entire team. Let t = the number of points scored by the entire team. Write an expression for the number of points Stacy scored.
15. The length of a rectangle is twice the width.
    a) Let w = the width. Write an expression for the length.
    b) Let l = the length. Write an expression for the width.
16. The base of a triangle is three times five more than the height. Let h = the height. Write an expression for the base.
17. The second angle of a triangle is three times as large as the first. The third angle measures 30° more than the first. Let x = the measure of the first angle. Write expressions for the other angles.
18. The second angle of a triangle is four times as large as the first. The third angle is 45° less than the sum of the other two angles. Let x = the measure of the first angle. Write expressions for the other angles.

Solve these problems. Even though you might find the answer quickly in some other way, practice using the five-step problem-solving process.

19. Three less than twice a number is 25. What is the number?

20. Two fewer than ten times a number is 118. What is the number?

21. Five times the sum of 3 and some number is 70. What is the number?

22. Twice the sum of 4 and some number is 34. What is the number?

23. When 18 is subtracted from six times a certain number, the result is 96. What is the number?

24. When 28 is subtracted from five times a certain number, the result is 232. What is the number?

25. If you double a number and then add 16, you get $\frac{2}{5}$ of the original number. What is the original number?

26. If you double a number and then add 85, you get $\frac{3}{4}$ of the original number. What is the original number?

27. If you add two fifths of a number to the number itself, you get 56. What is the number?

28. If you add one third of a number to the number itself, you get 48. What is the number?

29. A 180-m rope is cut into three pieces. The second piece is twice as long as the first. The third piece is three times as long as the second. How long is each piece of rope?

30. A 480-m wire is cut into three pieces. The second piece is three times as long as the first. The third piece is four times as long as the second. How long is each piece?

31. The sum of the page numbers on the facing pages of a book is 273. What are the page numbers?

32. The sum of the page numbers on the facing pages of a book is 281. What are the page numbers?

33. The sum of two consecutive even integers is 114. What are the integers?

34. The sum of two consecutive even integers is 106. What are the integers?

35. The sum of three consecutive integers is 108. What are the integers?

36. The sum of three consecutive integers is 126. What are the integers?

37. The sum of three consecutive odd integers is 189. What are the integers?

38. The sum of three consecutive even integers is 396. What are the integers?

39. The top of the John Hancock Building in Chicago is a rectangle whose length is 60 ft more than the width. The perimeter is 520 ft. Find the width and the length of the rectangle. Find the area of the rectangle.

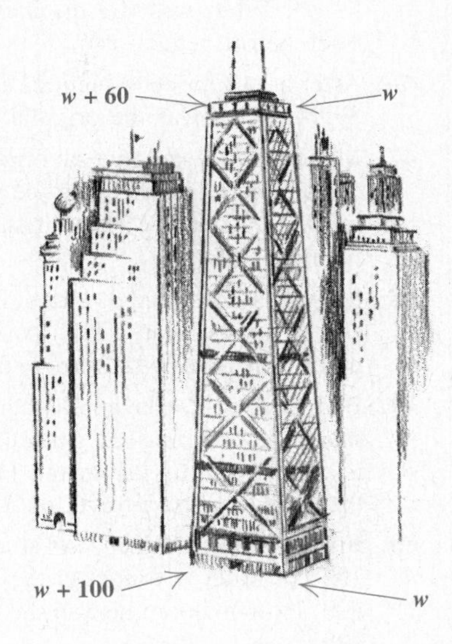

40. The ground floor of the John Hancock Building is a rectangle whose length is 100 ft more than the width. The perimeter is 860 ft. Find the width and the length of the rectangle. Find the area of the rectangle.

41. The perimeter of a standard-size piece of typewriter paper is 99 cm. The width is 6.3 cm less than the length. Find the length and the width.

42. The perimeter of the state of Wyoming is 1280 mi. The width is 90 mi less than the length. Find the width and the length.

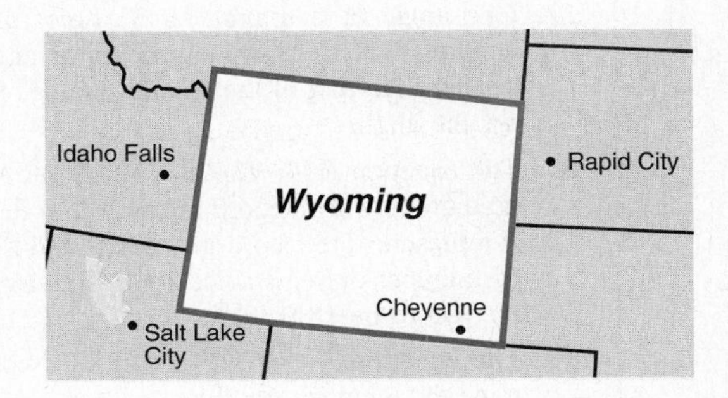

43. The second angle of a triangle is four times as large as the first. The third angle is 45° less than the sum of the other two angles. Find the measure of the first angle.

44. The second angle of a triangle is three times as large as the first. The third angle is 25° less than the sum of the other two angles. Find the measure of the first angle.

45. After a 40% price reduction, a shirt is on sale at \$9.60. What was the original price (that is, the price before reduction)?

46. After a 34% price reduction, a blouse is on sale at \$9.24. What was the original price?

47. Money is invested in a savings account at a rate of 6% simple interest. After one year, there is \$4664 in the account. How much was originally invested?

48. Money is borrowed at a rate of 10% simple interest. After one year, \$7194 pays off the loan. How much was originally borrowed?

49. Badger Rent-A-Car rents a compact car at a daily rate of \$34.95 plus 10¢ per mile. A businessperson is allotted \$80 for car rental. How many miles can the businessperson travel on the \$80 budget?

50. Badger rents midsized cars at a rate of \$43.95 plus 10¢ per mile. A tourist has a car-rental budget of \$90. How many miles can the tourist travel on the \$90?

51. The second angle of a triangle is three times as large as the first. The measure of the third angle is 40° greater than that of the first. How large are the angles?

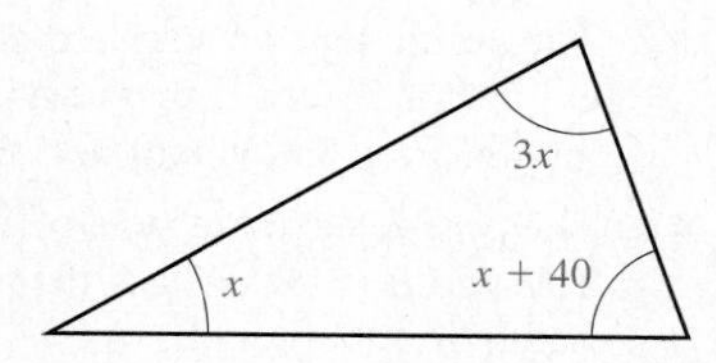

52. One angle of a triangle is 32 times as large as another. The measure of the third angle is 10° greater than that of the smallest angle. How large are the angles?

53. The equation $R = -0.028t + 20.8$ can be used to predict the world record in the 200-m dash, where R represents the record in seconds and t represents the number of years since 1920. In what year will the record be 18.0 sec?

54. The equation $F = \frac{1}{4}N + 40$ can be used to determine the temperature given the number of times a cricket chirps per minute, where F represents the temperature in degrees and N represents the number of chirps per minute. Determine the number of chirps per minute for a temperature of 80°.

Skill Maintenance

55. Factor: $3x - 12y + 60$.

56. Simplify: $5x - 3(9 - 2x) + 14$.

Synthesis

57. ◈ A student claims to be able to solve most of the problems of this section by guessing. Is there anything wrong with this approach? Why or why not?

58. ◈ Write a problem for a classmate to solve. Devise it so that the problem can be translated to the equation

$$x + (x + 2) + (x + 4) = 375.$$

59. Abraham Lincoln's 1863 Gettysburg Address refers to the year 1776 as "four *score* and seven years ago." Write an equation and find what a score is.

60. One number is 25% of another. The larger number is 12 more than the smaller. What are the numbers?

61. If the daily rental for a car is \$18.90 plus a certain price per mile and a person must drive 190 mi and stay within a \$55.00 budget, what is the highest price per mile that the person can afford?

62. Hal scored 78 on a test that had 4 fill-ins worth 7 points each and 24 multiple-choice questions worth 3 points each. He had one fill-in wrong. How many multiple-choice questions did Hal get right?

63. The width of a rectangle is three fourths of the length. The perimeter of the rectangle becomes 50 cm when the length and the width are each increased by 2 cm. Find the length and the width.

64. Apples are collected in a basket for six people. One third, one fourth, one eighth, and one fifth of the apples are given to four people, respectively. The fifth person gets ten apples, and one apple remains for the sixth person. Find the original number of apples in the basket.

65. In a basketball league, the Falcons won 15 of their first 20 games. In order to win 60% of the total number of games, how many more games will they have to play, assuming they win only half the remaining games?

66. Luke spends 12% of his weekly salary on dining out. Of the money spent dining out, 55%, or \$39.60, is spent on fast-food. What is Luke's weekly salary?

67. The area of a triangle is 2.9047 in^2. Find the height of the triangle if the base is 8 in.

68. Ella has an average score of 82 on three tests. Her average score on the first two tests is 85. What was the score on the third test?

69. A school purchases a piano and must choose between paying \$2000 at the time of purchase or \$2150 at the end of one year. Which option should the school select and why?

2.6 Solving Inequalities

Solutions of Inequalities • Graphs of Inequalities • Solving Inequalities Using the Addition Principle • Solving Inequalities Using the Multiplication Principle • Using the Principles Together

Many real-world situations can be translated to *inequalities*. For instance, a student might be interested in what test scores will assure *at least* a 90 average; an elevator might be designed to hold *at most* 2000 pounds; a tax credit might be allowable for families with incomes *less than* \$25,000; and so on. Before we can solve applications of this type, we must adapt our equation-solving principles to the solving of inequalities.

Solutions of Inequalities

In Section 1.4, we learned that an inequality is a number sentence with $>$ (is greater than), $<$ (is less than), $\geq$ (is greater than or equal to), or $\leq$ (is less than or equal to) as its verb. Inequalities like

$$-7 > x, \quad t < 5, \quad 5x - 2 \geq 9, \quad \text{and} \quad -3y + 8 \leq -7$$

are true for some replacements of the variable and false for others.

EXAMPLE 1

Determine whether the given number is a solution of $x < 2$: **(a)** -2; **(b)** 2.

Solution

a) Since $-2 < 2$ is true, -2 is a solution.
b) Since $2 < 2$ is false, 2 is not a solution. ❑

EXAMPLE 2

Determine whether the given number is a solution of $y \geq 6$: **(a)** 6; **(b)** -4.

a) Since $6 \geq 6$ is true, 6 is a solution.
b) Since $-4 \geq 6$ is false, -4 is not a solution. ❑

Graphs of Inequalities

Because the solutions of inequalities like $x < 2$ are too numerous to list, it is helpful to make a drawing that represents all the solutions. A **graph** of an inequality is just such a drawing. Graphs of inequalities in one variable can be drawn on a number line by shading all points that are solutions. Open dots are used to indicate endpoints that are *not* solutions and closed dots indicate endpoints that *are* solutions.

EXAMPLE 3

Graph each inequality: **(a)** $x < 2$; **(b)** $y \geq -3$; **(c)** $-2 < x \leq 3$.

Solution

a) The solutions of $x < 2$ are those numbers less than 2. They are shown on the graph by shading all points to the left of 2. The open dot at 2 indicates that 2 *is not* part of the graph, but numbers like 1.2 and 1.99 are.

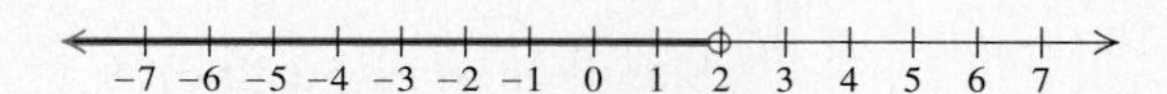

b) The solutions of $y \geq -3$ are shown on the number line by shading the point for -3 and all points to the right of -3. The closed dot at -3 indicates that -3 *is* part of the graph.

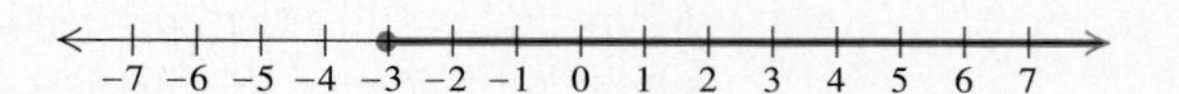

c) The inequality $-2 < x \leq 3$ is read "-2 is less than x *and* x is less than or equal to 3," or "x is greater than -2 *and* less than or equal to 3." To be a solution of this inequality, a number must be a solution of both $-2 < x$ *and* $x \leq 3$. The number 1 is a solution, as are -0.5, 2, 2.5, and 3. The solution set is graphed as follows:

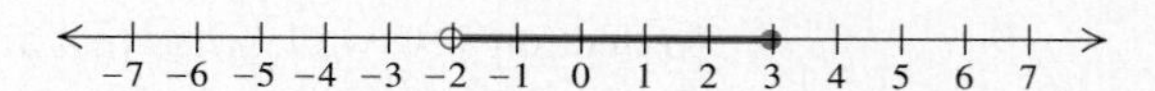

The open dot at -2 means that -2 is not part of the graph. The closed dot at 3 means that 3 is part of the graph. The other solutions are shaded. ❑

Solving Inequalities Using the Addition Principle

Consider a balance similar to one that appears in Section 2.1. Should one side of the balance weigh more than the other, the balance will tip in that direction. If equal amounts of weight are then added or subtracted on each side of the balance, the balance will remain tipped in the same direction.

The balance illustrates the idea that when a number, such as 2, is added (or subtracted) on both sides of an inequality, such as $3 < 7$, we get another true inequality:

$$3 + 2 < 7 + 2, \quad \text{or} \quad 5 < 9.$$

Similarly, if we add -3 on both sides, we get another true inequality:

$$3 + (-3) < 7 + (-3), \quad \text{or} \quad 0 < 4.$$

The Addition Principle for Inequalities

For any real numbers a, b, and c,

if $a < b$, then $a + c < b + c$; if $a > b$, then $a + c > b + c$;

if $a \leqslant b$, then $a + c \leqslant b + c$; if $a \geqslant b$, then $a + c \geqslant b + c$.

As with equation solving, when solving inequalities, our objective is to isolate a variable on one side.

EXAMPLE 4

Solve $x + 2 > 8$ and then graph the solution.

Solution We use the addition principle, subtracting 2 on both sides:

$$x + 2 - 2 > 8 - 2 \qquad \text{Subtracting 2 or adding } -2 \text{ on both sides}$$

$$x > 6.$$

Using the addition principle, we found an inequality, $x > 6$, from which we can determine the solutions easily.

Any number greater than 6 makes $x > 6$ true and is a solution of that inequality as well as the inequality $x + 2 > 8$. The graph is as follows:

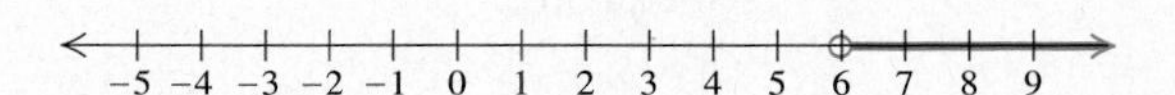

Because there are infinitely many solutions of most inequalities, we cannot possibly check them all. A partial check can be made using one of the possible solutions. In this case, we can check by substituting any number greater than 6, say 6.1, into the original inequality:

$x + 2 > 8$	
$6.1 + 2$	? 8
8.1	8 TRUE

Since $8.1 > 8$ is true, 6.1 is a solution. Any number greater than 6 is a solution. ❑

Although the inequality $x > 6$ is easy to solve (we merely replace x with numbers greater than 6), it is important to note that $x > 6$ is an *inequality,* not a *solution.* In fact, the solutions of $x > 6$ are numbers. To describe the set of all solutions, we will use **set-builder notation** to write the *solution set* of Example 4 as

$$\{x | \, x > 6\}.$$

This notation is read

"The set of all x such that x is greater than 6."

Thus a number is in $\{x|\ x > 6\}$ if that number is greater than 6. From now on, solutions of inequalities will be written using set-builder notation.

EXAMPLE 5

Solve $3x - 1 \leq 2x - 5$ and then graph the solution.

Solution

$$3x - 1 \leq 2x - 5$$

$$3x - 1 + 1 \leq 2x - 5 + 1 \quad \text{Adding 1}$$

$$3x \leq 2x - 4 \quad \text{Simplifying}$$

$$3x - 2x \leq 2x - 4 - 2x \quad \text{Subtracting } 2x$$

$$x \leq -4 \quad \text{Simplifying}$$

The graph is as follows:

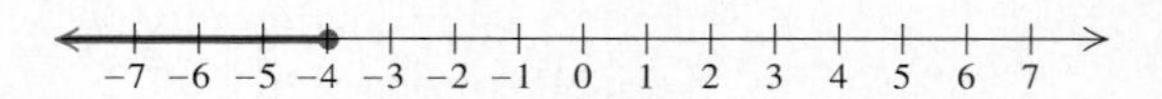

Any number less than or equal to -4 is a solution so the solution set is $\{x|\ x \leq -4\}$. ❑

Solving Inequalities Using the Multiplication Principle

There is a multiplication principle for inequalities similar to that for equations, but it must be modified when multiplying on both sides by a negative number. Consider the inequality

$$3 < 7.$$

If we multiply on both sides by a *positive* number like 2, we get another true inequality:

$$3 \cdot 2 < 7 \cdot 2, \quad \text{or} \quad 6 < 14. \qquad \text{TRUE}$$

If we multiply on both sides by a negative number like -2, we get a *false* inequality:

$$3 \cdot (-2) < 7 \cdot (-2), \quad \text{or} \quad -6 < -14. \qquad \text{FALSE}$$

The fact that $6 < 14$ is true, but $-6 < -14$ is false, stems from the fact that the negative numbers, in a sense, mirror the positive numbers. That is, whereas 14 is to the *right* of 6, the number -14 is to the *left* of -6. Thus if we reverse the inequality symbol in $-6 < -14$, we get a true inequality:

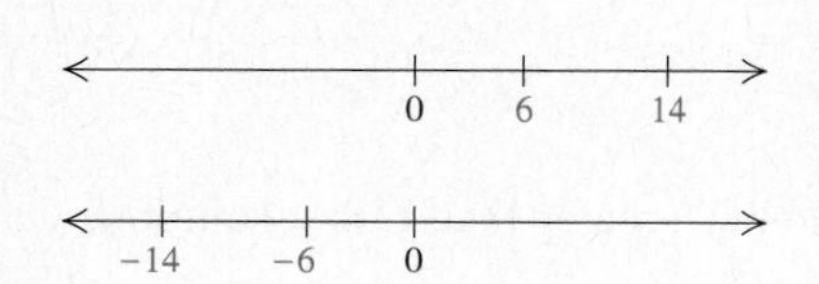

$$-6 > -14. \qquad \text{TRUE}$$

The Multiplication Principle for Inequalities

For any real numbers a and b, and for any *positive* number c,

if $a < b$, then $ac < bc$ and if $a > b$, then $ac > bc$.

For any real numbers a and b, and for any *negative* number c,

if $a < b$, then $ac > bc$ and if $a > b$, then $ac < bc$.

Similar statements hold for $\leq$ and $\geq$.

EXAMPLE 6

Solve and graph each inequality: **(a)** $\frac{1}{4}x < 7$; **(b)** $-2y < 18$.

Solution

a) $\frac{1}{4}x < 7$

$4 \cdot \frac{1}{4}x < 4 \cdot 7$ Multiplying by 4, the reciprocal of $\frac{1}{4}$

The symbol stays the same.

$x < 28$ Simplifying

The solution set is $\{x \mid x < 28\}$. The graph is as follows:

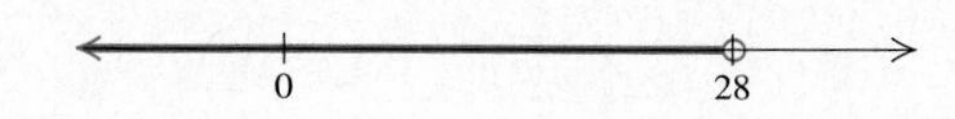

b) $-2y < 18$

$\frac{-2y}{-2} > \frac{18}{-2}$ Multiplying by $-\frac{1}{2}$ or dividing by -2

The symbol must be reversed!

$y > -9$ Simplifying

As a partial check we substitute a number greater than -9, say -8, into the original inequality:

$$\begin{array}{r|l} -2y & < 18 \\ \hline -2(-8) & ?\ 18 \\ 16 & 18 \quad \text{TRUE} \end{array}$$

The solution set is $\{y \mid y > -9\}$. The graph is as follows:

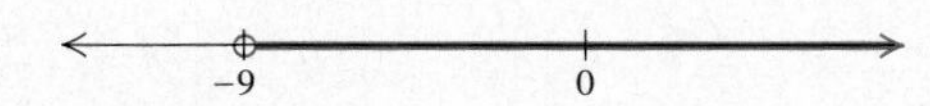

❑

Using the Principles Together

We use the addition and multiplication principles together to solve inequalities much as we did when solving equations. We generally use the addition principle first.

EXAMPLE 7

Solve: **(a)** $6 - 5y > 7$; **(b)** $2x - 9 \leq 7x + 1$.

Solution

a) $6 - 5y > 7$

$-6 + 6 - 5y > -6 + 7$ Adding -6

$-5y > 1$ Simplifying

$-\frac{1}{5} \cdot (-5y) < -\frac{1}{5} \cdot 1$ Multiplying by $-\frac{1}{5}$

The symbol must be reversed.

$y < -\frac{1}{5}$ Simplifying

As a check, we substitute a number smaller than $-\frac{1}{5}$, say -1, into the original inequality:

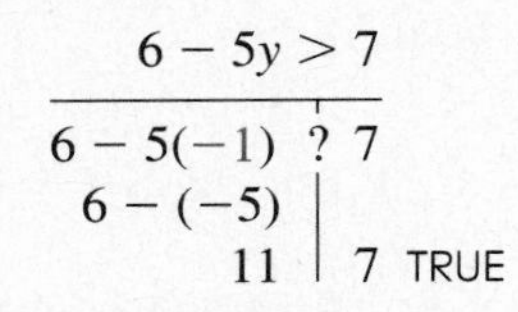

$$\begin{array}{c|l} \multicolumn{2}{c}{6 - 5y > 7} \\ \hline 6 - 5(-1) & ?\ 7 \\ 6 - (-5) & \\ 11 & 7 \text{ TRUE} \end{array}$$

The solution set is $\{y \mid y < -\frac{1}{5}\}$.

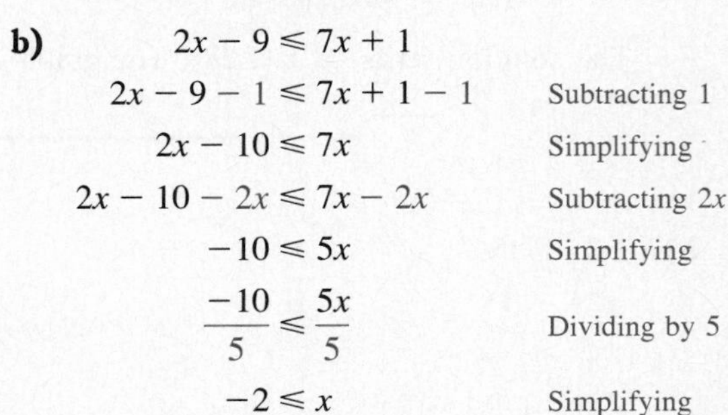

b)

$$\begin{aligned} 2x - 9 &\leq 7x + 1 & & \\ 2x - 9 - 1 &\leq 7x + 1 - 1 & &\text{Subtracting 1} \\ 2x - 10 &\leq 7x & &\text{Simplifying} \\ 2x - 10 - 2x &\leq 7x - 2x & &\text{Subtracting } 2x \\ -10 &\leq 5x & &\text{Simplifying} \\ \frac{-10}{5} &\leq \frac{5x}{5} & &\text{Dividing by 5} \\ -2 &\leq x & &\text{Simplifying} \end{aligned}$$

The solution set is $\{x \mid -2 \leq x\}$, or $\{x \mid x \geq -2\}$. ❑

All of the equation-solving techniques used in Sections 2.1 and 2.2 can be used with inequalities provided we remember to reverse the inequality symbol when multiplying or dividing on both sides by a negative number.

EXAMPLE 8

Solve: **(a)** $16.3 - 7.2p \leq -8.18$; **(b)** $3(x - 2) - 1 < 2 - 5(x + 6)$.

Solution

a) The greatest number of decimal places in any one number is *two*. Multiplying by 100, which has two 0's, will clear decimals. Then we proceed as before.

$$\begin{aligned} 16.3 - 7.2p &\leq -8.18 & & \\ 100(16.3 - 7.2p) &\leq 100(-8.18) & &\text{Multiplying by 100 on both sides} \\ 100(16.3) - 100(7.2p) &\leq 100(-8.18) & &\text{Using the distributive law} \\ 1630 - 720p &\leq -818 & &\text{Simplifying} \\ -720p &\leq -818 - 1630 & &\text{Subtracting 1630 on both sides} \\ -720p &\leq -2448 & &\text{Simplifying} \\ p &\geq \frac{-2448}{-720} & &\text{Multiplying by } -\frac{1}{720} \end{aligned}$$

The symbol must be reversed.

$$p \geq 3.4$$

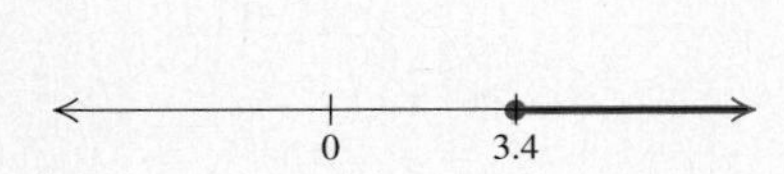

The solution set is $\{p \mid p \geq 3.4\}$.

b)

$$\begin{aligned} 3(x - 2) - 1 &< 2 - 5(x + 6) & & \\ 3x - 6 - 1 &< 2 - 5x - 30 & &\text{Using the distributive law to multiply and remove parentheses} \\ 3x - 7 &< -5x - 28 & &\text{Simplifying} \\ 3x + 5x &< -28 + 7 & &\text{Adding } 5x \text{ and also 7, to get all } x\text{-terms on one side and all other terms on the other side} \end{aligned}$$

and

$$8x < -21 \quad \text{Simplifying}$$

$$x < -\frac{21}{8} \quad \text{Multiplying by } \frac{1}{8}$$

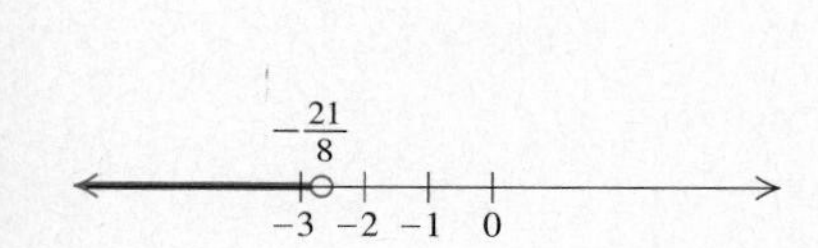

The solution set is $\{x \mid x < -\frac{21}{8}\}$. ❑

EXERCISE SET | 2.6

Determine whether each number is a solution of the given inequality.

1. $x > -4$
a) 4 **b)** 0 **c)** −4.1
d) −3.9 **e)** 5.6

2. $y < 5$
a) 0 **b)** 5 **c)** 4.99
d) −13 **e)** $7\frac{1}{4}$

3. $x \geq 6$
a) −6 **b)** 0 **c)** 6
d) 6.01 **e)** $-3\frac{1}{2}$

4. $x \leq 10$
a) 4 **b)** −10 **c)** 0
d) 10.2 **e)** −4.7

Graph on a number line.

5. $x > 4$ **6.** $y < 0$ **7.** $t < -3$
8. $y > 5$ **9.** $m \geq -1$ **10.** $p \leq 3$
11. $-3 < x \leq 4$ **12.** $-5 \leq x < 2$
13. $0 < x < 3$ **14.** $-5 \leq x \leq 0$

Describe the graph using set-builder notation.

15.

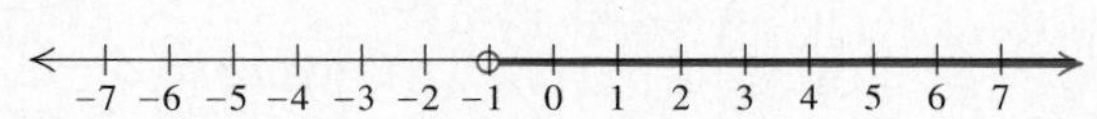

16.

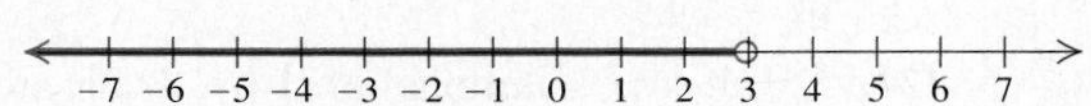

17.

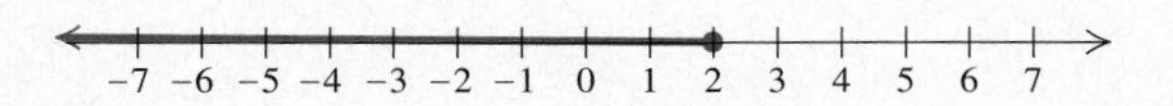

18.

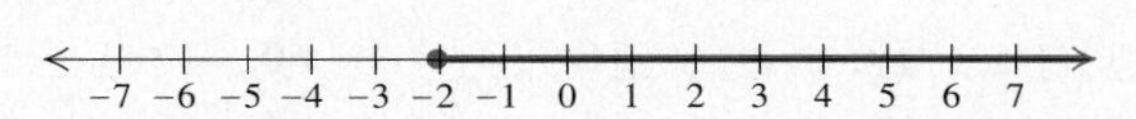

19.

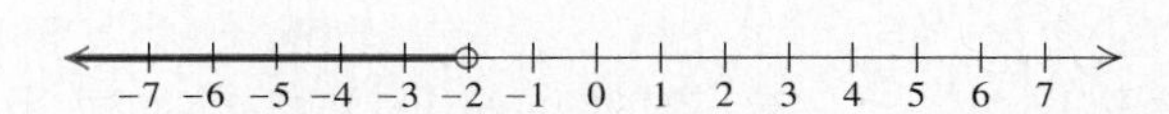

20.

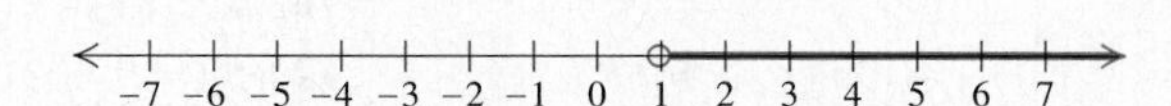

21.

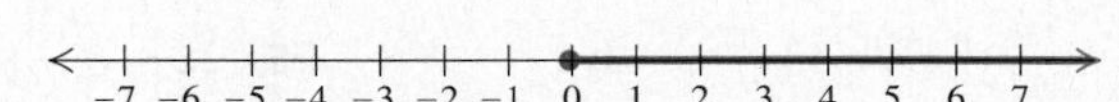

22.

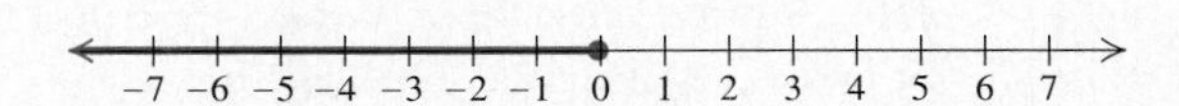

Solve using the addition principle. Graph and write set-builder notation for the answers.

23. $y + 5 > 8$ **24.** $y + 7 > 9$
25. $x + 8 \leq -10$ **26.** $x + 9 \leq -12$
27. $x - 7 < 9$ **28.** $x - 3 < 14$
29. $x - 6 \geq 2$ **30.** $x - 9 \geq 4$
31. $y - 7 > -12$ **32.** $y - 10 > -16$
33. $2x + 3 \leq x + 5$ **34.** $2x + 4 \leq x + 7$

Solve using the addition principle. Write the answers in set-builder notation.

35. $3x - 6 \geq 2x + 7$ **36.** $3x - 9 \geq 2x + 11$

37. $5x - 6 < 4x - 2$

38. $6x - 8 < 5x - 9$

39. $7 + c > 7$

40. $-9 + c > 9$

41. $y + \frac{1}{4} \leq \frac{1}{2}$

42. $y + \frac{1}{3} \leq \frac{5}{6}$

43. $x - \frac{1}{3} > \frac{1}{4}$

44. $x - \frac{1}{8} > \frac{1}{2}$

45. $-14x + 21 > 21 - 15x$

46. $-10x + 15 > 18 - 11x$

Solve using the multiplication principle. Graph and write set-builder notation for the answers.

47. $5x < 35$

48. $8x \geq 32$

49. $9y \leq 81$

50. $10x > 240$

51. $7x < 13$

52. $8y < 17$

53. $12x > -36$

54. $16x < -64$

Solve using the multiplication principle. Write the answers in set-builder notation.

55. $5y \geq -2$

56. $7x > -4$

57. $-2x \leq 12$

58. $-3y \leq 15$

59. $-4y \geq -16$

60. $-7x < -21$

61. $-3x < -17$

62. $-5y > -23$

63. $-2y > \frac{1}{7}$

64. $-4x \leq \frac{1}{9}$

65. $-\frac{6}{5} \leq -4x$

66. $-\frac{7}{8} > -56t$

Solve using the addition and multiplication principles.

67. $4 + 3x < 28$

68. $5 + 4y < 37$

69. $6 + 5y \geq 36$

70. $7 + 8x \geq 71$

71. $3x - 5 \leq 13$

72. $5y - 9 \leq 21$

73. $13x - 7 < -46$

74. $8y - 4 < -52$

75. $5x + 3 \geq -7$

76. $7y + 4 \geq -10$

77. $13 < 4 - 3y$

78. $22 < 6 - 8x$

79. $30 > 3 - 9x$

80. $40 > 5 - 7y$

81. $3 - 6y > 23$

82. $8 - 2y > 14$

83. $-3 < 8x + 7 - 7x$

84. $-5 < 9x + 8 - 8x$

85. $6 - 4y > 4 - 3y$

86. $7 - 8y > 5 - 7y$

87. $5 - 9y \leq 2 - 8y$

88. $6 - 13y \leq 4 - 12y$

89. $21 - 8y < 6y + 49$

90. $33 - 12x < 4x + 97$

91. $27 - 11x > 14x - 18$

92. $42 - 13y > 15y - 19$

93. $2.1x + 45.2 > 3.2 - 8.4x$

94. $0.96y - 0.79 \leq 0.21y + 0.46$

95. $0.7n - 15 + n \geq 2n - 8 - 0.4n$

96. $1.7t + 8 - 1.62t < 0.4t - 0.32 + 8$

97. $\frac{x}{3} - 2 \leq 1$

98. $\frac{2}{3} - \frac{x}{5} < \frac{4}{15}$

99. $\frac{y}{5} + 1 \leq \frac{2}{5}$

100. $\frac{3x}{5} \geq -15$

101. $3(2y - 3) < 27$

102. $4(2y - 3) > 28$

103. $5(d + 4) \leq 7(d - 2)$

104. $3(t - 2) \geq 9(t + 2)$

105. $8(2t + 1) > 4(7t + 7)$

106. $7(5x - 2) < 6(6x - 1)$

107. $3(r - 6) + 2 < 4(r + 2) - 21$

108. $5(t + 3) + 9 > 3(t - 2) + 6$

109. $\frac{2}{3}(2x - 1) \geq 10$

110. $\frac{4}{5}(3x + 4) \leq 20$

111. $\frac{3}{4}\left(3x - \frac{1}{2}\right) - \frac{2}{3} < \frac{1}{3}$

112. $\frac{2}{3}\left(\frac{7}{8} - 4x\right) - \frac{5}{8} < \frac{3}{8}$

Skill Maintenance

Simplify.

113. $10 \div 2 \cdot 5 - 3^2 + (-4)^2$

114. $7 - 3^2 + (8 - 3)^2 \cdot 4$

Synthesis

115. ◈ Are all solutions of $x > -5$ solutions of $-x < 5$? Why or why not?

116. ◈ Explain in your own words why it is necessary to reverse the inequality symbol when multiplying both sides of an inequality by a negative number.

Solve.

117. $2[4 - 2(3 - x)] - 1 \geq 4[2(4x - 3) + 7] - 25$

118. $5[3(7 - t) - 4(8 + 2t)] - 20 < -6[2(6 + 3t) - 4]$

Solve for x.

119. $-(x + 5) \geq 4a - 5$

120. $\frac{1}{2}(2x + 2b) > \frac{1}{3}(21 + 3b)$

121. $y < ax + b$ (Assume $a > 0$.)

122. $y < ax + b$ (Assume $a < 0$.)

123. Determine whether each number is a solution of the inequality $|x| < 3$.

a) 0
b) -2
c) -3
d) 4
e) 3
f) 1.7
g) -2.8

124. Graph the solutions of $|x| < 3$ on a number line.

2.7 Problem Solving Using Inequalities

Translating to Inequalities • Solving Problems

The five steps for problem solving can be used for problems involving inequalities.

Translating to Inequalities

Before solving problems that involve inequalities, we list some important phrases to look for. Sample translations are listed as well.

Important Words	Sample Sentence	Translation
is at least	Bill is at least 21 years old.	$b \geq 21$
is at most	At most 5 students dropped the course.	$n \leq 5$
cannot exceed	To qualify, earnings cannot exceed \$12,000.	$r \leq 12{,}000$
must exceed	The speed must exceed 15 mph.	$s > 15$
is less than	Spot's weight is less than 50 lb.	$w < 50$
is more than	Boston is more than 200 miles away.	$d > 200$
is between	The film was between 90 and 100 minutes long.	$90 < t < 100$

Solving Problems

EXAMPLE 1

Martha is taking an introductory algebra course in which four tests are to be given. To get an A, a student must average at least 90 on the four tests. Martha got scores of 96, 82, and 91 on the first three tests. Determine (in terms of an inequality) what scores on the last test will earn her an A.

Solution

1. FAMILIARIZE. Many students in Martha's situation might make a guess, like 87, and then compute the average of the four scores. The average of the four scores is their sum divided by the number of tests, 4:

$$\frac{96 + 82 + 91 + 87}{4} = 89.$$

Our work shows that an 87 will not earn Martha her A, since the average is not *at least* 90. Rather than make a second guess, let's translate to an inequality and solve. We let $x =$ Martha's score on the last test.

2. TRANSLATE. Having familiarized ourselves with the meaning of average,

we can reword and translate, using x as the fourth score:

Rewording: The average of the four scores must be at least 90

Translating: $\dfrac{96 + 82 + 91 + x}{4} \geqslant 90.$

3. **CARRY OUT.** To solve the inequality, we first multiply by 4 to clear fractions:

$$\frac{96 + 82 + 91 + x}{4} \geqslant 90$$

$$4\left(\frac{96 + 82 + 91 + x}{4}\right) \geqslant 4 \cdot 90 \quad \text{Multiplying by 4}$$

$$96 + 82 + 91 + x \geqslant 360$$

$$269 + x \geqslant 360 \quad \text{Simplifying}$$

$$x \geqslant 91.$$

The solution set is $\{x|\ x \geqslant 91\}$.

4. **CHECK.** We can obtain a partial check by substituting a number greater than or equal to 91. We leave it to the student to try 92 in a manner similar to what was done in the familiarization step.

5. **STATE.** A score of 91 or better on the last test will give Martha an A in the course. ❑

EXAMPLE 2

The women's volleyball team can spend at most $400 for its awards banquet at a local restaurant. If the restaurant charges a $45 set-up fee plus $12.50 per person, at most how many can attend?

Solution

1. **FAMILIARIZE.** Suppose 20 people were to attend the dinner. The cost would then be $45 + $12.50 · 20, or $295. This shows that more than 20 people could attend without exceeding $400. We could next make another guess, or subtract $295 from $400 and "use up" the difference. Instead, we translate to an inequality and solve. We let $n =$ the number of people in attendance.

2. **TRANSLATE.** The cost of the banquet will be $45 for the set-up fee plus $12.50 times the number of people attending. We can reword as follows:

Rewording: The set-up fee plus the cost of the dinners cannot exceed $400.

Translating: $45 + 12.50 \cdot n \leqslant 400.$

3. **CARRY OUT.** We solve the inequality:

$$45 + 12.50n \leqslant 400$$

$$12.5n \leqslant 355 \quad \text{Subtracting 45}$$

$$n \leqslant \frac{355}{12.5} \quad \text{Dividing by 12.5}$$

$$n \leqslant 28.4 \quad \text{Simplifying}$$

The solution set of the inequality is $\{n \mid n \leq 28.4\}$.

4. CHECK. Although the solution set of the inequality is all numbers less than or equal to 28.4, since n represents the number of people in attendance, we round down to 28. If 28 people attend, the cost will be $\$45 + \$12.50 \cdot 28$, or $395, and if 29 attend, the cost will exceed $400.

5. STATE. At most 28 people can attend the banquet. ❑

In the check to Example 2, an important point is made: Solutions of equations or inequalities do not always solve the problem from which the equation or inequality originates. In some cases, answers must be nonnegative, and in other cases, answers must be integers. Thus it is important to always check that the original problem has been solved.

EXERCISE SET 2.7

Translate to an inequality.

1. A number is greater than 4.

2. A number is less than 7.

3. A number is less than or equal to -6.

4. A number is greater than or equal to 13.

5. The temperature is at most 80°.

6. The bag weighs at least 2 pounds.

7. Between 75 and 100 people attended the concert.

8. The average speed was between 90 and 110 mph.

9. The number of people is at least 1200.

10. The cost is at most $3457.95.

11. The amount of acid is not to exceed 500 liters.

12. The cost of gasoline is no less than 99 cents per gallon.

13. Two more than 3 times a number is less than 13.

14. Five less than one half of a number is greater than 17.

Solve. Use the five steps for problem solving.

15. Your quiz grades are 73, 75, 89, and 91. What scores on a fifth quiz will make your average quiz grade at least 85? Use an inequality.

16. Alvin is taking a literature course in which four tests are to be given. To get a B, a student must average at least 80 on the four tests. Alvin got scores of 82, 76, and 78 on the first three tests. What scores on the last test will earn him at least a B? Use an inequality.

17. Ridem rents trucks at a daily rate of $42.95 plus $0.46 per mile. A family wants a one-day truck rental, but must stay within a budget of $200. What mileages will allow the family to stay within budget? Use an inequality and round to the nearest tenth of a mile.

18. Atlas rents a cargo van at a daily rate of $44.95 plus $0.39 per mile. A business has budgeted $250 for a one-day van rental. What mileages will allow the business to stay within budget? Use an inequality and round to the nearest tenth of a mile.

19. The width of a rectangle is fixed at 4 cm. For what lengths will the area be less than 86 cm²? Use an inequality.

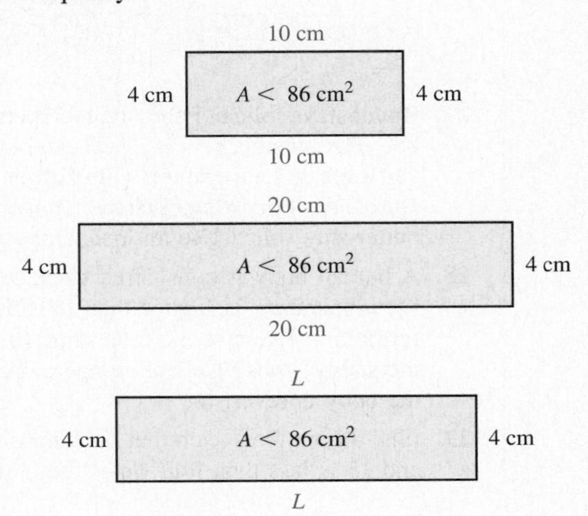

20. The width of a rectangle is fixed at 16 yd. For what lengths will the area be at least 264 yd^2? Use an inequality.

21. Laura is certain that every time she parks in the municipal garage it costs her at least \$2.20. If the garage charges 45¢ plus 25¢ for each half hour, how long is Laura's car parked? Use an inequality.

22. Simon claims that it costs him at least \$3.00 every time he calls a customer from a pay phone. If a typical call costs 75¢ plus 45¢ for each minute, how long do his calls last? Use an inequality.

23. The formula $R = -0.075t + 3.85$ can be used to predict the world record in the 1500-m run t years after 1930. Determine (in terms of an inequality) those years for which the world record will be less than 3.5 min.

24. The formula $R = -0.028t + 20.8$ can be used to predict the world record in the 200-m dash t years after 1920. Determine (in terms of an inequality) those years for which the world record will be less than 19.0 sec.

25. A 9-lb puppy is gaining weight at a rate of $\frac{3}{4}$ lb per week. When will the puppy's weight exceed $22\frac{1}{2}$ lb? Use an inequality.

26. On July 1, Garrett's Pond was 25 ft deep. Since that date, the water level has dropped $\frac{2}{3}$ ft per week. For what dates will the water level not exceed 21 ft? Use an inequality.

27. Butter stays solid at Fahrenheit temperatures below 88°. The formula $F = \frac{9}{5}C + 32$ can be used to convert Celsius temperatures C to Fahrenheit temperatures F. For what Celsius temperatures does butter stay solid? Use an inequality.

28. A human body is considered to be feverish when its temperature is higher than 98.6°F. Using the formula in Exercise 27, determine (in terms of an inequality) those Celsius temperatures for which the body is feverish.

29. Find all numbers such that the sum of the number and 15 is less than four times the number.

30. Find all numbers such that three times the number minus ten times the number is greater than or equal to eight times the number.

31. The length of a rectangle is fixed at 26 cm. What widths will make the perimeter greater than 80 cm?

32. The width of a rectangle is fixed at 8 ft. What lengths will make the perimeter at least 200 ft? at most 200 ft?

33. One side of a triangle is 2 cm shorter than the base. The other side is 3 cm longer than the base. What lengths of the base will allow the perimeter to be greater than 19 cm?

34. The perimeter of a rectangular swimming pool is not to exceed 70 ft. The length is to be twice the width. What widths will meet these conditions?

35. Dot's Electric made 17 customer calls last week and 22 calls this week. How many calls must be made next week in order to maintain an average of at least 20 for the three-week period?

36. George and Joan do volunteer work at a hospital. Joan worked 3 more hours than George, and together they worked more than 27 hours. What possible numbers of hours did each work?

37. Angelo is shopping for a new pair of jeans and two sweaters of the same kind. He is determined to spend no more than \$120.00 for the clothes. He buys jeans for \$21.95. What is the most that Angelo can spend for each sweater?

38. The medium-sized box of dog food weighs 1 lb more than the small size. The large size weighs 2 lb more than the small size. The total weight of the three boxes is at most 30 lb. What are the possible weights of the small box?

39. The width of a rectangle is 32 km. What lengths will make the area at least 2048 km^2?

40. The height of a triangle is 20 cm. What lengths of the base will make the area at most 40 cm^2?

41. The average price of a movie ticket can be estimated by the equation $P = 0.1522Y - 298.592$, where Y represents the year and P the average price, in dollars. The price is lower than what might be expected due to senior-citizen discounts, children's prices, and special volume discounts. For what years will the average price of a movie ticket be at least \$6?

42. The equation $y = 0.027x + 0.19$ can be used to determine the approximate cost y, in dollars, of driving x miles on the Indiana toll road. For what mileages x will the cost be at most \$6?

Skill Maintenance

Simplify.

43. $-3 + 2(-5)^2(-3) - 7$

44. $7 - a^2 - 9 + 5a^2$

45. $9x - 5 + 4x^2 - 2 - 13x$

46. $3x + 2[4 - 5(2x - 1)]$

Synthesis

47. The symbols $\not<$, $\not>$, $\not\leq$, and $\not\geq$ have not been discussed. Do you feel that this was an oversight? Why or why not?

48. Suppose that t = Todd's age and f = Frances's age. Write a sentence that would translate to the inequality

$$t \geq f + 10.$$

49. The area of a square can be no more than 64 cm^2. What lengths of a side will allow this?

50. The sum of two consecutive odd integers is less than 100. What is the largest possible pair of such integers?

51. Mack's Parking Garage charges \$4.00 for the first hour and \$2.50 for each additional hour. For how long has a car been parked when the charge exceeds \$16.50?

52. A salesperson can choose to be paid in one of two ways.

Plan A: A salary of \$600 per month, plus a commission of 4% of gross sales

Plan B: A salary of \$800 per month, plus a commission of 6% of gross sales over \$10,000

For what gross sales is plan A better than plan B, assuming that gross sales are always more than \$10,000?

53. When asked how much the parking charge is for a certain car (see Exercise 51), Mack replies "between 14 and 24 dollars." For how long has the car been parked?

54. Green ski wax works best at Fahrenheit temperatures between 5° and 15°. Determine those Celsius temperatures for which green ski wax works best. (See Exercise 27.)

55. Chassman and Bem Booksellers offers a preferred customer card for \$25. The card entitles a customer to a 10% discount on all purchases for a period of one year. Under what circumstances would an individual save money by purchasing a card?

56. After 9 quizzes, Jackie's average is 84. Is it possible for Jackie to improve her average two points with the next quiz? Why or why not?

SUMMARY AND REVIEW 2

KEY TERMS

Equivalent equations, p. 63
Coefficient, p. 66
Clearing fractions and decimals, p. 71
Formula, p. 75
Consecutive integers, p. 86
Graph of an inequality, p. 95
Solution set, p. 97
Set-builder notation, p. 97

IMPORTANT PROPERTIES AND FORMULAS

Solving Equations

The Addition Principle: For any real numbers a, b, and c, if $a = b$, then $a + c = b + c$.

The Multiplication Principle: For any real numbers a, b, and c, if $a = b$, then $ac = bc$.

Solving Inequalities

The Addition Principle: For any real numbers a, b, and c,
if $a < b$, then $a + c < b + c$, and
if $a > b$, then $a + c > b + c$.

The Multiplication Principle: For any real numbers a and b and any *positive* number c,
if $a < b$, then $ac < bc$, and
if $a > b$, then $ac > bc$.

For any real numbers a and b and any *negative* number c,
if $a < b$, then $ac > bc$, and
if $a > b$, then $ac < bc$.

An Equation-Solving Procedure

1. If necessary, use the distributive law to remove parentheses. Then collect like terms on each side.
2. Multiply on both sides to clear fractions or decimals. (This is optional, but it can ease computations.)
3. Get all terms with variables on one side and all constant terms on the other side, using the addition principle.
4. Collect like terms again, if necessary.
5. Multiply or divide to solve for the variable, using the multiplication principle.
6. Check all possible solutions in the original equation.

Percent Notation

$n\%$ means $n \times 0.01$, or $n \times \frac{1}{100}$, or $\frac{n}{100}$.

Five Steps for Problem Solving in Algebra

1. *Familiarize* yourself with the problem situation.
2. *Translate* to mathematical language. (This often means write an equation.)
3. *Carry out* some mathematical manipulation. (This often means *solve* an equation.)
4. *Check* your possible answer in the original problem.
5. *State* the answer clearly.

REVIEW EXERCISES

This chapter's review and test include Skill Maintenance exercises from Sections 1.1, 1.2, 1.7, and 1.8.

Solve.

1. $x + 5 = -17$

2. $-8x = -56$

3. $-\frac{x}{4} = 48$

4. $n - 7 = -6$

5. $15x = -35$

6. $x - 11 = 14$

7. $-\frac{2}{3} + x = -\frac{1}{6}$

8. $\frac{4}{5}y = -\frac{3}{16}$

9. $y - 0.9 = 9.09$

10. $5 - x = 13$

11. $5t + 9 = 3t - 1$

12. $7x - 6 = 25x$

13. $\frac{1}{4}x - \frac{5}{8} = \frac{3}{8}$

14. $14y = 23y - 17 - 10$

15. $0.22y - 0.6 = 0.12y + 3 - 0.8y$

16. $\frac{1}{4}x - \frac{1}{8}x = 3 - \frac{1}{16}x$

17. $4(x + 3) = 36$

18. $3(5x - 7) = -66$

19. $8(x - 2) = 5(x + 4)$

20. $-5x + 3(x + 8) = 16$

21. $C = \pi d$, for d

22. $V = \frac{1}{3}Bh$, for B

23. $A = \frac{a + b}{2}$, for a

24. Find decimal notation: 0.7%.

25. Find percent notation: $\frac{11}{25}$.

26. What percent of 60 is 12?

27. 198 is 55% of what number?

Determine whether the given number is a solution of the inequality $x \leq 4$.

28. -3

29. 7

30. 4

Graph on a number line.

31. $4x - 6 < x + 3$

32. $-2 < x \leq 5$

33. $y > 0$

Solve. Write the answers in set-builder notation.

34. $y + \frac{2}{3} \geq \frac{1}{6}$

35. $9x \geq 63$

36. $2 + 6y > 14$

37. $7 - 3y \geq 27 + 2y$

38. $3x + 5 < 2x - 6$

39. $-4y < 28$

40. $3 - 4x < 27$

41. $4 - 8x < 13 + 3x$

42. $-3y \geq -21$

43. $-4x \leq \frac{1}{3}$

44. A color television sold for \$629 in May. This was \$38 more than the cost in January. Find the cost in January.

45. Selma gets a \$4 commission for each appliance that she sells. One week she received \$108 in commissions. How many appliances did she sell?

46. An 8-m board is cut into two pieces. One piece is 2 m longer than the other. How long are the pieces?

47. If 14 is added to three times a certain number, the result is 41. Find the number.

48. The sum of two consecutive odd integers is 116. Find the integers.

49. The perimeter of a rectangle is 56 cm. The width is 6 cm less than the length. Find the width and the length.

50. After a 30% reduction, an item is on sale for \$154. What was the marked price (the price before reducing)?

51. A businessperson's salary is \$30,000, which is a 15% increase over the previous year's salary. What was the previous salary (to the nearest dollar)?

52. The measure of the second angle of a triangle is 50° more than that of the first. The measure of the third angle is 10° less than twice the first. Find the measures of the angles.

53. Steve's quiz grades are 71, 75, 82, and 86. What is the lowest grade that he can get on the next quiz and still have an average of at least 80?

54. The length of a rectangle is 43 cm. What widths will make the perimeter greater than 120 cm?

Skill Maintenance

55. Evaluate $\frac{x-y}{5}$ when $x = 27$ and $y = 2$.

56. Multiply: $4(3t + 2 + s)$.

57. Divide: $12.42 \div (-5.4)$.

58. Remove parentheses and simplify:

$$5x - 8(6x - y).$$

Synthesis

59. ◈ What is the difference between using the multiplication principle for solving equations and for solving inequalities?

60. ◈ Explain how checking the solutions of an equation differs from checking the solutions of an inequality.

61. The total length of the Nile and Amazon Rivers is 13,108 km. If the Amazon were 234 km longer, it would be as long as the Nile. Find the length of each river.

62. Consumer experts advise us never to pay the sticker price for a car. A rule of thumb is to pay the sticker price minus 20% of the sticker price, plus \$200. A car is purchased for \$11,520 using the rule. What was the sticker price?

Solve.

63. $2|n| + 4 = 50$

64. $|3n| = 60$

65. $y = 2a - ab + 3$, for a

CHAPTER TEST 2

Solve.

1. $x + 7 = 15$

2. $t - 9 = 17$

3. $3x = -18$

4. $-\frac{4}{7}x = -28$

5. $3t + 7 = 2t - 5$

6. $\frac{1}{2}x - \frac{3}{5} = \frac{2}{5}$

7. $8 - y = 16$

8. $-\frac{2}{5} + x = -\frac{3}{4}$

9. $3(x + 2) = 27$

10. $-3x + 6(x + 4) = 9$

11. $0.4p + 0.2 = 4.2p - 7.8 - 0.6p$

Solve. Write the answers in set-builder notation.

12. $x + 6 \leq 2$

13. $14x + 9 > 13x - 4$

14. $12x \leq 60$

15. $-2y \geq 26$

16. $-4y \leq -32$

17. $-5x \geq \frac{1}{4}$

18. $4 - 6x > 40$

19. $5 - 9x \geq 19 + 5x$

Solve the formula for the given letter.

20. $A = 2\pi rh$, for r

21. $w = \frac{P - 2l}{2}$, for l

22. Find decimal notation: 200%.

23. Find percent notation: 0.054.

24. What number is 42% of 50?

25. What percent of 75 is 33?

Graph on a number line.

26. $y < 9$

27. $-2 \leq x \leq 2$

Solve.

28. The perimeter of a rectangle is 36 cm. The length is 4 cm greater than the width. Find the width and the length.

29. If you triple a number and then subtract 14, you get two thirds of the original number. What is the original number?

30. The sum of three consecutive odd integers is 249. Find the integers.

31. Money is invested in a savings account at 6% simple interest. After 1 year, there is \$2650 in the account. How much was originally invested?

32. Find all numbers such that six times the number is greater than the number plus 30.

33. The width of a rectangle is 96 yd. Find all possible lengths so that the perimeter of the rectangle will be at least 540 yd.

Skill Maintenance

34. Translate to an algebraic expression: 10 less than x.

35. Factor: $3a + 24b + 12$.

36. Multiply: $-\frac{3}{8} \cdot \left(-\frac{4}{5}\right)$.

37. Simplify: $8 + (-10) \div 5 \cdot 2 + 1$.

Synthesis

38. Solve $c = \dfrac{1}{a - d}$, for d.

39. Solve: $3|w| - 8 = 37$.

40. A movie theater had a certain number of tickets to give away. Five people got the tickets. The first got one third of the tickets, the second got one fourth of the tickets, and the third got one fifth of the tickets. The fourth person got eight tickets, and there were five tickets left for the fifth person. Find the total number of tickets given away.

CHAPTER 3

Introduction to Graphing

AN APPLICATION

A smoker is 15 times more likely to die from lung cancer than a nonsmoker. An exsmoker who stopped smoking t years ago is w times more likely to die from lung cancer than a nonsmoker, where $t + w = 15$.

a) Sandy gave up smoking $2\frac{1}{2}$ years ago. How much more likely is she to die from lung cancer than Polly, who never smoked?
b) Graph the equation, using t as the first coordinate.

This problem appears as Exercise 21 in Section 3.4.

Don Kissack
DIRECTOR OF BENEFITS

"Corporations are increasingly looking at how factors like diet, exercise, smoking, alcohol, and seat belt use affect the welfare of employees and the health-care costs for a business.

"Algebra and graphing are important tools in human resources, especially in the areas of compensation and benefits."

We now begin our study of graphing. First we will examine graphs as they commonly appear in newspapers or magazines and develop some terminology. Following that, we will practice graphing equations whose graphs are lines. Finally we will practice using graphs as a problem-solving tool for certain applications.

In addition to material from this chapter, the review and test for Chapter 3 include material from Sections 1.3, 1.4, 2.2, and 2.3.

3.1 Ordered Pairs and Graphs

Problem Solving with Graphs • Points and Ordered Pairs • Quadrants • Finding Coordinates

Today's print and electronic media make almost constant use of graphs. This is due in part to the ease with which some graphs are prepared by computer, and in part to the large quantity of information that a graph can display. We first consider problem solving with bar graphs, line graphs, and circle graphs. Then we examine graphs that use a coordinate system.

Problem Solving with Graphs

Bar Graphs

A *bar graph* is convenient for showing comparisons. The bars can be either vertical or horizontal. Typically, certain units, such as body weight in the graph of Example 1, are shown horizontally. With each horizontal number, there is associated a vertical number, or unit. In Example 1, the vertical unit is the number of drinks required to raise the blood alcohol level to a point at which driving is illegal in all 50 states, 0.10%.

EXAMPLE 1

Although some states use a cut-off of 0.08%, in *all* states, a blood-alcohol level of 0.10% or higher indicates that an individual has consumed too much alcohol to drive. The following bar graph* shows the number of drinks that a person of a certain

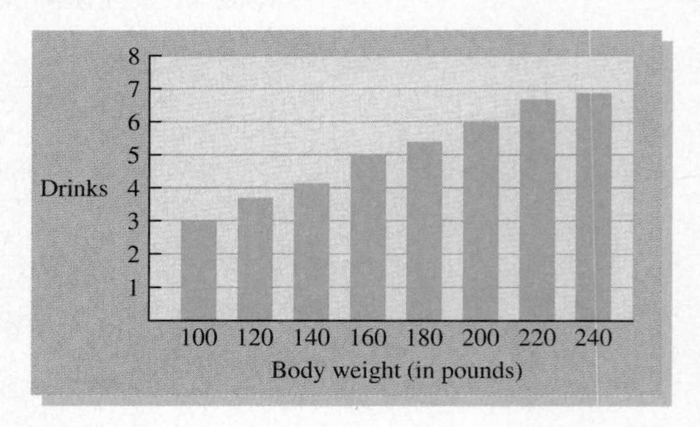

*Adapted from *Neighborhood Digest,* Vol. 7, No. 12.

weight would consume to achieve a blood-alcohol level of 0.10%. Note that a 12-ounce beer, a 5-ounce glass of wine, or a cocktail containing $1\frac{1}{2}$ ounces of distilled liquor all count as one drink.

a) Approximately how many drinks would a 200-pound person have consumed in order to have a blood-alcohol level of 0.10%?

b) At least how much would an individual have to weigh in order to consume 4 drinks without reaching a blood-alcohol level of 0.10%?

Solution

a) We go to the top of the bar that is above the body weight 200 lb. Then we move horizontally from the top of the bar to the vertical scale listing numbers of drinks. It appears that approximately 6 drinks will give a 200-lb person a blood-alcohol level of 0.10%.

b) By moving up the vertical scale to the number 4, and then moving horizontally, we see that the first bar to reach a height of 4 corresponds to a weight of 140 lb. An individual should weigh at least 140 lb if he or she wishes to consume 4 drinks without exceeding a blood-alcohol level of 0.10%. ❑

EXAMPLE 2

These reasons for dropping out of high school were given in a recent National Assessment of Educational Progress survey.

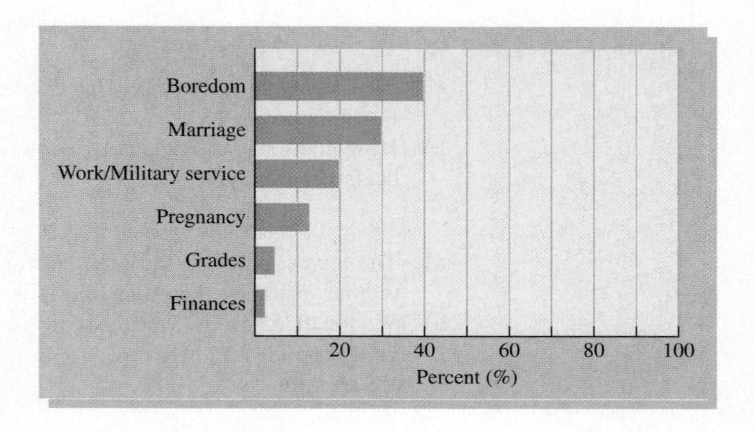

a) Approximately what percent of those surveyed dropped out because of pregnancy?

b) What reason for dropping out was given least often?

c) What reason for dropping out was given by about 30% of those surveyed?

Solution

a) We go to the right of the bar representing pregnancy and then look down to the percent scale. We find that approximately 12% dropped out because of pregnancy. Note that this is only an estimate.

b) The reason given least often is finances, since that is represented by the shortest bar.

c) We locate the 30% mark on the percent scale and then look up for a bar ending at approximately 30%.We then go across to the left and read the reason. The reason given by about 30% was marriage. ❑

Line Graphs

Line graphs are often used to show change over time. Certain points are drawn to represent given information. When segments are drawn to connect the points, a line graph is formed.

Sometimes it is impractical to begin the listing of horizontal or vertical values with zero. When this occurs, as in Example 3, the symbol ⌇ is used to indicate a break in the listing of values.

EXAMPLE 3

The following line graph shows the relationship between a person's resting pulse rate and months of regular exercise.*

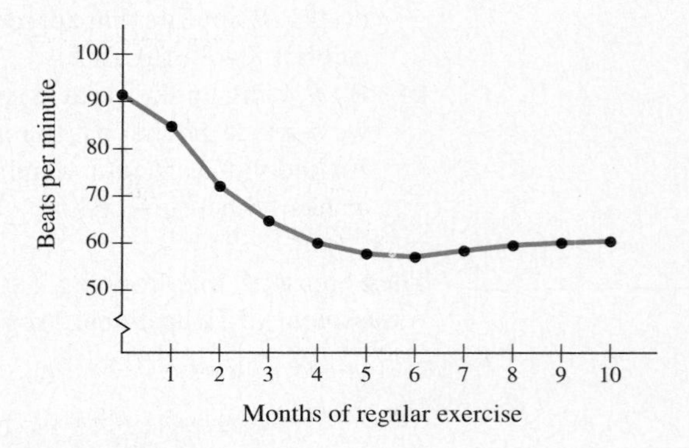

a) How many months of regular exercise are required to lower the pulse rate as much as possible?

b) How many months of regular exercise are needed to achieve a pulse rate of 65 beats per minute?

Solution

a) The lowest point on the graph occurs above the number 6. After 6 months of regular exercise, the pulse rate is lowered as much as possible.

b) We locate 65 on the vertical scale and then move right until the line is reached. At that point, we move down to the horizontal scale and read the information we are seeking.

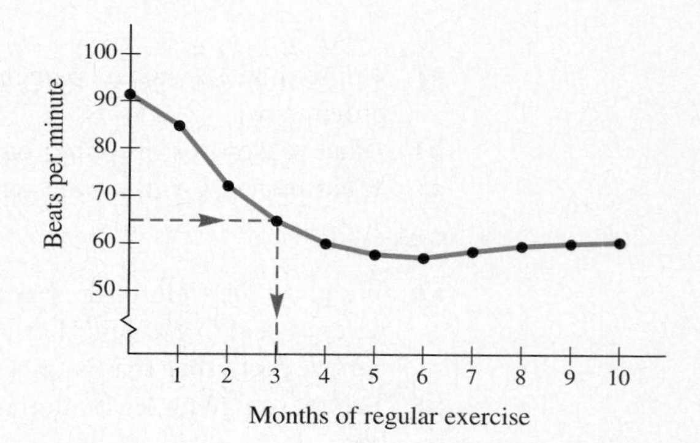

The pulse rate is 65 beats per minute after 3 months of regular exercise. ❑

*Data from *Body Clock* by Dr. Martin Hughes, p. 60. New York: Facts on File, Inc.

Circle Graphs

Circle graphs, or *pie charts,* are often used to show what percent of a whole each particular item in a group represents.

EXAMPLE 4

This pie chart shows expenses as a percent of income for a family of four, according to the Bureau of Labor Statistics.

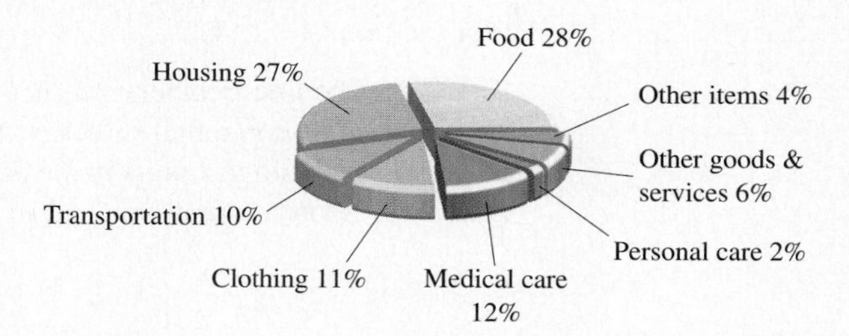

A family with a monthly income of \$2500 would typically spend how much on housing per month?

Solution

1. **FAMILIARIZE.** The chart tells us that housing is 27% of income. We let y = the amount spent on housing.
2. **TRANSLATE.** We reword and translate the problem as follows:

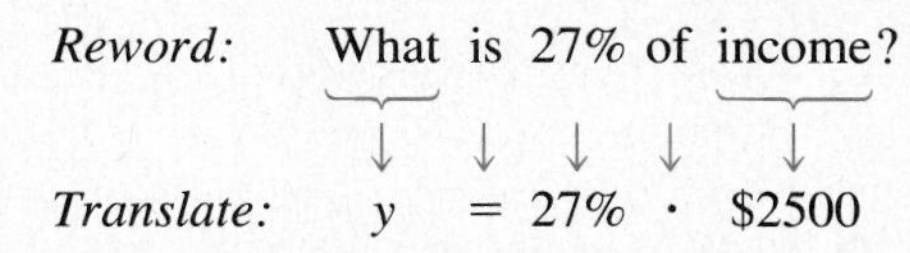

3. **CARRY OUT.** We solve by carrying out the computation:

   $y = 0.27 \cdot \$2500 = \$675.$

4. **CHECK.** We leave the check to the student.
5. **STATE.** The family would spend \$675 on housing. ❑

Points and Ordered Pairs

We have already graphed numbers on a number line. In order to graph an equation containing two variables, we must learn to graph pairs of numbers on a plane.

To graph pairs of numbers on a plane, we use two perpendicular number lines called **axes** (singular, **axis**). The axes cross at a point called the **origin.** Arrows on the axes show the positive directions.

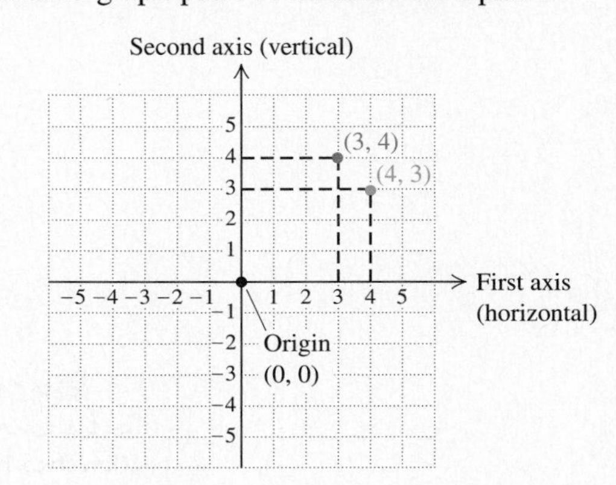

Consider the pair (3, 4). The numbers in such a pair are called **coordinates.** In (3, 4), the **first coordinate** is 3 and the **second coordinate** is 4. To plot (3, 4), we start at the origin and move horizontally to the 3. Then we move up vertically 4 units and make a ''dot.'' Note that (3, 4) is located above 3 on the first axis and to the right of 4 on the second axis.

The point (4, 3) is also plotted in the figure on the preceding page. Note that (3, 4) and (4, 3) give different points. They are called **ordered pairs** because the order in which the numbers appear is important.

EXAMPLE 5

Plot the point $(-3, 4)$.

Solution The first number, -3, is negative. Starting at the origin, we move 3 units in the negative horizontal direction (3 units to the left). The second number, 4, is positive, so we move 4 units in the positive vertical direction (up). The point $(-3, 4)$ is above -3 on the first axis and to the left of 4 on the second axis.

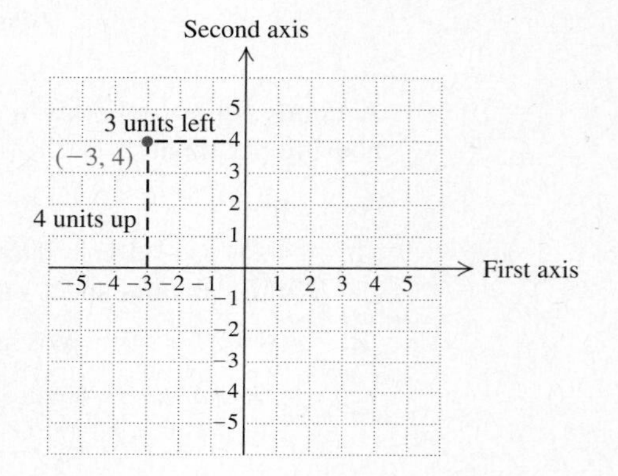

❑

Quadrants

The following figure shows some points and their coordinates. In region I (the *first quadrant*), both coordinates of any point are positive. In region II (the *second quadrant*), the first coordinate is negative and the second is positive. In region III (the *third quadrant*), both coordinates are negative. In region IV (the *fourth quadrant*), the first coordinate is positive and the second is negative.

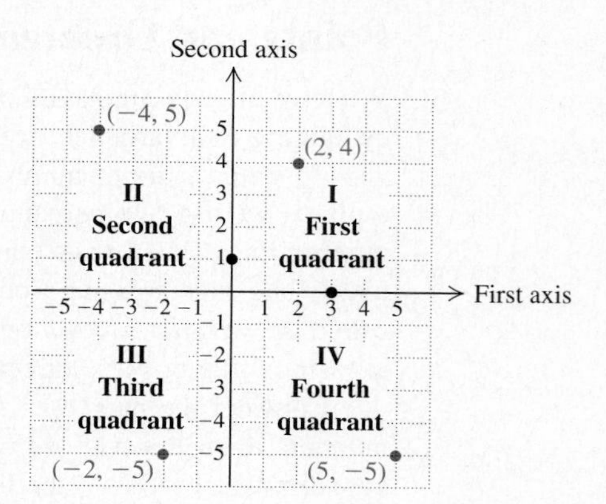

Note that the point $(-4, 5)$ is in the second quadrant and the point $(5, -5)$ is in the fourth quadrant. The points (3, 0) and (0, 1) are on the axes and are not considered to be in any quadrant.

Finding Coordinates

To find the coordinates of a point, we see how far to the right or left of the vertical axis it is and how far above or below the horizontal axis it is.

EXAMPLE 6

Find the coordinates of points A, B, C, D, E, F, and G.

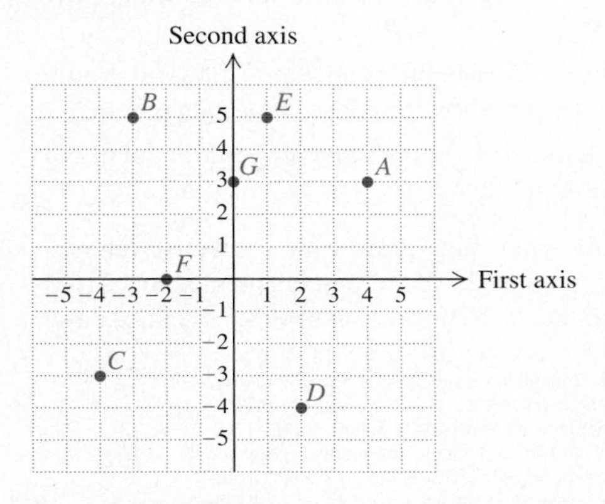

Solution Point A is 4 units to the right (horizontal direction) and 3 units up (vertical direction). Its coordinates are (4, 3). The coordinates of the other points are as follows:

B: $(-3, 5)$; C: $(-4, -3)$; D: $(2, -4)$;

E: $(1, 5)$; F: $(-2, 0)$; G: $(0, 3)$. ❑

EXERCISE SET 3.1

Use the bar graph in Example 1 to answer Exercises 1–4.

1. Approximately how many drinks would it take for a 100-lb person to reach a blood-alcohol level of 0.10%?
2. Approximately how many drinks would it take for a 160-lb person to reach a blood-alcohol level of 0.10%?
3. At least how much does an individual weigh if she has consumed $3\frac{1}{2}$ drinks without reaching a blood-alcohol level of 0.10%?
4. At least how much does an individual weigh if he has consumed $4\frac{1}{2}$ drinks without reaching a blood-alcohol level of 0.10%?

Use the bar graph in Example 2 to answer Exercises 5–8.

5. What reason was given most often for dropping out?
6. What reason was given by about 20% for dropping out?
7. Approximately what percent in the survey dropped out because of grades?
8. Approximately what percent in the survey dropped out because of boredom?

Use the line graph in Example 3 to answer Exercises 9–12.

9. Approximately how many months of regular exercise are needed to achieve a pulse rate of 85 beats per minute?
10. Approximately how many months of regular exercise are needed to achieve a pulse rate of 56 beats per minute?
11. What month caused the greatest drop in pulse rate?
12. During what month did the pulse rate first drop below 60?

Use the pie chart in Example 4 to answer Exercises 13–16.

13. What percent of income is spent on medical expenses?

14. A family with a monthly income of \$2000 would typically spend how much on clothing per month?

15. A family with a monthly income of \$2400 would typically spend how much on food per month?

16. What percent of the income is spent on food and housing combined?

Use the following line graphs to answer Exercises 17–26. The graphs show the percentages of the Gross National Product (GNP) spent on health, education, and defense over a period of 40 years.*

"Health vs. Education vs. Defense." Copyright 1992 by Consumers Union of U.S., Inc., Yonkers, NY 10703–1057. Adapted with permission from CONSUMER REPORTS, July 1992. Although this material originally appeared in CONSUMER REPORTS, the selective adaptation and resulting conclusions presented are those of the author(s) and are not sanctioned or endorsed in any way by Consumers Union, the publisher of CONSUMER REPORTS.

Percentage of GNP 1950-1990

Sources: U.S. National Center for Education Statistics, Health Care Financing Administration, U.S. Office of Management and Budget

17. Approximately what percent of the GNP was spent on public education in 1965?

18. Approximately what percent of the GNP was spent on defense in 1970?

19. In what year did health care costs represent about 10% of the GNP?

20. In what year did health care costs exceed 12% of the GNP?

21. In what three years were defense expenditures approximately 8% of the GNP?

22. In what year did public education expenditures (as a percent of GNP) peak?

23. In what year did defense expenditures (as a percent of GNP) peak?

24. In what two years were public education expenditures about $4\frac{1}{2}\%$?

25. Approximately how much did health care expenditures (as a percent of GNP) grow over the years 1970–1990?

26. Approximately how much did defense expenditures (as a percent of GNP) grow over the years 1950–1990?

Use the following circle graph to answer Exercises 27–32.

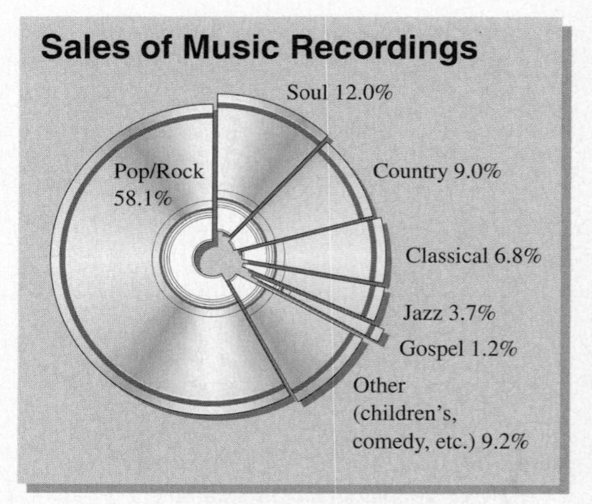

Source: National Association of Recording Merchandisers

27. What percent of all recordings sold are jazz?

28. What percent of all recordings sold are country?

29. Together, what percent of all recordings sold are either soul or pop/rock?

30. Together, what percent of all recordings sold are either classical or jazz?

31. A music store sells 3000 recordings a month. How many would you expect to be pop/rock? soul? country?

32. A music store sells 2500 recordings a month. How many would you expect to be pop/rock? classical? gospel?

33. Plot these points.

$(2, 5), (-1, 3), (3, -2), (-2, -4),$
$(0, 4), (0, -5), (5, 0), (-5, 0)$

*From *Consumer Reports,* July 1992, p. 439. Reprinted with permission.

34. Plot these points.

(4, 4), (−2, 4), (5, −3), (−5, −5),
(0, 4), (0, −4), (3, 0), (−4, 0)

In which quadrant is the point located?

35. (−5, 3)
36. (−12, 1)
37. (100, −1)
38. (35.6, −2.5)
39. (−6, −29)
40. (−3.6, −105.9)
41. (3.8, 9.2)
42. (1895, 1492)

43. In quadrant III, first coordinates are always ________ and second coordinates are always ________.

44. In quadrant II, ________ coordinates are always positive and ________ coordinates are always negative.

45. Find the coordinates of points *A*, *B*, *C*, *D*, and *E*.

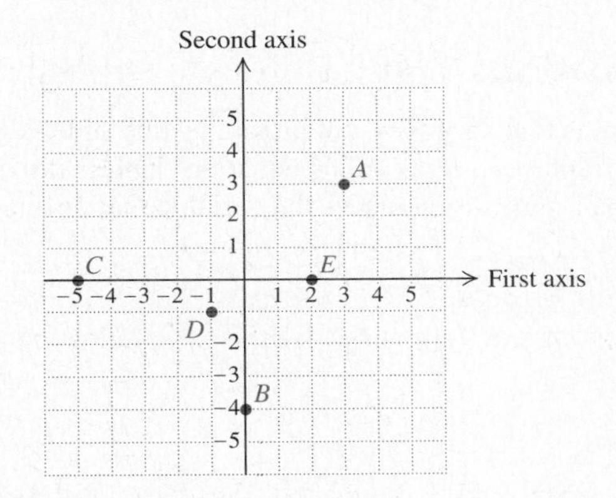

46. Find the coordinates of points *A*, *B*, *C*, *D*, and *E*.

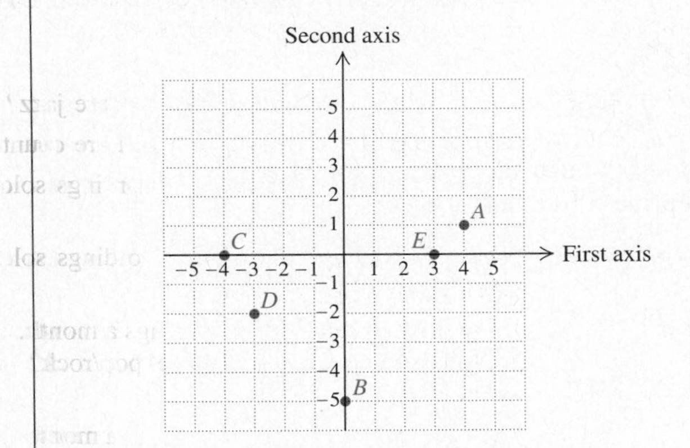

Skill Maintenance

Perform the indicated operation and simplify.

47. $\frac{3}{5} \cdot \frac{10}{9}$
48. $\frac{2}{3} + \frac{1}{5}$
49. $\frac{3}{7} - \frac{4}{5}$
50. $-\frac{2}{3} \div 5$

Synthesis

51. ◈ The graph accompanying Example 3 flattens out. Explain why the graph does not continue to slope downward.

52. ◈ Loreena and Phil's yearly income totals \$29,000. They are considering renting an apartment for \$700 per month. In light of Example 4, would you recommend that they rent the apartment? Why or why not?

In Exercises 53–56, tell in which quadrant(s) the given point could be located.

53. The first coordinate is positive.

54. The second coordinate is negative.

55. The first and second coordinates are equal.

56. The first coordinate is the additive inverse of the second coordinate.

57. The points (−1, 1), (4, 1), and (4, −5) are three vertices of a rectangle. Find the coordinates of the fourth vertex.

58. Three parallelograms share the vertices (−2, −3), (−1, 2), and (4, −3). Find the fourth vertex of each parallelogram.

59. Graph eight points such that the sum of the coordinates in each pair is 6.

60. Graph eight points such that the first coordinate minus the second coordinate is 1.

61. Find the perimeter of a rectangle whose vertices have coordinates (5, 3), (5, −2), (−3, −2), and (−3, 3).

62. Find the area of a triangle whose vertices have coordinates (0, 9), (0, −4), and (5, −4).

Coordinates on the globe. Coordinates can also be used to describe the location of three-dimensional objects: 0° latitude is the equator and 0° longitude is a line from the North Pole to the South Pole through France and Algeria. In the figure below, the hurricane Clara is at a point about 260 mi northwest of Bermuda near latitude 36.0° North, longitude 69.0° West.

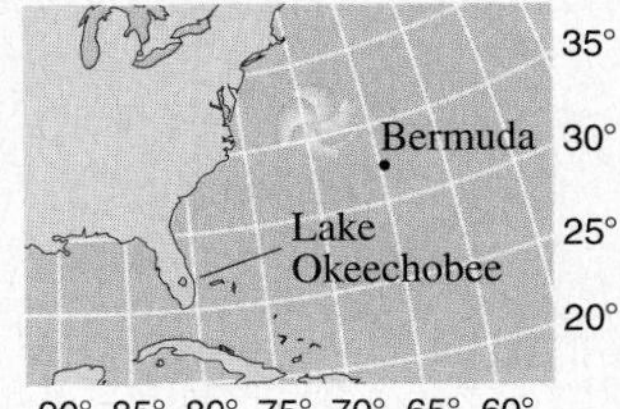

63. Approximate the latitude and the longitude of Bermuda.

64. Approximate the latitude and the longitude of Lake Okeechobee.

3.2 Graphing Linear Equations

Solutions of Equations • Graphing Equations of the Type $y = mx$ and $y = mx + b$

We have seen how bar, line, and circle graphs are used and how points are plotted using a coordinate system. Now we begin to learn how graphs can be used to represent solutions of equations.

Equations like $5x + 2y = 7$, $3y = 4 - 2x$, and $y = \frac{2}{3}x + 1$ are called **linear equations.** In general, any equation that can be written in the form $Ax + By = C$, where A, B, and C are real numbers with A and B not both zero, is a linear equation. We will find that when the solutions of a linear equation are graphed, the result is a straight line.

Solutions of Equations

When an equation contains two variables, solutions must be ordered pairs in which each number in the pair replaces a letter in the equation. Unless directed otherwise, the first number in the pair generally replaces the variable that occurs first alphabetically.

EXAMPLE 1

Determine whether each of the following pairs is a solution of $4q - 3p = 22$: **(a)** (2, 7); **(b)** (1, 6).

Solution

a) We substitute 2 for p and 7 for q (alphabetical order of variables):

$$\begin{array}{r|l} \multicolumn{2}{c}{4q - 3p = 22} \\ \hline 4 \cdot 7 - 3 \cdot 2 & ?\ 22 \\ 28 - 6 & \\ 22 & 22 \quad \text{TRUE} \end{array}$$

Since $22 = 22$ is *true,* the pair (2, 7) *is* a solution.

b) In this case, we replace p by 1 and q by 6:

$$\begin{array}{r|l} \multicolumn{2}{c}{4q - 3p = 22} \\ \hline 4 \cdot 6 - 3 \cdot 1 & ?\ 22 \\ 24 - 3 & \\ 21 & 22 \quad \text{FALSE} \end{array}$$

Since $21 = 22$ is *false,* the pair (1, 6) is *not* a solution. ❑

EXAMPLE 2

Show that the pairs (3, 7), (0, 1), and (−3, −5) are solutions of $y = 2x + 1$.

Solution We substitute, replacing x with the first coordinate and y with the second coordinate of each pair:

$y = 2x + 1$			$y = 2x + 1$			$y = 2x + 1$		
7 ?	$2 \cdot 3 + 1$		1 ?	$2 \cdot 0 + 1$		-5 ?	$2(-3) + 1$	
	$6 + 1$			$0 + 1$			$-6 + 1$	
7	7	TRUE	1	1	TRUE	-5	-5	TRUE

In each of the three cases, the substitution results in a true equation. Thus the pairs (3, 7), (0, 1), and (−3, −5) are all solutions. ❑

In Example 2, three pairs were shown to solve the equation $y = 2x + 1$. The pairs (5, 11), (10, 21), and (−5, −9) are also solutions. In fact, there are infinitely many pairs that are solutions of $y = 2x + 1$. Similarly, there are infinitely many solutions of the equation in Example 1, $4q - 3p = 22$.

For any linear equation, there are infinitely many solutions.

Graphing Equations of the Type $y = mx$ and $y = mx + b$

Because every linear equation has infinitely many solutions, we cannot possibly list them all. Instead, we can make a drawing that represents the solutions. Since all solutions are ordered pairs, each pair can be represented as a point on a plane. The drawing that represents all such points is called the **graph** of the equation.

To *graph* an equation means to make a drawing that represents its solutions.

In the following examples, equations of the form $y = mx$ or $y = mx + b$ are graphed. As with any type of linear equation, once a few points have been plotted, we will discover that the graph is a straight line.

EXAMPLE 3

Graph: $y = 2x$.

Solution Before a graph of the equation can be drawn, we need to find some ordered pairs that are solutions. Rather than attempt this by trial and error, we will *choose* a number for x, the first coordinate, and then find the associated value for y by substitution. For example,

if $x = 3$, then $y = 2 \cdot 3 = 6$; We get a solution: the ordered pair (3, 6).

if $x = 1$, then $y = 2 \cdot 1 = 2$; We get another solution: the ordered pair (1, 2).

if $x = 0$, then $y = 2 \cdot 0 = 0$; We get another solution: the ordered pair (0, 0).

if $x = -2$, then $y = 2(-2) = -4$. We get another solution: the ordered pair (−2, −4).

These results are often listed in a table, as shown below. The points corresponding to each pair are then plotted and we look for a pattern.

	y	
x	$y = 2x$	(x, y)
3	6	(3, 6)
1	2	(1, 2)
0	0	(0, 0)
−2	−4	(−2, −4)

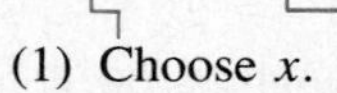

(1) Choose x.
(2) Compute y.
(3) Form the pair (x, y).
(4) Plot the points.

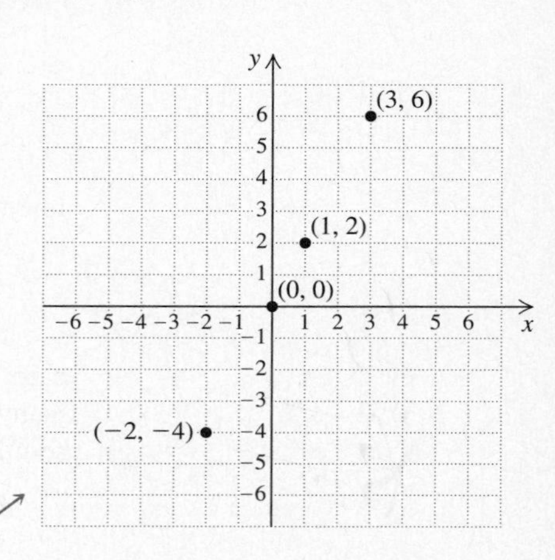

Note that the points resemble a straight line. We draw the line using a ruler or some other straightedge. Every point on the line (for instance, (1.5, 3)) represents a solution of $y = 2x$.

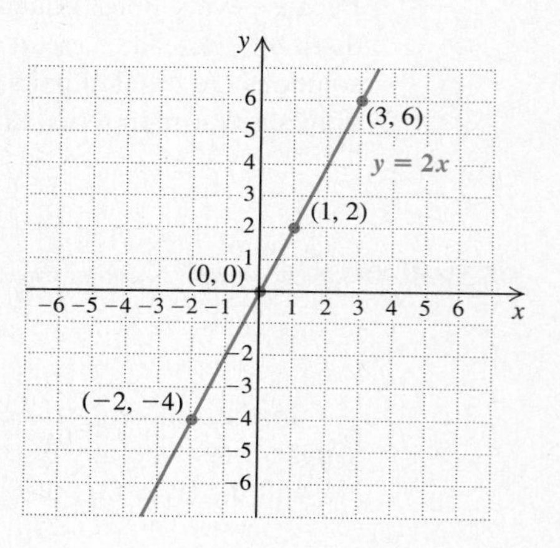

❑

EXAMPLE 4

Graph: $y = -3x$.

Solution Since two points determine a line, we need find only two ordered pairs that are solutions and plot the corresponding points. As a check against making a mistake, however, we generally plot *three* points. If all three points do not line up, we know that we have made a mistake.

We choose a number for x, the first coordinate, and calculate the associated y-value by substitution.

If $x = 1$, then $y = -3 \cdot 1 = -3$. We get the ordered pair $(1, -3)$.

If $x = -1$, then $y = -3(-1) = 3$. We get the ordered pair $(-1, 3)$.

If $x = 2$, then $y = -3 \cdot 2 = -6$. We get the ordered pair $(2, -6)$.

The following table lists these solutions. Next, we plot the points and see that they form a line. Finally, we draw and label the line.

x	y $y = -3x$	(x, y)
1	-3	$(1, -3)$
-1	3	$(-1, 3)$
2	-6	$(2, -6)$

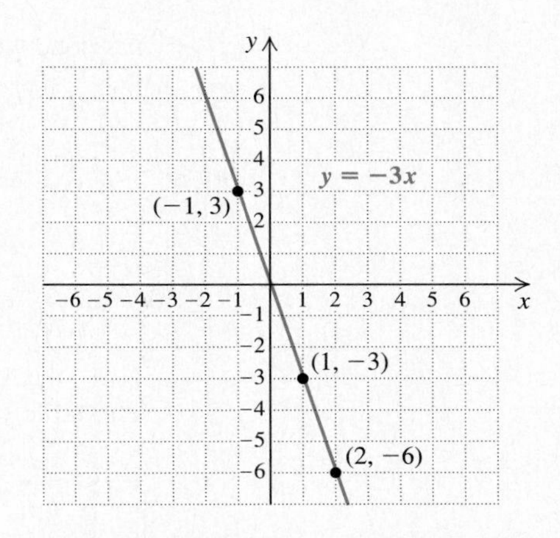

When choosing values for the x-coordinate, we try to avoid numbers that take us off the graph paper. If a fraction appears as the coefficient of x, we can choose x-coordinates that are multiples of the denominator and thereby avoid y-coordinates that are fractions.

EXAMPLE 5

Graph: $y = \frac{2}{3}x$.

Solution We make a table of solutions.

When $x = 3$, $y = \frac{2}{3} \cdot 3 = 2$.
When $x = -3$, $y = \frac{2}{3}(-3) = -2$.
When $x = 6$, $y = \frac{2}{3} \cdot 6 = 4$.
When $x = 1$, $y = \frac{2}{3} \cdot 1 = \frac{2}{3}$.

Note that when multiples of 3 are substituted for x, the y-coordinates are not fractions.

Next, we plot the points and complete the graph by drawing a line through them.

x	y $y = \frac{2}{3}x$	(x, y)
3	2	$(3, 2)$
-3	-2	$(-3, -2)$
6	4	$(6, 4)$
1	$\frac{2}{3}$	$\left(1, \frac{2}{3}\right)$

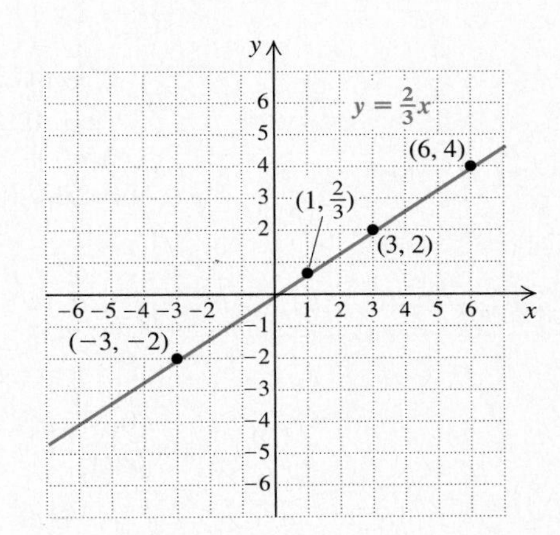

The lines in Examples 3–5 all contain the origin, (0, 0), because all three equations are of the form $y = mx$. What will happen if we add a number b on the right side to get an equation of the form $y = mx + b$?

EXAMPLE 6

Graph $y = 2x + 3$ and compare the graph with that of $y = 2x$.

Solution The equation $y = 2x$ was graphed on page 124. For the purpose of comparison, we graph $y = 2x + 3$ using the same choices for x. Note that:

if $x = 3$, then $y = 2 \cdot 3 + 3 = 9$; We get a solution: the ordered pair (3, 9).

if $x = 1$, then $y = 2 \cdot 1 + 3 = 5$; We get another solution: the ordered pair (1, 5).

if $x = 0$, then $y = 2 \cdot 0 + 3 = 3$; We get another solution: the ordered pair (0, 3).

if $x = -2$, then $y = 2(-2) + 3 = -1$. We get another solution: the ordered pair (−2, −1).

The table of values and the graph of $y = 2x + 3$ are shown below.

x	y $y = 2x + 3$	(x, y)
3	9	(3, 9)
1	5	(1, 5)
0	3	(0, 3)
−2	−1	(−2, −1)

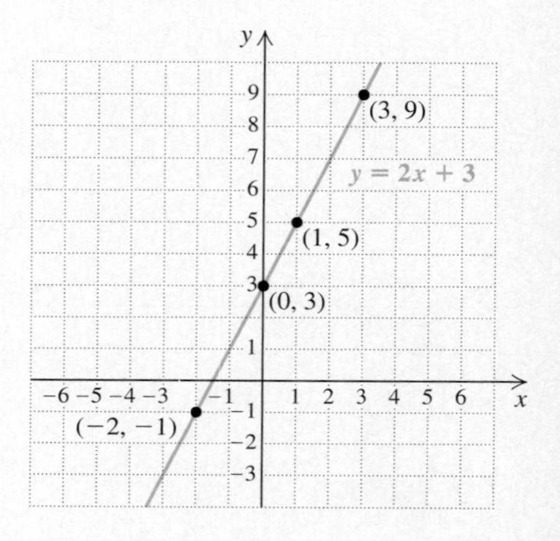

Note that the graph of $y = 2x + 3$ looks just like the graph of $y = 2x$ but shifted 3 units up. In particular, instead of crossing the y-axis at (0, 0), the graph now crosses the y-axis at (0, 3). ❑

In Example 6, when we replaced x with 0, the corresponding y-value was $2 \cdot 0 + 3$, or 3. In general, if $y = mx + b$ and x is replaced with 0, the corresponding y-value is $m \cdot 0 + b$, or b. Thus all graphs of equations of the form $y = mx + b$ will contain the point $(0, b)$.

EXAMPLE 7

Graph: $y = \frac{2}{5}x + 4$.

Solution Note that when $x = 0$, we have $y = \frac{2}{5} \cdot 0 + 4 = 4$ so the graph will contain (0, 4) as expected. We find two other pairs, using multiples of 5 to avoid fractions.

When $x = 5$, $y = \frac{2}{5} \cdot 5 + 4 = 2 + 4 = 6$.

When $x = -5$, $y = \frac{2}{5}(-5) + 4 = -2 + 4 = 2$.

x	y
0	4
5	6
−5	2

Now we can draw the graph of $y = \frac{2}{5}x + 4$.

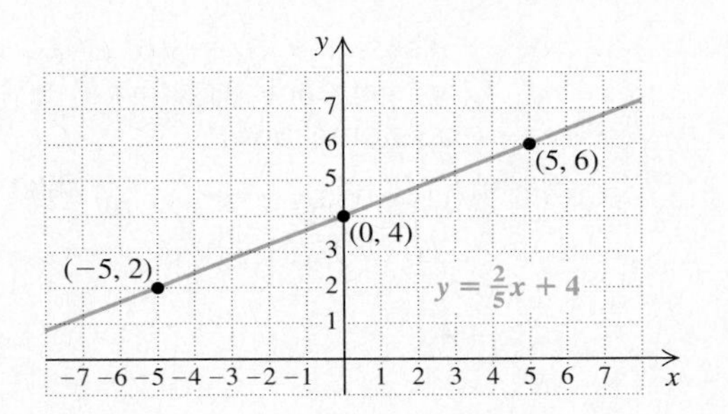

We have seen that any equation of the form $y = mx + b$ contains the point $(0, b)$. That point is called the **y-intercept** of the graph since it is the point at which the graph crosses the y-axis. Sometimes, for convenience, we may refer to b as the y-intercept rather than list both coordinates.

y-intercept

The graph of any equation $y = mx + b$ passes through the y-intercept $(0, b)$.

EXAMPLE 8

Graph: $x + 3y = -6$.

Solution To make use of what we've studied, we first solve for y to write an equivalent equation in the form $y = mx + b$:

$$x + 3y = -6$$
$$3y = -x - 6 \quad \text{Adding } -x \text{ on both sides}$$
$$y = \tfrac{1}{3}(-x - 6) \quad \text{Multiplying by } \tfrac{1}{3} \text{ on both sides}$$
$$y = -\tfrac{1}{3}x - 2. \quad \text{Using the distributive law}$$

TECHNOLOGY CONNECTION

Beginning in this chapter, we will include in some sections and exercise sets activities that utilize graphing calculators or computer graphing software. Such calculators and software will be referred to simply as *graphers*. Most activities will use only basic features common to virtually all graphers. All will be presented in a generic form—check with a user's manual or ask your instructor for more exact procedures.

All graphers have a *window*, the rectangular portion of the screen in which a graph appears. We will describe a window with four numbers, [L, R, B, T], that represent the *L*eft and *R*ight endpoints of the x-axis and the *B*ottom and *T*op endpoints of the y-axis. A *Range* feature is sometimes used to set these dimensions.

The primary use for graphers is to graph equations. For example, let's graph the equation $y = -\frac{4}{5}x + \frac{13}{5}$. Selecting the window $[-10, 10, -10, 10]$ results in the graph shown at the top of the next column.

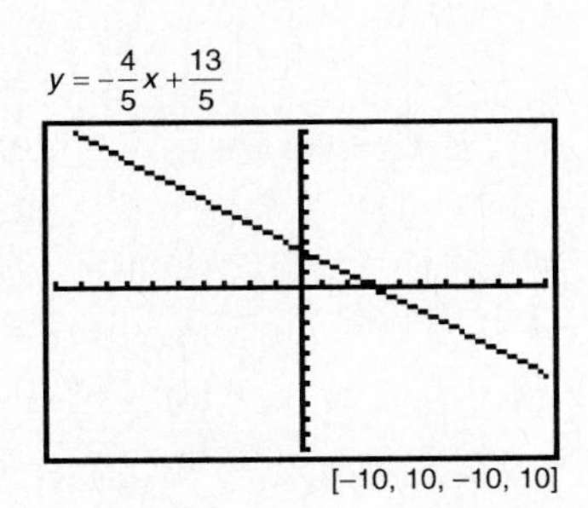

Use a grapher to draw each of the following lines. Select the window $[-10, 10, -10, 10]$ for each graph.

TC1. $y = -5x + 6.5$

TC2. $y = 3x - 4.5$

TC3. $y = \frac{4}{7}x - \frac{22}{7}$

TC4. $y = -\frac{11}{5}x - 4$

Thus, $x + 3y = -6$ is equivalent to $y = -\frac{1}{3}x - 2$, or $y = -\frac{1}{3}x + (-2)$. The y-intercept is therefore $(0, -2)$. We find two other pairs using multiples of 3 for x to avoid fractions.

When $x = 6$, $\quad y = -\frac{1}{3}\cdot 6 - 2 = -2 - 2 = -4$.
When $x = -6$, $\quad y = -\frac{1}{3}(-6) - 2 = 2 - 2 = 0$.

x	y
0	−2
6	−4
−6	0

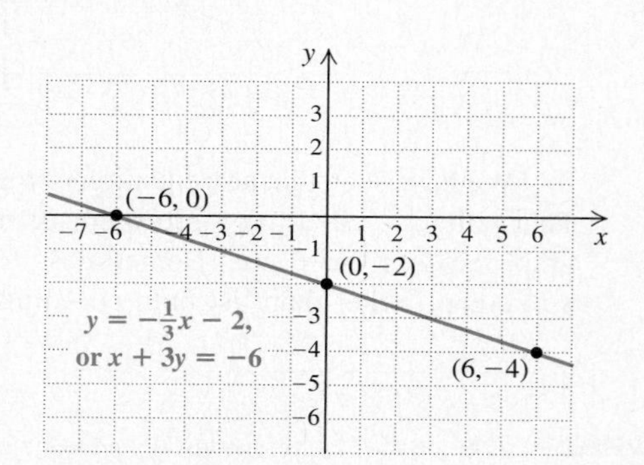

❑

EXERCISE SET 3.2

Determine whether the given point is a solution of the equation.

1. $(2, 5)$; $\quad y = 3x - 1$
2. $(1, 7)$; $\quad y = 2x + 5$
3. $(2, -3)$; $\quad 3x - y = 4$
4. $(-1, 4)$; $\quad 2x + y = 6$
5. $(-2, -1)$; $\quad 2c + 2d = -7$
6. $(0, -4)$; $\quad 4p + 2q = -9$

Graph.

7. $y = x$
8. $y = -x$
9. $y = -2x$
10. $y = -4x$
11. $y = \frac{1}{3}x$
12. $y = \frac{1}{2}x$
13. $y = -\frac{3}{2}x$
14. $y = -\frac{5}{4}x$
15. $y = x + 1$
16. $y = x + 3$
17. $y = 2x + 2$
18. $y = 3x - 2$
19. $y = \frac{1}{3}x - 1$
20. $y = \frac{1}{2}x + 1$
21. $y + x = -3$
22. $y + x = -2$
23. $y = \frac{5}{2}x + 3$
24. $y = \frac{5}{3}x - 2$
25. $y = -\frac{5}{2}x - 2$
26. $y = -\frac{5}{3}x - 2$
27. $y = \frac{1}{2}x - 5$
28. $y = \frac{3}{2}x - 6$
29. $2x + y = 3$
30. $5x + y = 7$
31. $y = x - \frac{1}{2}$
32. $y = x + \frac{2}{3}$
33. $x + 2y = -4$
34. $x + 2y = 8$
35. $6x - 3y = 9$
36. $8x - 4y = 12$
37. $6y + 2x = 8$
38. $8y + 2x = -4$

Skill Maintenance

Solve and check.

39. $3x - 7 = -34$
40. $2(x - 9) + 4 = 2 - 3x$
41. Solve $Ax + By = C$ for y.
42. Solve $A = \dfrac{T + Q}{2}$ for Q.

Synthesis

43. ◈ An equation is graphed by plotting first two points and then a third point as a check. If the three points form a line, is it still possible that the graph is incorrect? Why or why not?
44. ◈ Do all graphs of linear equations have y-intercepts? Why or why not?

45. Complete the following table for $y = x^2 + 1$. Plot the points on graph paper and draw the graph.

x	0	−1	1	−2	2	−3	3
y							

46. Find all whole-number solutions of
$x + y = 6$.

47. Find all whole-number solutions of
$x + 3y = 15$.

48. Translate to an equation: n nickels and d dimes total \$1.95. Find three solutions.

49. Translate to an equation: n nickels and q quarters total \$2.35. Find three solutions.

50. Find three solutions of $y = |x|$.

For Exercises 51–54, use a grapher to graph the equation. Use a $[-10, 10, -10, 10]$ window.

51. $y = -2.8x + 3.5$

52. $y = 4.5x + 2.1$

53. $y = \frac{2}{7}x - \frac{24}{5}$

54. $y = -\frac{33}{8}x - \frac{45}{7}$

55. Are the equations
$y = -\frac{2}{5}x + 3$ and $2x + 5y = 15$
equivalent? Why or why not?

3.3 More on Graphing Linear Equations

Graphing Using Intercepts • Graphing Horizontal or Vertical Lines

Although equations like $4x + 3y = 12$ can be rewritten in the form $y = mx + b$ and then graphed as in Section 3.2, there is a faster way to graph equations of the form $Ax + By = C$.

Graphing Using Intercepts

We have seen that the y-intercept of a graph occurs where the graph crosses the y-axis and that the first coordinate of the y-intercept is always 0.

The point at which a graph crosses the x-axis is called the ***x*-intercept.** Since x-intercepts are always on the x-axis, the second coordinate of the x-intercept will always be 0.

Intercepts are found as follows.

Intercepts

The y-intercept is $(0, b)$. To find b, let $x = 0$ and solve the original equation for y.

The x-intercept is $(a, 0)$. To find a, let $y = 0$ and solve the original equation for x.

EXAMPLE 1

Find the x- and y-intercepts of the graph of $4x + 3y = 12$.

Solution

To find the y-intercept, let $x = 0$. Then solve for y:

$$4 \cdot 0 + 3y = 12 \quad \text{Replacing } x \text{ with } 0$$

$$3y = 12$$

$$y = 4.$$

Thus, $(0, 4)$ is the y-intercept.

To find the x-intercept, let $y = 0$. Then solve for x:

$$4x + 3 \cdot 0 = 12 \quad \text{Replacing } y \text{ with } 0$$

$$4x = 12$$

$$x = 3.$$

Thus, $(3, 0)$ is the x-intercept. ❑

By plotting both intercepts and a third point as a check, we can quickly graph any linear equation of the form $Ax + By = C$.

EXAMPLE 2

Graph $4x + 3y = 12$ using intercepts.

Solution In Example 1, we found that the y-intercept is $(0, 4)$ and the x-intercept is $(3, 0)$. Before drawing a line, we plot a third point as a check. We substitute any convenient value for x and solve for y.

If we let $x = 1$, then

$$4 \cdot 1 + 3y = 12 \quad \text{Substituting 1 for } x$$

$$4 + 3y = 12$$

$$3y = 12 - 4 = 8$$

$$y = \tfrac{8}{3}, \quad \text{or } 2\tfrac{2}{3}. \quad \text{Solving for } y$$

Since the point $(1, 2\frac{2}{3})$ appears to line up with the intercepts, we draw and label the graph.

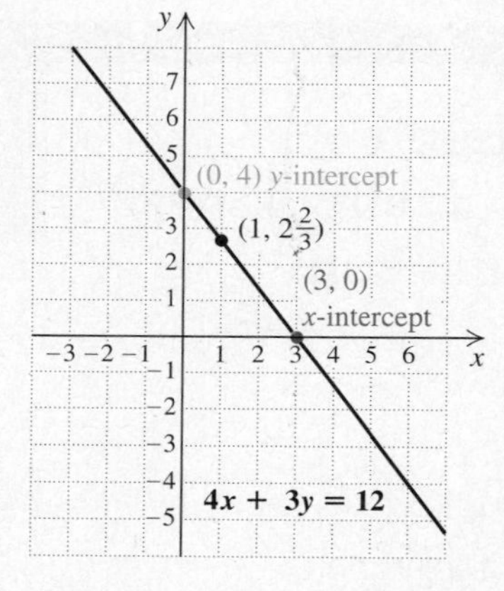

❑

In Example 1, the equation $4x + 3y = 12$ simplified to $3y = 12$ when we solved for the y-intercept. Thus, to find the y-intercept, we can simply "cover up" the x-term and solve the remaining equation.

In a similar manner, $4x + 3y = 12$ simplified to $4x = 12$ when we solved for the x-intercept. Thus, to find the x-intercept, we can simply "cover up" the y-term and solve the remaining equation.

EXAMPLE 3

Graph $3x - 2y = 6$ using intercepts.

Solution To find the y-intercept, let $x = 0$. This amounts to covering up the x-term and then solving:

$$-2y = 6 \quad \text{When } x \text{ is } 0,\ 3x - 2y \text{ is simply } -2y.$$

$$y = -3.$$

The y-intercept is $(0, -3)$.

To find the x-intercept, let $y = 0$. This is the same as covering up the y-term and then solving:

$3x = 6$ When y is 0, $3x - 2y$ is simply $3x$.

$x = 2$.

The x-intercept is (2, 0).

To find a third point, we replace x with 4 and solve for y:

$3 \cdot 4 - 2y = 6$ Other numbers besides 4 can be used for x.

$12 - 2y = 6$

$-2y = -6$

$y = 3$.

The point (4, 3) appears to line up with the intercepts so we draw the graph.

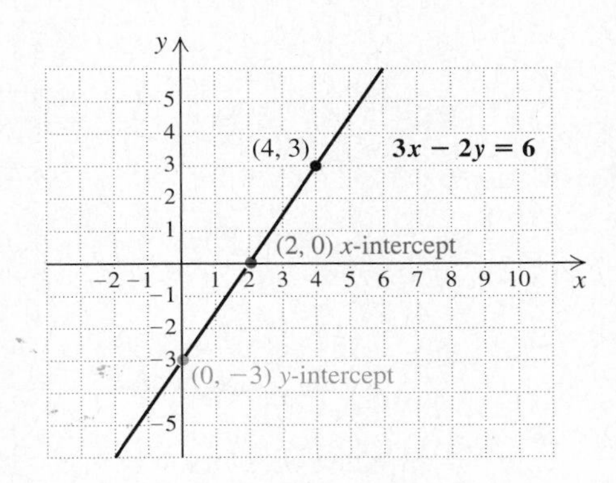

❑

Graphing Horizontal or Vertical Lines

Equations like $y = 3$ or $x = -4$ are in the form $Ax + By = C$, but with $A = 0$ or $B = 0$. The graph of such an equation will be either a horizontal or a vertical line. There will be a y-intercept or an x-intercept but never both.

EXAMPLE 4

Graph: $y = 3$.

Solution We can regard the equation $y = 3$ as $0 \cdot x + y = 3$. No matter what number we choose for x, we find that y must be 3 if the equation is to be solved. Consider the following table.

Choose any number for x. →

	y	
x	$y = 3$	(x, y)
−2	3	(−2, 3)
0	3	(0, 3)
4	3	(4, 3)

All pairs will have 3 as the y-coordinate.

y must be 3.

When we plot the ordered pairs $(-2, 3)$, $(0, 3)$, and $(4, 3)$ and connect the points, we obtain a horizontal line. Any ordered pair $(x, 3)$ is a solution. So the line is parallel to the x-axis with y-intercept $(0, 3)$.

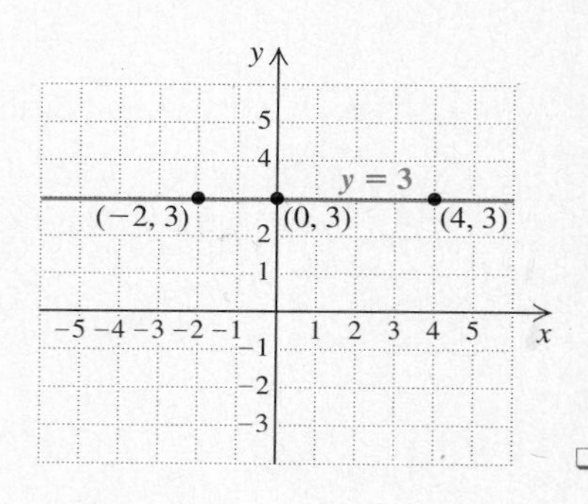

EXAMPLE 5

Graph: $x = -4$.

Solution We can regard the equation $x = -4$ as $x + 0 \cdot y = -4$. We make up a table with all -4's in the x-column.

x must be -4. →

$x = -4$	y	(x, y)
-4	-5	$(-4, -5)$
-4	1	$(-4, 1)$
-4	3	$(-4, 3)$

All pairs will have -4 as the x-coordinate.

Choose any number for y.

When we plot the ordered pairs $(-4, -5)$, $(-4, 1)$, and $(-4, 3)$ and connect them, we obtain a vertical line. Any ordered pair $(-4, y)$ is a solution. So the line is parallel to the y-axis with x-intercept $(-4, 0)$.

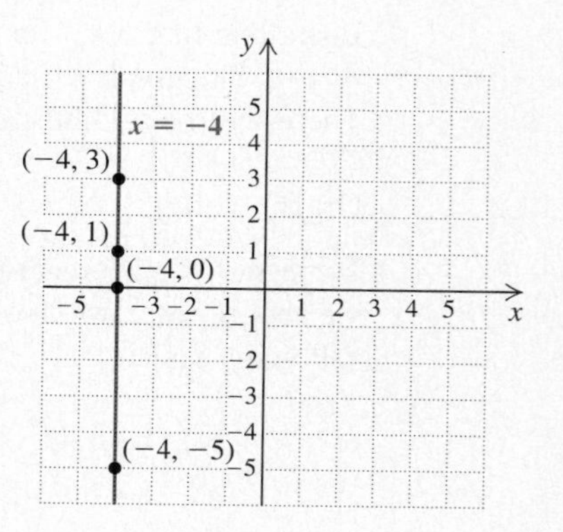

The graph of $y = b$ is a horizontal line, with y-intercept $(0, b)$.
The graph of $x = a$ is a vertical line, with x-intercept $(a, 0)$.

The following is a general procedure for graphing linear equations.

To Graph Linear Equations

1. If the equation is of the type $x = a$ or $y = b$, the graph will be a line parallel to an axis.

   Examples.

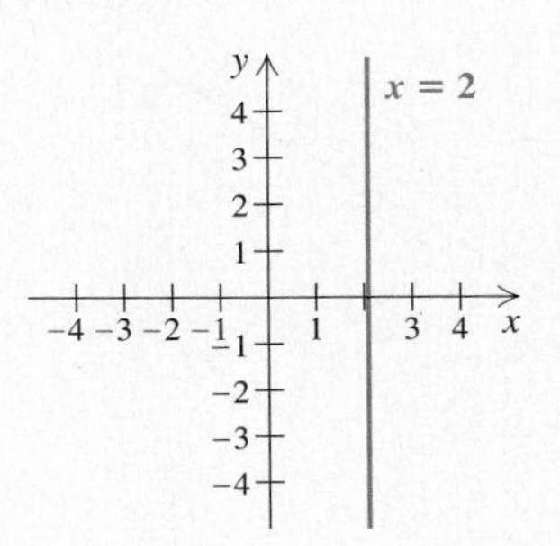

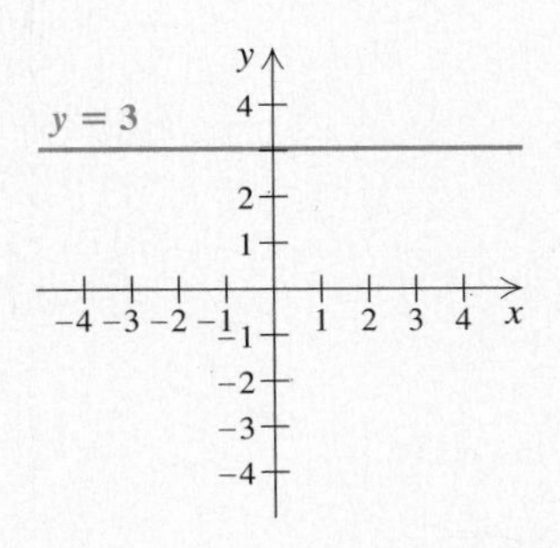

2. If the equation is of the type $y = mx$, both intercepts are the origin, (0, 0). Plot (0, 0) and one other point. A third point can be calculated as a check.

   Example.

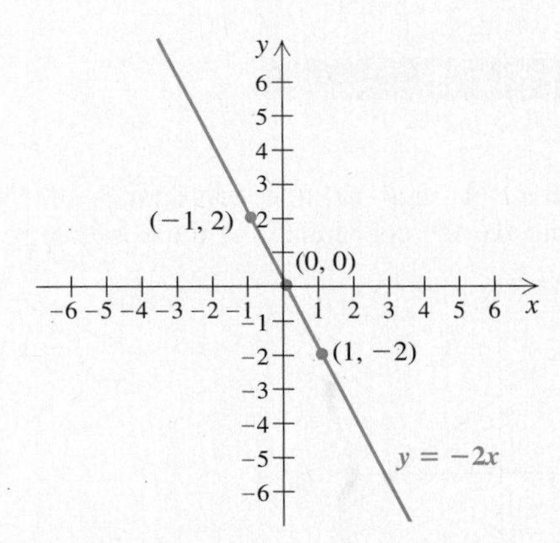

3. If the equation is of the type $y = mx + b$, the y-intercept is $(0, b)$. Plot $(0, b)$ and one other point. A third point can be calculated as a check.

   Example.

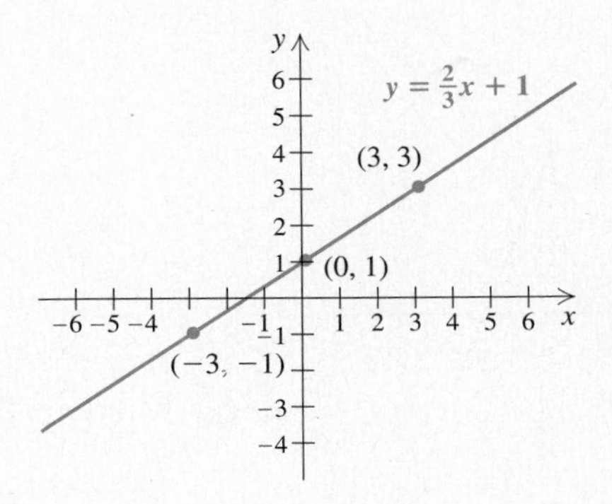

4. If the equation is of the type $Ax + By = C$, but not of the type $x = a$ or $y = b$, graph using intercepts. A third point can be used as a check.

Example.

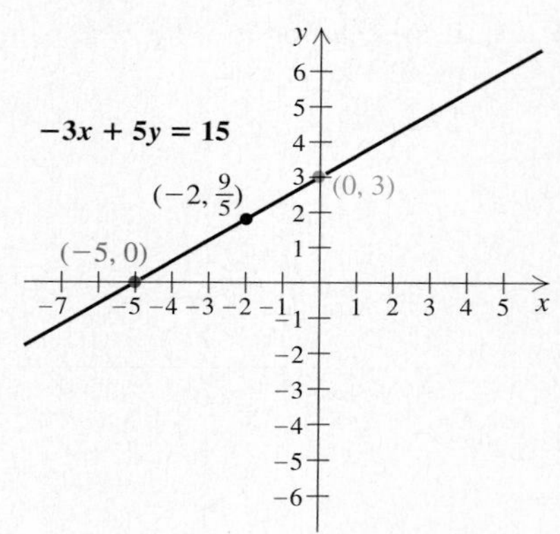

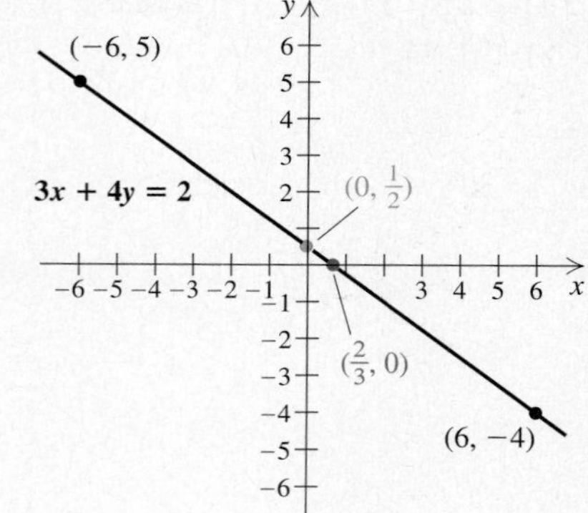

EXERCISE SET 3.3

For Exercises 1–4, find **(a)** the coordinates of the y-intercept and **(b)** the coordinates of the x-intercept.

1.

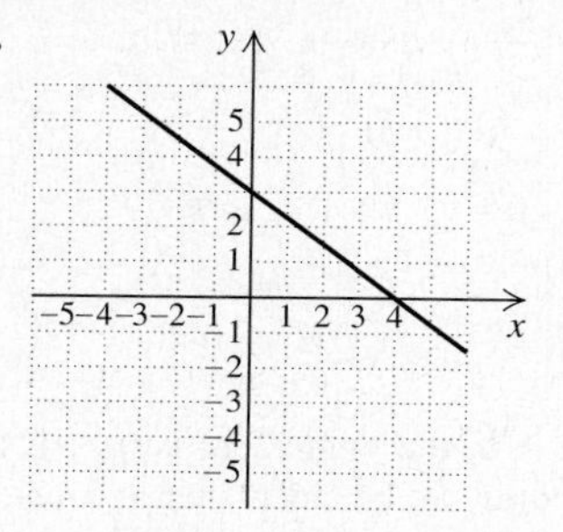

2.

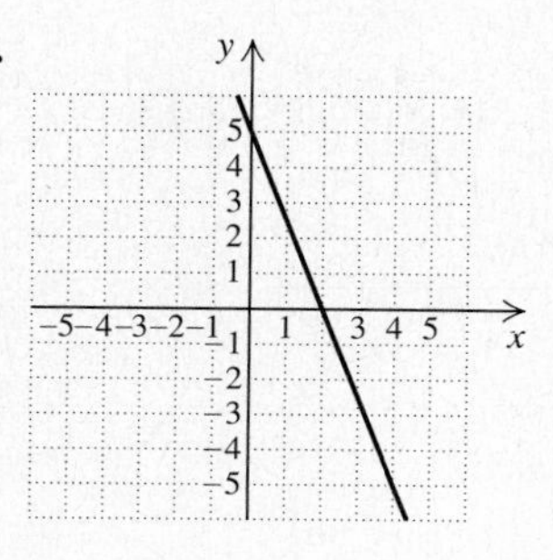

3.

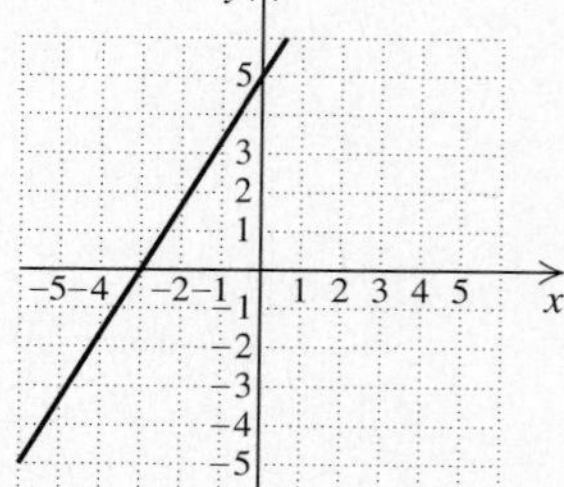

4.

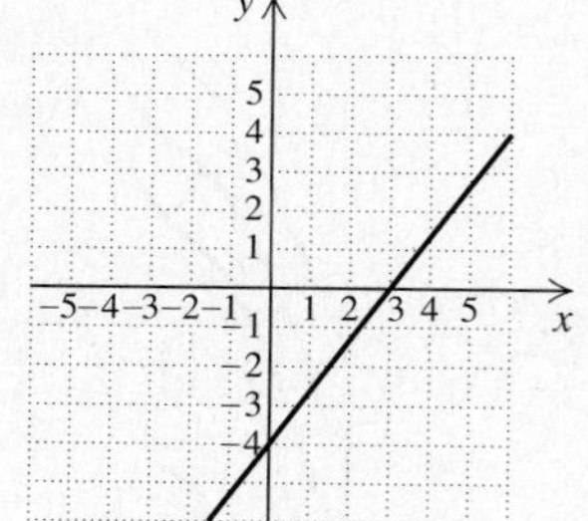

For Exercises 5–12, find **(a)** the coordinates of the y-intercept and **(b)** the coordinates of the x-intercept. Do not graph.

5. $2x + 5y = 20$

6. $5x + 3y = 15$

7. $4x - 3y = 24$
8. $2x - 7y = 28$
9. $-6x + y = 8$
10. $-8x + y = 10$
11. $2y - 4 = 6x$
12. $3y + 6 = 9x$

Find the intercepts. Then graph.

13. $3x + 2y = 12$
14. $2x + 4y = 16$
15. $x + 3y = 6$
16. $x + 2y = 8$
17. $-x + 2y = 4$
18. $-x + 3y = 9$
19. $3x + y = 9$
20. $2x + y = 6$
21. $2y - 2 = 6x$
22. $3y - 6 = 9x$
23. $3x - 9 = 3y$
24. $5x - 10 = 5y$
25. $2x - 3y = 6$
26. $2x - 5y = 10$
27. $4x + 5y = 20$
28. $2x + 6y = 12$
29. $2x + 3y = 8$
30. $x - 1 = y$
31. $x - 3 = y$
32. $2x - 1 = y$
33. $3x - 2 = y$
34. $4x - 3y = 12$
35. $6x - 2y = 18$
36. $7x + 2y = 6$
37. $3x + 4y = 5$
38. $y = -4 - 4x$
39. $y = -3 - 3x$
40. $-3x = 6y - 2$
41. $-4x = 8y - 5$
42. $3 = 2x - 5y$
43. $y - 3x = 0$
44. $x + 2y = 0$

Graph.

45. $x = -2$
46. $x = -1$
47. $y = 2$
48. $y = 4$
49. $x = 7$
50. $x = 3$
51. $y = 0$
52. $y = -1$
53. $x = \frac{3}{2}$
54. $x = -\frac{5}{2}$
55. $3y = -5$
56. $12y = 45$
57. $4x + 3 = 0$
58. $-3x + 12 = 0$
59. $18 - 3y = 0$
60. $63 + 7y = 0$

Skill Maintenance

Write the prime factorization of the number.

61. 98
62. 240

Simplify.

63. $\frac{36}{90}$
64. $\frac{12x}{84x}$

Synthesis

65. ◈ Explain in your own words why equations of the form $y = C$ have graphs that are horizontal lines.
66. ◈ Can any equation of the type $Ax + By = C$ be written in the form $y = mx + b$? Why or why not?
67. Write an equation for the y-axis.
68. Write an equation for the x-axis.
69. Find the coordinates of the point of intersection of the graphs of the equations $x = -3$ and $y = 6$.
70. Write an equation of a line parallel to the x-axis and 5 units below it.
71. Write an equation of a line parallel to the y-axis and 13 units to the right of it.
72. Write an equation of a line parallel to the x-axis and intersecting the y-axis at $(0, 2.8)$.
73. Find the value of m in the equation $y = mx + 3$ so that the x-intercept of its graph will be $(2, 0)$.
74. Find the value of b in the equation $2y = -5x + 3b$ so that the y-intercept of its graph will be $(0, -12)$.
75. ◈ For A and B nonzero, the graphs of $Ax + D = C$ and $By + D = C$ will be parallel to an axis. Explain why.

3.4 Graphs and Problem Solving

Translating to Equations • Graphing • Problem Solving

Suppose we are asked to find a pair of numbers whose sum is 5. There are infinitely many such pairs, so a graph would provide a convenient way of showing all solutions.

Translating to Equations

When a problem-solving situation involves pairs of numbers and many solutions, we can often translate it to an equation with two variables.

EXAMPLE 1

Translate to an equation.

a) The sum of two numbers is 5.
b) Harry is two years older than Jane.
c) The express train is twice as fast as the local.
d) Sondra earns $100 less than twice Jim's salary.

Solution

a) Let x and y represent the two numbers whose sum is 5. The translation follows immediately: $x + y = 5$.

b) Let h = Harry's age and j = Jane's age. We reword and then translate.

Reword: Harry's age is two more than Jane's age.

Translate: $h = j + 2$

c) Let s = the speed of the express train and l = the speed of the local train. Then we have

Reword: The speed of the express train is twice the speed of the local.

Translate: $s = 2l$

d) Let s = Sondra's salary and j = Jim's salary. Then we have

Reword: Sondra's salary is $100 less than twice Jim's salary.

Translate: $s = 2j - 100$ ❑

Graphing

When a problem-solving situation translates to a linear equation, we can label axes and draw a graph. The graph provides a visual representation of the situation.

EXAMPLE 2

The total number of calories C expended by a 150-lb person who is walking briskly for t hours can be estimated by the equation $C = 300t$.

a) Use the equation to estimate the energy expended in walking for 2 hr, 3.5 hr, and 5 hr.

b) Graph the equation. Assume that t is the first coordinate and C is the second.

Solution

a) To find the number of calories used to walk 2 hr, 3.5 hr, and 5 hr, we substitute and calculate as follows:

$C = 300(2) = 600$ calories; Walking 2 hr burns 600 Cal.

$C = 300(3.5) = 1050$ calories; Walking 3.5 hr burns 1050 Cal.

$C = 300(5) = 1500$ calories. Walking 5 hr burns 1500 Cal.

b) We use the values computed in part (a) and any others that we may calculate to make a table. Each pair of values represents a point, which we can plot using the horizontal axis for time and the vertical axis for the total number of calories burned.

Often, in problems such as this, it is impractical to count by 1's. In this case, we count by 100's on the vertical axis. By altering the *scale* in this way, we can plot all the pairs listed in the table.

Once the points are graphed, we draw a straight line through them.

Time (in Hours)	Calories Burned
2	600
3.5	1050
5	1500

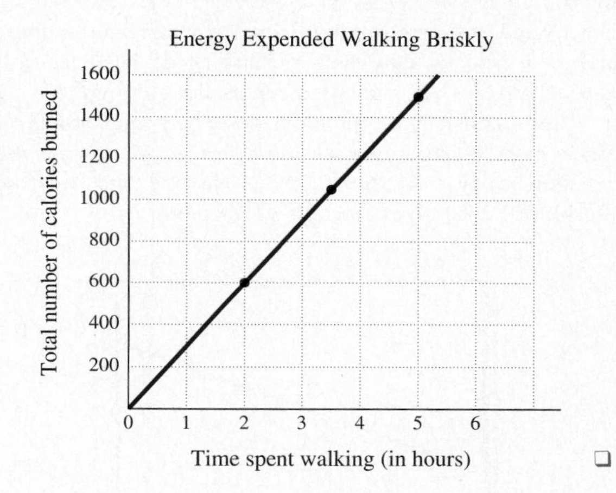

❑

Problem Solving

When a problem is translated and graphed, the graph often provides a quick and useful way of approximating values that could be calculated more precisely (but also more slowly) from a formula.

EXAMPLE 3

Ridem Trucks charges $49.95 per day plus 35¢ per mile for the rental of an 18-ft truck. To help customers predict the cost of a rental, the firm wishes to draw a graph in which mileage is measured on the horizontal axis and cost on the vertical axis. Using such a graph, predict the cost of renting an 18-ft truck for one day and driving 125 miles.

Solution

1. FAMILIARIZE. In Section 2.5, another truck rental problem was solved, so we already have some familiarity. Since the axes will represent mileage and cost, we let m = mileage and c = cost.

2. TRANSLATE. Since the cost of a rental is \$49.95 plus 35¢ for each mile, and since m miles are to be driven, we have the translation

$$c = 49.95 + 0.35m.$$

3. CARRY OUT. We make a table of values using some convenient choices for m.

When $m = 50$, $c = 49.95 + 0.35(50) = 67.45$.

When $m = 150$, $c = 49.95 + 0.35(150) = 102.45$.

When $m = 300$, $c = 49.95 + 0.35(300) = 154.95$.

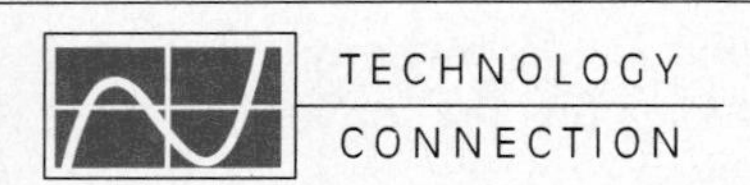

The *Trace* feature found on most graphers allows you to find the coordinates of a point on a line. When the Trace feature is activated, a cursor (often blinking) appears on the graph and its x- and y-coordinates are displayed. These coordinates change as you move the cursor along the graph.

Let's consider the problem described in Example 3. Since most graphers use x- and y-terms, we rewrite the equation as $y = 49.95 + 0.35x$. Selecting the window [0, 400, 0, 250] gives the following graph.

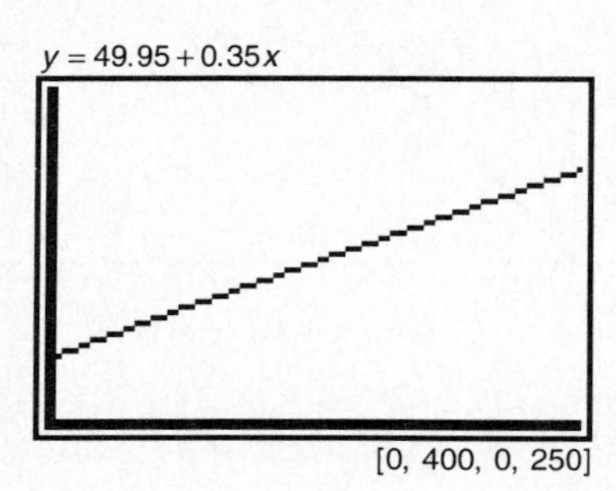

Note that the axes appear heavyset because the *scales* remain at 1. If this is bothersome, the *Range* feature can be used to adjust the scales for x and y. Using a scale of 50 on both axes results in this graph:

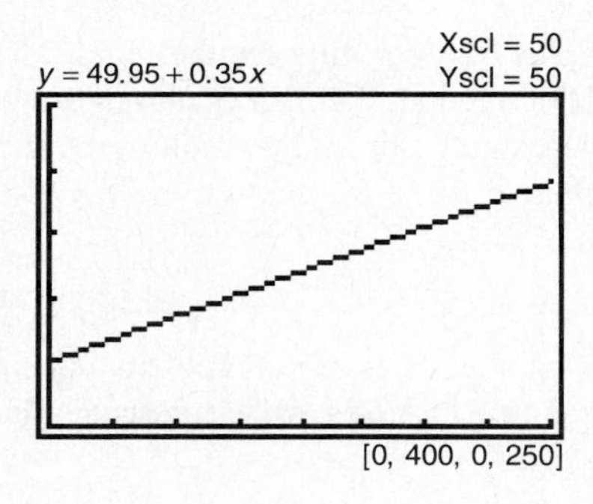

Let's use the graph to determine the cost of driving 210 mi. To do so, we activate the Trace feature and move the cursor until its x-coordinate is approximately 210. The displayed coordinates indicate a cost of approximately \$124.

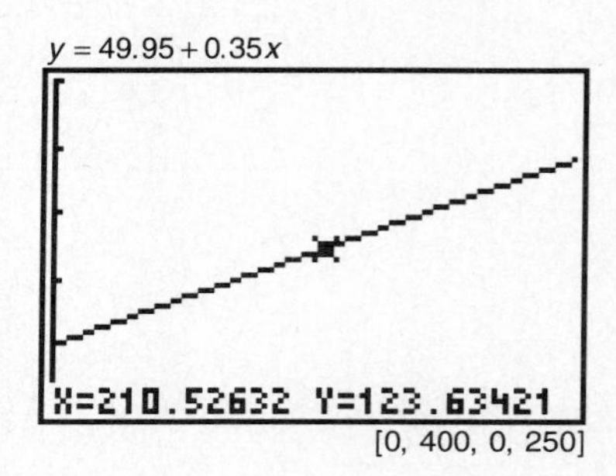

For more precision, we can zoom in on the point in question. To do so, we can shrink the window's dimensions or use a *Zoom* feature to magnify a portion of the graph. (Consult a manual or your instructor on how to best utilize the Zoom feature.) By zooming in or shrinking the window (say, to [209, 211, 122, 124]), we can use the Trace feature again for a better approximation: \$123.45.

Solve each of the following using a grapher.

TC1. Use the Trace feature and reset the range or zoom in to approximate to the nearest cent the cost of driving the rental truck 190 mi.

TC2. Starting at an altitude of 2000 ft, a pilot climbs at the rate of 3500 ft per minute. What is the plane's altitude after 4.5 min? (*Hint:* Use the window [0, 10, 2000, 40,000] with the y-scale 4000.)

TC3. Sally's Plumbing Service charges \$35 for each visit plus \$45 per hour. If a job lasts 6.5 hr, how much will it cost?

Mileage	Cost
50	$67.45
150	$102.45
300	$154.95

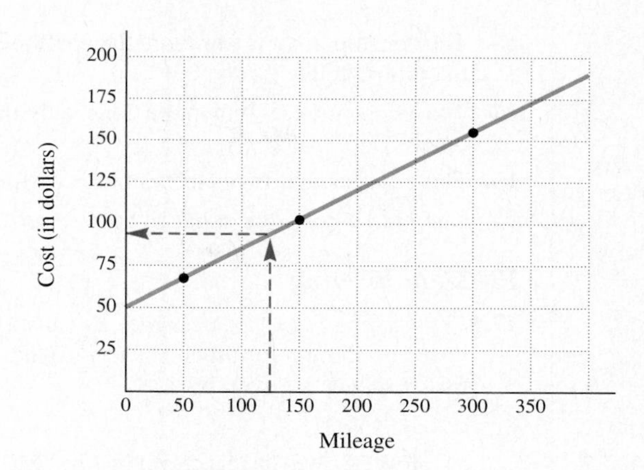

Being careful to label the axes correctly, we draw the graph by plotting the points listed in the table and then drawing a line through them.

To predict the cost of driving 125 mi, we locate 125 on the horizontal axis. From there we trace a path *up* to the line and then *left* from the line to the vertical axis. Since the predicted cost is closer to \$100 than to \$75, we estimate the cost of the rental at \$90.

4. CHECK. The three points that we graphed form a line, so the graph is probably correct. To check the predicted cost of \$90, we could calculate the cost precisely. Doing so may make the graph seem unnecessary, but the rental firm might still want the graph for other quick cost estimates:

$$\text{Cost of rental} = 49.95 + 0.35 \cdot (\text{miles driven}),$$
$$\text{Cost of driving } 125 \text{ mi} = 49.95 + 0.35(125) = \$93.70.$$

Our prediction, \$90, was close enough to serve as a good approximation.

5. STATE. Using the graph, we could predict that the cost of renting an 18-ft truck and driving it 125 mi is about \$90. ❑

EXERCISE SET | 3.4

Translate each sentence to an equation containing two variables. Be sure to state what each variable represents.

1. The sum of two numbers is 27.
2. The sum of two numbers is 53.
3. A number plus twice another number is 65.
4. The sum of a number and twice another number is 93.
5. One number is three times as great as another.
6. One number is half as great as another.
7. One number is 5 more than another.
8. One number is 7 less than another.
9. Hank's age plus 7 is twice Nanette's age.
10. Lisa's age is 5 less than twice Lou's.
11. Lois earns \$170 more than three times Roberta's salary.
12. Evelyn's salary plus \$200 is four times Eric's salary.

13. The nonstop flight is $\frac{3}{4}$ hr more than half the time of the direct flight.

14. The delay took 5 min more than half the flight time.

15. Three pizzas and two sandwiches cost $37.

16. One entree costs the same as two desserts.

Problem Solving

17. *Car traveling at a fixed speed.* A car travels at a speed of 55 mph for time t, in hours. The distance that it travels is given by

$$d = 55t.$$

a) How far does the car travel in 1 hr? in 2 hr? in 5 hr? in 10 hr?
b) Graph the equation. Assume that t is the first coordinate and d is the second.

18. *Hair growth.* Hair will normally grow at a rate of about $\frac{1}{2}$ in. per month. In t months, it will grow N in., where N is given by

$$N = \tfrac{1}{2}t.$$

a) How long will hair grow in 2 months? in 5 months? in 8 months? in 12 months?
b) Graph the equation. Assume that t is the first coordinate and N is the second.

19. *Women's shoe sizes in the United States and Britain.* Shoe sizes vary from country to country. The equation

$$y = x - 2$$

can be used to convert a women's shoe size x in the United States to its corresponding size y in Britain.

a) What women's shoe size in Britain corresponds to the U.S. size of 4? of 5? of 6? of 7? of 8?
b) Graph the equation.

20. *Toll road charges.* The equation

$$y = 0.027x + 0.19$$

can be used to determine the approximate cost y, in dollars, of driving x miles on the Indiana toll road.

a) Use the equation to predict the cost of driving 10 miles, 100 miles, and 150 miles.
b) Graph the equation.

21. *Increasing life expectancy.** A regular smoker is 15 times more likely to die from lung cancer than a nonsmoker. An exsmoker who stopped smoking t years ago is w times more likely to die from lung cancer than a nonsmoker, where

$$t + w = 15.$$

a) Sandy gave up smoking $2\frac{1}{2}$ years ago. How much more likely is she to die from lung cancer than Polly, who never smoked?
b) Graph the equation, using t as the first coordinate.

22. *Allocating resources.* Servemaster food services has $240 to spend on turkey at $4.00 per pound and/or roast beef at $6.00 per pound. If t pounds of turkey and r pounds of roast beef are bought, the equation

$$4t + 6r = 240$$

must be satisfied.

a) If 10 lb of turkey are bought, how much roast beef will be purchased?
b) Graph the equation, using r as the first coordinate.

23. *Truck rentals.* Iron Mike's Trucks charges $39.95 plus 55¢ per mile for the rental of a 20-ft truck. Draw a graph in which mileage is measured on the horizontal axis and cost on the vertical axis. Then use the graph to predict the cost of renting a 20-ft truck for one day and driving 180 mi.

24. *Van rentals.* Rent King charges $59.95 plus 45¢ per mile for the rental of its 20-ft van. Draw a graph in which mileage is measured on the horizontal axis and cost on the vertical axis. Then use the graph to predict the cost of renting the 20-ft van for one day and driving 340 mi.

25. *Wages and commissions.* Each salesperson at the Shoe Box is paid a weekly salary of $150 plus 4% of that person's sales. Draw a graph in which sales are measured on the horizontal axis and wages on the vertical axis. Then use the graph to estimate the wages paid when a salesperson sells $4500 in merchandise in one week.

26. *Wages and commissions.* Each salesperson at Grand Buy Appliances is paid a weekly salary of $200 plus 5% of that person's sales. Draw a graph in which sales are measured on the horizontal axis and wages on the vertical axis. Then use the graph to estimate the wages paid when a salesperson sells $3700 in appliances in one week.

27. *Cost of a road call.* Dave's Foreign Auto Village charges $35 for a road call plus $10 for each 15-min unit of time. Draw a graph that can be used to predict the cost of a service call. Use the horizontal axis for time and the vertical axis for cost. Then

---

**Source:* Data from *Body Clock* by Dr. Martin Hughes, p. 60. New York: Facts on File, Inc.

use the graph to estimate the cost of a $1\frac{1}{2}$-hr road call.

28. *Parking fees.* Karla's Parking charges $3.00 to park plus 50¢ for each 15-min unit of time. Draw a graph that can be used to predict the cost of parking at Karla's. Use the horizontal axis for time and the vertical axis for cost. Then use the graph to estimate how much it will cost to park for $3\frac{1}{2}$ hr.

29. *Food preparation.* Harriet's Catering believes that parties for more than 10 should include a 3-lb wheel of cheese and an additional $\frac{2}{9}$ lb for each person in excess of 10. Draw a graph that can be used to predict how much cheese should be purchased for parties of 10 or more. Use the horizontal axis for the number of people and the vertical axis for the number of pounds. Then use the graph to estimate the amount of cheese needed for a party of 21.

30. *Real-estate depreciation.* Because of wear and tear, rental property can be depreciated each year that it is in service. The depreciated value for some real estate is found by subtracting $\frac{1}{18}$ of the original value for each year that the property is rented. Draw a graph that can be used to find the depreciated value of a house that was valued at $150,000 when it was first rented. Use the horizontal axis for time and estimate the depreciated value after 8 yr of renting.

31. *Aviation.* Captain Hsu is landing a 747 from its cruising altitude of 32,000 ft. The jet descends at a rate of 3000 ft/min. Draw a graph in which the plane's altitude is measured on the vertical axis and use the graph to predict the altitude 8 min into the descent.

32. *Aviation.* Helga is landing a single-engine Tandem Taildragger from a cruising altitude of 6000 ft. Her plane is descending at a rate of 50 ft/min. Draw a graph in which the plane's altitude is measured on the vertical axis and use the graph to predict the altitude 17 min into the descent.

Skill Maintenance

33. Solve $s = vt + d$ for t.

34. Solve: $3(x - 4) + 7 = -2x + 6(x - 5)$.

Synthesis

35. ◈ Write a problem for a classmate to solve. Devise the problem so that a graph can be drawn to represent the situation and then predict a value (see Exercises 23–32).

36. ◈ Exercises 27 and 28 refer to units of time and Exercise 29 refers to the number of people at a party. Is it misleading to use solid lines rather than points in these graphs? Why or why not?

37. A Boeing 737 climbs from sea level to a cruising altitude of 34,000 ft at a rate of 6500 ft/min. After cruising for 3 min, the jet is forced to land, descending at a rate of 3500 ft/min. Draw a graph in which the plane's altitude is measured on the vertical axis and time on the horizontal axis.

38. Each salesperson at Mike's Bikes is paid $140 a week plus 13% of all sales up to $2000, and then 20% on any sales in excess of $2000. Draw a graph in which sales are measured on the horizontal axis and wages on the vertical axis. Then use the graph to estimate the wages paid when a salesperson sells $2700 in merchandise in one week.

39. Peggy earns $150 less than twice Paul's weekly salary. Paul's salary is $70 more than half of Jenna's. Draw a graph in which Jenna's salary is listed on the horizontal axis and Peggy's salary on the vertical axis. Then write an equation that relates Peggy's salary, p, to Jenna's salary, j.

40. Fast Eddie claims his truck rentals are always 5% less expensive than Rental Rick's. Rental Rick charges $40 plus 50¢ per mile for a 20-ft rental truck. Draw a graph that Rental Rick could use to calculate prices. Measure mileage on the horizontal axis and cost on the vertical axis. Then write an equation that relates Fast Eddie's cost, c, to the mileage driven, m.

Solve using a grapher.

41. Weekly pay at Bikes for Hikes is \$148 plus a 3.5% sales commission. If a salesperson sells \$3775 worth of merchandise, what is the weekly pay?

42. It costs Bert's Shirts \$38 plus \$2.35 a shirt to print tee-shirts for a day camp. Find Bert's cost for producing 144 shirts.

SUMMARY AND REVIEW 3

KEY TERMS

Bar graph, p. 114
Line graph, p. 116
Circle graph (or pie chart), p. 117
Axes (singular, axis), p. 117
Origin, p. 117
Coordinate, p. 117
Ordered pair, p. 118
Quadrant, p. 118
Linear equation, p. 122
Graph, p. 123
y-intercept, p. 127
x-intercept, p. 129

IMPORTANT PROPERTIES AND FORMULAS

To Graph Linear Equations

1. If the equation is of the type $x = a$ or $y = b$, the graph will be a line parallel to an axis.
2. If the equation is of the type $y = mx$, both intercepts are the origin, (0, 0). Plot (0, 0) and one other point. A third point can be calculated as a check.
3. If the equation is of the type $y = mx + b$, the y-intercept is $(0, b)$. Plot $(0, b)$ and one other point. A third point can be calculated as a check.
4. If the equation is of the type $Ax + By = C$, graph using intercepts. A third point can be used as a check.

REVIEW EXERCISES

This chapter's review and test include Skill Maintenance exercises from Sections 1.3, 1.4, 2.2, and 2.3.

This line graph shows the prime rate (the interest rate charged by banks to their best customers) in June for several years.

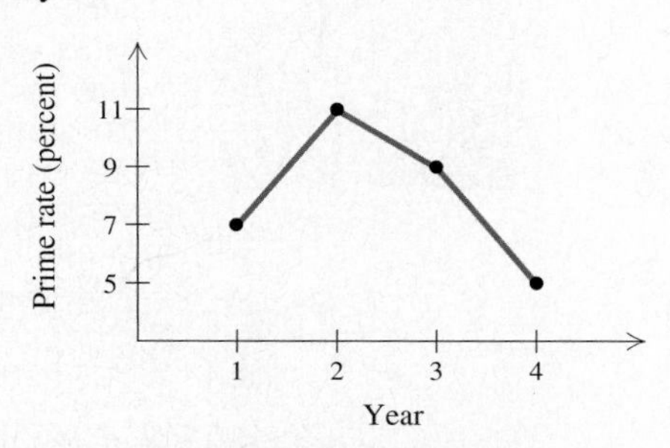

1. What was the highest prime rate?
2. Between what two consecutive years did the prime rate decrease the most?

Plot the point.

3. $(2, 5)$ 4. $(0, -3)$ 5. $(-4, -2)$

In which quadrant is the point located?

6. $(3, -8)$ 7. $(-20, -14)$ 8. $(4.9, 1.3)$

Find the coordinates of each point in the figure.

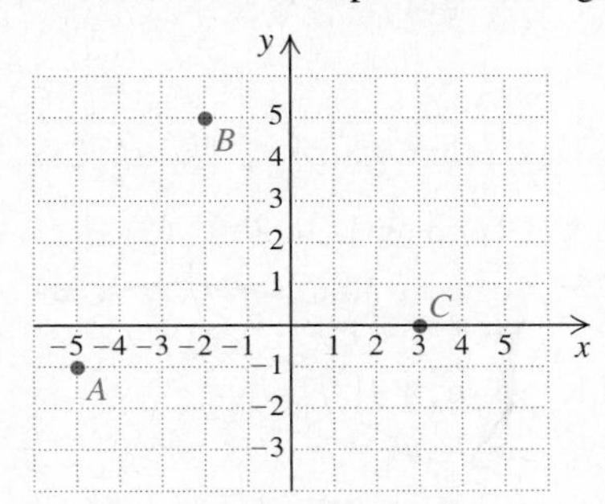

9. A 10. B 11. C

Determine whether the point is a solution of the equation $2y - x = 10$.

12. $(2, -6)$ 13. $(0, 5)$

Graph.

14. $y = 2x - 5$ 15. $y = -\frac{3}{4}x$
16. $y = -x + 4$ 17. $y = 3 - 4x$
18. $5x - 2y = 10$ 19. $4x + 3 = 0$

Find the x- and y-intercepts. Do not graph.

20. $x - 2y = 6$ 21. $2x + 3y = 27$

22. Translate to an equation containing two variables. State what each variable represents.

    One number is 3 less than twice another number.

23. *Impulses in nerve fibers.* Impulses in nerve fibers travel at a speed of 293 ft/sec. The distance d traveled in t seconds is given by

    $$d = 293t.$$

    a) How far will a nerve impulse travel in 0.01 sec? in 0.4 sec? in 1 sec?
    b) Graph the equation using decimal values for t from 0 to 1.

24. *Water rates.* Bay City Water Company charges each customer a \$6.40 monthly service charge, plus \$1.03 for each 100 cubic feet of water used. Draw a graph in which water usage is measured on the horizontal axis and the monthly bill on the vertical axis. Then use the graph to estimate the water bill for a month in which a customer used 800 cubic feet of water.

Skill Maintenance

25. Add and simplify: $\frac{3}{8} + \frac{5}{12}$.
26. Find decimal notation: $-\frac{7}{8}$.
27. Solve: $2(x - 3) = 10$.
28. Solve $A = \dfrac{m + n}{2}$ for m.

Synthesis

29. ◈ Describe two ways in which a small business might make use of graphs.
30. ◈ Explain why the first coordinate of the y-intercept is always 0.
31. Find the value of m in $y = mx + 3$ such that $(-2, 5)$ is on the graph.
32. Find the value of b in $y = -5x + b$ such that $(3, 4)$ is on the graph.
33. Find the area and the perimeter of a rectangle for which $(-2, 2)$, $(7, 2)$, and $(7, -3)$ are three of the vertices.
34. Find three solutions of $y = 4 - |x|$.

CHAPTER TEST 3

Consider the bar graph at right for Exercises 1–4.

1. What kind of degree was awarded most?
2. How many more bachelor's degrees than associate degrees were awarded?
3. How many more master's degrees than doctoral degrees were awarded?
4. In all, how many graduate degrees were awarded; that is, how many master's, doctoral, and professional degrees were awarded?

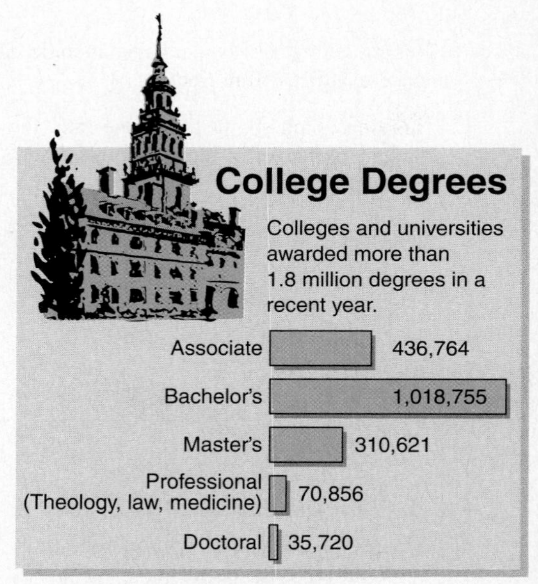

Source: National Center for Education Statistics, Digest of Education Statistics, 1992.

In which quadrant is the point located?

5. $\left(-\frac{1}{2}, 7\right)$
6. $(-5, -6)$

Find the coordinates of each point in the figure.

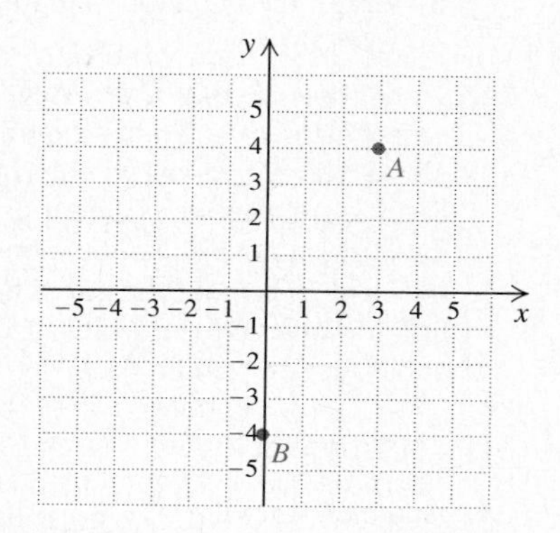

7. A
8. B
9. Determine whether the ordered pair $(2, -4)$ is a solution of the equation $y - 3x = -10$.

Graph.

10. $y = 2x - 1$
11. $2x - 4y = -8$
12. $y = 5$
13. $y = -\frac{3}{2}x$
14. $2x + 8 = 0$

Find the coordinates of the x- and y-intercepts. Do not graph.

15. $5x - 3y = 45$
16. $x = 10 - 4y$
17. Translate to an equation containing two variables. State what each variable represents.
    Greta earns \$50 more than twice Alice's salary.
18. Bright Electrical Service charges \$25 for a service call plus \$45 per hour. Draw a graph that could be used to predict the cost of a service call. Use the horizontal axis for time and the vertical axis for cost. Then use the graph to estimate the cost of a $\frac{1}{2}$-hr service call.

Skill Maintenance

19. Divide and simplify: $\frac{3}{5} \div \frac{3}{11}$.
20. Write a true sentence using either < or >: $-9.7 \;\blacksquare\; -11.3$.
21. Solve: $\frac{1}{3} + 2p = 4p - \frac{1}{5}$.
22. Solve $mx = b - nx$ for x.

Synthesis

23. A diagonal of a square connects the points $(-3, -1)$ and $(2, 4)$. Find the area and the perimeter of the square.
24. Write an equation of a line parallel to the x-axis and 3 units above it.

CUMULATIVE REVIEW 1–3

1. Evaluate $\frac{x}{2y}$ when $x = 10$ and $y = 2$.

2. Multiply: $3(4x - 5y + 7)$.

3. Factor: $3x + 9y + 15$.

4. Find the prime factorization of 42.

5. Find decimal notation: $\frac{9}{20}$.

6. Find the absolute value: $|-4|$.

7. Find the opposite of 5.

8. Find the reciprocal of 5.

9. Collect like terms: $2x - 5y + (-3x) + 4y$.

10. Find decimal notation: 78.5%.

Simplify.

11. $\frac{3}{5} - \frac{5}{12}$

12. $3.4 + (-0.8)$

13. $(-2)(-1.4)(2.6)$

14. $\frac{3}{8} \div \left(-\frac{9}{10}\right)$

15. $2 - [32 \div (4 + 2^2)]$

16. $-5 + 16 \div 2 \cdot 4$

17. $y - (3y + 7)$

18. $3(x - 1) - 2[x - (2x + 7)]$

Solve.

19. $1.5 = 2.7 + x$

20. $\frac{2}{7}x = -6$

21. $5x - 9 = 36$

22. $\frac{2}{3} = \frac{-m}{10}$

23. $5.4 - 1.9x = 0.8x$

24. $x - \frac{7}{8} = \frac{3}{4}$

25. $2(2 - 3x) = 3(5x + 7)$

26. $\frac{1}{4}x - \frac{2}{3} = \frac{3}{4} + \frac{1}{3}x$

27. $y + 5 - 3y = 5y - 9$

28. $x - 28 < 20 - 2x$

29. $2(x + 2) \geqslant 5(2x + 3)$

30. Solve $A = 2\pi rh + \pi r^2$, for h.

31. In which quadrant is the point $(3, -1)$ located?

32. Graph on a number line: $-1 < x \leqslant 2$.

Graph.

33. $y = -2$

34. $2x + 5y = 10$

35. $y = -2x + 1$

36. $y = \frac{2}{3}x$

Find the coordinates of the x- and y-intercepts. Do not graph.

37. $2x - 7y = 21$

38. $y = 4x + 5$

Solve.

39. A bill for a tie at Great Necks totaled \$14.28. How much did the tie cost if the sales tax is 5%?

40. If 25 is subtracted from a certain number, the result is 129. Find the number.

41. Sara and Becky purchased two dresses for a total of \$107. Sara paid \$17 more for her dress than Becky did. What did Becky pay?

42. A 143-m wire is cut into three pieces. The second is 3 m longer than the first. The third is four fifths as long as the first. How long is each piece?

43. Money is invested in a savings account at 4% simple interest. After 1 year, there is \$1560 in the account. How much was originally invested?

44. Sven's test grades are 75, 82, 86, and 79. Determine, in terms of an inequality, what scores on the last test will assure Sven an average test score of at least 80.

45. Translate to an equation containing two variables. Be sure to state what each variable represents.

> The steak cost \$5 more than three times the cost of the chicken.

46. Sparkle's Cleaning charges \$20 a visit plus \$15 an hour for commercial cleaning jobs. Draw a graph that can be used to predict the cost of a cleaning job. Use the horizontal axis for time and the vertical axis for cost. Then use the graph to estimate the cost of a $2\frac{1}{2}$-hr job.

Synthesis

47. Paula's salary at the end of a year is \$26,780. This reflects a 4% salary increase that preceded a 3% cost-of-living adjustment during the year. What was her salary at the beginning of the year?

Solve.

48. $4|x| - 13 = 3$

49. $4(x + 2) = 4(x - 2) + 16$

50. $0(x + 3) + 4 = 0$

51. $\frac{2 + 5x}{4} = \frac{11}{28} + \frac{8x + 3}{7}$

52. $5(7 + x) = (x + 7)5$

53. Solve $p = \frac{2}{m + Q}$, for Q.

CHAPTER 4

Polynomials

AN APPLICATION

During the first 13 seconds of a jump, the number of feet that a skydiver falls in t seconds is approximated by the polynomial

$$11.12t^2.$$

Approximately how far has a skydiver fallen 10 seconds after jumping from a plane?

This problem appears as Exercise 71 in Section 4.2.

Fran Strimenos
SKYDIVING INSTRUCTOR

"We use math every day in running our parachute company and school. Every jump requires calculations of wind velocity, the plane's speed, and the fall rate, and must take into consideration parachute design and the jumper's weight."

Algebraic expressions like $16t^2$, $5a^2 - 45$, and $3x^2 - 7x + 5$ are called polynomials. Polynomials occur frequently in applications and appear in most branches of mathematics. Thus learning to add, subtract, multiply, and divide polynomials is an important part of most courses in elementary algebra and is the focus of this chapter.

In addition to the material from this chapter, the review and test for Chapter 4 include material from Sections 1.5, 2.5, 2.6, and 3.1.

4.1 Exponents and Their Properties

Multiplying Powers with Like Bases • Dividing Powers with Like Bases • Zero as an Exponent • Raising a Power to a Power • Raising a Product or a Quotient to a Power

Because most of the polynomials with which we will work contain exponents, it is important that we develop some rules for simplifying exponential expressions.

Multiplying Powers with Like Bases

We know that an expression like a^3 means $a \cdot a \cdot a$ and that a^1 means a. Now consider multiplying powers with like bases:

$$a^3 \cdot a^2 = (a \cdot a \cdot a)(a \cdot a) \qquad \text{There are three factors in } a^3\text{; two factors in } a^2.$$

$$= a \cdot a \cdot a \cdot a \cdot a \qquad \text{Using an associative law}$$

$$= a^5.$$

Note that the exponent in a^5 is the sum of the exponents in $a^3 \cdot a^2$. That is, $3 + 2 = 5$. Similarly,

$$b^4 \cdot b^3 = (b \cdot b \cdot b \cdot b)(b \cdot b \cdot b) = b^7, \quad \text{where } 4 + 3 = 7.$$

Adding the exponents gives the correct result.

The Product Rule

For any number a and any positive integers m and n,

$$a^m \cdot a^n = a^{m+n}.$$

(When multiplying with exponential notation, if the bases are the same, keep the base and add the exponents.)

EXAMPLE 1

Multiply and simplify each of the following. (Here "simplify" means express the product as one base to a power whenever possible.)

a) $x^2 \cdot x^9$ **b)** $8^4 \cdot 8^3$ **c)** $x \cdot x^8$

d) $m^5 m^{10} m^3$ **e)** $(a^3 b^2)(a^3 b^5)$

Solution

a) $x^2 \cdot x^9 = x^{2+9}$ Adding exponents: $a^m \cdot a^n = a^{m+n}$
$= x^{11}$

b) $8^4 \cdot 8^3 = 8^{4+3}$
$= 8^7$

c) $x \cdot x^8 = x^1 \cdot x^8 = x^{1+8}$
$= x^9$

d) $m^5 m^{10} m^3 = m^{5+10+3}$
$= m^{18}$

e) $(a^3b^2)(a^3b^5) = a^3b^2a^3b^5$ Using an associative law
$= a^3a^3b^2b^5$ Using a commutative law
$= a^6b^7$ Adding exponents ❑

Dividing Powers with Like Bases

The following suggests a rule for dividing powers with like bases, such as a^5/a^2:

$$\frac{a^5}{a^2} = \frac{a \cdot a \cdot a \cdot a \cdot a}{a \cdot a} = \frac{a \cdot a \cdot a \cdot a \cdot a}{1 \cdot a \cdot a} = \frac{a \cdot a \cdot a}{1} \cdot \frac{a \cdot a}{a \cdot a} = \frac{a \cdot a \cdot a}{1} \cdot 1$$
$$= a \cdot a \cdot a = a^3.$$

Note that the exponent in a^3 is the difference of the exponents in a^5/a^2. Similarly,

$$\frac{x^4}{x^3} = \frac{x \cdot x \cdot x \cdot x}{x \cdot x \cdot x} = \frac{x}{1} \cdot \frac{x \cdot x \cdot x}{x \cdot x \cdot x} = \frac{x}{1} \cdot 1 = x^1, \quad \text{or } x.$$

Subtracting exponents gives the correct result.

The Quotient Rule

For any nonzero number a and any positive integers m and n for which $m > n$,

$$\frac{a^m}{a^n} = a^{m-n}.$$

(To divide with exponential notation, if the bases are the same, keep the base and subtract the exponent of the denominator from the exponent of the numerator.)

EXAMPLE 2

Divide and simplify. (Here "simplify" means express the quotient as one base to a power whenever possible.)

a) $\dfrac{x^8}{x^2}$ **b)** $\dfrac{6^5}{6^3}$ **c)** $\dfrac{t^{12}}{t}$ **d)** $\dfrac{p^5q^7}{p^2q^5}$

Solution

a) $\dfrac{x^8}{x^2} = x^{8-2}$ Subtracting exponents: $\dfrac{a^m}{a^n} = a^{m-n}$
$= x^6$

b) $\dfrac{6^5}{6^3} = 6^{5-3}$
$= 6^2$

c) $\dfrac{t^{12}}{t} = t^{12-1} = t^{11}$

d) $\dfrac{p^5q^7}{p^2q^5} = p^{5-2}q^{7-5} = p^3q^2$ ❑

Zero as an Exponent

The quotient rule can be used to help determine what 0 should mean when it appears as an exponent. Consider a^4/a^4, where a is nonzero. Since the numerator and the denominator are the same,

$$\frac{a^4}{a^4} = 1.$$

On the other hand, using the quotient rule gives us

$$\frac{a^4}{a^4} = a^{4-4} = a^0. \qquad \text{Subtracting exponents}$$

Since $a^0 = a^4/a^4 = 1$, it follows that $a^0 = 1$ for any nonzero value of a.

The Exponent Zero

For any real number a, $a \neq 0$,

$$a^0 = 1.$$

(Any nonzero number raised to the 0 power is 1.)

EXAMPLE 3

Simplify: **(a)** 1957^0; **(b)** $(-7)^0$; **(c)** $(-1)7^0$; **(d)** $(3x)^0$.

Solution

a) $1957^0 = 1$ Any nonzero number raised to the 0 power is 1.

b) $(-7)^0 = 1$ Any nonzero number raised to the 0 power is 1.

c) We have

$$(-1)7^0 = (-1)1 = -1.$$

Since multiplying by -1 is the same as finding the opposite, the expression $(-1)7^0$ could have been written as -7^0. Note that while $-7^0 = -1$, part (b) shows that $(-7)^0 = 1$.

d) The parentheses indicate that the base is $3x$. Thus,

$$(3x)^0 = 1 \quad \text{for any } x \neq 0.$$ ❑

To see why 0^0 is not defined, note that $0^0 = 0^{1-1} = 0^1/0^1 = 0/0$. As we saw in Section 1.7, 0/0 has not been defined. Thus, 0^0 is not defined either. Henceforth in this text we will assume that expressions like a^m do not represent 0^0.

Raising a Power to a Power

Consider an expression like $(5^2)^4$.

$$\begin{aligned}(5^2)^4 &= (5^2)(5^2)(5^2)(5^2) && \text{There are four factors of } 5^2. \\ &= (5 \cdot 5)(5 \cdot 5)(5 \cdot 5)(5 \cdot 5) \\ &= 5 \cdot 5 \cdot 5 \cdot 5 \cdot 5 \cdot 5 \cdot 5 \cdot 5 && \text{Using an associative law} \\ &= 5^8.\end{aligned}$$

Note that in this case we could have multiplied the exponents:

$$(5^2)^4 = 5^{2 \cdot 4} = 5^8.$$

Likewise, $(y^7)^3 = (y^7)(y^7)(y^7) = y^{21}$. Once again, we get the same result if we multiply the exponents:

$$(y^7)^3 = y^{7 \cdot 3} = y^{21}.$$

The Power Rule

For any number a and any whole numbers m and n,

$$(a^m)^n = a^{mn}.$$

(To raise a power to a power, multiply the exponents and leave the base unchanged.)

EXAMPLE 4

Simplify: **(a)** $(m^2)^5$; **(b)** $(3^5)^4$.

Solution

a) $(m^2)^5 = m^{2 \cdot 5}$ Multiplying exponents: $(a^m)^n = a^{mn}$
$= m^{10}$

b) $(3^5)^4 = 3^{5 \cdot 4}$
$= 3^{20}$ ❑

Raising a Product or a Quotient to a Power

When an expression inside parentheses is raised to a power, the inside expression is the base. Let us compare $2a^3$ and $(2a)^3$:

$2a^3 = 2 \cdot a \cdot a \cdot a$; The base is a.

$(2a)^3 = (2a)(2a)(2a)$ The base is $2a$.
$= (2 \cdot 2 \cdot 2)(a \cdot a \cdot a)$ Using an associative and a commutative law
$= 2^3a^3$
$= 8a^3$.

We see that $2a^3$ and $(2a)^3$ are *not* equivalent. Note too that $(2a)^3$ can be simplified by raising each factor to the power 3. This leads us to the following rule for raising a product to a power.

Raising a Product to a Power

For any numbers a and b and any whole number n,

$$(ab)^n = a^nb^n.$$

(To raise a product to the nth power, raise each factor to the nth power.)

EXAMPLE 5

Simplify: **(a)** $(4a)^3$; **(b)** $(5x^4)^2$; **(c)** $(-4a^5b^3)^3$.

Solution

a) $(4a)^3 = 4^3a^3 = 64a^3$ Raising each factor to the third power and simplifying

b) $(5x^4)^2 = 5^2(x^4)^2$ Raising each factor to the second power

$= 25x^8$ Simplifying 5^2; using the power rule

c) $(-4a^5b^3)^3 = (-4)^3(a^5)^3(b^3)^3$ Cubing each factor

$= -64a^{15}b^9$ A negative number raised to an odd power is negative; using the power rule. ❑

There is a similar rule for raising a quotient to a power.

Raising a Quotient to a Power

For any numbers a and b, $b \neq 0$, and any whole number n,

$$\left(\frac{a}{b}\right)^n = \frac{a^n}{b^n}.$$

(To raise a quotient to a power, raise the numerator to the power and divide by the denominator to the power.)

EXAMPLE 6

Simplify: **(a)** $\left(\frac{x}{5}\right)^2$; **(b)** $\left(\frac{5}{a^4}\right)^3$; **(c)** $\left(\frac{3a^4}{b^3}\right)^2$.

Solution

a) $\left(\frac{x}{5}\right)^2 = \frac{x^2}{5^2} = \frac{x^2}{25}$ Squaring the numerator and the denominator

b) $\left(\frac{5}{a^4}\right)^3 = \frac{5^3}{(a^4)^3} = \frac{125}{a^{4 \cdot 3}} = \frac{125}{a^{12}}$

c) $\left(\frac{3a^4}{b^3}\right)^2 = \frac{(3a^4)^2}{(b^3)^2} = \frac{3^2(a^4)^2}{b^{3 \cdot 2}} = \frac{9a^8}{b^6}$ ❑

In the following summary of definitions and rules, we assume that no denominators are 0 and that 0^0 is not considered.

For any whole numbers m and n,

1 as an exponent:	$a^1 = a$
0 as an exponent:	$a^0 = 1$
The Product Rule:	$a^m \cdot a^n = a^{m+n}$
The Quotient Rule:	$\frac{a^m}{a^n} = a^{m-n}$
The Power Rule:	$(a^m)^n = a^{mn}$
Raising a product to a power:	$(ab)^n = a^nb^n$
Raising a quotient to a power:	$\left(\frac{a}{b}\right)^n = \frac{a^n}{b^n}$

EXERCISE SET 4.1

Multiply and simplify.

1. $2^4 \cdot 2^3$
2. $3^5 \cdot 3^2$
3. $8^5 \cdot 8^9$
4. $n^3 \cdot n^{20}$
5. $x^4 \cdot x^3$
6. $y^7 \cdot y^9$
7. $9^{17} \cdot 9^{21}$
8. $t^0 \cdot t^{16}$
9. $(3y)^4(3y)^8$
10. $(2t)^8(2t)^{17}$
11. $(7y)^1(7y)^{16}$
12. $(8x)^0(8x)^1$
13. $(a^2b^7)(a^3b^2)$
14. $(m^5n^4)(m^6n^2)$
15. $(xy^9)(x^3y^5)$
16. $(a^8b^3)(a^4b)$
17. $r^3 \cdot r^7 \cdot r^2$
18. $s^4 \cdot s^5 \cdot s^2$
19. $x^3(xy^4)(xy)$
20. $a^4(a^3b)(ab)$

Divide and simplify.

21. $\dfrac{7^5}{7^2}$
22. $\dfrac{4^7}{4^3}$
23. $\dfrac{8^{12}}{8^6}$
24. $\dfrac{9^{14}}{9^2}$
25. $\dfrac{y^9}{y^5}$
26. $\dfrac{x^{12}}{x^{11}}$
27. $\dfrac{(5a)^7}{(5a)^6}$
28. $\dfrac{(3m)^9}{(3m)^8}$
29. $\dfrac{6^5x^8}{6^2x^3}$
30. $\dfrac{3^9a^7}{3^4a^5}$
31. $\dfrac{18m^5}{6m^2}$
32. $\dfrac{30n^7}{6n^3}$
33. $\dfrac{a^9b^7}{a^2b}$
34. $\dfrac{r^{10}s^7}{r^3s^0}$
35. $\dfrac{m^9n^8}{m^0n^4}$
36. $\dfrac{a^{10}b^{12}}{a^2b^3}$

Simplify.

37. x^0 when $x = -12$
38. y^0 when $y = 23$
39. $5x^0$ when $x = -4$
40. $7m^0$ when $m = 1.7$
41. n^0, $n \neq 0$
42. t^0, $t \neq 0$
43. 10^0
44. 9^0
45. $5^1 - 5^0$
46. $8^0 - 8^1$

Simplify. Answers should not contain parentheses.

47. $(x^3)^4$
48. $(a^4)^6$
49. $(2^3)^8$
50. $(5^7)^3$
51. $(m^7)^5$
52. $(n^9)^2$
53. $(a^{25})^3$
54. $(a^3)^{25}$
55. $(3x)^2$
56. $(5a)^2$
57. $(-2a)^3$
58. $(-3x)^3$
59. $(4m^3)^2$
60. $(5n^4)^2$
61. $(3a^2b)^3$
62. $(5xy^2)^3$
63. $(a^3b^2)^5$
64. $(m^4n^5)^6$
65. $(-5x^4y^5)^2$
66. $(-3a^5b^7)^4$
67. $\left(\dfrac{a}{4}\right)^3$
68. $\left(\dfrac{3}{x}\right)^4$
69. $\left(\dfrac{7}{5a}\right)^2$
70. $\left(\dfrac{4x}{3}\right)^3$
71. $\left(\dfrac{a^2}{b^3}\right)^4$
72. $\left(\dfrac{x^3}{y^4}\right)^5$
73. $\left(\dfrac{y^3}{2}\right)^2$
74. $\left(\dfrac{a^5}{3}\right)^3$
75. $\left(\dfrac{5x^2}{y^3}\right)^3$
76. $\left(\dfrac{7y^5}{x^4}\right)^2$
77. $\left(\dfrac{a^3}{-2b^5}\right)^4$
78. $\left(\dfrac{x^5}{-3y^3}\right)^4$
79. $\left(\dfrac{2a^2}{3b^4}\right)^3$
80. $\left(\dfrac{3x^5}{4y^3}\right)^2$
81. $\left(\dfrac{4x^3y^5}{3z^7}\right)^2$
82. $\left(\dfrac{5a^7}{2b^5c}\right)^3$

Skill Maintenance

83. Factor: $3s + 3t + 24$.
84. Factor: $-7x - 14$.
85. Collect like terms: $9x + 2y - 4x - 2y$.
86. 24 is what percent of 64?

Synthesis

87. ◈ Under what conditions does a^n represent a negative number? Why?
88. ◈ Using the quotient rule, explain why 9^0 is 1.
89. Solve for x: $\dfrac{w^{50}}{w^x} = w^x$.

Simplify.

90. $(y^{2x})(y^{3x})$
91. $a^{5k} \div a^{3k}$
92. $\dfrac{a^{6t}(a^{7t})}{a^{9t}}$
93. $\dfrac{\left(\frac{1}{2}\right)^4}{\left(\frac{1}{2}\right)^5}$
94. $\dfrac{(0.4)^5}{(0.4)^3(0.4)^2}$

Use $>$, $<$, or $=$ for ■ to write a true sentence.

95. 3^5 ■ 3^4
96. 4^2 ■ 4^3
97. 4^3 ■ 5^3
98. 4^3 ■ 3^4

Find a value of the variable that shows that the two expressions are *not* equivalent.

99. $3x^2$; $(3x)^2$
100. $(a + 3)^2$; $a^2 + 3^2$

101. $\dfrac{x+2}{2}$; x

102. $\dfrac{y^6}{y^3}$; y^2

Interest compounded annually. If a principal P is invested at interest rate r, compounded annually, in t years it will grow to an amount A given by

$$A = P(1 + r)^t.$$

103. Suppose \$10,400 is invested at 8.5% compounded annually. How much is in the account at the end of 5 years?

104. Suppose \$20,800 is invested at 4.5%, compounded annually. How much is in the account at the end of 6 years?

105. Find the area of a square if each side has length $5x$.

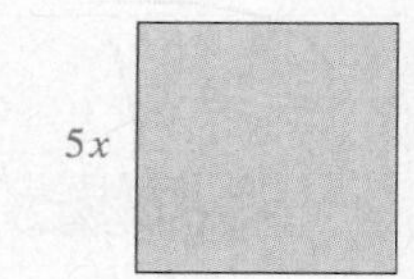

106. Find the volume of a cube if each side has length $7a$.

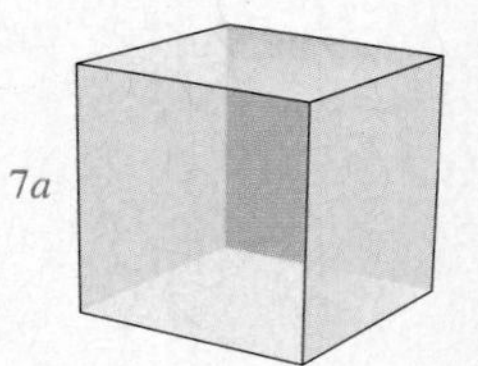

4.2 Polynomials

Terms • Collecting Like Terms • Evaluating Polynomials and Applications

We now begin our study of an important algebraic expression known as a *polynomial.* Certain polynomials have appeared earlier in this text so you already have some experience working with them.

One type of polynomial that we have already seen is a *monomial*. A **monomial** is a constant or the product of some constant and variable(s) raised to a whole-number power. Examples of monomials are

$$5, \quad x, \quad -7a, \quad 3m^2, \quad \text{and} \quad \tfrac{3}{4}x^2y^5.$$

Algebraic expressions like the following are **polynomials:**

$$\tfrac{3}{4}y^5, \quad -2, \quad 5y + 3, \quad 3x^2 + 2x - 5, \quad -7a^3 + 4ab, \quad 6x, \quad 37p^4, \quad x, \quad 0.$$

Polynomial

A *polynomial* is a monomial or a combination of sums and/or differences of monomials.

The following algebraic expressions are *not* polynomials:

$$\textbf{(1)}\ \frac{x+3}{x-4}, \qquad \textbf{(2)}\ 5x^3 - 2x^2 + \frac{1}{x}, \qquad \textbf{(3)}\ \frac{1}{x^3 - 2}.$$

Expressions (1) and (3) are not polynomials because they represent quotients, not sums or differences. Expression (2) is not a polynomial because $1/x$ is not a monomial.

A sum or difference of two monomials is called a **binomial.** When three monomials are added and/or subtracted, we have what is called a **trinomial.** Polynomials composed of four or more monomials have no special name.

Monomials	Binomials	Trinomials	No Special Name
$4x^2$	$2x + 4$	$3t^3 + 4t + 7$	$4x^3 - 5x^2 + x - 8$
9	$3a^5 + 6bc$	$6x^7 - 7x^2 + 4$	$z^5 + 2z^4 - z^3 + 7z + 3$
$-7a^{19}$	$-9x^7 - 6$	$4x^2 - 6x - \frac{1}{2}$	$4x^6 - 3x^5 + x^4 - x^3 + 2x - 1$

Terms

Although every monomial is a term, some terms (like $5/x^2$) are not monomials. To identify the terms of a polynomial, we look at the polynomial as a sum of monomials.

EXAMPLE 1

Identify the terms of the polynomial $3t^4 - 5t^6 - 4t + 2$.

Solution The terms are $3t^4$, $-5t^6$, $-4t$, and 2. If this is difficult to see, try rewriting all subtractions as additions of opposites:

$$3t^4 - 5t^6 - 4t + 2 = 3t^4 + (-5t^6) + (-4t) + 2.$$

$\uparrow \quad \uparrow \quad \uparrow \quad \uparrow$

These are the terms of the polynomial. ❑

The **coefficient** of a term is the number multiplying the variable(s) in the term. Thus the coefficient of the term $5x^3$ is 5.

EXAMPLE 2

Identify the coefficient of each term in the polynomial

$$4x^3 - 7x^2y + x - 8.$$

Solution

The coefficient of the first term is 4.

The coefficient of the second term is -7.

The coefficient of the third term is 1, since $x = 1x$.

The coefficient of the fourth term is -8. ❑

The **degree of a term** is the exponent of the variable.* The degree of the term $5x^3$ is 3.

EXAMPLE 3

Identify the degree of each term of $8x^4 + 3x + 7$.

Solution

The degree of $8x^4$ is 4.

The degree of $3x$ is 1. Recall that $x = x^1$.

The degree of 7 is 0. Think of 7 as $7x^0$. Recall that $x^0 = 1$. ❑

---

*A more detailed definition for terms with several variables is given in Section 4.6.

The **degree of a polynomial** is the largest of the degrees of its terms, unless it is the polynomial 0. We agree that 0 has *no* degree either as a term or as a polynomial. This is because we can express 0 as $0 = 0x^5 = 0x^7$, and so on, using any exponent we wish.

EXAMPLE 4

Identify the degree of the polynomial $5x^3 - 6x^4 + 7$.

Solution We have

$5x^3 - 6x^4 + 7$. The largest exponent is 4.

The degree of the polynomial is 4. ❑

Let us summarize the terminology we have learned for the polynomial

$3x^4 - 8x^3 + 5x^2 + 7x - 6$.

Term	Coefficient	Degree of the Term	Degree of the Polynomial
$3x^4$	3	4	4
$-8x^3$	-8	3	
$5x^2$	5	2	
$7x$	7	1	
-6	-6	0	

Collecting Like Terms

As we saw in Section 1.8, when a polynomial contains *like,* or *similar, terms*—that is, terms with the same variable(s) raised to the same power(s)—those terms can be *combined* or *collected.*

EXAMPLE 5

Identify the like terms in $4x^3 + 5x - 7x^2 + 2x^3 + x^2$.

Solution

Like terms: $4x^3$ and $2x^3$ Same variable and exponent

Like terms: $-7x^2$ and x^2 Same variable and exponent ❑

EXAMPLE 6

Collect like terms:

a) $2x^3 - 6x^3$

b) $5x^2 + 7 + 2x^4 + 4x^2 - 11 - 2x^4$

c) $7a^3 - 5a^2 + 9a^3 + a^2$

d) $\frac{2}{3}x^4 - x^3 - \frac{1}{6}x^4 + \frac{2}{5}x^3 - \frac{3}{10}x^3$

Solution

a) $2x^3 - 6x^3 = (2 - 6)x^3$ Using the distributive law

$= -4x^3$

b) $5x^2 + 7 + 2x^4 + 4x^2 - 11 - 2x^4$

$= (5 + 4)x^2 + (2 - 2)x^4 + (7 - 11)$

$= 9x^2 + 0x^4 + (-4)$

These steps are often done mentally.

$= 9x^2 - 4$

c) $7a^3 - 5a^2 + 9a^3 + a^2 = 7a^3 - 5a^2 + 9a^3 + 1a^2$

When a variable to a power appears without a coefficient, we can write in 1.

$$= 16a^3 - 4a^2$$

d)
$$\tfrac{2}{3}x^4 - x^3 - \tfrac{1}{6}x^4 + \tfrac{2}{5}x^3 - \tfrac{3}{10}x^3 = \left(\tfrac{2}{3} - \tfrac{1}{6}\right)x^4 + \left(-1 + \tfrac{2}{5} - \tfrac{3}{10}\right)x^3$$
$$= \left(\tfrac{4}{6} - \tfrac{1}{6}\right)x^4 + \left(-\tfrac{10}{10} + \tfrac{4}{10} - \tfrac{3}{10}\right)x^3$$
$$= \tfrac{3}{6}x^4 - \tfrac{9}{10}x^3$$
$$= \tfrac{1}{2}x^4 - \tfrac{9}{10}x^3$$

❑

Note in Examples 6(b), (c), and (d) that the solutions are written so that the term of highest degree appears first, followed by the term of next highest degree, and so on. This is known as *descending order* and is the form in which answers will normally appear.

Evaluating Polynomials and Applications

Recall from Chapter 1 that when we replace the variable in a polynomial with a number, the polynomial then represents a number, or *value*, that can be calculated using the rules for the order of operations.

EXAMPLE 7

Evaluate $-x^2 + 3x + 9$ when $x = -2$.

Solution When $x = -2$, we have

$$-x^2 + 3x + 9 = -(-2)^2 + 3(-2) + 9$$
$$= -4 + (-6) + 9$$
$$= -10 + 9 = -1.$$

❑

Polynomials occur in many real-world situations and are used in problem solving. The following examples are three such applications. Because the examples involve only the evaluation of a polynomial, we do not apply all five problem-solving steps.

EXAMPLE 8

Volume of a cube. The volume of a cube with sides of length x is given by the polynomial

$$x^3.$$

Find the volume of a cube with sides of length 5 cm.

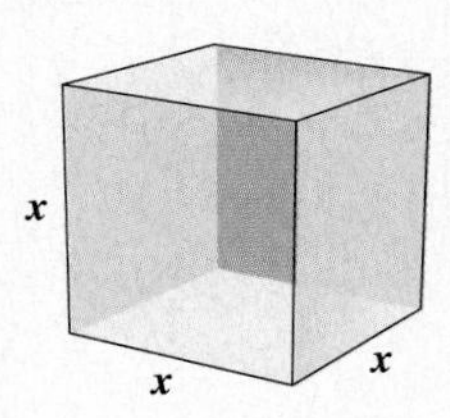

Solution We evaluate the polynomial for $x = 5$:

$$x^3 = 5^3 = 125.$$

The volume is 125 cm^3 (cubic centimeters).

❑

EXAMPLE 9

Games in a sports league. In a sports league of n teams in which each team plays every other team twice, the total number of games to be played is given by the polynomial

$$n^2 - n.$$

A women's softball league has 10 teams. What is the total number of games to be played?

Solution We evaluate the polynomial for $n = 10$:

$$\begin{aligned} n^2 - n &= 10^2 - 10 \\ &= 100 - 10 \\ &= 90. \end{aligned}$$

The league plays 90 games. ❑

EXAMPLE 10

Medical dosage. The concentration, in parts per million, of a certain medication in the bloodstream after t hours is given by the polynomial

$$-0.05t^2 + 2t + 2.$$

Find the concentration after 2 hr.

Solution To find the concentration after 2 hr, we evaluate the polynomial for $t = 2$:

$$\begin{aligned} -0.05t^2 + 2t + 2 &= -0.05(2)^2 + 2(2) + 2 \\ &= -0.05(4) + 2(2) + 2 \\ &= -0.2 + 4 + 2 \\ &= 5.8. \end{aligned}$$

The concentration after 2 hr is 5.8 parts per million. ❑

The polynomial in Example 10 can be graphed if we evaluate it for several values of t. Note that the concentration peaks at the 20-hr mark and after a bit more than 40 hr, the concentration is 0. Since neither time nor concentration can be negative, our graph uses only the first quadrant.

t	$-0.05t^2 + 2t + 2$
0	2
2	5.8
10	17
20	22
30	17

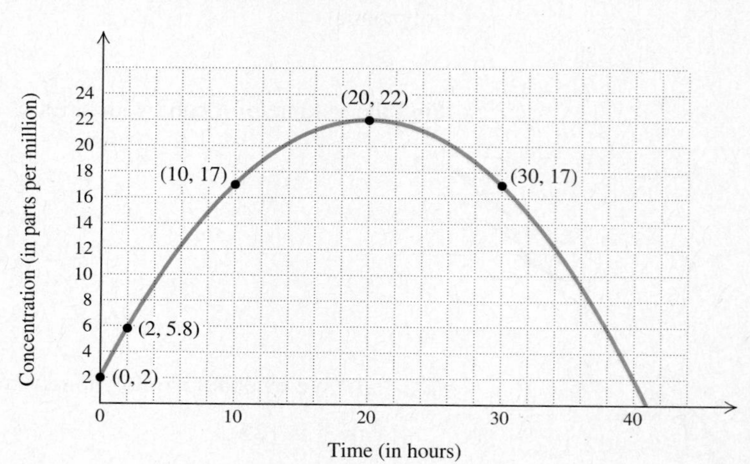

EXERCISE SET 4.2

Classify the polynomial as a monomial, binomial, trinomial, or none of these.

1. $x^2 - 10x + 25$ **2.** $-6x^4$

3. $x^3 - 7x^2 + 2x - 4$ **4.** $x^2 - 9$

5. $4x^2 - 25$

6. $2x^4 - 7x^3 + x^2 + x - 6$

7. $40x$ **8.** $4x^2 + 12x + 9$

Identify the terms of the polynomial.

9. $2 - 3x + x^2$ **10.** $2x^2 + 3x - 4$

Identify the like terms in the polynomial.

11. $5x^3 + 6x^2 - 3x^2$ **12.** $3x^2 + 4x^3 - 2x^2$

13. $2x^4 + 5x - 7x - 3x^4$ **14.** $-3t + t^3 - 2t - 5t^3$

Identify the coefficient of each term of the polynomial.

15. $-3x + 6$ **16.** $2x - 4$

17. $5x^2 + 3x + 3$ **18.** $3x^2 - 5x + 2$

19. $-7x^3 + 6x^2 + 3x + 7$

20. $5x^4 + x^2 - x + 2$

21. $-5x^4 + 6x^3 - 3x^2 + 8x - 2$

22. $7x^3 - 4x^2 - 4x + 5$

Determine the degree of each term of the polynomial and the degree of the polynomial.

23. $2x - 4$ **24.** $6 - 3x$

25. $3x^2 - 5x + 2$ **26.** $5x^3 - 2x^2 + 3$

27. $-7x^3 + 6x^2 + 3x + 7$ **28.** $5x^4 + x^2 - x + 2$

29. $x^2 - 3x + x^6 - 9x^4$ **30.** $8x - 3x^2 + 9 - 8x^3$

31. For the polynomial $-7x^4 + 6x^3 - 3x^2 + 8x - 2$, complete the following table.

Term	Coefficient	Degree of the Term	Degree of the Polynomial
$7x^4$	-7	4	4
$6x^3$	6	3	
$3x^2$	-3	2	
$8x$	8	1	
-2	-2	0	

32. For the polynomial $3x^2 + 8x^5 - 46x^3 + 6x - 2.4 - \frac{1}{2}x^4$, complete the following table.

Term	Coefficient	Degree of the Term	Degree of the Polynomial
		5	
$-\frac{1}{2}x^4$		4	
	-46		
$3x^2$		2	
	6		
-2.4			

Collect like terms. Write all answers in descending order.

33. $2x - 5x$ **34.** $2x^2 + 8x^2$

35. $x - 9x$ **36.** $x - 5x$

37. $5x^3 + 6x^3 + 4$ **38.** $6x^4 - 2x^4 + 5$

39. $5x^3 + 6x - 4x^3 - 7x$

40. $3a^4 - 2a + 2a + a^4$

41. $6b^5 + 3b^2 - 2b^5 - 3b^2$

42. $2x^2 - 6x + 3x + 4x^2$

43. $\frac{1}{4}x^5 - 5 + \frac{1}{2}x^5 - 2x - 37$

44. $\frac{1}{3}x^3 + 2x - \frac{1}{6}x^3 + 4 - 16$

45. $6x^2 + 2x^4 - 2x^2 - x^4 - 4x^2$

46. $8x^2 + 2x^3 - 3x^3 - 4x^2 - 4x^2$

47. $\frac{1}{4}x^3 - x^2 - \frac{1}{6}x^2 + \frac{3}{8}x^3 + \frac{5}{16}x^3$

48. $\frac{1}{5}x^4 + \frac{1}{5} - 2x^2 + \frac{1}{10} - \frac{3}{15}x^4 + 2x^2 - \frac{3}{10}$

49. $3x^4 - 5x^6 - 2x^4 + 6x^6$

50. $-1 + 5x^3 - 3 - 7x^3 + x^4 + 5$

51. $-2x + 4x^3 - 7x + 9x^3 + 8$

52. $-6x^2 + x - 5x + 7x^2 + 1$

53. $3x + 3x + 3x - x^2 - 4x^2$

54. $-2x - 2x - 2x + x^3 - 5x^3$

55. $-x + \frac{3}{4} + 15x^4 - x - \frac{1}{2} - 3x^4$

56. $2x - \frac{5}{6} + 4x^3 + x + \frac{1}{3} - 2x$

Evaluate the polynomial for $x = 4$.

57. $-5x + 2$ **58.** $-3x + 1$

59. $2x^2 - 5x + 7$

60. $3x^2 + x + 7$

61. $x^3 - 5x^2 + x$

62. $7 - x + 3x^2$

Evaluate the polynomial for $x = -1$.

63. $3x + 5$

64. $6 - 2x$

65. $x^2 - 2x + 1$

66. $5x - 6 + x^2$

67. $-3x^3 + 7x^2 - 3x - 2$

68. $-2x^3 - 5x^2 + 4x + 3$

Daily accidents. The average number of accidents per day involving drivers of age r can be approximated by the polynomial

$$0.4r^2 - 40r + 1039.$$

69. Evaluate the polynomial for $r = 18$ to find the daily number of accidents involving 18-year-old drivers.

70. Evaluate the polynomial for $r = 20$ to find the daily number of accidents involving 20-year-old drivers.

71. *Skydiving.* During the first 13 sec of a jump, the number of feet that a skydiver falls in t seconds is approximated by the polynomial

$$11.12t^2.$$

Approximately how far has a skydiver fallen 10 sec after jumping from a plane?

72. *Skydiving.* For jumps that exceed 13 sec, the polynomial $173t - 369$ can be used to approximate the distance, in feet, that a skydiver has fallen in t seconds. Approximately how far has a skydiver fallen 20 sec after jumping from a plane?

Total revenue. An electronics firm is marketing a new kind of stereo. Total revenue is the total amount of money taken in. The firm determines that when it sells x stereos, it will take in

$$280x - 0.4x^2 \text{ dollars.}$$

73. What is the total revenue from the sale of 75 stereos?

74. What is the total revenue from the sale of 100 stereos?

Total cost. The electronics firm determines that the total cost of producing x stereos is given by

$$5000 + 0.6x^2 \text{ dollars.}$$

75. What is the total cost of producing 500 stereos?

76. What is the total cost of producing 650 stereos?

Circumference. The circumference of a circle of radius r is given by the polynomial $2\pi r$,

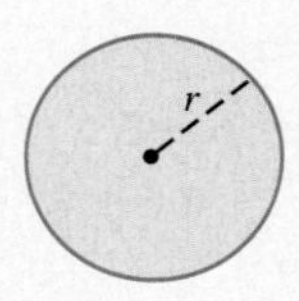

where π is an irrational number. For an approximation of π, use 3.14 unless indicated otherwise.

77. Find the circumference of a circle with radius 10 cm.

78. Find the circumference of a circle with radius 5 ft.

Area of a circle. The area of a circle of radius r is given by the polynomial πr^2.

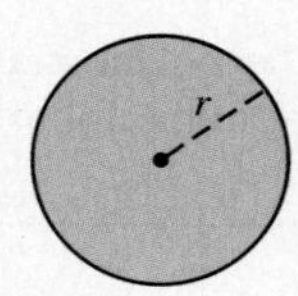

79. Find the area of a circle with radius 5 m. Use 3.14 for π.

80. Find the area of a circle with radius 10 in. Use 3.14 for π.

In Exercises 81 and 82, complete the table for the given choices of t. Then plot the points and connect them with a smooth curve representing the graph of the polynomial.

81.

t	$-t^2 + 6t - 4$
1	
2	
3	
4	
5	

82.

t	$-t^2 + 10t - 18$
3	
4	
5	
6	
7	

Skill Maintenance

83. The sum of the page numbers on the facing pages of a book is 549. What are the page numbers?

84. A family spent \$2011 to drive a car one year, during which the car was driven 7400 mi. The family spent \$972 for insurance and \$114 for a license registration fee. The only other cost was for gasoline. How much did gasoline cost per mile?

Synthesis

85. Explain why an understanding of the rules for order of operations is essential when evaluating polynomials.

86. Is it better to evaluate a polynomial before or after like terms have been combined? Why?

Simplify.

87. $\frac{9}{2}x^8 + \frac{1}{9}x^2 + \frac{1}{2}x^9 + \frac{9}{2}x + \frac{9}{2}x^9 + \frac{8}{9}x^2 + \frac{1}{2}x - \frac{1}{2}x^8$

88. $(3x^2)^3 + 4x^2 \cdot 4x^4 - x^4(2x)^2 + ((2x)^2)^3 - 100x^2(x^2)^2$

89. Evaluate both $s^2 - 50s + 675$ and $-s^2 + 50s - 675$ when $s = 18$, $s = 25$, and $s = 32$.

90. *Daily accidents.* The average number of accidents per day involving drivers of age r can be approximated by the polynomial

$$0.4r^2 - 40r + 1039.$$

For what age is the number of daily accidents smallest?

91. Construct a polynomial in x (meaning that x is the variable) of degree 5 with four terms and coefficients that are integers.

92. Construct a trinomial in y of degree 4 with coefficients that are rational numbers.

93. What is the degree of $(5m^5)^2$?

94. Construct three like terms of degree 4.

95. *Path of the Olympic arrow.* The Olympic flame at the 1992 Summer Olympics was lit by a flaming arrow. As the arrow moved d meters horizontally from the archer, its height, in meters, was approximated by the polynomial

$$-0.0064d^2 + 0.8d + 2.$$

Complete the table for the choices of d given. Then plot the points and draw a graph representing the path of the arrow.

d	$-0.0064d^2 + 0.8d + 2$
0	
30	
60	
90	
120	

96. A polynomial in x has degree 3. The coefficient of x^2 is 3 less than the coefficient of x^3. The coefficient of x is 3 times the coefficient of x^2. The remaining constant is 2 more than the coefficient of x^3. The sum of the coefficients is -4. Find the polynomial.

4.3 Addition and Subtraction of Polynomials

Addition of Polynomials • Opposites of Polynomials • Subtraction of Polynomials • Problem Solving

Addition of Polynomials

To add two polynomials, we write a plus sign between them and then collect like terms.

EXAMPLE 1

Add.

a) $(-3x^3 + 2x - 4) + (4x^3 + 3x^2 + 2)$
b) $\left(\frac{2}{3}x^4 + 3x^2 - 2x + \frac{1}{2}\right) + \left(-\frac{1}{3}x^4 + 5x^3 - 3x^2 + 3x - \frac{1}{2}\right)$

Solution

a) $(-3x^3 + 2x - 4) + (4x^3 + 3x^2 + 2)$

$= (-3 + 4)x^3 + 3x^2 + 2x + (-4 + 2)$ Collecting like terms; using the distributive law

$= x^3 + 3x^2 + 2x - 2$

b) $\left(\frac{2}{3}x^4 + 3x^2 - 2x + \frac{1}{2}\right) + \left(-\frac{1}{3}x^4 + 5x^3 - 3x^2 + 3x - \frac{1}{2}\right)$

$= \left(\frac{2}{3} - \frac{1}{3}\right)x^4 + 5x^3 + (3 - 3)x^2 + (-2 + 3)x + \left(\frac{1}{2} - \frac{1}{2}\right)$ Collecting like terms

$= \frac{1}{3}x^4 + 5x^3 + x$ ❑

After some practice, polynomial addition is often performed mentally.

EXAMPLE 2

Add: $(3x^2 - 2x + 2) + (5x^3 - 2x^2 + 3x - 4)$.

Solution

$(3x^2 - 2x + 2) + (5x^3 - 2x^2 + 3x - 4)$

$= 5x^3 + (3 - 2)x^2 + (-2 + 3)x + (2 - 4)$ You might do this step mentally.

$= 5x^3 + x^2 + x - 2$ Then you would write only this. ❑

We can also add polynomials by writing like terms in columns. Sometimes this makes like terms easier to see.

EXAMPLE 3

Add: $9x^5 - 2x^3 + 6x^2 + 3$ and $5x^4 - 7x^2 + 6$ and $3x^6 - 5x^5 + x^2 + 5$.

Solution We arrange the polynomials with like terms in columns.

$$\begin{array}{rrrrrrl}
 & 9x^5 & & -\ 2x^3 & +\ 6x^2 & +\ 3 & \\
 & & 5x^4 & & -\ 7x^2 & +\ 6 & \text{We leave spaces for missing terms.} \\
3x^6 & -\ 5x^5 & & & +\ 1x^2 & +\ 5 & \text{Writing } x^2 \text{ as } 1x^2 \\
\hline
3x^6 & +\ 4x^5 & +\ 5x^4 & -\ 2x^3 & & +\ 14 & \text{Adding}
\end{array}$$

We write the answer as $3x^6 + 4x^5 + 5x^4 - 2x^3 + 14$ without the extra space. ❑

Opposites of Polynomials

In Section 1.8, we used the property of negative one to determine that the opposite of a sum is the sum of the opposites. This idea can be extended to polynomials with any number of terms.

> To find an equivalent polynomial for the *opposite,* or *additive inverse,* of a polynomial, replace each term by its opposite—that is, *change the sign of every term.*

EXAMPLE 4

Find two equivalent expressions for the opposite of $4x^5 - 7x^3 - 8x + \frac{5}{6}$.

Solution

i) $-(4x^5 - 7x^3 - 8x + \frac{5}{6})$

ii) $-4x^5 + 7x^3 + 8x - \frac{5}{6}$ Changing the sign of every term

Thus, $-(4x^5 - 7x^3 - 8x + \frac{5}{6})$ is equivalent to $-4x^5 + 7x^3 + 8x - \frac{5}{6}$, and each is the opposite of the polynomial $4x^5 - 7x^3 - 8x + \frac{5}{6}$. ❑

EXAMPLE 5

Simplify: $-(-7x^4 - \frac{5}{9}x^3 + 8x^2 - x + 67)$.

Solution

$$-(-7x^4 - \tfrac{5}{9}x^3 + 8x^2 - x + 67) = 7x^4 + \tfrac{5}{9}x^3 - 8x^2 + x - 67$$ ❑

Subtraction of Polynomials

We can now subtract one polynomial from another by adding the opposite of the polynomial being subtracted.

EXAMPLE 6

Subtract.

a) $(9x^5 + x^3 - 2x^2 + 4) - (-2x^5 + x^4 - 4x^3 - 3x^2)$
b) $(7x^5 + x^3 - 9x) - (3x^5 - 4x^3 + 5)$

Solution

a) $(9x^5 + x^3 - 2x^2 + 4) - (-2x^5 + x^4 - 4x^3 - 3x^2)$

$= 9x^5 + x^3 - 2x^2 + 4 + 2x^5 - x^4 + 4x^3 + 3x^2$ Adding the opposite

$= 11x^5 - x^4 + 5x^3 + x^2 + 4$ Collecting like terms

b) $(7x^5 + x^3 - 9x) - (3x^5 - 4x^3 + 5)$

$= 7x^5 + x^3 - 9x + (-3x^5) + 4x^3 - 5$ Adding the opposite

$= 7x^5 + x^3 - 9x - 3x^5 + 4x^3 - 5$ Try to go directly to this step

$= 4x^5 + 5x^3 - 9x - 5$ Collecting like terms ❑

To use columns to subtract, we replace coefficients by their opposites and then add.

EXAMPLE 7

Write in columns and subtract: $(5x^2 - 3x + 6) - (9x^2 - 5x - 3)$.

Solution

i)
$$\begin{array}{r} 5x^2 - 3x + 6 \\ \underline{-(9x^2 - 5x - 3)} \end{array}$$
Writing similar terms in columns

ii)
$$\begin{array}{r} 5x^2 - 3x + 6 \\ \underline{-9x^2 + 5x + 3} \end{array}$$
Changing signs and removing parentheses

iii)
$$\begin{array}{r} 5x^2 - 3x + 6 \\ \underline{-9x^2 + 5x + 3} \\ -4x^2 + 2x + 9 \end{array}$$
Adding ❑

If you can do so without error, you can arrange the polynomials in columns and write just the answer.

EXAMPLE 8

Write in columns and subtract: $(x^3 + x^2 + 2x - 12) - (-2x^3 + x^2 - 3x)$.

Solution We have

$$\begin{array}{r} x^3 + x^2 + 2x - 12 \\ \underline{-(-2x^3 + x^2 - 3x \qquad)} \\ 3x^3 \qquad + 5x - 12. \end{array}$$
Leaving space for the missing term ❑

Problem Solving

EXAMPLE 9

Find a polynomial for the sum of the areas of these rectangles.

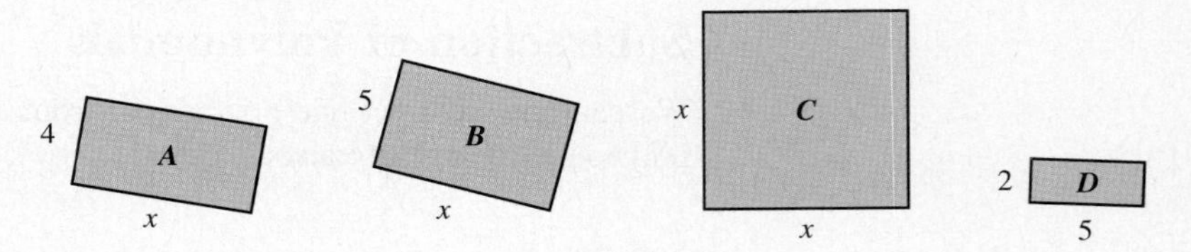

Solution

1. **FAMILIARIZE.** Recall that the area of a rectangle is the product of the length and the width.
2. **TRANSLATE.** We translate the problem to mathematical language. The sum of the areas is a sum of products. We find each product and then add:

$$\underbrace{\text{Area of } A}_{4x} \ \underbrace{\text{plus}}_{+} \ \underbrace{\text{area of } B}_{5x} \ \underbrace{\text{plus}}_{+} \ \underbrace{\text{area of } C}_{x \cdot x} \ \underbrace{\text{plus}}_{+} \ \underbrace{\text{area of } D}_{2 \cdot 5.}$$

3. **CARRY OUT.** We collect like terms:

$$4x + 5x + x^2 + 10 = x^2 + 9x + 10.$$

4. **CHECK.** We can check by going over our calculations. Another way to check is to replace x with a number, say 3. Then we find each of the areas and add the results:

$$4 \cdot 3 + 5 \cdot 3 + 3 \cdot 3 + 2 \cdot 5 = 12 + 15 + 9 + 10 = 46.$$

When we substitute 3 for x in the polynomial $x^2 + 9x + 10$, we should also get 46:

$$x^2 + 9x + 10 = 3^2 + 9(3) + 10 = 46.$$

Our check is only a partial check, since it is also possible for an incorrect answer to equal 46 when evaluated for $x = 3$. This would be very unlikely, especially if a second choice of x, say $x = 5$, also checks. We leave that check to the student.

5. **STATE.** A polynomial for the sum of the areas is $x^2 + 9x + 10$. ❑

EXAMPLE 10

A 4-ft by 4-ft sandbox is placed on a square lawn x ft on a side. Find a polynomial for the remaining area.

Solution

1. **FAMILIARIZE.** We draw a picture of the situation as follows.

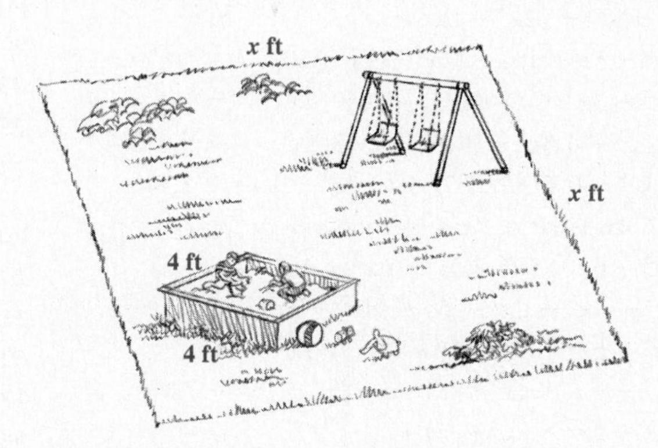

2. **TRANSLATE.** We reword the problem and translate as follows.

Reword:	Area of lawn	−	area of sandbox	=	area left over
	↓	↓	↓	↓	↓
Translate:	$x \cdot x$	−	$4 \cdot 4$	=	Area left over

3. **CARRY OUT.** We carry out the manipulation by multiplying the numbers:

$$x^2 - 16 = \text{Area left over}.$$

4. **CHECK.** We can perform a partial check by assigning some value to x—say, 10—and carrying out the computation of the area in two ways:

Area of lawn $= 10 \cdot 10 = 100\ \text{ft}^2$;

Area of sandbox $= 4 \cdot 4 = 16\ \text{ft}^2$;

Area left over $= 100 - 16 = 84\ \text{ft}^2$.

This is the same as substituting 10 for x in $x^2 - 16$:

$$10^2 - 16 = 100 - 16 = 84.$$

Thus our solution is probably correct.

5. **STATE.** The remaining area is $x^2 - 16$ ft^2. ❑

EXERCISE SET 4.3

Add.

1. $(3x + 2) + (-4x + 3)$
2. $(6x + 1) + (-7x + 2)$
3. $(-6x + 2) + (x^2 + x - 3)$
4. $(x^2 - 5x + 4) + (8x - 9)$
5. $(x^2 - 9) + (x^2 + 9)$
6. $(x^3 + x^2) + (2x^3 - 5x^2)$
7. $(3x^2 - 5x + 10) + (2x^2 + 8x - 40)$
8. $(6x^4 + 3x^3 - 1) + (4x^2 - 3x + 3)$
9. $(1.2x^3 + 4.5x^2 - 3.8x) + (-3.4x^3 - 4.7x^2 + 23)$
10. $(0.5x^4 - 0.6x^2 + 0.7) + (2.3x^4 + 1.8x - 3.9)$
11. $(1 + 4x + 6x^2 + 7x^3) + (5 - 4x + 6x^2 - 7x^3)$
12. $(3x^4 - 6x - 5x^2 + 5) + (6x^2 - 4x^3 - 1 + 7x)$
13. $(9x^8 - 7x^4 + 2x^2 + 5) + (8x^7 + 4x^4 - 2x)$
14. $(4x^5 - 6x^3 - 9x + 1) + (6x^3 + 9x^2 + 9x)$
15. $\left(\frac{1}{4}x^4 + \frac{2}{3}x^3 + \frac{5}{8}x^2 + 7\right) + \left(-\frac{3}{4}x^4 + \frac{3}{8}x^2 - 7\right)$
16. $\left(\frac{1}{3}x^9 + \frac{1}{5}x^5 - \frac{1}{2}x^2 + 7\right) + \left(-\frac{1}{5}x^9 + \frac{1}{4}x^4 - \frac{3}{5}x^5 + \frac{3}{4}x^2 + \frac{1}{2}\right)$
17. $(0.02x^5 - 0.2x^3 + x + 0.08) + (-0.01x^5 + x^4 - 0.8x - 0.02)$
18. $(0.03x^6 + 0.05x^3 + 0.22x + 0.05) + \left(\frac{7}{100}x^6 - \frac{3}{100}x^3 + 0.5\right)$
19. $$\begin{array}{r} -3x^4 + 6x^2 + 2x - 1 \\ \underline{-3x^2 + 2x + 1} \end{array}$$
20. $$\begin{array}{r} -4x^3 + 8x^2 + 3x - 2 \\ \underline{-4x^2 + 3x + 2} \end{array}$$
21. $$\begin{array}{rrrrr} 0.15x^4 & +\,0.10x^3 & -\,0.9x^2 & & \\ & -\,0.01x^3 & +\,0.01x^2 & +\,x & \\ 1.25x^4 & & +\,0.11x^2 & & +\,0.01 \\ & 0.27x^3 & & & +\,0.99 \\ -0.35x^4 & & +\quad 15x^2 & & -\,0.03 \\ \hline \end{array}$$
22. $$\begin{array}{rrrrr} 0.05x^4 & +\,0.12x^3 & -\,0.5x^2 & & \\ & -\,0.02x^3 & +\,0.02x^2 & +\,2x & \\ 1.5x^4 & & +\,0.01x^2 & & +\,0.15 \\ & 0.25x^3 & & & +\,0.85 \\ -0.25x^4 & & +\quad 10x^2 & & -\,0.04 \\ \hline \end{array}$$

Find two equivalent expressions for the opposite of the polynomial.

23. $-5x$
24. $x^2 - 3x$
25. $-x^2 + 10x - 2$
26. $-4x^3 - x^2 - x$
27. $12x^4 - 3x^3 + 3$
28. $4x^3 - 6x^2 - 8x + 1$

Simplify.

29. $-(3x - 7)$
30. $-(-2x + 4)$
31. $-(4x^2 - 3x + 2)$
32. $-(-6a^3 + 2a^2 - 9a + 1)$
33. $-(-4x^4 + 6x^2 + \frac{3}{4}x - 8)$
34. $-(-5x^4 + 4x^3 - x^2 + 0.9)$

Subtract.

35. $(3x + 2) - (-4x + 3)$
36. $(6x + 1) - (-7x + 2)$
37. $(-6x + 2) - (x^2 + x - 3)$
38. $(x^2 - 5x + 4) - (8x - 9)$
39. $(x^2 - 9) - (x^2 + 9)$
40. $(x^3 + x^2) - (2x^3 - 5x^2)$
41. $(6x^4 + 3x^3 - 1) - (4x^2 - 3x + 3)$
42. $(-4x^2 + 2x) - (3x^3 - 5x^2 + 3)$
43. $(1.2x^3 + 4.5x^2 - 3.8x) - (-3.4x^3 - 4.7x^2 + 23)$
44. $(0.5x^4 - 0.6x^2 + 0.7) - (2.3x^4 + 1.8x - 3.9)$
45. $(5x^2 + 6) - (3x^2 - 8)$
46. $(7x^3 - 2x^2 + 6) - (7x^2 + 2x - 4)$
47. $(6x^5 - 3x^4 + x + 1) - (8x^5 + 3x^4 - 1)$

48. $\left(\frac{1}{2}x^2 - \frac{3}{2}x + 2\right) - \left(\frac{3}{2}x^2 + \frac{1}{2}x - 2\right)$

49. $(6x^2 + 2x) - (-3x^2 - 7x + 8)$

50. $7x^3 - (-3x^2 - 2x + 1)$

51. $\left(\frac{5}{8}x^3 - \frac{1}{4}x - \frac{1}{3}\right) - \left(-\frac{1}{8}x^3 + \frac{1}{4}x - \frac{1}{3}\right)$

52. $\left(\frac{1}{5}x^3 + 2x^2 - 0.1\right) - \left(-\frac{2}{5}x^3 + 2x^2 + 0.01\right)$

53. $(0.08x^3 - 0.02x^2 + 0.01x) - (0.02x^3 + 0.03x^2 - 1)$

54. $(0.8x^4 + 0.2x - 1) - \left(\frac{7}{10}x^4 + \frac{1}{5}x - 0.1\right)$

55. $\begin{array}{l} x^2 + 5x + 6 \\ \underline{-(x^2 + 2x \qquad)} \end{array}$

56. $\begin{array}{l} x^3 \qquad\quad + 1 \\ \underline{-(x^3 + x^2 \qquad)} \end{array}$

57. $\begin{array}{l} \quad 5x^4 + 6x^3 - 9x^2 \\ \underline{-(-6x^4 - 6x^3 \qquad\quad + 8x + 9)} \end{array}$

58. $\begin{array}{l} 5x^4 \qquad + 6x^2 - 3x + 6 \\ \underline{-(\quad 6x^3 + 7x^2 - 8x - 9)} \end{array}$

59. $\begin{array}{l} \qquad\quad 3x^4 + 6x^2 + 8x - 1 \\ \underline{-(4x^5 - 6x^4 \qquad\quad - 8x - 7)} \end{array}$

60. $\begin{array}{l} \quad 6x^5 \qquad\quad + 3x^2 - 7x + 2 \\ \underline{-(10x^5 + 6x^3 - 5x^2 - 2x + 4)} \end{array}$

61. $\begin{array}{l} x^5 \qquad\qquad\qquad\qquad - 1 \\ \underline{-(x^5 - x^4 + x^3 - x^2 + x - 1)} \end{array}$

62. $\begin{array}{l} x^5 + x^4 - x^3 + x^2 - x + 2 \\ \underline{-(x^5 - x^4 + x^3 - x^2 - x + 2)} \end{array}$

Problem Solving

63. Solve.

a) Find a polynomial for the sum of the areas of these rectangles.

b) Find the sum of the areas when $x = 3$ and $x = 8$.

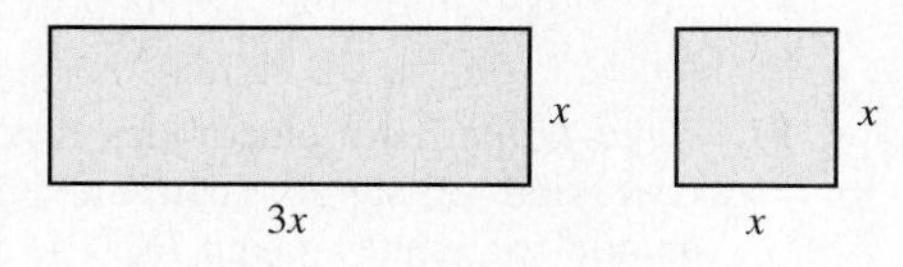

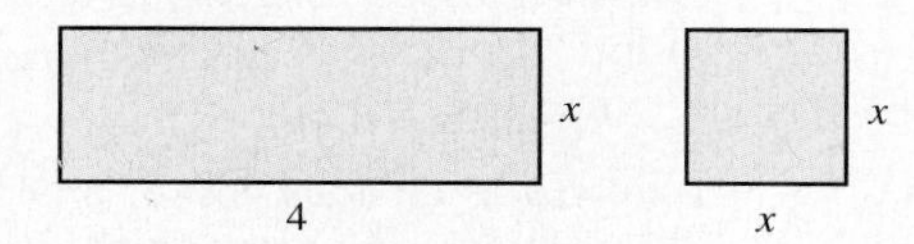

64. Solve.

a) Find a polynomial for the sum of the areas of these circles.

b) Find the sum of the areas when $r = 5$ and $r = 11.3$.

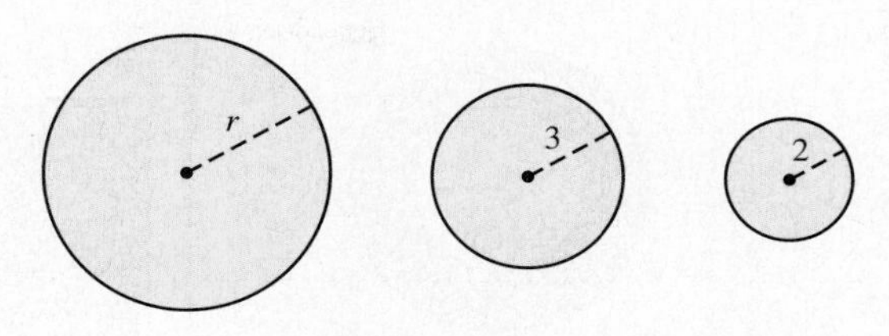

Find a polynomial for the perimeter of the figure.

65.

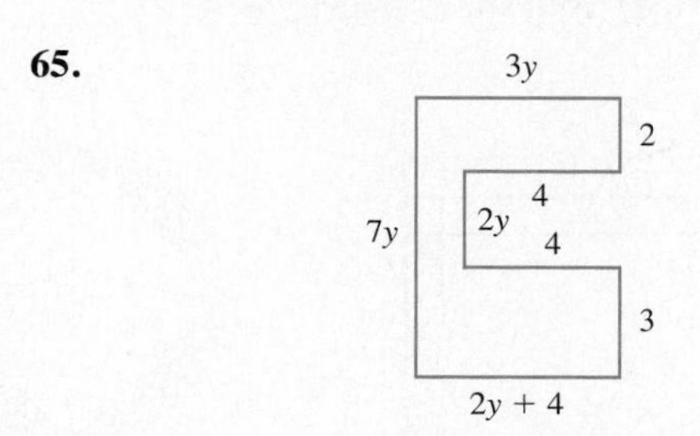

66.

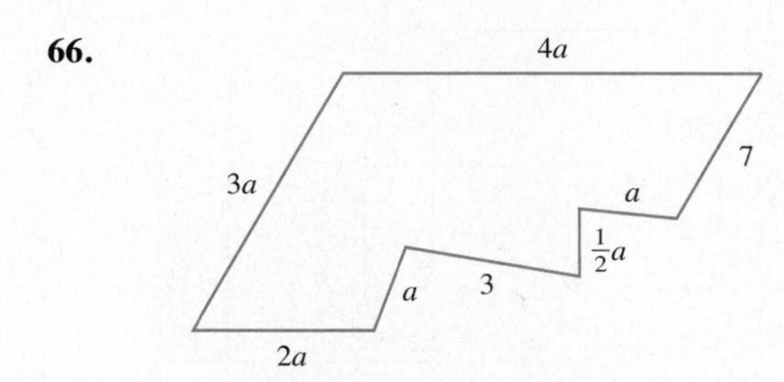

Find two algebraic expressions for the area of the figure. For one expression, view the figure as one large rectangle, and for the other, view the figure as a sum of four smaller rectangles.

67.

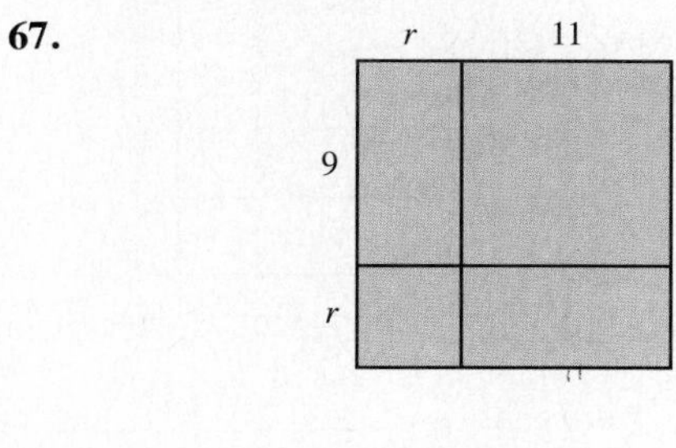

68.

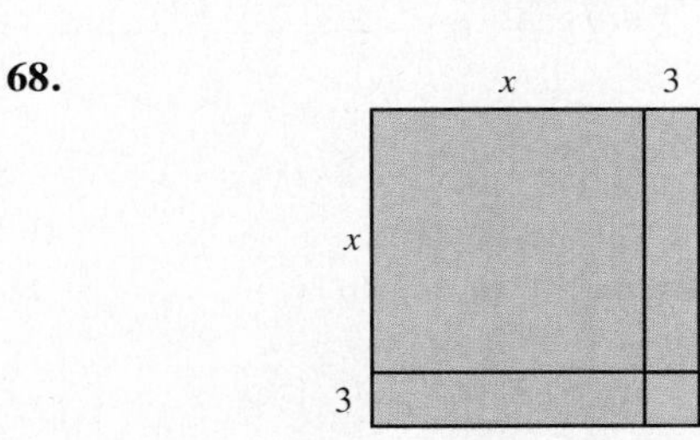

Find a polynomial for the shaded area of the figure.

69.

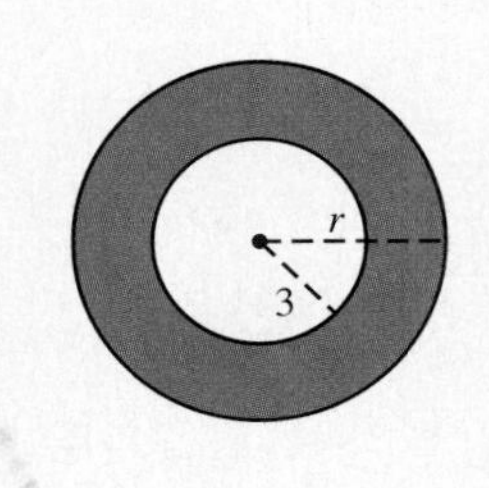

70.

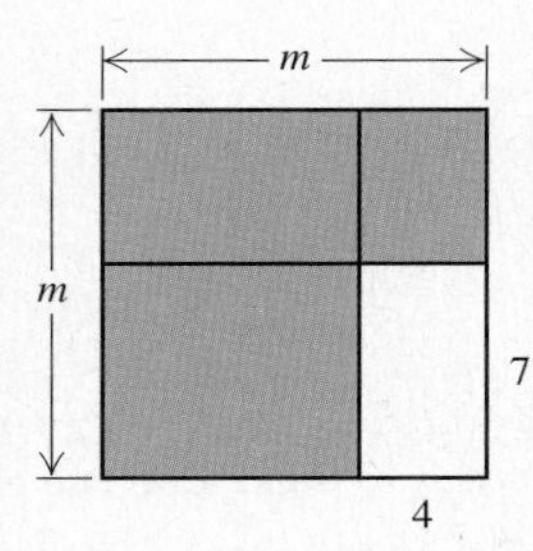

71.

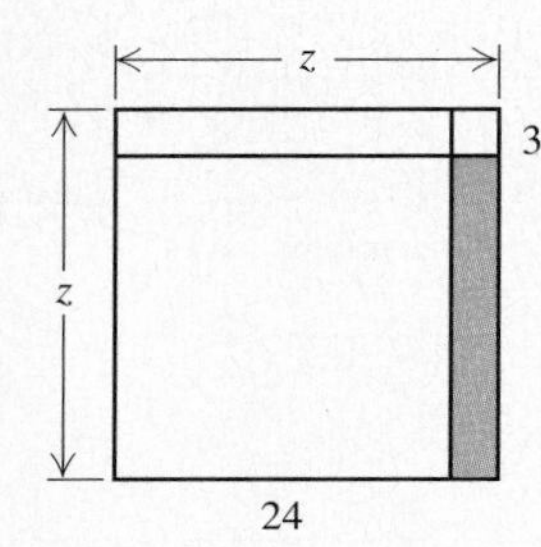

72.

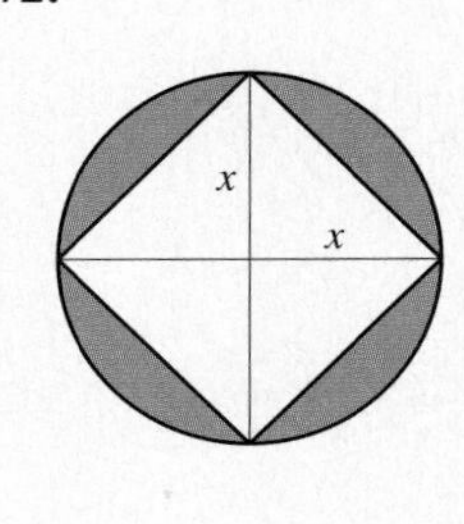

73.

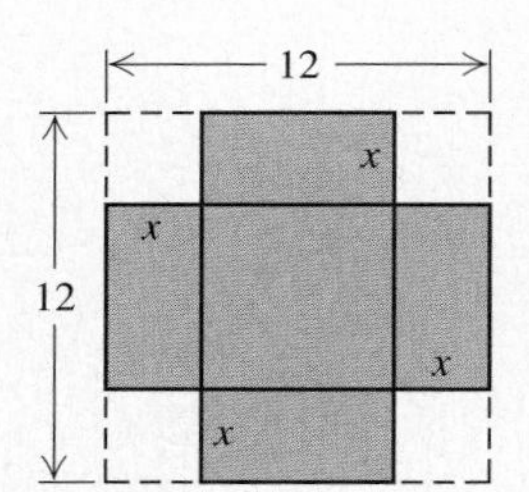

74. Find $(y - 2)^2$ using the four parts of this square.

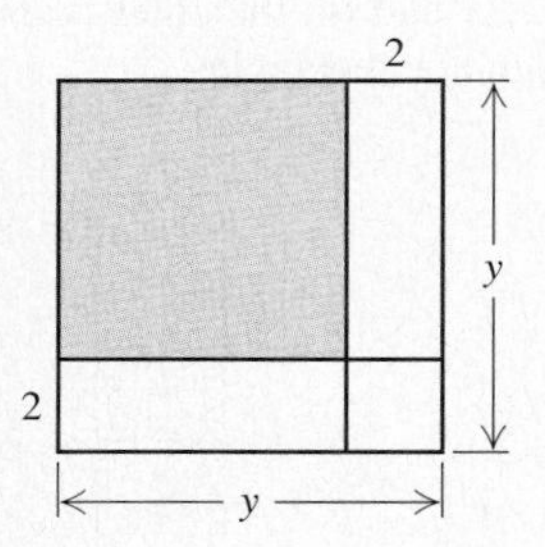

Skill Maintenance

Solve.

75. $1.5x - 2.7x = 23 - 5.6x$

76. $3x - 3 = -4x + 4$

77. $8(x - 2) = 16$

78. $4(x - 5) = 7(x + 8)$

Synthesis

79. ◈ Which, if any, of the commutative, associative, and distributive laws are needed for adding polynomials? Why?

80. ◈ Is the sum of two binomials ever a trinomial? Why or why not?

Simplify.

81. $(4a^2 - 3a) + (7a^2 - 9a - 13) - (6a - 9)$

82. $(3x^2 - 4x + 6) - (-2x^2 + 4) + (-5x - 3)$

83. $(-8y^2 - 4) - (3y + 6) - (2y^2 - y)$

84. $(5x^3 - 4x^2 + 6) - (2x^3 + x^2 - x) + (x^3 - x)$

85. $(-y^4 - 7y^3 + y^2) + (-2y^4 + 5y - 2) - (-6y^3 + y^2)$

86. $(-4 + x^2 + 2x^3) - (-6 - x + 3x^3) - (-x^2 - 5x^3)$

87. ▦ $(345.099x^3 - 6.178x) - (-224.508x^3 + 8.99x)$

Find a polynomial for the surface area of the right rectangular solid.

88.

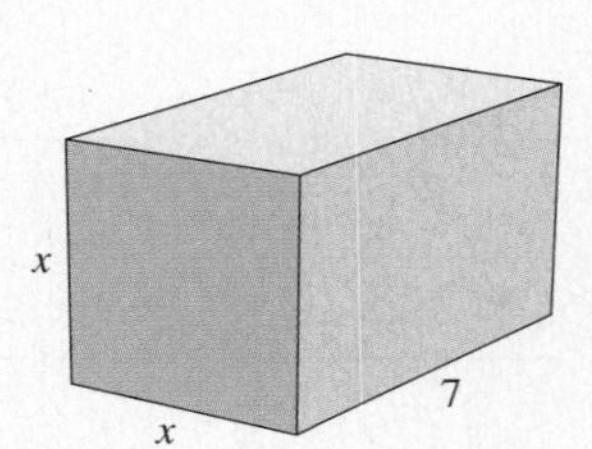

89.

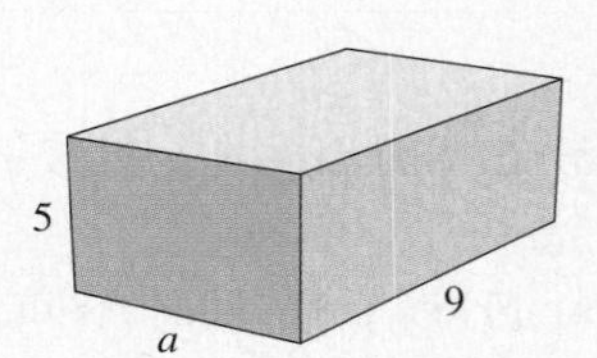

90. ◈ Does replacing each occurrence of the variable x in $5x^3 - 3x^2 + 2x$ with its opposite result in the opposite of the polynomial? Why or why not?

91. *Total profit.* An electronics firm is marketing a new kind of stereo. Total revenue is the total amount of money taken in. The firm determines that when it sells x stereos, its total revenue is given by

$$R = 280x - 0.4x^2.$$

Total cost is the total cost of producing x stereos. The electronics firm determines that the total cost of producing x stereos is given by

$$C = 5000 + 0.6x^2.$$

The total profit is

(Total Revenue) − (Total Cost) = $R - C$.

a) Find a polynomial for total profit.

b) What is the total profit on the production and sale of 75 stereos?

c) What is the total profit on the production and sale of 100 stereos?

4.4 Multiplication of Polynomials

Multiplying Monomials • Multiplying a Monomial and a Polynomial • Multiplying Two Binomials • Multiplying Any Polynomials

We now multiply polynomials using techniques based largely on the distributive, associative, and commutative laws and the rules for exponents.

Multiplying Monomials

Consider $(3x)(4x)$. We multiply as follows:

$(3x)(4x) = 3 \cdot x \cdot 4 \cdot x$ — By an associative law

$= 3 \cdot 4 \cdot x \cdot x$ — By a commutative law

$= (3 \cdot 4) \cdot x \cdot x$ — By an associative law

$= 12x^2.$

> To find an equivalent expression for the product of two monomials, multiply the coefficients and then multiply the variables using the product rule for exponents.

EXAMPLE 1

Multiply: **(a)** $(5x)(6x)$; **(b)** $(3a)(-a)$; **(c)** $(-7x^5)(4x^3)$.

Solution

a) $(5x)(6x) = (5 \cdot 6)(x \cdot x)$ — Multiplying the coefficients; multiplying the variables

$= 30x^2$ — Simplifying

b) $(3a)(-a) = (3a)(-1a)$ — Writing $-a$ as $-1a$ can ease calculations.

$= (3)(-1)(a \cdot a)$ — Using an associative and a commutative law

$= -3a^2$

c) $(-7x^5)(4x^3) = (-7 \cdot 4)(x^5 \cdot x^3)$

$= -28x^{5+3}$

$= -28x^8$ — Using the product rule for exponents ❑

After some practice, you can try writing only the answer.

Multiplying a Monomial and a Polynomial

To find an equivalent expression for the product of a monomial, such as $2x$, and a polynomial, such as $5x + 3$, we use the distributive law.

EXAMPLE 2

Multiply: **(a)** x and $x + 3$; **(b)** $5x(2x^2 - 3x + 4)$.

Solution

a) $x(x + 3) = x \cdot x + x \cdot 3$ Using the distributive law

$= x^2 + 3x$

b) $5x(2x^2 - 3x + 4) = (5x)(2x^2) - (5x)(3x) + (5x)(4)$ Using the distributive law

$= 10x^3 - 15x^2 + 20x$ Multiplying the three pairs of terms ❑

The product in Example 2(a) can be visualized as the area of a rectangle with width x and length $x + 3$.

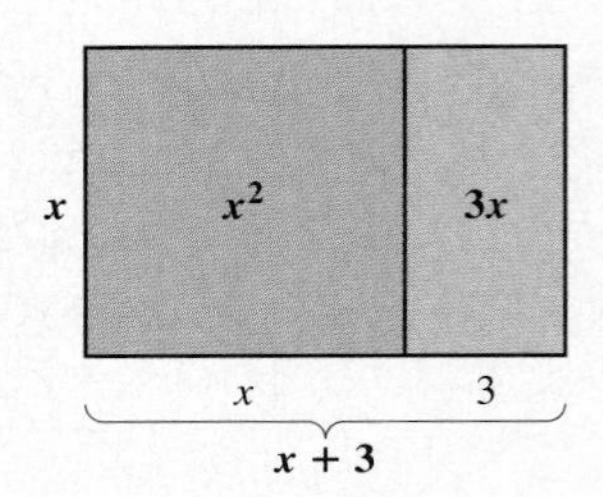

Note that the total area can be expressed as $x(x + 3)$ or, by adding the two smaller areas, $x^2 + 3x$.

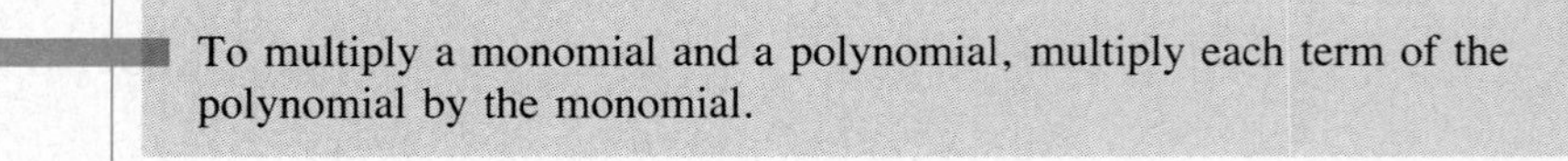
To multiply a monomial and a polynomial, multiply each term of the polynomial by the monomial.

Try to do this mentally, if possible.

EXAMPLE 3

Multiply: $2x^2(x^3 - 7x^2 + 10x - 4)$.

Solution

Think: $2x^2 \cdot x^3 - 2x^2 \cdot 7x^2 + 2x^2 \cdot 10x - 2x^2 \cdot 4$

$$2x^2(x^3 - 7x^2 + 10x - 4) = 2x^5 - 14x^4 + 20x^3 - 8x^2$$ ❑

Multiplying Two Binomials

To find an equivalent expression for the product of two binomials, we again begin by using the distributive law. This time, however, it is a *binomial* rather than a monomial that is being distributed.

EXAMPLE 4

Multiply: **(a)** $x + 5$ and $x + 4$; **(b)** $4x - 3$ and $x - 2$.

Solution

a) $(x + 5)(x + 4) = (x + 5)\,x + (x + 5)\,4$ — Using the distributive law

$= x(x + 5) + 4(x + 5)$ — Using the commutative law for multiplication

$= x \cdot x + x \cdot 5 + 4 \cdot x + 4 \cdot 5$ — Using the distributive law (twice)

$= x^2 + 5x + 4x + 20$ — Multiplying the monomials

$= x^2 + 9x + 20$ — Collecting like terms

b) $(4x - 3)(x - 2) = (4x - 3)\,x - (4x - 3)\,2$ — Using the distributive law

$= x(4x - 3) - 2(4x - 3)$ — Using the commutative law for multiplication. This step is often omitted.

$= x \cdot 4x - x \cdot 3 - 2 \cdot 4x - 2(-3)$ — Using the distributive law (twice)

$= 4x^2 - 3x - 8x + 6$ — Multiplying the monomials

$= 4x^2 - 11x + 6$ — Collecting like terms ❑

To visualize the product in Example 4(a), consider a rectangle of length $x + 5$ and width $x + 4$.

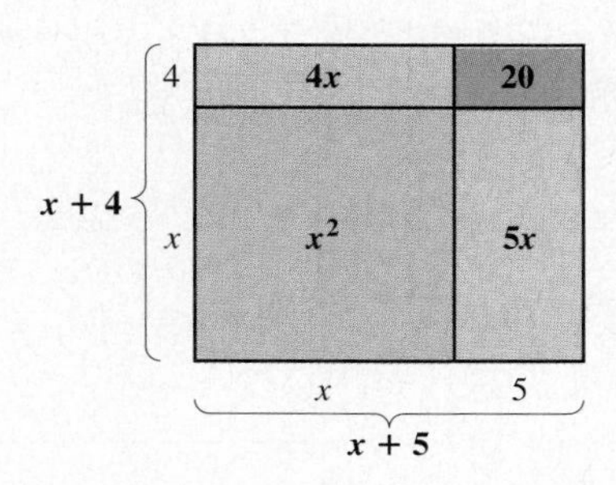

The total area can be expressed as $(x + 5)(x + 4)$ or, by adding the four smaller areas, $x^2 + 5x + 4x + 20$.

Multiplying Any Polynomials

Let us consider the product of a binomial and a trinomial. Again we make repeated use of the distributive law.

EXAMPLE 5

Multiply: $(x^2 + 2x - 3)(x + 4)$.

Solution

$$(x^2 + 2x - 3)(x + 4)$$

$$= (x^2 + 2x - 3)x + (x^2 + 2x - 3)4 \qquad \text{Using the distributive law}$$

$$= x(x^2 + 2x - 3) + 4(x^2 + 2x - 3) \qquad \text{Using a commutative law}$$

$$= x \cdot x^2 + x \cdot 2x - x \cdot 3 + 4 \cdot x^2 + 4 \cdot 2x - 4 \cdot 3 \qquad \text{Using the distributive law (twice)}$$

$$= x^3 + 2x^2 - 3x + 4x^2 + 8x - 12 \qquad \text{Multiplying the monomials}$$

$$= x^3 + 6x^2 + 5x - 12 \qquad \text{Collecting like terms}$$

❑

Perhaps you have discovered the following in the preceding examples.

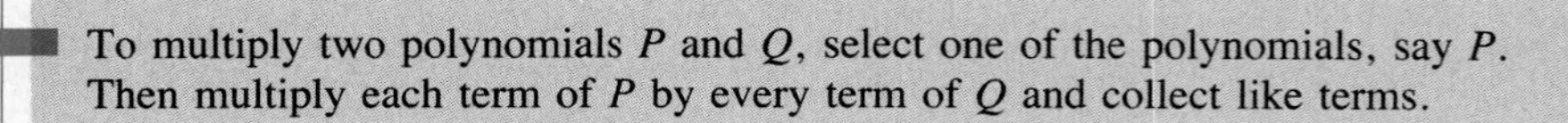

To multiply two polynomials P and Q, select one of the polynomials, say P. Then multiply each term of P by every term of Q and collect like terms.

To use columns for long multiplications, multiply each term at the top by every term at the bottom. We write like terms in columns, and then add the results. Such multiplication is like multiplying with whole numbers:

$$\begin{array}{rrrr} & 3 & 2 & 1 \\ \times & & 1 & 2 \\ \hline & 6 & 4 & 2 \\ 3 & 2 & 1 & \\ \hline 3 & 8 & 5 & 2 \end{array} \qquad \begin{array}{r} 300 + 20 + 1 \\ \times \qquad 10 + 2 \\ \hline 600 + 40 + 2 \\ 3000 + 200 + 10 \qquad \\ \hline 3000 + 800 + 50 + 2 \end{array}$$

EXAMPLE 6

Multiply: $(4x^3 - 2x^2 + 3x)(x^2 + 2x)$.

Solution

$$\begin{array}{rrrrl} & 4x^3 & -\,2x^2 & +\,3x & \\ & & x^2 & +\,2x & \\ \hline & 8x^4 & -\,4x^3 & +\,6x^2 & \text{Multiplying the top row by } 2x \\ 4x^5 & -\,2x^4 & +\,3x^3 & & \text{Multiplying the top row by } x^2 \\ \hline 4x^5 & +\,6x^4 & -\,x^3 & +\,6x^2 & \text{Collecting like terms} \end{array}$$

Line up like terms in columns.

❑

When terms are missing, it helps to leave space for them and align like terms as we multiply.

EXAMPLE 7

Multiply: $(-2x^2 - 3)(5x^3 - 3x + 4)$.

Solution

$$\begin{array}{rrrrrl} & 5x^3 & & -3x & +4 & \\ & -2x^2 & & & -3 & \\ \hline & -15x^3 & & +9x & -12 & \text{Multiplying by } -3 \\ -10x^5 & +6x^3 & -8x^2 & & & \text{Multiplying by } -2x^2 \\ \hline -10x^5 & -9x^3 & -8x^2 & +9x & -12 & \text{Collecting like terms} \end{array}$$

❑

With practice some steps can be skipped. Sometimes we can multiply horizontally, while still aligning like terms.

EXAMPLE 8

Multiply: $(2x^3 + 3x^2 - 4x + 6)(3x + 5)$.

Solution

$$\begin{aligned}(3x + 5)(2x^3 + 3x^2 - 4x + 6) &= 6x^4 + 9x^3 - 12x^2 + 18x && \text{Multiplying by } 3x \\ &\quad + 10x^3 + 15x^2 - 20x + 30 && \text{Multiplying by } 5 \\ &= 6x^4 + 19x^3 + 3x^2 - 2x + 30\end{aligned}$$

❑

EXERCISE SET 4.4

Multiply.

1. $(6x^2)(7)$
2. $(5x^2)(-2)$
3. $(-x^3)(-x)$
4. $(-x^4)(x^2)$
5. $(-x^5)(x^3)$
6. $(-x^6)(-x^2)$
7. $(3x^4)(2x^2)$
8. $(5x^3)(4x^5)$
9. $(7t^5)(4t^3)$
10. $(10a^2)(3a^2)$
11. $(-0.1x^6)(0.2x^4)$
12. $(0.3x^3)(-0.4x^6)$
13. $\left(-\frac{1}{5}x^3\right)\left(-\frac{1}{3}x\right)$
14. $\left(-\frac{1}{4}x^4\right)\left(\frac{1}{5}x^8\right)$
15. $(-4x^2)(0)$
16. $(-4m^5)(-1)$
17. $(3x^2)(-4x^3)(2x^6)$
18. $(-2y^5)(10y^4)(-3y^3)$
19. $3x(-x + 5)$
20. $2x(4x - 6)$
21. $4x(x + 1)$
22. $3x(x + 2)$
23. $(x + 7)5x$
24. $(x - 6)3x$
25. $x^2(x^3 + 1)$
26. $-2x^3(x^2 - 1)$
27. $3x(2x^2 - 6x + 1)$
28. $-4x(2x^3 - 6x^2 - 5x + 1)$
29. $4x^2(3x + 6)$
30. $5x^2(-2x + 1)$
31. $-6x^2(x^2 + x)$
32. $-4x^2(x^2 - x)$
33. $3y^2(6y^4 + 8y^3)$
34. $4y^4(y^3 - 6y^2)$
35. $3x^4(14x^{50} + 20x^{11} + 6x^{57} + 60x^{15})$
36. $5x^6(4x^{32} - 10x^{19} + 5x^8)$
37. $(x + 6)(x + 3)$
38. $(x + 5)(x + 2)$
39. $(x + 5)(x - 2)$
40. $(x + 6)(x - 2)$
41. $(x - 4)(x - 3)$
42. $(x - 7)(x - 3)$
43. $(x + 3)(x - 3)$
44. $(x + 6)(x - 6)$
45. $(5 - x)(5 - 2x)$
46. $(3 + x)(6 + 2x)$
47. $(2x + 5)(2x + 5)$

48. $(3x - 4)(3x - 4)$
49. $(3y - 4)(3y + 4)$
50. $(2y + 1)(2y - 1)$
51. $\left(x - \frac{5}{2}\right)\left(x + \frac{2}{5}\right)$
52. $\left(x + \frac{4}{3}\right)\left(x + \frac{3}{2}\right)$
53. $(x^2 + x + 1)(x - 1)$
54. $(x^2 - x + 2)(x + 2)$
55. $(2x + 1)(2x^2 + 6x + 1)$
56. $(3x - 1)(4x^2 - 2x - 1)$
57. $(y^2 - 3)(3y^2 - 6y + 2)$
58. $(3y^2 - 3)(y^2 + 6y + 1)$
59. $(x^3 + x^2)(x^3 + x^2 - x)$
60. $(x^3 - x^2)(x^3 - x^2 + x)$
61. $(-5x^3 - 7x^2 + 1)(2x^2 - x)$
62. $(-4x^3 + 5x^2 - 2)(5x^2 + 1)$
63. $(1 + x + x^2)(-1 - x + x^2)$
64. $(1 - x + x^2)(1 - x + x^2)$
65. $(2x^2 + 3x - 4)(2x^2 + x - 2)$
66. $(2x^2 - x - 3)(2x^2 - 5x - 2)$
67. $(x + 1)(x^3 + 7x^2 + 5x + 4)$
68. $(x + 2)(x^3 + 5x^2 + 9x + 3)$
69. $(2x^2 + x - 2)(-2x^2 + 4x - 5)$
70. $(3x^2 - 8x + 1)(-2x^2 - 4x + 2)$
71. $(2x + 1)(x^3 - 4x^2 + 3x - 2)$
72. $(4x + 3)(x^3 - 2x^2 + 5x - 1)$
73. $(x^3 + x^2 + x + 1)(x - 1)$
74. $(x^3 - x^2 + x - 2)(x - 2)$
75. $(x^3 + x^2 - x - 3)(x - 3)$
76. $(x^3 - x^2 - x + 4)(x + 4)$

Skill Maintenance

77. Subtract: $-\frac{1}{4} - \frac{1}{2}$.
78. Factor: $16x - 24y + 36$.

Synthesis

79. ◈ The polynomials

$$(a + b + c + d) \quad \text{and} \quad (r + s + m + p)$$

are multiplied. Without performing the multiplication, determine how many terms the product will contain. Provide a justification for your answer.

80. ◈ Is it possible to understand polynomial multiplication without first understanding the distributive law? Why or why not?

Find a polynomial for the shaded area of the figure.

81.

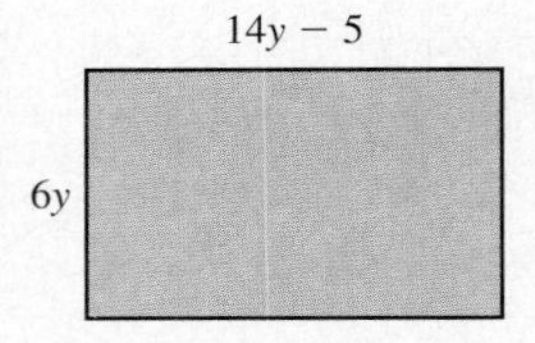

82.

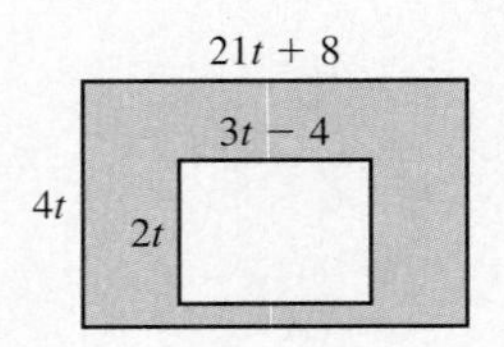

83. A box with a square bottom is to be made from a 12-in.-square piece of cardboard. Squares with side x are cut out of the corners and the sides are folded up. Find polynomials for the volume and the outside surface area of the box.

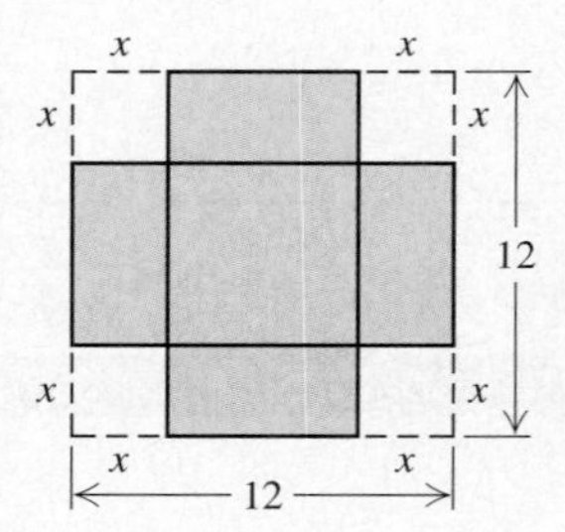

84. An open wooden box is a cube with side x cm. The wood from which the box is made is 1 cm thick. Find a polynomial for the interior volume of the cube.

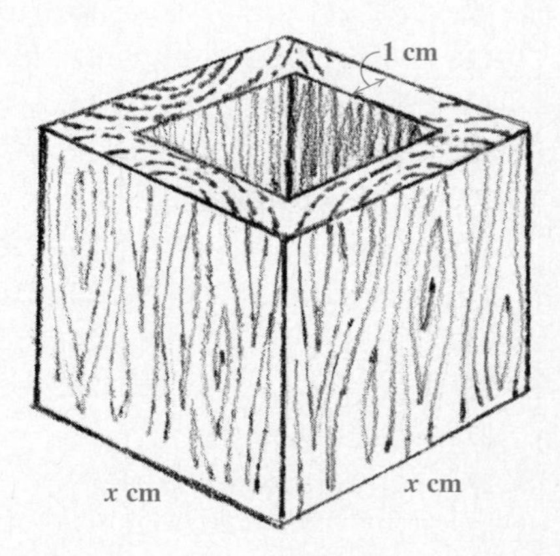

85. The height of a triangle is 4 ft longer than its base. Find a polynomial for the area.

86. A rectangular garden is twice as long as it is wide (see the figure at right). It is surrounded by a sidewalk that is 4 ft wide. The area of the garden and the sidewalk together is 256 ft^2 more than the area of the garden alone. Find the dimensions of the garden.

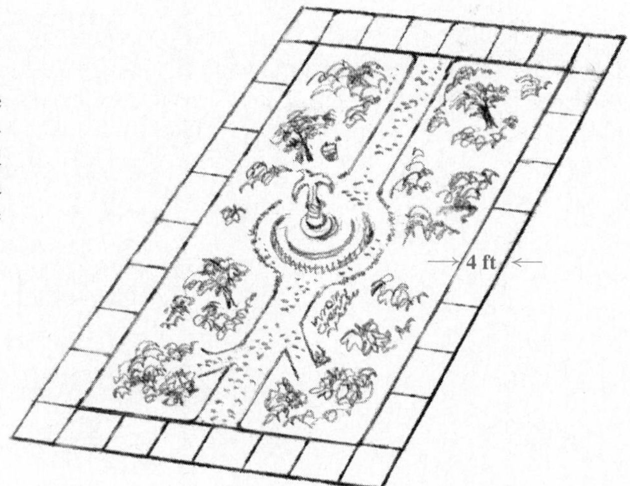

Compute and simplify.

87. $(x + 3)(x + 6) + (x + 3)(x + 6)$

88. $(x - 2)(x - 7) + (x - 2)(x - 7)$

89. $(x + 5)^2 - (x - 3)^2$

90. $(x - 6)^2 + (4 - x)^2$

4.5 Special Products

Products of Two Binomials • Multiplying Sums and Differences of Two Terms • Squaring Binomials • Multiplications of Various Types

Certain products of two binomials occur so often that it is helpful to be able to compute them quickly. In this section, we develop methods for computing "special" products more quickly than we were able to in Section 4.4.

Products of Two Binomials

To multiply two binomials, we can select one binomial and multiply each term of that binomial by every term of the other. Then we collect like terms. Consider the product $(x + 5)(x + 4)$:

$$\begin{aligned}(x + 5)(x + 4) &= x \cdot x + x \cdot 4 + 5 \cdot x + 5 \cdot 4 \\ &= x^2 + 4x + 5x + 20 \\ &= x^2 + 9x + 20.\end{aligned}$$

Note that the product $x \cdot x$ is found by multiplying the *First* terms of each binomial, $x \cdot 4$ is found by multiplying the *Outside* terms of the two binomials, $5 \cdot x$ is the product of the *Inside* terms of the two binomials, and $5 \cdot 4$ is the product of the *Last* terms of each binomial:

$$(x + 5)(x + 4) = \underbrace{x \cdot x}_{\text{First terms}} + \underbrace{4 \cdot x}_{\text{Outside terms}} + \underbrace{5 \cdot x}_{\text{Inside terms}} + \underbrace{5 \cdot 4}_{\text{Last terms}}.$$

To remember this method of multiplying, use the initials **FOIL.**

The FOIL Method

To multiply two binomials, $A + B$ and $C + D$, multiply the First terms AC, the Outside terms AD, the Inside terms BC, and then the Last terms BD. Then collect like terms, if possible.

$$(A + B)(C + D) = AC + AD + BC + BD$$

1. Multiply First terms: AC.
2. Multiply Outside terms: AD.
3. Multiply Inside terms: BC.
4. Multiply Last terms: BD.

↓

FOIL

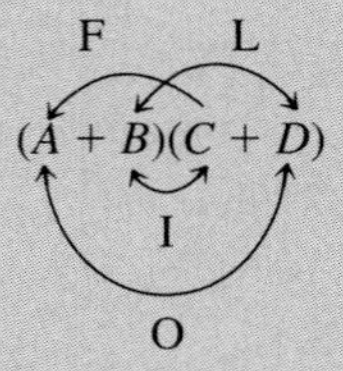

EXAMPLE 1

Multiply: $(x + 8)(x^2 + 5)$.

Solution

F O I L

$$(x + 8)(x^2 + 5) = x^3 + 5x + 8x^2 + 40$$
$$= x^3 + 8x^2 + 5x + 40$$

□

After multiplying, any like terms should be collected.

EXAMPLE 2

Multiply.

a) $(x + 7)(x + 4)$ **b)** $(y + 3)(y - 2)$
c) $(4t^3 + 5t)(3t^2 - 2)$ **d)** $(3 - 4x)(7 - 5x^3)$

Solution

a) $(x + 7)(x + 4) = x^2 + 4x + 7x + 28$ Using FOIL
$= x^2 + 11x + 28$ Collecting like terms

b) $(y + 3)(y - 2) = y^2 - 2y + 3y - 6$
$= y^2 + y - 6$

c) $(4t^3 + 5t)(3t^2 - 2) = 12t^5 - 8t^3 + 15t^3 - 10t$ Remember to add exponents when multiplying terms with the same base.
$= 12t^5 + 7t^3 - 10t$

d) $(3 - 4x)(7 - 5x^3) = 21 - 15x^3 - 28x + 20x^4$
$= 21 - 28x - 15x^3 + 20x^4$ Because the original binomials are in *ascending* order, we write the answer that way. □

Multiplying Sums and Differences of Two Terms

Consider the product of the sum and difference of the same two terms, such as

$$(x + 2)(x - 2).$$

Since this is the product of two binomials, we can use FOIL. This product occurs so often, however, that it will be even quicker to use another method. To find a faster way to compute such a product, look for a pattern in the following:

a) $(x + 2)(x - 2) = x^2 - 2x + 2x - 4$
$= x^2 - 4;$

b) $(3a - 5)(3a + 5) = 9a^2 + 15a - 15a - 25$
$= 9a^2 - 25;$

c) $\left(x^3 + \frac{2}{7}\right)\left(x^3 - \frac{2}{7}\right) = x^6 - \frac{2}{7}x^3 + \frac{2}{7}x^3 - \frac{4}{49}$ — $x^3 \cdot x^3 = (x^3)^2 = x^6$
$= x^6 - \frac{4}{49}.$

Perhaps you discovered in each case that when we multiply the two binomials, the "outer" and "inner" products add to 0 and "drop out."

The Product of a Sum and Difference

The product of the sum and difference of the same two terms is the square of the first term minus the square of the second term:

$$(A + B)(A - B) = A^2 - B^2.$$

This is called a *difference of squares*.

EXAMPLE 3

Multiply.

a) $(x + 4)(x - 4)$

b) $(5 + 2w)(5 - 2w)$

c) $(3a^4 - 5)(3a^4 + 5)$

d) $(-4x - 10)(-4x + 10)$

Solution

$(A + B)(A - B) = A^2 - B^2$ — Saying the words can help.

a) $(x + 4)(x - 4) = x^2 - 4^2$ — "The square of the first term, x^2, minus the square of the second, 4^2."
$= x^2 - 16$ — Simplifying

b) $(5 + 2w)(5 - 2w) = 5^2 - (2w)^2$
$= 25 - 4w^2$ — Squaring both 2 and w

c) $(3a^4 - 5)(3a^4 + 5) = (3a^4)^2 - 5^2$
$= 9a^8 - 25$ — Using the rules for exponents. Remember to multiply exponents when raising a power to a power.

d) $(-4x - 10)(-4x + 10) = (-4x)^2 - 10^2$
$= 16x^2 - 100$ — Squaring both -4 and x ❑

Squaring Binomials

Consider the square of a binomial, such as $(x + 3)^2$. This can be expressed as $(x + 3)(x + 3)$. Since this is the product of two binomials, we can use FOIL. But again, this product occurs so often that it will speed up our work to be able to use an even faster method. Look for a pattern in the following:

a) $(x + 3)^2 = (x + 3)(x + 3)$
$= x^2 + 3x + 3x + 9$
$= x^2 + 6x + 9;$

b) $(5 + 3p)^2 = (5 + 3p)(5 + 3p)$
$= 25 + 15p + 15p + 9p^2$
$= 25 + 30p + 9p^2;$

c) $(a^3 - 1)^2 = (a^3 - 1)(a^3 - 1)$
$= a^6 - a^3 - a^3 + 1$
$= a^6 - 2a^3 + 1;$

d) $(7x - 2)^2 = (7x - 2)(7x - 2)$
$= 49x^2 - 14x - 14x + 4$
$= 49x^2 - 28x + 4.$

Perhaps you noticed that in each product the "outer" and "inner" products are identical. The other two terms, the "first" and "last" products, are squares.

The Square of a Binomial

The square of a binomial is the square of the first term, plus twice the product of the two terms, plus the square of the last term:

$$(A + B)^2 = A^2 + 2AB + B^2;$$
$$(A - B)^2 = A^2 - 2AB + B^2.$$

EXAMPLE 4

Multiply: **(a)** $(x + 7)^2$; **(b)** $(t - 5)^2$; **(c)** $(3a + 0.4)^2$; **(d)** $(5x - 3x^4)^2$.

Solution

$(A + B)^2 = A^2 + 2 \cdot A \cdot B + B^2$ — Saying the words can help.

a) $(x + 7)^2 = x^2 + 2 \cdot x \cdot 7 + 7^2$ — "The square of the first term, x^2, plus twice the product of the terms, $2 \cdot 7x$, plus the square of the second term, 7^2."

$= x^2 + 14x + 49$

b) $(t - 5)^2 = t^2 - 2 \cdot t \cdot 5 + 5^2$
$= t^2 - 10t + 25$

c) $(3a + 0.4)^2 = (3a)^2 + 2 \cdot 3a \cdot 0.4 + 0.4^2$
$= 9a^2 + 2.4a + 0.16$

d) $(5x - 3x^4)^2 = (5x)^2 - 2 \cdot 5x \cdot 3x^4 + (3x^4)^2$
$= 25x^2 - 30x^5 + 9x^8$ Using the rules for exponents ❑

CAUTION! Note carefully in Example 4 that the square of a sum is *not* the sum of the squares:

The middle term $2AB$ is missing.

$$(A + B)^2 \neq A^2 + B^2.$$

To confirm this inequality, note that

$$(20 + 5)^2 = 25^2 = 625,$$

whereas

$$20^2 + 5^2 = 400 + 25 = 425, \quad \text{and } 425 \neq 625.$$

Geometrically, $(A + B)^2$ can be viewed as follows.

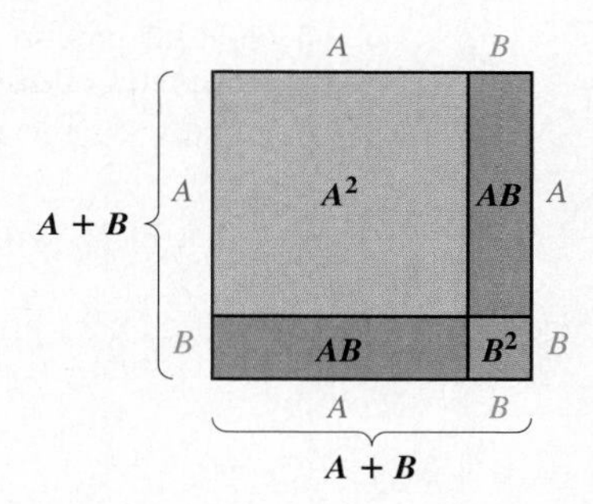

The area of the large square is

$$(A + B)(A + B) = (A + B)^2.$$

This is equal to the sum of the areas of the smaller rectangles:

$$A^2 + AB + AB + B^2 = A^2 + 2AB + B^2.$$

Thus,

$$(A + B)^2 = A^2 + 2AB + B^2.$$

Multiplications of Various Types

Let us now try a variety of multiplications mixed together so that we can learn to sort them out. First try to determine what kind of product you have. Then use the best method. The formulas you should know and the questions you should ask yourself are as follows.

Multiplying Two Polynomials

1. Is the product the square of a binomial? If so, use the following:

$$(A + B)(A + B) = (A + B)^2 = A^2 + 2AB + B^2,$$

or

$$(A - B)(A - B) = (A - B)^2 = A^2 - 2AB + B^2.$$

The square of a binomial is the square of the first term, plus *twice* the product of the two terms, plus the square of the last term.

[The answer has 3 terms.]

2. Is it the product of the sum and difference of the *same* two terms? If so, use the following:

$$(A + B)(A - B) = A^2 - B^2.$$

The product of the sum and difference of the same two terms is a difference of squares.

[The answer has 2 terms.]

3. Is it the product of two binomials other than those above? If so, use FOIL.

[The answer will have 3 or 4 terms.]

4. Is it the product of a monomial and a polynomial? If so, multiply each term of the polynomial by the monomial.
5. Is it the product of two polynomials other than those above? If so, multiply each term of one by every term of the other. Use columns if you wish.

Note that FOIL will work instead of either of the first two methods, but not as quickly.

EXAMPLE 5

Multiply.

a) $(x + 3)(x - 3)$ **b)** $(t + 7)(t - 5)$ **c)** $(x + 7)(x + 7)$

d) $2x^3(9x^2 + x - 7)$ **e)** $(5x^3 - 7x)^2$ **f)** $(p + 3)(p^2 + 2p - 1)$

g) $\left(3x + \frac{1}{4}\right)^2$

Solution

a) $(x + 3)(x - 3) = x^2 - 9$ The product of the sum and difference of the same two terms

b) $(t + 7)(t - 5) = t^2 - 5t + 7t - 35$ Using FOIL

$= t^2 + 2t - 35$ Try to go directly to this step.

c) $(x + 7)(x + 7) = x^2 + 14x + 49$ The product is the square of a binomial.

d) $2x^3(9x^2 + x - 7) = 18x^5 + 2x^4 - 14x^3$ Multiplying each term of the trinomial by the monomial

e) $(5x^3 - 7x)^2 = 25x^6 - 2(5x^3)(7x) + 49x^2$ Squaring a binomial; remember the rules for powers.

$= 25x^6 - 70x^4 + 49x^2$

f)
$$\begin{array}{r} p^2 + 2p - 1 \\ p + 3 \\ \hline 3p^2 + 6p - 3 \\ p^3 + 2p^2 - p \phantom{{}-3} \\ \hline p^3 + 5p^2 + 5p - 3 \end{array}$$

Using columns to multiply a binomial and a trinomial

Multiplying by 3

Multiplying by p

g) $\left(3x + \frac{1}{4}\right)^2 = 9x^2 + 2(3x)\left(\frac{1}{4}\right) + \frac{1}{16}$ Squaring a binomial

$= 9x^2 + \frac{3}{2}x + \frac{1}{16}$ ❑

EXERCISE SET 4.5

Multiply. Try to write only the answer. If you need more steps, by all means use them.

1. $(x + 1)(x^2 + 3)$ **2.** $(x^2 - 3)(x - 1)$

3. $(x^3 + 2)(x + 1)$ **4.** $(x^4 + 2)(x + 12)$

5. $(y + 2)(y - 3)$ **6.** $(a + 2)(a + 2)$

7. $(3x + 2)(3x + 3)$ **8.** $(4x + 1)(2x + 2)$

9. $(5x - 6)(x + 2)$ **10.** $(x - 8)(x + 8)$

11. $(3t - 1)(3t + 1)$ **12.** $(2m + 3)(2m + 3)$

13. $(4x - 2)(x - 1)$ **14.** $(2x - 1)(3x + 1)$

15. $\left(p - \frac{1}{4}\right)\left(p + \frac{1}{4}\right)$ **16.** $\left(q + \frac{3}{4}\right)\left(q + \frac{3}{4}\right)$

17. $(x - 0.1)(x + 0.1)$ **18.** $(x + 0.3)(x - 0.4)$

19. $(2x^2 + 6)(x + 1)$ **20.** $(2x^2 + 3)(2x - 1)$

21. $(-2x + 1)(x + 6)$ **22.** $(3x + 4)(2x - 4)$

23. $(a + 7)(a + 7)$ **24.** $(2y + 5)(2y + 5)$

25. $(1 + 2x)(1 - 3x)$ **26.** $(-3x - 2)(x + 1)$

27. $(x^2 + 3)(x^3 - 1)$ **28.** $(x^4 - 3)(2x + 1)$

29. $(3x^2 - 2)(x^4 - 2)$ **30.** $(x^{10} + 3)(x^{10} - 3)$

31. $(3x^5 + 2)(2x^2 + 6)$ **32.** $(1 - 2x)(1 + 3x^2)$

33. $(8x^3 + 1)(x^3 + 8)$ **34.** $(4 - 2x)(5 - 2x^2)$

35. $(4x^2 + 3)(x - 3)$ **36.** $(7x - 2)(2x - 7)$

37. $(4y^4 + y^2)(y^2 + y)$

38. $(5y^6 + 3y^3)(2y^6 + 2y^3)$

Multiply mentally, writing only the answer if possible. If you need extra steps, by all means use them.

39. $(x + 4)(x - 4)$ **40.** $(x + 1)(x - 1)$

41. $(2x + 1)(2x - 1)$ **42.** $(x^2 + 1)(x^2 - 1)$

43. $(5m - 2)(5m + 2)$ **44.** $(3x^4 + 2)(3x^4 - 2)$

45. $(2x^2 + 3)(2x^2 - 3)$ **46.** $(6x^5 - 5)(6x^5 + 5)$

47. $(3x^4 - 4)(3x^4 + 4)$ **48.** $(t^2 - 0.2)(t^2 + 0.2)$

49. $(x^6 - x^2)(x^6 + x^2)$

50. $(2x^3 - 0.3)(2x^3 + 0.3)$

51. $(x^4 + 3x)(x^4 - 3x)$ **52.** $\left(\frac{3}{4} + 2x^3\right)\left(\frac{3}{4} - 2x^3\right)$

53. $(x^{12} - 3)(x^{12} + 3)$

54. $(12 - 3x^2)(12 + 3x^2)$

55. $(2y^8 + 3)(2y^8 - 3)$ **56.** $\left(m - \frac{2}{3}\right)\left(m + \frac{2}{3}\right)$

57. $(x + 2)^2$ **58.** $(2x - 1)^2$

59. $(3x^2 + 1)^2$ **60.** $\left(3x + \frac{3}{4}\right)^2$

61. $\left(a - \frac{1}{2}\right)^2$ **62.** $\left(2a - \frac{1}{5}\right)^2$

63. $(3 + x)^2$ **64.** $(x^3 - 1)^2$

65. $(x^2 + 1)^2$ **66.** $(8x - x^2)^2$

67. $(2 - 3x^4)^2$ **68.** $(6x^3 - 2)^2$

69. $(5 + 6t^2)^2$ **70.** $(3p^2 - p)^2$

71. $(7x - 0.3)^2$ **72.** $(4a - 0.6)^2$

73. $5a^3(2a^2 - 1)$

74. $(a - 3)(a^2 + 2a - 4)$

75. $(x^2 - 5)(x^2 + x - 1)$

76. $9x^4(3x^2 - x)$ **77.** $(3 - 2x^3)^2$

78. $(x - 4x^3)^2$ **79.** $4x(x^2 + 6x - 3)$

80. $8x(-x^5 + 6x^2 + 9)$ **81.** $\left(2x^2 - \frac{1}{2}\right)\left(2x^2 - \frac{1}{2}\right)$

82. $(-x^2 + 1)^2$ **83.** $(-1 + 3p)(1 + 3p)$

84. $(-3q + 2)(3q + 2)$ **85.** $3t^2(5t^3 - t^2 + t)$

86. $-6x^2(x^3 + 8x - 9)$ **87.** $(6x^4 + 4)^2$

88. $(8a + 5)^2$ **89.** $(3x + 2)(4x^2 + 5)$

90. $(2x^2 - 7)(3x^2 + 9)$ **91.** $(8 - 6x^4)^2$

92. $\left(\frac{1}{5}x^2 + 9\right)\left(\frac{3}{5}x^2 - 7\right)$

93. $(t - 1)(t^2 + t + 1)$

94. $(y + 5)(y^2 - 5y + 25)$

Compute and compare.

95. $3^2 + 4^2$; $(3 + 4)^2$

96. $6^2 + 7^2$; $(6 + 7)^2$

97. $9^2 - 5^2$; $(9 - 5)^2$

98. $11^2 - 4^2$; $(11 - 4)^2$

Find the total shaded area.

99.

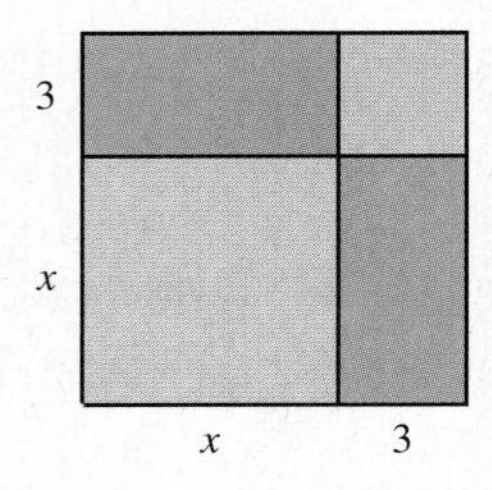

100.

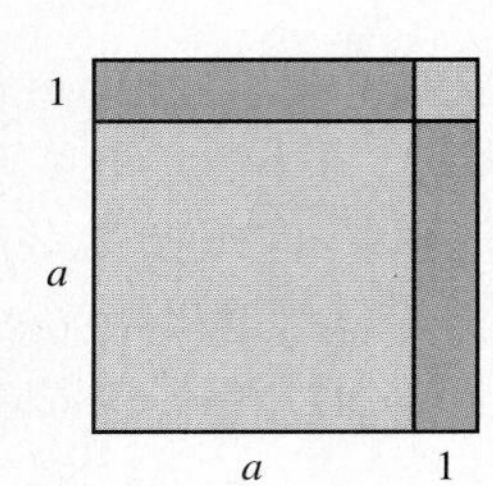

101.

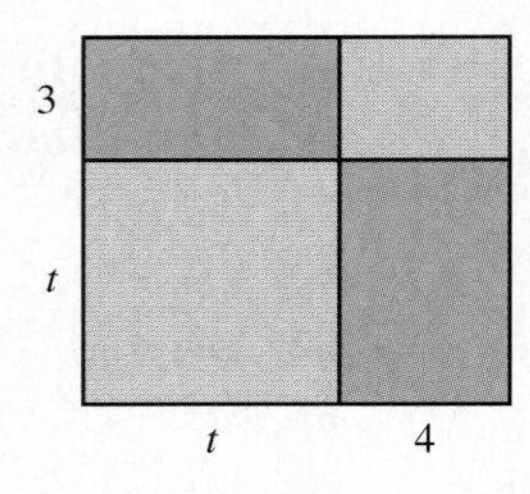

102.

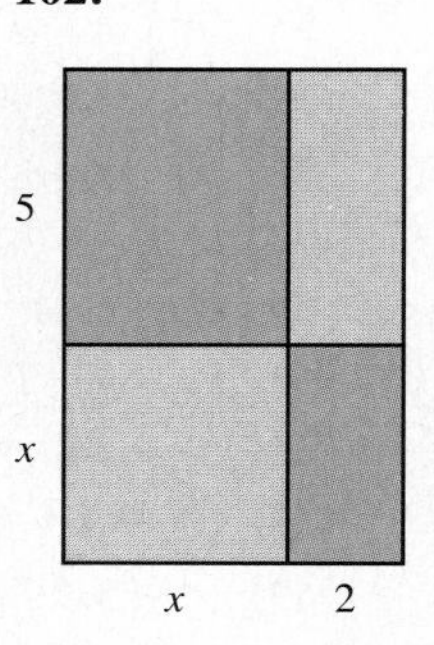

103.

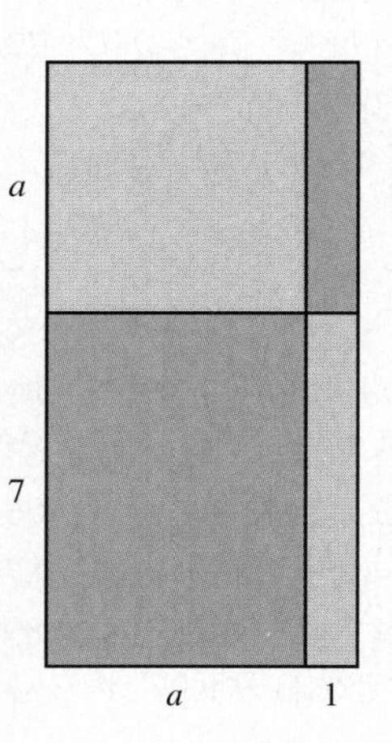

104.

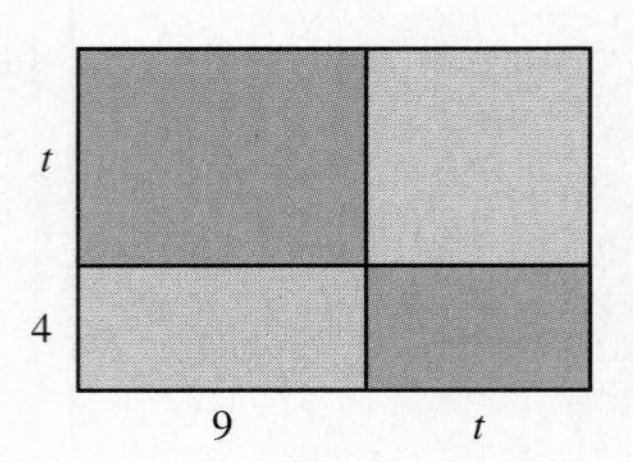

Skill Maintenance

105. In an apartment, lamps, an air conditioner, and a television set are all operating at the same time. The lamps take 10 times as many watts as the television set, and the air conditioner takes 40 times as many watts as the television set. The total wattage used in the apartment is 2550 watts. How many watts are used by each appliance?

106. Solve: $3x - 8x = 4(7 - 8x)$.

Synthesis

107. ◈ The product $(A + B)^2$ can be regarded as the sum of the areas of four regions (as shown following Example 4). How might one visually represent $(A + B)^3$? Why?

108. ◈ Anais claims that by writing $19 \cdot 21$ as $(20 - 1)(20 + 1)$ she can find the product mentally. How is this possible?

Multiply.

109. $4y(y + 5)(2y + 8)$

110. $8x(2x - 3)(5x + 9)$

111. $[(3x - 2)(3x + 2)](9x^2 + 4)$

112. $[(2x - 1)(2x + 1)](4x^2 + 1)$

113. $(5t^3 - 3)^2(5t^3 + 3)^2$
[*Hint:* Regroup as two pairs of binomials.]

114. $[3a - (2a - 3)][3a + (2a - 3)]$
[*Hint:* Do not collect like terms before multiplying.]

115. ▦ $(67.58x + 3.225)^2$

116. $[(x + 3) - 7]^2$

Calculate as the difference of squares.

117. 18×22 [*Hint:* $(20 - 2)(20 + 2)$.]

118. 93×107

Solve.

119. $(x + 2)(x - 5) = (x + 1)(x - 3)$

120. $(2x + 5)(x - 4) = (x + 5)(2x - 4)$

The height of a box is one more than its length l, and the length is one more than its width w. Find a polynomial for the volume V in terms of the following.

121. The width w

122. The length l

123. The height h

Find two expressions for the total shaded area.

124.

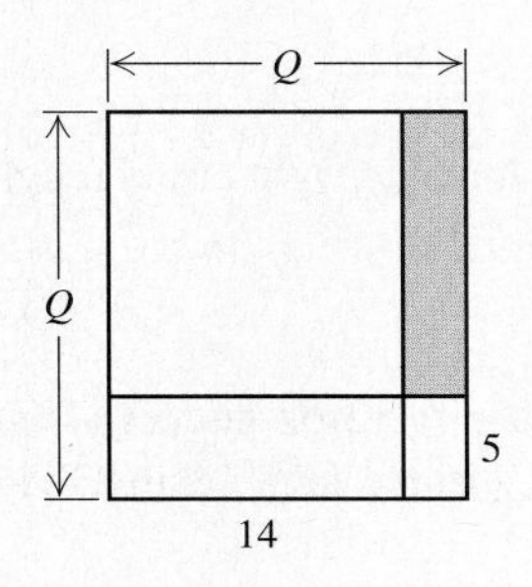

125.

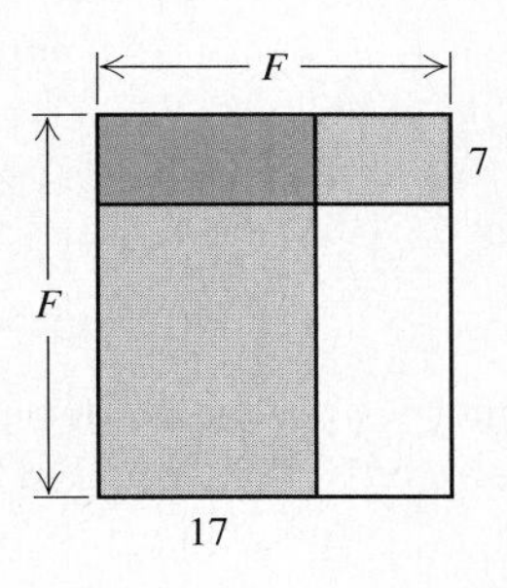

126.

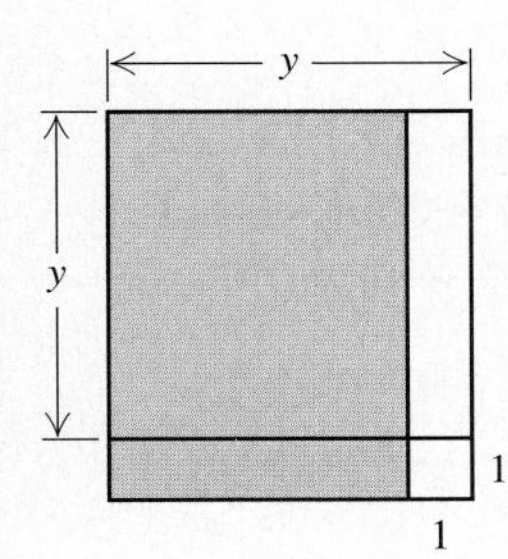

127. A polynomial for the shaded area in this rectangle is $(A + B)(A - B)$.

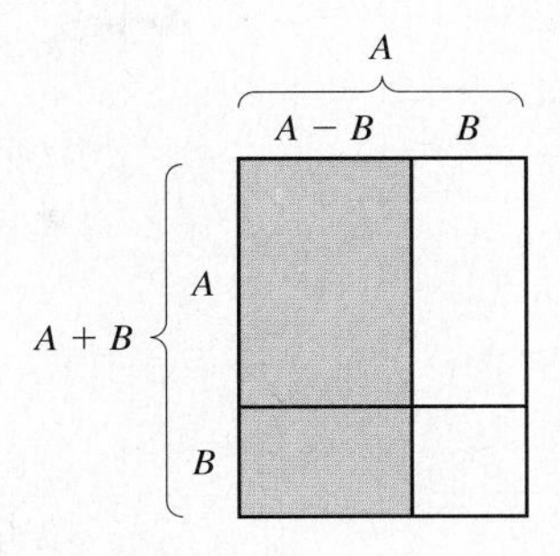

a) Find a polynomial for the area of the entire rectangle.

b) Find a polynomial for the sum of the areas of the two small unshaded rectangles.

c) Find a polynomial for the area in part (a) minus the area in part (b).

d) Find a polynomial for the area of the shaded region and compare this with the polynomial found in part (c).

128. Find three consecutive integers for which the sum of the squares is 65 more than three times the square of the smallest integer.

129. Find $(10x + 5)^2$. Use your result to show how to mentally square any two-digit number ending in 5.

4.6 Polynomials in Several Variables

Evaluating Polynomials • Coefficients and Degrees • Collecting Like Terms • Addition and Subtraction • Multiplication

The polynomials that we have been studying have only one variable. When a polynomial contains two or more variables, it is referred to as a **polynomial in several variables.** Here are some examples:

$$3a + ab^2 + 5b + 4, \qquad 8xy^2z - 2x^3z - 13x^4y^2 + 15, \qquad 9m^2 - 4n^2.$$

In this section, we will find that polynomials in several variables can be evaluated, added, subtracted, and multiplied much like polynomials in one variable.

Evaluating Polynomials

To evaluate a polynomial in two or more variables, we substitute numbers for the variables. Then we compute, using the rules for order of operations.

EXAMPLE 1

Evaluate the polynomial $4 + 3x + xy^2 + 8x^3y^3$ when $x = -2$ and $y = 5$.

Solution We replace x by -2 and y by 5:

$$4 + 3x + xy^2 + 8x^3y^3 = 4 + 3(-2) + (-2)\cdot 5^2 + 8(-2)^3\cdot 5^3$$
$$= 4 - 6 - 50 - 8000 = -8052.$$

❑

EXAMPLE 2

Surface area of a right circular cylinder. The surface area of a right circular cylinder is given by the polynomial

$$2\pi rh + 2\pi r^2,$$

where h is the height and r is the radius of the base. A 12-oz beverage can has a height of 4.7 in. and a radius of 1.2 in. Find its surface area, using 3.14 as an approximation for π.*

Solution We evaluate the polynomial for $h = 4.7$, $r = 1.2$, and $\pi \approx 3.14$:

$$2\pi rh + 2\pi r^2 \approx 2(3.14)(1.2)(4.7) + 2(3.14)(1.2)^2$$
$$= 2(3.14)(1.2)(4.7) + 2(3.14)(1.44)$$
$$= 35.4192 + 9.0432 = 44.4624.$$

The surface area is about 44.4624 in^2 (square inches). ❑

Coefficients and Degrees

The **degree of a term** is the sum of the exponents of the variables. The **degree of a polynomial** is the degree of the term of highest degree.

EXAMPLE 3

Identify the coefficient and the degree of each term and the degree of the polynomial

$$9x^2y^3 - 14xy^2z^3 + xy + 4y + 5x^2 + 7.$$

Solution

Term	Coefficient	Degree	Degree of the Polynomial
$9x^2y^3$	9	5	6
$-14xy^2z^3$	-14	6	
xy	1	2	
$4y$	4	1	
$5x^2$	5	2	
7	7	0	

❑

*Many pocket calculators have a key that gives a better approximation of π. Often the label, π, is printed above the key itself, meaning that another key, often labeled SHIFT or 2ND FCN, must be pressed before the key labeled π. Pressing SHIFT π, we have $\pi \approx 3.141592654$.

Collecting Like Terms

Like terms (or **similar terms**) have exactly the same variables with exactly the same exponents. For example,

$3x^2y^3$ and $-7x^2y^3$ are like terms;

and

$9a^4b^7$ and $12a^4b^7$ are like terms.

On the other hand,

$13xy^5$ and $-2x^2y^5$ are *not* like terms, because the x-factors have different exponents;

and

$3abc^2$ and $4ab$ are *not* like terms, because the factor c^2 does not appear in the second expression.

As always, collecting like terms is based on the distributive law.

EXAMPLE 4

Collect like terms.

a) $9x^2y + 3xy^2 - 5x^2y - xy^2$
b) $7ab - 5ab^2 + 3ab^2 + 6a^3 + 9ab - 11a^3 + b - 1$

Solution

a) $9x^2y + 3xy^2 - 5x^2y - xy^2 = (9 - 5)x^2y + (3 - 1)xy^2$

$= 4x^2y + 2xy^2$ Try to go directly to this step.

b) $7ab - 5ab^2 + 3ab^2 + 6a^3 + 9ab - 11a^3 + b - 1$

$= -2ab^2 + 16ab - 5a^3 + b - 1$ ❑

Addition and Subtraction

The procedure used for adding polynomials in one variable is used to add polynomials in several variables.

EXAMPLE 5

Add.

a) $(-5x^3 + 3y - 5y^2) + (8x^3 + 4x^2 + 7y^2)$
b) $(5ab^2 - 4a^2b + 5a^3 + 2) + (3ab^2 - 2a^2b + 3a^3b - 5)$

Solution

a) $(-5x^3 + 3y - 5y^2) + (8x^3 + 4x^2 + 7y^2)$

$= (-5 + 8)x^3 + 4x^2 + 3y + (-5 + 7)y^2$ Try to do this step mentally.

$= 3x^3 + 4x^2 + 3y + 2y^2$

b) $(5ab^2 - 4a^2b + 5a^3 + 2) + (3ab^2 - 2a^2b + 3a^3b - 5)$

$= 8ab^2 - 6a^2b + 5a^3 + 3a^3b - 3$ ❑

When subtracting a polynomial, remember to find the opposite of each term in that polynomial and then add.

EXAMPLE 6

Subtract: $(4x^2y + x^3y^2 + 3x^2y^3 + 6y) - (4x^2y - 6x^3y^2 + x^2y^2 - 5y)$.

Solution

$$(4x^2y + x^3y^2 + 3x^2y^3 + 6y) - (4x^2y - 6x^3y^2 + x^2y^2 - 5y)$$
$$= 4x^2y + x^3y^2 + 3x^2y^3 + 6y - 4x^2y + 6x^3y^2 - x^2y^2 + 5y \quad \text{Adding the opposite}$$
$$= 7x^3y^2 + 3x^2y^3 - x^2y^2 + 11y \quad \text{Collecting like terms}$$

❑

Multiplication

To multiply polynomials in several variables, multiply each term of one polynomial by every term of the other, much as we did in Sections 4.4 and 4.5.

EXAMPLE 7

Multiply: $(3x^2y - 2xy + 3y)(xy + 2y)$.

Solution

$$
\begin{array}{rl}
3x^2y - 2xy + 3y & \\
xy + 2y & \\
\hline
6x^2y^2 - 4xy^2 + 6y^2 & \text{Multiplying by } 2y \\
3x^3y^2 - 2x^2y^2 + 3xy^2 \quad\quad\quad & \text{Multiplying by } xy \\
\hline
3x^3y^2 + 4x^2y^2 - xy^2 + 6y^2 & \text{Adding}
\end{array}
$$

❑

The special products that we have studied can be used to speed up our multiplication of polynomials in several variables.

EXAMPLE 8

Multiply.

a) $(p + 5q)(2p - 3q)$
b) $(3x + 2y)^2$
c) $(a^3 - 7a^2b)^2$
d) $(3x^2y + 2y)(3x^2y - 2y)$
e) $(-2x^3y^2 + 5t)(2x^3y^2 + 5t)$
f) $(2x + 3 - 2y)(2x + 3 + 2y)$

Solution

a) $$(p + 5q)(2p - 3q) = \overset{F}{2p^2} - \overset{O}{3pq} + \overset{I}{10pq} - \overset{L}{15q^2}$$
$$= 2p^2 + 7pq - 15q^2 \quad \text{Collecting like terms}$$

$$(A + B)^2 = A^2 + 2 \cdot A \cdot B + B^2$$

b) $$(3x + 2y)^2 = (3x)^2 + 2(3x)(2y) + (2y)^2 \quad \text{Squaring a binomial}$$
$$= 9x^2 + 12xy + 4y^2$$

$$(A - B)^2 = A^2 - 2 \cdot A \cdot B + B^2$$

c) $$(a^3 - 7a^2b)^2 = (a^3)^2 - 2(a^3)(7a^2b) + (7a^2b)^2 \quad \text{Squaring a binomial}$$
$$= a^6 - 14a^5b + 49a^4b^2 \quad \text{Using the rules for exponents}$$

$$(A + B)(A - B) = A^2 - B^2$$

d) $(3x^2y + 2y)(3x^2y - 2y) = (3x^2y)^2 - (2y)^2$ — Multiplying the sum and the difference of the same two terms

$= 9x^4y^2 - 4y^2$

e) $(-2x^3y^2 + 5t)(2x^3y^2 + 5t) = (5t - 2x^3y^2)(5t + 2x^3y^2)$ — Using the commutative law for addition twice

$= (5t)^2 - (2x^3y^2)^2$ — Multiplying the sum and the difference of the same two terms

$= 25t^2 - 4x^6y^4$

$$(A - B)(A + B) = A^2 - B^2$$

f) $(2x + 3 - 2y)(2x + 3 + 2y) = (2x + 3)^2 - (2y)^2$ — Multiplying a sum and a difference

$= 4x^2 + 12x + 9 - 4y^2$ — Squaring a binomial

❑

Note that in Example 8 we recognized patterns that might have evaded some students, particularly in parts (e) and (f). In part (e), we could have used FOIL, and in part (f), we could have used long multiplication, but doing so would have been slower. By carefully inspecting a problem before "jumping in," we can often save ourselves considerable work.

EXERCISE SET 4.6

Evaluate the polynomial when $x = 3$ and $y = -2$.

1. $x^2 - y^2 + xy$

2. $x^2 + y^2 - xy$

Evaluate the polynomial when $x = 2$, $y = -3$, and $z = -1$.

3. $xyz^2 + z$

4. $xy - xz + yz$

Interest compounded annually for two years. An amount of money P is invested at interest rate r. In 2 years, it will grow to an amount given by the polynomial

$$A = P(1 + r)^2.$$

5. Evaluate the polynomial when $P = 10{,}000$ and $r = 0.08$ to find the amount to which \$10,000 will grow at 8% interest for 2 years.

6. Evaluate the polynomial when $P = 10{,}000$ and $r = 0.07$ to find the amount to which \$10,000 will grow at 7% interest for 2 years.

Interest compounded annually for three years. An amount of money P is invested at interest rate r. In 3 years, it will grow to an amount given by the polynomial

$$A = P(1 + r)^3.$$

7. Evaluate the polynomial when $P = 10{,}000$ and $r = 0.08$ to find the amount to which \$10,000 will grow at 8% interest for 3 years.

8. Evaluate the polynomial when $P = 10{,}000$ and $r = 0.07$ to find the amount to which \$10,000 will grow at 7% interest for 3 years.

Surface area of a silo. A silo is a structure that is shaped like a right circular cylinder with a half sphere on top. The surface area of a silo of height h and radius r (including the area of the base) is given by the polynomial

$$2\pi rh + \pi r^2.$$

9. A $1\frac{1}{2}$-oz bottle of roll-on deodorant has a height of 4 in. and a radius of $\frac{3}{4}$ in. Find the surface area of the bottle if the bottle is shaped like a silo. Use 3.14 for π.

10. A container of tennis balls is silo-shaped, with a height of $7\frac{1}{2}$ in. and a radius of $1\frac{1}{4}$ in. Find the surface area of the container. Use 3.14 for π.

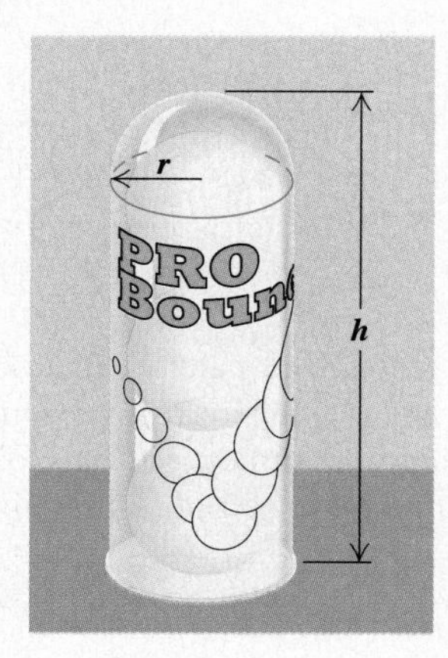

Identify the coefficient and the degree of each term of the polynomial. Then find the degree of the polynomial.

11. $x^3y - 2xy + 3x^2 - 5$

12. $5y^3 - y^2 + 15y + 1$

13. $17x^2y^3 - 3x^3yz - 7$

14. $6 - xy + 8x^2y^2 - y^5$

Collect like terms.

15. $a + b - 2a - 3b$

16. $y^2 - 1 + y - 6 - y^2$

17. $3x^2y - 2xy^2 + x^2$

18. $m^3 + 2m^2n - 3m^2 + 3mn^2$

19. $2u^2v - 3uv^2 + 6u^2v - 2uv^2$

20. $3x^2 + 6xy + 3y^2 - 5x^2 - 10xy - 5y^2$

21. $6au + 3av + 14au + 7av$

22. $3x^2y - 2z^2y + 3xy^2 + 5z^2y$

Add or subtract, as indicated.

23. $(2x^2 - xy + y^2) + (-x^2 - 3xy + 2y^2)$

24. $(2z - z^2 + 5) + (z^2 - 3z + 1)$

25. $(r^3 + 3rs - 5s^2) - (5r^3 + rs + 4s^2)$

26. $(7a^4 - 5ab + 6ab^2) - (9a^4 + 3ab - ab^2)$

27. $(r - 2s + 3) + (2r + 3s - 7)$

28. $(b^3a^2 - 2b^2a^3 + 3ba + 4) + (b^2a^3 - 4b^3a^2 + 2ba - 1)$

29. $(2x^2 - 3xy + y^2) + (-4x^2 - 6xy - y^2) + (x^2 + xy - y^2)$

30. $(x^3 - y^3) - (-2x^3 + x^2y - xy^2 + 2y^3)$

31. $(xy - ab) - (xy - 3ab)$

32. $(3y^4x^2 + 2y^3x - 3y) - (2y^4x^2 + 2y^3x - 4y - 2x)$

33. $(-2a + 7b - c) + (-3b + 4c - 8d)$

34. $(5a^2b + 7ab) + (9a^2b - 5ab) + (a^2b - 6ab)$

35. Subtract $7x + 3y$ from the sum of $4x + 5y$ and $-5x + 6y$.

36. Subtract $5a + 2b$ from the sum of $2a + b$ and $3a - 4b$.

Multiply.

37. $(3z - u)(2z + 3u)$

38. $(a - b)(a^2 + b^2 + 2ab)$

39. $(a^2b - 2)(a^2b - 5)$

40. $(xy + 7)(xy - 4)$

41. $(a^3 + bc)(a^3 - bc)$

42. $(m^2 + n^2 - mn)(m^2 + mn + n^2)$

43. $(y^4x + y^2 + 1)(y^2 + 1)$

44. $(a - b)(a^2 + ab + b^2)$

45. $(3xy - 1)(4xy + 2)$

46. $(m^3n + 8)(m^3n - 6)$

47. $(3 - c^2d^2)(4 + c^2d^2)$

48. $(6x - 2y)(5x - 3y)$

49. $(m^2 - n^2)(m + n)$

50. $(pq + 0.2)(0.4pq - 0.1)$

51. $(xy + x^5y^5)(x^4y^4 - xy)$

52. $(x - y^3)(2y^3 + x)$

53. $(x + h)^2$

54. $(3a + 2b)^2$

55. $(r^3t^2 - 4)^2$

56. $(3a^2b - b^2)^2$

57. $(p^4 + m^2n^2)^2$

58. $(ab + cd)^2$

59. $(2a - b)(2a + b)$

60. $(x - y)(x + y)$

61. $(c^2 - d)(c^2 + d)$

62. $(p^3 - 5q)(p^3 + 5q)$

63. $(ab + cd^2)(ab - cd^2)$

64. $(xy + pq)(xy - pq)$

65. $(x + y - 3)(x + y + 3)$

66. $(p + q + 4)(p + q - 4)$
67. $[x + y + z][x - (y + z)]$
68. $[a + b + c][a - (b + c)]$
69. $(a + b + c)(a - b - c)$
70. $(3x + 2 - 5y)(3x + 2 + 5y)$

Find a polynomial for the shaded area.

71.

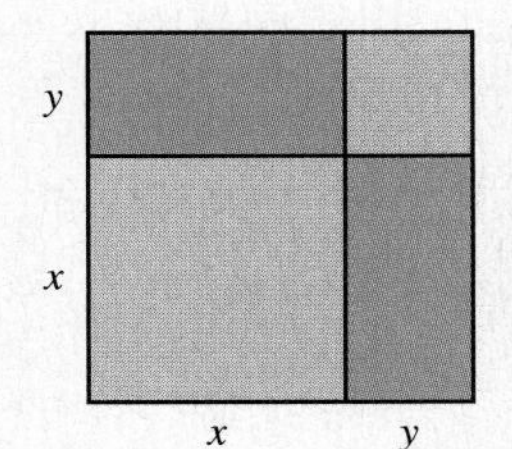

72.

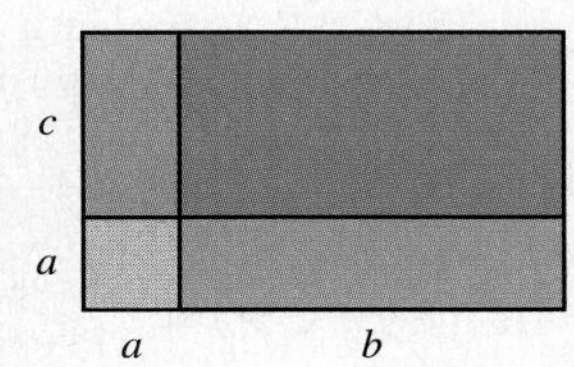

73.

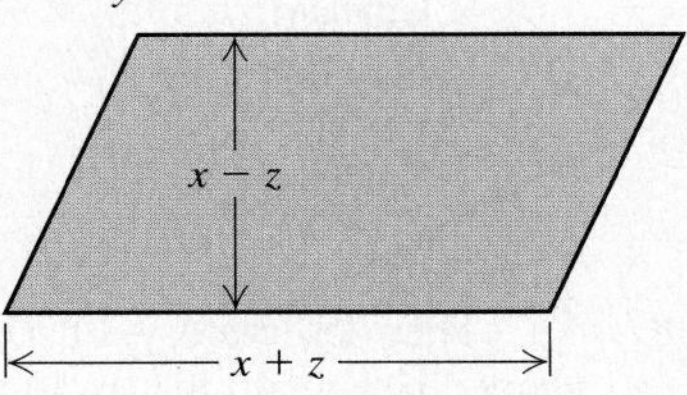

74.

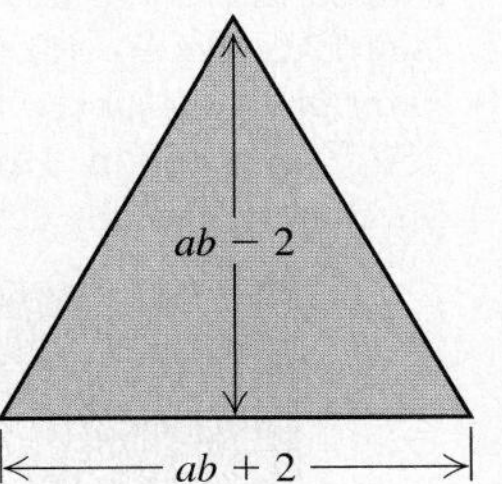

Skill Maintenance

The graph at right* shows the prices paid for a ton of white office paper and for a ton of newsprint by recyclers in the Boston market.

75. How much was being paid for white office paper in December 1989?
76. At what date did recyclers switch from paying for newsprint to charging for collecting it?
77. When did the value of newsprint peak?
78. When did the price that was paid for white office paper first drop to $20 per ton?

*From Burlington Free Press, 7/13/92, Burlington, VT. Reprinted with permission.

Synthesis

79. Is it possible for a polynomial in 4 variables to have a degree less than 4? Why or why not?
80. Explain how the formulas for the surface area of a right circular cylinder (Example 2) and a silo (Exercise 9) can be used to find a formula for the surface area of a sphere.

Find a polynomial for the shaded area. (Leave results in terms of π where appropriate.)

81.

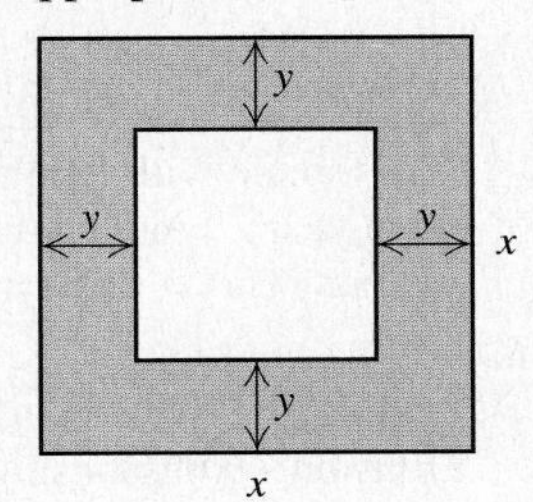

82.

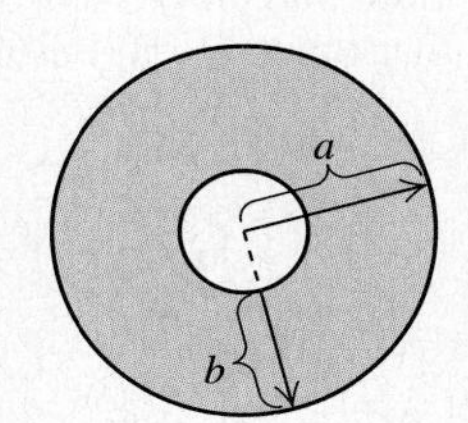

83.

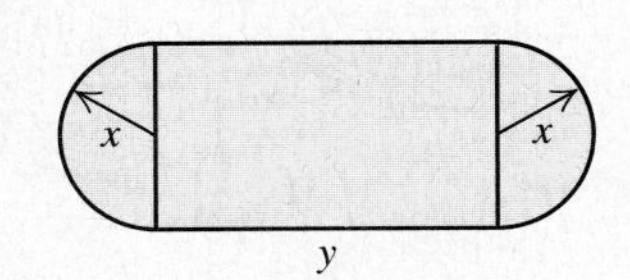

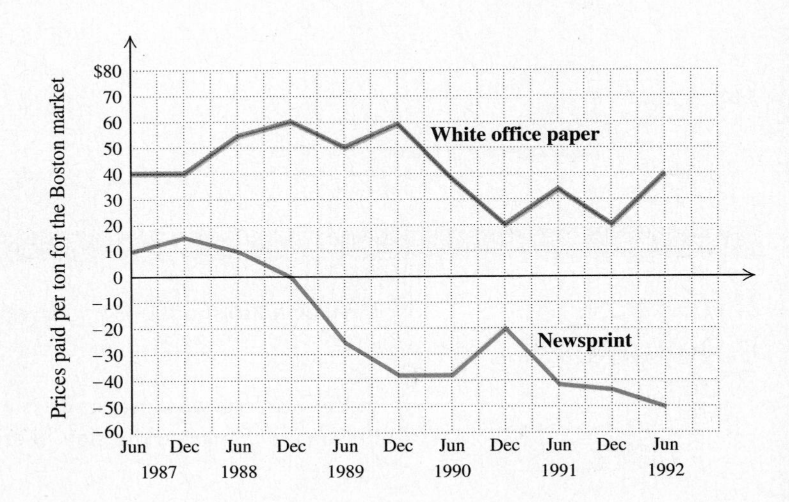

84.

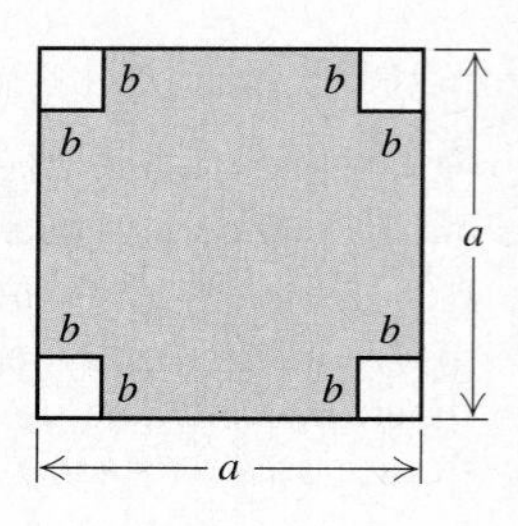

85. *Lung capacity.* The polynomial

$$0.041h - 0.018A - 2.69$$

can be used to estimate the lung capacity, in liters, of a female with height h, in centimeters, and age A, in years. Find the lung capacity of a 30-year-old woman who is 165 cm tall.

86. *The magic number.* The Boston Red Sox are leading the New York Yankees for the Eastern Division championship of the American League. The magic number is 8. This means that any combination of Red Sox wins and Yankee losses that totals 8 will ensure the championship for the Red Sox. The magic number is given by the polynomial

$$G - P - L + 1,$$

where G is the number of games in the season, P is the number of games that the leading team has played, and L is the number of games by which the leading team is ahead of the second-place team in the loss column.

Given the situation shown in the table and assuming a 162-game season, what is the magic number for the Philadelphia Phillies?

Magic number = $G - P - L + 1$

	W	L
Philadelphia	77	40
Pittsburgh	65	53
New York	61	60
Chicago	55	67
St. Louis	51	65
Florida	46	68
Montreal	41	73

87. ◈ The observatory at Danville University is shaped like a silo that is 40 ft high and 30 ft wide (see Exercise 9). Rick and Annie are to paint the exterior of the observatory using paint that covers 250 ft^2 per gallon. How many gallons should they purchase? Explain your reasoning.

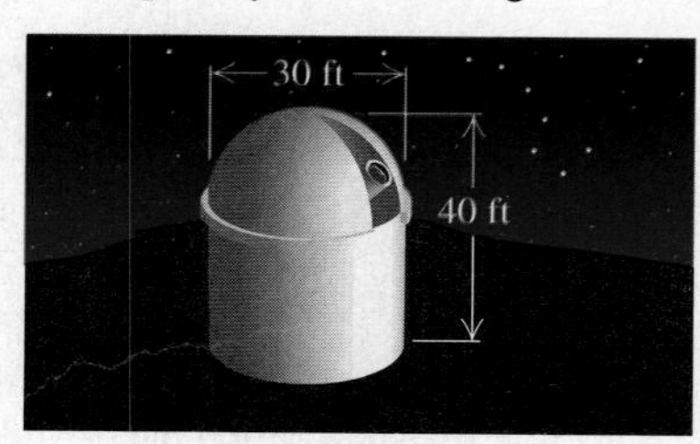

4.7 Division of Polynomials

Divisor a Monomial • Divisor a Binomial

In this section, we consider division of polynomials. You will see that polynomial division is similar to division in arithmetic.

Divisor a Monomial

We first consider division by a monomial. When dividing a monomial by a monomial, we use the quotient rule of Section 4.1 to subtract exponents when bases are the same. For example,

$$\begin{aligned}\frac{15x^{10}}{3x^4} &= 5x^{10-4}\\ &= 5x^6;\end{aligned}$$

and

$$\begin{aligned}\frac{42a^2b^5}{-3ab^2} &= \frac{42}{-3}a^{2-1}b^{5-2}\\ &= -14ab^3.\end{aligned}$$

When we are dividing a monomial into a polynomial, we use the rule for addition using fractional notation. That is, since

$$\frac{A}{C}+\frac{B}{C}=\frac{A+B}{C},$$

we know that

$$\frac{A+B}{C}=\frac{A}{C}+\frac{B}{C}.$$

You have actually used this rule for divisions. Consider $86 \div 2$: Although we might simply write

$$\frac{86}{2}=43,$$

we are really saying

$$\frac{80+6}{2}=\frac{80}{2}+\frac{6}{2}=40+3.$$

EXAMPLE 1

Divide $9x^8 + 12x^6$ by $3x^2$.

Solution This is equivalent to

$$\frac{9x^8}{3x^2}+\frac{12x^6}{3x^2}.$$ To see this, add and get the original expression.

We can now perform the separate divisions. *Caution*! The coefficients are *divided*, but the exponents are *subtracted*.

$$\begin{aligned}\frac{9x^8}{3x^2}+\frac{12x^6}{3x^2} &= \frac{9}{3}x^{8-2}+\frac{12}{3}x^{6-2}\\ &= 3x^6+4x^4.\end{aligned}$$

To check, we multiply the quotient by $3x^2$:

$$(3x^6+4x^4)3x^2 = 9x^8+12x^6.$$ The answer checks. ❑

EXAMPLE 2

Divide and check.

a) $(x^3 + 10x^2 - 8x) \div 2x$

b) $(10a^5b^4 - 2a^3b^2 + 6a^2b) \div 2a^2b$

Solution

a)
$$\frac{x^3 + 10x^2 - 8x}{2x} = \frac{x^3}{2x} + \frac{10x^2}{2x} - \frac{8x}{2x}$$
$$= \frac{1}{2}x^{3-1} + \frac{10}{2}x^{2-1} - \frac{8}{2}x^{1-1}$$
Dividing coefficients and subtracting exponents
$$= \frac{1}{2}x^2 + 5x - 4$$

Check: We check by multiplying the quotient by $2x$:

$$\begin{array}{r} \frac{1}{2}x^2 + 5x - 4 \\ 2x \\ \hline x^3 + 10x^2 - 8x \end{array}$$

Multiplying

The answer checks.

b)
$$\frac{10a^5b^4 - 2a^3b^2 + 6a^2b}{2a^2b} = \frac{10a^5b^4}{2a^2b} - \frac{2a^3b^2}{2a^2b} + \frac{6a^2b}{2a^2b}$$
$$= \frac{10}{2}a^{5-2}b^{4-1} - \frac{2}{2}a^{3-2}b^{2-1} + \frac{6}{2}$$
$$= 5a^3b^3 - ab + 3$$

Check:

$$\begin{array}{r} 5a^3b^3 - ab + 3 \\ 2a^2b \\ \hline 10a^5b^4 - 2a^3b^2 + 6a^2b \end{array}$$

Multiplying

The answer checks. ❑

> To divide a polynomial by a monomial, divide each term by the monomial.

Divisor a Binomial

When the divisor has more than one term, we use long division very much as we do in arithmetic. We write polynomials in descending order and write in any missing terms.

EXAMPLE 3

Divide $x^2 + 5x + 6$ by $x + 2$.

Solution We have

$$\begin{array}{r} x \\ x + 2\overline{)x^2 + 5x + 6} \\ x^2 + 2x \\ \hline 3x \end{array}$$

- x ← Divide the first term, x^2, by the first term in the divisor: $x^2/x = x$. Ignore the term 2.
- $x^2 + 2x$ ← Multiply x above by the divisor, $x + 2$.
- $3x$ ← Subtract: $(x^2 + 5x) - (x^2 + 2x) = x^2 + 5x - x^2 - 2x = 3x$.

Now we "bring down" the next term—in this case, 6—and repeat the procedure:

$$\begin{array}{r} x + 3 \\ x+2\overline{)x^2 + 5x + 6} \\ \underline{x^2 + 2x} \\ 3x + 6 \\ \underline{3x + 6} \\ 0 \end{array}$$

- $x + 3$ ← Consider the "remainder" $3x + 6$. Divide its first term by the first term of the divisor: $3x/x = 3$.
- $3x + 6$ ← The 6 has been "brought down."
- $3x + 6$ ← Multiply 3 by the divisor, $x + 2$.
- 0 ← Subtract: $(3x + 6) - (3x + 6) = 0$.

The quotient is $x + 3$. The remainder is 0, usually expressed as R0. A remainder of 0 is generally not listed in an answer.

To check, we multiply the quotient by the divisor and add the remainder, if any, to see if we get the dividend:

$$\underbrace{(x+2)}_{\text{Divisor}}\ \underbrace{(x+3)}_{\text{Quotient}} + \underbrace{0}_{\text{Remainder}} = \underbrace{x^2 + 5x + 6}_{\text{Dividend}}.$$

The division checks. ❑

EXAMPLE 4

Divide and check: $(x^2 + 2x - 12) \div (x - 3)$.

Solution We have

$$\begin{array}{r} x \\ x-3\overline{)x^2 + 2x - 12} \\ \underline{x^2 - 3x} \\ 5x \end{array}$$

- x ← Divide the first term by the first term: $x^2/x = x$.
- $x^2 - 3x$ ← Multiply x above by the divisor, $x - 3$.
- $5x$ ← Subtract: $(x^2 + 2x) - (x^2 - 3x) = x^2 + 2x - x^2 + 3x = 5x$.

Next we bring down the next term of the dividend, -12.

$$\begin{array}{r} x + 5 \\ x-3\overline{)x^2 + 2x - 12} \\ \underline{x^2 - 3x} \\ 5x - 12 \\ \underline{5x - 15} \\ 3 \end{array}$$

- $x + 5$ ← Divide the first term of $5x - 12$ by the first term of the divisor: $5x/x = 5$.
- $5x - 12$ ← The -12 has been "brought down."
- $5x - 15$ ← Multiply 5 by the divisor, $x - 3$.
- 3 ← Subtract: $(5x - 12) - (5x - 15) = 5x - 12 - 5x + 15 = 3$.

The answer is $x + 5$ with R3, or

$$\underbrace{x + 5}_{\text{Quotient}} + \frac{3 \rightarrow \text{Remainder}}{\underbrace{x - 3}_{\text{Divisor}}}$$

(This is the way answers will be given at the back of the book.)

Check: In arithmetic, to check that $9 \div 4 = 2\frac{1}{4}$, we can multiply and add: $4 \cdot 2 + 1$. A similar procedure, used to check that

$$(x^2 + 2x - 12) \div (x - 3) = x + 5 + \frac{3}{x - 3},$$

is as follows:

$$(x - 3)(x + 5) + 3 = x^2 + 2x - 15 + 3 = x^2 + 2x - 12.$$ ❑

Our division procedure ends when the degree of the remainder is less than the degree of the divisor. Check this for Example 4.

EXAMPLE 5

Divide: **(a)** $(x^3 + 1) \div (x + 1)$; **(b)** $(x^4 - 3x^2 + 1) \div (x - 4)$.

Solution

a)

$$\begin{array}{r l}
& x^2 - x + 1 \\
x + 1 \overline{)\, x^3 + 0x^2 + 0x + 1} & \longleftarrow \text{Fill in the missing terms.} \\
\underline{x^3 + x^2} \qquad\qquad & \\
-x^2 + 0x \qquad & \longleftarrow \text{Subtracting } x^3 + x^2 \text{ from } x^3 + 0x^2 \text{ and bringing down the } 0x \\
\underline{-x^2 - x} \qquad & \\
x + 1 & \longleftarrow \text{Subtracting } -x^2 - x \text{ from } -x^2 + 0x \text{ and bringing down the 1} \\
\underline{x + 1} & \\
0 &
\end{array}$$

The answer is $x^2 - x + 1$.

Check: $(x + 1)(x^2 - x + 1) + 0 = x^3 - x^2 + x + x^2 - x + 1 + 0 = x^3 + 1$.

b)

$$\begin{array}{r l}
& x^3 + 4x^2 + 13x + 52 \\
x - 4 \overline{)\, x^4 + 0x^3 - 3x^2 + 0x + 1} & \longleftarrow \text{Fill in the missing terms.} \\
\underline{x^4 - 4x^3} \qquad\qquad\qquad\qquad & \\
4x^3 - 3x^2 \qquad\qquad & x^4 + 0x^3 - (x^4 - 4x^3) = 4x^3 \\
\underline{4x^3 - 16x^2} \qquad\qquad & \\
13x^2 + 0x \qquad & (4x^3 - 3x^2) - (4x^3 - 16x^2) = 13x^2 \\
\underline{13x^2 - 52x} \qquad & \\
52x + 1 & \\
\underline{52x - 208} & \\
209 &
\end{array}$$

The answer is $x^3 + 4x^2 + 13x + 52$, with R209, or

$$x^3 + 4x^2 + 13x + 52 + \frac{209}{x - 4}.$$

Check:

$$\begin{aligned}
&(x - 4)(x^3 + 4x^2 + 13x + 52) + 209 \\
&= x^4 + 4x^3 + 13x^2 + 52x - 4x^3 - 16x^2 - 52x - 208 + 209 \\
&= x^4 - 3x^2 + 1
\end{aligned}$$

❑

EXERCISE SET 4.7

Divide and check.

1. $\dfrac{24x^4 - 4x^3}{8}$

2. $\dfrac{12a^4 - 3a^2}{6}$

3. $\dfrac{u - 2u^2 - u^5}{u}$

4. $\dfrac{50x^5 - 7x^4 + x^2}{x}$

5. $(15t^3 + 24t^2 - 6t) \div 3t$

6. $(25t^3 + 15t^2 - 30t) \div 5t$

7. $(20x^6 - 20x^4 - 5x^2) \div (-5x^2)$

8. $(24x^6 + 32x^5 - 8x^2) \div (-8x^2)$

9. $(24x^5 - 40x^4 + 6x^3) \div (4x^3)$

10. $(18x^6 - 27x^5 - 3x^3) \div (9x^3)$

11. $\dfrac{8x^2 - 3x + 1}{2}$

12. $\dfrac{6x^2 + 3x - 2}{3}$

13. $\dfrac{2x^3 + 6x^2 + 4x}{2x}$

14. $\dfrac{2x^4 - 3x^3 + 5x^2}{x^2}$

15. $\dfrac{9r^2s^2 + 3r^2s - 6rs^2}{-3rs}$

16. $\dfrac{4x^4y - 8x^6y^2 + 12x^8y^6}{4x^4y}$

17. $(x^2 + 4x + 4) \div (x + 2)$

18. $(x^2 - 6x + 9) \div (x - 3)$

19. $(x^2 - 10x - 25) \div (x - 5)$

20. $(x^2 + 8x - 16) \div (x + 4)$

21. $(x^2 + 4x - 14) \div (x + 6)$

22. $(x^2 + 5x - 9) \div (x - 2)$

23. $\dfrac{x^2 - 9}{x + 3}$

24. $\dfrac{x^2 - 25}{x + 5}$

25. $\dfrac{x^5 + 1}{x + 1}$

26. $\dfrac{x^5 - 1}{x - 1}$

27. $\dfrac{8x^3 - 22x^2 - 5x + 12}{4x + 3}$

28. $\dfrac{2x^3 - 9x^2 + 11x - 3}{2x - 3}$

29. $(x^6 - 13x^3 + 42) \div (x^3 - 7)$

30. $(x^6 + 5x^3 - 24) \div (x^3 - 3)$

31. $(x^4 - 16) \div (x - 2)$

32. $(x^4 - 81) \div (x - 3)$

33. $(t^3 - t^2 + t - 1) \div (t - 1)$

34. $(t^3 - t^2 + t - 1) \div (t + 1)$

Skill Maintenance

35. The perimeter of a rectangle is 640 ft. The length is 15 ft greater than the width. Find the area of the rectangle.

36. Solve: $2x > 12 + 7x$.

37. Plot the points $(4, -1)$, $(0, 5)$, $(-2, 3)$, and $(-3, 0)$.

38. In which quadrant are both coordinates negative?

Synthesis

39. ◈ On an assignment, a student writes

$$\frac{12x^3 - 6x}{3x} = 4x^2 - 6x.$$

What mistake is the student making and how might you convince the person that a mistake has been made?

40. ◈ Explain how you might divide $x^2 - 49$ by $x - 7$ without doing long division.

Divide.

41. $(x^4 + 9x^2 + 20) \div (x^2 + 4)$

42. $(y^4 + a^2) \div (y + a)$

43. $(5a^3 + 8a^2 - 23a - 1) \div (5a^2 - 7a - 2)$

44. $(15y^3 - 30y + 7 - 19y^2) \div (3y^2 - 2 - 5y)$

45. Divide the sum of $4x^5 - 14x^3 - x^2 + 3$ and $2x^5 + 3x^4 + x^3 - 3x^2 + 5x$ by $3x^3 - 2x - 1$.

46. Divide $5x^7 - 3x^4 + 2x^2 - 10x + 2$ by the sum of $(x - 3)^2$ and $5x - 8$.

47. Divide $6a^{3h} + 13a^{2h} - 4a^h - 15$ by $2a^h + 3$.

If the remainder is 0 when one polynomial is divided by another, the divisor is a *factor* of the dividend. Find the value(s) of c for which $x - 1$ is a factor of the polynomial.

48. $x^2 + 4x + c$

49. $2x^2 + 3cx - 8$

50. $c^2x^2 - 2cx + 1$

4.8 Negative Exponents and Scientific Notation

Negative Integers as Exponents • Scientific Notation • Multiplying and Dividing Using Scientific Notation • Problem Solving with Scientific Notation

We now develop a definition of negative exponents. Once we are able to consider both positive and negative powers, we will study a method of writing numbers known as *scientific notation*.

Negative Integers as Exponents

No meaning has yet been attached to negative exponents. By defining negative exponents in a certain way, however, all the rules that apply to whole-number exponents will hold for integer exponents as well.

Consider $5^3/5^7$ and first simplify by removing a factor of 1:

$$\frac{5^3}{5^7} = \frac{5 \cdot 5 \cdot 5}{5 \cdot 5 \cdot 5 \cdot 5 \cdot 5 \cdot 5 \cdot 5}$$
$$= \frac{5 \cdot 5 \cdot 5}{5 \cdot 5 \cdot 5} \cdot \frac{1}{5 \cdot 5 \cdot 5 \cdot 5} = \frac{1}{5^4}.$$

Next, simplify the same expression using the quotient rule:

$$\frac{5^3}{5^7} = 5^{3-7} = 5^{-4}.$$

From these two expressions for $5^3/5^7$, it follows that

$$5^{-4} = \frac{1}{5^4}.$$

This leads to our definition of negative exponents.

Negative Exponents

For any real number a that is nonzero and any integer n,

$$a^{-n} = \frac{1}{a^n}.$$

(The numbers a^{-n} and a^n are reciprocals.)

EXAMPLE 1

Express using positive exponents, and then simplify: **(a)** m^{-3}; **(b)** 4^{-2}; **(c)** $(-3)^{-2}$; **(d)** ab^{-1}; **(e)** $3c^{-5}$; **(f)** $1/x^{-3}$.

Solution

a) $m^{-3} = \dfrac{1}{m^3}$ — m^{-3} is the reciprocal of m^3.

b) $4^{-2} = \dfrac{1}{4^2} = \dfrac{1}{16}$ — 4^{-2} is the reciprocal of 4^2. Note that $4^{-2} \neq 4(-2)$.

c) $(-3)^{-2} = \dfrac{1}{(-3)^2} = \dfrac{1}{(-3)(-3)} = \dfrac{1}{9}$ — $(-3)^{-2}$ is the reciprocal of $(-3)^2$. Note that $(-3)^{-2} \neq -\dfrac{1}{3^2}$.

d) $ab^{-1} = a\left(\dfrac{1}{b^1}\right) = a\left(\dfrac{1}{b}\right) = \dfrac{a}{b}$ — b^{-1} is the reciprocal of b^1.

e) $3c^{-5} = 3\left(\dfrac{1}{c^5}\right) = \dfrac{3}{c^5}$

f) $\dfrac{1}{x^{-3}} = x^{-(-3)} = x^3$ — The reciprocal of x^{-3} is $x^{-(-3)}$, or x^3. ❑

Notice in Examples 1(b) and 1(c) that a negative exponent does not, in itself, indicate that an expression represents a negative number.

The following is another way to understand why negative exponents are defined as they are:

On this side, we divide by 5 at each step.		On this side, the exponents decrease by 1.
	$5 \cdot 5 \cdot 5 = 5^3$	
	$5 \cdot 5 = 5^2$	
	$5 = 5^1$	
	$1 = 5^0$	
	$\frac{1}{5} = 5^?$	
↓	$\frac{1}{25} = 5^?$	↓

To continue the pattern, it should follow that

$$\frac{1}{5} = \frac{1}{5^1} = 5^{-1} \quad \text{and} \quad \frac{1}{25} = \frac{1}{5^2} = 5^{-2}.$$

The rules for powers still hold when exponents are negative.

EXAMPLE 2

Simplify: **(a)** $a^5 \cdot a^{-2}$; **(b)** x/x^7; **(c)** b^{-4}/b^{-5}; **(d)** $(y^{-5})^{-7}$; **(e)** $(5x^2y^{-3})^4$.

Solution

a) $a^5 \cdot a^{-2} = a^{5+(-2)}$ Adding exponents

$= a^3$

b) $\frac{x}{x^7} = x^{1-7}$ Subtracting exponents

$= x^{-6}$, or $\frac{1}{x^6}$

c) $\frac{b^{-4}}{b^{-5}} = b^{-4-(-5)} = b^1 = b$ We subtract exponents even if the exponent in the denominator is negative.

d) $(y^{-5})^{-7} = y^{(-5)(-7)}$ Multiplying exponents

$= y^{35}$

e) $(5x^2y^{-3})^4 = 5^4(x^2)^4(y^{-3})^4$ Raising each factor to the fourth power

$= 625x^8y^{-12}$, or $\frac{625x^8}{y^{12}}$ ❑

Some manipulations with negative exponents can be performed quickly when certain patterns are discovered. For example, since $m^{-5} = 1/m^5$ and $1/x^{-3} = x^3$, we have

$$\frac{m^{-5}}{x^{-3}} = m^{-5} \cdot \frac{1}{x^{-3}} = \frac{1}{m^5} \cdot x^3 = \frac{x^3}{m^5}.$$

Note how the signs of the exponents change.

EXAMPLE 3

Simplify: $\left(\dfrac{y^3}{5}\right)^{-2}$.

Solution

$$\left(\frac{y^3}{5}\right)^{-2} = \frac{(y^3)^{-2}}{5^{-2}} \quad \text{Raising a quotient to a power}$$

$$= \frac{y^{-6}}{5^{-2}} \quad \text{Using the power rule}$$

$$= \frac{5^2}{y^6} \quad \text{Rewriting with positive exponents}$$

$$= \frac{25}{y^6}$$

❑

The following summary of definitions and rules assumes that no denominators are 0 and that 0^0 is not considered.

Definitions and Rules for Exponents

For any integers m and n,

1 as an exponent: $a^1 = a$

0 as an exponent: $a^0 = 1$

Negative exponents: $a^{-n} = \dfrac{1}{a^n}$

The Product Rule: $a^m \cdot a^n = a^{m+n}$

The Quotient Rule: $\dfrac{a^m}{a^n} = a^{m-n}$

The Power Rule: $(a^m)^n = a^{mn}$

Raising a product to a power: $(ab)^n = a^n b^n$

Raising a quotient to a power: $\left(\dfrac{a}{b}\right)^n = \dfrac{a^n}{b^n}$

Scientific Notation

When working with the very large or very small numbers that frequently arise in science, **scientific notation** is an especially useful way of writing numbers. The following are examples of scientific notation.

The distance from the earth to the sun:

9.3×10^7 mi = 93,000,000 mi

The mass of a hydrogen atom:

1.7×10^{-24} gm = 0.0000000000000000000000017 gm

Scientific Notation

Scientific notation for a number is an expression of the type

$$N \times 10^n,$$

where N is at least 1 but less than 10 ($1 \leq N < 10$), N is expressed in decimal notation, and n is an integer.

Converting from scientific to decimal notation involves multiplying by a power of 10. Consider the following.

Scientific Notation $N \times 10^n$	*Multiplication*	*Decimal Notation*
4.52×10^2	4.52×100	452.
4.52×10^1	4.52×10	45.2
4.52×10^0	4.52×1	4.52
4.52×10^{-1}	4.52×0.1	0.452
4.52×10^{-2}	4.52×0.01	0.0452

Note that when n, the power of 10, is positive, the decimal point moves right n places in decimal notation. When n is negative, the decimal point moves left n places. We generally try to perform this multiplication mentally.

EXAMPLE 4

Convert to decimal notation: **(a)** 7.893×10^5; **(b)** 4.7×10^{-8}.

Solution

a) Since the exponent is positive, the decimal point moves to the right:

7.89300. (5 places) $\quad 7.893 \times 10^5 = 789{,}300$ The decimal point moves 5 places to the right.

b) Since the exponent is negative, the decimal point moves to the left:

0.00000004.7 (8 places) $\quad 4.7 \times 10^{-8} = 0.000000047$ The decimal point moves 8 places to the left. ❑

To convert from decimal to scientific notation, we reverse the above procedure.

EXAMPLE 5

Write in scientific notation: **(a)** 7800; **(b)** 0.0549.

Solution

a) We must have $7800 = N \times 10^n$, where $1 \leq N < 10$. Because multiplication by 10^n moves only the decimal point, we must have $N = 7.8$:

$$7800 = 7.8 \times 10^n.$$

Multiplying 7.8 by 10^3 moves the decimal point 3 places to the right. Thus, n is 3 and

$$7800 = 7.8 \times 10^3.$$

b) In scientific notation, 0.0549 is written as 5.49×10^n. Multiplying 5.49 by 10^{-2} moves the decimal point 2 places to the left. Thus, n is -2 and

$$0.0549 = 5.49 \times 10^{-2}.$$

❑

You should try to make conversions to scientific notation mentally as much as possible. Remember that positive exponents are used when representing large numbers and negative exponents are used when representing numbers between 0 and 1.

Multiplying and Dividing Using Scientific Notation

Products and quotients of numbers written in scientific notation are found using the rules for exponents.

EXAMPLE 6

Simplify: **(a)** $(1.8 \times 10^6) \cdot (2.3 \times 10^{-4})$; **(b)** $(3.41 \times 10^5) \div (1.1 \times 10^{-3})$.

Solution

a)
$$\begin{aligned}(1.8 \times 10^6) \cdot (2.3 \times 10^{-4}) &= (1.8 \cdot 2.3) \times (10^6 \cdot 10^{-4}) \\ &= 4.14 \times 10^{6+(-4)} \qquad \text{Adding exponents} \\ &= 4.14 \times 10^2\end{aligned}$$

b)
$$\begin{aligned}(3.41 \times 10^5) \div (1.1 \times 10^{-3}) &= \frac{3.41 \times 10^5}{1.1 \times 10^{-3}} \\ &= \frac{3.41}{1.1} \times \frac{10^5}{10^{-3}} \\ &= 3.1 \times 10^{5-(-3)} \qquad \text{Subtracting exponents} \\ &= 3.1 \times 10^8\end{aligned}$$

❑

When a problem is stated using scientific notation, it is customary to use scientific notation for the answer.

EXAMPLE 7

Simplify: **(a)** $(3.1 \times 10^5) \cdot (4.5 \times 10^{-3})$; **(b)** $(7.2 \times 10^{-7}) \div (8.0 \times 10^6)$.

Solution

a) We have

$$\begin{aligned}(3.1 \times 10^5) \cdot (4.5 \times 10^{-3}) &= (3.1 \times 4.5)(10^5 \cdot 10^{-3}) \\ &= 13.95 \times 10^2.\end{aligned}$$

Our answer is not yet in scientific notation because 13.95 is not between 1 and 10. We convert to scientific notation as follows:

$$\begin{aligned}13.95 \times 10^2 &= 1.395 \times 10^1 \times 10^2 \qquad \text{Substituting } 1.395 \times 10^1 \text{ for } 13.95 \\ &= 1.395 \times 10^3 \qquad \text{Adding exponents}\end{aligned}$$

b)
$$\begin{aligned}(7.2 \times 10^{-7}) \div (8.0 \times 10^6) &= \frac{7.2 \times 10^{-7}}{8.0 \times 10^6} = \frac{7.2}{8.0} \times \frac{10^{-7}}{10^6} \\ &= 0.9 \times 10^{-13} \\ &= 9.0 \times 10^{-1} \times 10^{-13} \qquad \text{Substituting } 9.0 \times 10^{-1} \text{ for } 0.9 \\ &= 9.0 \times 10^{-14} \qquad \text{Adding exponents}\end{aligned}$$

❑

Problem Solving with Scientific Notation

EXAMPLE 8

Light traveling at a rate of 300,000 kilometers per second (km/s) takes 499 seconds to reach the earth from the sun. Find the distance, expressed in scientific notation, from the sun to the earth.

Solution

1. FAMILIARIZE. The time t that it takes for light to reach the earth from the sun is 4.99×10^2 sec (s). The speed r is 3.0×10^5 km/s. Recall that distance can be expressed in terms of speed and time:

$$d = rt$$

$$\text{Distance} = \text{Speed} \times \text{Time}.$$

(If you did not know this formula, you might look it up in a reference book.)

2. TRANSLATE. We translate the problem to mathematical language by substituting 3.0×10^5 for r and 4.99×10^2 for t:

$$\begin{aligned} d &= rt \\ &= (3.0 \times 10^5)(4.99 \times 10^2). \end{aligned}$$

3. CARRY OUT. We carry out the computation and express the results using scientific notation for the answer:

$$\begin{aligned} d &= (3.0 \times 10^5)(4.99 \times 10^2) \\ &= 14.97 \times 10^7 \\ &= 1.497 \times 10^8 \text{ km.} \end{aligned}$$

Converting to scientific notation

4. CHECK. We can check by reviewing our computations. Note too that our answer seems reasonable since it far exceeds the time or the rate alone.

5. STATE. The distance from the earth to the sun is 1.497×10^8 km. ❑

EXERCISE SET | 4.8

Express using positive exponents. Then simplify.

1. 3^{-2} **2.** 2^{-3} **3.** 10^{-4}

4. 5^{-6} **5.** 7^{-3} **6.** 5^{-2}

7. a^{-3} **8.** x^{-2} **9.** $\frac{1}{y^{-4}}$

10. $\frac{1}{t^{-7}}$ **11.** $\frac{1}{z^{-n}}$ **12.** $\frac{1}{h^{-m}}$

13. 2^{-1} **14.** $\left(\frac{2}{3}\right)^{-1}$

15. $\left(\frac{1}{4}\right)^{-2}$ **16.** $\left(\frac{4}{5}\right)^{-2}$

Express using negative exponents.

17. $\frac{1}{4^3}$ **18.** $\frac{1}{5^2}$ **19.** $\frac{1}{x^3}$

20. $\frac{1}{y^2}$ **21.** $\frac{1}{a^4}$ **22.** $\frac{1}{t^5}$

23. $\frac{1}{p^n}$ **24.** $\frac{1}{m^n}$ **25.** $\frac{1}{5}$

26. $\frac{1}{8}$ **27.** $\frac{1}{t}$ **28.** $\frac{1}{m}$

Simplify.

29. $3^{-5} \cdot 3^8$
30. $5^{-8} \cdot 5^9$
31. $x^{-2} \cdot x$
32. $x \cdot x^{-1}$
33. $x^{-7} \cdot x^{-6}$
34. $y^{-5} \cdot y^{-8}$
35. $\frac{m^6}{m^{12}}$
36. $\frac{p^4}{p^5}$
37. $\frac{(8x)^6}{(8x)^{10}}$
38. $\frac{(9t)^4}{(9t)^{11}}$
39. $\frac{18^9}{18^9}$
40. $\frac{(6y)^7}{(6y)^7}$
41. $(a^{-3}b^{-5})(a^{-4}b^{-6})$
42. $(x^{-2}y^{-7})(x^{-3}y^{-2})$
43. $\frac{x^7}{x^{-2}}$
44. $\frac{t^8}{t^{-3}}$
45. $\frac{z^{-6}}{z^{-2}}$
46. $\frac{y^{-7}}{y^{-3}}$
47. $\frac{x^{-5}}{x^{-8}}$
48. $\frac{y^{-4}}{y^{-9}}$
49. $\frac{x}{x^{-1}}$
50. $\frac{x^6}{x}$
51. $(a^{-3})^5$
52. $(x^{-5})^6$
53. $(5^2)^{-3}$
54. $(9^3)^{-4}$
55. $(x^{-3})^{-4}$
56. $(a^{-5})^{-6}$
57. $(m^{-3})^7$
58. $(n^{-2})^8$
59. $(ab)^{-3}$
60. $(mn)^{-5}$
61. $(5ab)^{-2}$
62. $(4xy)^{-2}$
63. $(6x^{-5})^2$
64. $(3a^{-4})^4$
65. $(x^4y^5)^{-3}$
66. $(t^5x^3)^{-4}$
67. $(x^{-6}y^{-2})^{-4}$
68. $(x^{-2}y^{-7})^{-5}$
69. $(3x^3y^{-8}z^{-3})^2$
70. $(2a^2y^{-4}z^{-5})^3$
71. $(x^3y^{-4}z^{-5})(x^{-4}y^{-2}z^9)$
72. $(a^{-5}b^7c^{-2})(a^{-3}b^{-2}c^6)$
73. $(m^{-4}n^7p^3)(m^9n^{-2}p^{-10})$
74. $(t^{-9}p^{10}m^8)(t^{-5}p^{-7}m^{-2})$
75. $\left(\frac{y^2}{2}\right)^{-3}$
76. $\left(\frac{a^4}{3}\right)^{-2}$
77. $\left(\frac{3}{a^2}\right)^3$
78. $\left(\frac{7}{x^7}\right)^2$
79. $\left(\frac{x^2y}{z}\right)^3$
80. $\left(\frac{m}{n^4p}\right)^3$
81. $\left(\frac{a^2b}{cd^3}\right)^{-2}$
82. $\left(\frac{2a^2}{3b^4}\right)^{-3}$

Convert to decimal notation.

83. 2.14×10^3
84. 8.92×10^2
85. 6.92×10^{-3}
86. 7.26×10^{-4}
87. 7.84×10^8
88. 1.35×10^7
89. 8.764×10^{-10}
90. 9.043×10^{-3}
91. 10^8
92. 10^4
93. 10^{-4}
94. 10^{-7}

Convert to scientific notation.

95. 25,000
96. 71,500
97. 0.00371
98. 0.0814
99. 78,000,000,000
100. 3,700,000,000,000
101. 907,000,000,000,000,000
102. 168,000,000,000,000
103. 0.00000374
104. 0.000000000275
105. 0.000000018
106. 0.00000000002
107. 10,000,000
108. 100,000,000,000
109. 0.000000001
110. 0.0000001

Multiply or divide, and write scientific notation for the result.

111. $(3 \times 10^4)(2 \times 10^5)$
112. $(1.9 \times 10^8)(3.4 \times 10^{-3})$
113. $(5.2 \times 10^5)(6.5 \times 10^{-2})$
114. $(7.1 \times 10^{-7})(8.6 \times 10^{-5})$
115. $(9.9 \times 10^{-6})(8.23 \times 10^{-8})$
116. $(1.123 \times 10^4) \times 10^{-9}$
117. $\frac{8.5 \times 10^8}{3.4 \times 10^{-5}}$
118. $\frac{5.6 \times 10^{-2}}{2.5 \times 10^5}$
119. $(3.0 \times 10^6) \div (6.0 \times 10^9)$
120. $(1.5 \times 10^{-3}) \div (1.6 \times 10^{-6})$
121. $\frac{7.5 \times 10^{-9}}{2.5 \times 10^{12}}$
122. $\frac{4.0 \times 10^{-3}}{8.0 \times 10^{20}}$

Problem Solving

Write scientific notation for each answer.

123. There are 3064 members of the Professional Bowlers Association. There are 249 million people in the United States. What fractional part of the population are members of the Professional Bowlers Association?

124. There are 300,000 words in the English language. The average person knows about 10,000 of them. What fractional part of the total number of words does the average person know?

125. Americans eat 6.5 million gal of popcorn each day. How much popcorn do they eat in one year?

126. Americans drink 3 million gal of orange juice each day. How much orange juice do Americans consume in one year?

127. The average discharge at the mouth of the Amazon River is 4,200,000 cubic feet per second. How much water is discharged from the Amazon River in one hour? in one year?

128. There are 300,000 words in the English language. The exceptional person knows about 20,000 of them. What fractional part of the total number of words does the exceptional person know?

Skill Maintenance

Collect like terms.

129. $-9a + 17a$

130. $-12x + (-5x)$

131. Plot the points $(-4, 1)$, $(-3, -2)$, $(5, 2)$, and $(-1, 4)$.

132. In which two quadrants is the first coordinate positive?

Synthesis

Carry out the indicated operations. Write scientific notation for the result.

133. $\dfrac{(5.2 \times 10^6)(6.1 \times 10^{-11})}{1.28 \times 10^{-3}}$

134. $\dfrac{3.9 \times 10^{15}}{(8.0 \times 10^{-12})(3.2 \times 10^{19})}$

135. Perform the indicated operations. Express the result in scientific notation.

$$\{2.1 \times 10^6[(2.5 \times 10^{-3}) \div (5.0 \times 10^{-5})]\} \div (3.0 \times 10^{17})$$

136. Find the reciprocal. Express in scientific notation.

a) 6.25×10^{-3}
b) 4.0×10^{10}

137. Write $4^{-3} \cdot 8 \cdot 16$ as a power of 2.

138. Write $2^8 \cdot 16^{-3} \cdot 64$ as a power of 4.

Simplify.

139. $(5^{-12})^2 \cdot 5^{25}$

140. $49^{18} \cdot 7^{-35}$

141. $\left(\dfrac{1}{a}\right)^{-n}$

142. $\dfrac{(0.4)^5}{[(0.4)^3]^2}$

Determine whether each of the following is true for any pairs of integers m and n and any positive numbers x and y.

143. $x^m \cdot y^n = (xy)^{mn}$

144. $x^m \cdot y^m = (xy)^{2m}$

145. $(x - y)^m = x^m - y^m$

SUMMARY AND REVIEW 4

KEY TERMS

Polynomial, p. 154
Monomial, p. 154
Binomial, p. 155
Trinomial, p. 155
Coefficient, p. 155
Degree, pp. 155, 184
Descending order, p. 157
Opposite of a polynomial, p. 163
FOIL, p. 176
Difference of squares, p. 177
Polynomial in several variables, p. 183
Like terms, p. 185
Scientific notation, p. 198

IMPORTANT PROPERTIES AND FORMULAS

Definitions and Rules for Exponents

Assuming that no denominator is 0 and that 0^0 is not considered, for any integers m and n,

1 as an exponent: $a^1 = a$

0 as an exponent: $a^0 = 1$

Negative exponents: $a^{-n} = \frac{1}{a^n}$

The Product Rule: $a^m \cdot a^n = a^{m+n}$

The Quotient Rule: $\frac{a^m}{a^n} = a^{m-n}$

The Power Rule: $(a^m)^n = a^{mn}$

Raising a product to a power: $(ab)^n = a^n b^n$

Raising a quotient to a power: $\left(\frac{a}{b}\right)^n = \frac{a^n}{b^n}$

Special Products of Polynomials

$(A + B)(A - B) = A^2 - B^2$

$(A + B)(A + B) = A^2 + 2AB + B^2$

$(A - B)(A - B) = A^2 - 2AB + B^2$

Scientific Notation: $N \times 10^n$, where $1 \leq N < 10$

REVIEW EXERCISES

This chapter's review and test include Skill Maintenance exercises from Sections 1.5, 2.5, 2.6, and 3.1.

Simplify.

1. $y^7 \cdot y^3 \cdot y$ **2.** $(3x)^5 \cdot (3x)^9$ **3.** $t^8 \cdot t^0$

4. $\frac{4^5}{4^2}$ **5.** $\frac{(7x)^4}{(7x)^4}$ **6.** $\left(\frac{3t^4}{2s^3}\right)^2$

7. $(-2xy^2)^3$ **8.** $(2x^3)^2(-3x)^2$

9. $2(x^3)^2(-3x)^2$

Identify the terms of the polynomial.

10. $3x^2 + 6x + \frac{1}{2}$ **11.** $-4y^5 + 7y^2 - 3y - 2$

Identify the coefficient of each term of the polynomial.

12. $6x^2 + 17$ **13.** $4x^3 + 6x^2 - 5x + \frac{5}{3}$

Determine the degree of each term and the degree of the polynomial.

14. $x^3 + 4x - 6$

15. $3 - 2x^4 + 3x^9 + x^6 - \frac{3}{4}x^3$

Classify the polynomial as a monomial, a binomial, a trinomial, or none of these.

16. $4x^3 - 1$

17. $4 - 9t^3 - 7t^4 + 10t^2$

18. $7y^2$

Collect like terms and write in descending order.

19. $5x - x^2 + 4x$

20. $\frac{3}{4}x^3 + 4x^2 - x^3 + 7$

21. $-2x^4 + 16 + 2x^4 + 9 - 3x^5$

22. $3x^2 - 2x + 3 - 5x^2 - 1 - x$

23. $-x + \frac{1}{2} + 14x^4 - 7x^2 - 1 - 4x^4$

Evaluate the polynomial when $x = -1$.

24. $7x - 10$ **25.** $x^2 - 3x + 6$

Add or subtract.

26. $(3x^4 - x^3 + x - 4) + (x^5 + 7x^3 - 3x^2 - 5) + (-5x^4 + 6x^2 - x)$

27. $(3x^5 - 4x^4 + x^3 - 3) + (3x^4 - 5x^3 + 3x^2) + (4x^5 + 4x^3) + (-5x^5 - 5x^2) + (-5x^4 + 2x^3 + 5)$

28. $(5x^2 - 4x + 1) - (3x^2 + 7)$

29. $(3x^5 - 4x^4 + 2x^2 + 3) - (2x^5 - 4x^4 + 3x^3 + 4x^2 - 5)$

30.
$$\begin{array}{rrrrr} -\frac{3}{4}x^4 & +\frac{1}{2}x^3 & & & +\frac{7}{8} \\ & -\frac{1}{4}x^3 & -x^2 & -\frac{7}{4}x & \\ +\frac{3}{2}x^4 & & +\frac{2}{3}x^2 & & -\frac{1}{2} \\ \hline \end{array}$$

31.
$$\begin{array}{r} 2x^5 \qquad - x^3 \qquad + x + 3 \\ -(3x^5 - x^4 + 4x^3 + 2x^2 - x + 3) \\ \hline \end{array}$$

32. The length of a rectangle is 4 m greater than its width.

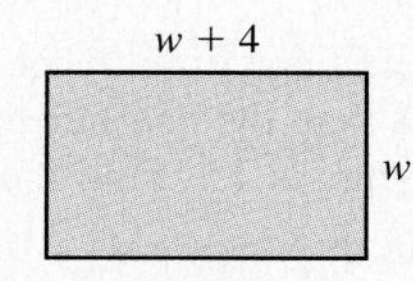

a) Find a polynomial for the perimeter.
b) Find a polynomial for the area.

Multiply.

33. $3x(-4x^2)$

34. $(7x + 1)^2$

35. $\left(x + \frac{2}{3}\right)\left(x + \frac{1}{2}\right)$

36. $(1.5x - 6.5)(0.2x + 1.3)$

37. $(4x^2 - 5x + 1)(3x - 2)$

38. $(x - 9)^2$

39. $5x^4(3x^3 - 8x^2 + 10x + 2)$

40. $(x + 4)(x - 7)$

41. $(x - 0.3)(x - 0.75)$

42. $(x^4 - 2x + 3)(x^3 + x - 1)$

43. $(3y^2 - 2y)^2$

44. $(2t^2 + 3)(t^2 - 7)$

45. $(x^3 - 2x + 3)(4x^2 - 5x)$

46. $(3x^2 + 4)(3x^2 - 4)$

47. $(2 - x)(2 + x)$

48. $(13x - 3)(x - 13)$

49. Evaluate $2 - 5xy + y^2 - 4xy^3 + x^6$ when $x = -1$ and $y = 2$.

Identify the coefficient and the degree of each term of the polynomial. Then find the degree of the polynomial.

50. $x^5y - 7xy + 9x^2 - 8$

51. $x^2y^5z^9 - y^{40} + x^{13}z^{10}$

Collect like terms.

52. $y + w - 2y + 8w - 5$

53. $m^6 - 2m^2n + m^2n^2 + n^2m - 6m^3 + m^2n^2 + 7n^2m$

Add or subtract.

54. $(5x^2 - 7xy + y^2) + (-6x^2 - 3xy - y^2) + (x^2 + xy - 2y^2)$

55. $(6x^3y^2 - 4x^2y - 6x) - (-5x^3y^2 + 4x^2y + 6x^2 - 6)$

Multiply.

56. $(p - q)(p^2 + pq + q^2)$

57. $(3a^4 - \frac{1}{3}b^3)^2$

Divide.

58. $(10x^3 - x^2 + 6x) \div 2x$

59. $(6x^3 - 5x^2 - 13x + 13) \div (2x + 3)$

60. $\dfrac{t^4 + t^3 + 2t^2 - t - 3}{t + 1}$

61. $\dfrac{2x^5 + x^3 - x^2 + 1}{x^3 - 1}$

62. Express using a positive exponent: y^{-4}.

63. Express using a negative exponent: $\dfrac{1}{t^5}$.

Simplify.

64. $7^2 \cdot 7^{-4}$

65. $\dfrac{a^{-5}b}{a^8b^8}$

66. $(x^3)^{-4}$

67. $(2x^{-3}y)^{-2}$

68. $\left(\dfrac{2x}{y}\right)^{-3}$

69. Convert to decimal notation: 8.3×10^6.

70. Convert to scientific notation: 0.0000328.

Multiply or divide and write scientific notation for the result.

71. $(3.8 \times 10^4)(5.5 \times 10^{-1})$

72. $\dfrac{1.28 \times 10^{-8}}{2.5 \times 10^{-4}}$

73. Each day Americans eat 170 million eggs. How many eggs do Americans eat in one year? Write scientific notation for the answer.

Skill Maintenance

74. Find the perimeter of the figure.

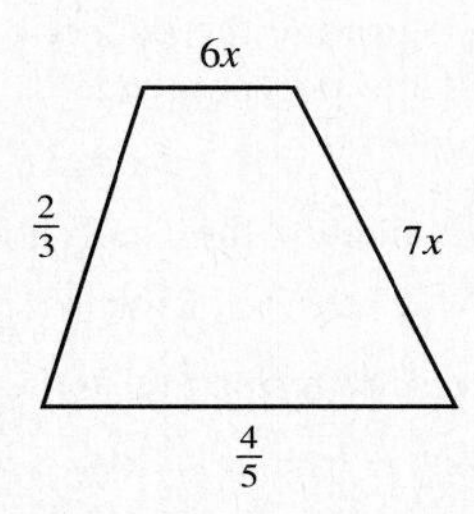

75. The perimeter of a rectangle is 540 m. The width is 19 m less than the length. Find the width and the length.

76. Solve: $3 - 2x \leq 7$.

77. In which quadrant is the point $(3, -1)$ located?

Synthesis

78. ◈ Explain why $5x^3$ and $(5x)^3$ are not equivalent expressions.

79. ◈ If two polynomials of degree n are added, is the sum also of degree n? Why or why not?

80. If a and b are positive, how many terms are there in each of the following?

a) $(x - a)(x - b) + (x - a)(x - b)$
b) $(x + a)(x - b) + (x - a)(x + b)$

81. Collect like terms:

$$-3x^5 \cdot 3x^3 - x^6(2x)^2 + (3x^4)^2 + (2x^2)^4 - 40x^2(x^3)^2.$$

82. A polynomial has degree 4. The x^2-term is missing. The coefficient of x^4 is two times the coefficient of x^3. The coefficient of x is 3 less than the coefficient of x^4. The remaining coefficient is 7 less than the coefficient of x. The sum of the coefficients is 15. Find the polynomial.

83. Multiply: $[(x - 4) - x^3][(x + 4) + 4x^3]$.

84. Solve: $(x - 7)(x + 10) = (x - 4)(x - 6)$.

CHAPTER TEST 4

Simplify.

1. $x^6 \cdot x^2 \cdot x$ **2.** $(4a)^3 \cdot (4a)^8$ **3.** $\dfrac{3^5}{3^2}$

4. $\dfrac{(2x)^5}{(2x)^5}$ **5.** $(x^3)^2$ **6.** $(-3y^2)^3$

7. $(3x^2)^3(-2x^5)^3$ **8.** $3(x^2)^3(-2x^5)^3$

9. Classify the polynomial as a monomial, a binomial, a trinomial, or none of these:

$7 - x$.

10. Identify the coefficient of each term of the polynomial:

$\frac{1}{3}x^5 - x + 7$.

11. Determine the degree of each term and the degree of the polynomial:

$2x^3 - 4 + 5x + 3x^6$.

12. Evaluate the polynomial $x^2 + 5x - 1$ when $x = -2$.

Collect like terms and write in descending order.

13. $4a^2 - 6 + a^2$ **14.** $y^2 - 3y - y + \frac{3}{4}y^2$

15. $3 - x^2 + 2x^3 + 5x^2 - 6x - 2x + x^5$

Add or subtract.

16. $(3x^5 + 5x^3 - 5x^2 - 3) + (x^5 + x^4 - 3x^3 - 3x^2 + 2x - 4)$

17. $\left(x^4 + \frac{2}{3}x + 5\right) + \left(4x^4 + 5x^2 + \frac{1}{3}x\right)$

18. $(2x^4 + x^3 - 8x^2 - 6x - 3) - (6x^4 - 8x^2 + 2x)$

19. $(x^3 - 0.4x^2 - 12) - (x^5 + 0.3x^3 + 0.4x^2 + 9)$

Multiply.

20. $-3x^2(4x^2 - 3x - 5)$ **21.** $\left(x - \frac{1}{3}\right)^2$

22. $(3x + 10)(3x - 10)$ **23.** $(3b + 5)(b - 3)$

24. $(x^6 - 4)(x^8 + 4)$ **25.** $(8 - y)(6 + 5y)$

26. $(2x + 1)(3x^2 - 5x - 3)$

27. $(5t + 2)^2$

28. Collect like terms:

$x^3y - y^3 + xy^3 + 8 - 6x^3y - x^2y^2 + 11$.

29. Subtract:

$(8a^2b^2 - ab + b^3) - (-6ab^2 - 7ab - ab^3 + 5b^3)$.

30. Multiply: $(3x^5 - 4y^5)(3x^5 + 4y^5)$.

Divide.

31. $(12x^4 + 9x^3 - 15x^2) \div 3x^2$

32. $(6x^3 - 8x^2 - 14x + 13) \div (3x + 2)$

33. Express using a positive exponent: 5^{-3}.

34. Express using a negative exponent: $\dfrac{1}{y^8}$.

Simplify.

35. $6^{-2} \cdot 6^{-3}$ **36.** $\dfrac{x^3y^2}{x^8y^{-3}}$

37. $(2a^3b^{-1})^{-4}$ **38.** $\left(\dfrac{ab}{c}\right)^{-3}$

39. Convert to scientific notation: 3,900,000,000.

40. Convert to decimal notation: 5×10^{-8}.

Multiply or divide and write scientific notation for the result.

41. $\dfrac{5.6 \times 10^6}{3.2 \times 10^{-11}}$

42. $(2.4 \times 10^5)(5.4 \times 10^{16})$

43. Each day Americans eat 170 million eggs. There are 249 million people in this country. How many eggs does the average person eat in one year? Write scientific notation for the answer.

Skill Maintenance

44. Solve: $7x - 4x - 2 > 37$.

45. Plot the point $(-1, 5)$.

46. Add: $\frac{2}{5} + \left(-\frac{3}{4}\right)$.

47. The first angle of a triangle is four times as large as the second. The measure of the third angle is 30° greater than that of the second. How large are the angles?

Synthesis

48. The height of a box is 1 less than its length, and the length is 2 more than its width. Find the volume in terms of the length.

49. Solve: $x^2 + (x - 7)(x + 4) = 2(x - 6)^2$.

CHAPTER 5

Polynomials and Factoring

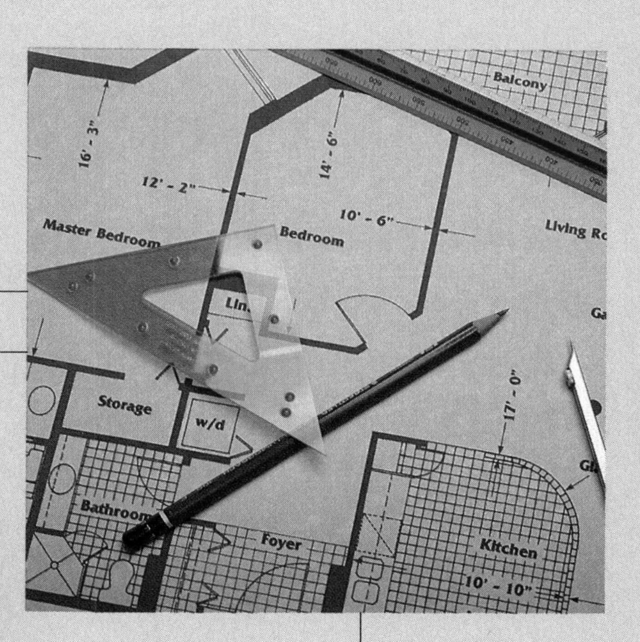

AN APPLICATION

An architect has allocated a rectangular space of 264 ft^2 for a square dining room and a 10-ft wide kitchen. Determine the dimensions of each room.

This problem appears as Exercise 31 in Section 5.7.

Paula Haynes
ARCHITECT

"In general, people think of architecture as art only. But art is only one portion of the picture. Architecture integrates both art and science. A solid background in mathematics is essential for anyone pursuing a future in architecture."

In Chapter 1, we learned that *factoring* is multiplying reversed. To *factor* a polynomial is to find an equivalent expression that is a product. Factoring polynomials requires a solid command of the multiplication methods studied in Chapter 4.

One important reason for studying factoring is that it can be used to solve certain "new" equations that occur in problem-solving situations. Factoring is also very important elsewhere in algebra, as we will see in the remaining chapters of this text.

In addition to the material from this chapter, the review and test for Chapter 5 include material from Sections 1.6, 2.7, 3.2, and 4.1.

5.1 Introduction to Factoring

Factoring Monomials • Factoring When Terms Have a Common Factor • Factoring by Grouping

Just as a number, like 15, can be factored as $3 \cdot 5$, a polynomial, like $x^2 + 7x$, can be factored as $x(x + 7)$. In both cases, we ask ourselves, "What was multiplied to obtain the given result?" The situation is much like a popular television game show in which an "answer" is given and participants must find a "question" to which the answer corresponds.

Factoring

To *factor* a polynomial is to find an equivalent expression that is a product.

Factoring Monomials

To factor a monomial, we find two monomials whose product is equivalent to the original monomial. For example, $20x^2$ can be factored as $2 \cdot 10x^2$, $4x \cdot 5x$, or $10x \cdot 2x$ in addition to other ways that we haven't listed.

EXAMPLE 1

Find three factorizations of $15x^3$.

Solution

a) $15x^3 = (3 \cdot 5)(x \cdot x^2)$
$= (3x)(5x^2)$ The factors here are $3x$ and $5x^2$.

b) $15x^3 = (3 \cdot 5)(x^2 \cdot x)$
$= (3x^2)(5x)$ The factors here are $3x^2$ and $5x$.

c) $15x^3 = ((-5)(-3))x^3$
$= (-5)(-3x^3)$ The factors here are -5 and $-3x^3$. ❑

Note that each part of a factorization is referred to as a *factor*, so the word "factor" operates as a verb or a noun, depending on the context in which it appears.

Factoring When Terms Have a Common Factor

To multiply a polynomial of two or more terms by a monomial, we multiply each term by the monomial, using the distributive law $a(b + c) = ab + ac$. To factor, we do the reverse. We express a polynomial as a product, using the same law, read from right to left: $ab + ac = a(b + c)$. Consider the following:

Multiply	*Factor*
$3x(x^2 + 2x - 4)$	$3x^3 + 6x^2 - 12x$
$= 3x \cdot x^2 + 3x \cdot 2x - 3x \cdot 4$	$= 3x \cdot x^2 + 3x \cdot 2x - 3x \cdot 4$
$= 3x^3 + 6x^2 - 12x;$	$= 3x(x^2 + 2x - 4).$

In the factorization on the right, note that since $3x$ appears as a factor of $3x^3$, $6x^2$, and $-12x$, it is a *common factor* for all the terms of the trinomial $3x^3 + 6x^2 - 12x$.

To factor a polynomial with two or more terms, always try to first find a factor common to all terms. In some cases, there may not be a common factor (other than 1). If a common factor *does* exist, we generally use the common factor with the largest possible coefficient and the largest possible exponent. Such a factor is called the *largest common factor*.

EXAMPLE 2

Factor: $5x^2 + 15$.

Solution We have

$$5x^2 + 15 = 5 \cdot x^2 + 5 \cdot 3 \quad \text{Factoring each term}$$
$$= 5(x^2 + 3). \quad \text{Factoring out the common factor, 5}$$

To check, we multiply: $5(x^2 + 3) = 5 \cdot x^2 + 5 \cdot 3 = 5x^2 + 15$. ❑

CAUTION! $5 \cdot x^2 + 5 \cdot 3$ is a factorization of the *terms* of $5x^2 + 15$, but not the polynomial itself. The factorization of $5x^2 + 15$ is $5(x^2 + 3)$.

When all terms in a polynomial contain the same letter raised to various powers, we factor out the largest power possible.

EXAMPLE 3

Factor: $24x^3 + 30x^2$.

Solution The largest factor common to 24 and 30 is 6. The largest power of x common to x^3 and x^2 is x^2. (To see this, write x^3 as $x^2 \cdot x$.) Thus the largest common factor of $24x^3$ and $30x^2$ is $6x^2$. We factor as follows:

$$24x^3 + 30x^2 = 6x^2 \cdot 4x + 6x^2 \cdot 5 \quad \text{Factoring each term}$$
$$= 6x^2(4x + 5). \quad \text{Factoring out } 6x^2$$

❑

Suppose in Example 3 that you had not recognized the *largest* common factor and removed only part of it, as follows:

$$24x^3 + 30x^2 = 2x^2 \cdot 12x + 2x^2 \cdot 15$$
$$= 2x^2(12x + 15). \quad 12x + 15 \text{ still has a common factor.}$$

Note that $12x + 15$ still has a common factor, 3. To find the largest common factor, continue factoring out common factors, as follows, until no more exist:

$$= 2x^2[3(4x + 5)]$$
$$= 6x^2(4x + 5).$$ Using an associative law

EXAMPLE 4

Factor: $15x^5 - 12x^4 + 27x^3 - 3x^2$.

Solution

$$15x^5 - 12x^4 + 27x^3 - 3x^2$$
$$= 3x^2 \cdot 5x^3 - 3x^2 \cdot 4x^2 + 3x^2 \cdot 9x - 3x^2 \cdot 1$$ Try to do this mentally.
$$= 3x^2(5x^3 - 4x^2 + 9x - 1)$$ Factoring out $3x^2$

CAUTION! Don't forget the term -1.

If you can spot the largest common factor without writing out a factorization of each term, you can write the answer in one step.

EXAMPLE 5

Factor: **(a)** $8m^3 - 16m$; **(b)** $14p^2y^3 - 8py^2 + 2py$; **(c)** $\frac{4}{5}x^2 + \frac{1}{5}x + \frac{2}{5}$.

Solution

a) $8m^3 - 16m = 8m(m^2 - 2)$
b) $14p^2y^3 - 8py^2 + 2py = 2py(7py^2 - 4y + 1)$
c) $\frac{4}{5}x^2 + \frac{1}{5}x + \frac{2}{5} = \frac{1}{5}(4x^2 + x + 2)$

Determine the largest common factor by inspection; then carefully fill in the parentheses.

Below are two of the most important points to keep in mind as we study this chapter.

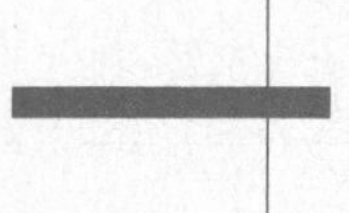

1. Before doing any other kind of factoring, first try to factor out the largest common factor.
2. You can always check your factoring by multiplying.

Factoring by Grouping

Sometimes algebraic expressions contain a common factor that is a polynomial with two or more terms.

EXAMPLE 6

Factor: $x^2(x + 1) + 2(x + 1)$.

Solution The binomial $x + 1$ is a factor of both $x^2(x + 1)$ and $2(x + 1)$. Thus, $x + 1$ is a common factor:

$$x^2(x + 1) + 2(x + 1) = (x + 1)x^2 + (x + 1)2$$ Using a commutative law twice
$$= (x + 1)(x^2 + 2).$$ Factoring out the common factor, $x + 1$

The factorization is $(x + 1)(x^2 + 2)$.

Next consider a four-term polynomial like $x^3 + x^2 + 2x + 2$. Although there is no factor, other than 1, common to all terms, we can factor $x^3 + x^2$ and $2x + 2$ separately:

$$x^3 + x^2 = x^2(x + 1) \quad \text{and} \quad 2x + 2 = 2(x + 1).$$

Note that $x^3 + x^2$ and $2x + 2$ share a common factor of $x + 1$.

When a polynomial can be split into two groups of terms that share a common factor, we can factor out that common factor. This gives a factorization of the original polynomial:

$$\begin{aligned} x^3 + x^2 + 2x + 2 &= (x^3 + x^2) + (2x + 2) && \text{Using an associative law} \\ &= x^2(x + 1) + 2(x + 1) && \text{Factoring each binomial} \\ &= (x + 1)(x^2 + 2). && \text{Using Example 6} \end{aligned}$$

This method is called **factoring by grouping.** Factoring by grouping can be tried on any polynomial with four terms.

EXAMPLE 7

Factor by grouping.

a) $6x^3 - 9x^2 + 4x - 6$

b) $12x^5 + 20x^2 - 21x^3 - 35$

Solution

a) $$\begin{aligned} 6x^3 - 9x^2 + 4x - 6 &= (6x^3 - 9x^2) + (4x - 6) \\ &= 3x^2(2x - 3) + 2(2x - 3) && \text{Factoring each binomial} \\ &= (2x - 3)(3x^2 + 2) && \text{Factoring out the common factor, } 2x - 3 \end{aligned}$$

b) $$\begin{aligned} 12x^5 + 20x^2 - 21x^3 - 35 &= 4x^2(3x^3 + 5) - 7(3x^3 + 5) && \text{Factoring two binomials. Using } -7 \text{ gives a common binomial factor.} \\ &= (3x^3 + 5)(4x^2 - 7) \end{aligned}$$

❑

Although factoring by grouping is a very useful skill, many polynomials, like $x^3 + x^2 + 2x - 2$, cannot be factored by grouping:

$$x^3 + x^2 + 2x - 2 = x^2(x + 1) + 2(x - 1). \qquad \text{There is no common factor.}$$

EXERCISE SET 5.1

Find three factorizations for the monomial.

1. $6x^3$ **2.** $9x^4$ **3.** $-9x^5$

4. $-12x^6$ **5.** $24x^4$ **6.** $15x^5$

Factor. Remember to use the largest common factor and to check by multiplying.

7. $x^2 - 4x$ **8.** $x^2 + 8x$

9. $2x^2 + 6x$ **10.** $3x^2 - 3x$

11. $x^3 + 6x^2$ **12.** $4x^4 + x^2$

13. $8x^4 - 24x^2$ **14.** $5x^5 + 10x^3$

15. $2x^2 + 2x - 8$ **16.** $6x^2 + 3x - 15$

17. $17x^5y^3 + 34x^3y^2 + 51xy$

18. $16x^6y^4 - 32x^5y^3 - 48xy^2$

19. $6x^4 - 10x^3 + 3x^2$
20. $5x^5 + 10x^2 - 8x$
21. $x^5y^5 + x^4y^3 + x^3y^3 - x^2y^2$
22. $x^9y^6 - x^7y^5 + x^4y^4 + x^3y^3$
23. $2x^7 - 2x^6 - 64x^5 + 4x^3$
24. $10x^3 + 25x^2 + 15x - 20$
25. $1.6x^4 - 2.4x^3 + 3.2x^2 + 6.4x$
26. $2.5x^6 - 0.5x^4 + 5x^3 + 10x^2$
27. $\frac{5}{3}x^6 + \frac{4}{3}x^5 + \frac{1}{3}x^4 + \frac{1}{3}x^3$
28. $\frac{5}{7}x^7 + \frac{3}{7}x^5 - \frac{6}{7}x^3 - \frac{1}{7}x$

Factor.

29. $y(y + 3) + 4(y + 3)$
30. $b(b - 5) - 3(b - 5)$
31. $x^2(x + 3) + 2(x + 3)$
32. $3z^2(2z + 1) + (2z + 1)$
33. $y^2(y + 8) + (y + 8)$
34. $x^2(x - 7) - 3(x - 7)$

Factor by grouping, if possible.

35. $x^3 + 3x^2 + 2x + 6$
36. $6z^3 + 3z^2 + 2z + 1$
37. $2x^3 + 6x^2 + x + 3$
38. $3x^3 + 2x^2 + 3x + 2$
39. $8x^3 - 12x^2 + 6x - 9$
40. $10x^3 - 25x^2 + 4x - 10$
41. $12x^3 - 16x^2 + 3x - 4$
42. $18x^3 - 21x^2 + 30x - 35$
43. $x^3 + 8x^2 - 3x - 24$
44. $2x^3 + 12x^2 - 5x - 30$
45. $w^3 - 7w^2 + 4w - 28$
46. $y^3 + 8y^2 - 2y - 16$
47. $x^3 - x^2 - 2x + 5$
48. $p^3 + p^2 - 3p + 10$
49. $2x^3 - 8x^2 - 9x + 36$
50. $20g^3 - 4g^2 - 25g + 5$

Skill Maintenance

51. Graph: $y = x - 6$.
52. Solve: $4x - 8x + 16 \geq 6(x - 2)$.
53. Subtract: $-13 - (-25)$.
54. Solve $A = \dfrac{p + q}{2}$ for p.

Multiply.

55. $(y + 5)(y + 7)$
56. $(y + 7)^2$
57. $(y + 7)(y - 7)$
58. $(y - 7)^2$

Synthesis

59. ◈ Write a two-sentence paragraph in which the word "factor" is used at least once as a noun and once as a verb.

60. ◈ In answering a factoring problem, Taylor says the largest common factor is $-5x^2$ and Natasha says the largest common factor is $5x^2$. Can they both be correct? Why or why not?

Factor, if possible.

61. $4x^5 + 6x^3 + 6x^2 + 9$
62. $x^6 + x^4 + x^2 + 1$
63. $x^{12} + x^7 + x^5 + 1$
64. $x^3 + x^2 - 2x + 2$
65. $p^3 - p^2 + 3p + 3$
66. $ax^2 + 2ax + 3a + x^2 + 2x + 3$
67. ◈ Explain how to construct a four-term polynomial that can be factored by grouping.
68. ◈ Explain how to construct a polynomial of degree 9 for which $5x^3y^2$ is the largest common factor.

5.2 Factoring Trinomials of the Type $x^2 + bx + c$

Constant Term Positive • Constant Term Negative

We now learn how to factor trinomials like

$$x^2 + 5x + 4 \quad \text{or} \quad x^2 + 3x - 10,$$

for which no common factor exists. We will limit our attention to trinomials of the type $ax^2 + bx + c$, where $a = 1$. The coefficient a is often called the **leading coefficient.**

Constant Term Positive

Recall the FOIL method of multiplying two binomials:

$$\begin{aligned}
& \quad\ \ \text{F} \quad \text{O} \quad \text{I} \quad \text{L} \\
(x+2)(x+5) &= x^2 + \underbrace{5x + 2x}_{\downarrow} + 10 \\
&= x^2 + \quad 7x \quad + 10.
\end{aligned}$$

To factor $x^2 + 7x + 10$, we think of FOIL in reverse. We multiplied x times x to get the first term of the trinomial, so we know that the first term of each binomial factor is x. Next we look for numbers p and q such that

$$x^2 + 7x + 10 = (x + p)(x + q).$$

To get the middle term and the last term of the trinomial, we look for two numbers p and q whose product is 10 and whose sum is 7. Those numbers are 2 and 5. Thus the factorization is

$$(x + 2)(x + 5).$$

EXAMPLE 1

Factor: $x^2 + 5x + 6$.

Solution Think of FOIL in reverse. The first term of each factor is x:

$$(x + p)(x + q).$$

We then look for two numbers p and q whose product is 6 and whose sum is 5.

Pairs of Factors of 6	Sums of Factors
1, 6	7
2, 3	5 ← The numbers we seek are 2 and 3.
−1, −6	−7
−2, −3	−5

Since $2 \cdot 3 = 6$ and $2 + 3 = 5$, the factorization of $x^2 + 5x + 6$ is $(x + 2)(x + 3)$. To check, we simply multiply the two binomials to see whether we get the original trinomial.

Check: $(x + 2)(x + 3) = x^2 + 3x + 2x + 6 = x^2 + 5x + 6$.

Note that since 5 and 6 are both positive, when factoring $x^2 + 5x + 6$ we need not consider negative factors of 6. Note too that changing the signs of the factors changes the sign of the sum. ❑

We began this section examining the product of two sums: $(x + 2)(x + 5)$. Note that a product of two differences, like $(x - 2)(x - 5)$, also contains a positive constant term:

$$\begin{aligned}
& \quad\ \ \text{F} \quad \text{O} \quad \text{I} \quad \text{L} \\
(x-2)(x-5) &= x^2 - \underbrace{5x - 2x}_{\downarrow} + 10 \\
&= x^2 - \quad 7x \quad + 10.
\end{aligned}$$

When the constant term of a trinomial is positive, we look for two numbers with the same sign. The sign is that of the middle term:

$$(x^2 - 7x + 10) = (x - 2)(x - 5).$$

EXAMPLE 2

Factor: $y^2 - 8y + 12$.

Solution Since the constant term is positive and the coefficient of the middle term is negative, we look for a factorization of 12 in which both factors are negative. Their sum must be -8.

Pairs of Factors of 12	Sums of Factors
$-1, -12$	-13
$-2, -6$	-8 ← The numbers we need are -2 and -6.
$-3, -4$	-7

The factorization is $(y - 2)(y - 6)$. The check is left for the student. ❑

Constant Term Negative

Sometimes when we use FOIL, the product has a negative constant term. Consider these multiplications:

$$\text{a) } (x - 5)(x + 2) = \overset{F}{x^2} + \overset{O}{2x} - \overset{I}{5x} - \overset{L}{10} = x^2 - 3x - 10;$$

$$\text{b) } (x + 5)(x - 2) = \overset{F}{x^2} - \overset{O}{2x} + \overset{I}{5x} - \overset{L}{10} = x^2 + 3x - 10.$$

Reversing the signs of -5 and 2 changes only the sign of the middle term.

When the constant term of a trinomial is negative, we look for two numbers whose product is negative. One of them must be positive and the other negative. Their sum must be the coefficient of the middle term.

EXAMPLE 3

Factor: $x^2 - 8x - 20$.

Solution The factorization of the constant term, -20, must have one factor positive and one factor negative. The sum must be -8, so the negative factor must have the larger absolute value. Thus we consider only pairs of factors in which the negative factor has the larger absolute value.

Pairs of Factors of -20	Sums of Factors
1, -20	-19
2, -10	-8
4, -5	-1
5, -4	1
10, -2	8
20, -1	19

The numbers we need are 2 and -10.

Since the positive factor in each of these pairs has the larger absolute value, the sums are all positive. For this problem, we can disregard these pairs. Note that changing the signs of the factors changes the sign of the sum.

The numbers we need are 2 and -10. Thus the factorization is $(x + 2)(x - 10)$.

Check: $(x + 2)(x - 10) = x^2 - 10x + 2x - 20 = x^2 - 8x - 20.$ ❑

EXAMPLE 4

Factor: $t^2 - 24 + 5t$.

Solution It helps to first write the trinomial in descending order: $t^2 + 5t - 24$. The factorization of the constant term, -24, must have one factor positive and one factor negative. The sum must be 5, so the positive factor must have the larger absolute value. Thus we consider only pairs of factors in which the positive factor has the larger absolute value.

Pairs of Factors of -24	Sums of Factors
-1, 24	23
-2, 12	10
-3, 8	5
-4, 6	2

The numbers we need are -3 and 8.

The factorization is $(t - 3)(t + 8)$. The check is left for the student. ❑

Polynomials in two or more variables, such as $a^2 + 4ab - 21b^2$, can be factored in a similar manner.

EXAMPLE 5

Factor: $a^2 + 4ab - 21b^2$.

Solution It may help to write the trinomial in the equivalent form

$$a^2 + 4ba - 21b^2.$$

This way we think of $-21b^2$ as the "constant" term and $4b$ as the "coefficient" of the middle term. Then we try to express $-21b^2$ as a product of two factors whose sum is $4b$. Those factors are $-3b$ and $7b$. Thus the factorization is

$$(a - 3b)(a + 7b).$$

Check: $(a - 3b)(a + 7b) = a^2 + 7ab - 3ba - 21b^2 = a^2 + 4ab - 21b^2.$ ❑

EXAMPLE 6

Factor: $x^2 - x + 5$.

Solution Since 5 has very few factors, we can easily check all possibilities.

Pairs of Factors of 5	Sums of Factors
5, 1	6
−5, −1	−6

Since there are no factors whose sum is −1, the polynomial is *not* factorable into binomials. ❑

A polynomial like $x^2 - x + 5$ that cannot be factored further is said to be **prime.**

Often factoring requires two or more steps. In general, when told to factor, we should *factor completely*. This means that the final factorization should not contain any factors that can be factored further.

EXAMPLE 7

Factor: $2x^3 - 20x^2 + 50x$.

Solution *Always* look first for a common factor. This time there is one, $2x$, which we factor out first:

$$2x^3 - 20x^2 + 50x = 2x(x^2 - 10x + 25).$$

Now consider $x^2 - 10x + 25$. Since the constant term is positive and the coefficient of the middle term is negative, we look for a factorization of 25 in which both factors are negative. Their sum must be −10.

Pairs of Factors of 25	Sums of Factors
−25, −1	−26
−5, −5	−10 ← The numbers we need are −5 and −5.

The factorization of

$$x^2 - 10x + 25$$

is

$$(x - 5)(x - 5), \quad \text{or} \quad (x - 5)^2,$$

but we must not forget the common factor, $2x$. The factorization of

$$2x^3 - 20x^2 + 50x$$

is

$$2x(x - 5)(x - 5), \quad \text{or} \quad 2x(x - 5)^2.$$

Check: $2x(x - 5)(x - 5) = 2x[x^2 - 10x + 25]$ Multiplying binomials

$= 2x^3 - 20x^2 + 50x.$ Using the distributive law ❑

Once any common factors are factored out, the following summary can be used to factor $x^2 + bx + c$.

To factor $x^2 + bx + c$:

1. First arrange in descending order. Use a trial-and-error process to express c as a product of two factors whose sum is b.
   a) If c is positive, the signs of the factors are the same as the sign of b.
   b) If c is negative, one factor is positive and the other is negative. Select the factors so that the factor with the larger absolute value is the factor with the same sign as b.
2. Check by multiplying.

EXERCISE SET 5.2

Factor completely. Remember that you can check by multiplying.

1. $x^2 + 8x + 15$
2. $x^2 + 5x + 6$
3. $x^2 + 7x + 12$
4. $x^2 + 9x + 8$
5. $x^2 - 6x + 9$
6. $y^2 + 11y + 28$
7. $x^2 + 9x + 14$
8. $a^2 + 11a + 30$
9. $b^2 + 5b + 4$
10. $x^2 - \frac{2}{5}x + \frac{1}{25}$
11. $x^2 + \frac{2}{3}x + \frac{1}{9}$
12. $z^2 - 8z + 7$
13. $d^2 - 7d + 10$
14. $x^2 - 8x + 15$
15. $y^2 - 11y + 10$
16. $x^2 - 2x - 15$
17. $x^2 + x - 42$
18. $x^2 + 2x - 15$
19. $2x^2 - 14x - 36$
20. $3y^2 - 9y - 84$
21. $x^3 - 6x^2 - 16x$
22. $x^3 - x^2 - 42x$
23. $y^2 - 4y - 45$
24. $x^2 - 7x - 60$
25. $-2x - 99 + x^2$
26. $x^2 - 72 + 6x$
27. $c^4 + c^3 - 56c^2$
28. $5b^2 + 25b - 120$
29. $2a^2 + 4a - 70$
30. $x^5 + x^4 - 2x^3$
31. $x^2 + x + 1$
32. $x^2 + 2x + 3$
33. $7 - 2p + p^2$
34. $11 - 3w + w^2$
35. $x^2 + 20x + 100$
36. $x^2 + 20x + 99$
37. $3x^3 - 63x^2 - 300x$
38. $2x^3 - 40x^2 + 192x$
39. $x^2 - 21x - 72$
40. $4x^2 + 40x + 100$
41. $x^2 - 25x + 144$
42. $y^2 - 21y + 108$
43. $a^4 + a^3 - 132a^2$
44. $a^6 + 9a^5 - 90a^4$
45. $120 - 23x + x^2$
46. $96 + 22d + d^2$
47. $108 - 3x - x^2$
48. $112 + 9y - y^2$
49. $y^2 - 0.2y - 0.08$
50. $t^2 - 0.3t - 0.10$
51. $p^2 + 3pq - 10q^2$
52. $a^2 - 2ab - 3b^2$
53. $m^2 + 5mn + 5n^2$
54. $x^2 - 11xy + 24y^2$
55. $s^2 - 2st - 15t^2$
56. $b^2 + 8bc - 20c^2$
57. $2x^3 - 10x^2 + 12x$
58. $3a^6 - 24a^5 + 36a^4$
59. $7a^9 - 28a^8 - 35a^7$
60. $6x^{10} - 30x^9 - 84x^8$

Skill Maintenance

Multiply.

61. $(x + 6)(3x + 4)$
62. $(7w + 6)^2$
63. In a recent year, 29,090 people were arrested for counterfeiting. This figure was down 1.2% from the year before. How many people were arrested the year before?
64. The first angle of a triangle is four times as large as the second. The measure of the third angle is 30° greater than that of the second. How large are the angles?

Synthesis

65. ◈ Without multiplying $(x - 17)(x - 18)$, explain why it cannot possibly be a factorization of $x^2 + 35x + 306$.
66. ◈ A student factors $x^3 - 8x^2 + 15x$ as $(x^2 - 5x)(x - 3)$. Is the student wrong? Why or why not? What advice would you offer the student?
67. Find all integers m for which $y^2 + my + 50$ can be factored.
68. Find all integers b for which $a^2 + ba - 50$ can be factored.

Factor completely.

69. $x^2 - \frac{1}{2}x - \frac{3}{16}$

70. $x^2 - \frac{1}{4}x - \frac{1}{8}$

71. $x^2 + \frac{30}{7}x - \frac{25}{7}$

72. $\frac{1}{3}x^3 + \frac{1}{3}x^2 - 2x$

73. $b^{2n} + 7b^n + 10$

74. $a^{2m} - 11a^m + 28$

75. $(x + 1)a^2 + (x + 1)3a + (x + 1)2$

76. $ax^2 - 5x^2 + 8ax - 40x - (a - 5)9$
(*Hint:* See Exercise 75.)

Find a polynomial in factored form for the shaded area in the figure. (Leave answers in terms of π.)

77.

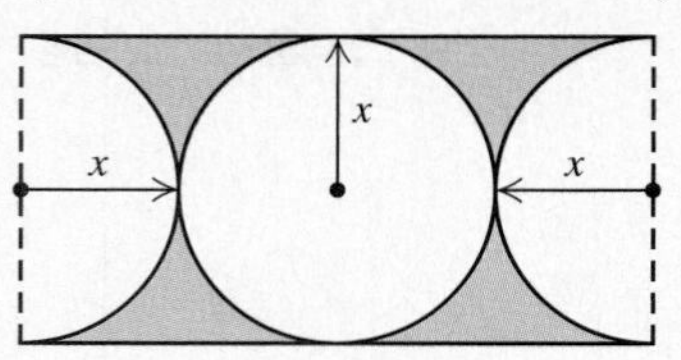

78.

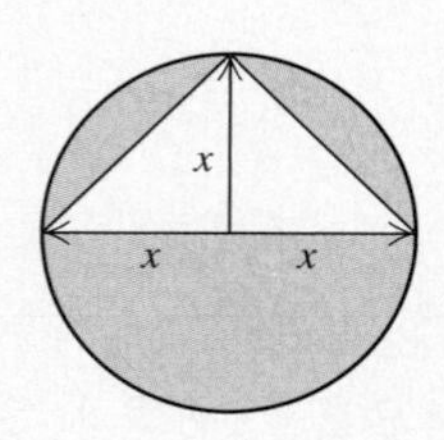

5.3 Factoring Trinomials of the Type $ax^2 + bx + c$, $a \neq 1$

Factoring with FOIL • The Grouping Method

In Section 5.2, we learned a trial-and-error method for factoring trinomials of the type $x^2 + bx + c$. Now we learn to factor trinomials in which the leading, or x^2, coefficient is not 1. First we will study the standard trial-and-error method and then we will consider an alternative method that involves factoring by grouping.

Factoring with FOIL

We want to factor trinomials of the type $ax^2 + bx + c$. Consider the following multiplication:

$$\begin{aligned} & \quad\ \ \text{F} \quad\ \text{O} \quad\ \ \text{I} \quad\ \text{L} \\ (2x + 5)(3x + 4) &= 6x^2 + 8x + 15x + 20 \\ &= 6x^2 + \quad 23x \quad + 20 \end{aligned}$$

To factor $6x^2 + 23x + 20$, we reverse the above multiplication and look for two binomials whose product is this trinomial. The product of the First terms must be $6x^2$. The product of the Outside terms plus the product of the Inside terms must be $23x$. The product of the Last terms must be 20. We know from the preceding discussion that the answer is

$$(2x + 5)(3x + 4).$$

Generally, however, finding such an answer is a trial-and-error process. It turns out that $(-2x - 5)(-3x - 4)$ is also a correct answer, but we usually choose an answer in which the first coefficients are positive.

We will use the following trial-and-error method.

To factor $ax^2 + bx + c$, $a \neq 1$, using the FOIL method:

1. Factor out the largest common factor, if one exists.
2. Find two First terms whose product is ax^2:

$$(\blacksquare x + \quad)(\blacksquare x + \quad) = ax^2 + bx + c.$$

FOIL

3. Find two Last terms whose product is c:

$$(\quad x + \blacksquare)(\quad x + \blacksquare) = ax^2 + bx + c.$$

FOIL

4. Redo steps (2) and (3) until a combination is found for which the sum of the Outer and Inner products is bx:

$$(\blacksquare x + \blacksquare)(\blacksquare x + \blacksquare) = ax^2 + bx + c.$$

I O FOIL

EXAMPLE 1

Factor: $3x^2 - 10x - 8$.

Solution

1) First, factor out a common factor, if any. There is none (other than 1 or -1).

2) Find two **First** terms whose product is $3x^2$.

The only possibilities for the **First** terms are $3x$ and x, so any factorization must be of the form

$$(3x + \quad)(x + \quad).$$

3) Find two **Last** terms whose product is -8.

Possible factorizations of -8 are

$$(-8)\cdot 1, \quad 8\cdot(-1), \quad (-2)\cdot 4, \quad \text{and} \quad 2\cdot(-4).$$

Since the First terms are not identical, we must also consider

$$1\cdot(-8), \quad (-1)\cdot 8, \quad 4\cdot(-2), \quad \text{and} \quad (-4)\cdot 2.$$

4) Inspect the **Outer** and **Inner** products resulting from steps (2) and (3). Look for a combination in which the sum of the products is the middle term, $-10x$:

Trial	*Product*	
$(3x - 8)(x + 1)$	$3x^2 + 3x - 8x - 8$	
	$= 3x^2 - 5x - 8$	← Wrong middle term
$(3x + 8)(x - 1)$	$3x^2 - 3x + 8x - 8$	
	$= 3x^2 + 5x - 8$	← Wrong middle term
$(3x - 2)(x + 4)$	$3x^2 + 12x - 2x - 8$	
	$= 3x^2 + 10x - 8$	← Wrong middle term

(*continued*)

$(3x + 2)(x - 4)$	$3x^2 - 12x + 2x - 8$	
	$= 3x^2 - 10x - 8$	← Correct middle term!
$(3x + 1)(x - 8)$	$3x^2 - 24x + x - 8$	
	$= 3x^2 - 23x - 8$	← Wrong middle term
$(3x - 1)(x + 8)$	$3x^2 + 24x - x - 8$	
	$= 3x^2 + 23x - 8$	← Wrong middle term
$(3x + 4)(x - 2)$	$3x^2 - 6x + 4x - 8$	
	$= 3x^2 - 2x - 8$	← Wrong middle term
$(3x - 4)(x + 2)$	$3x^2 + 6x - 4x - 8$	
	$= 3x^2 + 2x - 8$	← Wrong middle term

The correct factorization is $(3x + 2)(x - 4)$. ❑

Two observations can be made from Example 1. First, we listed all possible trials even though we could have stopped after finding the correct factorization. We did this to show that each trial differs only in the middle term of the product. Second, note that as in Section 5.2, only the sign of the middle term changes when the signs in the binomials are reversed.

> If a trial produces a middle term that is the opposite of the term we want, we need only change the signs in the two binomials.

EXAMPLE 2

Factor: $24x^2 - 76x + 40$.

Solution

1) First we factor out the largest common factor, 4:

$$4(6x^2 - 19x + 10).$$

Now we factor the trinomial $6x^2 - 19x + 10$.

2) Because $6x^2$ can be factored as $3x \cdot 2x$ or $6x \cdot x$, we have these possibilities for factorizations:

$$(3x + \quad)(2x + \quad) \quad \text{or} \quad (6x + \quad)(x + \quad).$$

3) There are four pairs of factors of 10 and they each can be listed in two ways:

$$10, 1 \qquad -10, -1 \qquad 5, 2 \qquad -5, -2$$

and

$$1, 10 \qquad -1, -10 \qquad 2, 5 \qquad -2, -5.$$

4) The two possibilities from step (2) and the eight possibilities from step (3) give $2 \cdot 8$, or 16 possibilities for factorizations. We look for **O**uter and **I**nner products resulting from steps (2) and (3) for which the sum is the middle term, $-19x$. Since the sign of the middle term is negative, but the sign of the last term, 10, is positive, the signs of both factors of the last term, 10, must be negative. This means only four pairings from step (3) need be considered. We first try these factors with $(3x + \quad)(2x + \quad)$. If none gives the correct factorization, we will consider $(6x + \quad)(x + \quad)$.

Trial	*Product*	
$(3x - 10)(2x - 1)$	$6x^2 - 3x - 20x + 10$	
	$= 6x^2 - 23x + 10$	← Wrong middle term
$(3x - 1)(2x - 10)$	$6x^2 - 30x - 2x + 10$	
	$= 6x^2 - 32x + 10$	← Wrong middle term
$(3x - 5)(2x - 2)$	$6x^2 - 6x - 10x + 10$	
	$= 6x^2 - 16x + 10$	← Wrong middle term
$(3x - 2)(2x - 5)$	$6x^2 - 15x - 4x + 10$	
	$= 6x^2 - 19x + 10$	← Correct middle term!

Since we have a correct factorization, we need not consider

$$(6x + \quad)(x + \quad).$$

Look again at the possibility $(3x - 5)(2x - 2)$. Without multiplying, we can reject such a possibility. To see why, consider the following:

$$(3x - 5)(2x - 2) = 2(3x - 5)(x - 1).$$

The expression $2x - 2$ has a common factor, 2. But we removed the *largest* common factor before we began. If this expression were a factorization, then 2 would have to be a common factor in addition to the original 4. Thus, $(2x - 2)$ cannot be part of the factorization of the original trinomial.

> Given that we factored out the largest common factor at the outset, we can eliminate factorizations that have a common factor.

The factorization of $6x^2 - 19x + 10$ is $(3x - 2)(2x - 5)$, but *do not forget the common factor*! We must include it in order to factor the original trinomial:

$$24x^2 - 76x + 40 = 4(6x^2 - 19x + 10)$$
$$= 4(3x - 2)(2x - 5).$$

❑

EXAMPLE 3

Factor: $10x^2 + 37x + 7$.

Solution

1) There is no common factor (other than 1 or -1).

2) Because $10x^2$ factors as $10x \cdot x$ or $5x \cdot 2x$, we have these possibilities for factorizations:

$$(10x + \quad)(x + \quad)$$

and

$$(5x + \quad)(2x + \quad).$$

3) There are two pairs of factors of 7 and they each can be listed in two ways:

$$1, 7 \qquad -1, -7$$

and

$$7, 1 \qquad -7, -1.$$

4) From steps (2) and (3), we see that there are 8 possibilities for factorizations. Look for **O**uter and **I**nner products for which the sum is the middle term. Because all coefficients in $10x^2 + 37x + 7$ are positive, we need consider only positive factors of 7. The possibilities are

$$(10x + 1)(x + 7) = 10x^2 + 71x + 7,$$
$$(10x + 7)(x + 1) = 10x^2 + 17x + 7,$$
$$(5x + 7)(2x + 1) = 10x^2 + 19x + 7,$$
$$(5x + 1)(2x + 7) = 10x^2 + 37x + 7.$$

The factorization is $(5x + 1)(2x + 7)$. ❑

Tips for factoring $ax^2 + bx + c$, $a \neq 1$:

1. Reversing the signs in the binomials reverses the sign of the middle term.
2. If the largest common factor has been factored out of the original trinomial, then no binomial factor can have a common factor (other than 1 or -1).
3. If c is positive, then the signs in both binomial factors must match the sign of b.
4. Be systematic about your trials. Keep track of those you have tried and those you have not.

Keep in mind that this method of factoring trinomials of the type $ax^2 + bx + c$ involves trial and error. As you practice, you will find that you can make fewer trials. Don't forget: When factoring any polynomial, always look first for a common factor. Failure to do so is such a common error that this caution bears repeating.

EXAMPLE 4

Factor: $6p^2 - 13pq - 28q^2$.

Solution Since no common factor exists, we examine the first term, $6p^2$. We have these possibilities:

$$(2p + \quad)(3p + \quad) \quad \text{and} \quad (6p + \quad)(p + \quad).$$

The last term, $-28q^2$, has the following pairs of factors:

$$28q, -q \qquad 14q, -2q \qquad 7q, -4q$$

and

$$-28q, q \qquad -14q, 2q \qquad -7q, 4q,$$

as well as each of the pairings reversed.

Some trials, like $(2p + 28q)(3p - q)$ and $(2p + 14q)(3p - 2q)$, cannot be correct because they contain a common factor, 2. We try $(2p + 7q)(3p - 4q)$:

$$\begin{aligned}(2p + 7q)(3p - 4q) &= 6p^2 - 8pq + 21pq - 28q^2 \\ &= 6p^2 + 13pq - 28q^2.\end{aligned}$$

Our trial is incorrect, but only because of the sign of the middle term. The correct factorization is found by changing the signs in the binomials:

$$(2p - 7q)(3p + 4q).$$

Check: $(2p - 7q)(3p + 4q) = 6p^2 - 13pq - 28q^2$. ❑

The Grouping Method

Another method of factoring trinomials of the type $ax^2 + bx + c$ is known as the *grouping method*. The grouping method relies on finding two numbers, p and q, for which $p + q = b$ and $ax^2 + px + qx + c$ can be factored by grouping. The method is outlined as follows.

To factor $ax^2 + bx + c$, $a \neq 1$, using the grouping method:

1. Factor out the largest common factor, if one exists.
2. Multiply the leading coefficient a and the constant c.
3. List pairs of factors of ac until you find two numbers whose product is ac and whose sum is b.
4. Rewrite the middle term. That is, write it as a sum or difference using the factors found in step (3).
5. Then factor by grouping.

EXAMPLE 5

Factor: $3x^2 - 10x - 8$.

Solution

1) First note that there is no common factor (other than 1 or -1).

2) We multiply the leading coefficient, 3, and the constant, -8:

$$3(-8) = -24.$$

3) We then look for a factorization of -24 in which the sum of the factors is the coefficient of the middle term, -10.

Pairs of Factors of −24	Sums of Factors
1, −24	−23
−1, 24	23
2, −12	−10 ← $2 + (-12) = -10$
−2, 12	10
3, −8	−5
−3, 8	5
4, −6	−2
−4, 6	2

We generally stop listing pairs of factors once we have found the one we are after.

4) Next, we express the middle term as a sum or difference using the factors found in step (3):

$$-10x = -12x + 2x.$$

5) We now factor by grouping as follows:

$$3x^2 - 10x - 8 = 3x^2 - 12x + 2x - 8$$ Substituting $-12x + 2x$ for $-10x$

$$= 3x(x - 4) + 2(x - 4)$$ Factoring by grouping; see Section 5.1

$$= (x - 4)(3x + 2).$$ Factoring out the common factor, $x - 4$

Check: $(x - 4)(3x + 2) = 3x^2 - 10x - 8$. ❑

EXAMPLE 6

Factor: $8x^3 + 22x^2 - 6x$.

Solution

1) We factor out the largest common factor, $2x$:

$$8x^3 + 22x^2 - 6x = 2x(4x^2 + 11x - 3).$$

2) To factor $4x^2 + 11x - 3$ by grouping, we multiply the leading coefficient, 4, and the constant term, -3:

$$4(-3) = -12.$$

3) We next look for factors of -12 that add to 11.

Pairs of Factors of -12	Sums of Factors
$1, -12$	-11
$-1, 12$	11 ← Since $-1 + 12 = 11$, there is no need to list other pairs of factors.
⋮	⋮

4) We then rewrite the middle term, $11x$, as follows:

$$11x = 12x - 1x.$$

5) Next, we factor by grouping:

$$4x^2 + 11x - 3 = 4x^2 + 12x - 1x - 3$$ Rewriting the middle term

$$= 4x(x + 3) - 1(x + 3)$$ Factoring by grouping. Removing -1 gives a common factor of $x + 3$.

$$= (x + 3)(4x - 1).$$ Factoring out the common factor

The factorization of $4x^2 + 11x - 3$ is $(x + 3)(4x - 1)$. But don't forget the common factor, $2x$, when giving the factorization of the original trinomial:

$$8x^3 + 22x^2 - 6x = 2x(x + 3)(4x - 1).$$ ❑

EXERCISE SET 5.3

Factor completely. If a polynomial is prime, state so.

1. $2x^2 - 7x - 4$
2. $3x^2 - x - 4$
3. $5x^2 + x - 18$
4. $3x^2 - 4x - 15$
5. $6x^2 + 23x + 7$
6. $6x^2 + 13x + 6$
7. $3x^2 + 4x + 1$
8. $7x^2 + 15x + 2$
9. $4x^2 + 4x - 15$
10. $9x^2 + 6x - 8$
11. $2x^2 - x - 1$
12. $15x^2 - 19x - 10$
13. $9x^2 + 18x - 16$
14. $2x^2 + 5x + 2$
15. $3x^2 - 5x - 2$
16. $18x^2 - 3x - 10$
17. $12x^2 + 31x + 20$
18. $15x^2 + 19x - 10$
19. $14x^2 + 19x - 3$
20. $35x^2 + 34x + 8$
21. $9x^2 + 18x + 8$
22. $4 - 13x + 6x^2$
23. $49 - 42x + 9x^2$
24. $25x^2 + 40x + 16$
25. $24x^2 + 47x - 2$
26. $16a^2 + 78a + 27$
27. $35x^2 - 57x - 44$
28. $9a^2 + 12a - 5$
29. $2x^2 - 6x - 19$
30. $2x^2 - x - 15$
31. $12x^2 + 28x - 24$
32. $6x^2 + 33x + 15$
33. $30x^2 - 24x - 54$
34. $20x^2 - 25x + 5$

35. $4x + 6x^2 - 10$
36. $-9 + 18x^2 - 21x$
37. $3x^2 - 4x + 1$
38. $6x^2 - 13x + 6$
39. $12x^2 - 28x - 24$
40. $6x^2 - 33x + 15$
41. $-1 + 2x^2 - x$
42. $-19x + 15x^2 + 6$
43. $9x^2 - 18x - 16$
44. $14x^2 + 35x + 14$
45. $15x^2 - 25x - 10$
46. $18x^2 + 3x - 10$
47. $12x^3 + 31x^2 + 20x$
48. $15x^3 + 19x^2 - 10x$
49. $14x^4 + 19x^3 - 3x^2$
50. $70x^4 + 68x^3 + 16x^2$
51. $168x^3 - 45x^2 + 3x$
52. $144x^5 + 168x^4 + 48x^3$
53. $15x^2 - 19x + 6$
54. $9x^2 + 18x + 8$
55. $25t^2 + 80t + 64$
56. $9x^2 - 42x + 49$
57. $6x^3 + 4x^2 - 10x$
58. $18x^3 - 21x^2 - 9x$
59. $25x^2 + 89x + 64$
60. $9y^2 - 42y + 47$
61. $x^2 + 3x - 7$
62. $x^2 + 13x - 12$
63. $12m^2 + mn - 20n^2$
64. $12a^2 + 17ab + 6b^2$
65. $6a^2 - ab - 15b^2$
66. $3p^2 - 16pq - 12q^2$
67. $9a^2 + 18ab + 8b^2$
68. $10s^2 + 4st - 6t^2$
69. $35p^2 + 34pq + 8q^2$
70. $30a^2 + 87ab + 30b^2$
71. $18x^2 - 6xy - 24y^2$
72. $15a^2 - 5ab - 20b^2$

Factor. Use factoring by grouping even though it would seem reasonable to first collect like terms.

73. $y^2 + 4y + y + 4$
74. $x^2 + 5x + 2x + 10$
75. $x^2 - 4x - x + 4$
76. $a^2 + 5a - 2a - 10$
77. $6x^2 + 4x + 9x + 6$
78. $3x^2 - 2x + 3x - 2$
79. $3x^2 - 4x - 12x + 16$
80. $24 - 18y - 20y + 15y^2$
81. $35x^2 - 40x + 21x - 24$
82. $8x^2 - 6x - 28x + 21$
83. $4x^2 + 6x - 6x - 9$
84. $2x^4 - 6x^2 - 5x^2 + 15$

Factor by grouping. If a polynomial is prime, state so.

85. $2x^2 - 7x - 4$
86. $3x^2 - x - 4$
87. $5x^2 + x - 18$
88. $3x^2 - 4x - 15$
89. $6x^2 + 23x + 7$
90. $6x^2 + 13x + 6$
91. $3x^2 + 4x + 1$
92. $7x^2 + 15x + 2$
93. $4x^2 + 4x - 15$
94. $9x^2 + 6x - 8$
95. $2x^2 - x - 1$
96. $15x^2 - 19x - 10$
97. $9x^2 + 18x - 16$
98. $2x^2 + 5x + 2$
99. $3x^2 - 5x - 2$
100. $18x^2 - 3x - 10$
101. $12x^2 + 31x + 20$
102. $15x^2 + 19x - 10$
103. $14x^2 + 19x - 3$
104. $35x^2 + 34x + 8$
105. $9x^2 + 18x + 8$
106. $6 - 13x + 6x^2$
107. $49 - 42x + 9x^2$
108. $25x^2 + 40x + 16$

Skill Maintenance

109. The earth is a sphere (or ball) that is about 40,000 km in circumference. Find the radius of the earth, in kilometers and in miles. Use 3.14 for π. (*Hint:* 1 km $\approx$ 0.62 mi.)

110. The second angle of a triangle is 10° less than twice the first. The third angle is 15° more than four times the first. Find the measure of the second angle.

111. Graph: $y = \frac{2}{5}x - 1$.

112. Divide: $\frac{y^{12}}{y^4}$.

Synthesis

113. ◈ A student presents the following work:

$$4x^2 + 28x + 48 = (2x + 6)(2x + 8)$$
$$= 2(x + 3)(x + 4).$$

Is this correct? Explain.

114. ◈ If a trinomial's leading coefficient and constant term are both prime numbers, at most how many trials can be made when factoring the trinomial? Why?

Factor.

115. $9x^{10} - 12x^5 + 4$
116. $16x^{10} + 8x^5 + 1$
117. $20x^{2n} + 16x^n + 3$
118. $-15x^{2m} + 26x^m - 8$
119. $3x^{6a} - 2x^{3a} - 1$
120. $x^{2n+1} - 2x^{n+1} + x$
121. $3(a + 1)^{n+1}(a + 3)^2 - 5(a + 1)^n(a + 3)^3$

5.4 Factoring Trinomial Squares and Differences of Squares

Recognizing Trinomial Squares • Factoring Trinomial Squares • Recognizing Differences of Squares • Factoring Differences of Squares • Factoring Completely

In Chapter 4, we studied some shortcuts for finding certain products of binomials. We now reverse these procedures to discover shortcuts for factoring certain polynomials.

Recognizing Trinomial Squares

Some trinomials are squares of binomials. For example, the trinomial $x^2 + 10x + 25$ is the square of the binomial $x + 5$. To see this, we can calculate $(x + 5)^2$. It is $x^2 + 2 \cdot x \cdot 5 + 5^2$, or $x^2 + 10x + 25$. A trinomial that is the square of a binomial is called a **trinomial square.**

In Chapter 4, we considered squaring binomials as a special-product rule:

$$(A + B)^2 = A^2 + 2AB + B^2;$$
$$(A - B)^2 = A^2 - 2AB + B^2.$$

Written from right to left, these equations can be used to factor trinomial squares. Note that in order for a trinomial to be the square of a binomial, we must have the following:

A. Two terms, A^2 and B^2, must be squares, such as

$$4, \quad x^2, \quad 81m^2, \quad 16t^2.$$

B. There must be no minus sign before A^2 or B^2.

C. If we multiply A and B (the square roots of A^2 and B^2) and double the result, we get the remaining term, $2 \cdot A \cdot B$, or its opposite, $-2 \cdot A \cdot B$.

EXAMPLE 1

Determine whether $x^2 + 6x + 9$ is a trinomial square.

Solution

A. We know that x^2 and 9 are squares.

B. There is no minus sign before x^2 or 9.

C. If we multiply the square roots, x and 3, and double the product, we get the remaining term: $2 \cdot x \cdot 3 = 6x$.

Thus, $x^2 + 6x + 9$ is a trinomial square. ❑

EXAMPLE 2

Determine whether $x^2 + 6x + 11$ is a trinomial square.

Solution The answer is no, because only one term is a square. ❑

EXAMPLE 3

Determine whether $16x^2 + 49 - 56x$ is a trinomial square.

Solution It helps to first write the trinomial in descending order:

$$16x^2 - 56x + 49.$$

A. We know that $16x^2$ and 49 are squares.
B. There is no minus sign before $16x^2$ or 49.
C. If we multiply the square roots, $4x$ and 7, and double the product, we get

$$2 \cdot 4x \cdot 7, \quad \text{or} \quad 56x;$$

$56x$ is the opposite of the remaining term.

Thus, $16x^2 + 49 - 56x$ is a trinomial square. □

Factoring Trinomial Squares

We can use the trial-and-error or grouping methods from Sections 5.2 and 5.3 to factor trinomial squares, but a faster method uses the following equations:

$$A^2 + 2AB + B^2 = (A + B)^2;$$
$$A^2 - 2AB + B^2 = (A - B)^2.$$

The factorization uses the square roots of the squared terms and the sign of the remaining term.

EXAMPLE 4

Factor: **(a)** $x^2 + 6x + 9$; **(b)** $x^2 + 49 - 14x$; **(c)** $16x^2 - 40x + 25$.

Solution

a) $x^2 + 6x + 9 = x^2 + 2 \cdot x \cdot 3 + 3^2 = (x + 3)^2$ The sign of the middle term is positive.

$$\mathbf{A^2 + 2\ A\ B + B^2 = (A + B)^2}$$

b) $x^2 + 49 - 14x = x^2 - 14x + 49$ Using a commutative law to write descending order

$= x^2 - 2 \cdot x \cdot 7 + 7^2$ The sign of the middle term is negative.

$= (x - 7)^2$ Factoring the trinomial square

c) $16x^2 - 40x + 25 = (4x)^2 - 2 \cdot 4x \cdot 5 + 5^2 = (4x - 5)^2$

$$\mathbf{A^2 - 2\ A\ B + B^2 = (A - B)^2}$$ □

With practice, you will be able to spot trinomial squares whenever they occur and factor them quickly.

EXAMPLE 5

Factor: **(a)** $4p^2 - 12pq + 9q^2$; **(b)** $75m^3 + 60m^2 + 12m$.

Solution

a) $4p^2 - 12pq + 9q^2 = (2p)^2 - 2(2p)(3q) + (3q)^2$ Recognizing the trinomial square

$= (2p - 3q)^2$ The sign of the middle term is negative.

Check: $(2p - 3q)(2p - 3q) = 4p^2 - 12pq + 9q^2.$

b) *Always* look first for a common factor. This time there is one, $3m$:

$$75m^3 + 60m^2 + 12m = 3m[25m^2 + 20m + 4]$$ Factoring out the largest common factor

$$= 3m[(5m)^2 + 2(5m)(2) + 2^2]$$ Recognizing the trinomial square. Try to do this mentally.

$$= 3m(5m + 2)^2.$$

Check: $3m(5m + 2)^2 = 3m(5m + 2)(5m + 2)$
$= 3m(25m^2 + 20m + 4)$
$= 75m^3 + 60m^2 + 12m.$ ❑

Recognizing Differences of Squares

Some binomials represent the difference of two squares. For example, the binomial $16x^2 - 9$ is a difference of two expressions, $16x^2$ and 9, that are squares. To see this, note that $16x^2 = (4x)^2$ and $9 = 3^2$.

Any expression, like $16x^2 - 9$, that can be written in the form $A^2 - B^2$ is called a **difference of squares.** Note that in order for a binomial to be a difference of squares, we must have the following:

A. There must be two expressions, both squares, such as

$$4x^2, \quad 9, \quad 4x^2y^2, \quad 1, \quad x^6, \quad 49y^8.$$

B. The terms in the binomial must have different signs.

Note that in order for a term to be a square, its coefficient must be a perfect square and the power(s) of the variable(s) must be even.

EXAMPLE 6

Is $9x^2 - 64$ a difference of squares?

Solution

A. The first expression is a square: $9x^2 = (3x)^2$.
The second expression is a square: $64 = 8^2$.
B. The terms have different signs.

Thus we have a difference of squares, $(3x)^2 - 8^2$. ❑

EXAMPLE 7

Is $25 - t^3$ a difference of squares?

Solution

A. The expression t^3 is not a square.

Thus, $25 - t^3$ is not a difference of squares. ❑

EXAMPLE 8

Is $-4x^2 + 16$ a difference of squares?

Solution

A. The expressions $4x^2$ and 16 are squares: $4x^2 = (2x)^2$ and $16 = 4^2$.
B. The terms have different signs.

Thus we have a difference of squares. We can also see this by rewriting in the equivalent form: $16 - 4x^2$. ❑

Factoring Differences of Squares

To factor a difference of squares, we use an equation that first appeared, written from right to left, in Chapter 4:

$$A^2 - B^2 = (A + B)(A - B).$$

To factor a difference of squares $A^2 - B^2$, we first find the square roots A and B. Then we use A and B to form two factors. One factor is $A + B$, and the other is $A - B$.

EXAMPLE 9

Factor: **(a)** $x^2 - 4$; **(b)** $m^2 - 9p^2$.

Solution

a) $x^2 - 4 = x^2 - 2^2 = (x + 2)(x - 2)$

$$A^2 - B^2 = (A + B)(A - B)$$

b) $m^2 - 9p^2 = m^2 - (3p)^2 = (m + 3p)(m - 3p)$

$$A^2 - B^2 = (A + B)(A - B)$$ ❑

When powers larger than 2 arise, it is important to remember that $(a^r)^s = a^{r \cdot s}$ and $a^n \cdot a^m = a^{n+m}$.

EXAMPLE 10

Factor: **(a)** $9 - 16t^{10}$; **(b)** $18x^2 - 50x^6$; **(c)** $49x^4 - 9x^6$.

Solution

a) $9 - 16t^{10} = 3^2 - (4t^5)^2$ — Using the rules for powers

$$A^2 - B^2$$

$= (3 + 4t^5)(3 - 4t^5)$ — Try to go directly to this step.

$$(A + B)(A - B)$$

b) *Always* look first for a common factor. This time there is one, $2x^2$:

$18x^2 - 50x^6 = 2x^2(9 - 25x^4)$ — Factoring out the largest common factor

$= 2x^2[3^2 - (5x^2)^2]$ — Recognizing $A^2 - B^2$. Try to do this mentally.

$= 2x^2(3 + 5x^2)(3 - 5x^2)$ — Factoring the difference of squares

Check: $2x^2(3 + 5x^2)(3 - 5x^2) = 2x^2(9 - 25x^4) = 18x^2 - 50x^6$.

c) $49x^4 - 9x^6 = x^4(49 - 9x^2)$ — Factoring out the largest common factor

$= x^4(7 + 3x)(7 - 3x)$ — Factoring the difference of squares

Check: $x^4(7 + 3x)(7 - 3x) = x^4(49 - 9x^2) = 49x^4 - 9x^6$. ❑

CAUTION! Note carefully in these examples that a difference of squares is *not* the square of the difference; that is,

$$A^2 - B^2 \neq (A - B)^2.$$

For example,

$$8^2 - 3^2 = 64 - 9 = 55,$$

but

$$(8 - 3)^2 = 5^2 = 25.$$

Factoring Completely

If a factor with more than one term can still be factored, you should do so. When no factor can be factored further, you have factored completely. Always factor completely whenever you are asked to factor.

EXAMPLE 11

Factor: $p^4 - 16$.

Solution

$p^4 - 16 = (p^2)^2 - 4^2$

$= (p^2 + 4)(p^2 - 4)$ — Factoring a difference of squares

$= (p^2 + 4)(p + 2)(p - 2)$ — Factoring further. The factor $p^2 - 4$ is itself a difference of squares.

Check: $(p^2 + 4)(p + 2)(p - 2) = (p^2 + 4)(p^2 - 4) = p^4 - 16$. ❑

Observe in Example 11 that the factor $p^2 + 4$ is a *sum* of squares that cannot be factored further.

CAUTION! If the largest common factor has been removed, then you cannot factor a sum of squares further. In particular,

$$A^2 + B^2 \neq (A + B)^2.$$

Consider $25x^2 + 100$. This is a case in which we have a sum of squares with a common factor, 25. Factoring, we get $25(x^2 + 4)$, where $x^2 + 4$ is prime.

As you proceed through the exercises, these suggestions may prove helpful.

1. Always look first for a common factor! If there is one, factor it out.
2. Be alert for trinomial squares and differences of squares. Once recognized, they can be factored without trial and error.
3. Always factor completely.
4. Check by multiplying.

EXERCISE SET 5.4

Determine whether each of the following is a trinomial square.

1. $x^2 - 14x + 49$
2. $x^2 - 16x + 64$
3. $x^2 + 16x - 64$
4. $x^2 - 14x - 49$
5. $x^2 - 3x + 9$
6. $x^2 + 2x + 4$
7. $9x^2 - 36x + 24$
8. $36x^2 - 24x + 16$

Factor completely. Remember to look first for a common factor and to check by multiplying.

9. $x^2 - 14x + 49$
10. $x^2 - 16x + 64$
11. $x^2 + 16x + 64$
12. $x^2 + 14x + 49$
13. $x^2 - 2x + 1$
14. $x^2 + 2x + 1$
15. $4 + 4x + x^2$
16. $4 + x^2 - 4x$
17. $9x^2 + 6x + 1$
18. $25x^2 - 10x + 1$
19. $49 - 56y + 16y^2$
20. $120m + 75 + 48m^2$
21. $2x^2 - 4x + 2$
22. $2x^2 - 40x + 200$
23. $x^3 - 18x^2 + 81x$
24. $x^3 + 24x^2 + 144x$
25. $20x^2 + 100x + 125$
26. $12x^2 + 36x + 27$
27. $49 - 42x + 9x^2$
28. $64 - 112x + 49x^2$
29. $5y^2 + 10y + 5$
30. $2a^2 + 28a + 98$
31. $2 + 20x + 50x^2$
32. $7 - 14a + 7a^2$
33. $4p^2 + 12pq + 9q^2$
34. $25m^2 + 20mn + 4n^2$
35. $a^2 - 14ab + 49b^2$
36. $x^2 - 6xy + 9y^2$
37. $64m^2 + 16mn + n^2$
38. $81p^2 - 18pq + q^2$
39. $16s^2 - 40st + 25t^2$
40. $36a^2 + 96ab + 64b^2$

Determine whether each of the following is a difference of squares.

41. $x^2 - 4$
42. $x^2 - 36$
43. $x^2 + 36$
44. $x^2 + 4$
45. $x^2 - 35$
46. $x^2 - 50y^2$
47. $16x^2 - 25y^2$
48. $-1 + 36x^2$

Factor completely. Remember to look first for a common factor.

49. $y^2 - 4$
50. $x^2 - 36$
51. $p^2 - 9$
52. $q^2 - 1$
53. $-49 + t^2$
54. $-64 + m^2$
55. $a^2 - b^2$
56. $p^2 - q^2$
57. $25t^2 - m^2$
58. $w^2 - 49z^2$
59. $100 - k^2$
60. $81 - w^2$
61. $16a^2 - 9$
62. $25x^2 - 4$
63. $4x^2 - 25y^2$
64. $9a^2 - 16b^2$
65. $8x^2 - 98$
66. $24x^2 - 54$
67. $36x - 49x^3$
68. $16x - 81x^3$
69. $49a^4 - 81$
70. $25a^4 - 9$
71. $x^4 - 1$
72. $x^4 - 16$
73. $4x^4 - 64$
74. $5x^4 - 80$
75. $1 - y^8$
76. $x^8 - 1$
77. $3x^3 - 24x^2 + 48x$
78. $2a^4 - 36a^3 + 162a^2$
79. $x^{12} - 16$
80. $x^8 - 81$
81. $y^2 - \frac{1}{16}$
82. $x^2 - \frac{1}{25}$
83. $a^8 - 2a^7 + a^6$
84. $x^8 - 8x^7 + 16x^6$
85. $25 - \frac{1}{49}x^2$
86. $4 - \frac{1}{9}y^2$
87. $16m^4 - t^4$
88. $1 - a^4b^4$

Skill Maintenance

89. Bonnie is taking an astronomy course. To get an A, a student must average at least 90 after four hour exams. Bonnie scored 96, 98, and 89 on the first three tests. Determine (in terms of an inequality) what scores on the last test will earn her an A.

90. About 5 L of oxygen can be dissolved in 100 L of water at 0°C. This is 1.6 times the amount that can be dissolved in the same volume of water at 20°C. How much oxygen can be dissolved at 20°C?

Simplify.

91. $(x^3y^5)(x^9y^7)$
92. $(5a^2b^3)^2$

Synthesis

93. ◈ A student concludes that since $x^2 - 9 = (x - 3)(x + 3)$, it must follow that $x^2 + 9 = (x + 3)(x - 3)$. What mistake(s) is the student making?

94. ◈ Explain in your own words how to determine if a polynomial is a trinomial square.

Factor completely. If a polynomial is prime, state so.

95. $49x^2 - 216$
96. $x^2 - 5x + 25$
97. $18x^3 + 12x^2 + 2x$
98. $162x^2 - 82$
99. $x^8 - 2^8$
100. $4x^4 - 4x^2$

101. $3x^5 - 12x^3$

102. $3x^2 - \frac{1}{3}$

103. $18x^3 - \frac{8}{25}x$

104. $x^2 - 2.25$

105. $0.49p - p^3$

106. $0.64x^2 - 1.21$

107. $(x + 3)^2 - 9$

108. $(y - 5)^2 - 36q^2$

109. $x^2 - \left(\frac{1}{x}\right)^2$

110. $a^{2n} - 49b^{2n}$

111. $81 - b^{4k}$

112. $x^4 - 8x^2 - 9$

113. $9b^{2n} + 12b^n + 4$

114. $16x^4 - 96x^2 + 144$

115. $(y + 3)^2 + 2(y + 3) + 1$

116. $49(x + 1)^2 - 42(x + 1) + 9$

117. $27x^3 - 63x^2 - 147x + 343$

118. Subtract $(x^2 + 1)^2$ from $x^2(x + 1)^2$ and factor the result.

Factor by grouping. Look for a grouping of three terms that is a trinomial square.

119. $a^2 + 2a + 1 - 9$

120. $y^2 + 6y + 9 - x^2 - 8x - 16$

Find c so that the polynomial will be the square of a binomial.

121. $cy^2 + 6y + 1$

122. $cy^2 - 24y + 9$

123. Show that the difference of the squares of two consecutive integers is the sum of the integers. (*Hint:* Use x for the smaller number.)

124. Find the value of a if $x^2 + a^2x + a^2$ factors into $(x + a)^2$.

5.5 Factoring: A General Strategy

Choosing the Right Method • Checking by Evaluating

We now combine all of our factoring techniques and consider a general strategy for factoring polynomials. Here we will encounter polynomials of all the types we have considered, in random order, so you will have to determine which method to use.

To factor a polynomial:

A. Always look first for a common factor. If there is one, factor out the largest common factor. Be sure to include it in your final answer.

B. Then look at the number of terms.

Two terms: If you have a difference of squares, factor accordingly. Do not try to factor a sum of squares: $A^2 + B^2$.

Three terms: Determine whether the trinomial is a square. If so, factor accordingly. If not, try trial and error, using the standard method or grouping.

Four terms: Try factoring by grouping.

C. Always *factor completely*. If a factor with more than one term can still be factored, you should do so.

D. Check by multiplying.

EXAMPLE 1

Factor: $5t^4 - 80$.

Solution

A. We look for a common factor:

$$5t^4 - 80 = 5(t^4 - 16).$$

B. The factor $t^4 - 16$ has only two terms. It is a difference of squares: $(t^2)^2 - 4^2$. We factor it, being careful to rewrite the common factor:

$$5t^4 - 80 = 5(t^2 + 4)(t^2 - 4).$$

C. We see that one of the factors is again a difference of squares. We factor it:

$$5t^4 - 80 = 5(t^2 + 4)(t - 2)(t + 2).$$

↑ This is a sum of squares. It cannot be factored!

We have factored completely because no factor with more than one term can be factored further.

D. *Check:* $5(t^2 + 4)(t - 2)(t + 2) = 5(t^2 + 4)(t^2 - 4)$
$= 5(t^4 - 16) = 5t^4 - 80.$ ❑

EXAMPLE 2

Factor: $2x^3 + 10x^2 + x + 5$.

Solution

A. We look for a common factor. There isn't one.

B. There are four terms. We try factoring by grouping:

$2x^3 + 10x^2 + x + 5$

$= (2x^3 + 10x^2) + (x + 5)$ — Separating into two binomials

$= 2x^2(x + 5) + 1(x + 5)$ — Factoring out the largest common factor from each binomial

$= (x + 5)(2x^2 + 1)$ — Factoring out the common factor, $x + 5$

C. No factor with more than one term can be factored further, so we have factored completely.

D. *Check:* $(x + 5)(2x^2 + 1) = 2x^3 + x + 10x^2 + 5 = 2x^3 + 10x^2 + x + 5.$ ❑

EXAMPLE 3

Factor: $x^5 - 2x^4 - 35x^3$.

Solution

A. We look first for a common factor. This time there is one, x^3:

$$x^5 - 2x^4 - 35x^3 = x^3(x^2 - 2x - 35).$$

B. The factor $x^2 - 2x - 35$ has three terms, but it is not a trinomial square. We factor it using trial and error:

$x^5 - 2x^4 - 35x^3 = x^3(x^2 - 2x - 35)$

$= x^3(x - 7)(x + 5).$ — Don't forget to rewrite the common factor.

C. No factor with more than one term can be factored further, so we have factored completely.

D. *Check:* $x^3(x - 7)(x + 5) = x^3(x^2 - 2x - 35) = x^5 - 2x^4 - 35x^3.$ ❑

EXAMPLE 4

Factor: $x^2 - 20x + 100$.

Solution

A. We look first for a common factor. There isn't one.

B. There are three terms. This polynomial is a trinomial square, so we factor it accordingly:

$$x^2 - 20x + 100 = x^2 - 2 \cdot x \cdot 10 + 10^2 \quad \text{Try to do this step mentally.}$$
$$= (x - 10)^2.$$

C. No factor with more than one term can be factored further, so we have factored completely.

D. *Check:* $(x - 10)(x - 10) = x^2 - 20x + 100$. ❑

EXAMPLE 5

Factor: $6x^2y^4 - 21x^3y^5 + 3x^2y^6$.

Solution

A. We look first for a common factor:

$$6x^2y^4 - 21x^3y^5 + 3x^2y^6 = 3x^2y^4(2 - 7xy + y^2).$$

B. There are three terms in $2 - 7xy + y^2$. Since only y^2 is a square, we do not have a trinomial square. Can the trinomial be factored by trial and error? A key to the answer is that x is only in the term $-7xy$. If the polynomial factored into a form like $(1 - y)(2 - y)$, there would be no x in the middle term. Thus, $2 - 7xy + y^2$ cannot be factored.

C. Have we factored completely? Yes, because no factor with more than one term can be factored further.

D. *Check:* $3x^2y^4(2 - 7xy + y^2) = 6x^2y^4 - 21x^3y^5 + 3x^2y^6$. ❑

EXAMPLE 6

Factor: $(p + q)(x + 2) + (p + q)(x + y)$.

Solution

A. We look for a common factor:

$$(p + q)(x + 2) + (p + q)(x + y) = (p + q)[(x + 2) + (x + y)]$$
$$= (p + q)(2x + y + 2). \quad \text{Collecting like terms}$$

B. There are three terms in $2x + y + 2$, but this trinomial cannot be factored further.

C. No factor with more than one term can be factored further, so we have factored completely.

D. To check, the student can reverse the steps in part (A). ❑

EXAMPLE 7

Factor: $px + py + qx + qy$.

Solution

A. We look first for a common factor. There isn't one.

B. There are four terms. We try factoring by grouping:

$$px + py + qx + qy = p(x + y) + q(x + y)$$
$$= (x + y)(p + q).$$

C. Since no factor with more than one term can be factored further, we have factored completely.

D. *Check:* $(x+y)(p+q) = xp + xq + yp + yq = px + py + qx + qy$. ❑

EXAMPLE 8

Factor: $25x^2 + 20xy + 4y^2$.

Solution

A. We look first for a common factor. There isn't one.

B. There are three terms. We determine whether the trinomial is a square. The first term and the last term are squares:

$$25x^2 = (5x)^2 \quad \text{and} \quad 4y^2 = (2y)^2.$$

Since twice the product of $5x$ and $2y$ is the other term,

$$2 \cdot 5x \cdot 2y = 20xy,$$

the trinomial is a square.

We factor by writing a binomial squared. We determine the binomial by writing the square roots of the square terms and the sign of the middle term:

$$25x^2 + 20xy + 4y^2 = (5x + 2y)^2.$$

C. No factor with more than one term can be factored further, so we have factored completely.

D. *Check:* $(5x+2y)(5x+2y) = 25x^2 + 20xy + 4y^2$. ❑

EXAMPLE 9

Factor: $p^2q^2 + 7pq + 12$.

Solution

A. We look first for a common factor. There isn't one.

B. There are three terms. Since the first term is a square, but neither of the other terms is, we do not have a trinomial square. We use trial and error, thinking of the product pq as a single variable. The binomials will then have the following form:

$$(pq + \quad)(pq + \quad).$$

We factor the last term, 12. All the signs are positive, so we consider only positive factors. Possibilities are 1, 12 and 2, 6 and 3, 4. The pair 3, 4 gives a sum of 7 for the coefficient of the middle term. Thus,

$$p^2q^2 + 7pq + 12 = (pq + 3)(pq + 4).$$

C. No factor with more than one term can be factored further, so we have factored completely.

D. The check is left to the student. ❑

EXAMPLE 10

Factor: $a^4 - 16b^4$.

Solution

A. We look first for a common factor. There isn't one.

B. There are two terms. Since $a^4 = (a^2)^2$ and $16b^4 = (4b^2)^2$, we see that we do have a difference of squares. Thus,

$$a^4 - 16b^4 = (a^2 + 4b^2)(a^2 - 4b^2).$$

C. The last factor in $(a^2 + 4b^2)(a^2 - b^2)$ can be factored further. It too is a difference of squares. Thus,

$$a^4 - 16b^4 = (a^2 + 4b^2)(a + 2b)(a - 2b).$$

D. *Check:* $(a^2 + 4b^2)(a + 2b)(a - 2b) = (a^2 + 4b^2)(a^2 - 4b^2) = a^4 - 16b^4.$ ❑

Checking by Evaluating

Multiplication is but one way of checking a factorization. Another method, which serves as a partial check, is to evaluate the original polynomial and the proposed factorization using the same replacement(s). If the factorization is correct, the polynomial and the factorization will have the same value for any replacement(s) of the variable(s).

EXAMPLE 11

Check the factorization of Example 3, $x^5 - 2x^4 - 35x^3 = x^3(x - 7)(x + 5)$, by evaluating.

Solution We choose a convenient value for x, say 1, and evaluate both expressions:

$$\begin{aligned} x^5 - 2x^4 - 35x^3 &= 1^5 - 2 \cdot 1^4 - 35 \cdot 1^3 \\ &= 1 - 2 - 35 \\ &= -36; \end{aligned} \qquad \begin{aligned} x^3(x - 7)(x + 5) &= 1^3(1 - 7)(1 + 5) \\ &= 1(-6)(6) \\ &= -36. \end{aligned}$$

Since the value of both expressions is -36, the factorization is probably correct. ❑

Evaluating both a polynomial and its factorization is a quick way to check without multiplying out the factorization. Because checking by evaluating is not foolproof, however, it is a good idea to use this method as only a partial check.

EXERCISE SET 5.5

Factor completely. If a polynomial is prime, state so.

1. $2x^2 - 128$
2. $3t^2 - 27$
3. $a^2 + 25 - 10a$
4. $y^2 + 49 + 14y$
5. $2x^2 - 11x + 12$
6. $8y^2 - 18y - 5$
7. $x^3 + 24x^2 + 144x$
8. $x^3 - 18x^2 + 81x$
9. $x^3 + 3x^2 - 4x - 12$
10. $x^3 - 5x^2 - 25x + 125$
11. $24x^2 - 54$
12. $8x^2 - 98$
13. $20x^3 - 4x^2 - 72x$
14. $9x^3 + 12x^2 - 45x$
15. $x^2 + 4$
16. $t^2 + 25$
17. $x^4 + 7x^2 - 3x^3 - 21x$
18. $m^4 + 8m^3 + 8m^2 + 64m$
19. $x^5 - 14x^4 + 49x^3$
20. $2x^6 + 8x^5 + 8x^4$
21. $20 - 6x - 2x^2$
22. $45 - 3x - 6x^2$
23. $x^2 + 3x + 1$
24. $x^2 + 5x + 2$
25. $4x^4 - 64$
26. $5x^5 - 80x$
27. $t^8 - 1$
28. $1 - n^8$
29. $x^5 - 4x^4 + 3x^3$
30. $x^6 - 2x^5 + 7x^4$
31. $x^2 - y^2$
32. $p^2q^2 - r^2$
33. $12n^2 + 24n^3$
34. $ax^2 + ay^2$
35. $9x^2y^2 - 36xy$
36. $x^2y - xy^2$
37. $2\pi rh + 2\pi r^2$

38. $10p^4q^4 + 35p^3q^3 + 10p^2q^2$
39. $(a + b)(x - 3) + (a + b)(x + 4)$
40. $5c(a^3 + b) - (a^3 + b)$
41. $(x - 1)(x + 1) - y(x + 1)$
42. $x^2 + x + xy + y$
43. $n^2 + 2n + np + 2p$
44. $a^2 - 3a + ay - 3y$
45. $2x^2 - 4x + xz - 2z$
46. $6y^2 - 3y + 2py - p$
47. $x^2 + y^2 - 2xy$
48. $4b^2 + a^2 - 4ab$
49. $9c^2 + 6cd + d^2$
50. $16x^2 + 24xy + 9y^2$
51. $7p^4 - 7q^4$
52. $4x^2y^2 + 12xyz + 9z^2$
53. $25z^2 + 10zy + y^2$
54. $a^4b^4 - 16$
55. $a^5 + 4a^4b - 5a^3b^2$
56. $4p^2q + pq^2 + 4p^3$
57. $a^2 - ab - 2b^2$
58. $3b^2 - 17ab - 6a^2$
59. $2mn - 360n^2 + m^2$
60. $15 + x^2y^2 + 8xy$
61. $m^2n^2 - 4mn - 32$
62. $p^2q^2 + 7pq + 6$
63. $a^5b^2 + 3a^4b - 10a^3$
64. $m^2n^6 + 4mn^5 - 32n^4$
65. $49m^2 - 112mn + 64n^2$
66. $2s^6t^2 + 10s^3t^3 + 12t^4$
67. $x^6 + x^5y - 2x^4y^2$
68. $a^2 + 2a^2bc + a^2b^2c^2$
69. $36a^2 - 15a + \frac{25}{16}$
70. $\frac{1}{81}x^2 - \frac{8}{27}x + \frac{16}{9}$
71. $\frac{1}{4}a^2 + \frac{1}{3}ab + \frac{1}{9}b^2$
72. $0.01x^2 - 0.1xy + 0.25y^2$
73. $81a^4 - b^4$
74. $1 - 16x^{12}y^{12}$
75. $w^3 - 7w^2 - 4w + 28$
76. $y^3 + 8y^2 - y - 8$

Skill Maintenance

77. Show that the pairs $(-1, 11)$, $(0, 7)$, and $(3, -5)$ are solutions of $y = -4x + 7$.
78. Graph: $y = -\frac{1}{2}x + 4$.
79. Solve $A = aX + bX - 7$ for X.
80. Solve: $4(x - 9) - 2(x + 7) < 14$.

Synthesis

81. ◈ In your own words, describe a strategy that can be used to factor polynomials.
82. ◈ Kelly factored $16 - 8x + x^2$ as $(x - 4)^2$, while Tony factored it as $(4 - x)^2$. Evaluate each expression for several values of x. Then explain why $(x - 4)^2$ and $(4 - x)^2$ are equivalent.

Check that the factorization is most likely correct by evaluating with the values given.

83. $6x^2 - xy - 15y^2 = (2x + 3y)(3x - 5y)$; $x = 1,\ y = 1$
84. $6x^2y^2 - 23xy + 20 = (2xy - 5)(3xy - 4)$; $x = 1,\ y = -1$

Factor.

85. $18 + y^3 - 9y - 2y^2$
86. $-(x^4 - 7x^2 - 18)$
87. $a^3 + 4a^2 + a + 4$
88. $x^3 + x^2 - (4x + 4)$
89. $x^4 - 7x^2 - 18$
90. $3x^4 - 15x^2 + 12$
91. $x^3 - x^2 - 4x + 4$
92. $y^2(y + 1) - 4y(y + 1) - 21(y + 1)$
93. $y^2(y - 1) - 2y(y - 1) + (y - 1)$
94. $6(x - 1)^2 + 7y(x - 1) - 3y^2$
95. $(y + 4)^2 + 2x(y + 4) + x^2$
96. $2(a + 3)^2 - (a + 3)(b - 2) - (b - 2)^2$
97. Factor $x^{2k} - 2^{2k}$ when $k = 4$.
98. ◈ At most how many factors can a seventh-degree polynomial in x have? Why?

5.6 Solving Quadratic Equations by Factoring

The Principle of Zero Products • Factoring to Solve Equations • Graphing and Quadratic Equations

In this section, we will use the factoring skills we have just learned to solve equations like $x^2 - 8x = -16$ and $x^2 + x - 156 = 0$. Second-degree equations of this type are said to be **quadratic.**

Quadratic Equation

A *quadratic equation* is an equation equivalent to one of the form

$$ax^2 + bx + c = 0, \quad \text{where } a \neq 0.$$

The Principle of Zero Products

The product of two numbers is 0 if one or both of the numbers is 0. Furthermore, *if any product is* 0, *then at least one of the factors must be* 0. For example, if $7x = 0$, then we can conclude that x must be 0. If $x(2x - 9) = 0$, we can conclude that $x = 0$ and/or $2x - 9 = 0$. If $(x + 3)(x - 2) = 0$, we can conclude that $x + 3 = 0$ and/or $x - 2 = 0$. In a product like $ab = 24$, we cannot conclude that either factor must be a specific value.

EXAMPLE 1

Solve: $(x + 3)(x - 2) = 0$.

Solution We have a product of 0. This equation will be true when either factor is 0. Hence it is true when

$$x + 3 = 0 \quad or \quad x - 2 = 0.$$

Here we have two simple equations that we know how to solve:

$$x = -3 \quad or \quad x = 2.$$

Each of the numbers -3 and 2 is a solution of the original equation, as we can see in the following checks.

Check: For -3:

$$\begin{array}{r|l} (x + 3)(x - 2) & = 0 \\ \hline (-3 + 3)(-3 - 2) & ?\ 0 \\ 0(-5) & \\ 0 & 0 \quad \text{TRUE} \end{array}$$

For 2:

$$\begin{array}{r|l} (x + 3)(x - 2) & = 0 \\ \hline (2 + 3)(2 - 2) & ?\ 0 \\ 5(0) & \\ 0 & 0 \quad \text{TRUE} \end{array}$$

❑

We now have a principle to help in solving quadratic equations.

The Principle of Zero Products

An equation $ab = 0$ is true if and only if $a = 0$ or $b = 0$, or both. (A product is 0 if and only if at least one factor is 0.)

EXAMPLE 2

Solve: $(5x + 1)(x - 7) = 0$.

Solution

$$(5x + 1)(x - 7) = 0$$

$$5x + 1 = 0 \quad or \quad x - 7 = 0 \qquad \text{Using the principle of zero products}$$

$$5x = -1 \quad or \quad x = 7 \qquad \text{Solving the two equations separately}$$

$$x = -\tfrac{1}{5} \quad or \quad x = 7$$

Check: For $-\frac{1}{5}$:

$$\begin{array}{c|l} (5x+1)(x-7) & = 0 \\ \hline (5(-\frac{1}{5})+1)(-\frac{1}{5}-7) & ?\ 0 \\ (-1+1)(-7\frac{1}{5}) & \\ 0(-7\frac{1}{5}) & \\ 0 & 0 \text{ TRUE} \end{array}$$

For 7:

$$\begin{array}{c|l} (5x+1)(x-7) & = 0 \\ \hline (5(7)+1)(7-7) & ?\ 0 \\ (35+1)0 & \\ 36 \cdot 0 & \\ 0 & 0 \text{ TRUE} \end{array}$$

The solutions are $-\frac{1}{5}$ and 7. ❑

The principle of zero products can be used any time a product equals 0—even if a factor has only one term.

EXAMPLE 3

Solve: $x(2x - 9) = 0$.

Solution

$$x(2x - 9) = 0$$

$$x = 0 \quad or \quad 2x - 9 = 0 \qquad \text{Using the principle of zero products}$$

$$x = 0 \quad or \quad 2x = 9$$

$$x = 0 \quad or \quad x = \tfrac{9}{2}$$

The solutions are 0 and $\frac{9}{2}$. The check is left to the student. ❑

Factoring to Solve Equations

By factoring and using the principle of zero products, we can now solve quadratic equations.

EXAMPLE 4

Solve: $x^2 + 5x + 6 = 0$.

Solution This equation differs from those that we solved in Chapter 2. There are no like terms to collect, and we have a squared term. We first factor the polynomial. Then we use the principle of zero products:

$$x^2 + 5x + 6 = 0$$

$$(x + 2)(x + 3) = 0 \qquad \text{Factoring}$$

$$x + 2 = 0 \quad or \quad x + 3 = 0 \qquad \text{Using the principle of zero products}$$

$$x = -2 \quad or \quad x = -3.$$

Check: For -2:

$$\begin{array}{c|l} x^2+5x+6 & = 0 \\ \hline (-2)^2+5(-2)+6 & ?\ 0 \\ 4-10+6 & \\ -6+6 & \\ 0 & 0 \text{ TRUE} \end{array}$$

For -3:

$$\begin{array}{c|l} x^2+5x+6 & = 0 \\ \hline (-3)^2+5(-3)+6 & ?\ 0 \\ 9-15+6 & \\ -6+6 & \\ 0 & 0 \text{ TRUE} \end{array}$$

The solutions are -2 and -3. ❑

CAUTION! We *must* have 0 on one side before using the principle of zero products. Get all nonzero terms on one side and 0 on the other.

EXAMPLE 5

Solve: **(a)** $x^2 - 8x = -16$; **(b)** $x^2 + 5x = 0$; **(c)** $4x^2 = 25$.

Solution

a) We first add 16 to get 0 on one side:

$$x^2 - 8x = -16$$
$$x^2 - 8x + 16 = 0 \qquad \text{Adding 16 on both sides}$$
$$(x - 4)(x - 4) = 0 \qquad \text{Factoring}$$
$$x - 4 = 0 \quad or \quad x - 4 = 0 \qquad \text{Using the principle of zero products}$$
$$x = 4 \quad or \quad x = 4.$$

There is only one solution, 4. The check is left to the student.

b) $x^2 + 5x = 0$

$$x(x + 5) = 0 \qquad \text{Factoring out a common factor}$$
$$x = 0 \quad or \quad x + 5 = 0 \qquad \text{Using the principle of zero products}$$
$$x = 0 \quad or \quad x = -5$$

The solutions are 0 and -5. The check is left to the student.

c)

$$4x^2 = 25$$
$$4x^2 - 25 = 0 \qquad \text{Subtracting 25 on both sides to get 0 on one side}$$
$$(2x - 5)(2x + 5) = 0 \qquad \text{Factoring a difference of squares}$$
$$2x - 5 = 0 \quad or \quad 2x + 5 = 0$$
$$2x = 5 \quad or \quad 2x = -5$$
$$x = \tfrac{5}{2} \quad or \quad x = -\tfrac{5}{2}$$

The solutions are $\frac{5}{2}$ and $-\frac{5}{2}$. The check is left to the student. ❑

EXAMPLE 6

Solve: $(x + 3)(2x - 1) = 9$.

Solution Be careful with an equation like this! Remember that since we must have 0 on one side, we multiply out the product on the left and then subtract 9 from both sides.

$$(x + 3)(2x - 1) = 9$$
$$2x^2 + 5x - 3 = 9 \qquad \text{Multiplying on the left}$$
$$2x^2 + 5x - 3 - 9 = 9 - 9 \qquad \text{Subtracting 9 on both sides}$$
$$2x^2 + 5x - 12 = 0$$
$$(2x - 3)(x + 4) = 0 \qquad \text{Factoring}$$
$$2x - 3 = 0 \quad or \quad x + 4 = 0 \qquad \text{Using the principle of zero products}$$
$$2x = 3 \quad or \quad x = -4$$
$$x = \tfrac{3}{2} \quad or \quad x = -4$$

Check: For $\frac{3}{2}$:

$$\begin{array}{c|l} (x + 3)(2x - 1) & = 9 \\ \hline (\tfrac{3}{2} + 3)(2 \cdot \tfrac{3}{2} - 1) & ?\ 9 \\ (\tfrac{9}{2})(2) & \\ 9 & 9 \quad \text{TRUE} \end{array}$$

For -4:

$$\begin{array}{c|l} (x + 3)(2x - 1) & = 9 \\ \hline (-4 + 3)(2(-4) - 1) & ?\ 9 \\ (-1)(-9) & \\ 9 & 9 \quad \text{TRUE} \end{array}$$

The solutions are $\frac{3}{2}$ and -4. ❑

Graphing and Quadratic Equations

In Chapter 3, we graphed linear equations of the form $Ax + By = C$ and $y = mx + b$. Recall that to find the x-intercept, we replaced y with 0 and solved for x. This same procedure can be used to find the x-intercepts when an equation of the form $y = ax^2 + bx + c$ $(a \neq 0)$ is graphed. Equations like this are graphed in Chapter 10. Their graphs are shaped like the following curves:

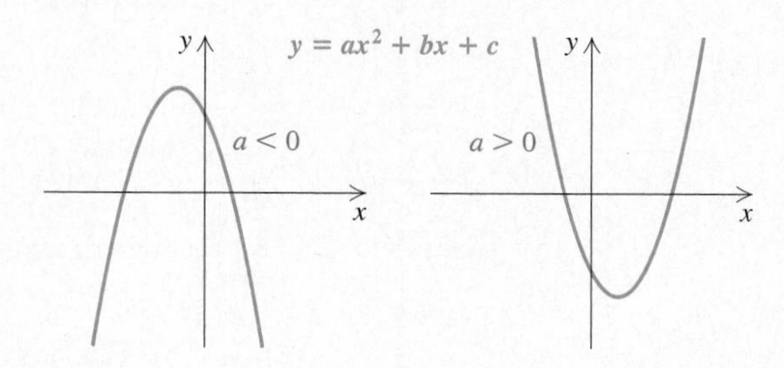

A grapher can help us solve quadratic equations by zooming in on any x-intercepts that may exist. This technique works whether the equation is factorable or not. As an example, consider the quadratic equation $y = x^2 - 3x - 5$. Graphing this equation in a $[-10, 10, -10, 10]$ window gives the following graph:

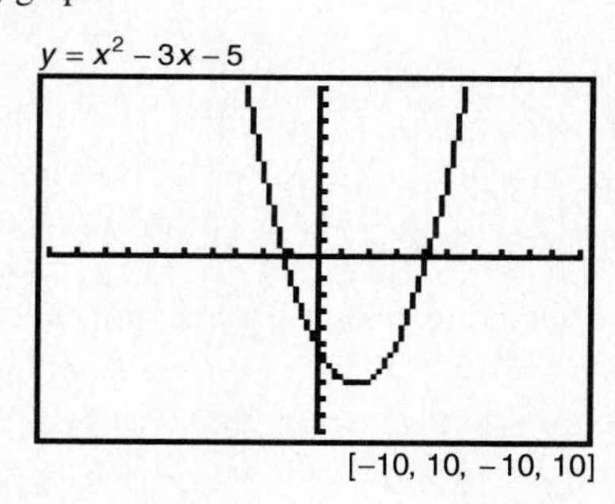

$[-10, 10, -10, 10]$

There appears to be an x-intercept between -2 and -1 and another one between 4 and 5. Let's first examine the negative intercept using a $[-2, -1, -1, 1]$ window or the Zoom feature:

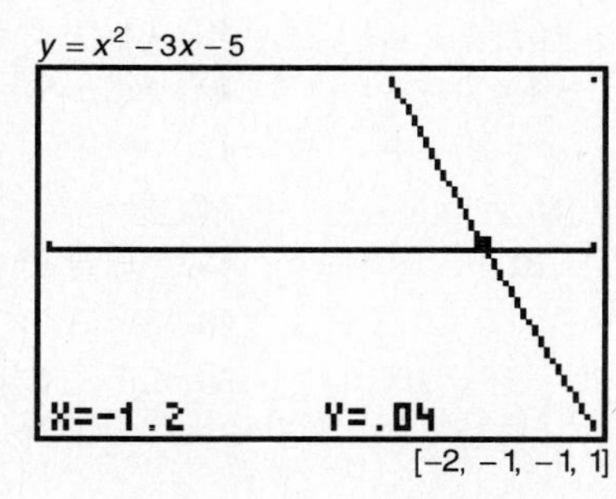

$[-2, -1, -1, 1]$

Using the Trace feature, we can see that the intercept is close to -1.2. If we change the window dimensions to $[-1.22, -1.18, -0.1, 0.1]$ or zoom in again, it becomes clear that the intercept is very close to -1.19:

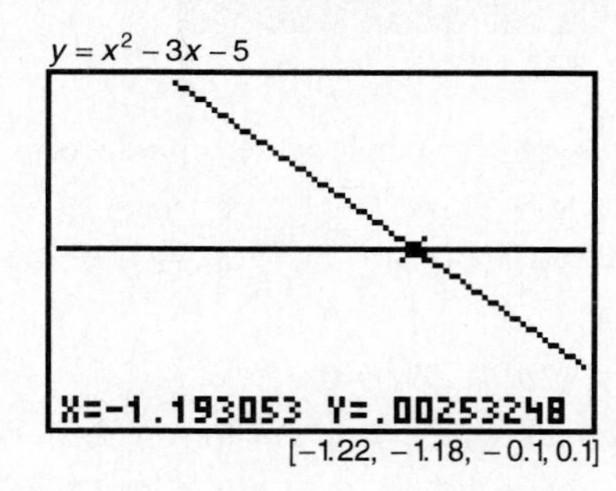

$[-1.22, -1.18, -0.1, 0.1]$

By similarly zooming in on the positive intercept, we can conclude that the other solution of $x^2 - 3x - 5 = 0$ is close to 4.19.

Use a grapher to find the solutions, if they exist, accurate to two decimal places.

TC1. $x^2 + 4x - 3 = 0$

TC2. $x^2 - 5x - 2 = 0$

TC3. $x^2 + 13.54x + 40.95 = 0$

TC4. $x^2 - 4.43x + 6.32 = 0$

EXAMPLE 7 Find the x-intercepts for the graph of the equation shown.

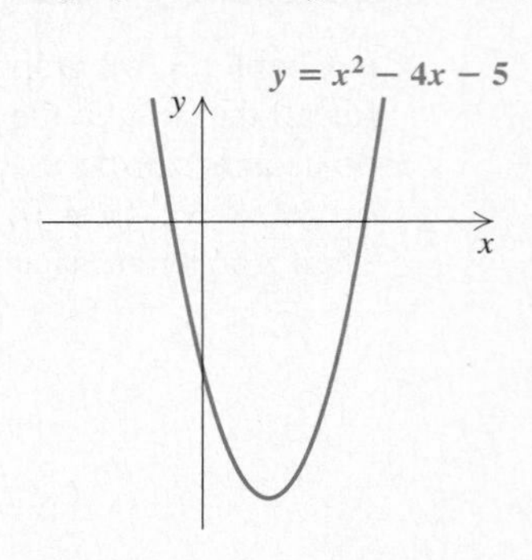

Solution To find the x-intercepts, we let $y = 0$ and solve for x:

$0 = x^2 - 4x - 5$	Substituting 0 for y
$0 = (x - 5)(x + 1)$	Factoring
$x - 5 = 0 \quad or \quad x + 1 = 0$	Using the principle of zero products
$x = 5 \quad or \quad x = -1.$	Solving for x

The x-intercepts are $(5, 0)$ and $(-1, 0)$. ❑

EXERCISE SET 5.6

Solve using the principle of zero products.

1. $(x + 8)(x + 6) = 0$
2. $(x + 3)(x + 2) = 0$
3. $(x - 3)(x + 5) = 0$
4. $(x + 9)(x - 3) = 0$
5. $(x + 12)(x - 11) = 0$
6. $(x - 13)(x + 53) = 0$
7. $x(x + 5) = 0$
8. $y(y + 7) = 0$
9. $0 = y(y + 10)$
10. $0 = x(x - 21)$
11. $(2x + 5)(x + 4) = 0$
12. $(2x + 9)(x + 8) = 0$
13. $(5x + 1)(4x - 12) = 0$
14. $(4x + 9)(14x - 7) = 0$
15. $(7x - 28)(28x - 7) = 0$
16. $(12x - 11)(8x - 5) = 0$
17. $2x(3x - 2) = 0$
18. $75x(8x - 9) = 0$
19. $\frac{1}{2}x\left(\frac{2}{3}x - 12\right) = 0$
20. $\frac{5}{7}x\left(\frac{3}{4}x - 6\right) = 0$
21. $\left(\frac{1}{5} + 2x\right)\left(\frac{1}{9} - 3x\right) = 0$
22. $\left(\frac{7}{4}x - \frac{1}{12}\right)\left(\frac{2}{3}x - \frac{12}{11}\right) = 0$
23. $(0.3x - 0.1)(0.05x - 1) = 0$
24. $(0.1x - 0.3)(0.4x - 20) = 0$
25. $9x(3x - 2)(2x - 1) = 0$
26. $(x - 5)(x + 55)(5x - 1) = 0$

Solve by factoring and using the principle of zero products.

27. $x^2 + 6x + 5 = 0$
28. $x^2 + 7x + 6 = 0$
29. $x^2 + 7x - 18 = 0$
30. $x^2 + 4x - 21 = 0$
31. $x^2 - 8x + 15 = 0$
32. $x^2 - 9x + 14 = 0$
33. $x^2 - 8x = 0$
34. $x^2 - 3x = 0$
35. $x^2 + 19x = 0$
36. $x^2 + 12x = 0$
37. $x^2 = 16$
38. $100 = x^2$
39. $9x^2 - 4 = 0$
40. $4x^2 - 9 = 0$
41. $0 = 6x + x^2 + 9$
42. $0 = 25 + x^2 + 10x$
43. $x^2 + 16 = 8x$
44. $1 + x^2 = 2x$
45. $5x^2 = 6x$
46. $7x^2 = 8x$
47. $6x^2 - 4x = 10$
48. $3x^2 - 7x = 20$
49. $12y^2 - 5y = 2$
50. $2y^2 + 12y = -10$

51. $x(x - 5) = 14$
52. $t(3t + 1) = 2$
53. $64m^2 - 25 = 56$
54. $100t^2 - 9 = 40$
55. $3x^2 + 8x = 9 + 2x$
56. $x^2 - 5x = 18 + 2x$
57. $(3x + 5)(x + 3) = 7$
58. $(5x + 4)(x - 1) = 2$

Find the x-intercepts for the graph of each equation.

59. $y = x^2 - x - 6$

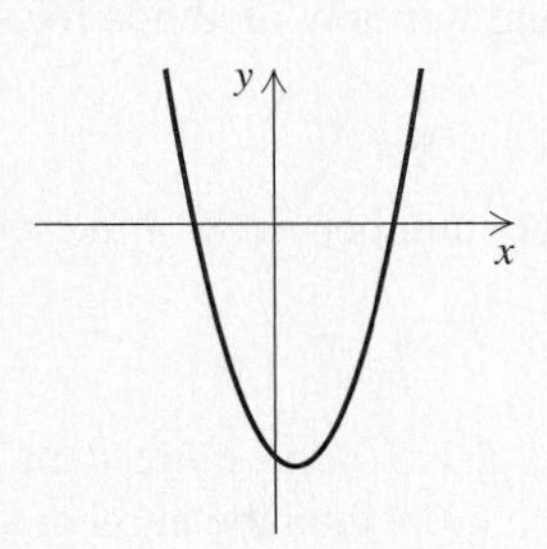

60. $y = x^2 + 3x - 4$

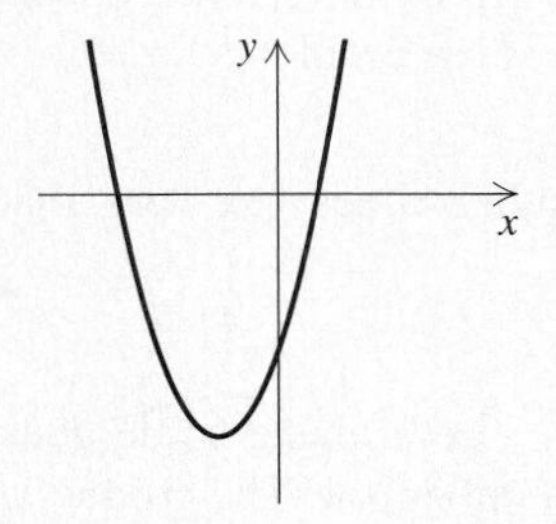

61. $y = x^2 + 2x - 8$

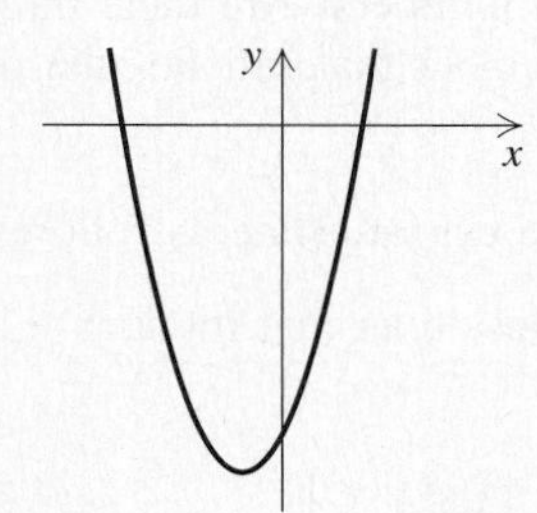

62. $y = x^2 - 2x - 15$

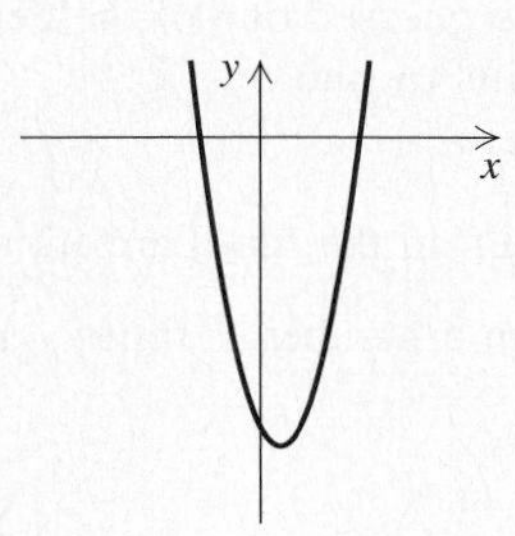

63. $y = 2x^2 + 3x - 9$

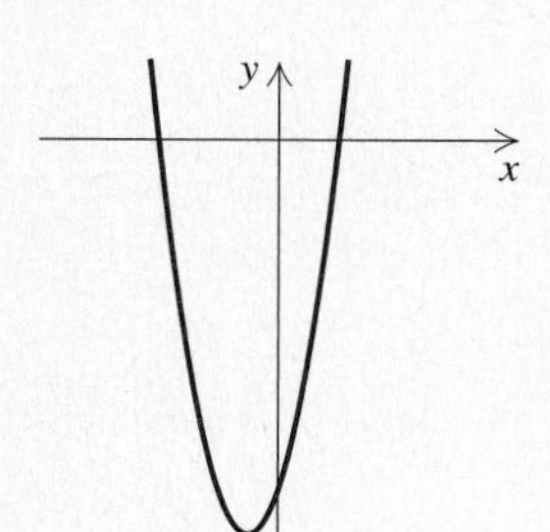

64. $y = 2x^2 + x - 10$

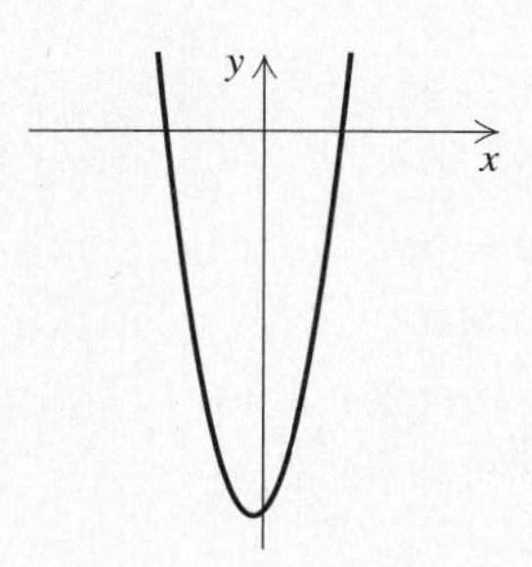

Skill Maintenance

Translate to an algebraic expression.

65. The square of the sum of a and b
66. The sum of the squares of a and b

Translate to an inequality.

67. 5 more than twice a number is less than 19.
68. 7 less than half of a number exceeds 24.

Synthesis

69. What is the difference between a trinomial and a quadratic equation?

70. The equation $x^2 + 1$ has no real-number solutions. What implications does this have for the graph of $y = x^2 + 1$?

71. What is wrong with solving $x^2 = 3x$ by dividing both sides of the equation by x?

72. When the principle of zero products is used to solve an equation, will there always be two solutions? Why or why not?

73. Find an equation with integer coefficients that has the given numbers as solutions. For example, 3 and -2 are solutions to $x^2 - x - 6 = 0$.

a) $-3, 4$ **b)** $-3, -4$ **c)** $\frac{1}{2}, \frac{1}{2}$
d) $5, -5$ **e)** $0, 0.1, \frac{1}{4}$

Solve.

74. $b(b + 9) = 4(5 + 2b)$
75. $y(y + 8) = 16(y - 1)$
76. $(t - 5)^2 = 2(5 - t)$
77. $x^2 - \frac{1}{64} = 0$
78. $x^2 - \frac{25}{36} = 0$
79. $\frac{5}{16}x^2 = 5$
80. $\frac{27}{25}x^2 = \frac{1}{3}$

81. For each equation on the left, find an equivalent equation on the right.

a) $3x^2 - 4x + 8 = 0$ $\quad 4x^2 + 8x + 36 = 0$
b) $(x - 6)(x + 3) = 0$ $\quad (2x + 8)(2x - 5) = 0$
c) $x^2 + 2x + 9 = 0$ $\quad 9x^2 - 12x + 24 = 0$
d) $(2x - 5)(x + 4) = 0$ $\quad (x + 1)(5x - 5) = 0$
e) $5x^2 - 5 = 0$ $\quad x^2 - 3x - 18 = 0$
f) $x^2 + 10x - 2 = 0$ $\quad 2x^2 + 20x - 4 = 0$

82. Explain how to construct an equation that has seven solutions.

83. Explain how the graph in Exercise 59 can be used to visualize the solutions of

$$x^2 - x - 6 = -4.$$

Use a grapher to find the solutions of the equation, accurate to two decimal places.

84. $x^2 - 1.80x - 5.69 = 0$
85. $x^2 + 9.10x + 15.77 = 0$
86. $-x^2 + 0.63x + 0.22 = 0$
87. $x^2 + 13.74x + 42.00 = 0$
88. $6.4x^2 - 8.45x - 94.06 = 0$
89. $-0.25x^2 - 2.50x - 5.48 = 0$

5.7 Problem Solving

Applications • The Pythagorean Theorem

We can use our five-step problem-solving process and our new methods for solving quadratic equations to solve problems.

EXAMPLE 1

One more than a number times one less than the number is 8. Find all such numbers.

Solution

1. FAMILIARIZE. Let's make some guesses. Try 5. One more than 5 is 6. One less than the number is 4. The product of one more than the number and one less than the number is $6 \cdot 4$, or 24, which is too large. Let's try 3. One more than 3 is 4. One less than 3 is 2. The product of these numbers is $4 \cdot 2$, or 8. This checks and we have guessed one of the desired numbers. Are there more? We use our algebra skills to find out. Let x = a number that satisfies the requirements of the problem.

2. TRANSLATE. From the familiarization, we can translate as follows:

$$\underbrace{\text{One more than a number}}_{(x+1)} \quad \underbrace{\text{times}}_{\cdot} \quad \underbrace{\text{one less than that number}}_{(x-1)} \quad \underbrace{\text{is}}_{=} \quad \underbrace{8.}_{8}$$

3. CARRY OUT. We solve the equation as follows:

$$
\begin{aligned}
(x+1)(x-1) &= 8 \\
x^2 - 1 &= 8 && \text{Multiplying} \\
x^2 - 1 - 8 &= 0 && \text{Subtracting 8 to get 0 on one side} \\
x^2 - 9 &= 0 \\
(x-3)(x+3) &= 0 && \text{Factoring} \\
x - 3 = 0 \quad &or \quad x + 3 = 0 && \text{Using the principle of zero products} \\
x = 3 \quad &or \quad x = -3.
\end{aligned}
$$

4. CHECK. We already guessed and checked one of the solutions, 3, in the *Familiarize* step. To check -3, note that one more than -3 is -2 and one less than -3 is -4. The product of -2 and -4 is 8. Thus, -3 also checks.

5. STATE. There are two such numbers, 3 and -3. ❑

EXAMPLE 2

A vacant square-shaped lot is being turned into a community garden. Because a path 2 m wide is needed at one end, only 48 m^2 of the lot will be garden space. Find the dimensions of the lot.

Solution

1. **FAMILIARIZE.** We first make a drawing. Recall that the area of any rectangle, including a square, is length · width. We let $x =$ the length (or width) of the square lot. Note that the path is a rectangle 2 m wide and x m long.

2. **TRANSLATE.** It helps to reword this problem before translating:

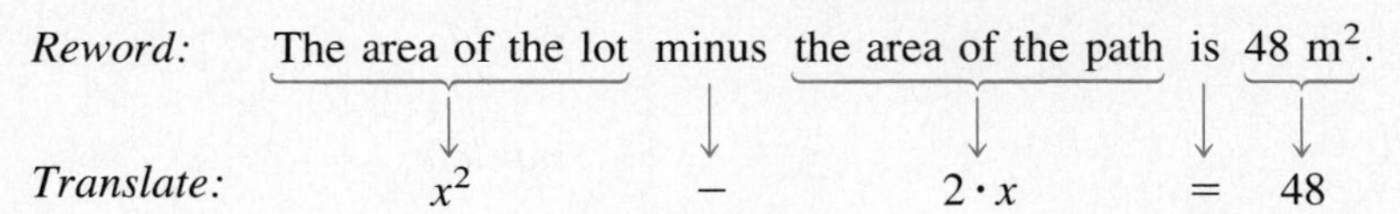

3. **CARRY OUT.** We solve the equation as follows:

$$x^2 - 2x = 48$$

$$x^2 - 2x - 48 = 0 \quad \text{Subtracting 48 to get 0 on one side}$$

$$(x - 8)(x + 6) = 0 \quad \text{Factoring}$$

$$x - 8 = 0 \quad or \quad x + 6 = 0 \quad \text{Using the principle of zero products}$$

$$x = 8 \quad or \quad x = -6.$$

4. **CHECK.** Since measurements cannot be negative, we must disregard -6 as a solution of the original problem. If $x = 8$, the area of the lot is $8 \cdot 8 = 64$ m^2 and the area of the path is $2 \cdot 8 = 16$ m^2, leaving $64 - 16 = 48$ m^2 for the garden. Thus, 8 checks. Another approach to this problem is to express the garden's area as $x(x - 2)$ and set this equal to 48. This provides a second check, since $8(8 - 2) = 8 \cdot 6 = 48$.

5. **STATE.** The lot is 8 m long and 8 m wide. ❑

EXAMPLE 3

The height of a triangular sail is 7 ft more than the base. The area of the triangle is 30 ft^2. Find the height and the base.

Solution

1. **FAMILIARIZE.** We first make a drawing. The formula for the area of a triangle is Area $= \frac{1}{2} \cdot (\text{base}) \cdot (\text{height})$. We let $b =$ the length, in feet, of the triangle's base.

2. **TRANSLATE.** We reword the problem and translate:

Reword:	$\frac{1}{2}$	times	the base	times	the base plus 7	is	30.
	↓	↓	↓	↓	↓	↓	↓
Translate:	$\frac{1}{2}$	$\cdot$	b	$\cdot$	$(b + 7)$	$=$	30

3. **CARRY OUT.** We solve the equation as follows:

$$\frac{1}{2} \cdot b \cdot (b + 7) = 30$$

$$\frac{1}{2}(b^2 + 7b) = 30 \qquad \text{Multiplying}$$

$$b^2 + 7b = 60 \qquad \text{Multiplying by 2 to clear fractions}$$

$$b^2 + 7b - 60 = 0 \qquad \text{Subtracting 60 to get 0 on one side}$$

$$(b + 12)(b - 5) = 0 \qquad \text{Factoring}$$

$$b + 12 = 0 \quad or \quad b - 5 = 0 \qquad \text{Using the principle of zero products}$$

$$b = -12 \quad or \quad b = 5.$$

4. **CHECK.** The base of a triangle cannot have a negative length, so -12 cannot be a solution. Suppose the base is 5 ft. Then the height is 7 ft more than the base, so the height is 12 ft and the area is $\frac{1}{2}(5)(12)$, or 30 ft^2. These numbers check in the original problem.

5. **STATE.** The height is 12 ft and the base is 5 ft. ❑

EXAMPLE 4

In a sports league of n teams in which each team plays every other team twice, the total number N of games to be played is given by

$$n^2 - n = N.$$

If a basketball league plays a total of 240 games, how many teams are in the league?

Solution

1. FAMILIARIZE. To familiarize yourself with this equation, reread Example 9 in Section 4.2, where we first considered it.

2. TRANSLATE. We are trying to find the number of teams n in a league in which 240 games are played. We substitute 240 for N in order to solve for n:

$$n^2 - n = 240. \quad \text{Substituting 240 for } N$$

3. CARRY OUT. We solve the equation as follows:

$$n^2 - n = 240$$
$$n^2 - n - 240 = 0 \quad \text{Subtracting 240 to get 0 on one side}$$
$$(n - 16)(n + 15) = 0 \quad \text{Factoring}$$
$$n - 16 = 0 \quad \text{or} \quad n + 15 = 0 \quad \text{Using the principle of zero products}$$
$$n = 16 \quad \text{or} \quad n = -15.$$

4. CHECK. The solutions of the equation are 16 and -15. Since the number of teams cannot be negative, -15 cannot be a solution. However, 16 checks, since $16^2 - 16 = 256 - 16 = 240$.

5. STATE. There are 16 teams in the league. ❑

EXAMPLE 5

The product of the page numbers on two consecutive pages of a book is 156. Find the page numbers.

Solution

1. FAMILIARIZE. Recall that consecutive page numbers are one apart, like 49 and 50. Let $x =$ the first page number; then $x + 1 =$ the next page number.

2. TRANSLATE. We reword the problem before translating:

Reword:	The first page number	times	the next page number	is	156.
	↓	↓	↓	↓	↓
Translate:	x	$\cdot$	$(x + 1)$	$=$	156

3. CARRY OUT. We solve the equation as follows:

$$x(x + 1) = 156$$
$$x^2 + x = 156 \quad \text{Multiplying}$$
$$x^2 + x - 156 = 0 \quad \text{Subtracting 156 to get 0 on one side}$$
$$(x - 12)(x + 13) = 0 \quad \text{Factoring}$$
$$x - 12 = 0 \quad \text{or} \quad x + 13 = 0 \quad \text{Using the principle of zero products}$$
$$x = 12 \quad \text{or} \quad x = -13.$$

4. CHECK. The solutions of the equation are 12 and -13. Since page numbers cannot be negative, -13 can't be a solution. On the other hand, if x is 12, then $x + 1$ is 13 and $12 \cdot 13 = 156$. Thus, 12 checks.

5. STATE. The pair of page numbers is 12 and 13. ❑

The following problem involves the Pythagorean theorem, which relates the lengths of the sides of a right triangle. A **right triangle** has a 90° angle. The side opposite the 90° angle is called the **hypotenuse.** The other sides are called **legs.**

The Pythagorean Theorem

The sum of the squares of the legs of a right triangle is equal to the square of the hypotenuse:

$$a^2 + b^2 = c^2.$$

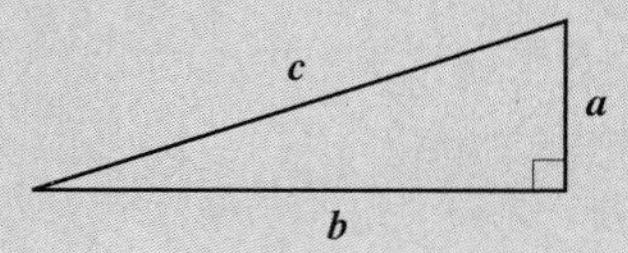

EXAMPLE 6

The length of one leg of a right triangle is 7 ft longer than the other. The length of the hypotenuse is 13 ft. Find the lengths of the legs.

Solution

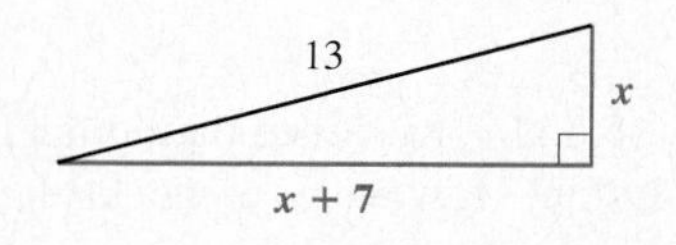

1. **FAMILIARIZE.** We make a drawing and let $x =$ the length of one leg. Since the other leg is 7 ft longer, we know that $x + 7 =$ the length of the other leg. The hypotenuse has length 13 ft.

2. **TRANSLATE.** Applying the Pythagorean theorem, we obtain the following translation:

$$a^2 + b^2 = c^2$$
$$x^2 + (x + 7)^2 = 13^2. \quad \text{Substituting}$$

3. **CARRY OUT.** We solve the equation as follows:

$$x^2 + (x^2 + 14x + 49) = 169 \quad \text{Squaring the binomial and 13}$$
$$2x^2 + 14x + 49 = 169 \quad \text{Collecting like terms}$$
$$2x^2 + 14x - 120 = 0 \quad \text{Subtracting 169 to get 0 on side}$$
$$2(x^2 + 7x - 60) = 0 \quad \text{Factoring out a common factor}$$
$$2(x + 12)(x - 5) = 0 \quad \text{Factoring}$$
$$x + 12 = 0 \quad or \quad x - 5 = 0 \quad \text{Using the principle of zero products}$$
$$x = -12 \quad or \quad x = 5.$$

4. **CHECK.** The integer -12 cannot be a length of a side because it is negative. When $x = 5$, $x + 7 = 12$, and $5^2 + 12^2 = 13^2$. So 5 checks.

5. **STATE.** The lengths of the legs are 5 ft and 12 ft. ❑

EXERCISE SET 5.7

Solve.

1. If you subtract a number from four times its square, the result is 3. Find all such numbers.

2. If 7 is added to the square of a number, the result is 32. Find all such numbers.

3. Eight more than the square of a number is six times the number. Find all such numbers.

4. Fifteen more than the square of a number is eight times the number. Find all such numbers.

5. The product of the page numbers on two facing pages of a book is 210. Find the page numbers.
6. The product of the page numbers on two facing pages of a book is 110. Find the page numbers.
7. The product of two consecutive even integers is 168. Find the integers.
8. The product of two consecutive even integers is 224. Find the integers.
9. The product of two consecutive odd integers is 255. Find the integers.
10. The product of two consecutive odd integers is 143. Find the integers.
11. The length of a rectangular garden is 4 m greater than the width. The area of the rectangle is 96 m^2. Find the length and the width.

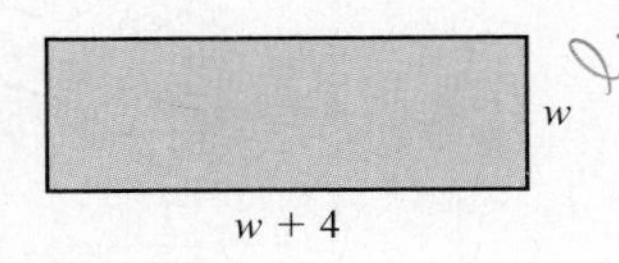

12. The length of a rectangular calculator is 5 cm greater than the width. The area of the rectangle is 84 cm^2. Find the length and the width.

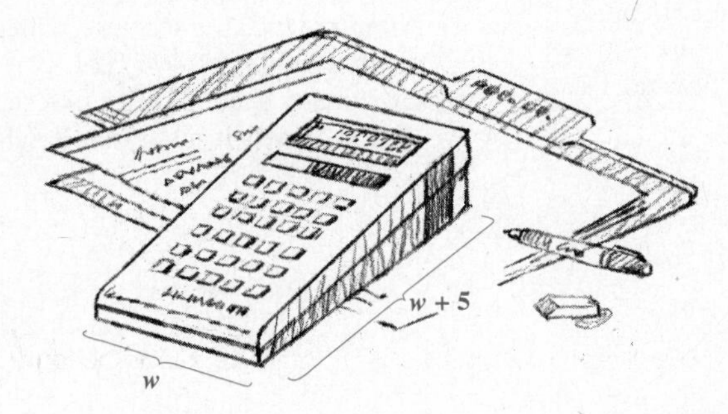

13. The area of a square bookcase is 5 more than the perimeter. Find the length of a side.
14. The perimeter of a square porch is 3 more than the area. Find the length of a side.
15. The base of a triangle is 10 cm greater than the height. The area is 28 cm^2. Find the height and the base.

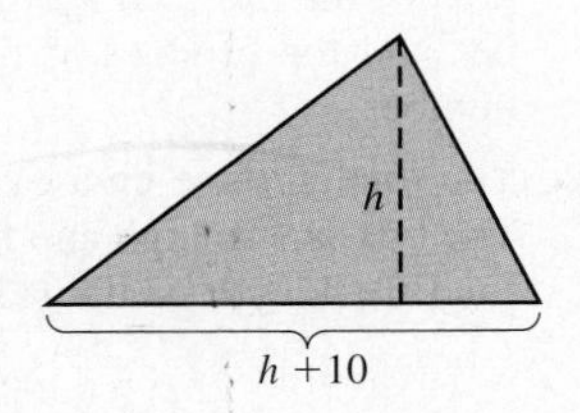

16. The height of a triangle is 8 m less than the base. The area is 10 m^2. Find the height and the base.
17. If the sides of a square are lengthened by 3 m, the area becomes 81 m^2. Find the length of a side of the original square.
18. If the sides of a square are lengthened by 7 km, the area becomes 121 km^2. Find the length of a side of the original square.
19. The sum of the squares of two consecutive odd whole numbers is 74. Find the numbers.
20. The sum of the squares of two consecutive odd whole numbers is 130. Find the numbers.

Use the formula from Example 4, $n^2 - n = N$, for Exercises 21–24.

21. A women's volleyball league has 23 teams. What is the total number of games to be played?
22. A chess league has 14 teams. What is the total number of games to be played?
23. A women's softball league plays a total of 132 games. How many teams are in the league?

24. A basketball league plays a total of 90 games. How many teams are in the league?

The number of possible handshakes within a group of n people is given by $N = \frac{1}{2}(n^2 - n)$.

25. At a meeting, there are 40 people. How many handshakes are possible?
26. At a party, there are 100 people. How many handshakes are possible?
27. During a toast at a party, there were 190 "clicks" of glasses. How many people took part in the toast?
28. After winning the championship, all members of a team exchanged "high fives." Altogether, there were 300 high fives. How many people were on the team?

29. The length of one leg of a right triangle is 8 ft. The length of the hypotenuse is 2 ft longer than the other leg. Find the lengths of the hypotenuse and the other leg.

30. The length of one leg of a right triangle is 24 ft. The length of the other leg is 16 ft shorter than the hypotenuse. Find the lengths of the hypotenuse and the other leg.

31. An architect has allocated a rectangular space of 264 ft^2 for a square dining room and a 10-ft wide kitchen. Find the dimensions of each room.

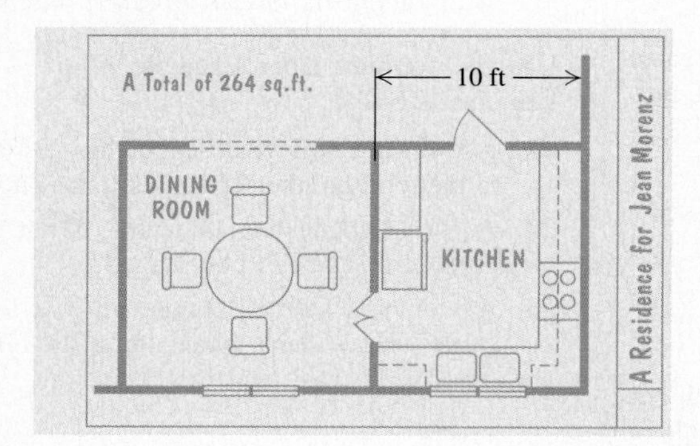

32. A television antenna's guy wire is 1 m longer than the height of the antenna. If the guy wire is anchored 3 m from the foot of the antenna, how tall is the antenna?

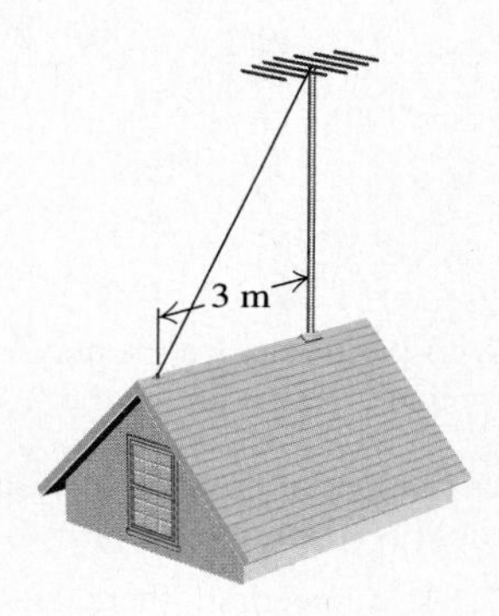

Skill Maintenance

Graph.

33. $y = -\frac{2}{3}x + 1$

34. $y = \frac{3}{5}x - 1$

Simplify.

35. $7x^0$ when $x = -4$

36. $\frac{m^7n^9}{mn^3}$

Synthesis

37. ◈ Write a problem in which a quadratic equation must be solved in order to solve the problem.

38. ◈ Write a problem for a classmate to solve so that only one of two solutions of a quadratic equation can be used as an answer.

39. A cement walk of uniform width is built around a 20-ft by 40-ft rectangular pool. The total area of the pool and the walk is 1500 ft^2. Find the width of the walk.

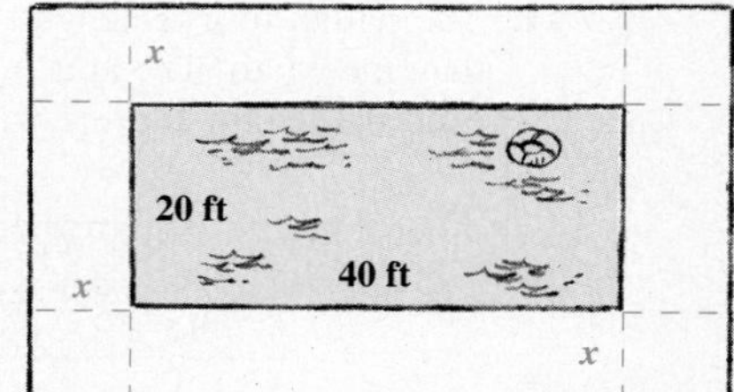

40. A model rocket is launched with an initial velocity of 180 ft/sec. Its height h, in feet, after t seconds is given by the formula $h = 180t - 16t^2$.

a) After how many seconds will the rocket first reach a height of 464 ft?

b) How many seconds after first reaching a height of 464 ft will it be at that height again?

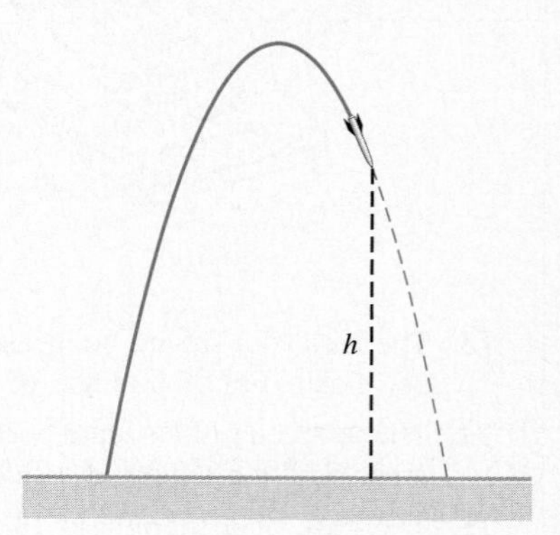

41. The one's digit of a number less than 100 is four greater than the ten's digit. The sum of the number and the product of the digits is 58. Find the number.

42. The total surface area of a closed box is 350 m^2. The box is 9 m high and has a square base and lid. Find the length of the side of the base.

43. A rectangular piece of cardboard is twice as long as it is wide. A 4-cm square is cut out of each corner, and the sides are turned up to make a box with an open top. The volume of the box is 616 cm^3. Find the original dimensions of the cardboard.

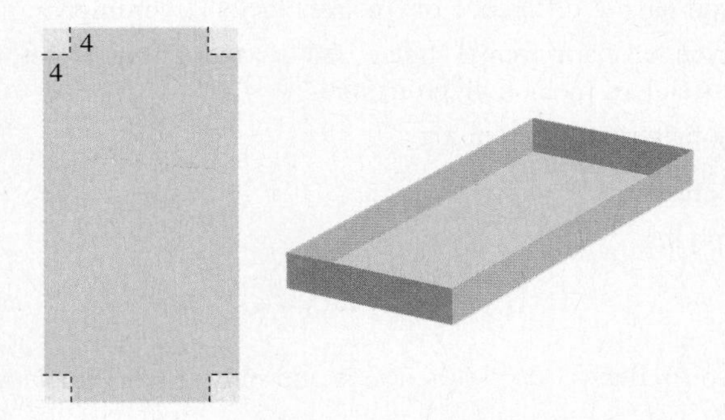

44. An open rectangular gutter is made by turning up the sides of a piece of metal 20 in. wide. The area of the cross-section of the gutter is 50 in^2. Find the depth of the gutter.

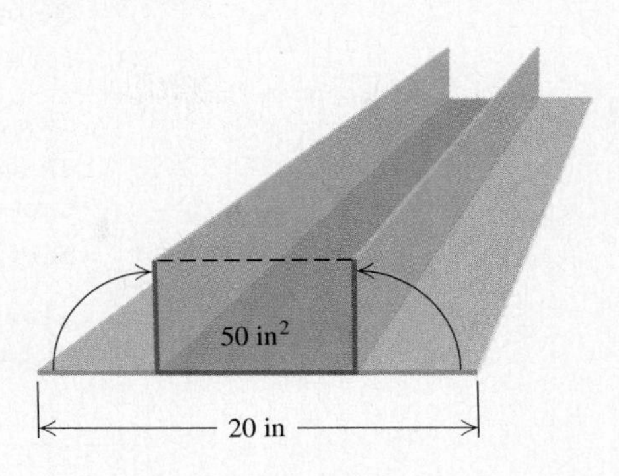

45. The length of each side of a square is increased by 5 cm to form a new square. The area of the new square is $2\frac{1}{4}$ times the area of the original square. Find the area of each square.

SUMMARY AND REVIEW 5

KEY TERMS

Factor, p. 210
Common factor, p. 211
Factoring by grouping, p. 213
Leading coefficient, p. 214
Prime polynomial, p. 218
Factor completely, p. 218
Trinomial square, p. 228
Difference of squares, p. 230
Quadratic equation, p. 239
Right triangle, p. 249
Hypotenuse, p. 249
Leg, p. 249

IMPORTANT PROPERTIES AND FORMULAS

Factoring Formulas

$$A^2 + 2AB + B^2 = (A + B)^2,$$
$$A^2 - 2AB + B^2 = (A - B)^2,$$
$$A^2 - B^2 = (A + B)(A - B)$$

To factor a polynomial:

A. Look first for a common factor. If there is one, factor out the largest common factor. Be sure to include it in your final answer.

B. Look at the number of terms.

Two terms: If you have a difference of squares, factor accordingly.

Three terms: If you have a trinomial square, factor accordingly. If not, try trial and error, using the standard method or grouping.

Four terms: Try factoring by grouping.

C. Always factor completely.

D. Check by multiplying.

The Principle of Zero Products: $ab = 0$ is true if and only if $a = 0$ or $b = 0$, or both.

The Pythagorean theorem: $a^2 + b^2 = c^2$

REVIEW EXERCISES

This chapter's review and test include Skill Maintenance exercises from Sections 1.6, 2.7, 3.2, and 4.1.

Find three factorizations of the monomial.

1. $-10x^2$

2. $36x^5$

Factor completely. If a polynomial is prime, state so.

3. $5 - 20x^6$

4. $x^2 - 3x$

5. $9x^2 - 4$

6. $x^2 + 4x - 12$

7. $x^2 + 14x + 49$

8. $6x^3 + 12x^2 + 3x$

9. $x^3 + x^2 + 3x + 3$

10. $6x^2 - 5x + 1$

11. $x^4 - 81$

12. $9x^3 + 12x^2 - 45x$

13. $2x^2 - 50$

14. $x^4 + 4x^3 - 2x - 8$

15. $16x^4 - 1$

16. $8x^6 - 32x^5 + 4x^4$

17. $75 + 12x^2 + 60x$

18. $x^2 + 9$

19. $x^3 - x^2 - 30x$

20. $4x^2 - 25$

21. $9x^2 + 25 - 30x$

22. $6x^2 - 28x - 48$

23. $x^2 - 6x + 9$

24. $2x^2 - 7x - 4$

25. $18x^2 - 12x + 2$

26. $3x^2 - 27$

27. $15 - 8x + x^2$

28. $25x^2 - 20x + 4$

29. $x^2y^2 + xy - 12$

30. $12a^2 + 84ab + 147b^2$

31. $m^2 + 5m + mt + 5t$

32. $32x^4 - 128y^4z^4$

Solve.

33. $(x - 1)(x + 3) = 0$

34. $x^2 + 2x - 35 = 0$

35. $x^2 + x - 12 = 0$

36. $3x^2 + 2 = 5x$

37. $2x^2 + 5x = 12$

38. $16 = x(x - 6)$

39. The square of a number is 6 more than the number. Find the number.

40. Find the x-intercepts for the graph of $y = 2x^2 - 3x - 5$.

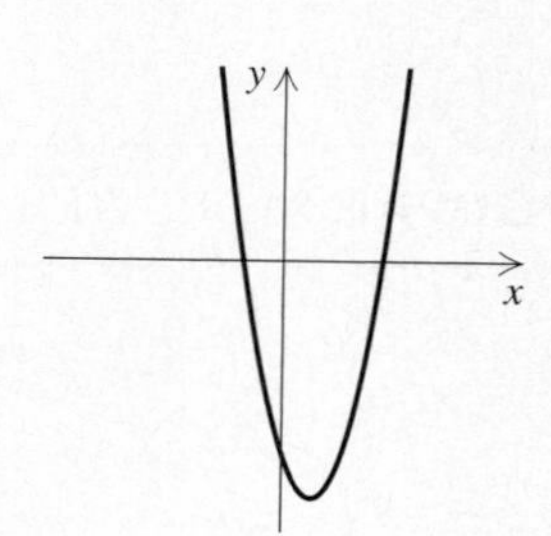

41. The product of two consecutive odd integers is 323. Find the integers.

42. Twice the square of a number is 10 more than the number. Find the number.

Skill Maintenance

43. Subtract: $\frac{2}{5} - \left(-\frac{1}{10}\right)$.

44. Graph: $y = -\frac{3}{4}x$.

45. Divide and simplify: $\dfrac{m^3n^{10}}{mn^5}$.

46. Translate to an inequality: 2 less than half a number is no less than 10.

Synthesis

47. ◈ Compare the two methods of checking factorizations that were discussed.

48. ◈ How do the equations solved in this chapter differ from those solved in previous chapters?

Solve.

49. The pages of a book measure 15 cm by 20 cm. Margins of equal width surround the printing on each page and constitute one half of the area of the page. Find the width of the margins.

50. The cube of a number is the same as twice the square of the number. Find the number.

51. The length of a rectangle is two times its width. When the length is increased by 20 and the width is decreased by 1, the area is 160. Find the original length and width.

Solve.

52. $x^2 + 25 = 0$

53. $(x - 2)(x + 3)(2x - 5) = 0$

CHAPTER TEST 5

1. Find three factorizations of $4x^3$.

Factor completely.

2. $x^2 - 7x + 10$

3. $x^2 + 25 - 10x$

4. $6y^2 - 8y^3 + 4y^4$

5. $x^3 + x^2 + 2x + 2$

6. $x^2 - 5x$

7. $x^3 + 2x^2 - 3x$

8. $28x - 48 + 10x^2$

9. $4x^2 - 9$

10. $x^2 - x - 12$

11. $6m^3 + 9m^2 + 3m$

12. $3w^2 - 75$

13. $60x + 45x^2 + 20$

14. $3x^4 - 48$

15. $49x^2 - 84x + 36$

16. $5x^2 - 26x + 5$

17. $x^4 + 2x^3 - 3x - 6$

18. $80 - 5x^4$

19. $4x^2 - 4x - 15$

20. $6t^3 + 9t^2 - 15t$

21. $3m^2 - 9mn - 30n^2$

Solve.

22. $x^2 - x - 20 = 0$

23. $2x^2 + 7x = 15$

24. $x(x - 3) = 28$

25. The square of a number is 24 more than five times the number. Find the number.

26. The length of a rectangle is 6 m more than the width. The area of the rectangle is 40 m^2. Find the length and the width.

Skill Maintenance

27. Simplify: $-2.8 - 3.5 + 4.2 - (-1.7)$.

28. The width of a rectangle must be 8 cm. For what lengths will the area be less than 104 cm^2? Use an inequality.

29. Graph: $y = \frac{3}{4}x + 1$.

30. Simplify: $(-7a^3b^5)^2$.

Synthesis

31. The length of a rectangle is five times its width. When the length is decreased by 3 and the width is increased by 2, the area of the new rectangle is 60. Find the original length and width.

32. Factor: $(a + 3)^2 - 2(a + 3) - 35$.

CHAPTER 6

Rational Expressions and Equations

AN APPLICATION

To determine the number of fish in a lake, a park ranger catches 225 fish, tags them, and throws them back into the lake. Later, 108 fish are caught, and 15 are found to be tagged. Estimate how many fish are in the lake.

This problem appears as Example 6 in Section 6.8.

Norman A. Bishop
RESEARCH INTERPRETER

"At Yellowstone National Park, we all collect data and use and interpret statistics. In the 1960s, the native trout population of Yellowstone Lake was in precipitous decline. Mathematical census methods helped us turn the situation around. As a result, many forms of wildlife have an abundant source of food—trout from the lake and its tributaries."

Just as fractional notation helps us to solve arithmetic problems, rational expressions similar to those in the following pages will help us solve algebra problems. We now learn how to simplify, as well as add, subtract, multiply, and divide, rational expressions. These skills are important for solving problems like the one on the preceding page.

In addition to the material from this chapter, the review and test for Chapter 6 include material from Sections 2.1, 3.3, 4.3, and 5.2.

6.1 Rational Expressions

Simplifying Rational Expressions • Factors That Are Opposites

Whereas a rational number is a quotient of two integers, a **rational expression** is a quotient of two polynomials. The following are rational expressions:

$$\frac{7}{3}, \quad \frac{5}{x+6}, \quad \frac{t^2-5t+6}{4t^2-7}.$$

Rational expressions are examples of what are sometimes called *algebraic fractions* or *fractional expressions*.

Because rational expressions indicate division, we must be careful to avoid denominators that are 0. When a variable is replaced by a number that produces a denominator of 0, the rational expression is undefined. For example, in the expression

$$\frac{x+3}{x-7},$$

when x is replaced with 7, the denominator is 0, and the expression is undefined:

$$\frac{x+3}{x-7} = \frac{7+3}{7-7} = \frac{10}{0} \leftarrow \text{Undefined.}$$

When x is replaced with a number other than 7, like 6, the expression *is* defined because the denominator is nonzero:

$$\frac{x+3}{x-7} = \frac{6+3}{6-7} = \frac{9}{-1} = -9.$$

EXAMPLE 1

Find all numbers for which the rational expression $\dfrac{x+4}{x^2-3x-10}$ is undefined.

Solution To determine which numbers make the rational expression undefined, we set the denominator equal to 0 and solve:

$$x^2 - 3x - 10 = 0$$

$$(x-5)(x+2) = 0 \qquad \text{Factoring}$$

$$x - 5 = 0 \quad \text{or} \quad x + 2 = 0 \qquad \text{Using the principle of zero products}$$

$$x = 5 \quad \text{or} \quad x = -2.$$

Check:

For $x = 5$:

$$\frac{x+4}{x^2-3x-10} = \frac{5+4}{5^2-3\cdot 5-10} = \frac{9}{25-15-10} = \frac{9}{0},$$

which is undefined.

For $x = -2$:

$$\frac{x+4}{x^2-3x-10} = \frac{-2+4}{(-2)^2-3(-2)-10} = \frac{2}{4+6-10} = \frac{2}{0},$$

which is undefined.

Thus, $\dfrac{x+4}{x^2-3x-10}$ is undefined for $x = 5$ and $x = -2$. ❑

Simplifying Rational Expressions

Simplifying rational expressions is similar to simplifying the fractional expressions studied in Section 1.3. We saw, for example, that an expression like $\frac{15}{40}$ could be simplified as follows:

$\dfrac{15}{40} = \dfrac{3\cdot 5}{8\cdot 5}$ — Factoring the numerator and the denominator. Note the common factor of 5.

$= \dfrac{3}{8}\cdot\dfrac{5}{5}$ — Rewriting as a product of two fractions

$= \dfrac{3}{8}\cdot 1$ — $\dfrac{5}{5} = 1$

$= \dfrac{3}{8}.$ — Using the identity property of 1. We call this "removing a factor of 1."

The same steps are followed when simplifying rational expressions: We factor and then remove a factor of 1.

EXAMPLE 2

Simplify: $\dfrac{8x^2}{24x}$.

Solution

$\dfrac{8x^2}{24x} = \dfrac{8\cdot x\cdot x}{3\cdot 8\cdot x}$ — Factoring the numerator and the denominator. Note the common factor of $8\cdot x$.

$= \dfrac{x}{3}\cdot\dfrac{8x}{8x}$ — Rewriting as a product of two rational expressions

$= \dfrac{x}{3}\cdot 1$ — $\dfrac{8x}{8x} = 1$

$= \dfrac{x}{3}$ — Removing a factor of 1 ❑

When two or more terms appear in a numerator or a denominator, we factor as we did in Chapter 5. Then we try to remove a factor of 1.

EXAMPLE 3

Simplify: **(a)** $\dfrac{5a + 15}{10}$; **(b)** $\dfrac{6x + 12}{7x + 14}$; **(c)** $\dfrac{6a^2 + 4a}{2a^2 + 2a}$; **(d)** $\dfrac{x^2 + 3x + 2}{x^2 - 1}$.

Solution

a) $\dfrac{5a + 15}{10} = \dfrac{5(a + 3)}{5 \cdot 2}$ Factoring the numerator and the denominator

$= \dfrac{5}{5} \cdot \dfrac{a + 3}{2}$ Rewriting as a product of two rational expressions

$= 1 \cdot \dfrac{a + 3}{2}$

$= \dfrac{a + 3}{2}$

Removing a factor of 1: $\dfrac{5}{5} = 1$

b) $\dfrac{6x + 12}{7x + 14} = \dfrac{6(x + 2)}{7(x + 2)}$ Factoring the numerator and the denominator

$= \dfrac{6}{7} \cdot \dfrac{x + 2}{x + 2}$ Rewriting as a product of two rational expressions

$= \dfrac{6}{7} \cdot 1$

$= \dfrac{6}{7}$

Removing a factor of 1: $\dfrac{x + 2}{x + 2} = 1$

c) $\dfrac{6a^2 + 4a}{2a^2 + 2a} = \dfrac{2a(3a + 2)}{2a(a + 1)}$ Factoring the numerator and the denominator

$= \dfrac{2a}{2a} \cdot \dfrac{3a + 2}{a + 1}$ Rewriting as a product of two rational expressions

$= 1 \cdot \dfrac{3a + 2}{a + 1}$ $\dfrac{2a}{2a} = 1$

$= \dfrac{3a + 2}{a + 1}$ Removing a factor of 1. Note in this step that you *cannot* remove the a's because they are not factors of the entire numerator and the entire denominator.

d) $\dfrac{x^2 + 3x + 2}{x^2 - 1} = \dfrac{(x + 2)(x + 1)}{(x + 1)(x - 1)}$ Factoring

$= \dfrac{x + 1}{x + 1} \cdot \dfrac{x + 2}{x - 1}$ Rewriting as a product of two rational expressions

$= 1 \cdot \dfrac{x + 2}{x - 1}$

$= \dfrac{x + 2}{x - 1}$ ❑

Canceling

Canceling is a shortcut that can be used—and easily *misused*—when working with rational expressions. As we stated in Section 1.3, canceling must be done with care and understanding. Essentially, canceling streamlines the steps in which a rational expression is simplified by removing a factor of 1. Example 3(d) could have been done faster as follows:

$$\frac{x^2 + 3x + 2}{x^2 - 1} = \frac{(x+2)\cancel{(x+1)}}{\cancel{(x+1)}(x-1)}$$ When a factor of 1 is noted, it is "canceled": $\frac{x+1}{x+1} = 1$.

$$= \frac{x+2}{x-1}.$$ Simplifying

> CAUTION! Canceling is often used incorrectly. The following cancellations are *incorrect*:
>
> $$\frac{\cancel{x}+2}{\cancel{x}+3}; \quad \frac{a^2-\cancel{5}}{\cancel{5}}; \quad \frac{\cancel{6x^2}+5\cancel{x}+1}{\cancel{4x^2}-3\cancel{x}}.$$
>
> Wrong! Wrong! Wrong!
>
> None of the above cancellations represents removing a factor of 1. Factors are parts of products. For example, in $x \cdot 2$, x and 2 are factors, but in $x + 2$, x and 2 are *not* factors. If you can't factor, you can't cancel! When in doubt, don't cancel!

EXAMPLE 4

Simplify: $\dfrac{3x^2 - 2x - 1}{x^2 - 3x + 2}$.

Solution We factor the numerator and the denominator and look for common factors:

$$\frac{3x^2 - 2x - 1}{x^2 - 3x + 2} = \frac{(3x+1)\cancel{(x-1)}}{(x-2)\cancel{(x-1)}}$$ Try to visualize this as $\frac{3x+1}{x-2} \cdot \frac{x-1}{x-1}$.

$$= \frac{3x+1}{x-2}.$$ Removing a factor of 1: $\frac{x-1}{x-1} = 1$ ❑

Checking

When a rational expression is simplified, the result is an equivalent expression. Example 3(a) says that

$$\frac{5a+15}{10} \quad \text{is equivalent to} \quad \frac{a+3}{2}.$$

This result can be partially checked using a value of a for which both expressions are defined. For instance, if $a = 2$, then

$$\frac{5a+15}{10} = \frac{5 \cdot 2 + 15}{10} = \frac{25}{10} = \frac{5}{2} \quad \text{and} \quad \frac{a+3}{2} = \frac{2+3}{2} = \frac{5}{2}.$$

Had we evaluated both expressions and obtained differing results, we would have known that a mistake had been made.

For instance, had $(5a + 15)/10$ been incorrectly simplified as $(a + 15)/2$ and we had evaluated using $a = 2$, we would have found that

$$\frac{5a+15}{10} = \frac{5 \cdot 2 + 15}{10} = \frac{5}{2}, \quad \text{whereas} \quad \frac{a+15}{2} = \frac{2+15}{2} = \frac{17}{2}.$$

Factors That Are Opposites

Consider

$$\frac{x-4}{4-x}.$$

At first glance the numerator and the denominator do not appear to have any common factors other than 1. But $x - 4$ and $4 - x$ are opposites, or additive inverses, of each other. Thus we can find a common factor by factoring out -1 in one expression.

EXAMPLE 5

Simplify: $\frac{x-4}{4-x}$.

Solution

$$\frac{x-4}{4-x} = \frac{x-4}{-1(-4+x)}$$ Factoring out -1 in the denominator

$$= \frac{x-4}{-1(x-4)}$$ $-4 + x = x + (-4) = x - 4$

$$= \frac{1}{-1} \cdot \frac{x-4}{x-4}$$ Rewriting as a product. It helps to write the 1 in the numerator.

$$= -1$$

As a partial check, note that for any choice of x other than 4, the value of the rational expression is -1. For instance, if $x = 6$, then

$$\frac{x-4}{4-x} = \frac{6-4}{4-6} = \frac{2}{-2}$$

$$= -1.$$

❑

EXERCISE SET 6.1

List all numbers for which the rational expression is undefined.

1. $\frac{-5}{2x}$
2. $\frac{14}{-5y}$
3. $\frac{a+7}{a-8}$
4. $\frac{a-8}{a+7}$
5. $\frac{3}{2y+5}$
6. $\frac{x^2-9}{4x-12}$
7. $\frac{x^2+11}{x^2-3x-28}$
8. $\frac{p^2-9}{p^2-7p+10}$
9. $\frac{m^3-2m}{m^2-25}$
10. $\frac{7-3x+x^2}{49-x^2}$

Simplify by removing a factor of 1. Show all steps.

11. $\frac{10a^3b}{30ab^2}$
12. $\frac{45x^3y^2}{9x^5y}$
13. $\frac{35x^2y}{14x^3y^5}$
14. $\frac{12a^5b^6}{18a^3b}$
15. $\frac{9x+15}{6x+10}$
16. $\frac{14x-7}{10x-5}$
17. $\frac{a^2-25}{a^2+6a+5}$
18. $\frac{a^2+5a+6}{a^2-9}$

Simplify, if possible. Then check by evaluating.

19. $\frac{48x^4}{18x^6}$
20. $\frac{76a^5}{24a^3}$

21. $\frac{4x - 12}{4x}$

22. $\frac{-2y + 6}{-4y}$

23. $\frac{3m^2 + 3m}{6m^2 + 9m}$

24. $\frac{4y^2 - 2y}{5y^2 - 5y}$

25. $\frac{a^2 - 9}{a^2 + 5a + 6}$

26. $\frac{t^2 - 25}{t^2 + t - 20}$

27. $\frac{2t^2 + 6t + 4}{4t^2 - 12t - 16}$

28. $\frac{3a^2 - 9a - 12}{6a^2 + 30a + 24}$

29. $\frac{x^2 - 25}{x^2 - 10x + 25}$

30. $\frac{x^2 + 8x + 16}{x^2 - 16}$

31. $\frac{a^2 - 1}{a - 1}$

32. $\frac{t^2 - 1}{t + 1}$

33. $\frac{x^2 + 1}{x + 1}$

34. $\frac{y^2 + 4}{y + 2}$

35. $\frac{6x^2 - 54}{4x^2 - 36}$

36. $\frac{8x^2 - 32}{4x^2 - 16}$

37. $\frac{6t + 12}{t^2 - t - 6}$

38. $\frac{5y + 5}{y^2 + 7y + 6}$

39. $\frac{a^2 - 10a + 21}{a^2 - 11a + 28}$

40. $\frac{y^2 - 3y - 18}{y^2 - 2y - 15}$

41. $\frac{t^2 - 4}{(t + 2)^2}$

42. $\frac{(a - 3)^2}{a^2 - 9}$

43. $\frac{6 - x}{x - 6}$

44. $\frac{x - 8}{8 - x}$

45. $\frac{a - b}{b - a}$

46. $\frac{q - p}{-p + q}$

47. $\frac{6t - 12}{2 - t}$

48. $\frac{5a - 15}{3 - a}$

49. $\frac{a^2 - 1}{1 - a}$

50. $\frac{a^2 - b^2}{b^2 - a^2}$

Skill Maintenance

Factor.

51. $x^2 + 8x + 7$

52. $x^2 - 9x + 14$

Find the intercepts. Then graph.

53. $5x + 2y = 20$

54. $2x - 4y = 8$

Synthesis

55. ◈ How is canceling related to the identity property of 1?

56. ◈ Explain why evaluating is not a foolproof check when simplifying rational expressions.

Simplify.

57. $\frac{x^4 - 16y^4}{(x^2 + 4y^2)(x - 2y)}$

58. $\frac{(a - b)^2}{b^2 - a^2}$

59. $\frac{(t^4 - 1)(t^2 - 9)(t - 9)^2}{(t^4 - 81)(t^2 + 1)(t + 1)^2}$

60. $\frac{(t + 2)^3(t^2 + 2t + 1)(t + 1)}{(t + 1)^3(t^2 + 4t + 4)(t + 2)}$

61. $\frac{(x^2 - y^2)(x^2 - 2xy + y^2)}{(x - y)^2(x^2 - 4xy - 5y^2)}$

62. $\frac{(x - 1)(x^4 - 1)(x^2 - 1)}{(x^2 + 1)(x - 1)^2(x^4 - 2x^2 + 1)}$

63. ◈ Select any number x, multiply by 2, add 5, multiply by 5, subtract 25, and divide by 10. What do you get? Explain how this procedure can be used for a number trick.

6.2 Multiplication and Division

Multiplication • Division

Multiplication and division of rational expressions is similar to multiplication and division with fractional notation.

Multiplication

Recall that to multiply fractions, we simply multiply their numerators and multiply their denominators. Rational expressions are multiplied the same way.

To multiply rational expressions, multiply numerators and multiply denominators:

$$\frac{A}{B} \cdot \frac{C}{D} = \frac{AC}{BD}.$$

Then factor and simplify the result if possible.

For example,

$$\frac{3}{5} \cdot \frac{8}{11} = \frac{24}{55}$$

and

$$\frac{x}{3} \cdot \frac{x+2}{y} = \frac{x(x+2)}{3y}.$$

Fraction bars are grouping symbols, so parentheses are needed when writing some products. Because we normally simplify, it is best to leave parentheses in the product. There is no need to multiply further.

EXAMPLE 1

Multiply and simplify.

a) $\dfrac{5a^3}{4} \cdot \dfrac{2}{5a}$

b) $\dfrac{x^2+6x+9}{x^2-4} \cdot \dfrac{x-2}{x+3}$

c) $\dfrac{x^2+x-2}{15} \cdot \dfrac{5}{2x^2-3x+1}$

Solution

a) $\dfrac{5a^3}{4} \cdot \dfrac{2}{5a} = \dfrac{5a^3(2)}{4(5a)}$ Multiplying the numerators and the denominators

$= \dfrac{2 \cdot 5 \cdot a \cdot a \cdot a}{2 \cdot 2 \cdot 5 \cdot a}$ Factoring the numerator and the denominator

$= \dfrac{\cancel{2} \cdot \cancel{5} \cdot \cancel{a} \cdot a \cdot a}{\cancel{2} \cdot 2 \cdot \cancel{5} \cdot \cancel{a}}$

$= \dfrac{a^2}{2}$ Removing a factor of 1: $\dfrac{2 \cdot 5 \cdot a}{2 \cdot 5 \cdot a} = 1$

b) $\dfrac{x^2+6x+9}{x^2-4} \cdot \dfrac{x-2}{x+3} = \dfrac{(x^2+6x+9)(x-2)}{(x^2-4)(x+3)}$ Multiplying the numerators and the denominators

$= \dfrac{(x+3)(x+3)(x-2)}{(x+2)(x-2)(x+3)}$ Factoring the numerator and the denominator

$= \dfrac{\cancel{(x+3)}(x+3)\cancel{(x-2)}}{(x+2)\cancel{(x-2)}\cancel{(x+3)}}$

$= \dfrac{x+3}{x+2}$ Removing a factor of 1: $\dfrac{(x+3)(x-2)}{(x+3)(x-2)} = 1$

c) $$\frac{x^2 + x - 2}{15} \cdot \frac{5}{2x^2 - 3x + 1} = \frac{(x^2 + x - 2)5}{15(2x^2 - 3x + 1)}$$ Multiplying the numerators and the denominators

$$= \frac{(x + 2)(x - 1)5}{5(3)(x - 1)(2x - 1)}$$ Factoring the numerator and the denominator. Try to go directly to this step.

$$= \frac{(x + 2)\cancel{(x - 1)}\cancel{5}}{\cancel{5}(3)\cancel{(x - 1)}(2x - 1)}$$

$$= \frac{x + 2}{3(2x - 1)}$$ Removing a factor of 1: $\frac{5(x - 1)}{5(x - 1)} = 1$

You need not carry out this multiplication.

Division

As with fractions, reciprocals of rational expressions are found by interchanging the numerator and the denominator. For example,

the reciprocal of $\frac{2}{7}$ is $\frac{7}{2}$, and the reciprocal of $\frac{3x}{x + 5}$ is $\frac{x + 5}{3x}$.

To divide by a rational expression, multiply by its reciprocal:

$$\frac{A}{B} \div \frac{C}{D} = \frac{A}{B} \cdot \frac{D}{C} = \frac{AD}{BC}.$$

Then factor and simplify if possible.

EXAMPLE 2

Divide: **(a)** $\frac{x}{5} \div \frac{7}{y}$; **(b)** $(x + 1) \div \frac{x - 1}{x + 3}$.

Solution

a) $$\frac{x}{5} \div \frac{7}{y} = \frac{x}{5} \cdot \frac{y}{7}$$ Multiplying by the reciprocal of the divisor

$$= \frac{xy}{35}$$ Multiplying rational expressions

b) $$(x + 1) \div \frac{x - 1}{x + 3} = \frac{x + 1}{1} \cdot \frac{x + 3}{x - 1}$$ Multiplying by the reciprocal of the divisor. Writing $x + 1$ as $\frac{x + 1}{1}$ can be helpful.

$$= \frac{(x + 1)(x + 3)}{x - 1}$$

As usual, we should simplify when possible.

EXAMPLE 3

Divide and simplify.

a) $\dfrac{x+1}{x^2-1} \div \dfrac{x+1}{x^2-2x+1}$

b) $\dfrac{a^2+3a+2}{a^2+4} \div (5a^2+10a)$

c) $\dfrac{x^2-2x-3}{x^2-4} \div \dfrac{x+1}{x+5}$

Solution

a) $\dfrac{x+1}{x^2-1} \div \dfrac{x+1}{x^2-2x+1} = \dfrac{x+1}{x^2-1} \cdot \dfrac{x^2-2x+1}{x+1}$ Multiplying by the reciprocal

$= \dfrac{(x+1)(x-1)(x-1)}{(x+1)(x-1)(x+1)}$ Multiplying rational expressions and factoring numerators and denominators

$= \dfrac{\cancel{(x+1)}\cancel{(x-1)}(x-1)}{\cancel{(x+1)}\cancel{(x-1)}(x+1)}$

$= \dfrac{x-1}{x+1}$

Removing a factor of 1: $\dfrac{(x+1)(x-1)}{(x+1)(x-1)} = 1$

b) $\dfrac{a^2+3a+2}{a^2+4} \div (5a^2+10a) = \dfrac{a^2+3a+2}{a^2+4} \cdot \dfrac{1}{5a^2+10a}$ Multiplying by the reciprocal

$= \dfrac{(a+2)(a+1)}{(a^2+4)5a(a+2)}$ Multiplying rational expressions and factoring

$= \dfrac{\cancel{(a+2)}(a+1)}{(a^2+4)5a\cancel{(a+2)}}$

$= \dfrac{a+1}{(a^2+4)5a}$

Removing a factor of 1: $\dfrac{a+2}{a+2} = 1$

c) $\dfrac{x^2-2x-3}{x^2-4} \div \dfrac{x+1}{x+5} = \dfrac{x^2-2x-3}{x^2-4} \cdot \dfrac{x+5}{x+1}$ Multiplying by the reciprocal

$= \dfrac{(x-3)(x+1)(x+5)}{(x-2)(x+2)(x+1)}$ Multiplying rational expressions and factoring

$= \dfrac{(x-3)\cancel{(x+1)}(x+5)}{(x-2)(x+2)\cancel{(x+1)}}$

$= \dfrac{(x-3)(x+5)}{(x-2)(x+2)}$

Removing a factor of 1: $\dfrac{x+1}{x+1} = 1$ ❑

EXERCISE SET 6.2

Multiply. Leave parentheses in the product.

1. $\dfrac{3x}{2} \cdot \dfrac{x+4}{x-1}$
2. $\dfrac{4x}{5} \cdot \dfrac{x-3}{x+2}$
3. $\dfrac{x-1}{x+2} \cdot \dfrac{x+1}{x+2}$
4. $\dfrac{x-2}{x-5} \cdot \dfrac{x-2}{x+5}$

5. $\dfrac{2x+3}{4} \cdot \dfrac{x+1}{x-5}$

6. $\dfrac{-5}{3x-4} \cdot \dfrac{-6}{5x+6}$

7. $\dfrac{a-5}{a^2+1} \cdot \dfrac{a+2}{a^2-1}$

8. $\dfrac{t+3}{t^2-2} \cdot \dfrac{t+3}{t^2-2}$

9. $\dfrac{x+1}{2+x} \cdot \dfrac{x-1}{x+1}$

10. $\dfrac{m^2+5}{m+8} \cdot \dfrac{m^2-4}{m^2-4}$

Multiply and simplify, if possible.

11. $\dfrac{4x^3}{3x} \cdot \dfrac{14}{x}$

12. $\dfrac{32}{b^4} \cdot \dfrac{3b^2}{8}$

13. $\dfrac{3c}{d^2} \cdot \dfrac{4d}{6c^3}$

14. $\dfrac{3x^2y}{2} \cdot \dfrac{4}{xy^3}$

15. $\dfrac{x^2-3x-10}{(x-2)^2} \cdot \dfrac{x-2}{x-5}$

16. $\dfrac{t^2}{t^2-4} \cdot \dfrac{t^2-5t+6}{t^2-3t}$

17. $\dfrac{a^2-25}{a^2-4a+3} \cdot \dfrac{2a-5}{2a+5}$

18. $\dfrac{x+3}{x^2+9} \cdot \dfrac{x^2+5x+4}{x+9}$

19. $\dfrac{a^2-9}{a^2} \cdot \dfrac{a^2-3a}{a^2+a-12}$

20. $\dfrac{x^2+10x-11}{x^2-1} \cdot \dfrac{x+1}{x+11}$

21. $\dfrac{4a^2}{3a^2-12a+12} \cdot \dfrac{3a-6}{2a}$

22. $\dfrac{5v+5}{v-2} \cdot \dfrac{v^2-4v+4}{v^2-1}$

23. $\dfrac{t^2+2t-3}{t^2+4t-5} \cdot \dfrac{t^2-3t-10}{t^2+5t+6}$

24. $\dfrac{x^2+5x+4}{x^2-6x+8} \cdot \dfrac{x^2+5x-14}{x^2+8x+7}$

25. $\dfrac{5a^2-180}{10a^2-10} \cdot \dfrac{20a+20}{2a-12}$

26. $\dfrac{2t^2-98}{4t^2-4} \cdot \dfrac{8t+8}{16t-112}$

27. $\dfrac{x^2-1}{x^2-9} \cdot \dfrac{(x-3)^4}{(x+1)^2}$

28. $\dfrac{(x+2)^5}{(x-1)^3} \cdot \dfrac{x^2-1}{x^2+5x+6}$

29. $\dfrac{a^2-4}{a^2+2a+1} \cdot \dfrac{a-1}{a^4+1}$

30. $\dfrac{a^2+4}{a^2-6a+9} \cdot \dfrac{a^2+6a+9}{a^4+16}$

31. $\dfrac{(t-2)^3}{(t-1)^3} \cdot \dfrac{t^2-2t+1}{t^2-4t+4}$

32. $\dfrac{(y+4)^3}{(y+2)^3} \cdot \dfrac{y^2+4y+4}{y^2+8y+16}$

Find the reciprocal.

33. $\dfrac{4}{x}$

34. $\dfrac{a+3}{a-1}$

35. x^2-y^2

36. $\dfrac{1}{a+b}$

37. $\dfrac{x^2+2x-5}{x^2-4x+7}$

38. $\dfrac{x^2-3xy+y^2}{x^2+7xy-y^2}$

Divide and simplify, if possible.

39. $\dfrac{2}{5} \div \dfrac{4}{3}$

40. $\dfrac{5}{6} \div \dfrac{2}{3}$

41. $\dfrac{2}{x} \div \dfrac{8}{x}$

42. $\dfrac{x}{2} \div \dfrac{3}{x}$

43. $\dfrac{x^2}{y} \div \dfrac{x^3}{y^3}$

44. $\dfrac{a}{b^2} \div \dfrac{a^2}{b^3}$

45. $\dfrac{a+2}{a-3} \div \dfrac{a-1}{a+3}$

46. $\dfrac{y+2}{4} \div \dfrac{y}{2}$

47. $\dfrac{x^2-1}{x} \div \dfrac{x+1}{x-1}$

48. $\dfrac{4y-8}{y+2} \div \dfrac{y-2}{y^2-4}$

49. $\dfrac{x+1}{6} \div \dfrac{x+1}{3}$

50. $\dfrac{a}{a-b} \div \dfrac{b}{a-b}$

51. $(y^2-9) \div \dfrac{y^2-2y-3}{y^2+1}$

52. $(x^2-5x-6) \div \dfrac{x^2-1}{x+6}$

53. $\dfrac{5x-5}{16} \div \dfrac{x-1}{6}$

54. $\dfrac{-4+2x}{8} \div \dfrac{x-2}{2}$

55. $\dfrac{-6+3x}{5} \div \dfrac{4x-8}{25}$

56. $\dfrac{-12+4x}{4} \div \dfrac{-6+2x}{6}$

57. $\dfrac{a+2}{a-1} \div \dfrac{3a+6}{a-5}$

58. $\dfrac{t-3}{t+2} \div \dfrac{4t-12}{t+1}$

59. $(x-5) \div \dfrac{2x^2-11x+5}{4x^2-1}$

60. $(a+7) \div \dfrac{3a^2+14a-49}{a^2+8a+7}$

61. $\dfrac{x^2-4}{x} \div \dfrac{x-2}{x+2}$

62. $\dfrac{x+y}{x-y} \div \dfrac{x^2+y}{x^2-y^2}$

63. $\dfrac{x^2-9}{4x+12} \div \dfrac{x-3}{6}$

64. $\dfrac{x-b}{2x} \div \dfrac{x^2-b^2}{5x^2}$

65. $\dfrac{c^2+3c}{c^2+2c-3} \div \dfrac{c}{c+1}$

66. $\dfrac{x-5}{2x} \div \dfrac{x^2-25}{4x^2}$

67. $\dfrac{2y^2-7y+3}{2y^2+3y-2} \div \dfrac{6y^2-5y+1}{3y^2+5y-2}$

68. $\dfrac{x^2-x-20}{x^2+7x+12} \div \dfrac{x^2-10x+25}{x^2+6x+9}$

69. $\dfrac{c^2+10c+21}{c^2-2c-15} \div (c^2+2c-35)$

70. $\dfrac{1-z}{1+2z-z^2} \div (1-z)$

71. $\dfrac{(t+5)^3}{(t-5)^3} \div \dfrac{(t+5)^2}{(t-5)^2}$

72. $\dfrac{(y-3)^3}{(y+3)^3} \div \dfrac{(y-3)^2}{(y+3)^2}$

Skill Maintenance

73. Sixteen more than the square of a number is eight times the number. Find the number.

Subtract.

74. $(6x^2+7)-(4x^2-9)$

75. $(8x^3-3x^2+7)-(8x^2+3x-5)$

76. $(0.08y^3-0.04y^2+0.01y)-(0.02y^3+0.05y^2+1)$

Synthesis

77. ◈ Explain why the quotient

$$\frac{x+3}{x-5} \div \frac{x-7}{x+1}$$

is undefined for $x=5$, $x=-1$, and $x=7$.

78. ◈ A student claims to be able to divide, but not multiply, rational expressions. Why is this claim difficult to believe?

Simplify.

79. $\dfrac{2a^2-5ab}{c-3d} \div (4a^2-25b^2)$

80. $(x-2a) \div \dfrac{a^2x^2-4a^4}{a^2x+2a^3}$

81. $\dfrac{3a^2-5ab-12b^2}{3ab+4b^2} \div (3b^2-ab)$

82. $\dfrac{3x^2-2xy-y^2}{x^2-y^2} \div (3x^2+4xy+y^2)$

83. $xy \cdot \dfrac{y^2-4xy}{y-x} \div \dfrac{16x^2y^2-y^4}{4x^2-3xy-y^2}$

84. $\dfrac{z^2-8z+16}{z^2+8z+16} \div \dfrac{(z-4)^5}{(z+4)^5}$

85. $\dfrac{x^2-x+xy-y}{x^2+6x-7} \div \dfrac{x^2+2xy+y^2}{4x+4y}$

86. $\dfrac{3x+3y+3}{9x} \div \dfrac{x^2+2xy+y^2-1}{x^4+x^2}$

87. $\dfrac{t^4-1}{t^4-81} \cdot \dfrac{t^2-9}{t^2+1} \div \dfrac{(t+1)^2}{(t-9)^2}$

88. $\dfrac{(t+2)^3}{(t+1)^3} \div \dfrac{t^2+4t+4}{t^2+2t+1} \cdot \dfrac{t+1}{t+2}$

89. $\left(\dfrac{y^2+5y+6}{y^2} \cdot \dfrac{3y^3+6y^2}{y^2-y-12}\right) \div \dfrac{y^2-y}{y^2-2y-8}$

90. $\dfrac{a^4-81b^4}{a^2c-6abc+9b^2c} \cdot \dfrac{a+3b}{a^2+9b^2} \div \dfrac{a^2+6ab+9b^2}{(a-3b)^2}$

6.3 Addition and Subtraction

Addition When Denominators Are the Same • Subtraction When Denominators Are the Same • Addition and Subtraction When Denominators Are Opposites

Addition When Denominators Are the Same

Recall that to add fractions having the same denominator, like $\frac{2}{7}$ and $\frac{3}{7}$, we add the numerators: $\frac{2}{7}+\frac{3}{7}=\frac{5}{7}$. The same procedure is used to add rational expressions.

To add when the denominators are the same, add the numerators and keep the same denominator:

$$\frac{A}{B} + \frac{C}{B} = \frac{A + C}{B}.$$

Whenever possible, we will simplify the final result.

EXAMPLE 1

Add.

a) $\dfrac{4}{a} + \dfrac{3 + a}{a}$ **b)** $\dfrac{3x}{x - 5} + \dfrac{2x + 1}{x - 5}$

c) $\dfrac{2x^2 + 3x - 7}{2x + 1} + \dfrac{x^2 + x - 8}{2x + 1}$ **d)** $\dfrac{x - 5}{x^2 - 9} + \dfrac{2}{x^2 - 9}$

Solution

a) $\dfrac{4}{a} + \dfrac{3 + a}{a} = \dfrac{7 + a}{a}$ When the denominators are alike, add the numerators.

b) $\dfrac{3x}{x - 5} + \dfrac{2x + 1}{x - 5} = \dfrac{5x + 1}{x - 5}$ Adding the numerators

c) $\dfrac{2x^2 + 3x - 7}{2x + 1} + \dfrac{x^2 + x - 8}{2x + 1} = \dfrac{(2x^2 + 3x - 7) + (x^2 + x - 8)}{2x + 1}$

$= \dfrac{3x^2 + 4x - 15}{2x + 1}$ Collecting like terms in the numerator

d) $\dfrac{x - 5}{x^2 - 9} + \dfrac{2}{x^2 - 9} = \dfrac{(x - 5) + 2}{x^2 - 9}$

$= \dfrac{x - 3}{x^2 - 9}$ Collecting like terms in the numerator

$= \dfrac{x - 3}{(x - 3)(x + 3)}$ Factoring

$= \dfrac{\cancel{x - 3}}{(\cancel{x - 3})(x + 3)}$

$= \dfrac{1}{x + 3}$ Removing a factor of 1: $\dfrac{x - 3}{x - 3} = 1$ ❑

Subtraction When Denominators Are the Same

Recall that to subtract fractions having the same denominator, we subtract the numerators—for example, $\frac{5}{7} - \frac{2}{7} = \frac{3}{7}$. The same procedure is used to subtract rational expressions.

To subtract when the denominators are the same, subtract the numerators and keep the same denominator:

$$\frac{A}{B} - \frac{C}{B} = \frac{A - C}{B}.$$

CAUTION! Keep in mind that a fraction bar is a grouping symbol. When a numerator is being subtracted, remember to subtract *every* term in that expression.

EXAMPLE 2

Subtract: **(a)** $\frac{3x}{x+2} - \frac{x-5}{x+2}$; **(b)** $\frac{x^2}{x-4} - \frac{x+12}{x-4}$.

Solution

a) $\frac{3x}{x+2} - \frac{x-5}{x+2} = \frac{3x - (x-5)}{x+2}$ The parentheses are needed to make sure that we subtract both terms.

$= \frac{3x - x + 5}{x+2}$ Removing the parentheses and changing signs

$= \frac{2x+5}{x+2}$ Collecting like terms

b) $\frac{x^2}{x-4} - \frac{x+12}{x-4} = \frac{x^2 - (x+12)}{x-4}$ Remember the parentheses!

$= \frac{x^2 - x - 12}{x-4}$ Removing parentheses

$= \frac{(x-4)(x+3)}{x-4}$

$= \frac{\cancel{(x-4)}(x+3)}{\cancel{x-4}}$ Simplifying by factoring and removing a factor of 1: $\frac{x-4}{x-4} = 1$

$= x + 3$

❑

Addition and Subtraction When Denominators Are Opposites

Recall from Chapter 1 that

$$\frac{a}{-b} = \frac{-a}{b} = -\frac{a}{b} \quad \text{and} \quad m - (-n) = m + n.$$

These results can be used to add or subtract when denominators are opposites.

EXAMPLE 3

Perform the indicated operation: **(a)** $\frac{x}{2} + \frac{3}{-2}$; **(b)** $\frac{x}{5} - \frac{3x-4}{-5}$.

Solution

a) We rewrite $\frac{3}{-2}$ as $\frac{-3}{2}$ to obtain a common denominator of 2. Then we add and, if possible, simplify:

$$\frac{x}{2} + \frac{3}{-2} = \frac{x}{2} + \frac{-3}{2} \qquad \text{Since } \frac{a}{-b} = \frac{-a}{b}$$

$$= \frac{x + (-3)}{2} = \frac{x - 3}{2}.$$

b)
$$\frac{x}{5} - \frac{3x - 4}{-5} = \frac{x}{5} - \left(-\frac{3x - 4}{5}\right) \qquad \text{Since } \frac{a}{-b} = -\frac{a}{b}$$

$$= \frac{x}{5} + \frac{3x - 4}{5} \qquad \text{Since } m - (-n) = m + n$$

$$= \frac{x + (3x - 4)}{5}$$

$$= \frac{4x - 4}{5}$$

❑

Sometimes denominators are of the form $x - a$ and $a - x$. As we saw in Section 6.1, expressions of this form are opposites of each other.

EXAMPLE 4

Perform the indicated operation: **(a)** $\frac{5}{x - 2} - \frac{3}{2 - x}$; **(b)** $\frac{3y}{y - 5} + \frac{y + 1}{5 - y}$.

Solution

a) Note that $x - 2$ and $2 - x$ are opposites. Thus,

$$\frac{5}{x - 2} - \frac{3}{2 - x} = \frac{5}{x - 2} - \frac{3}{-(x - 2)} \qquad \text{Since } 2 - x \text{ is the opposite of } x - 2$$

$$= \frac{5}{x - 2} - \left(-\frac{3}{x - 2}\right) \qquad \frac{a}{-b} = -\frac{a}{b}$$

$$= \frac{5}{x - 2} + \frac{3}{x - 2} \qquad m - (-n) = m + n$$

$$= \frac{8}{x - 2}.$$

b)
$$\frac{3y}{y - 5} + \frac{y + 1}{5 - y} = \frac{3y}{y - 5} + \frac{y + 1}{-(y - 5)} \qquad \text{Since } 5 - y \text{ is the opposite of } y - 5$$

$$= \frac{3y}{y - 5} + \frac{-(y + 1)}{y - 5} \qquad \frac{a}{-b} = \frac{-a}{b}$$

$$= \frac{3y}{y - 5} + \frac{-y - 1}{y - 5} \qquad \text{The opposite of } y + 1 \text{ is } -y - 1.$$

$$= \frac{3y + (-y - 1)}{y - 5} \qquad \text{Adding rational expressions}$$

$$= \frac{2y - 1}{y - 5}$$

❑

Note that the second step of Example 4(b) could have been written as

$$\frac{3y}{y-5} + \left(-\frac{y+1}{y-5}\right) \quad \text{or} \quad \frac{3y}{y-5} - \frac{y+1}{y-5}.$$

Either expression would have led to the same result.

EXERCISE SET 6.3

Perform the indicated operation. Simplify, if possible.

1. $\frac{3}{x} + \frac{5}{x}$
2. $\frac{4}{a^2} + \frac{9}{a^2}$
3. $\frac{x}{15} + \frac{2x+1}{15}$
4. $\frac{a}{7} + \frac{3a-4}{7}$
5. $\frac{2}{a+3} + \frac{4}{a+3}$
6. $\frac{5}{x+2} + \frac{8}{x+2}$
7. $\frac{9}{a+6} - \frac{5}{a+6}$
8. $\frac{8}{x+7} - \frac{2}{x+7}$
9. $\frac{3y+9}{2y} - \frac{y+1}{2y}$
10. $\frac{5+3t}{4t} - \frac{2t+1}{4t}$
11. $\frac{9x+5}{x+1} + \frac{2x+3}{x+1}$
12. $\frac{3a+2}{a+4} + \frac{2a+7}{a+4}$
13. $\frac{9x+5}{x+1} - \frac{2x+3}{x+1}$
14. $\frac{3a+2}{a+4} - \frac{2a+7}{a+4}$
15. $\frac{a^2}{a-4} + \frac{a-20}{a-4}$
16. $\frac{x^2}{x+5} + \frac{7x+10}{x+5}$
17. $\frac{x^2}{x-2} - \frac{6x-8}{x-2}$
18. $\frac{a^2}{a+3} - \frac{2a+15}{a+3}$
19. $\frac{t^2+4t}{t-1} + \frac{2t-7}{t-1}$
20. $\frac{y^2+6y}{y+2} + \frac{2y+12}{y+2}$
21. $\frac{x+1}{x^2+5x+6} + \frac{2}{x^2+5x+6}$
22. $\frac{-7}{x^2-4x+3} + \frac{x+4}{x^2-4x+3}$
23. $\frac{a^2+3}{a^2+5a-6} - \frac{4}{a^2+5a-6}$
24. $\frac{a^2-3}{a^2-7a+12} - \frac{6}{a^2-7a+12}$
25. $\frac{t^2-3t}{t^2+6t+9} + \frac{2t-12}{t^2+6t+9}$
26. $\frac{y^2-7y}{y^2+8y+16} + \frac{6y-20}{y^2+8y+16}$
27. $\frac{2x^2+3}{x^2-6x+5} - \frac{x^2-5x+9}{x^2-6x+5}$
28. $\frac{2x^2+x}{x^2-8x+12} - \frac{x^2-2x+10}{x^2-8x+12}$
29. $\frac{3x}{8} + \frac{x}{-8}$
30. $\frac{5a}{6} + \frac{a}{-6}$
31. $\frac{3}{t} + \frac{4}{-t}$
32. $\frac{5}{-a} + \frac{8}{a}$
33. $\frac{2x+7}{x-6} + \frac{3x}{6-x}$
34. $\frac{3x-2}{4x-3} + \frac{2x-5}{3-4x}$
35. $\frac{a}{6} - \frac{7a}{-6}$
36. $\frac{x}{8} - \frac{5x}{-8}$
37. $\frac{5}{a} - \frac{8}{-a}$
38. $\frac{3}{t} - \frac{4}{-t}$
39. $\frac{x}{4} - \frac{3x-5}{-4}$
40. $\frac{2}{x-1} - \frac{2}{1-x}$
41. $\frac{y^2}{y-3} + \frac{9}{3-y}$
42. $\frac{t^2}{t-2} + \frac{4}{2-t}$
43. $\frac{b-7}{b^2-16} + \frac{7-b}{16-b^2}$
44. $\frac{a-3}{a^2-25} + \frac{a-3}{25-a^2}$
45. $\frac{3-2t}{t^2-5t+4} + \frac{2-3t}{t^2-5t+4}$
46. $\frac{7-2x}{x^2-6x+8} + \frac{3-3x}{x^2-6x+8}$
47. $\frac{3-x}{x-7} - \frac{2x-5}{7-x}$
48. $\frac{t^2}{t-2} - \frac{4}{2-t}$
49. $\frac{x-8}{x^2-16} - \frac{x-8}{16-x^2}$
50. $\frac{x-2}{x^2-25} - \frac{6-x}{25-x^2}$
51. $\frac{4-x}{x-9} - \frac{3x-8}{9-x}$
52. $\frac{3-x}{x-7} - \frac{2x-5}{7-x}$
53. $\frac{5-3x}{x^2-2x+1} - \frac{x+1}{x^2-2x+1}$

54. $\dfrac{x-7}{x^2+3x-4} - \dfrac{2x-3}{x^2+3x-4}$

Skill Maintenance

Graph.

55. $y = -1$

56. $x = 4$

57. $y = x - 1$

58. $y = \frac{1}{2}x - 1$

Synthesis

59. ◈ Are parentheses as important for adding rational expressions as they are for subtracting rational expressions? Why or why not?

60. ◈ Explain in your own words why the expressions

$$\frac{1}{3-x} \quad \text{and} \quad \frac{1}{x-3}$$

are opposites of each other.

Perform the indicated operations and simplify.

61. $\dfrac{3(2x+5)}{x-1} - \dfrac{3(2x-3)}{1-x} + \dfrac{6x-1}{x-1}$

62. $\dfrac{2x-y}{x-y} + \dfrac{x-2y}{y-x} - \dfrac{3x-3y}{x-y}$

63. $\dfrac{x-y}{x^2-y^2} + \dfrac{x+y}{x^2-y^2} - \dfrac{2x}{x^2-y^2}$

64. $\dfrac{x+y}{2(x-y)} - \dfrac{2x-2y}{2(x-y)} + \dfrac{x-3y}{2(y-x)}$

65. $\dfrac{10}{2y-1} - \dfrac{6}{1-2y} + \dfrac{y}{2y-1} + \dfrac{y-4}{1-2y}$

66. $\dfrac{(x+3)(2x-1)}{(2x-3)(x-3)} - \dfrac{(x-3)(x+1)}{(3-x)(3-2x)} + \dfrac{(2x+1)(x+3)}{(3-2x)(x-3)}$

67. $\dfrac{x}{(x-y)(y-z)} - \dfrac{x}{(y-x)(z-y)}$

68. $\dfrac{x}{x-y} + \dfrac{y}{y-x} + \dfrac{x+y}{x-y} + \dfrac{x-y}{y-x}$

69. $\dfrac{x^2}{3x^2-5x-2} - \dfrac{2x}{3x+1}\cdot\dfrac{1}{x-2}$

70. $\dfrac{3}{x+4}\cdot\dfrac{2x+11}{x-3} - \dfrac{-1}{4+x}\cdot\dfrac{6x+3}{3-x}$

71. ◈ Explain how evaluating can be used to perform a partial check on the result of Example 1(d):

$$\frac{x-5}{x^2-9} + \frac{2}{x^2-9} = \frac{1}{x+3}.$$

6.4 Least Common Multiples and Denominators

Least Common Multiples • Equivalent Expressions and Least Common Denominators

Like fractions, rational expressions must have common denominators before they can be added or subtracted. Our work will be easier if we use the smallest, or *least*, common denominator.

Least Common Multiples

To add fractions like $\frac{5}{12}$ and $\frac{7}{30}$, we first look for the smallest number that contains both 12 and 30 as factors. Such a number, the **least common multiple**, or **LCM**, of 12 and 30, will then be used as the **least common denominator**, or **LCD**.

Let's find the LCM of 12 and 30 using a method that can also be used with polynomials. We begin by writing the prime factorization of 12:

$$12 = 2 \cdot 2 \cdot 3.$$

Two factors of 2 and a factor of 3 must appear in the LCM if 12 is to be a factor of the LCM.

Next we write the prime factorization of 30:

$$30 = 2 \cdot 3 \cdot 5.$$

The factors 2, 3, and 5 must appear in the LCM if 30 is to be a factor of the LCM.

Note that the prime factorization of 12 already includes a 2 and a 3, but lacks a 5. The smallest product that contains both 12 and 30 as factors, 60, is found by multiplying $2 \cdot 2 \cdot 3$ by 5:

$$\text{LCM} = 2 \cdot 2 \cdot 3 \cdot 5$$

12 is a factor of the LCM. ($2 \cdot 2 \cdot 3$)

30 is a factor of the LCM. ($2 \cdot 3 \cdot 5$)

The factors common to the two denominators—in this case, 2 and 3—are used the greatest number of times that they appear in either of the individual factorizations.

To find the least common denominator (LCD):

1. Write the prime factorization of each denominator.
2. Select one of the factorizations and inspect it to see if it contains the other.
   a) If it does, it represents the LCD.
   b) If it does not, multiply that factorization by any factors of the other denominator that it lacks. The final product is the LCD.
3. The LCD should include each factor the greatest number of times that it occurs in any one factorization.

Let's finish adding $\frac{5}{12}$ and $\frac{7}{30}$:

$$\frac{5}{12} + \frac{7}{30} = \frac{5}{2 \cdot 2 \cdot 3} + \frac{7}{2 \cdot 3 \cdot 5}.$$

The least common denominator, or LCD, is $2 \cdot 2 \cdot 3 \cdot 5$. To get the LCD in the first denominator, we need a 5. To get the LCD in the second denominator, we need another 2. We get these numbers by multiplying by 1:

$$\frac{5}{12} + \frac{7}{30} = \frac{5}{2 \cdot 2 \cdot 3} \cdot \frac{5}{5} + \frac{7}{2 \cdot 3 \cdot 5} \cdot \frac{2}{2}$$ $\frac{5}{5} = 1$ and $\frac{2}{2} = 1$

$$= \frac{25}{2 \cdot 2 \cdot 3 \cdot 5} + \frac{14}{2 \cdot 3 \cdot 5 \cdot 2}$$ The denominators are now the LCD.

$$= \frac{39}{2 \cdot 2 \cdot 3 \cdot 5}$$ Adding the numerators and keeping the LCD

$$= \frac{\cancel{3} \cdot 13}{2 \cdot 2 \cdot \cancel{3} \cdot 5}$$

$$= \frac{13}{20}.$$

Simplifying by removing a factor of 1: $\frac{3}{3} = 1$

Rational expressions, like $7/(24x)$ and $5/(36x^2)$, can be added in much the same way that we added $\frac{5}{12}$ and $\frac{7}{30}$.

EXAMPLE 1

Find the LCM of $24x$ and $36x^2$.

Solution

1. We begin by writing the prime factorizations of $24x$ and $36x^2$:

$$24x = 2 \cdot 2 \cdot 2 \cdot 3 \cdot x,$$
$$36x^2 = 2 \cdot 2 \cdot 3 \cdot 3 \cdot x \cdot x.$$

2. Note that the factorization of $36x^2$ contains the entire factorization of $24x$ except for a third factor of 2. To find the smallest product that contains both $24x$ and $36x^2$ as factors, we need to add a third factor of 2 to $2 \cdot 2 \cdot 3 \cdot 3 \cdot x \cdot x$:

$$\text{LCM} = 2 \cdot 2 \cdot 3 \cdot 3 \cdot x \cdot x \cdot 2.$$

$36x^2$ is a factor.

$24x$ is a factor.

The LCM is thus $2^3 \cdot 3^2 \cdot x^2$, or $72x^2$. ❑

Now let's finish adding $\dfrac{7}{24x}$ and $\dfrac{5}{36x^2}$:

$$\frac{7}{24x} + \frac{5}{36x^2} = \frac{7}{2 \cdot 2 \cdot 2 \cdot 3 \cdot x} + \frac{5}{2 \cdot 2 \cdot 3 \cdot 3 \cdot x \cdot x}.$$

In Example 1, we found that the LCD is $2 \cdot 2 \cdot 2 \cdot 3 \cdot 3 \cdot x \cdot x$. To obtain equivalent expressions with this LCD, we multiply each expression by 1, using the missing factors of the LCD to write 1:

$$\frac{7}{24x} + \frac{5}{36x^2} = \frac{7}{2 \cdot 2 \cdot 2 \cdot 3 \cdot x} \cdot \frac{3 \cdot x}{3 \cdot x} + \frac{5}{2 \cdot 2 \cdot 3 \cdot 3 \cdot x \cdot x} \cdot \frac{2}{2}$$

The LCD requires additional factors of 3 and x.

The LCD requires another factor of 2.

$$= \frac{21x}{2 \cdot 2 \cdot 2 \cdot 3 \cdot x \cdot 3 \cdot x} + \frac{10}{2 \cdot 2 \cdot 3 \cdot 3 \cdot x \cdot x \cdot 2}$$

Both denominators are now the LCD.

$$= \frac{21x + 10}{72x^2}.$$

You now have the "big" picture of how LCMs are used with rational expressions. For the remainder of this section, we will practice finding LCMs and rewriting rational expressions so that they have the LCD as the denominator. In Section 6.5, we will carry out the actual addition and subtraction of rational expressions.

EXAMPLE 2

For each pair of polynomials, find the least common multiple.

a) $15a$ and $35b$

b) $21x^3y^6$ and $7x^5y^2$

c) $x^2 + 5x - 6$ and $x^2 - 1$

Solution

a) We write the prime factorizations and then construct the LCM:

$$15a = 3 \cdot 5 \cdot a$$
$$35b = 5 \cdot 7 \cdot b$$

15*a* is a factor of the LCM.

$$\text{LCM} = 3 \cdot 5 \cdot a \cdot 7 \cdot b$$

35*b* is a factor of the LCM.

The LCM is $3 \cdot 5 \cdot a \cdot 7 \cdot b$, or $105ab$.

b) $21x^3y^6 = 3 \cdot 7 \cdot x \cdot x \cdot x \cdot y \cdot y \cdot y \cdot y \cdot y \cdot y$
$7x^5y^2 = 7 \cdot x \cdot x \cdot x \cdot x \cdot x \cdot y \cdot y$

Try to visualize the factors of x and y mentally.

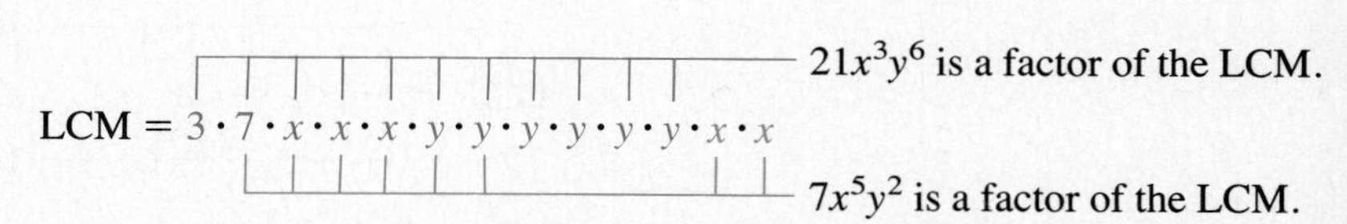

Note that we used the highest power of each factor in $3 \cdot 7 \cdot x^3y^6$ and $7x^5y^2$. The LCM is $21x^5y^6$.

c) $x^2 + 5x - 6 = (x - 1)(x + 6)$
$x^2 - 1 = (x - 1)(x + 1)$

$x^2 + 5x - 6$ is a factor of the LCM.

$$\text{LCM} = (x - 1)(x + 6)(x + 1)$$

$x^2 - 1$ is a factor of the LCM.

The LCM is $(x - 1)(x + 6)(x + 1)$. There is no need to multiply this out. ❑

The above procedure can be used to find the LCM of three polynomials as well. We factor each polynomial and then construct the LCM using each factor the greatest number of times that it appears in any one factorization.

EXAMPLE 3

For each group of polynomials, find the LCM.

a) $12x$, $16y$, and $8xyz$ **b)** $x^2 + 4$, $x + 1$, and 5

Solution

a) $12x = 2 \cdot 2 \cdot 3 \cdot x$
$16y = 2 \cdot 2 \cdot 2 \cdot 2 \cdot y$
$8xyz = 2 \cdot 2 \cdot 2 \cdot x \cdot y \cdot z$

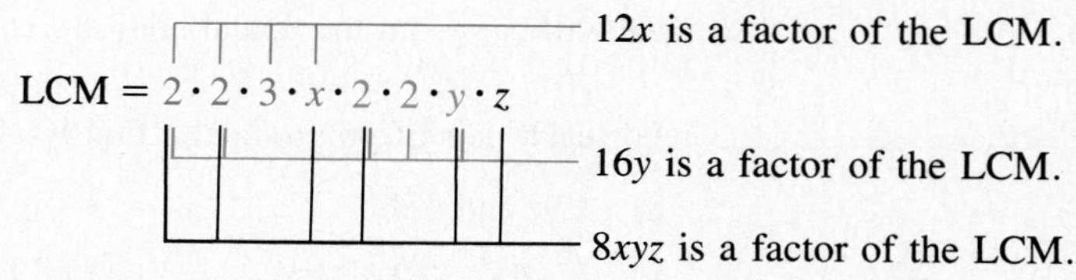

The LCM is $2^4 \cdot 3 \cdot xyz$, or $48xyz$.

b) Since $x^2 + 4$, $x + 1$, and 5 are not factorable, the LCM is their product: $5(x^2 + 4)(x + 1)$. ❑

When two or more rational expressions have different denominators, it is important to be able to write equivalent expressions that have the LCD.

EXAMPLE 4

Find equivalent expressions that have the LCD:

$$\frac{x+3}{x^2+5x-6}, \quad \frac{x+7}{x^2-1}.$$

Solution Look at Example 2(c). Note that the LCD is $(x + 6)(x - 1)(x + 1)$. Since $x^2 + 5x - 6 = (x + 6)(x - 1)$, the factor of the LCD that is missing from the first denominator is $x + 1$. We multiply by 1 using $(x + 1)/(x + 1)$:

$$\frac{x+3}{x^2+5x-6} = \frac{x+3}{(x+6)(x-1)} \cdot \frac{x+1}{x+1}$$
$$= \frac{(x+3)(x+1)}{(x+6)(x-1)(x+1)}.$$

For the second expression, we have $x^2 - 1 = (x + 1)(x - 1)$. The factor of the LCD that is missing is $x + 6$. We multiply by 1 using $(x + 6)/(x + 6)$:

$$\frac{x+7}{x^2-1} = \frac{x+7}{(x+1)(x-1)} \cdot \frac{x+6}{x+6}$$
$$= \frac{(x+7)(x+6)}{(x+1)(x-1)(x+6)}.$$

We leave the answers in factored form. ❑

EXERCISE SET 6.4

Find the LCM.

1. 12, 27
2. 10, 15
3. 8, 9
4. 12, 15
5. 6, 9, 21
6. 8, 36, 40
7. 24, 36, 40
8. 3, 4, 5
9. 28, 42, 60
10. 10, 100, 500

Add, first finding the LCM of the denominators. Simplify, if possible.

11. $\frac{7}{24} + \frac{11}{18}$
12. $\frac{7}{60} + \frac{6}{75}$
13. $\frac{1}{6} + \frac{3}{40} + \frac{2}{75}$
14. $\frac{5}{24} + \frac{3}{20} + \frac{7}{30}$
15. $\frac{2}{15} + \frac{5}{9} + \frac{3}{20}$
16. $\frac{1}{20} + \frac{1}{30} + \frac{2}{45}$

Find the LCM.

17. $6x^2$, $12x^3$
18. $2a^2b$, $8ab^2$
19. $2x^2$, $6xy$, $18y^2$
20. c^2d, cd^2, c^3d
21. $2(y - 3)$, $6(y - 3)$
22. $4(x - 1)$, $8(x - 1)$
23. t, $t + 2$, $t - 2$
24. x, $x + 3$, $x - 3$
25. $x^2 - 4$, $x^2 + 5x + 6$
26. $x^2 + 3x + 2$, $x^2 - 4$
27. $t^3 + 4t^2 + 4t$, $t^2 - 4t$
28. $y^3 - y^2$, $y^4 - y^2$
29. $9a^5b^2$, $6ab^6$
30. $10a^4b^7$, $15ab^8$
31. $10x^2y$, $6y^2z$, $5xz^3$
32. $8x^3z$, $12xy^2$, $4y^5z^2$
33. $a + 1$, $(a - 1)^2$, $a^2 - 1$
34. $x^2 - y^2$, $2x + 2y$, $x^2 + 2xy + y^2$
35. $m^2 - 5m + 6$, $m^2 - 4m + 4$
36. $2x^2 + 5x + 2$, $2x^2 - x - 1$
37. $2 + 3x$, $4 - 9x^2$, $2 - 3x$
38. $3 - 2x$, $9 - 4x^2$, $3 + 2x$

39. $10v^2 + 30v,\ 5v^2 + 35v + 60$

40. $12a^2 + 24a,\ 4a^2 + 20a + 24$

41. $9x^3 - 9x^2 - 18x,\ 6x^5 - 24x^4 + 24x^3$

42. $x^5 - 4x^3,\ x^3 + 4x^2 + 4x$

43. $x^5 + 4x^4 + 4x^3,\ 3x^2 - 12,\ 2x + 4$

44. $x^5 + 2x^4 + x^3,\ 2x^3 - 2x,\ 5x - 5$

Find equivalent expressions that have the LCD.

45. $\frac{7}{6x^5}, \frac{y}{12x^3}$

46. $\frac{3}{10a^3}, \frac{b}{5a^6}$

47. $\frac{3}{2a^2b}, \frac{5}{8ab^2}$

48. $\frac{7}{3x^4y^2}, \frac{4}{9xy^3}$

49. $\frac{x+1}{x^2-4}, \frac{x-2}{x^2+5x+6}$

50. $\frac{x-4}{x^2-9}, \frac{x+2}{x^2+11x+24}$

51. $\frac{3}{t}, \frac{4}{t+2}, \frac{t}{t-2}$

52. $\frac{1}{x}, \frac{-2}{x+3}, \frac{x^2}{x-3}$

53. $\frac{x+1}{2x-3}, \frac{x-2}{4x^2-9}, \frac{x+1}{2x+3}$

54. $\frac{x}{x^5+4x^4+4x^3}, \frac{3}{3x^2-12}$
(*Hint:* Simplify first.)

Skill Maintenance

Factor.

55. $x^2 - 19x + 60$

56. $x^2 + 9x - 36$

Find a polynomial that can represent the shaded area of the figure.

57.

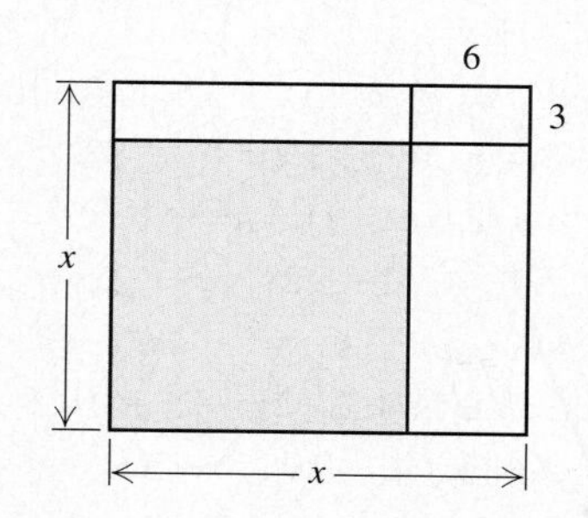

58.

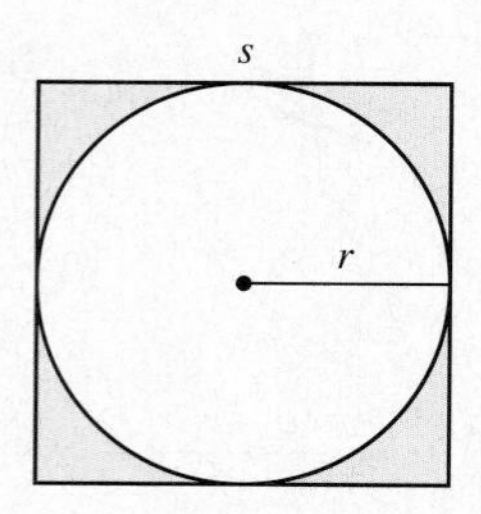

Synthesis

59. ◈ Explain why the product of two numbers is not always their LCM.

60. ◈ Is every LCD an LCM? Why or why not?

Find the LCM.

61. 72, 90, 96

62. $8x^2 - 8,\ 6x^2 - 12x + 6,\ 10 - 10x$

63. Two joggers leave the starting point of a fitness loop at the same time. One jogger completes a lap in 6 min and the second jogger in 8 min. Assuming they continue to run at the same pace, when will they next meet at the starting place?

64. ◈ If the LCM of two expressions is the same as one of the expressions, what relationship exists between the two expressions?

6.5

Addition and Subtraction with Unlike Denominators

Using LCDs for Addition and Subtraction • Combined Additions and Subtractions

We now know how to rewrite any two rational expressions in an equivalent form that uses the LCD. Once rational expressions share a common denominator, they can be added or subtracted just as they were in Section 6.3. The procedure is as follows.

To add or subtract rational expressions that have different denominators:

1. Find the LCD.
2. Multiply each rational expression by an expression for 1 made up of the factors of the LCD missing from that expression's denominator.
3. Add or subtract the numerators, as indicated. Write the sum or difference over the LCD.
4. Simplify, if possible.

EXAMPLE 1

Add: $\dfrac{5x^2}{8} + \dfrac{7x}{12}$.

Solution First, we find the LCD:

$$\left.\begin{aligned} 8 &= 2\cdot 2\cdot 2 \\ 12 &= 2\cdot 2\cdot 3 \end{aligned}\right\} \qquad \text{LCD} = 2\cdot 2\cdot 2\cdot 3, \text{ or } 24.$$

The denominator 8 needs to be multiplied by 3 to obtain the LCD. The denominator 12 needs to be multiplied by 2 to obtain the LCD. Thus we multiply by $\frac{3}{3}$ and $\frac{2}{2}$ to get the LCD. Then we add and, if possible, simplify.

$$\begin{aligned} \frac{5x^2}{8} + \frac{7x}{12} &= \frac{5x^2}{2\cdot 2\cdot 2} + \frac{7x}{2\cdot 2\cdot 3} \\ &= \frac{5x^2}{2\cdot 2\cdot 2}\cdot\frac{3}{3} + \frac{7x}{2\cdot 2\cdot 3}\cdot\frac{2}{2} \qquad \text{Multiplying twice by an expression for 1 to get the LCD} \\ &= \frac{15x^2}{24} + \frac{14x}{24} \\ &= \frac{15x^2 + 14x}{24} \end{aligned}$$

❑

Subtraction is performed in much the same way.

EXAMPLE 2

Subtract: $\frac{7}{8x} - \frac{5}{12x^2}$.

Solution First, we find the LCD:

$$\left.\begin{aligned} 8x &= 2\cdot 2\cdot 2\cdot x \\ 12x^2 &= 2\cdot 2\cdot 3\cdot x\cdot x \end{aligned}\right\} \quad \text{LCD} = 2\cdot 2\cdot 3\cdot x\cdot x\cdot 2, \text{ or } 24x^2.$$

The denominator $8x$ must be multiplied by $3x$ to obtain the LCD. The denominator $12x^2$ must be multiplied by 2 to obtain the LCD. Thus we multiply by $3x/3x$ and $2/2$ to get the LCD. Then we subtract and, if possible, simplify.

$$\frac{7}{8x} - \frac{5}{12x^2} = \frac{7}{8x}\cdot\frac{3\cdot x}{3\cdot x} - \frac{5}{12x^2}\cdot\frac{2}{2}$$

$$= \frac{21x}{24x^2} - \frac{10}{24x^2}$$

CAUTION! Do not simplify *these* rational expressions or you will lose the LCD.

$$= \frac{21x - 10}{24x^2}$$

❑

When denominators contain polynomials with two or more terms, the same steps are used.

EXAMPLE 3

Perform the indicated operations.

a) $\frac{2a}{a^2 - 1} + \frac{1}{a^2 + a}$

b) $\frac{x + 4}{x - 2} - \frac{x - 7}{x + 5}$

c) $\frac{x}{x^2 + 11x + 30} + \frac{-5}{x^2 + 9x + 20}$

d) $\frac{x}{x^2 + 5x + 6} - \frac{2}{x^2 + 3x + 2}$

Solution

a) First, we find the LCD:

$$\left.\begin{aligned} a^2 - 1 &= (a - 1)(a + 1) \\ a^2 + a &= a(a + 1) \end{aligned}\right\} \quad \text{LCD} = (a - 1)(a + 1)a.$$

We multiply by 1 to get the LCD in each expression. Then we add and, if possible, simplify.

$$\frac{2a}{a^2 - 1} + \frac{1}{a^2 + a} = \frac{2a}{(a - 1)(a + 1)}\cdot\frac{a}{a} + \frac{1}{a(a + 1)}\cdot\frac{a - 1}{a - 1}$$

Writing 1 as $\frac{a}{a}$ and as $\frac{a-1}{a-1}$ to get the LCD

$$= \frac{2a^2}{(a - 1)(a + 1)a} + \frac{a - 1}{(a - 1)(a + 1)a}$$

Then

$$\frac{2a}{a^2-1}+\frac{1}{a^2+a}=\frac{2a^2+a-1}{(a-1)(a+1)a}$$

$$=\frac{(2a-1)(a+1)}{(a-1)(a+1)a}$$

$$=\frac{2a-1}{(a-1)a}$$

Simplifying by factoring and removing a factor of 1: $\frac{a+1}{a+1}=1$

b) First, we find the LCD. It is just the product of the denominators:

$$\text{LCD}=(x-2)(x+5).$$

We multiply by 1 to get the LCD in each expression. Then we subtract and, if possible, simplify.

$$\frac{x+4}{x-2}-\frac{x-7}{x+5}=\frac{x+4}{x-2}\cdot\frac{x+5}{x+5}-\frac{x-7}{x+5}\cdot\frac{x-2}{x-2}$$

$$=\frac{x^2+9x+20}{(x-2)(x+5)}-\frac{x^2-9x+14}{(x-2)(x+5)}$$

Multiplying out numerators (but not denominators)

$$=\frac{x^2+9x+20-(x^2-9x+14)}{(x-2)(x+5)}$$

When we are subtracting a numerator with more than one term, parentheses are important.

$$=\frac{x^2+9x+20-x^2+9x-14}{(x-2)(x+5)}$$

Removing parentheses in the numerator

$$=\frac{18x+6}{(x-2)(x+5)}$$

Although $18x + 6$ can be factored as $6(3x + 1)$, doing so will not enable us to simplify our result.

c) $$\frac{x}{x^2+11x+30}+\frac{-5}{x^2+9x+20}$$

$$=\frac{x}{(x+5)(x+6)}+\frac{-5}{(x+5)(x+4)}$$

Factoring the denominators in order to find the LCD. The LCD is $(x + 5)(x + 6)(x + 4)$.

$$=\frac{x}{(x+5)(x+6)}\cdot\frac{x+4}{x+4}+\frac{-5}{(x+5)(x+4)}\cdot\frac{x+6}{x+6}$$

Multiplying to get the LCD

$$=\frac{x^2+4x}{(x+5)(x+6)(x+4)}+\frac{-5x-30}{(x+5)(x+6)(x+4)}$$

Multiplying in each numerator

$$=\frac{x^2+4x-5x-30}{(x+5)(x+6)(x+4)}$$

Adding numerators

$$=\frac{x^2-x-30}{(x+5)(x+6)(x+4)}$$

$$=\frac{(x+5)(x-6)}{(x+5)(x+6)(x+4)}$$

$$=\frac{x-6}{(x+6)(x+4)}$$

Always simplify the result, if possible, by removing a factor of 1. (Here $(x + 5)/(x + 5) = 1$.)

d) $\dfrac{x}{x^2 + 5x + 6} - \dfrac{2}{x^2 + 3x + 2}$

$$= \frac{x}{(x+2)(x+3)} - \frac{2}{(x+2)(x+1)}$$

Factoring denominators. The LCD is $(x + 2)(x + 3)(x + 1)$.

$$= \frac{x}{(x+2)(x+3)} \cdot \frac{x+1}{x+1} - \frac{2}{(x+2)(x+1)} \cdot \frac{x+3}{x+3}$$

$$= \frac{x^2 + x}{(x+2)(x+3)(x+1)} - \frac{2x+6}{(x+2)(x+3)(x+1)}$$

$$= \frac{x^2 + x - (2x + 6)}{(x+2)(x+3)(x+1)}$$

Don't forget the parentheses!

$$= \frac{x^2 + x - 2x - 6}{(x+2)(x+3)(x+1)}$$

$$= \frac{x^2 - x - 6}{(x+2)(x+3)(x+1)}$$

$$= \frac{(x+2)(x-3)}{(x+2)(x+3)(x+1)}$$

$$= \frac{x-3}{(x+3)(x+1)}$$

Simplifying; $\dfrac{x+2}{x+2} = 1$ ❑

Suppose that after factoring to find the LCD, we find a factor in one denominator that is the opposite of a factor in the other denominator. When this happens, we reuse a "trick" from Section 6.3: $b - a$ can be rewritten as $-(a - b)$ since $b - a$ is the opposite of $a - b$.

EXAMPLE 4

Add: $\dfrac{x}{x^2 - 25} + \dfrac{3}{5 - x}$.

Solution

$$\frac{x}{x^2 - 25} + \frac{3}{5 - x} = \frac{x}{(x-5)(x+5)} + \frac{3}{-(x-5)}$$

Since $b - a = -(a - b)$

$$= \frac{x}{(x-5)(x+5)} + \frac{-3}{(x-5)}$$

Since $\dfrac{a}{-b} = \dfrac{-a}{b}$

$$= \frac{x}{(x-5)(x+5)} + \frac{-3}{(x-5)} \cdot \frac{x+5}{x+5}$$

The LCD is $(x - 5)(x + 5)$.

$$= \frac{x}{(x-5)(x+5)} + \frac{-3x - 15}{(x-5)(x+5)}$$

$$= \frac{-2x - 15}{(x-5)(x+5)}$$

❑

Combined Additions and Subtractions

EXAMPLE 5

Perform the indicated operations and simplify.

a) $\dfrac{x+9}{x^2 - 4} + \dfrac{5 - x}{4 - x^2} - \dfrac{2 + x}{x^2 - 4}$ **b)** $\dfrac{1}{x} - \dfrac{1}{x^2} + \dfrac{2}{x+1}$

Solution

a) We have

$$\frac{x+9}{x^2-4}+\frac{5-x}{4-x^2}-\frac{2+x}{x^2-4}$$

$$=\frac{x+9}{x^2-4}+\frac{5-x}{-(x^2-4)}-\frac{2+x}{x^2-4}$$ Recognizing the middle denominator as the opposite of the others

$$=\frac{x+9}{x^2-4}+\frac{-(5-x)}{x^2-4}-\frac{2+x}{x^2-4}$$ Using $\frac{a}{-b}=\frac{-a}{b}$

$$=\frac{x+9-5+x-(2+x)}{x^2-4}$$

$$=\frac{x+9-5+x-2-x}{x^2-4}$$ Be careful with parentheses.

$$=\frac{x+2}{x^2-4}$$

$$=\frac{\cancel{(x+2)}\cdot 1}{\cancel{(x+2)}(x-2)}$$ Simplifying

$$=\frac{1}{x-2}.$$

b) The LCD $= x\cdot x\cdot(x+1)$, or $x^2(x+1)$.

$$\frac{1}{x}-\frac{1}{x^2}+\frac{2}{x+1}=\frac{1}{x}\cdot\frac{x(x+1)}{x(x+1)}-\frac{1}{x^2}\cdot\frac{(x+1)}{(x+1)}+\frac{2}{x+1}\cdot\frac{x^2}{x^2}$$

$$=\frac{x(x+1)}{x^2(x+1)}-\frac{x+1}{x^2(x+1)}+\frac{2x^2}{x^2(x+1)}$$

$$=\frac{x^2+x-(x+1)+2x^2}{x^2(x+1)}$$ Subtract this numerator. Don't forget the parentheses.

$$=\frac{x^2+x-x-1+2x^2}{x^2(x+1)}$$

$$=\frac{3x^2-1}{x^2(x+1)}$$ ❑

EXERCISE SET 6.5

Perform the indicated operation. Simplify, if possible.

1. $\frac{2}{x}+\frac{5}{x^2}$
2. $\frac{4}{x}+\frac{8}{x^2}$
3. $\frac{5}{6r}-\frac{7}{8r}$
4. $\frac{2}{9t}-\frac{11}{6t}$
5. $\frac{4}{xy^2}+\frac{6}{x^2y}$
6. $\frac{2}{c^2d}+\frac{7}{cd^3}$
7. $\frac{2}{9t^3}-\frac{1}{6t^2}$
8. $\frac{-2}{3xy^2}-\frac{6}{x^2y^3}$
9. $\frac{x+5}{8}+\frac{x-3}{12}$
10. $\frac{x-4}{9}+\frac{x+5}{6}$

11. $\frac{x-2}{6}-\frac{x+1}{3}$

12. $\frac{a+2}{2}-\frac{a-4}{4}$

13. $\frac{a+4}{16a}+\frac{3a+4}{4a^2}$

14. $\frac{2a-1}{3a^2}+\frac{5a+1}{9a}$

15. $\frac{4z-9}{3z}-\frac{3z-8}{4z}$

16. $\frac{x-1}{4x}-\frac{2x+3}{x}$

17. $\frac{x+y}{xy^2}+\frac{3x+y}{x^2y}$

18. $\frac{2c-d}{c^2d}+\frac{c+d}{cd^2}$

19. $\frac{4x+2t}{3xt^2}-\frac{5x-3t}{x^2t}$

20. $\frac{5x+3y}{2x^2y}-\frac{3x+4y}{xy^2}$

21. $\frac{3}{x-2}+\frac{3}{x+2}$

22. $\frac{2}{x-1}+\frac{2}{x+1}$

23. $\frac{5}{x+5}-\frac{3}{x-5}$

24. $\frac{2z}{z-1}-\frac{3z}{z+1}$

25. $\frac{3}{x+1}+\frac{2}{3x}$

26. $\frac{2}{x+5}+\frac{3}{4x}$

27. $\frac{3}{2t^2-2t}-\frac{5}{2t-2}$

28. $\frac{8}{x^2-4}-\frac{3}{x+2}$

29. $\frac{2x}{x^2-16}+\frac{x}{x-4}$

30. $\frac{4x}{x^2-25}+\frac{x}{x+5}$

31. $\frac{6}{z+4}-\frac{2}{3z+12}$

32. $\frac{t}{t-3}-\frac{5}{4t-12}$

33. $\frac{3}{x-1}+\frac{2}{(x-1)^2}$

34. $\frac{2}{x+3}+\frac{4}{(x+3)^2}$

35. $\frac{2t}{t^2-9}-\frac{3}{t-3}$

36. $\frac{2}{5x^2+5x}-\frac{4}{3x+3}$

37. $\frac{4a}{5a-10}+\frac{3a}{10a-20}$

38. $\frac{3a}{4a-20}+\frac{9a}{6a-30}$

39. $\frac{a}{x+a}-\frac{a}{x-a}$

40. $\frac{t}{y-t}-\frac{y}{y+t}$

41. $\frac{x+4}{x}+\frac{x}{x+4}$

42. $\frac{x}{x-5}+\frac{x-5}{x}$

43. $\frac{x}{x^2+5x+6}-\frac{2}{x^2+3x+2}$

44. $\frac{x}{x^2+11x+30}-\frac{5}{x^2+9x+20}$

45. $\frac{x}{x^2+2x+1}+\frac{1}{x^2+5x+4}$

46. $\frac{7}{a^2+a-2}+\frac{5}{a^2-4a+3}$

47. $\frac{x}{x^2+15x+56}-\frac{6}{x^2+13x+42}$

48. $\frac{-5}{x^2+17x+16}-\frac{3}{x^2+9x+8}$

49. $\frac{10}{x^2+x-6}+\frac{3x}{x^2-4x+4}$

50. $\frac{2}{z^2-z-6}+\frac{3}{z^2-9}$

51. $\frac{y+2}{y-7}+\frac{3-y}{49-y^2}$

52. $\frac{4-p}{25-p^2}+\frac{p+1}{p-5}$

53. $\frac{8x}{16-x^2}-\frac{5}{x-4}$

54. $\frac{5x}{x^2-9}-\frac{4}{3-x}$

55. $\frac{a}{a^2-1}+\frac{2a}{a-a^2}$

56. $\frac{3x+2}{3x+6}+\frac{x}{4-x^2}$

57. $\frac{4x}{x^2-y^2}-\frac{6}{y-x}$

58. $\frac{4-a^2}{a^2-9}-\frac{a-2}{3-a}$

Simplify.

59. $\frac{4y}{y^2-1}-\frac{2}{y}-\frac{2}{y+1}$

60. $\frac{x+6}{4-x^2}-\frac{x+3}{x+2}+\frac{x-3}{2-x}$

61. $\frac{2z}{1-2z}+\frac{3z}{2z+1}-\frac{3}{4z^2-1}$

62. $\frac{1}{x+y}+\frac{1}{x-y}-\frac{2x}{x^2-y^2}$

63. $\frac{5}{3-2x}+\frac{3}{2x-3}-\frac{x-3}{2x^2-x-3}$

64. $\frac{2r}{r^2-s^2}+\frac{1}{r+s}-\frac{1}{r-s}$

65. $\frac{3}{2c-1}-\frac{1}{c+2}-\frac{5}{2c^2+3c-2}$

66. $\frac{3y-1}{2y^2+y-3}-\frac{2-y}{y-1}$

67. $\frac{1}{x+y}-\frac{1}{x-y}+\frac{2x}{x^2-y^2}$

68. $\frac{1}{a-b}-\frac{1}{a+b}+\frac{2b}{a^2-b^2}$

Skill Maintenance

Graph.

69. $y=\frac{1}{2}x-5$

70. $y=-\frac{1}{2}x-5$

71. $y=3$

72. $x=-5$

Synthesis

73. ◈ In your own words, describe the procedure for adding any two rational expressions.

74. ◈ Under what circumstances is the product of the denominators the LCD?

Write expressions for the perimeter and the area of the rectangle.

75.

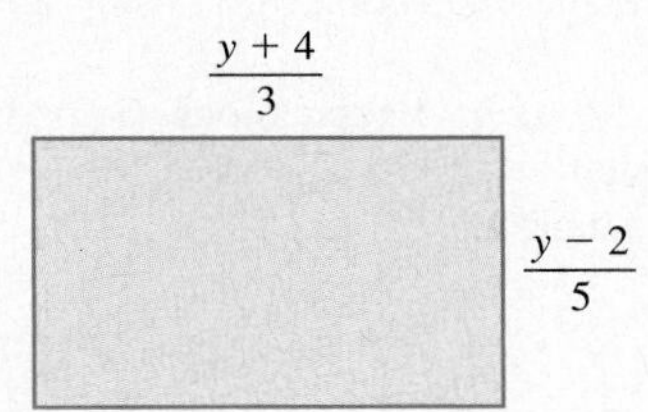

76.

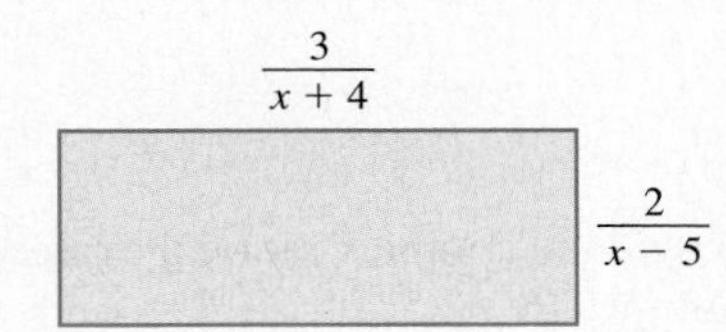

Add or subtract as indicated. Simplify, if possible.

77. $\dfrac{5}{z+2} + \dfrac{4z}{z^2-4} + 2$

78. $\dfrac{3z^2}{z^4-4} + \dfrac{5z^2-3}{2z^4+z^2-6}$

79. $\dfrac{1}{2xy-6x+ay-3a} - \dfrac{ay+xy}{(a^2-4x^2)(y^2-6y+9)}$

80. $\dfrac{x}{x^4-y^4} - \dfrac{1}{x^2+2xy+y^2}$

81. Express

$$\frac{a-3b}{a-b}$$

as a sum of two rational expressions with denominators that are opposites of each other. Answers may vary.

6.6 Complex Rational Expressions

Multiplying by the LCD • Simplifying by Adding or Subtracting

A **complex rational expression,** or **complex fractional expression,** is a rational expression that has one or more rational expressions within its numerator or denominator. Here are some examples:

$$\frac{1+\dfrac{2}{x}}{3}, \quad \frac{\dfrac{x+y}{2}}{\dfrac{2x}{x+1}}, \quad \frac{\dfrac{1}{3}+\dfrac{1}{5}}{\dfrac{2}{x}-\dfrac{x}{y}}.$$

These are rational expressions within the complex rational expression.

There are two methods to simplify complex rational expressions. We will consider them both. Use the one that works best for you or the one that your instructor directs you to use.

Multiplying by the LCD (Method 1)

Our first method of simplifying complex rational expressions relies on multiplying by an expression for 1.

To simplify a complex rational expression by using the LCD:

1. Find the LCD of all expressions *within* the complex rational expression.
2. Multiply the complex rational expression by 1, using the LCD to construct the 1.
3. Distribute and simplify. No rational expressions should remain within the complex rational expression.
4. Factor and, if possible, simplify.

EXAMPLE 1

Simplify: $\dfrac{\dfrac{1}{2}+\dfrac{3}{4}}{\dfrac{5}{6}-\dfrac{3}{8}}$.

Solution The denominators *within* the complex rational expression are 2, 4, 6, and 8. Their LCD is 24, so we multiply by $\frac{24}{24}$:

$$\frac{\dfrac{1}{2}+\dfrac{3}{4}}{\dfrac{5}{6}-\dfrac{3}{8}} = \frac{\dfrac{1}{2}+\dfrac{3}{4}}{\dfrac{5}{6}-\dfrac{3}{8}} \cdot \frac{24}{24}$$ Multiplying by 1, using the LCD: $\frac{24}{24} = 1$

$$= \frac{\left(\dfrac{1}{2}+\dfrac{3}{4}\right)24}{\left(\dfrac{5}{6}-\dfrac{3}{8}\right)24}.$$

Multiplying the numerator by 24
Don't forget the parentheses!
Multiplying the denominator by 24

Using the distributive law, we carry out the multiplications:

$$= \frac{\dfrac{1}{2}(24)+\dfrac{3}{4}(24)}{\dfrac{5}{6}(24)-\dfrac{3}{8}(24)}$$

$$= \frac{12+18}{20-9}$$ Simplifying

$$= \frac{30}{11}.$$ ❑

Multiplying in this manner has the effect of clearing fractions in both the top and bottom of the complex rational expression.

EXAMPLE 2

Simplify: **(a)** $\dfrac{\dfrac{3}{x}+\dfrac{1}{2x}}{\dfrac{1}{3x}-\dfrac{3}{4x}}$; **(b)** $\dfrac{1-\dfrac{1}{x}}{1-\dfrac{1}{x^2}}$.

Solution

a) The denominators within the complex expression are x, $2x$, $3x$, and $4x$, so the LCD is $12x$. We multiply by 1 using $(12x)/(12x)$:

$$\frac{\frac{3}{x}+\frac{1}{2x}}{\frac{1}{3x}-\frac{3}{4x}} = \frac{\frac{3}{x}+\frac{1}{2x}}{\frac{1}{3x}-\frac{3}{4x}} \cdot \frac{12x}{12x}$$

$$= \frac{\frac{3}{x}(12x)+\frac{1}{2x}(12x)}{\frac{1}{3x}(12x)-\frac{3}{4x}(12x)}$$ Using the distributive law

$$= \frac{36+6}{4-9}$$ Simplifying. All fractions have been cleared in both the numerator and the denominator.

$$= -\frac{42}{5}.$$

b) $$\frac{1-\frac{1}{x}}{1-\frac{1}{x^2}} = \frac{1-\frac{1}{x}}{1-\frac{1}{x^2}} \cdot \frac{x^2}{x^2}$$ The LCD is x^2 so we multiply by 1 using x^2/x^2.

$$= \frac{1 \cdot x^2 - \frac{1}{x} \cdot x^2}{1 \cdot x^2 - \frac{1}{x^2} \cdot x^2}$$ Using the distributive law

$$= \frac{x^2 - x}{x^2 - 1}$$ All fractions have been cleared within the complex rational expression.

$$= \frac{x(x-1)}{(x+1)(x-1)}$$

$$= \frac{x}{x+1}$$ Factoring and simplifying; $\frac{x-1}{x-1} = 1$ ❑

Simplifying by Adding or Subtracting (Method 2)

A second method of simplifying complex rational expressions involves rewriting the expression as a quotient of two rational expressions.

To simplify a complex rational expression by first adding or subtracting:

1. Add or subtract, as indicated, to get a single rational expression in the numerator.
2. Add or subtract, as indicated, to get a single rational expression in the denominator.
3. Divide the numerator by the denominator (invert and multiply).
4. If possible, simplify by removing a factor of 1.

We will redo Examples 1 and 2 using this method.

EXAMPLE 3

Simplify.

a) $\dfrac{\frac{1}{2}+\frac{3}{4}}{\frac{5}{6}-\frac{3}{8}}$ b) $\dfrac{\frac{3}{x}+\frac{1}{2x}}{\frac{1}{3x}-\frac{3}{4x}}$ c) $\dfrac{1-\frac{1}{x}}{1-\frac{1}{x^2}}$

Solution

a) $\dfrac{\frac{1}{2}+\frac{3}{4}}{\frac{5}{6}-\frac{3}{8}} = \dfrac{\frac{1}{2}\cdot\frac{2}{2}+\frac{3}{4}}{\frac{5}{6}\cdot\frac{4}{4}-\frac{3}{8}\cdot\frac{3}{3}}$ ← Multiplying by 1 to get the LCD, 4, in the numerator; ← Multiplying by 1 to get the LCD, 24, in the denominator

$= \dfrac{\frac{2}{4}+\frac{3}{4}}{\frac{20}{24}-\frac{9}{24}} = \dfrac{\frac{5}{4}}{\frac{11}{24}}$ Adding in the numerator; subtracting in the denominator

$= \dfrac{5}{4}\cdot\dfrac{24}{11}$ Multiplying by the reciprocal of the divisor

$= \dfrac{5}{2\cdot 2}\cdot\dfrac{3\cdot 2\cdot 2\cdot 2}{11}$ Factoring

$= \dfrac{5}{\not{2}\cdot\not{2}}\cdot\dfrac{3\cdot 2\cdot\not{2}\cdot\not{2}}{11}$ Removing a factor of 1: $\dfrac{2\cdot 2}{2\cdot 2}=1$

$= \dfrac{30}{11}$

b) $\dfrac{\frac{3}{x}+\frac{1}{2x}}{\frac{1}{3x}-\frac{3}{4x}} = \dfrac{\frac{3}{x}\cdot\frac{2}{2}+\frac{1}{2x}}{\frac{1}{3x}\cdot\frac{4}{4}-\frac{3}{4x}\cdot\frac{3}{3}}$ ← Multiplying by 1 to get the LCD, $2x$, in the numerator; ← Multiplying by 1 to get the LCD, $12x$, in the denominator

$= \dfrac{\frac{6}{2x}+\frac{1}{2x}}{\frac{4}{12x}-\frac{9}{12x}} = \dfrac{\frac{7}{2x}}{\frac{-5}{12x}}$ Adding in the numerator; subtracting in the denominator

$= \dfrac{7}{2x}\cdot\dfrac{12x}{-5}$ Multiplying by the reciprocal of the divisor

$= \dfrac{7}{2x}\cdot\dfrac{6(2x)}{-5}$ Factoring

$= \dfrac{7}{\not{2x}}\cdot\dfrac{6(\not{2x})}{-5}$ Removing a factor of 1: $\dfrac{2x}{2x}=1$

$= \dfrac{42}{-5}$

$= -\dfrac{42}{5}$

c) $$\frac{1-\frac{1}{x}}{1-\frac{1}{x^2}} = \frac{\frac{x}{x}-\frac{1}{x}}{\frac{x^2}{x^2}-\frac{1}{x^2}}$$ ← Rewriting the numerator using the LCD; ← Rewriting the denominator using the LCD

$$= \frac{\frac{x-1}{x}}{\frac{x^2-1}{x^2}}$$ Subtracting in the numerator; subtracting in the denominator

$$= \frac{x-1}{x}\cdot\frac{x^2}{x^2-1}$$ Multiplying by the reciprocal of the divisor

$$= \frac{(x-1)x\cdot x}{x(x-1)(x+1)}$$ Factoring

$$= \frac{\cancel{(x-1)}\cancel{x}\cdot x}{\cancel{x}\cancel{(x-1)}(x+1)}$$ Removing a factor of 1: $\frac{x(x-1)}{x(x-1)} = 1$

$$= \frac{x}{x+1}$$ ❑

EXERCISE SET | 6.6

Simplify.

1. $\dfrac{1+\frac{9}{16}}{1-\frac{3}{4}}$

2. $\dfrac{9-\frac{1}{4}}{3+\frac{1}{2}}$

3. $\dfrac{1-\frac{3}{5}}{1+\frac{1}{5}}$

4. $\dfrac{\frac{5}{27}-5}{\frac{1}{3}+1}$

5. $\dfrac{\frac{1}{x}+3}{\frac{1}{x}-5}$

6. $\dfrac{\frac{3}{s}+s}{\frac{s}{3}+s}$

7. $\dfrac{\frac{1}{2}+\frac{3}{4}}{\frac{5}{8}-\frac{5}{6}}$

8. $\dfrac{\frac{2}{3}-\frac{5}{6}}{\frac{3}{4}+\frac{7}{8}}$

9. $\dfrac{\frac{2}{y}+\frac{1}{2y}}{y+\frac{y}{2}}$

10. $\dfrac{4-\frac{1}{x^2}}{2-\frac{1}{x}}$

11. $\dfrac{8+\frac{8}{d}}{1+\frac{1}{d}}$

12. $\dfrac{2-\frac{3}{b}}{2-\frac{b}{3}}$

13. $\dfrac{\frac{x}{8}-\frac{8}{x}}{\frac{1}{8}+\frac{1}{x}}$

14. $\dfrac{\frac{2}{m}+\frac{m}{2}}{\frac{m}{2}-\frac{2}{m}}$

15. $\dfrac{1+\frac{1}{y}}{1-\frac{1}{y^2}}$

16. $\dfrac{\frac{1}{q^2}-1}{\frac{1}{q}+1}$

17. $\dfrac{\frac{1}{5}-\frac{1}{a}}{\frac{5-a}{5}}$

18. $\dfrac{2-\frac{1}{x}}{\frac{2}{x}}$

19. $\dfrac{\frac{x}{x-y}}{\frac{x^2}{x^2-y^2}}$

20. $\dfrac{\frac{x}{y}-\frac{y}{x}}{\frac{1}{y}+\frac{1}{x}}$

21. $\dfrac{\frac{3}{m}+\frac{2}{m^3}}{\frac{4}{m^2}-\frac{3}{m}}$

22. $\dfrac{\frac{a}{a^2-b^2}}{\frac{a^2}{a+b}}$

23. $\dfrac{\frac{5}{4x^3}-\frac{3}{8x}}{\frac{3}{2x}+\frac{3}{4x^3}}$

24. $\dfrac{\frac{2}{7a^4}-\frac{1}{14a}}{\frac{3}{5a^2}+\frac{2}{15a}}$

25. $\dfrac{\frac{a}{6b^3}+\frac{4}{9b^2}}{\frac{5}{6b}-\frac{1}{9b^3}}$

26. $\dfrac{\frac{x}{5y^3}-\frac{3}{10y}}{\frac{x}{10y}+\frac{3}{y^4}}$

27. $\dfrac{\dfrac{2}{x^2y} + \dfrac{3}{xy^2}}{\dfrac{2}{xy^3} + \dfrac{1}{x^2y}}$

28. $\dfrac{\dfrac{5}{ab^4} + \dfrac{2}{a^3b}}{\dfrac{5}{a^3b} - \dfrac{3}{ab}}$

29. $\dfrac{3 - \dfrac{2}{a^4}}{2 + \dfrac{3}{a^3}}$

30. $\dfrac{2 - \dfrac{3}{x^2}}{2 + \dfrac{3}{x^4}}$

31. $\dfrac{x + \dfrac{3}{x}}{x - \dfrac{2}{x}}$

32. $\dfrac{t - \dfrac{2}{t}}{t + \dfrac{5}{t}}$

33. $\dfrac{5 + \dfrac{3}{x^2y}}{\dfrac{3 + x}{x^3y}}$

34. $\dfrac{7 - \dfrac{5}{ab^3}}{\dfrac{4 + a}{a^2b}}$

35. $\dfrac{\dfrac{x + 5}{x^2}}{\dfrac{2}{x} - \dfrac{3}{x^2}}$

36. $\dfrac{\dfrac{a - 7}{a^3}}{\dfrac{3}{a^2} + \dfrac{2}{a}}$

37. $\dfrac{x - 3 + \dfrac{2}{x}}{x - 4 + \dfrac{3}{x}}$

38. $\dfrac{1 + \dfrac{a}{b - a}}{\dfrac{a}{a + b} - 1}$

39. $\dfrac{a + 5 - \dfrac{3}{a}}{a - 3 + \dfrac{5}{a}}$

40. $\dfrac{x - 2 + \dfrac{x}{3}}{x + 7 - \dfrac{4}{5x}}$

Skill Maintenance

41. Subtract:

$$(5x^4 - 6x^3 + 23x^2 - 79x + 24) - (-18x^4 - 56x^3 + 84x - 17).$$

42. The length of a rectangle is 3 yd greater than the width. The area of the rectangle is 10 yd^2. Find the perimeter.

Synthesis

43. ◈ Which of the two methods presented would you use to simplify Exercise 22? Why?

44. ◈ Which of the two methods presented would you use to simplify Exercise 28? Why?

45. Find simplified form for the reciprocal of

$$\frac{2}{x - 1} - \frac{1}{3x - 2}.$$

Simplify.

46. $\dfrac{\dfrac{a}{b} + \dfrac{c}{d}}{\dfrac{b}{a} + \dfrac{d}{c}}$

47. $\dfrac{\dfrac{a}{b} - \dfrac{c}{d}}{\dfrac{b}{a} - \dfrac{d}{c}}$

48. $\left[\dfrac{\dfrac{x + 1}{x - 1} + 1}{\dfrac{x + 1}{x - 1} - 1}\right]^5$

49. $1 + \dfrac{1}{1 + \dfrac{1}{1 + \dfrac{1}{x}}}$

50. $\dfrac{\dfrac{z}{1 - \dfrac{z}{2 + 2z}} - 2z}{\dfrac{2z}{5z - 2} - 3}$

6.7 Solving Rational Equations

A New Type of Equation • **A Visual Interpretation**

In Chapters 1 and 2, we first distinguished between *equivalent expressions*, like $x + x$ and $2x$, and *equivalent equations*, like $3(x + 2) = 18$ and $3x + 6 = 18$. Recall that equivalent equations have the same solution sets. In Sections 6.1–6.6, we saw how to write equivalent expressions but nowhere did we solve an equation. We now begin to solve equations and, in so doing, we will use the equation-solving principles discussed in Sections 2.1 and 5.6.

EXERCISE SET | 6.7

Solve.

1. $\frac{3}{8}+\frac{4}{5}=\frac{x}{20}$

2. $\frac{3}{5}+\frac{2}{3}=\frac{x}{9}$

3. $\frac{2}{3}-\frac{5}{6}=\frac{1}{x}$

4. $\frac{1}{8}-\frac{3}{5}=\frac{1}{x}$

5. $\frac{1}{6}+\frac{1}{8}=\frac{1}{t}$

6. $\frac{1}{8}+\frac{1}{10}=\frac{1}{t}$

7. $x+\frac{4}{x}=-5$

8. $x+\frac{3}{x}=-4$

9. $\frac{x}{4}-\frac{4}{x}=0$

10. $\frac{x}{5}-\frac{5}{x}=0$

11. $\frac{5}{x}=\frac{6}{x}-\frac{1}{3}$

12. $\frac{4}{x}=\frac{5}{x}-\frac{1}{2}$

13. $\frac{5}{3x}+\frac{3}{x}=1$

14. $\frac{3}{4x}+\frac{5}{x}=1$

15. $\frac{x-7}{x+2}=\frac{1}{4}$

16. $\frac{a-2}{a+3}=\frac{3}{8}$

17. $\frac{2}{x+1}=\frac{1}{x-2}$

18. $\frac{5}{x-1}=\frac{3}{x+2}$

19. $\frac{x}{6}-\frac{x}{10}=\frac{1}{6}$

20. $\frac{x}{8}-\frac{x}{12}=\frac{1}{8}$

21. $\frac{x+1}{3}-1=\frac{x-1}{2}$

22. $\frac{x+2}{5}-1=\frac{x-2}{4}$

23. $\frac{a-3}{3a+2}=\frac{1}{5}$

24. $\frac{x-1}{2x+5}=\frac{1}{4}$

25. $\frac{x-1}{x-5}=\frac{4}{x-5}$

26. $\frac{x-7}{x-9}=\frac{2}{x-9}$

27. $\frac{2}{x+3}=\frac{5}{x}$

28. $\frac{3}{x+4}=\frac{4}{x}$

29. $\frac{x-2}{x-3}=\frac{x-1}{x+1}$

30. $\frac{2b-3}{3b+2}=\frac{2b+1}{3b-2}$

31. $\frac{1}{x+3}+\frac{1}{x-3}=\frac{1}{x^2-9}$

32. $\frac{4}{x-3}+\frac{2x}{x^2-9}=\frac{1}{x+3}$

33. $\frac{x}{x+4}-\frac{4}{x-4}=\frac{x^2+16}{x^2-16}$

34. $\frac{5}{y-3}-\frac{30}{y^2-9}=1$

35. $\frac{-3}{y-7}=\frac{-10-y}{7-y}$

36. $\frac{4-m}{8-m}=\frac{4}{m-8}$

Skill Maintenance

Simplify.

37. $(a^2b^5)^{-3}$

38. $(x^{-2}y^{-3})^{-4}$

39. $\left(\frac{2x}{t^2}\right)^4$

40. $\left(\frac{y^3}{w^2}\right)^{-2}$

Synthesis

41. ◈ Explain the difference between adding rational expressions and solving rational equations.

42. ◈ Without multiplying by the LCD and solving, explain why the rational equation

$$\frac{x}{x+2}=\frac{-2}{x+2}$$

cannot have a solution. (*Hint:* Examine both numerators and denominators carefully.)

Solve.

43. $\frac{4}{y-2}-\frac{2y-3}{y^2-4}=\frac{5}{y+2}$

44. $\frac{x}{x^2+3x-4}+\frac{x+1}{x^2+6x+8}=\frac{2x}{x^2+x-2}$

45. $\frac{12-6x}{x^2-4}=\frac{3x}{x+2}-\frac{2x-3}{x-2}$

46. $\frac{x^2}{x^2-4}=\frac{x}{x+2}-\frac{2x}{2-x}$

47. $4a-3=\frac{a+13}{a+1}$

48. $\frac{3x-9}{x-3}=\frac{5x-4}{2}$

49. $\frac{y^2-4}{y+3}=2-\frac{y-2}{y+3}$

50. $\frac{3a-5}{a^2+4a+3}+\frac{2a+2}{a+3}=\frac{a-3}{a+1}$

51. Use a grapher to check the solutions to Examples 1–2(b). Be sure to use the Zoom and Trace features carefully.

52. Use a grapher to confirm the answers to Exercises 9, 25, and 43.

6.8 Problem Solving: Rational Equations and Proportions

Problem Solving • Problems Involving Work • Problems Involving Motion • Problems Involving Proportions

In many areas of study, applications involving rates, proportions, or reciprocals translate to rational equations. By using the five steps for problem solving and the lessons of Section 6.7, we can now solve such problems.

Problem Solving

EXAMPLE 1

A number plus three times its reciprocal is -4. Find the number.

Solution

1. FAMILIARIZE. Let's try to guess the number. Try 2: $2 + 3 \cdot \frac{1}{2} = \frac{7}{2}$. Although $\frac{7}{2} \neq -4$, the guess helps us to better understand how the problem can be translated. We can also see that a positive number cannot be a solution of the problem. Let x = the number.

2. TRANSLATE. From the familiarization step, we can translate directly:

A number plus three times its reciprocal is -4.

$$x + 3 \cdot \frac{1}{x} = -4$$

3. CARRY OUT. We solve the equation:

$$x + 3 \cdot \frac{1}{x} = -4$$

$$x\left(x + \frac{3}{x}\right) = x(-4) \quad \text{Multiplying by the LCD, } x\text{, on both sides}$$

$$x \cdot x + x \cdot \frac{3}{x} = -4x \quad \text{Using the distributive law}$$

$$x^2 + 3 = -4x \quad \text{Simplifying}$$

$$x^2 + 4x + 3 = 0$$

$$(x + 3)(x + 1) = 0$$

$$x + 3 = 0 \quad or \quad x + 1 = 0$$

(Using the principle of zero products)

$$x = -3 \quad or \quad x = -1.$$

4. CHECK. Three times the reciprocal of -3 is $3 \cdot \frac{1}{-3}$, or -1. Since $-3 + (-1) = -4$, the number -3 is a solution.

Three times the reciprocal of -1 is $3 \cdot \frac{1}{-1}$, or -3. Since $-1 + (-3) = -4$, the number -1 is a solution.

5. STATE. The solutions are -3 and -1. ❑

Problems Involving Work

EXAMPLE 2

Cecilia and Aaron work as volunteers at a town's recycling depot. Cecilia can sort a day's accumulation of recyclables in 4 hr, while Aaron requires 6 hr to do the same job. How long would it take them, working together, to sort the recyclables?

Solution

1. FAMILIARIZE. We familiarize ourselves with the problem by considering two *incorrect* ways of translating the problem to mathematical language.

   a) A common incorrect way to translate the problem is to just add the two times:

   $$4 \text{ hr} + 6 \text{ hr} = 10 \text{ hr}.$$

   Let's think about this. If Cecilia can do the sorting alone in 4 hr, then Cecilia and Aaron together should take *less* than 4 hr. Thus we reject 10 hr as a solution and reason that the answer must be less than 4 hr.

   b) Another incorrect way to translate the problem is to assume that Cecilia does half the sorting and Aaron does the other half. Then

   Cecilia sorts $\frac{1}{2}$ of the accumulation in $\frac{1}{2}(4 \text{ hr})$, or 2 hr, and
   Aaron sorts $\frac{1}{2}$ of the accumulation in $\frac{1}{2}(6 \text{ hr})$, or 3 hr.

   This would waste time since Cecilia would finish 1 hr earlier than Aaron. If Cecilia helps Aaron after completing her half, the entire job should take between 2 hr and 3 hr.

We proceed to a translation by considering how much of the sorting is finished in 1 hr, 2 hr, 3 hr, and so on. It takes Cecilia 4 hr to sort the recyclables alone. Then, in 1 hr, she can do $\frac{1}{4}$ of the job. It takes Aaron 6 hr to do the sorting alone. Then, in 1 hr, he can do $\frac{1}{6}$ of the job. Working together, they can complete

$$\frac{1}{4} + \frac{1}{6}, \quad \text{or} \quad \frac{5}{12} \text{ of the sorting in 1 hr.}$$

In 2 hr, Cecilia can do $2(\frac{1}{4})$ of the sorting and Aaron can do $2(\frac{1}{6})$ of the sorting. Working together, they can complete

$$2\left(\frac{1}{4}\right) + 2\left(\frac{1}{6}\right), \quad \text{or } \frac{5}{6} \text{ of the sorting in 2 hr.}$$

Continuing this reasoning, we can form a table like the following one.

	Fraction of the Sorting Completed		
Time	**Cecilia**	**Aaron**	**Together**
1 hr	$\frac{1}{4}$	$\frac{1}{6}$	$\frac{1}{4} + \frac{1}{6}$, or $\frac{5}{12}$
2 hr	$2\left(\frac{1}{4}\right)$	$2\left(\frac{1}{6}\right)$	$2\left(\frac{1}{4}\right) + 2\left(\frac{1}{6}\right)$, or $\frac{5}{6}$
3 hr	$3\left(\frac{1}{4}\right)$	$3\left(\frac{1}{6}\right)$	$3\left(\frac{1}{4}\right) + 3\left(\frac{1}{6}\right)$, or $1\frac{1}{4}$
t hr	$t\left(\frac{1}{4}\right)$	$t\left(\frac{1}{6}\right)$	$t\left(\frac{1}{4}\right) + t\left(\frac{1}{6}\right)$

From the table, we see that if they work 3 hr, the fraction of the sorting that they complete is $1\frac{1}{4}$, which is more of the job than needs to be done. We also see that the answer is somewhere between 2 hr and 3 hr. What we want is a number t such that the fraction of the sorting that is completed in t hours is 1; that is, the job is just completed—not more and not less.

2. TRANSLATE. From the table, we see that the time we want is some number t for which

Portion of work done by Cecilia in t hr; Portion of work done by Aaron in t hr

$$t\left(\frac{1}{4}\right) + t\left(\frac{1}{6}\right) = 1,$$

or

$$\frac{t}{4} + \frac{t}{6} = 1.$$

3. CARRY OUT. We solve the equation:

$$\frac{t}{4} + \frac{t}{6} = 1$$

$$12\left(\frac{t}{4} + \frac{t}{6}\right) = 12 \cdot 1 \qquad \text{The LCD is } 2 \cdot 2 \cdot 3, \text{ or } 12.$$

$$12 \cdot \frac{t}{4} + 12 \cdot \frac{t}{6} = 12$$

$$3t + 2t = 12$$

$$5t = 12$$

$$t = \frac{12}{5}, \quad \text{or } 2\frac{2}{5} \text{ hr.}$$

4. CHECK. The check can be done following the pattern used in the table of the *Familiarize* step above:

$$\frac{12}{5}\left(\frac{1}{4}\right) + \frac{12}{5}\left(\frac{1}{6}\right) = \frac{3}{5} + \frac{2}{5} = \frac{5}{5} = 1.$$

We also have a partial check in that we expected the answer to be between 2 hr and 3 hr.

5. STATE. It takes $2\frac{2}{5}$ hr for them to complete the sorting working together. ❑

The Work Principle

Suppose that a = the time it takes A to complete a task, b = the time it takes B to complete the same task, and t = the time it takes them to complete the task working together. Then

$$t\left(\frac{1}{a}\right) + t\left(\frac{1}{b}\right) = 1, \quad \text{or} \quad \frac{t}{a} + \frac{t}{b} = 1.$$

Problems Involving Motion

Problems that deal with distance, speed (or rate), and time are called **motion problems.** Translation of these problems involves the distance formula, $d = r \cdot t$, and/or the equivalent formulas $r = d/t$ and $t = d/r$.

EXAMPLE 3

One car travels 20 km/h faster than another. In the same time that one car travels 240 km, the other travels 160 km. Find their speeds.

Solution

1. FAMILIARIZE. Let's guess that the slow car is moving 30 km/h. The fast car would then be traveling 30 + 20, or 50 km/h. Thus, if r represents the slow car's speed in kilometers per hour, then the fast car's speed can be represented by $r + 20$.

If the fast car were traveling 50 km/h, it would travel 240 km in 240/50, or $4\frac{4}{5}$ hr. At 30 km/h, the slow car would travel 160 km in 160/30, or $5\frac{1}{3}$ hr. Because we are told that both cars spend the same amount of time traveling, and because

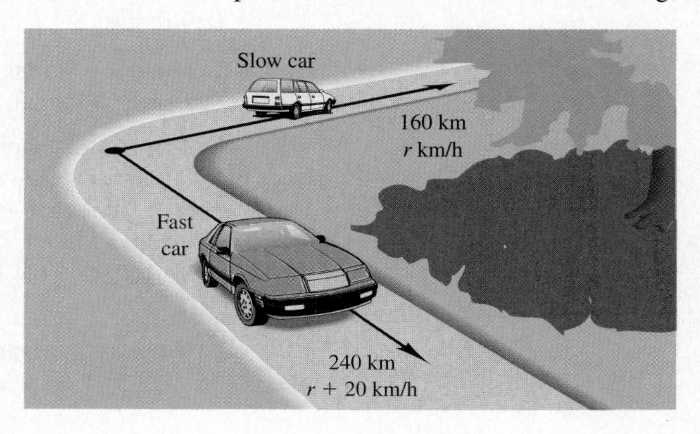

$4\frac{4}{5}$ hr $\neq 5\frac{1}{3}$ hr, we see that our guess of 30 km/h is incorrect. We let $t =$ the time, in hours, that the cars spend traveling and organize the given information in a table.

$$d \quad = \quad r \quad \cdot \quad t$$

	Distance	Speed	Time
Slow Car	160	r	t
Fast Car	240	$r + 20$	t

2. TRANSLATE. Examine how we checked our guess. We found, and then compared, the times of the two cars. To find the times, we divided the distances, 240 km and 160 km, by the rates, 50 km/h and 30 km/h, respectively. Thus the t's in the above table can be replaced, using the formula $t = d/r$.

	Distance	Speed	Time
Slow Car	160	r	$160/r$
Fast Car	240	$r + 20$	$240/(r + 20)$

The times are the same.

Since the times are the same for both cars, we have the equation

$$\frac{160}{r} = \frac{240}{r + 20}.$$

3. CARRY OUT. To solve the equation, we first multiply on both sides by the LCD, $r(r + 20)$:

$$r(r + 20) \cdot \frac{160}{r} = r(r + 20) \cdot \frac{240}{r + 20} \quad \text{Multiplying on both sides by the LCD, } r(r + 20)$$

$$160(r + 20) = 240r \quad \text{Simplifying}$$

$$160r + 3200 = 240r \quad \text{Removing parentheses}$$

$$3200 = 80r \quad \text{Subtracting } 160r$$

$$\frac{3200}{80} = r \quad \text{Dividing by 80}$$

$$40 = r.$$

We now have a possible solution. The speed of the slow car is 40 km/h, and the speed of the fast car is 40 + 20, or 60 km/h.

4. CHECK. We first reread the problem to see what we were to find. We check the speeds of 40 km/h for the slow car and 60 km/h for the fast car. The fast car does travel 20 km/h faster than the slow car. If the fast car goes 240 km at 60 km/h, it has traveled for 240/60, or 4 hr. If the slow car goes 160 km at 40 km/h, it has traveled for 160/40, or 4 hr. Since the times are the same, the speeds check.

5. STATE. The slow car has a speed of 40 km/h, and the fast car has a speed of 60 km/h. ❑

Problems Involving Proportions

A **ratio** of two quantities is their quotient. For example, 37% is the ratio of 37 to 100, $\frac{37}{100}$. A **proportion** is an equation stating that two ratios are equal.

Proportion

An equality of ratios, $A/B = C/D$, is called a *proportion*. The numbers named in a proportion are said to be *proportional* to each other.

Proportions can be used to solve applied problems by expressing a single ratio in two ways.

EXAMPLE 4

A car travels 135 mi on 6 gal of gas. Find the amount of gas required for a 360-mi trip.

Solution By assuming that the car always burns gas at the same rate, we can form a proportion in which the ratio of miles to gallons is expressed in two ways:

$$\begin{array}{l} \text{Miles} \rightarrow \\ \text{Gas} \rightarrow \end{array} \frac{135}{6} = \frac{360}{x}. \begin{array}{l} \leftarrow \text{Miles} \\ \leftarrow \text{Gas} \end{array}$$

To solve for x, we multiply both sides of the equation by the LCD, $6x$:

$$6x \cdot \frac{135}{6} = 6x \cdot \frac{360}{x}$$

$$\cancel{6} \cdot \frac{135x}{\cancel{6}} = \cancel{x} \cdot \frac{6 \cdot 360}{\cancel{x}}. \qquad \text{Removing factors of 1: } \frac{6}{6} = 1 \text{ and } \frac{x}{x} = 1$$

Note that the equation $135x = 6 \cdot 360$ could have been obtained from the original equation by *cross-multiplying:*

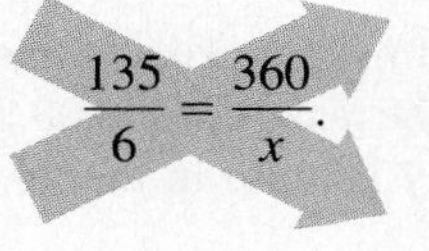

$$\frac{135}{6} = \frac{360}{x}. \qquad 135x \text{ and } 6 \cdot 360 \text{ are called } \textit{cross-products}.$$

We complete the problem as follows:

$$135x = 6 \cdot 360$$

$$x = \frac{6 \cdot 360}{135} \qquad \text{Dividing by 135 on both sides}$$

$$x = 16. \qquad \text{Simplifying}$$

The trip will require 16 gal of gas. ❑

Proportions occur in geometry when we are studying *similar triangles*. Two triangles are said to be **similar** when corresponding angles have the same measure and corresponding sides are proportional. To illustrate, if triangle ABC is similar to triangle RST, the measure of angle A = the measure of angle R, the measure of angle B = the measure of angle S, the measure of angle C = the measure of angle T, and

$$\frac{a}{r} = \frac{b}{s} = \frac{c}{t}.$$

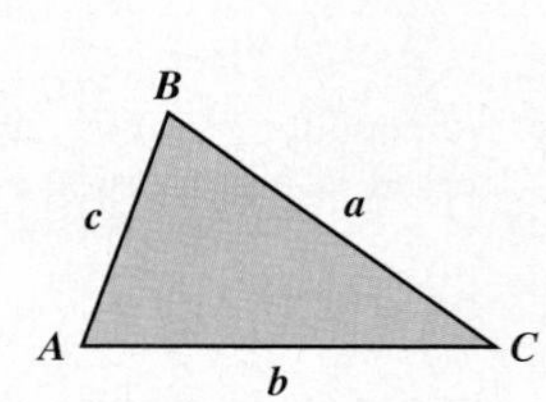

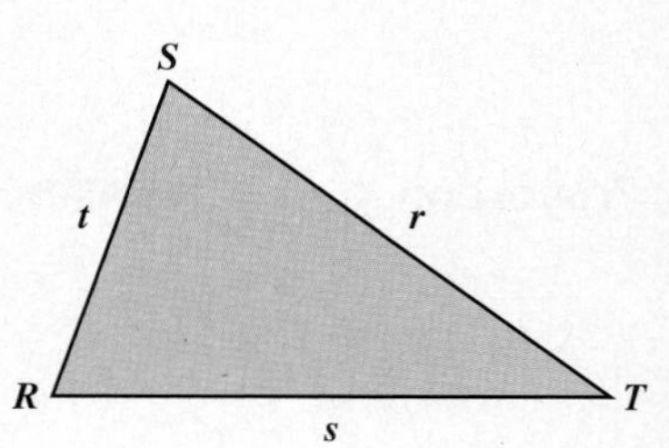

EXAMPLE 5

Triangles ABC and XYZ are similar. Solve for z if $x = 10$, $a = 8$, and $c = 5$.

Solution We make a sketch, write a proportion, and then solve. Note that side a is always opposite angle A, side x is always opposite angle X, and so on.

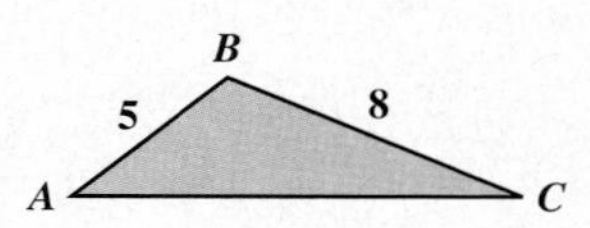

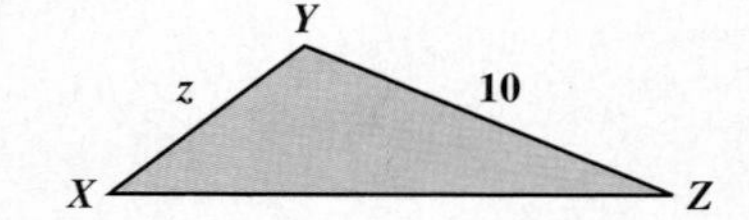

We have

$$\frac{z}{5} = \frac{10}{8}$$ The proportions $\frac{5}{z} = \frac{8}{10}$, $\frac{5}{8} = \frac{z}{10}$, or $\frac{8}{5} = \frac{10}{z}$ could also be used.

$$z = \frac{10}{8} \cdot 5$$ Multiplying by 5 on both sides

$$z = \frac{50}{8}, \quad \text{or } 6.25.$$ ❑

EXAMPLE 6

To determine the number of fish in a lake, a park ranger catches 225 fish, tags them, and throws them back into the lake. Later, 108 fish are caught, and 15 are found to be tagged. Estimate how many fish are in the lake.

Solution

1. **FAMILIARIZE.** If we knew that the 225 tagged fish constituted, say, 10% of the fish population, we could easily calculate the total fish population from the proportion

$$\frac{225}{F} = \frac{10}{100},$$

where F is the fish population. Unfortunately, we are *not* told the percentage of fish that were tagged. We must reread the problem, looking for numbers that could be used to approximate the percentage of the total fish population that was tagged.

2. **TRANSLATE.** Since 15 of 108 fish that were later caught had tags, we can use the ratio 15/108 to estimate the percentage of fish that had tags. Then we can

translate to a proportion.

$$\text{Fish tagged originally} \rightarrow \frac{225}{F} = \frac{15}{108} \leftarrow \text{Tagged fish caught later}$$
$$\text{Fish in lake} \rightarrow F \qquad 108 \leftarrow \text{Fish caught later}$$

3. CARRY OUT. To solve the proportion, we multiply by the LCD, $108F$:

$$108F \cdot \frac{225}{F} = 108F \cdot \frac{15}{108} \qquad \text{Multiplying by } 108F$$
$$108 \cdot 225 = F \cdot 15$$
$$\frac{108 \cdot 225}{15} = F \quad \text{or} \quad F = 1620.$$

4. CHECK. We leave the check to the student.

5. STATE. There are about 1620 fish in the lake. □

EXERCISE SET 6.8

Solve.

1. A number minus twice its reciprocal is 1. Find the number.
2. A number minus four times its reciprocal is 3. Find the number.
3. The sum of a number and its reciprocal is 2. Find the number.
4. The sum of a number and five times its reciprocal is 6. Find the number.
5. It takes David 4 hr to paint a certain area of a house. It takes Sierra 5 hr to do the same job. How long would it take them, working together, to complete the painting job?

6. By checking work records, a carpenter finds that Juanita can build a certain type of garage in 12 hr. Antoine can do the same job in 16 hr. How long would it take if they worked together?
7. Vern can shovel his driveway in 45 min after a snowfall. Nina can do the same job in 60 min. How long would it take Nina and Vern to shovel the driveway if they worked together?
8. Zoë can rake her yard in 4 hr. Steffi does the same job in 3 hr. How long would it take the two of them, working together, to rake the yard?
9. By checking work records, a plumber finds that Rory can do a certain job in 12 hr. Mira can do the same job in 9 hr. How long would it take if they worked together?
10. A tank can be filled in 18 hr by pipe A alone and in 24 hr by pipe B alone. How long would it take to fill the tank if both pipes were working?
11. By checking work records, a contractor finds that it takes Red Bryck 6 hr to construct a wall of a certain size. It takes Lotta Mudd 8 hr to construct the same wall. How long would it take if they worked together?

12. Bobbi can pick a quart of raspberries in 20 min. Blanche can pick a quart in 25 min. How long would it take if Bobbi and Blanche worked together?

13. One car travels 40 km/h faster than another. In the same time that one travels 150 km, the other goes 350 km. Find their speeds.

Complete the tables as part of the familiarization. Do not use t's in the second table.

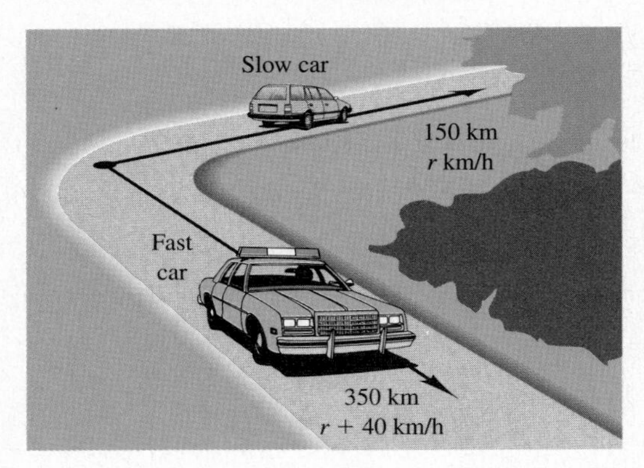

$$d = r \cdot t$$

	Distance	Speed	Time
Slow car	150	r	
Fast car	350		t

	Distance	Speed	Time
Slow car	150	r	$\frac{150}{r}$
Fast car	350		

14. One car travels 30 km/h faster than another. In the same time that one goes 250 km, the other goes 400 km. Find their speeds.

15. The speed of a freight train is 14 km/h slower than the speed of a passenger train. The freight train travels 330 km in the same time that it takes the passenger train to travel 400 km. Find the speed of each train.

Complete the tables as part of the familiarization. Do not use t's in the second table.

$$d = r \cdot t$$

	Distance	Speed	Time
Freight	330		t
Passenger	400	r	

	Distance	Speed	Time
Freight	330		
Passenger	400	r	

16. The speed of a freight train is 15 km/h slower than the speed of a passenger train. The freight train travels 390 km in the same time that it takes the passenger train to travel 480 km. Find the speed of each train.

17. Gail and Dexter bicycle at the same rate. After 16 mi, Dexter stopped for a repair. Gail continued for another 3 hr, biking a total of 50 mi that day. How many hours did Dexter ride before stopping?

18. Tucker and Jasmine's snowmobiles travel at the same rate. After 50 km, Tucker stops to fish, while Jasmine continues for another 4 hr. If Jasmine rides a total of 150 km, how many hours did Tucker travel before stopping?

19. Manley's tractor is just as fast as Caledonia's. It takes Manley 1 hr more than it takes Caledonia to drive to town. If Manley is 20 mi from town and Caledonia is 15 mi from town, how long does it take Caledonia to drive to town?

20. Tory and Emilio's motorboats both travel at the same speed. Tory pilots her boat 40 km before docking. Emilio continues for another 2 hr, traveling a total of 100 km before docking. How long did it take Tory to navigate the 40 km?

Find the ratio of the following. Simplify, if possible.

21. 54 days, 6 days

22. 800 mi, 50 gal

Solve.

23. A black racer snake travels 4.6 km in 2 hr. What is its speed in kilometers per hour?

24. Light travels 558,000 mi in 3 sec. What is its speed in miles per second?

25. The coffee beans from 14 trees are needed to produce 7.7 kg of coffee. (This is the average that each person in the United States consumes each year.) How many trees are needed to produce 320 kg of coffee?

26. Last season a minor-league baseball player got 240 hits in 600 times at bat. This season, his ratio of hits to number of times at bat is the same. He batted 500 times. How many hits has he had?

27. Wanda walked 234 km in 14 days. At this rate, how far would she walk in 42 days?

28. In a potato bread recipe, the ratio of milk to flour is $\frac{3}{13}$. If 5 cups of milk are used, how many cups of flour are used?

29. A normal 10-cc specimen of human blood contains 1.2 g of hemoglobin. How many grams would 16 cc of the same blood contain?

30. The winner of an election for class president won by a vote of 3 to 2, with 324 votes. How many votes did the loser get?

For each pair of similar triangles, find the value of the indicated letter.

31. b

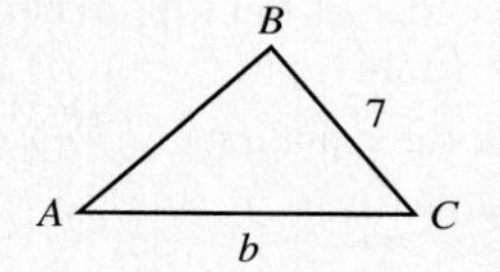

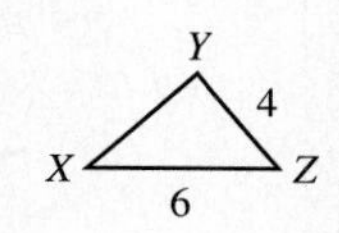

32. a

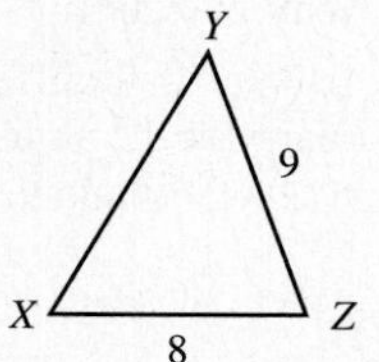

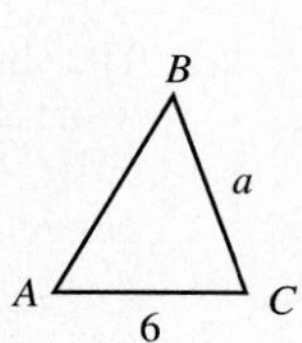

33. f

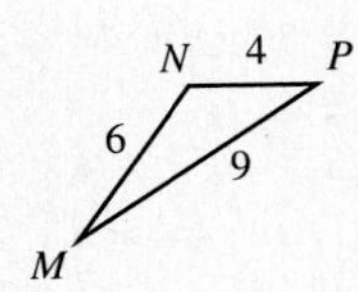

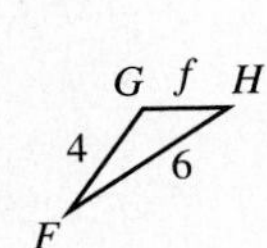

34. r

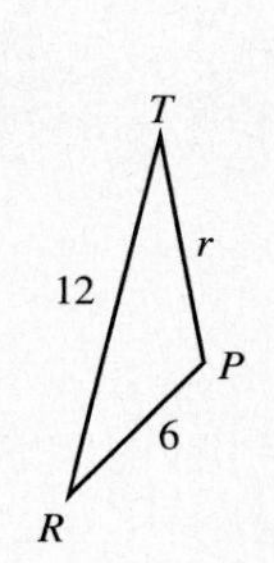

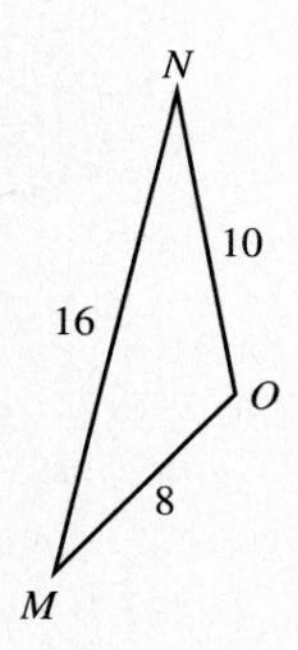

35. n

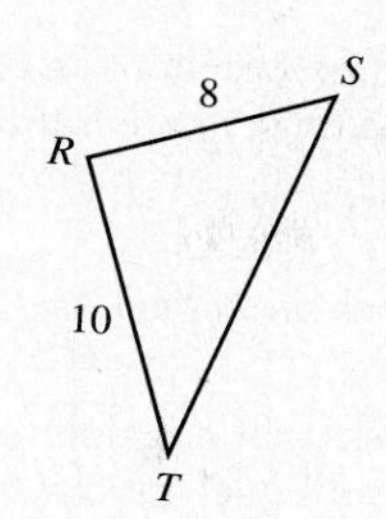

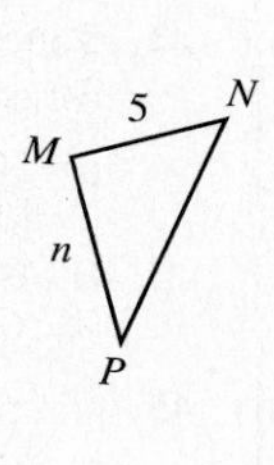

36. h

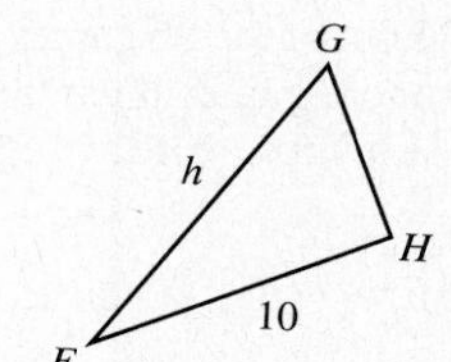

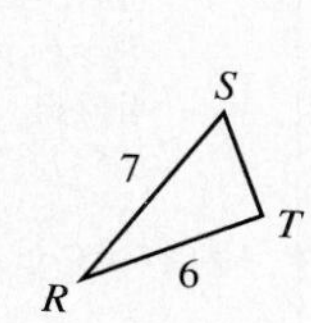

Solve.

37. To determine the number of trout in a lake, a naturalist catches 112 trout, tags them, and throws them back into the lake. Later, 82 trout are caught; 32 of them have tags. Estimate the number of trout in the lake.

38. To determine the number of deer in a game preserve, a game warden catches 318 deer, tags them, and lets them loose. Later, 168 deer are caught; 56 of them have tags. Estimate the number of deer in the preserve.

39. To determine the number of deer in a forest, a game warden catches 612 deer, tags them, and lets them loose. Later, 244 deer are caught and 72 of them have tags. Estimate how many deer are in the forest.

40. A sample of 184 light bulbs contained 6 defective bulbs. How many defective bulbs would you expect in a sample of 1288 bulbs?

41. A sample of 144 firecrackers contained 9 "duds." How many duds would you expect in a sample of 320 firecrackers?

42. To determine the number of moose in a park, a naturalist catches, tags, and then releases 25 moose. Later, 36 moose are caught; 4 of them have tags. Estimate the moose population of the park.

43. The ratio of the weight of an object on the moon to the weight of an object on earth is 0.16 to 1.

a) How much would a 12-ton rocket weigh on the moon?

b) How much would a 180-lb astronaut weigh on the moon?

44. The ratio of the weight of an object on Mars to the weight of an object on earth is 0.4 to 1.

a) How much would a 12-ton rocket weigh on Mars?

b) How much would a 120-lb astronaut weigh on Mars?

45. Simplest fractional notation for a rational number is $\frac{9}{17}$. Find an equal ratio where the sum of the numerator and the denominator is 104.

46. A baseball team has 12 more games to play. They have won 25 out of the 36 games they have played. How many more games must they win in order to finish with a 0.750 record?

Skill Maintenance

Subtract.

47. $(x + 2) - (x + 1)$

48. $(x^2 + x) - (x + 1)$

49. $(4y^3 - 5y^2 + 7y - 24) - (-9y^3 + 9y^2 - 5y + 49)$

50. The perimeter of a rectangle is 642 ft. The length is 15 ft greater than the width. Find the area of the rectangle.

Synthesis

51. ◈ Write a problem similar to Example 2 for a classmate to solve. Design the problem so that the translation step is

$$\frac{t}{7} + \frac{t}{5} = 1.$$

52. ◈ Write a problem similar to Example 3 for a classmate to solve. Design the problem so that the translation step is

$$\frac{30}{r + 4} = \frac{18}{r}.$$

53. The denominator of a fraction is 1 more than the numerator. If 2 is subtracted from both the numerator and the denominator, the resulting fraction is $\frac{1}{2}$. Find the original fraction.

54. Ann and Betty work together and complete a job in 4 hr. Working alone, it would take Betty 6 hr longer to complete the job than it would Ann. How long would it take each of them to complete the job working alone?

55. The speed of a boat in still water is 10 mph. It travels 24 mi upstream and 24 mi downstream in a total time of 5 hr. What is the speed of the current?

56. Express 100 as the sum of two numbers for which the ratio of one number, increased by 5, to the other number, decreased by 5, is 4.

57. Given that

$$\frac{A}{B} = \frac{C}{D},$$

write three different proportions using A, B, C, and D.

58. How soon after 5 o'clock will the hands on a clock first be together?

59. Rosina, Ng, and Oscar can write a computer program in 3 days. Rosina can write the program in 8 days and Ng can do it in 10 days. How many days will it take Oscar to write the program?

60. Together, Michelle, Sal, and Kristen can wax a car in 1 hr and 20 min. To complete the job alone, Michelle needs twice the time that Sal needs and 2 hr more than Kristen. How long would it take each to wax the car working alone?

61. To reach an appointment 50 mi away, Dr. Wright allowed 1 hr. After driving 30 mi, she realized that her speed would have to be increased 15 mph for the remainder of the trip. What was her speed for the first 30 mi?

62. ◈ Are the equations

$$\frac{A + B}{B} = \frac{C + D}{D} \quad \text{and} \quad \frac{A}{B} = \frac{C}{D}$$

equivalent? Why or why not?

63. The shadow from a 40-ft cliff just reaches across a water-filled quarry at the same time that a 6-ft tall diver casts a 10-ft shadow. How wide is the pond?

6.9 Formulas

Formulas from Applications • Tips for Solving Formulas

Formulas arise frequently in the natural and social sciences, business, and engineering. When a formula takes the form of a rational equation, we can use our equation-solving techniques to solve for any specified letter.

EXAMPLE 1

Gravitational force. The gravitational force f between objects of mass M and m, at a distance d from each other, is given by

$$f = \frac{kMm}{d^2},$$

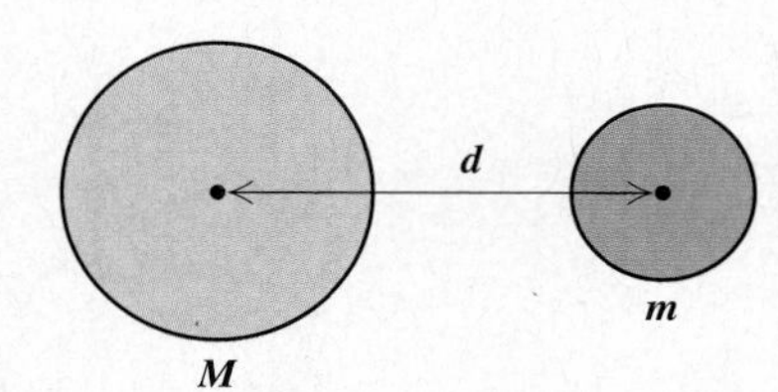

where k represents a fixed number constant. Solve for m.

Solution

$$f = \frac{kMm}{d^2}$$

$$fd^2 = kMm \quad \text{Multiplying by the LCD, } d^2$$

$$\frac{fd^2}{kM} = m \quad \text{Dividing by } kM$$

The procedure we follow is similar to the one used in Section 2.3.

To solve a formula for a given letter:

1. Identify the letter being solved for and multiply on both sides to clear fractions or decimals, if necessary.
2. Multiply, if necessary, to remove parentheses.
3. Get all terms with the letter to be solved for on one side of the equation and all other terms on the other side, using the addition principle.
4. Factor out the letter being solved for if it appears in more than one term.
5. Multiply or divide to solve for the letter in question.

EXAMPLE 2

The area of a trapezoid. The area A of a trapezoid is half the product of the height h and the sum of the lengths a and b of the parallel sides:

$$A = \tfrac{1}{2}(a + b)h.$$

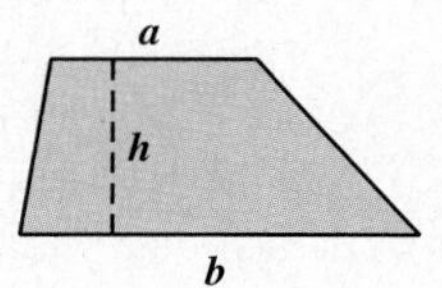

Solve for b.

Solution

$$A = \frac{1}{2}(a + b)h$$

$$2A = (a + b)h \qquad \text{Multiplying by 2 to clear fractions}$$

$$2A = ah + bh \qquad \text{Using the distributive law to remove parentheses}$$

$$2A - ah = bh \qquad \text{Subtracting } ah \text{ to get the } b\text{-term alone}$$

$$\frac{2A - ah}{h} = b \qquad \text{Dividing by } h$$

❑

EXAMPLE 3

A work formula. The formula $t/a + t/b = 1$ was used in Section 6.8. Solve it for t.

Solution

$$\frac{t}{a} + \frac{t}{b} = 1$$

$$ab\left(\frac{t}{a} + \frac{t}{b}\right) = ab \cdot 1 \qquad \text{Multiplying by the LCD, } ab\text{, to clear fractions}$$

$$\frac{\not{a}bt}{\not{a}} + \frac{a\not{b}t}{\not{b}} = ab \qquad \text{Using the distributive law to remove parentheses}$$

$$bt + at = ab \qquad \text{Removing factors of 1: } \frac{a}{a} = 1 \text{ and } \frac{b}{b} = 1$$

$$(b + a)t = ab \qquad \text{Factoring out } t\text{, the letter for which we are solving}$$

$$t = \frac{ab}{b + a} \qquad \text{Dividing by } b + a$$

❑

The answer to Example 3 can be used to find solutions to problems such as Example 2 in Section 6.8:

$$t = \frac{4 \cdot 6}{6 + 4} = \frac{24}{10} = 2\frac{2}{5}.$$

In Examples 1 and 2, the letter for which we solved was on the right side of the equation. In Example 3, the letter was on the left. The location of the letter is unimportant, since all equations are reversible.

Recall from Section 2.3 that the variable to be solved for should be alone on one side of the equation, with *no* occurrence of that variable on the other side.

EXAMPLE 4

Solve $y = \dfrac{x + y}{a}$ for y.

Solution

$$y = \frac{x + y}{a}$$

$$a \cdot y = a \cdot \frac{x + y}{a} \qquad \text{Multiplying by } a \text{ to clear fractions}$$

$$ay = x + y \qquad \text{Simplifying}$$

> CAUTION! If we next divide by a, we will not isolate y since y would still appear on both sides of the equation.

$$ay - y = x \quad \text{Subtracting } y \text{ to get all terms involving } y \text{ on one side}$$

$$y(a - 1) = x \quad \text{Factoring out } y$$

$$y = \frac{x}{a - 1} \quad \text{Dividing by } a - 1$$

❑

EXERCISE SET 6.9

Solve.

1. $S = 2\pi rh$, for r
2. $A = P(1 + rt)$, for t
   (An interest formula)
3. $A = \frac{1}{2}bh$, for b
   (The area of a triangle)

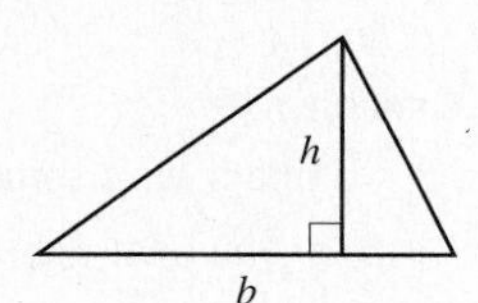

4. $s = \frac{1}{2}gt^2$, for g
   (A physics formula for distance)

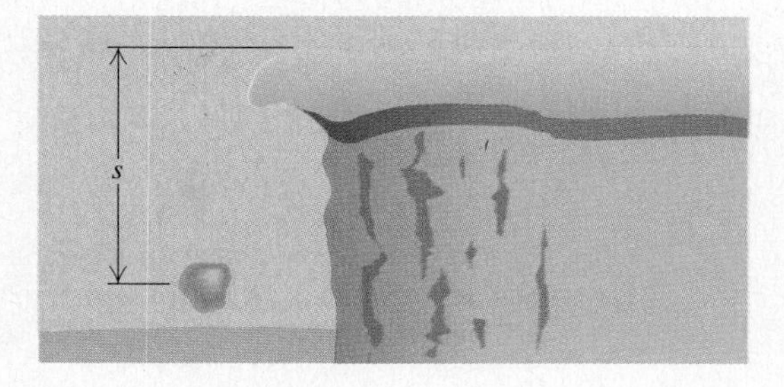

5. $\dfrac{1}{180} = \dfrac{n - 2}{s}$, for n
6. $S = \dfrac{n}{2}(a + l)$, for a
7. $V = \frac{1}{3}k(B + b + 4M)$, for b
8. $A = P + Prt$, for P
   (*Hint:* Factor the right-hand side.)
9. $rl - rS = L$, for r
10. $T = mg - mf$, for m
    (*Hint:* Factor the right-hand side.)
11. $A = \frac{1}{2}h(b_1 + b_2)$, for h
12. $S = 2\pi r(r + h)$, for h
    (The surface area of a right circular cylinder)

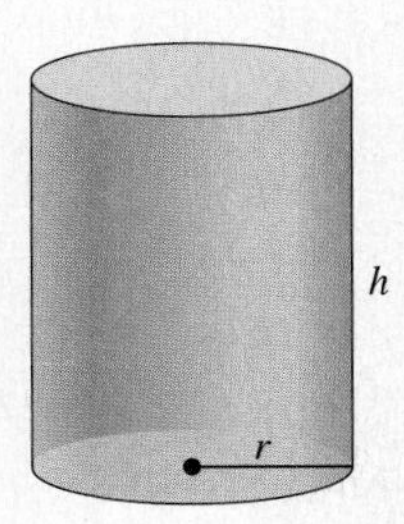

13. $ab = ac + d$, for a
14. $mn + p = np$, for n
15. $\dfrac{r}{p} = q$, for p
16. $n = \dfrac{m}{v}$, for v
17. $a + b = \dfrac{c}{d}$, for d
18. $\dfrac{m}{n} = p - q$, for n
19. $\dfrac{x - y}{z} = p + q$, for z

20. $\dfrac{M-g}{t} = r + s$, for t

21. $\dfrac{1}{p} + \dfrac{1}{q} = \dfrac{1}{f}$, for f

(An optics formula)

22. $\dfrac{1}{R} = \dfrac{1}{r_1} + \dfrac{1}{r_2}$, for R

(An electricity formula)

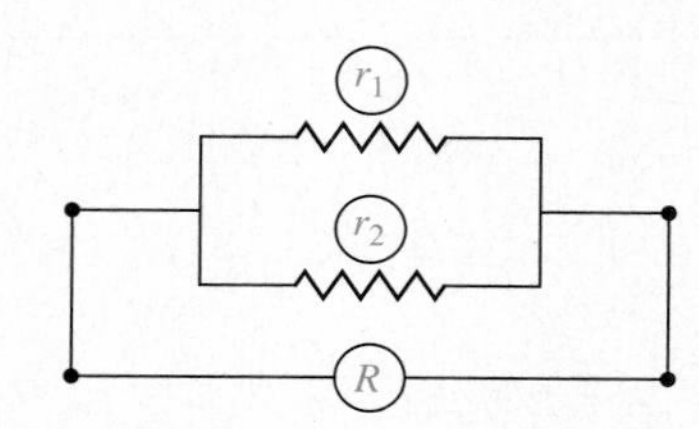

23. $r = \dfrac{v^2pL}{a}$, for p

24. $P = 2(l + w)$, for l

25. $\dfrac{a}{c} = n + bn$, for n

26. $ab - ac = \dfrac{Q}{M}$, for a

27. $S = \dfrac{a + 2b}{3b}$, for b

28. $C = \dfrac{Ka - b}{a}$, for a

29. $C = \frac{5}{9}(F - 32)$, for F

(A temperature conversion formula)

30. $V = \frac{4}{3}\pi r^3$, for r^3

(The volume of a sphere)

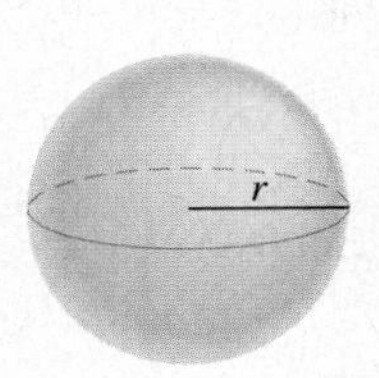

31. $f = \dfrac{gm - t}{m}$, for g

32. $S = \dfrac{rl - a}{r - l}$, for a

33. $f = \dfrac{gm - t}{m}$, for m

34. $S = \dfrac{rl - a}{r - l}$, for r

35. $C = \dfrac{a + Kb}{a}$, for a

36. $Q = \dfrac{t + Ph}{t}$, for t

Skill Maintenance

37. Solve: $-\frac{3}{5}x = \frac{9}{20}$.

38. Find the intercepts and then graph $3x + 4y = 24$.

39. Factor: $x^2 - 13x - 30$.

40. Subtract: $(5x^3 - 7x^2 + 9) - (8x^3 - 2x^2 + 4)$.

Synthesis

41. ◈ Is it easier to solve

$$\frac{1}{25} + \frac{1}{23} = \frac{1}{x} \quad \text{for } x,$$

or to solve

$$\frac{1}{p} + \frac{1}{q} = \frac{1}{f} \quad \text{for } f?$$

Explain why.

42. ◈ Explain why someone might want to solve

$A = \frac{1}{2}bh$ for b.

(See Exercise 3.)

Solve.

43. $u = -F\left(E - \dfrac{P}{T}\right)$, for T

44. $\dfrac{n_1}{p_1} + \dfrac{n_2}{p_2} = \dfrac{n_2 - n_1}{R}$, for n_2

45. The formula

$$C = \tfrac{5}{9}(F - 32)$$

is used to convert Fahrenheit temperatures to Celsius temperatures. At what temperature are the Fahrenheit and Celsius readings the same?

46. In

$$N = \frac{a}{c},$$

what is the effect on N when c increases? when c decreases? Assume that a, c, and N are positive.

SUMMARY AND REVIEW 6

KEY TERMS

IMPORTANT PROPERTIES AND FORMULAS

To add, subtract, multiply, and divide rational expressions:

$$\frac{A}{B}\cdot\frac{C}{D}=\frac{AC}{BD};\quad \frac{A}{B}\div\frac{C}{D}=\frac{A}{B}\cdot\frac{D}{C}=\frac{AD}{BC};$$

$$\frac{A}{B}+\frac{C}{B}=\frac{A+C}{B};\quad \frac{A}{B}-\frac{C}{B}=\frac{A-C}{B}.$$

The Work Principle: $t\left(\frac{1}{a}\right)+t\left(\frac{1}{b}\right)=1$, or $\frac{t}{a}+\frac{t}{b}=1$

To find the least common denominator (LCD):

1. Write the prime factorization of each denominator.
2. Select one of the factorizations and inspect it to see if it contains the other.
   a) If it does, it represents the LCD.
   b) If it does not, multiply that factorization by any factors of the other denominator that it lacks. The final product is the LCD.
3. The LCD should include each factor the greatest number of times that it occurs in any one factorization.

To add or subtract rational expressions that have different denominators:

1. Find the LCD.
2. Multiply each rational expression by an expression for 1 made up of the factors of the LCD missing from that expression's denominator.
3. Add or subtract the numerators, as indicated. Write the sum or difference over the LCD.
4. Simplify, if possible.

To simplify a complex rational expression by using the LCD:

1. Find the LCD of all expressions *within* the complex rational expression.
2. Multiply the complex rational expression by 1, using the LCD to construct the 1.
3. Distribute and simplify. No rational expressions should remain within the complex rational expression.
4. Factor and, if possible, simplify.

To simplify a complex rational expression by first adding or subtracting:

1. Add or subtract, as indicated, to get a single rational expression in the numerator.
2. Add or subtract, as indicated, to get a single rational expression in the denominator.
3. Divide the numerator by the denominator (invert and multiply).
4. If possible, simplify by removing a factor of 1.

To solve a rational equation:

1. Clear the equation of fractions by multiplying on both sides by the LCD of all rational expressions in the equation.
2. Solve the resulting equation using the addition principle, the multiplication principle, and the principle of zero products.
3. Check the possible solution(s) in the original equation.

REVIEW EXERCISES

This chapter's review and test include Skill Maintenance exercises from Sections 2.1, 3.3, 4.3, and 5.2.

List all numbers for which the expression is undefined.

1. $\dfrac{3}{x}$

2. $\dfrac{4}{x-6}$

3. $\dfrac{x+5}{x^2-36}$

4. $\dfrac{x^2-3x+2}{x^2+x-30}$

5. $\dfrac{-4}{(x+2)^2}$

6. $\dfrac{x-5}{x^3-8x^2+15x}$

Simplify.

7. $\dfrac{4x^2-8x}{4x^2+4x}$

8. $\dfrac{14x^2-x-3}{2x^2-7x+3}$

9. $\dfrac{(y-5)^2}{y^2-25}$

Multiply or divide and simplify, if possible.

10. $\dfrac{a^2-36}{10a}\cdot\dfrac{2a}{a+6}$

11. $\dfrac{6t-6}{2t^2+t-1}\cdot\dfrac{t^2-1}{t^2-2t+1}$

12. $\dfrac{10 - 5t}{3} \div \dfrac{t - 2}{12t}$

13. $\dfrac{4x^4}{x^2 - 1} \div \dfrac{2x^3}{x^2 - 2x + 1}$

14. $\dfrac{x^2 + 1}{x - 2} \cdot \dfrac{2x + 1}{x + 1}$

15. $(t^2 + 3t - 4) \div \dfrac{t^2 - 1}{t + 4}$

Find the LCM.

16. $3x^2,\ 10xy,\ 15y^2$

17. $x^2 - x,\ x^5 - x^3,\ x^4$

18. $y^2 - y - 2,\ y^2 - 4$

Add or subtract and simplify, if possible.

19. $\dfrac{x + 8}{x + 7} + \dfrac{10 - 4x}{x + 7}$

20. $\dfrac{3}{3x - 9} + \dfrac{x - 2}{3 - x}$

21. $\dfrac{6x - 3}{x^2 - x - 12} - \dfrac{2x - 15}{x^2 - x - 12}$

22. $\dfrac{3x - 1}{2x} - \dfrac{x - 3}{x}$

23. $\dfrac{x + 3}{x - 2} - \dfrac{x}{2 - x}$

24. $\dfrac{2a}{a + 1} - \dfrac{4a}{1 - a^2}$

25. $\dfrac{d^2}{d - c} + \dfrac{c^2}{c - d}$

26. $\dfrac{1}{x^2 - 25} - \dfrac{x - 5}{x^2 - 4x - 5}$

27. $\dfrac{3x}{x + 2} - \dfrac{x}{x - 2} + \dfrac{8}{x^2 - 4}$

28. $\dfrac{3}{2x} + \dfrac{1}{2x + 1}$

Simplify.

29. $\dfrac{\dfrac{1}{z} + 1}{\dfrac{1}{z^2} - 1}$

30. $\dfrac{2 + \dfrac{1}{xy^2}}{\dfrac{1 + x}{x^4y}}$

31. $\dfrac{\dfrac{c}{d} - \dfrac{d}{c}}{\dfrac{1}{c} + \dfrac{1}{d}}$

Solve.

32. $\dfrac{3}{y} - \dfrac{1}{4} = \dfrac{1}{y}$

33. $\dfrac{5}{x + 3} = \dfrac{3}{x + 2}$

34. $\dfrac{15}{x} - \dfrac{15}{x + 2} = 2$

Problem Solving

35. In checking records, a contractor finds that Sean can build a deck in 9 hr. Shane can do the same job in 12 hr. How long would it take if they worked together?

36. A lab is testing two high-speed trains. One train travels 40 km/h faster than the other. In the same time that one train travels 70 km, the other travels 60 km. Find the speed of each train.

37. The reciprocal of 1 more than a number is twice the reciprocal of the number itself. What is the number?

38. A sample of 250 batteries contained 8 defective batteries. How many defective batteries would you expect among 5000 batteries?

39. Triangles ABC and XYZ are similar. Find the value of x.

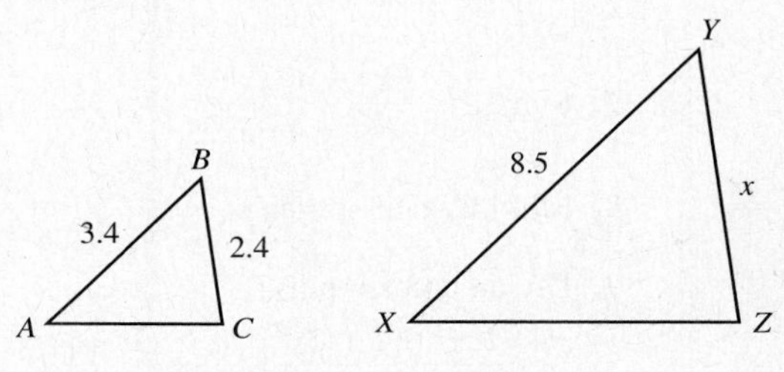

Solve.

40. $\dfrac{1}{r} + \dfrac{1}{s} = \dfrac{1}{t}$, for s

41. $F = \dfrac{9C + 160}{5}$, for C

Skill Maintenance

42. Solve: $-3 + x = 8$.

43. Find the intercepts and graph: $2x - y = 6$.

44. Factor: $x^2 + 8x - 48$.

45. Subtract:

$(5x^3 - 4x^2 + 3x - 4) - (7x^3 - 7x^2 - 9x + 14)$.

Synthesis

46. Why is factoring an important skill to master before beginning a study of rational equations?

47. A student insists on finding a common denominator by always multiplying the denominators of the expressions being added. How could the student's approach be improved?

Simplify.

48. $\dfrac{2a^2 + 5a - 3}{a^2} \cdot \dfrac{5a^3 + 30a^2}{2a^2 + 7a - 4} \div \dfrac{a^2 + 6a}{a^2 + 7a + 12}$

49. $\dfrac{12a}{(a - b)(b - c)} - \dfrac{2a}{(b - a)(c - b)}$

CHAPTER TEST 6

List all numbers for which the expression is undefined.

1. $\dfrac{8}{2x}$

2. $\dfrac{5}{x+8}$

3. $\dfrac{x-7}{x^2-49}$

4. $\dfrac{x^2+x-30}{x^2-3x+2}$

5. $\dfrac{11}{(x-1)^2}$

6. $\dfrac{x+2}{x^3+8x^2+15x}$

7. Simplify: $\dfrac{6x^2+17x+7}{2x^2+7x+3}$.

8. Multiply and simplify: $\dfrac{a^2-25}{6a} \cdot \dfrac{3a}{a-5}$.

9. Divide and simplify:
$$\frac{25x^2-1}{9x^2-6x} \div \frac{5x^2+9x-2}{3x^2+x-2}.$$

10. Find the LCM:
$$y^2-9,\ y^2+10y+21,\ y^2+4y-21.$$

Add or subtract. Simplify, if possible.

11. $\dfrac{16+x}{x^3}+\dfrac{7-4x}{x^3}$

12. $\dfrac{5-t}{t^2+1}-\dfrac{t-3}{t^2+1}$

13. $\dfrac{x-4}{x-3}+\dfrac{x-1}{3-x}$

14. $\dfrac{x-4}{x-3}-\dfrac{x-1}{3-x}$

15. $\dfrac{5}{t-1}+\dfrac{3}{t}$

16. $\dfrac{1}{x^2-16}-\dfrac{x+4}{x^2-3x-4}$

17. $\dfrac{1}{x-1}+\dfrac{4}{x^2-1}-\dfrac{2}{x^2-2x+1}$

Simplify.

18. $\dfrac{9-\dfrac{1}{y^2}}{3-\dfrac{1}{y}}$

19. $\dfrac{\dfrac{3}{a^2b}-\dfrac{2}{ab^3}}{\dfrac{1}{ab}+\dfrac{2}{a^4b}}$

Solve.

20. $\dfrac{7}{y}-\dfrac{1}{3}=\dfrac{1}{4}$

21. $\dfrac{15}{x}-\dfrac{15}{x-2}=-2$

Problem Solving

22. The reciprocal of 3 less than a number is four times the reciprocal of the number itself. What is the number?

23. A sample of 125 spark plugs contained 4 defective spark plugs. How many defective spark plugs would you expect among 500 spark plugs?

24. One car travels 20 km/h faster than another. In the same time that one goes 225 km, the other goes 325 km. Find the speed of each car.

25. Solve $d = rt + wt$ for t.

Skill Maintenance

26. Solve: $-3y = \frac{9}{7}$.

27. Find the intercepts and graph: $2x + 5y = 20$.

28. Factor: $x^2 - 4x - 45$.

29. Subtract:
$$(5x^2-19x+34)-(-8x^2+10x-42).$$

Synthesis

30. Reggie and Rema work together to mulch the flower beds around an office complex in $2\frac{6}{7}$ hr. Working alone, it would take Reggie 6 hr more than it would take Rema. How long would it take each of them to complete the landscaping working alone?

31. Simplify: $1+\dfrac{1}{1+\dfrac{1}{1+\dfrac{1}{a}}}$.

CUMULATIVE REVIEW 1–6

1. Use the commutative law of addition to write an expression equivalent to $a + 2b$.
2. Write a true sentence using either $<$ or $>$:
$-3.1 \blacksquare -3.15$.
3. Evaluate $(y - 1)^2$ when $y = -6$.
4. Remove parentheses and simplify:
$-4[2(x - 3) - 1]$.

Simplify.

5. $-\frac{1}{2} + \frac{3}{8} + (-6) + \frac{3}{4}$
6. $-\frac{72}{108} \div \left(-\frac{2}{3}\right)$
7. $-6.262 \div 1.01$
8. $4 \div (-2) \cdot 2 + 3 \cdot 4$

Solve.

9. $3(x - 2) = 24$
10. $6y + 3 = -15$
11. $-4x = -18$
12. $5x + 7 = -3x - 9$
13. $4(y - 5) = -2(y + 2)$
14. $-6x - 2(x - 4) = 10$
15. $\frac{1}{3}x - \frac{2}{9} = \frac{2}{3} + \frac{4}{9}x$
16. $-\frac{5}{6} = x - \frac{1}{3}$
17. $3 - y \geqslant 2y + 5$
18. $(3x - 4)(2x + 5) = 0$
19. $2x^2 + 7x = 4$
20. $16 = x^2$
21. $\dfrac{x^2}{x + 2} = \dfrac{4}{x + 2}$
22. $x + \dfrac{1}{x} = 2$
23. $\dfrac{2}{x^2 - 9} + \dfrac{5}{x - 3} = \dfrac{3}{x + 3}$

Solve the formula.

24. $A = \dfrac{4b}{t}$, for t
25. $\dfrac{1}{t} = \dfrac{1}{m} - \dfrac{1}{n}$, for n
26. $r = \dfrac{a - b}{c}$, for c

Collect like terms.

27. $x + 2y - 2z + \frac{1}{2}x - z$
28. $2x^3 - 7 + \frac{3}{7}x^2 - 6x^3 - \frac{4}{7}x^2 + 5$

Graph.

29. $y = 1 - \frac{1}{2}x$
30. $x = -3$
31. $x - 6y = 6$

Simplify.

32. $x^8 \cdot x^2$
33. $\dfrac{z^4}{z^{-7}}$
34. $-(3x^2y)^3$
35. Subtract:
$(-8y^2 - y + 2) - (y^3 - 6y^2 + y - 5)$.

Multiply.

36. $4(3x + 4y + z)$
37. $(2.5a + 7.5)(0.4a - 1.2)$
38. $(2x^2 - 1)(x^3 + x - 3)$
39. $(6x - 5y)^2$
40. $(2x^5 + 3)(3x^2 - 6)$
41. $(2x^3 + 1)(2x^3 - 1)$

Factor.

42. $6x - 2x^2 - 24x^4$
43. $16x^2 - 81$
44. $x^2 - 10x + 24$
45. $8x^2 + 10x + 3$
46. $6x^2 - 28x + 16$
47. $2x^2 - 18$
48. $16x^2 + 40x + 25$
49. $3x^2 + 10x - 8$
50. $x^4 + 2x^3 - 3x - 6$

Simplify.

51. $\dfrac{y^2 - 36}{2y + 8} \cdot \dfrac{y + 4}{y + 6}$
52. $\dfrac{x^2 - 1}{x^2 - x - 2} \div \dfrac{x - 1}{x - 2}$
53. $\dfrac{5ab}{a^2 - b^2} + \dfrac{a + b}{a - b}$
54. $\dfrac{x + 2}{4 - x} - \dfrac{x + 3}{x - 4}$
55. $\dfrac{1 + \frac{2}{x}}{1 - \frac{4}{x^2}}$
56. $\dfrac{\frac{1}{t} + 2t}{t - \frac{2}{t^2}}$

Divide.

57. $\dfrac{15x^4 - 12x^3 + 6x^2 + 2x + 18}{3x^2}$
58. $(15x^4 - 12x^3 + 6x^2 + 2x + 18) \div (x + 3)$

Problem Solving

59. The sum of two consecutive even integers is -554. Find the integers.
60. What number is 96% of 567?
61. If you double a number and then add 20, you get $\frac{2}{3}$ of the original number. What is the original number?

62. Linnae has \$36 budgeted for stationery. Engraved stationery costs \$20 for the first 25 sheets and \$0.08 for each additional sheet. Use an inequality to express the amounts of stationery that will enable Linnae to stay within her budget.

63. If the sides of a square are increased by 2 ft, the sum of the areas of the two squares is 452. Find the length of a side of the original square.

64. One car travels 10 km/h faster than another. In the same time it takes one car to go 120 km, the other car goes 150 km. Find the speed of each car.

65. By checking work records, a contractor finds that it takes Rita 6 hr to construct a wall of a certain size. It takes Lotta 8 hr to construct the same wall. How long would it take if they worked together?

Synthesis

66. Simplify:

$$(x + 7)(x - 4) - (x + 8)(x - 5).$$

67. Solve: $\frac{1}{3}|n| + 8 = 56.$

68. Multiply:

$$[4y^3 - (y^2 - 3)][4y^3 + (y^2 - 3)].$$

69. Factor: $2a^{32} - 13{,}122b^{40}.$

70. Solve: $(x - 4)(x + 7)(x - 12) = 0.$

71. Simplify: $-|0.875 - \left(-\frac{1}{8}\right) - 8|.$

CHAPTER 7

Graphs and Slope

AN APPLICATION

The grade of a road tells us how steep a road up a hill is. For example, a 3% grade means that the road rises 3 ft for every 100 ft that it runs horizontally. If a road rises 50 ft vertically over a horizontal distance of 1250 ft, find the grade of the road.

This problem appears as Example 5 in Section 7.1.

Cheryl Adams
ROADWAY DESIGNER

"I use math daily in evaluating designs that are submitted by other engineers in the Department of Transportation. We must use math for everything from setting the grade to identifying the quantities of material and the costs of a project."

The basics of graphing were first introduced in Chapter 3. There we learned how to graph equations by using tables or intercepts. The treatment of graphing in this chapter focuses on a topic that is of great importance in mathematics, science, and business: slope. Once slope is understood, we can develop faster and easier ways of graphing certain equations. Slope will also prove useful when we learn to graph linear inequalities in Section 7.4.

In addition to the material from this chapter, the review and test for Chapter 7 include material from Sections 4.4, 4.5, 5.1, and 6.1.

7.1 Slope

Rate and Slope • Horizontal and Vertical Lines • Applications

Rate and Slope

Suppose that a car manufacturer is operating two plants: one in Michigan and one in Pennsylvania. If we know that the Michigan plant produces 3 cars every 2 hours and the Pennsylvania plant produces 5 cars every 4 hours, we can set up tables listing the number of cars produced after various amounts of time.

Michigan Plant	
Hours Elapsed	Cars Produced
0	0
2	3
4	6
6	9
8	12

Pennsylvania Plant	
Hours Elapsed	Cars Produced
0	0
4	5
8	10
12	15
16	20

By comparing the number of cars produced at each plant over a specified period of time, we can compare the **rates** at which the plants produce cars. For example, the Michigan plant produces 3 cars every 2 hours, so its *rate* is $3 \div 2 = 1\frac{1}{2}$, or $\frac{3}{2}$ cars per hour. Since the Pennsylvania plant produces 5 cars every 4 hours, its rate is $5 \div 4 = 1\frac{1}{4}$, or $\frac{5}{4}$ cars per hour.

Let's now graph the pairs of numbers listed in the tables, using the horizontal axis for time and the vertical axis for the number of cars produced.

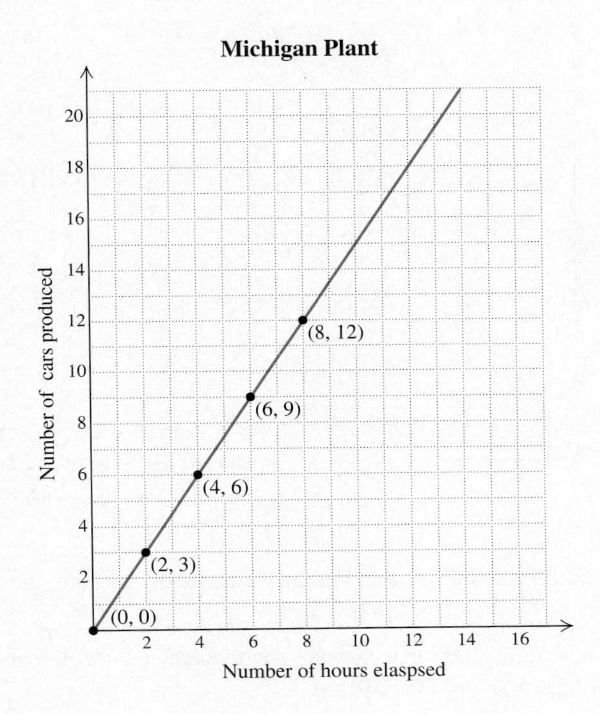

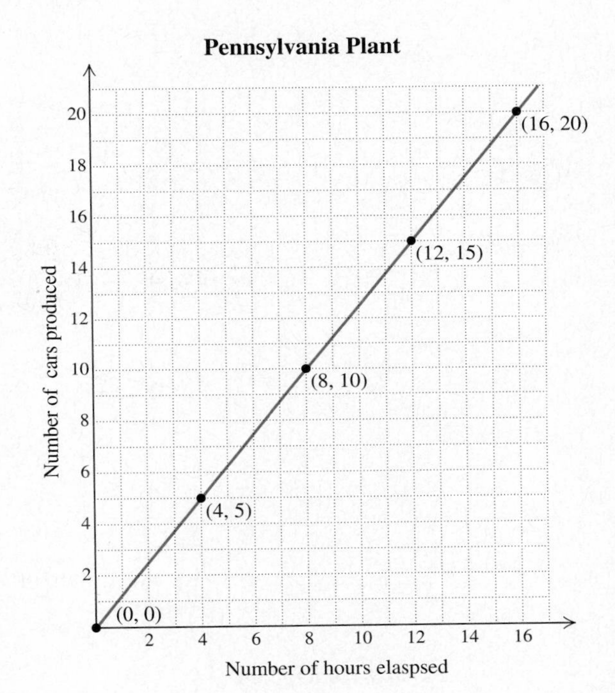

The rates of $\frac{3}{2}$ and $\frac{5}{4}$ can also be found using the coordinates of any two points that are on the line. For example, we can use the points (6, 9) and (8, 12) to find the production rate for the Michigan plant. To do so, remember that these coordinates tell us that after 6 hr, 9 cars have been produced, and after 8 hr, 12 cars have been produced. In the 2 hr between the 6-hr and 8-hr points, $12 - 9$, or 3 cars were produced. Thus,

$$\begin{aligned}\text{Michigan rate} &= \frac{\text{change in number of cars produced}}{\text{corresponding change in time}}\\ &= \frac{12 - 9 \text{ cars}}{8 - 6 \text{ hr}}\\ &= \frac{3 \text{ cars}}{2 \text{ hr}} = \frac{3}{2} \text{ cars per hour.}\end{aligned}$$

The same rate can be found using other points on that line, such as (0, 0) and (4, 6):

$$\text{Michigan rate} = \frac{6 - 0 \text{ cars}}{4 - 0 \text{ hr}} = \frac{6 \text{ cars}}{4 \text{ hr}} = \frac{3}{2} \text{ cars per hour.}$$

Note that the rate is always the vertical change divided by the associated horizontal change.

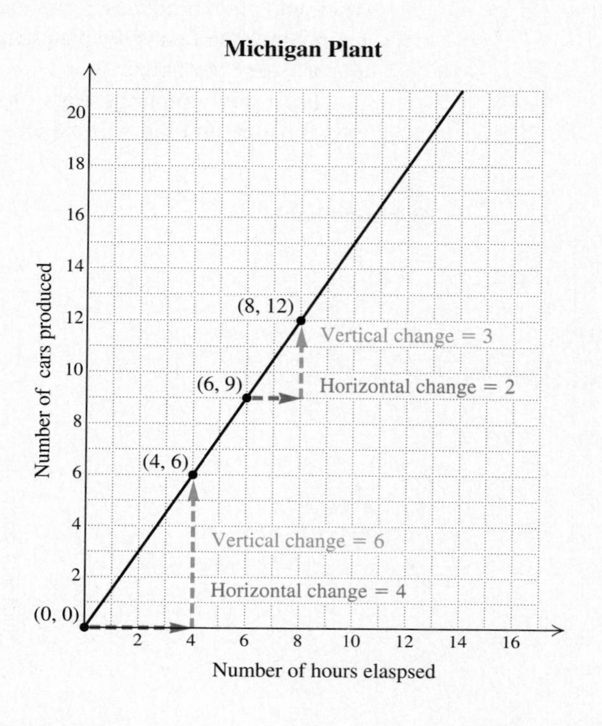

EXAMPLE 1

Use the graph of car production at the Pennsylvania plant to find the rate of production.

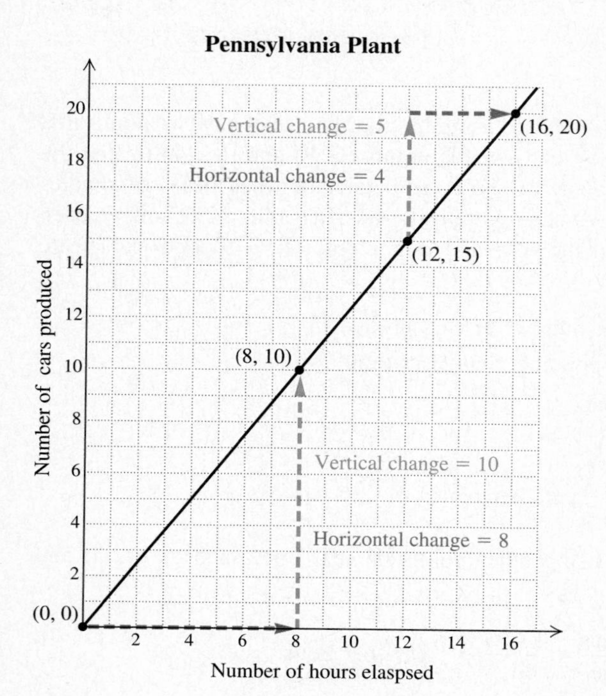

Solution We can use any two points on the line, such as (12, 15) and (16, 20):

$$\text{Pennsylvania rate} = \frac{\text{change in number of cars produced}}{\text{corresponding change in time}}$$

$$= \frac{20 - 15 \text{ cars}}{16 - 12 \text{ hr}}$$

$$= \frac{5 \text{ cars}}{4 \text{ hr}}$$

$$= \frac{5}{4} \text{ cars per hour.}$$

As a check, we can use another pair of points, like (0, 0) and (8, 10):

$$\text{Pennsylvania rate} = \frac{10 - 0 \text{ cars}}{8 - 0 \text{ hr}}$$

$$= \frac{10 \text{ cars}}{8 \text{ hr}}$$

$$= \frac{5}{4} \text{ cars per hour.}$$

❑

Even when a graph's axes are labeled simply x and y, it is useful to form the ratio of vertical change to horizontal change. This ratio gives a measure of a line's slant, or *slope*.

Consider a line passing through the points (2, 3) and (6, 5), as shown. We find the ratio of vertical change, or *rise,* to horizontal change, or *run,* as follows:

$$\text{Ratio of vertical change to horizontal change} = \frac{\text{change in } y}{\text{change in } x} = \frac{\text{rise}}{\text{run}}$$

$$= \frac{5-3}{6-2}$$

$$= \frac{2}{4}, \quad \text{or } \frac{1}{2}.$$

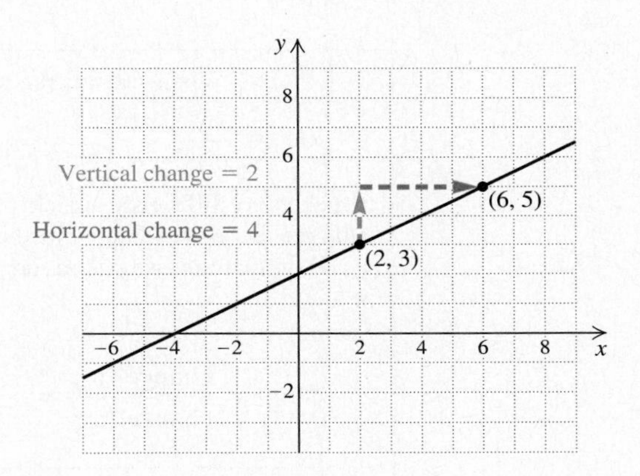

Thus the y-coordinate of a point on the line increases 2 units for every 4-unit increase in x, 1 unit for every 2-unit increase in x, and $\frac{1}{2}$ unit for every 1-unit increase in x. The slope of the line is $\frac{1}{2}$.

Slope

The *slope* of a line containing points (x_1, y_1) and (x_2, y_2) is given by

$$m = \frac{\text{change in } y}{\text{change in } x} = \frac{\text{rise}}{\text{run}} = \frac{y_2 - y_1}{x_2 - x_1}.$$

EXAMPLE 2

Graph the line containing the points $(-4, 3)$ and $(2, -6)$ and find the slope.

Solution The graph is on the next page. From $(-4, 3)$ to $(2, -6)$, the change in y, or rise, is $-6 - 3$, or -9. The change in x, or run, is $2 - (-4)$, or 6.

Thus,

$$\text{Slope} = \frac{\text{change in } y}{\text{change in } x} = \frac{\text{rise}}{\text{run}}$$
$$= \frac{-6-3}{2-(-4)}$$
$$= \frac{-9}{6}$$
$$= -\frac{9}{6}, \quad \text{or } -\frac{3}{2}.$$

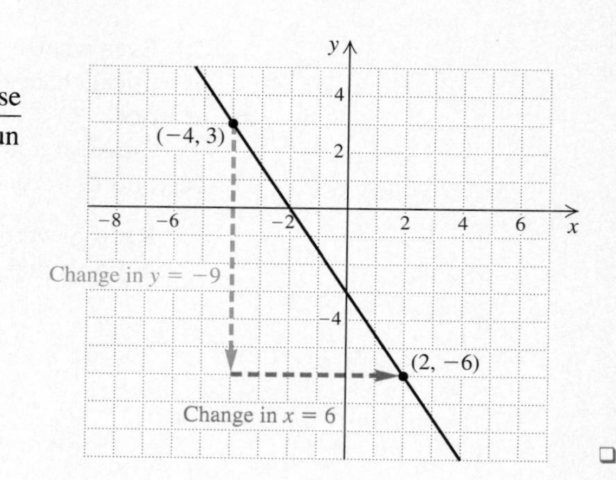

❑

> CAUTION! When we use the formula
>
> $$m = \frac{y_2 - y_1}{x_2 - x_1},$$
>
> it makes no difference which of the two points is considered (x_1, y_1) so long as we subtract the y-coordinates in the same order that we subtract the x-coordinates.
>
> To illustrate, we can reverse the subtractions in Example 2 and still obtain a slope of $-\frac{3}{2}$:
>
> $$\text{Slope} = \frac{\text{change in } y}{\text{change in } x} = \frac{3-(-6)}{-4-2} = \frac{9}{-6} = -\frac{3}{2}.$$

If a line has a positive slope, it slants up from left to right. The larger the slope, the steeper the slant. A line with negative slope slants downward from left to right.

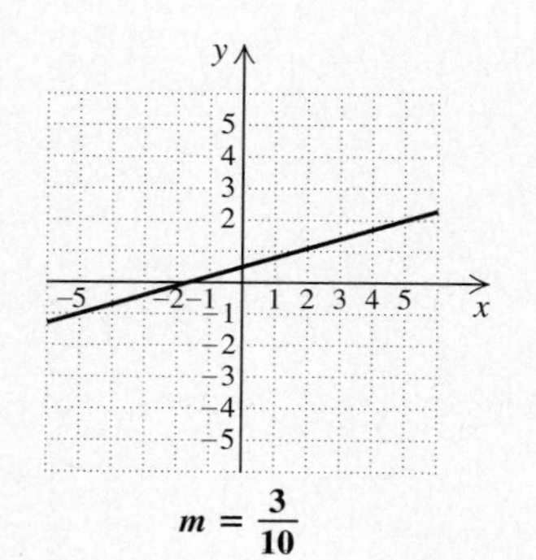

$m = \frac{3}{10}$

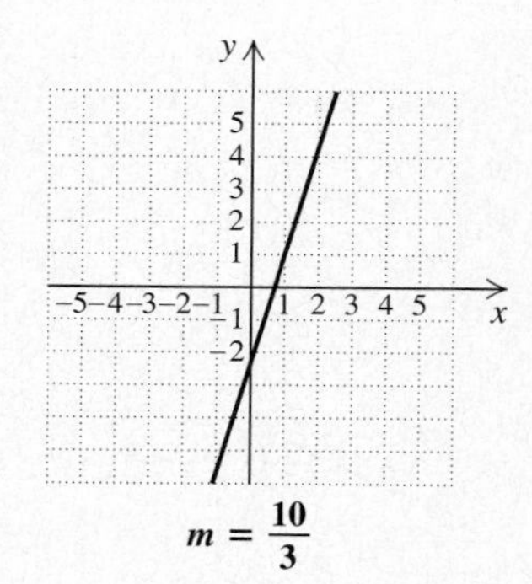

$m = \frac{10}{3}$

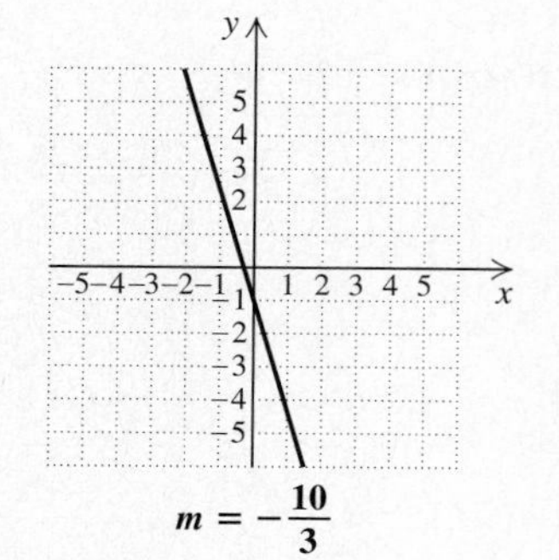

$m = -\frac{10}{3}$

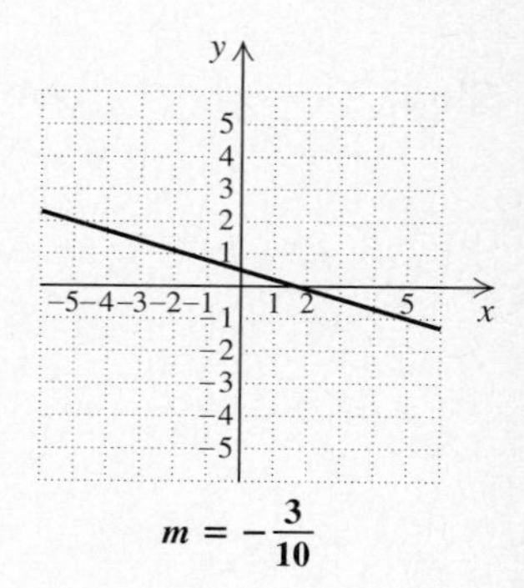

$m = -\frac{3}{10}$

Horizontal and Vertical Lines

What about the slope of a horizontal or a vertical line?

EXAMPLE 3

Find the slope of the line $y = 4$.

Solution Consider the points $(-3, 4)$ and $(2, 4)$, which are on the line.

The change in $y = 4 - 4$, or 0.
The change in $x = -3 - 2$, or -5.

$$m = \frac{4 - 4}{-3 - 2} = \frac{0}{-5} = 0$$

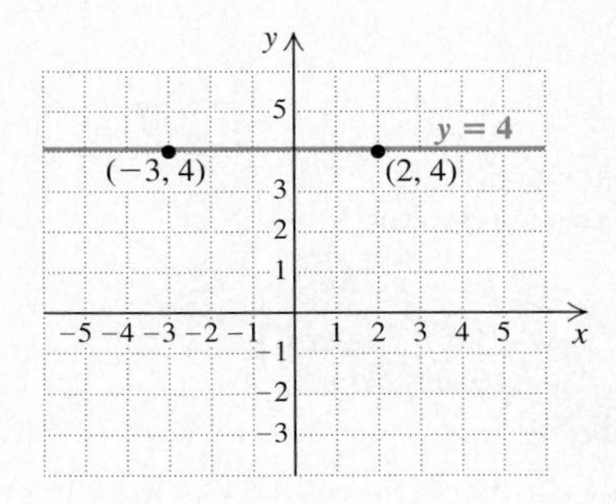

Any two points on a horizontal line have the same y-coordinate. Thus the change in y is 0, so the slope is 0. ❑

A horizontal line has slope 0.

EXAMPLE 4

Find the slope of the line $x = -3$.

Solution Consider the points $(-3, 4)$ and $(-3, -2)$, which are on the line.

The change in $y = 4 - (-2)$, or 6.
The change in $x = -3 - (-3)$, or 0.

$$m = \frac{4 - (-2)}{-3 - (-3)} = \frac{6}{0} \quad \text{(undefined)}$$

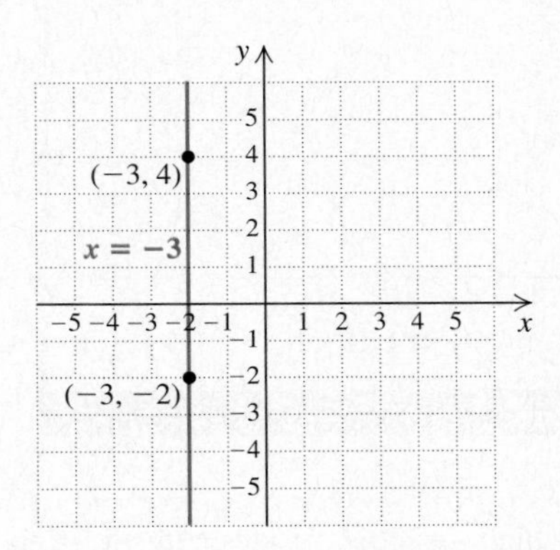

Since division by 0 is not defined, the slope of this line is not defined. The answer to this example is "The slope of this line is undefined." ❑

The slope of a vertical line is undefined.

Applications

Slope has many real-world applications. For example, numbers like 2%, 3%, and 6% are often used to represent the **grade** of a road, a measure of how steep a road on a hill or mountain is. For example, a 3% grade means that for every horizontal distance of 100 ft, the road rises or drops 3 ft. The concept of grade also occurs in fitness training when a person runs on a treadmill. A trainer may change the slope or grade of a treadmill to measure its effect on heartbeat. Another application of slope occurs when engineering a dam. A river's force or strength depends on how much the river drops over a specified horizontal distance.

EXAMPLE 5

A road rises 50 ft vertically over a horizontal distance of 1250 ft. Find the grade of the road.

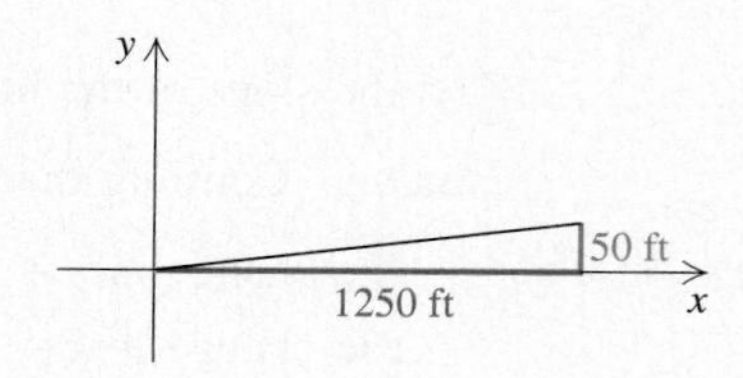

Solution The grade of the road is the slope of the line:

$$m = \frac{50}{1250} = 0.04 = 4\%.$$

EXERCISE SET 7.1

Find the slope, if it is defined, of the line.

1.

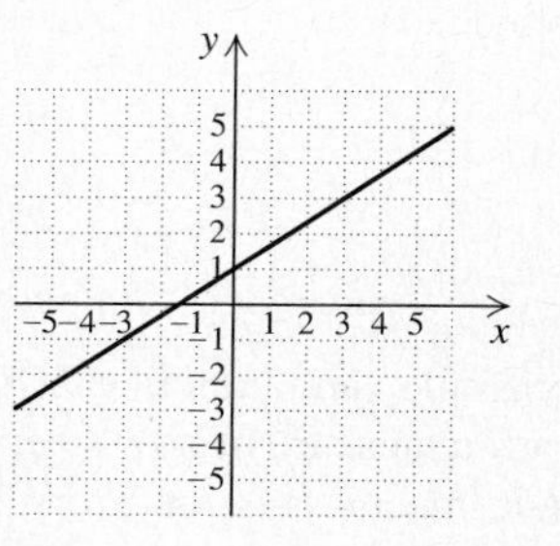

2.

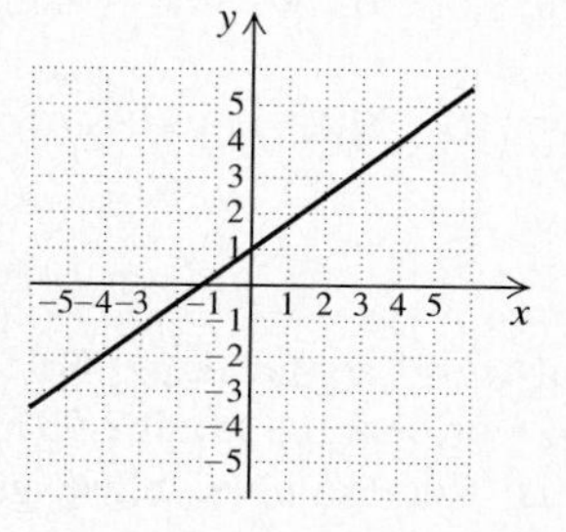

3.

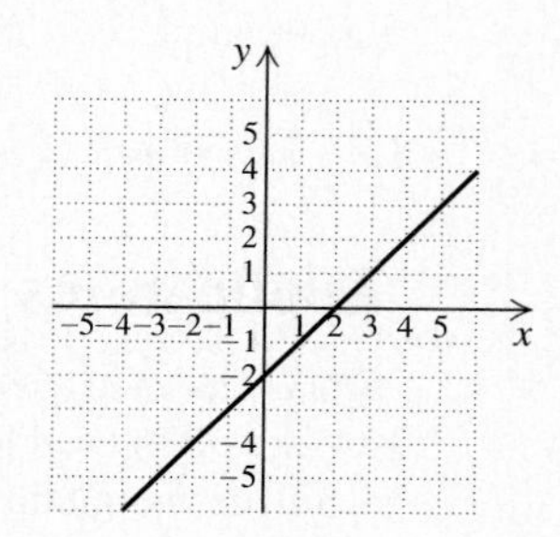

4.

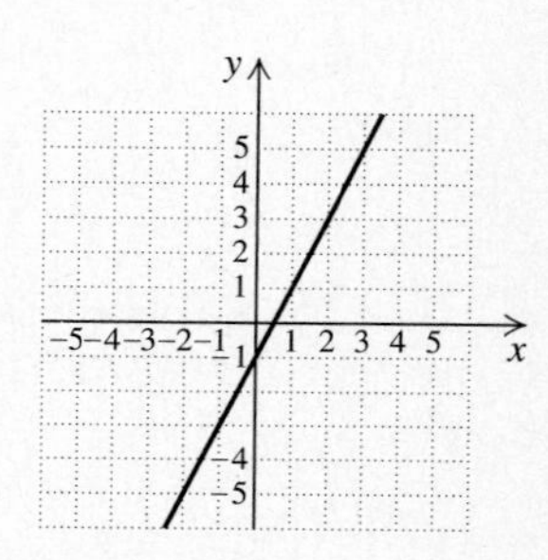

5.

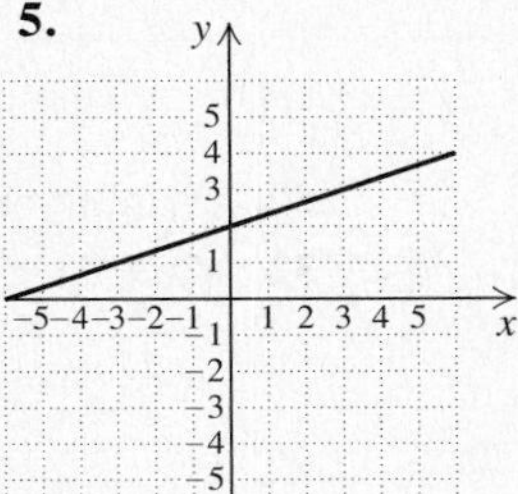

6.

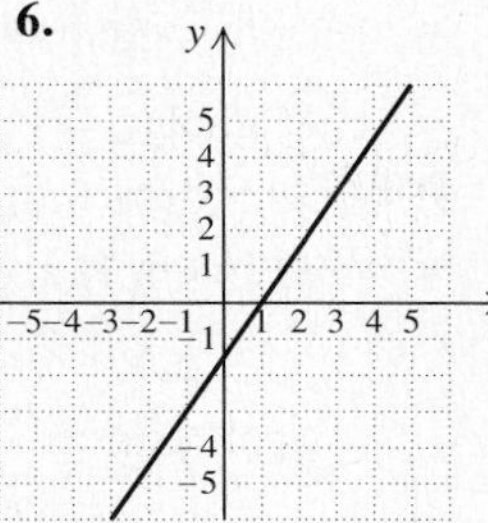

15.

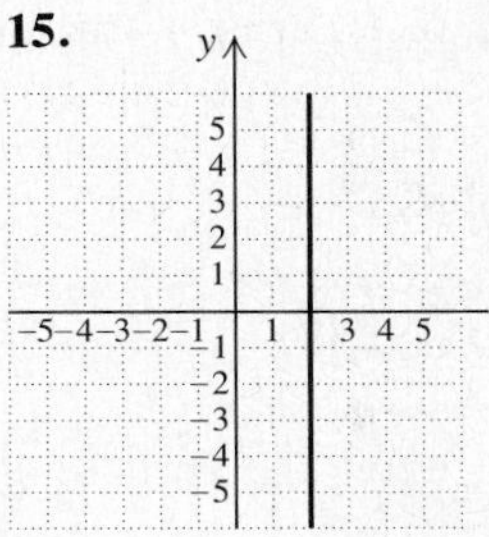

16.

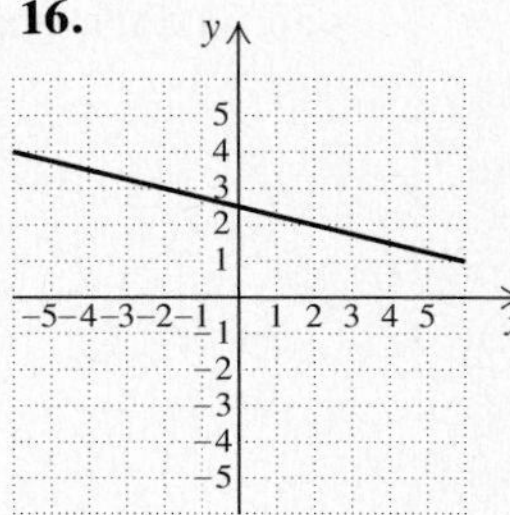

7.

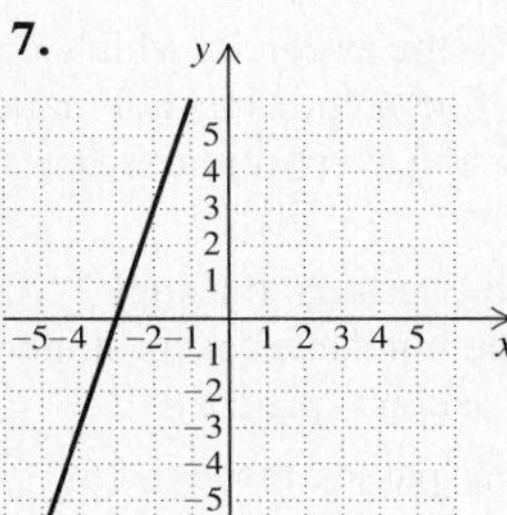

8.

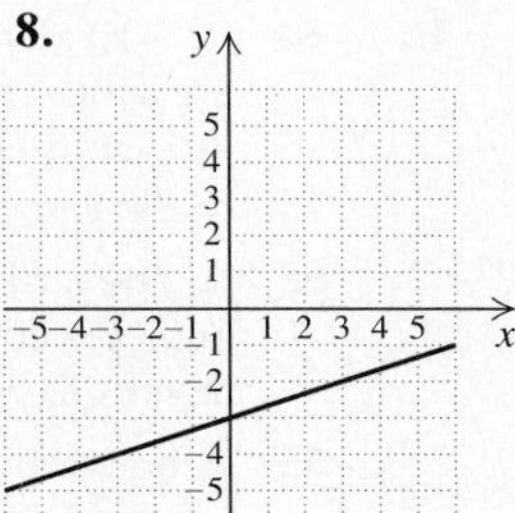

9.

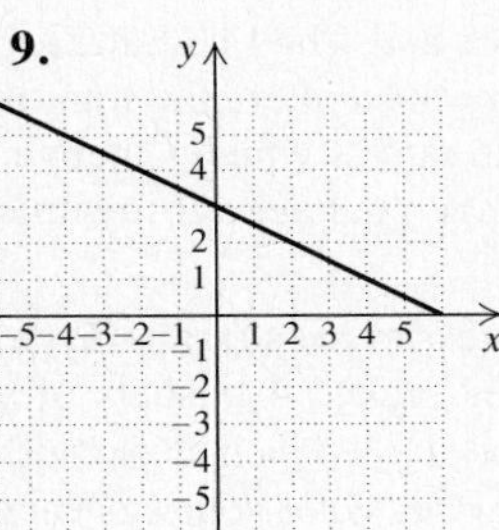

10.

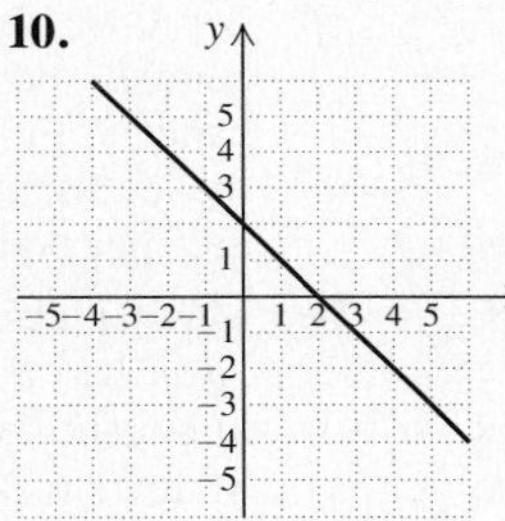

11.

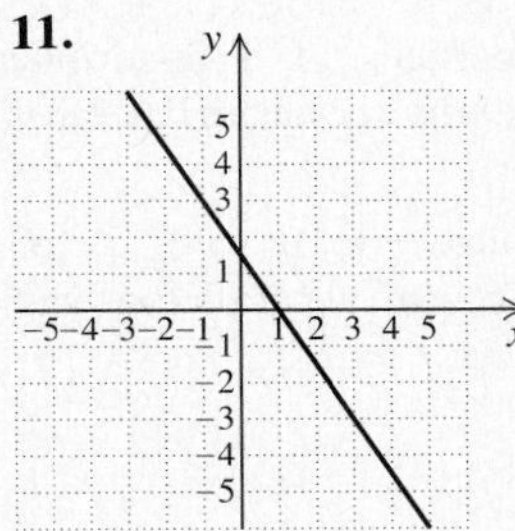

12.

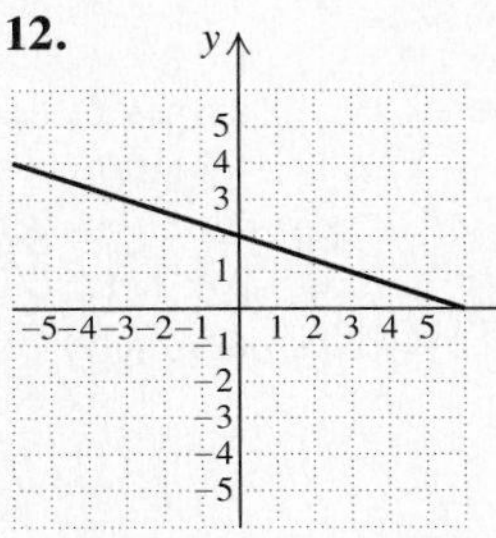

13.

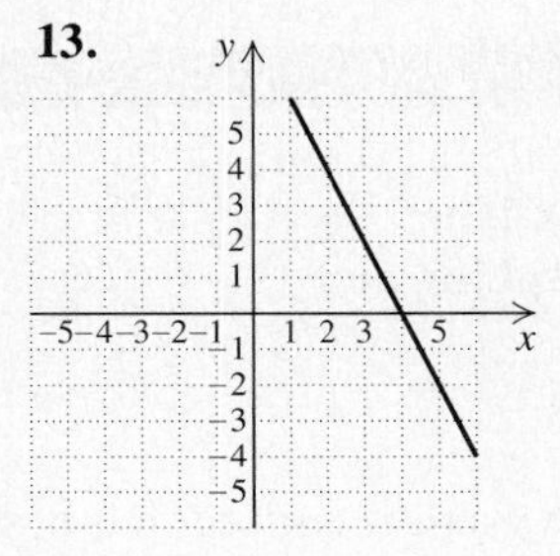

14.

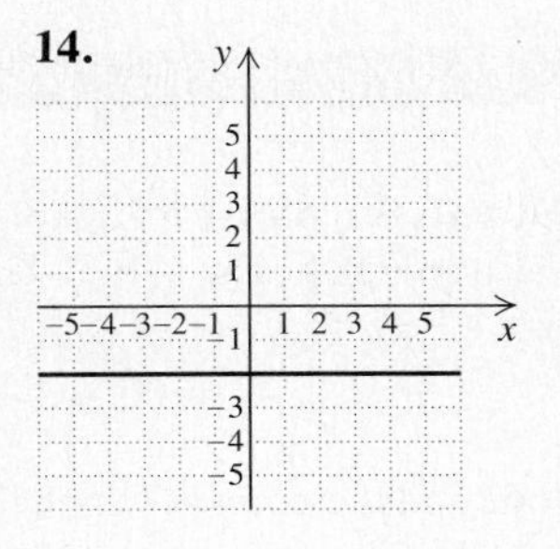

Find the slope of the line containing the given pair of points.

17. (3, 2) and (−1, 5)
18. (4, 1) and (−2, −3)
19. (−2, 4) and (3, 0)
20. (−4, 2) and (2, −3)
21. (4, 0) and (5, 7)
22. (3, 0) and (6, 2)
23. (0, 8) and (−3, 10)
24. (0, 9) and (4, 7)
25. (3, −2) and (5, −6)
26. (−2, 4) and (6, −7)
27. $\left(-2, \frac{1}{2}\right)$ and $\left(-5, \frac{1}{2}\right)$
28. (8, −3) and (10, −3)
29. (9, −4) and (9, −7)
30. (−10, 3) and (−10, 4)
31. (−1, 5) and (4, 5)
32. (−4, −2) and (−4, 7)

Find the slope of the line.

33. $x = -8$
34. $x = -4$
35. $y = 2$
36. $y = 17$
37. $x = 9$
38. $x = 6$
39. $y = -9$
40. $y = -4$

41. Find the grade of the road.

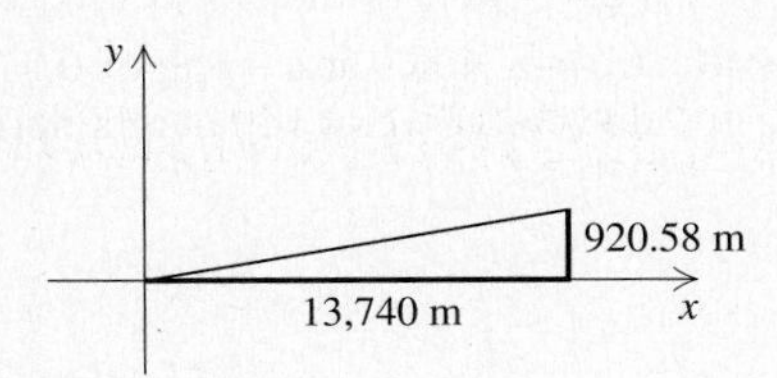

42. Find the slope (or pitch) of the roof.

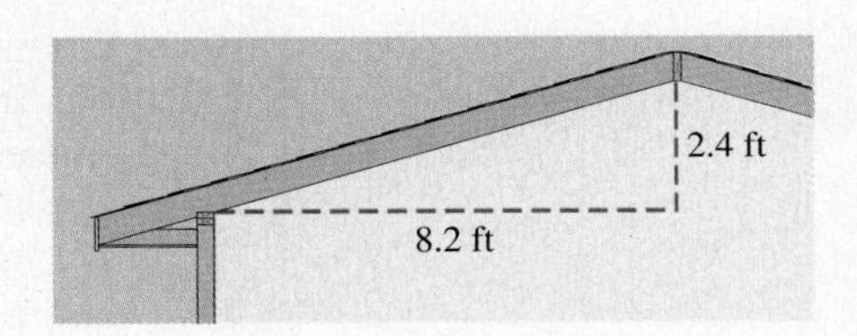

43. Find the slope (or grade) of the treadmill.

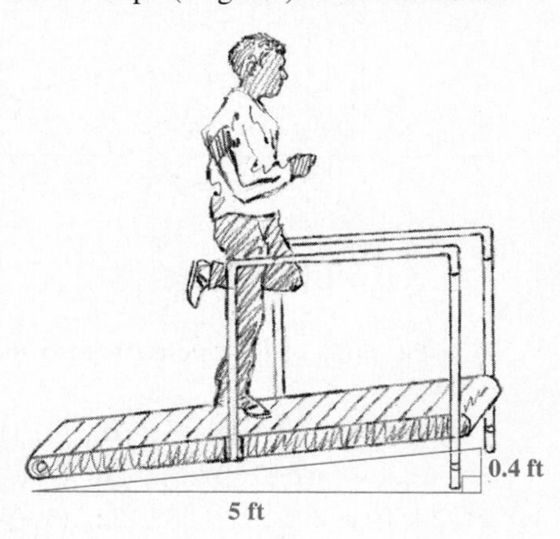

44. Find the slope of the river.

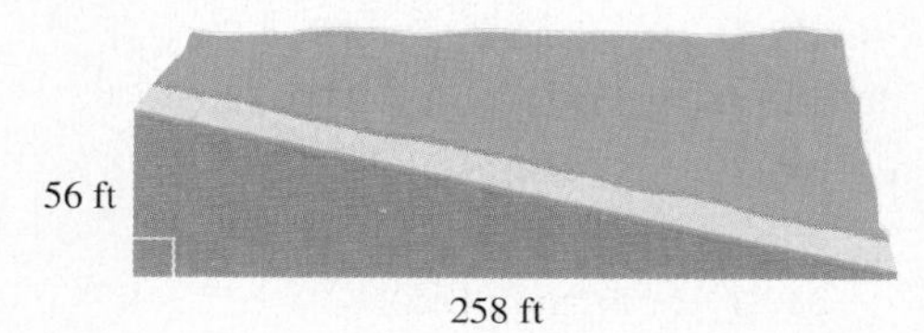

45. A road drops 158.4 ft vertically over a horizontal distance of 5280 ft. What is the grade of the road?

46. A river drops 55.71 ft vertically over a horizontal distance of 1238 ft. What is the slope of the river?

47. A treadmill that casts a shadow 5 ft long is set at a 12% grade when a heart arrhythmia occurs. How far off the floor is the end of the treadmill if the light source is directly overhead?

48. A river flows at a slope of 0.12. How many feet does it fall vertically over a horizontal distance of 250 ft?

Skill Maintenance

Multiply.

49. $5x(9x - 3)$

50. $(x - 2)(x^2 + 3x - 5)$

51. $(x - 7)(x + 7)$

52. $(5x + 2)^2$

Synthesis

53. ◈ If one line has a slope of -3 and another has a slope of 2, which line is steeper? Why?

54. ◈ Explain why the order in which coordinates are subtracted to find slope does not matter so long as y-coordinates and x-coordinates are subtracted in the same order.

55. A nonvertical line passes through (3, 4). What numbers could the line have for its slope if the line never enters the second quadrant?

56. A nonvertical line passes through $(5, -6)$. What numbers could the line have for its slope if the line never enters the first quadrant?

57. By 3:00, Catanya and Chad had already made 46 candles. Forty minutes later, the total reached 64 candles. Find the rate at which Catanya and Chad made candles. Give your answer in number of candles per hour.

58. Marcy picks apples twice as fast as Ryan. By 4:30, Ryan had already picked 4 bushels of apples; 50 minutes later, his total reached $5\frac{1}{2}$ bushels. Find Marcy's picking rate. Give your answer in number of bushels per hour.

59. ◈ The points $(-4, -3)$, $(1, 4)$, $(4, 2)$, and $(-1, -5)$ are vertices of a quadrilateral. Use slopes to explain why the quadrilateral is a parallelogram.

60. ◈ Can the points $(-4, 0)$, $(-1, 5)$, $(6, 2)$, and $(2, -3)$ be vertices of a parallelogram? Why or why not?

7.2 Slope–Intercept Form

Using the y-intercept and the Slope to Graph a Line • Equations in Slope–Intercept Form • Graphing and Slope–Intercept Form

If we know a line's slope and the point at which the y-axis is crossed, it is possible to draw a graph of the line. In this section, we will learn how to find a line's slope and

y-intercept from its equation. We will then be able to graph certain equations quite easily.

Using the y-intercept and the Slope to Graph a Line

Let's return to the car production situation described at the beginning of Section 7.1. Now suppose that as a new workshift begins, 4 cars have already been produced. At the Michigan plant, 3 cars were being produced every 2 hours, a rate of $\frac{3}{2}$ cars per hour. If this rate remains the same regardless of how many cars have already been produced, the following table and graph can be made.

Michigan Plant	
Hours Elapsed	**Cars Produced**
0	4
2	7
4	10
6	13
8	16

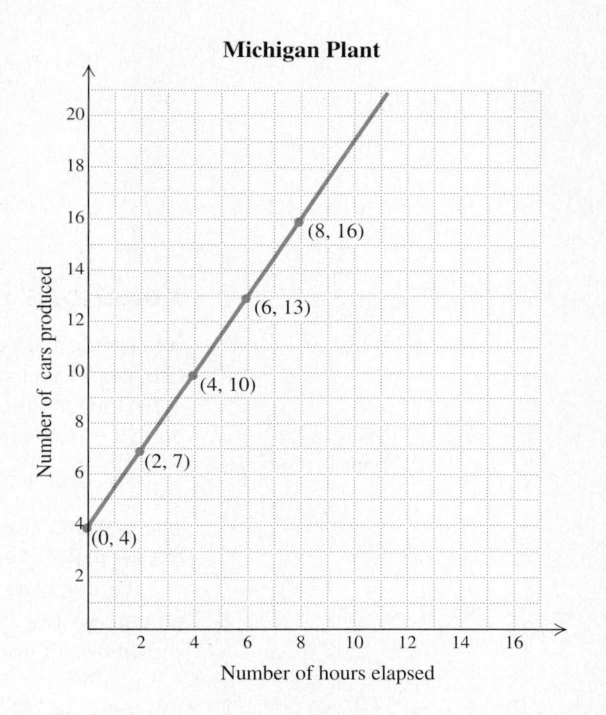

To confirm that the production rate is still $\frac{3}{2}$, we compute the slope using any pair of points on the graph. We choose (0, 4) and (2, 7):

$$\text{Slope} = \frac{\text{change in } y}{\text{change in } x}$$

$$= \frac{7-4}{2-0} = \frac{3}{2}.$$

Note that had we simply plotted the point (0, 4) and from there moved *up* 3 units and *to the right* 2 units, we could have located the point (2, 7) without consulting the table. Using (0, 4) and (2, 7), we then could have drawn the line. This is the method used in the following example.

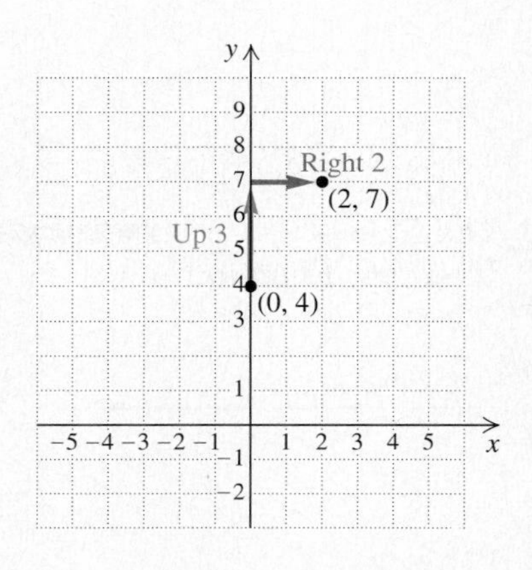

EXAMPLE 1

Draw a line that has slope $\frac{1}{4}$ and y-intercept $(0, 2)$.

Solution We plot $(0, 2)$ and from there move *up* 1 unit and *to the right* 4 units. This locates the point $(4, 3)$. We plot $(4, 3)$ and draw a line passing through $(0, 2)$ and $(4, 3)$, as shown on the right below.

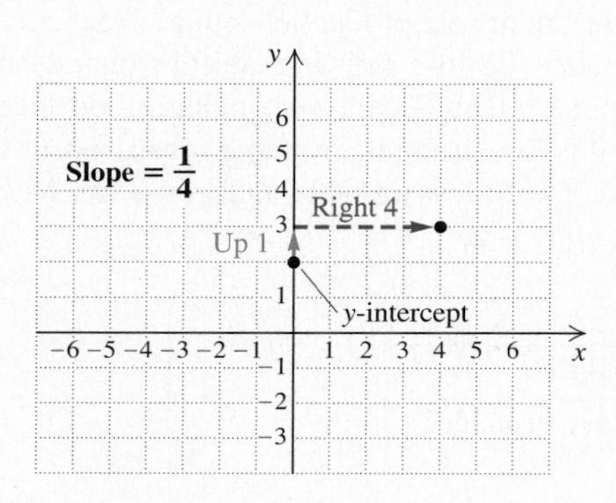

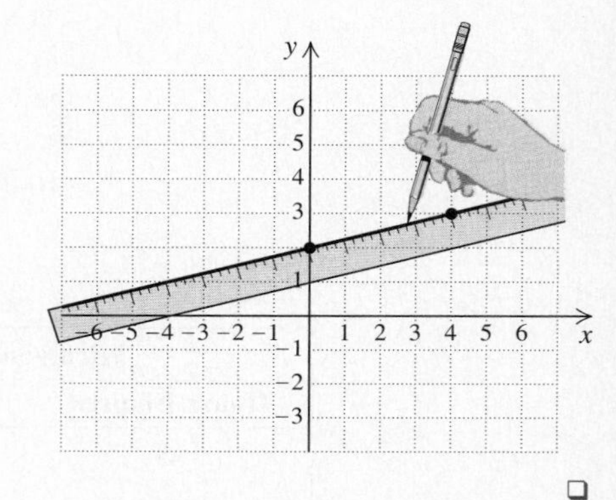

Equations in Slope–Intercept Form

It is not difficult to find a line's slope and y-intercept from its equation. Recall from Chapter 3 that to find the y-intercept of an equation's graph, we replace x with 0 and solve the resulting equation for y. For example, to find the y-intercept of the graph of $y = 2x + 3$, we replace x with 0 and solve as follows:

$$y = 2 \cdot 0 + 3 = 0 + 3 = 3.$$

The y-intercept of the graph of $y = 2x + 3$ is $(0, 3)$. In a similar manner, it can be shown that the y-intercept of the graph of any equation $y = mx + b$ is $(0, b)$.

To calculate the slope of the graph of $y = 2x + 3$, we need two points. The y-intercept $(0, 3)$ is one point and a second point, $(1, 5)$, can be found easily by substituting 1 for x. We then have

$$\begin{aligned} \text{Slope} &= \frac{\text{change in } y}{\text{change in } x} \\ &= \frac{5 - 3}{1 - 0} = \frac{2}{1} = 2. \end{aligned}$$

Note that the slope, 2, is also the coefficient of x in $y = 2x + 3$. In a similar manner, it can be shown that the slope of the graph of any equation $y = mx + b$ is m (see Exercise 73).

The Slope–Intercept Equation

The equation $y = mx + b$ is called the *slope–intercept equation*. The equation represents a line of slope m with y-intercept $(0, b)$.

EXAMPLE 2

Find the slope and the y-intercept of the line: **(a)** $y = \frac{4}{5}x - 8$; **(b)** $2x + y = 5$; **(c)** $3x + 4y = 7$.

Solution

a) We rewrite $y = \frac{4}{5}x - 8$ as $y = \frac{4}{5}x + (-8)$. Now we simply read the slope and the y-intercept from the equation:

$$y = \frac{4}{5}x + (-8).$$

The slope is $\frac{4}{5}$. The y-intercept is $(0, -8)$.

b) We first solve for y to rewrite the equation in the form $y = mx + b$:

$$2x + y = 5$$

$$y = -2x + 5. \qquad \text{Adding } -2x \text{ on both sides}$$

The slope is -2. The y-intercept is $(0, 5)$.

c) We rewrite the equation in the form $y = mx + b$:

$$3x + 4y = 7$$

$$4y = -3x + 7 \qquad \text{Adding } -3x \text{ on both sides}$$

$$y = \frac{1}{4}(-3x + 7) \qquad \text{Multiplying by } \frac{1}{4} \text{ on both sides}$$

$$y = -\frac{3}{4}x + \frac{7}{4} \qquad \text{Using the distributive law}$$

The slope is $-\frac{3}{4}$. The y-intercept is $\left(0, \frac{7}{4}\right)$. ❑

EXAMPLE 3

A line has slope $-\frac{12}{5}$ and y-intercept $(0, 11)$. Find an equation of the line.

Solution We use the slope–intercept equation, substituting $-\frac{12}{5}$ for m and 11 for b:

$$y = mx + b$$

$$y = -\frac{12}{5}x + 11.$$ ❑

Recall that *parallel* lines extend indefinitely without intersecting. When two lines have the same slope but different y-intercepts, they are parallel.

EXAMPLE 4

Determine whether the graphs of $y = -3x + 4$ and $6x + 2y = -10$ are parallel.

Solution The equation $y = -3x + 4$ represents a line with slope -3 and y-intercept $(0, 4)$. To find the slope of the second line, we rewrite $6x + 2y = -10$ in slope–intercept form:

$$6x + 2y = -10$$

$$2y = -6x - 10$$

$$y = -3x - 5. \qquad \text{The slope is } -3 \text{ and the } y\text{-intercept is } (0, -5).$$

Since both lines have slope -3 but different y-intercepts, the graphs are parallel. ❑

Graphing and Slope–Intercept Form

Our work in Examples 1 and 2 can be easily combined.

EXAMPLE 5

Graph: **(a)** $y = \frac{2}{5}x + 4$; **(b)** $2x + 3y = 3$.

Solution

a) First we plot the y-intercept, $(0, 4)$. We then consider the slope, $\frac{2}{5}$. Starting at the y-intercept and using the slope, we plot a second point by moving *up* 2 units (since the numerator is *positive* and corresponds to the change in y) and *to the right* 5 units (since the denominator is *positive* and corresponds to the change in x). We reach a new point, $(5, 6)$.

We can also rewrite the slope as $\frac{-2}{-5}$. We again start at the y-intercept, $(0, 4)$, but move *down* 2 units (since the numerator is *negative* and corresponds to the change in y) and *to the left* 5 units (since the denominator is *negative* and corresponds to the change in x). We reach another new point, $(-5, 2)$. Once two or three points have been plotted, the line can be drawn.

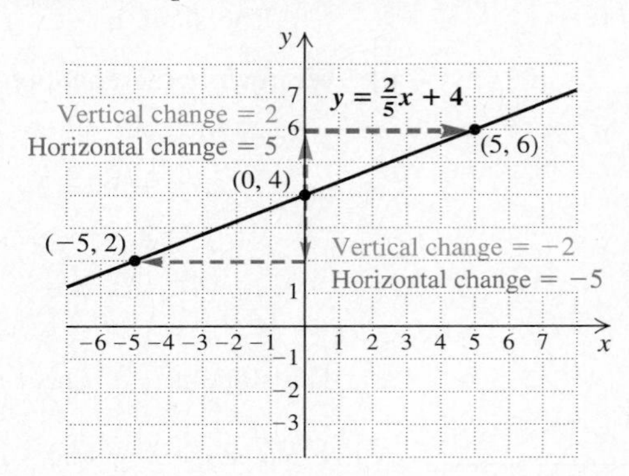

b) We graph the equation $2x + 3y = 3$ by first rewriting it in slope–intercept form:

$$2x + 3y = 3$$

$$3y = -2x + 3 \quad \text{Adding } -2x \text{ on both sides}$$

$$y = \tfrac{1}{3}(-2x + 3) \quad \text{Multiplying by } \tfrac{1}{3} \text{ on both sides}$$

$$y = -\tfrac{2}{3}x + 1. \quad \text{Using the distributive law}$$

TECHNOLOGY CONNECTION

Using a standard $[-10, 10, -10, 10]$ window, graph the equations $y_1 = \frac{2}{3}x + 1$, $y_2 = \frac{3}{8}x + 1$, $y_3 = \frac{2}{3}x + 5$, and $y_4 = \frac{3}{8}x + 5$. If you can, use your grapher in the *Mode* that will graph equations *simultaneously*. Once the four lines have been drawn, try to decide which equation corresponds to each line. After matching equations with lines, you can check your matches by graphing the same equations in a *sequence* mode, if your grapher has one. In the sequence mode, equation y_1 is drawn first, y_2 is drawn next, and so on.

TC1. Graph the equations $y_1 = -\frac{3}{4}x - 2$, $y_2 = -\frac{1}{5}x - 2$, $y_3 = -\frac{3}{4}x - 5$, and $y_4 = -\frac{1}{5}x - 5$ using a grapher. If possible, use the simultaneous mode. Then match each line with the corresponding equation. Check using the sequence mode.

TC2. Write four different slope–intercept equations, two of which have the same slope but different y-intercepts, and two of which have the same y-intercepts but different slopes. Then use a grapher to draw all four lines and ask a classmate to match each equation with the appropriate line.

To graph $y = -\frac{2}{3}x + 1$, we first plot the y-intercept, (0, 1). We can think of the slope as $\frac{-2}{3}$. Starting at the y-intercept and using the slope, we find a second point by moving *down* 2 units (since the numerator is *negative*) and *to the right* 3 units (since the denominator is *positive*). We plot the new point, (3, −1). In a similar manner, we can move from the point (3, −1) to locate a third point, (6, −3). The line can then be drawn.

If the slope is thought of as $\frac{2}{-3}$, we can start at (0, 1), but this time move *up* 2 units (since the numerator is *positive*) and *to the left* 3 units (since the denominator is *negative*). We get another point on the graph, (−3, 3).

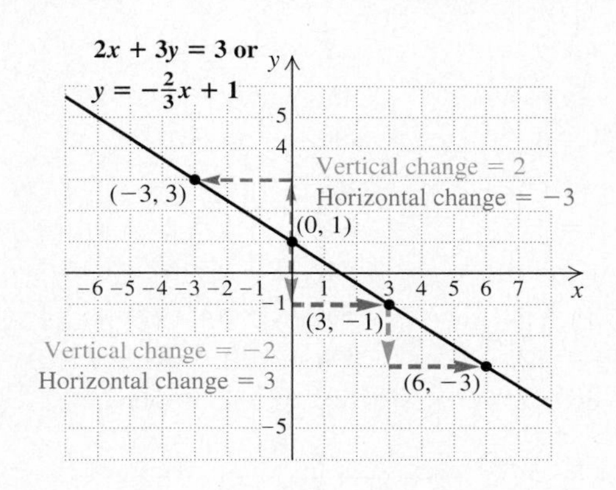

□

EXERCISE SET 7.2

Draw a line that has the given slope and y-intercept.

1. Slope $\frac{2}{5}$; y-intercept (0, 1)
2. Slope $\frac{5}{3}$; y-intercept (0, −1)
3. Slope $\frac{5}{2}$; y-intercept (0, −3)
4. Slope $\frac{3}{5}$; y-intercept (0, 2)
5. Slope $-\frac{3}{4}$; y-intercept (0, 5)
6. Slope $-\frac{4}{5}$; y-intercept (0, 6)
7. Slope 2; y-intercept (0, −4)
8. Slope −2; y-intercept (0, −3)
9. Slope −3; y-intercept (0, 2)
10. Slope 3; y-intercept (0, 4)

Find the slope and the y-intercept of the line.

11. $y = \frac{3}{7}x + 6$
12. $y = -\frac{3}{8}x + 7$
13. $y = -\frac{5}{6}x + 2$
14. $y = \frac{7}{2}x + 4$
15. $y = \frac{9}{4}x - 7$
16. $y = \frac{2}{9}x - 1$
17. $y = -\frac{2}{5}x$
18. $y = \frac{4}{3}x$
19. $-2x + y = 4$
20. $-5x + y = 5$
21. $4x - 3y = -12$
22. $x - 2y = 9$
23. $x - 3y = -2$
24. $x + y = 7$
25. $-2x + 4y = 8$
26. $-5x + 7y = 2$
27. $y = 5$
28. $y + 2 = 6$

Find the slope–intercept equation for the line with the indicated slope and y-intercept.

29. Slope 5; y-intercept (0, 6)
30. Slope −4; y-intercept (0, −2)
31. Slope $\frac{7}{8}$; y-intercept (0, −1)
32. Slope $\frac{5}{7}$; y-intercept (0, 4)
33. Slope $-\frac{5}{3}$; y-intercept (0, −8)
34. Slope $\frac{3}{4}$; y-intercept (0, 23)
35. Slope −2; y-intercept (0, 3)
36. Slope 7; y-intercept (0, −6)

Determine whether the pair of equations represents parallel lines.

37. $y = \frac{2}{3}x + 7,$
$y = \frac{2}{3}x - 5$

38. $y = -\frac{5}{4}x + 1,$
$y = \frac{5}{4}x + 3$

39. $y = 2x - 5,$
$4x + 2y = 9$

40. $y = -3x + 1,$
$6x + 2y = 8$

41. $3x + 4y = 8,$
$7 - 12y = 9x$

42. $3x = 5y - 2,$
$10y = 4 - 6x$

Graph.

43. $y = \frac{3}{5}x + 2$

44. $y = -\frac{3}{5}x - 1$

45. $y = -\frac{3}{5}x + 4$

46. $y = \frac{3}{5}x - 2$

47. $y = \frac{5}{3}x + 3$

48. $y = \frac{5}{3}x - 2$

49. $y = -\frac{3}{2}x - 2$

50. $y = -\frac{4}{3}x + 3$

51. $2x + y = 1$

52. $3x + y = 2$

53. $3x - y = 4$

54. $2x - y = 5$

55. $2x + 3y = 9$

56. $4x + 5y = 15$

57. $x - 4y = 12$

58. $x + 5y = 20$

59. $5x - 6y = 24$

60. $6x - 7y = 56$

Skill Maintenance

61. Solve: $2x^2 + 6x = 0$.

62. Factor: $x^3 + 5x^2 - 14x$.

63. The product of two consecutive odd integers is 195. Find the integers.

64. Eleven less than the square of a number is ten times the number. Find the number.

Synthesis

65. ◈ Under what circumstances might you draw an incorrect graph of a line even though three points were plotted and lined up with each other before the line was drawn?

66. ◈ Can an equation of a horizontal line be written in slope–intercept form? Why or why not?

Two lines are *perpendicular* if either the product of their slopes is -1, or one line is vertical and the other horizontal. For Exercises 67–72, determine whether each pair of equations represents perpendicular lines.

67. $3y = 5x - 3,$
$3x + 5y = 10$

68. $y + 3x = 10,$
$2x - 6y = 18$

69. $3x + 5y = 10,$
$15x + 9y = 18$

70. $10 - 4y = 7x,$
$7y + 21 = 4x$

71. $x = 5,$
$y = \frac{1}{2}$

72. $y = -2,$
$x = 2$

73. Show that the slope of the line given by $y = mx + b$ is m. (*Hint:* Substitute both 0 and 1 for x to find two pairs of coordinates. Then use the formula Slope = change in y/change in x.)

74. Find an equation that can be used to predict the total number of cars N produced after t hours by the Michigan plant discussed in this section.

75. Find an equation of the line with the same slope as the line $3x - 2y = 8$ and the same y-intercept as the line $2y + 3x = -4$.

7.3 Point–Slope Form

Writing Equations in Point–Slope Form • Graphing and Point–Slope Form

We now learn how to write an equation of a line using the line's slope and any one point through which the line passes.

Writing Equations in Point–Slope Form

Consider a line with slope 2 passing through the point (4, 1), as shown in the figure. In order for a point (x, y) to be on the line, the coordinates x and y must solve the slope equation

$$\frac{y-1}{x-4} = 2.$$

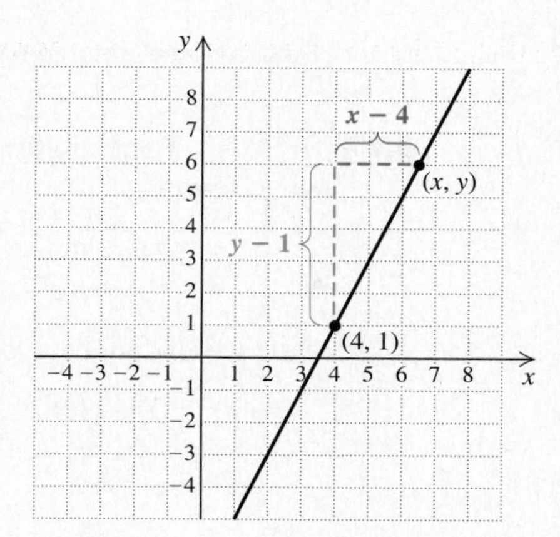

Multiplying on both sides by $x - 4$, we have

$$y - 1 = 2(x - 4).$$

This is considered **point–slope form** for the line shown.

To generalize, a line with slope m passing through the point (x_1, y_1) will include a point (x, y) if the coordinates x and y solve the slope equation

$$\frac{y - y_1}{x - x_1} = m.$$

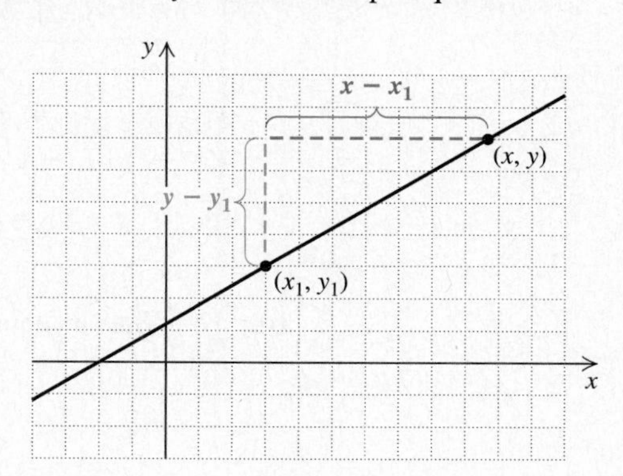

Multiplying on both sides by $x - x_1$ gives us the *point–slope equation*

$$y - y_1 = m(x - x_1).$$

The Point–Slope Equation

The equation $y - y_1 = m(x - x_1)$ is called the *point–slope equation* for the line with slope m that contains the point (x_1, y_1).

EXAMPLE 1

Find a point–slope equation for the line with slope $\frac{1}{5}$ that contains the point $(-2, -3)$.

Solution We substitute $\frac{1}{5}$ for m, -2 for x_1, and -3 for y_1:

$y - y_1 = m(x - x_1)$	Using the point–slope equation
$y - (-3) = \frac{1}{5}(x - (-2)).$	Substituting

❑

EXAMPLE 2

Find a slope–intercept equation for the line with slope 3 that contains the point (1, −5).

Solution There are two parts to this solution. First we write an equation in point–slope form:

$$y - y_1 = m(x - x_1)$$
$$y - (-5) = 3(x - 1). \quad \text{Substituting}$$

Next we find an equivalent equation of the form $y = mx + b$:

$$y - (-5) = 3(x - 1)$$
$$y + 5 = 3x - 3 \quad \text{Simplifying the subtraction and using the distributive law}$$
$$y = 3x - 8. \quad \text{This is in slope–intercept form.}$$

❑

EXAMPLE 3

A line passes through the points (3, −5) and (−4, 9). Find an equation for the line **(a)** in point–slope form and **(b)** in slope–intercept form.

Solution

a) To find a point–slope equation, we first compute the slope:

$$m = \frac{9 - (-5)}{-4 - 3} = \frac{14}{-7} = -2.$$

Next we use the point–slope equation and substitute, using −2 for m and either (3, −5) or (−4, 9) as (x_1, y_1):

$$y - y_1 = m(x - x_1)$$
$$y - (-5) = -2(x - 3). \quad \text{Using } (3, -5) \text{ for } (x_1, y_1)$$

b) To write an equation in slope–intercept form, we use the result of part (a) above:

$$y - (-5) = -2(x - 3)$$
$$y + 5 = -2x + 6 \quad \text{Simplifying the subtraction and using the distributive law}$$
$$y = -2x + 1. \quad \text{This is in slope–intercept form.}$$

Had we used (−4, 9) as (x_1, y_1) in part (a), we would have obtained the same slope–intercept equation:

$$y - 9 = -2(x - (-4)) \quad \text{Using } (-4, 9) \text{ for } (x_1, y_1)$$
$$y - 9 = -2(x + 4)$$
$$y - 9 = -2x - 8$$
$$y = -2x + 1.$$

❑

Graphing and Point–Slope Form

Equations written in point–slope form are easily graphed.

EXAMPLE 4

Graph: $y - 2 = 3(x - 4)$.

Solution Since $y - 2 = 3(x - 4)$ is in point–slope form, we know that the line has slope 3, or $\frac{3}{1}$, and passes through the point (4, 2). We plot (4, 2) and then find a second point by moving *up* 3 units and *to the right* 1 unit. The line can then be drawn, as shown below.

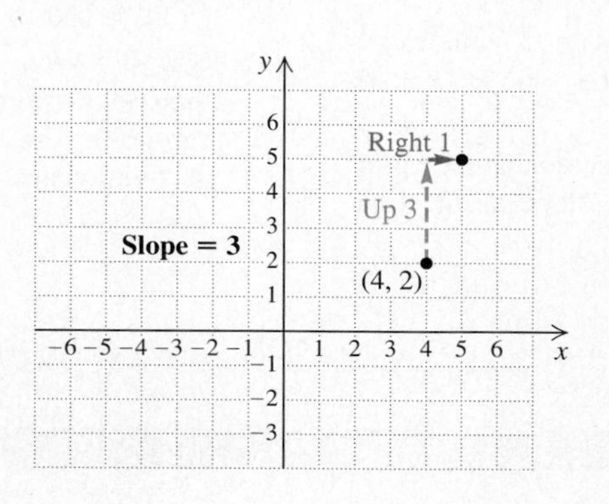

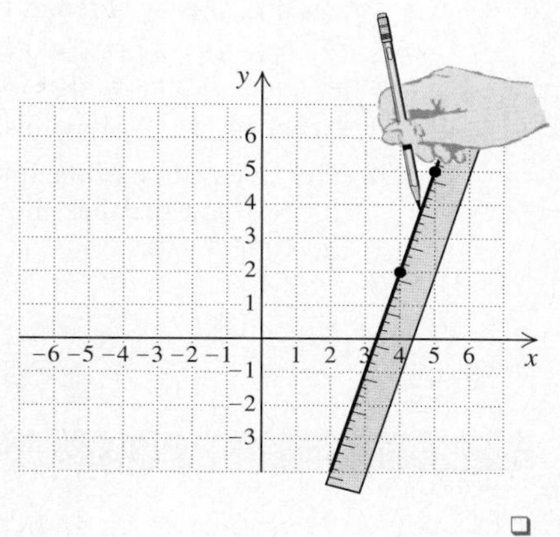

EXERCISE SET 7.3

Find a point–slope equation for the line containing the given point and having the given slope.

1. (2, 5), $m = 5$
2. (−3, 0), $m = -2$
3. (2, 4), $m = \frac{3}{4}$
4. $(\frac{1}{2}, 2)$, $m = -1$
5. (2, −6), $m = 1$
6. (4, −2), $m = 6$
7. (−3, 0), $m = -3$
8. (0, 3), $m = -3$
9. (5, 6), $m = \frac{2}{3}$
10. (2, 7), $m = \frac{5}{6}$

Find the slope–intercept equation for the line containing the given point and having the given slope.

11. (3, 7), $m = 2$
12. (1, 5), $m = 4$
13. (4, 5), $m = -1$
14. (2, −3), $m = 1$
15. (−2, 3), $m = \frac{1}{2}$
16. (6, −4), $m = -\frac{1}{2}$
17. (−6, −5), $m = -\frac{1}{3}$
18. (−5, 7), $m = \frac{1}{5}$
19. (4, −3), $m = \frac{5}{4}$
20. (−3, 8), $m = \frac{4}{3}$

Find the slope–intercept equation for the line containing the given pair of points. (*Hint:* First use point–slope form.)

21. (−6, 1) and (2, 3)
22. (12, 16) and (1, 5)
23. (0, 4) and (4, 2)
24. (0, 0) and (4, 2)
25. (3, 2) and (1, 5)
26. (−4, 1) and (−1, 4)
27. (5, 0) and (0, −2)
28. (−2, −2) and (1, 3)
29. (−2, −4) and (2, −1)
30. (−3, 5) and (−1, −3)

Graph.

31. $y - 5 = \frac{1}{2}(x - 3)$
32. $y - 2 = \frac{1}{3}(x - 5)$
33. $y - 3 = -\frac{1}{2}(x - 5)$
34. $y - 1 = -\frac{1}{4}(x - 3)$
35. $y + 5 = \frac{1}{2}(x - 3)$
36. $y - 2 = \frac{1}{3}(x + 5)$
37. $y + 2 = 3(x + 1)$
38. $y + 4 = 2(x + 1)$
39. $y - 4 = -2(x + 1)$
40. $y + 3 = -1(x - 4)$
41. $y + 3 = -(x + 2)$
42. $y + 4 = 2(x + 2)$

Skill Maintenance

Factor.

43. $7x^3y^2 + 35x^2y^6$
44. $5x^2(3x - 1) + (3x - 1)$

Simplify.

45. $\dfrac{5x^2 + 5x}{10x^3 - 10x^2}$
46. $\dfrac{x^2 - 7x + 10}{x^2 - 4}$

Synthesis

47. ◈ In your own words, describe a procedure that can be used to write a slope–intercept equation for any line passing through two given points.

48. ◈ The graph of $(y - 1)/(x - 4) = 2$ includes all the points found in the graph of $y - 1 = 2(x - 4)$ except (4, 1). Why is this?

49. Find an equation of the line that contains the point (2, −3) and that has the same slope as the line $3x - y + 4 = 0$.

50. Find an equation of the line that has the same y-intercept as the line $x - 3y = 6$ and contains the point (5, −1).

51. Find an equation of the line that has x-intercept (−2, 0) and is parallel to $4x - 8y = 12$.

52. ◈ Why is slope–intercept form more useful than point–slope form when using a grapher? How can point–slope form be modified so that it better accommodates graphers?

7.4 Linear Inequalities in Two Variables

Graphing Linear Inequalities • Test Points

Just as we can graph solutions of linear equations like $5x + 4y = 13$ or $y = \frac{1}{2}x + 1$, we can graph solutions of *linear inequalities* like $5x + 4y < 13$ or $y > \frac{1}{2}x + 1$.

Graphing Linear Inequalities

In Section 2.6, we found that the solution of an inequality like $5x + 9 \leq 4x + 3$ can be represented by a shaded portion of a number line. When a solution included an endpoint, we drew a solid dot, and when the endpoint was excluded, we drew an open dot. To graph inequalities like $y > \frac{1}{2}x + 1$ or $2x + 3y \leq 6$, we will shade a region of a plane. That region will be either above or below the graph of a "boundary line" (in this case, $y = \frac{1}{2}x + 1$ or $2x + 3y = 6$). When a solution includes the boundary line, we will draw that line solid. If the boundary line is excluded, we will draw it dashed.

EXAMPLE 1

Graph: $y > \frac{1}{2}x + 1$.

Solution We begin by graphing the line $y = \frac{1}{2}x + 1$. The slope is $\frac{1}{2}$ and the y-intercept is (0, 1). This line is drawn dashed since points on the line, like (2, 2), are *not* solutions of $y > \frac{1}{2}x + 1$:

$$\begin{array}{c|l} y > \frac{1}{2}x + 1 & \\ \hline 2 \ ? & \frac{1}{2}\cdot 2 + 1 \\ & 1 + 1 \\ 2 & 2 \end{array} \quad \text{FALSE}$$

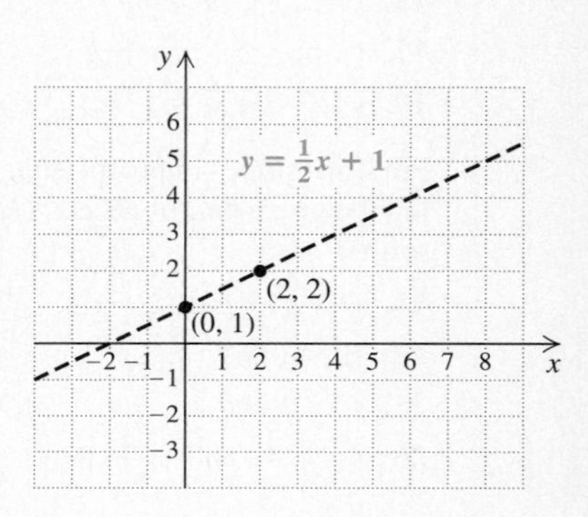

Note that the plane is now split into two regions. If we consider the coordinates of a few points above the line, we will find that all are solutions of $y > \frac{1}{2}x + 1$.

Here's a check for the point (2, 3):

$$\begin{array}{l|l} \multicolumn{2}{c}{y > \frac{1}{2}x + 1} \\ \hline 3 \; ? & \frac{1}{2}\cdot 2 + 1 \\ & 1 + 1 \\ 3 & 2 \end{array} \quad \text{TRUE}$$

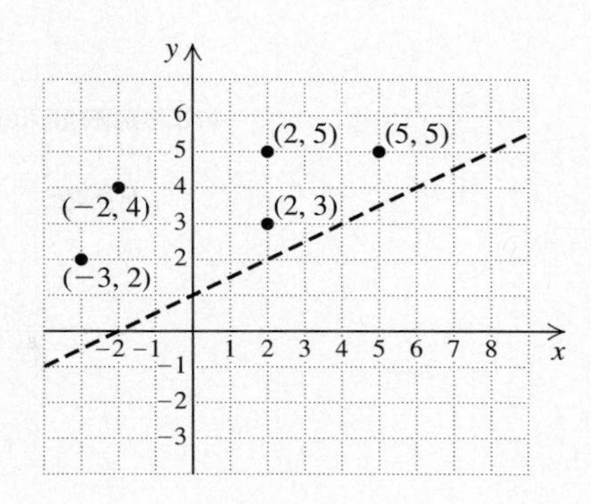

The student can check that *any* point on the same side of the dashed line as (2, 3) is a solution. Thus, if one point in a region solves an inequality, then *all* points in that region are solutions. The graph of

$$y > \tfrac{1}{2}x + 1$$

is shown below. The solution set consists of all points in the shaded region.

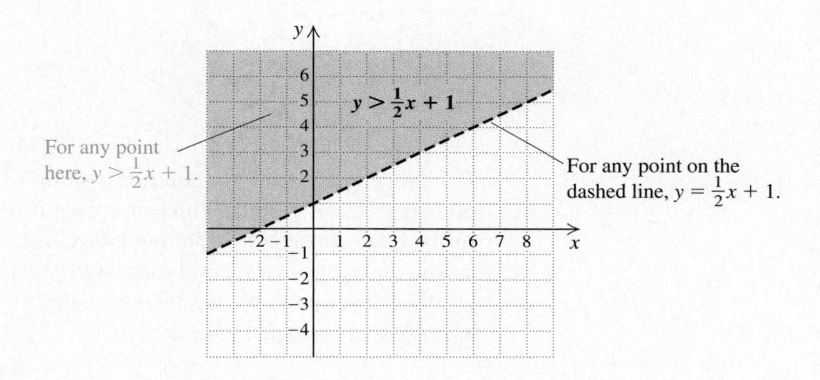

❑

EXAMPLE 2

Graph: $2x + 3y \leqslant 6$.

Solution First we graph the line $2x + 3y = 6$. This can be done by either using the intercepts, (0, 2) and (3, 0), or finding slope–intercept form, $y = -\frac{2}{3}x + 2$. Since the inequality contains the symbol $\leqslant$, we draw the boundary line solid to indicate that any pair on the line is a solution. The solution of the inequality will include the line and the region either above it or below it. By using a "test point" that is clearly above or below the line, we can determine which region to shade. The origin, (0, 0), is often a convenient test point:

$$\begin{array}{r|l} \multicolumn{2}{c}{2x + 3y \leqslant 6} \\ \hline 2\cdot 0 + 3\cdot 0 & ? \; 6 \\ 0 & 6 \end{array} \quad \text{TRUE}$$

The point (0, 0) is a solution and it appears in the region below the boundary line. Thus this region, along with the line itself, represents the solution.

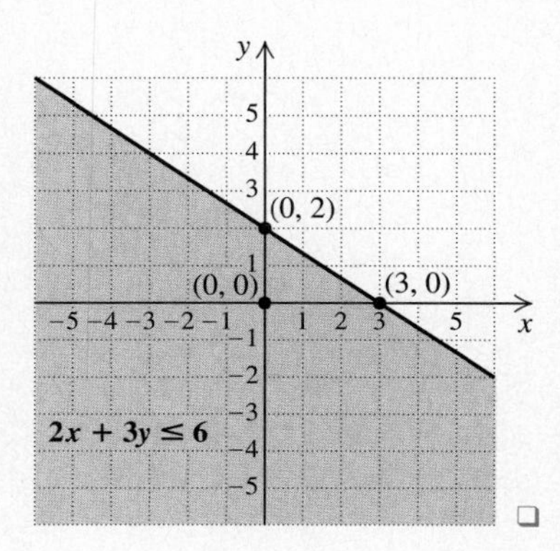

❑

To graph a linear inequality:

1. Draw the boundary line by replacing the inequality symbol with an equals sign and graphing the resulting equation. If the inequality symbol is $<$ or $>$, draw the line dashed. If the symbol is $\leq$ or $\geq$, draw the line solid.
2. Shade the region on one side of the boundary line. To determine which side, select a point not on the line as a test point. If that point solves the inequality, shade the region containing the point. If not, shade the other region.

EXAMPLE 3

Graph $y \leq -2$ on a plane.

Solution We graph $y = -2$ as a solid line to indicate that all points on the line are solutions. Next we use (0, 0) as a test point. It may help to write the inequality $y \leq -2$ as $0 \cdot x + y \leq -2$:

$$\begin{array}{r|l} 0 \cdot x + y & \leq -2 \\ \hline 0 \cdot 0 + 0 & ?\ -2 \\ 0 & -2 \quad \text{FALSE} \end{array}$$

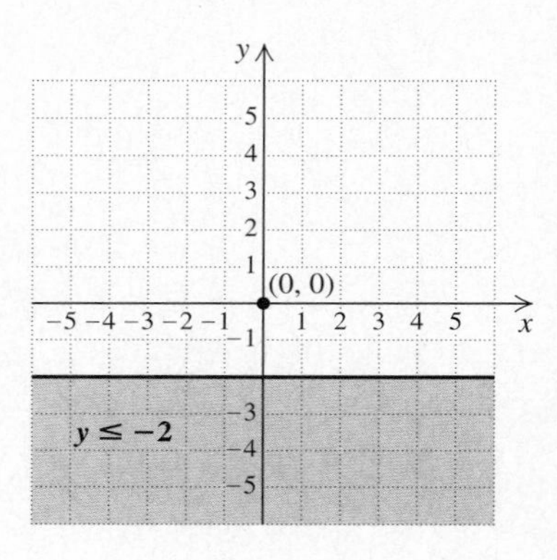

Since (0, 0) is *not* a solution, we do not shade the region in which it appears. Instead, we shade below the boundary line as shown. Note that the solution consists of all ordered pairs whose y-coordinates are less than or equal to -2. ❑

EXAMPLE 4

Graph $x < 3$ on a plane.

Solution We graph the equation $x = 3$ using a dashed line. To determine which region to shade, we consider a test point, (0, 0). It helps to write the inequality $x < 3$ as $x + 0 \cdot y < 3$:

$$\begin{array}{r|l} x + 0 \cdot y & < 3 \\ \hline 0 + 0 \cdot 0 & ?\ 3 \\ 0 & 3 \end{array}$$

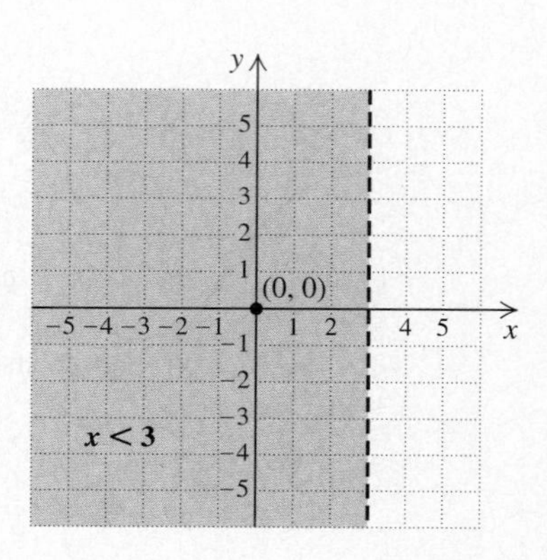

Since (0, 0) is a solution, we shade left of the boundary line, as shown. Note that the solution consists of all ordered pairs whose first coordinate is less than 3. ❑

EXERCISE SET 7.4

1. Determine whether $(-3, -5)$ is a solution of $-x - 3y < 18$.
2. Determine whether $(5, -3)$ is a solution of $-2x + 4y \leq -2$.
3. Determine whether $\left(\frac{1}{2}, -\frac{1}{4}\right)$ is a solution of $7y - 9x > -3$.
4. Determine whether $(-6, 5)$ is a solution of $x + 0 \cdot y < 3$.

Graph on a plane.

5. $y > -3x$
6. $y < -5x$
7. $y \leq x - 3$
8. $y \leq x - 5$
9. $y < x + 1$
10. $y < x + 4$
11. $y \geq x - 2$
12. $y \geq x - 1$
13. $y \leq 2x - 1$
14. $y \leq 3x + 2$
15. $x + y \leq 3$
16. $x + y \leq 4$
17. $x - y > 7$
18. $x - y > -2$
19. $x - 3y < 6$
20. $x - y < -10$
21. $2x + 3y \leq 12$
22. $5x + 4y \geq 20$
23. $y \geq 1 - 2x$
24. $y - 2x \leq -1$
25. $y + 4x > 0$
26. $y - x < 0$
27. $x > 2y$
28. $x > 3y$
29. $x < 4$
30. $x \leq 0$
31. $x \geq 3$
32. $x > -4$
33. $y \leq 3$
34. $y > -1$
35. $y \geq -5$
36. $y < 0$

Skill Maintenance

Multiply.

37. $(2x^2 - x - 1)(x + 3)$
38. $(3x - 5)(3x + 5)$
39. Factor: $3a^3 + 18a^2 - 4a - 24$.
40. Simplify: $\dfrac{x^2 - 9x + 14}{x^2 + 3x - 10}$.

Synthesis

41. ◈ Examine the solution of Example 2. Why is the point $(4.5, -1)$ *not* a good choice for a test point?
42. ◈ Describe a procedure that could be used to graph any inequality of the form $Ax + By < C$.
43. *Elevators.* Many elevators have a capacity of 1 metric ton (1000 kg). Suppose c children, each weighing 35 kg, and a adults, each weighing 75 kg, are on an elevator. Find and graph an inequality that asserts that the elevator is overloaded.
44. *Hockey wins and losses.* A hockey team figures that it needs at least 60 points for the season in order to make the playoffs. A win w is worth 2 points and a tie t is worth 1 point. Find and graph an inequality that describes the situation.

Find an inequality whose graph is as shown.

45.

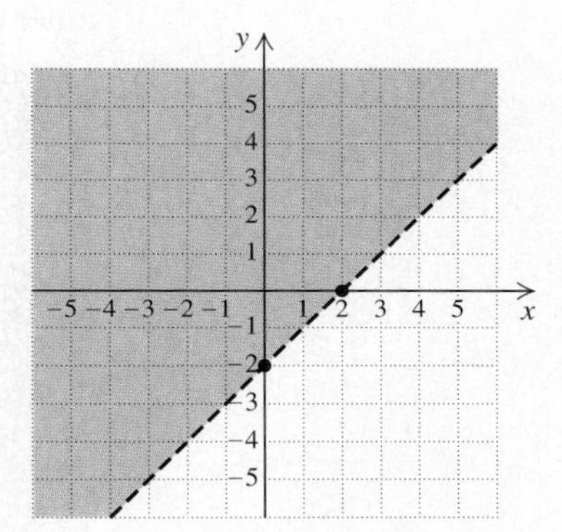

46.

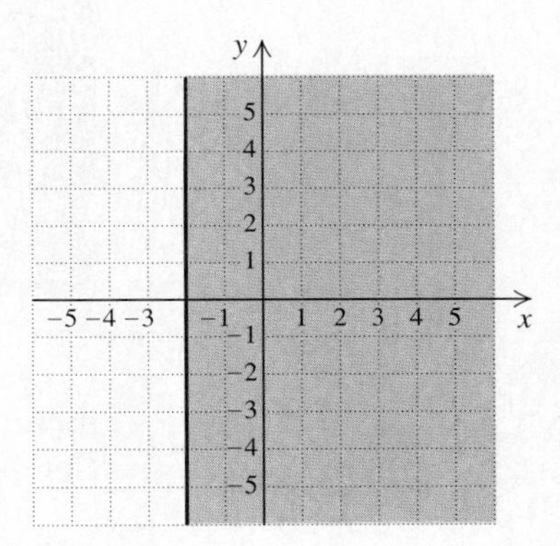

Graph on a plane. (*Hint:* Use several test points.)

47. $xy \leq 0$
48. $xy \geq 0$

7.5 Direct and Inverse Variation

Equations of Direct Variation • Problem Solving with Direct Variation • Equations of Inverse Variation • Problem Solving with Inverse Variation

We conclude our study of graphing by looking at some problems that lead to equations of the form $y = kx$ or $y = k/x$, for some constant k. Such equations are called *equations of variation*. Their graphs are sometimes nonlinear.

Equations of Direct Variation

A bicycle is traveling at a speed of 15 km/h. In 1 hr, it goes 15 km. In 2 hr, it goes 30 km. In 3 hr, it goes 45 km, and so on. We will use the number of hours as the first coordinate and the number of kilometers traveled as the second coordinate: (1, 15), (2, 30), (3, 45), (4, 60), and so on. Note that the second coordinate is always 15 times the first.

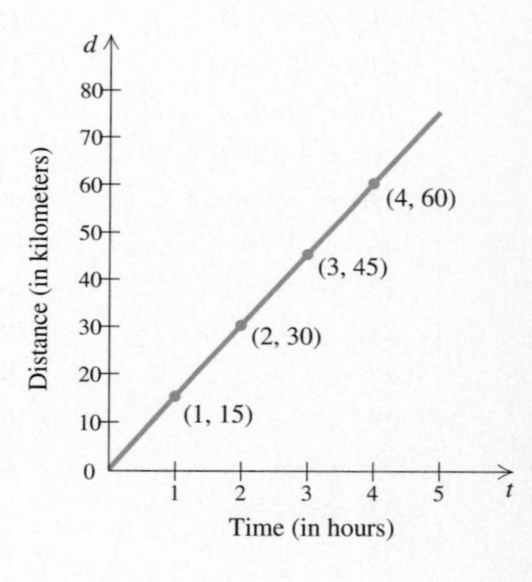

In this example, distance is a constant multiple of time, so we say that there is **direct variation** and that distance **varies directly** as time. The **equation of variation** is $d = 15t$.

Direct Variation

When a situation translates to an equation described by $y = kx$, where k is a constant, $y = kx$ is called an *equation of direct variation*, and k is called the *variation constant*. We say that y *varies directly* as x.

The terminologies

"y varies as x,"

"y is directly proportional to x," and

"y is proportional to x"

also imply direct variation and are used in many situations. The constant k is often referred to as a **constant of proportionality.** It can be found if one pair of values of x and y is known. Once k is known, other pairs can be determined.

EXAMPLE 1

If y varies directly as x and $y = 7$ when $x = 25$, find the equation of variation.

Solution We substitute to find k:

$$y = kx$$
$$7 = k \cdot 25$$
$$\frac{7}{25} = k, \quad \text{or } k = 0.28.$$

Thus the equation of variation is $y = 0.28x$. ❑

EXAMPLE 2

Find an equation in which s varies directly as t and $s = 10$ when $t = 15$. Then find the value of s when $t = 32$.

Solution We have

$s = kt$ We know that s varies directly as t.

$10 = k \cdot 15$ Substituting 10 for s and 15 for t

$\frac{10}{15} = k, \quad$ or $k = \frac{2}{3}$.

Thus the equation of variation is $s = \frac{2}{3}t$.

When $t = 32$,

$s = \frac{2}{3}t$

$s = \frac{2}{3} \cdot 32$ Substituting 32 for t in the equation of variation

$s = \frac{64}{3}, \quad$ or $21\frac{1}{3}$.

The value of s is $21\frac{1}{3}$ when $t = 32$. ❑

Problem Solving with Direct Variation

In applications, it is often necessary to find an equation of variation and then use it to find other values, much as we did in Example 2.

EXAMPLE 3

The karat rating R of a gold object varies directly as the actual percentage P of gold in the object. A 14-karat gold ring is 58.25% gold. What is the percentage of gold in a 24-karat gold ring?

Solution

1., 2. FAMILIARIZE and TRANSLATE. The problem states that we have direct variation between the variables R and P. Thus an equation $R = kP$ applies.

3. CARRY OUT. We find an equation of variation:

$$R = kP$$

$$14 = k(0.5825) \quad \text{Substituting 14 for } R \text{ and 58.25\%, or 0.5825, for } P$$

$$\frac{14}{0.5825} = k$$

$$24.03 \approx k. \quad \text{Dividing and rounding to the nearest hundredth}$$

The equation of variation is $R = 24.03P$.

When $R = 24$,

$$R = 24.03P$$

$$24 = 24.03P \quad \text{Substituting 24 for } R$$

$$\frac{24}{24.03} = P \quad \text{Solving for } P$$

$$0.999 \approx P$$

$$99.9\% \approx P.$$

4. CHECK. The check might be done by repeating the computations. You might also do some reasoning about the answer. The karat rating increased from 14 to 24. Similarly, the percentage increased from 58.25% to 99.9%. The ratios 14/0.5825 and 24/0.999 are the same value: about 24.03.

5. STATE. A 24-karat gold ring is 99.9% gold. ❑

Equations of Inverse Variation

A car is traveling a distance of 20 mi. At a speed of 5 mph, the trip will take 4 hr. At 20 mph, it will take 1 hr. At 40 mph, it will take $\frac{1}{2}$ hr, and so on. This determines a set of pairs of numbers:

$$(5, 4), \quad (20, 1), \quad \left(40, \tfrac{1}{2}\right), \quad \text{and so on.}$$

Note that the product of speed and time for each of these pairs is 20. The pairs can be plotted and connected with a curve. Although you are not expected to draw such graphs yet, doing so can help you visualize the situation.

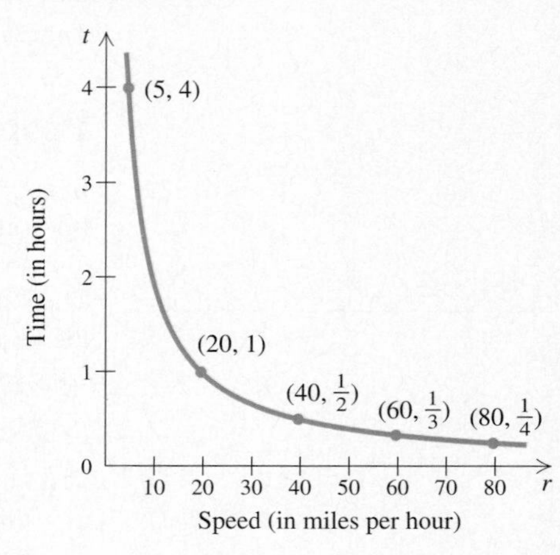

In this example, the product of speed and time is constant so we say there is **inverse variation** and that time **varies inversely** as speed. The equation of variation is

$$rt = 20(\text{a constant}), \quad \text{or } t = \frac{20}{r}.$$

Inverse Variation

When a situation translates to an equation described by $y = k/x$, where k is a constant, then $y = k/x$ is called an *equation of inverse variation*. We say that y *varies inversely* as x.

The terminology

"y is inversely proportional to x"

also implies inverse variation and is used in some situations. The constant k is again referred to as a **variation constant.**

EXAMPLE 4

If y varies inversely as x and $y = 145$ when $x = 0.8$, find the equation of variation.

Solution We substitute to find k:

$$y = \frac{k}{x}$$

$$145 = \frac{k}{0.8}$$

$$(0.8)145 = k$$

$$116 = k.$$

The equation of variation is $y = 116/x$. ❑

Problem Solving with Inverse Variation

Often in an applied situation, we must decide what kind of variation, if any, might apply to the problem.

EXAMPLE 5

It takes 4 hr for 20 people to raise a barn. How long would it take 25 people to complete the job?

Solution

1. FAMILIARIZE. Think about the situation in the problem. What kind of variation would be used? It seems reasonable that the greater the number of people working on the job, the less time it will take to finish. (One might argue that too many people in a crowded area would be counterproductive, but we will disregard that possibility.) Thus we assume that inverse variation applies. We let T = the time to complete the job, in hours, and N = the number of people.

2. TRANSLATE. Since inverse variation applies, we write the equation

$$T = \frac{k}{N}.$$

3. CARRY OUT. We find an equation of variation:

$$T = \frac{k}{N}$$

$$4 = \frac{k}{20} \quad \text{Substituting 4 for } T \text{ and 20 for } N$$

$$20 \cdot 4 = k$$

$$80 = k.$$

The equation of variation is $T = 80/N$.
When $N = 25$,

$$T = \frac{80}{25} \quad \text{Substituting 25 for } N$$

$$T = 3.2.$$

4. CHECK. A check might be done by repeating the computations or by noting that the products (3.2)(25) and (4)(20) are both 80. We might also note that as the number of people increased from 20 to 25, the time needed to complete the job decreased, as expected.

5. STATE. It should take 3.2 hr for 25 people to raise a barn. ❑

EXERCISE SET 7.5

Find an equation of variation in which y varies directly as x and the following are true.

1. $y = 28$, when $x = 7$

2. $y = 30$, when $x = 8$

3. $y = 0.7$, when $x = 0.4$

4. $y = 0.8$, when $x = 0.5$

5. $y = 400$, when $x = 125$

6. $y = 630$, when $x = 175$

7. $y = 200$, when $x = 300$

8. $y = 500$, when $x = 60$

Find an equation of variation in which y varies inversely as x and the following are true.

9. $y = 25$, when $x = 3$

10. $y = 45$, when $x = 2$

11. $y = 8$, when $x = 10$

12. $y = 7$, when $x = 10$

13. $y = 0.125$, when $x = 8$

14. $y = 6.25$, when $x = 0.16$

15. $y = 42$, when $x = 25$

16. $y = 42$, when $x = 50$

17. $y = 0.2$, when $x = 0.3$

18. $y = 0.4$, when $x = 0.6$

Problem Solving

19. A production line produces 15 compact-disc players in 8 hr. How many players can it produce in 37 hr?

a) What kind of variation might apply to this situation?

b) Solve the problem.

20. A person works for 15 hr and makes $93.75. How much would the person make by working 35 hr?

a) What kind of variation might apply to this situation?

b) Solve the problem.

21. It takes 16 hr for 2 people to resurface a gym floor. How long would it take 6 people to do the job?

a) What kind of variation might apply to this situation?

b) Solve the problem.

22. It takes 4 hr for 9 cooks to prepare a school lunch. How long would it take 8 cooks to prepare the lunch?

a) What kind of variation might apply to this situation?

b) Solve the problem.

23. A person's paycheck P varies directly as the number of hours worked H. For 15 hr of work, the pay is $78.75. Find the pay for 35 hr of work.

24. The number of bolts B that a machine can make varies directly as the time T that it operates. It can make 6578 bolts in 2 hr. How many can it make in 5 hr?

25. The number of servings S of meat that can be obtained from a turkey varies directly as its weight W. From a turkey weighing 14 kg, one can get 40 servings of meat. How many servings can be obtained from an 8-kg turkey?

26. The number of servings S of meat that can be obtained from round steak varies directly as the weight W. From 9 kg of round steak, one can get 70 servings of meat. How many servings can one get from 12 kg of round steak?

27. The volume V of a gas varies inversely as the pressure P on it. The volume of a gas is 200 cm^3 (cubic centimeters) under a pressure of 32 kg/cm^2. What will be its volume under a pressure of 20 kg/cm^2?

28. The current I in an electrical conductor varies inversely as the resistance R of the conductor. The current is 2 amperes when the resistance is 960 ohms. What is the current when the resistance is 540 ohms?

29. The weight M of an object on the moon varies directly as its weight E on earth. A person who weighs 171.6 lb on earth weighs 28.6 lb on the moon. How much would a 110-lb person weigh on the moon?

30. The weight M of an object on Mars varies directly as its weight E on earth. A person who weighs 209 lb on earth weighs 79.42 lb on Mars. How much would a 176-lb person weigh on Mars?

31. The pitch P of a musical tone varies inversely as its wavelength W. One tone has a pitch of 660 vibrations per second and a wavelength of 1.6 ft. Find the wavelength of another tone that has a pitch of 440 vibrations per second.

32. The time t required to empty a tank varies inversely as the rate r of pumping. A pump can empty a tank in 90 min at the rate of 1200 L/min. How long will it take the pump to empty the tank at the rate of 2000 L/min?

33. The cost c of operating a television varies directly as the number of hours n that it is in operation. It costs $14.00 to operate a standard-size color television continuously for 30 days. At this rate, how much would it cost to operate the television for 1 day? for 1 hr?

34. The amount C that a family spends on car expenses varies directly as its income I. A family making $21,760 a year will spend $3264 a year on car expenses. How much will a family making $30,000 a year spend on car expenses?

35. The number of minutes m that a student should allow for each question on a quiz is inversely proportional to the number of questions n on the quiz. If a 16-question quiz means that students have 2.5 minutes per question, how many questions would appear on a quiz in which students have 4 minutes per question?

36. The cost per person c of a chartered fishing boat is inversely proportional to the number of people n who are chartering the boat. If it costs $17.50 per person when 9 people charter a boat, how many people would be going fishing if the cost were $31.50 per person?

Synthesis

State whether the situation represents direct variation, inverse variation, or neither. Give reasons for your answers.

37. ◈ The cost of mailing a letter in the United States and the distance that it travels

38. ◈ A runner's speed in a race and the time it takes to run the race

39. ◈ The weight of a turkey and the cooking time

40. ◈ The number of plays it takes to go 80 yd for a touchdown and the average gain per play

Write an equation of direct variation to describe the situation. If possible, give a value for k and graph the equation.

41. The perimeter P of an equilateral octagon varies directly as the length S of a side.

42. The circumference C of a circle varies directly as the radius r.

43. The number of bags of peanuts B sold at a baseball game varies directly as the number of people N in attendance.

44. The cost C of building a new house varies directly as the area A of the floor space of the house.

45. Show that if p varies directly as q, then q varies directly as p.

46. The area of a circle varies directly as the square of the length of the radius. What is the variation constant?

Write an equation of variation to describe the situation.

47. In a stream, the amount of salt S carried varies directly as the sixth power of the speed of the stream v.

48. The square of the pitch P of a vibrating string varies directly as the tension t on the string.

49. The intensity of illumination I from a light source varies inversely as the square of the distance d from the source.

50. The density D of a given mass varies inversely as its volume V.

51. The volume V of a sphere varies directly as the cube of the radius r.

52. The power P in a windmill varies directly as the cube of the wind speed v.

53. ◈ If a varies directly as b and b varies directly as c, does it follow that a varies directly as c? Why or why not?

SUMMARY AND REVIEW 7

KEY TERMS

Rate, p. 319
Slope, p. 321
Rise, p. 321
Run, p. 321
Grade, p. 323
Slope–intercept equation, p. 328
Parallel lines, p. 329
Point–slope equation, p. 333
Linear inequality, pp. 340, 343
Direct variation, p. 340
Variation constant, pp. 340, 343
Inverse variation, p. 343

IMPORTANT PROPERTIES AND FORMULAS

$$\text{Slope} = m = \frac{y_2 - y_1}{x_2 - x_1} = \frac{\text{change in } y}{\text{change in } x} = \frac{\text{rise}}{\text{run}}$$

Horizontal line:	Slope 0
Vertical line:	Slope undefined
Slope–intercept equation:	$y = mx + b$
Parallel lines:	Slopes equal, y-intercepts different
Point–slope equation:	$y - y_1 = m(x - x_1)$
Direct variation:	$y = kx$
Inverse variation:	$y = k/x$

To graph a linear inequality:

1. Draw the boundary line by replacing the inequality symbol with an equals sign and graphing the resulting equation. If the inequality symbol is $<$ or $>$, draw the line dashed. If the symbol is $\leqslant$ or $\geqslant$, draw the line solid.
2. Shade the region on one side of the boundary line. To determine which side, select a point not on the line as a test point. If that point solves the inequality, shade the region containing the point. If not, shade the other region.

REVIEW EXERCISES

This chapter's review and test include Skill Maintenance exercises from Sections 4.4, 4.5, 5.1, and 6.1.

Find the slope of the line.

1.

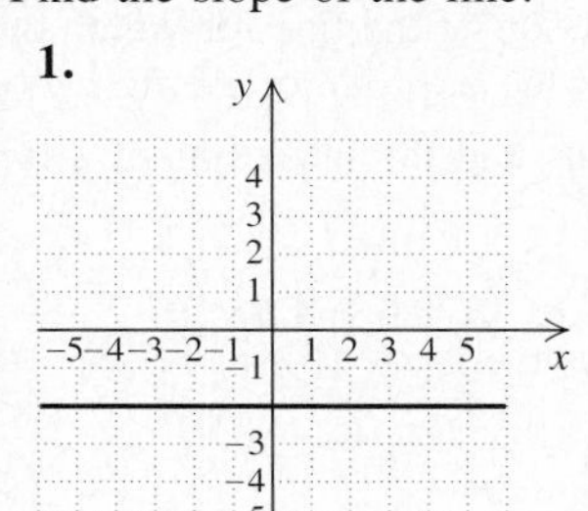

2.

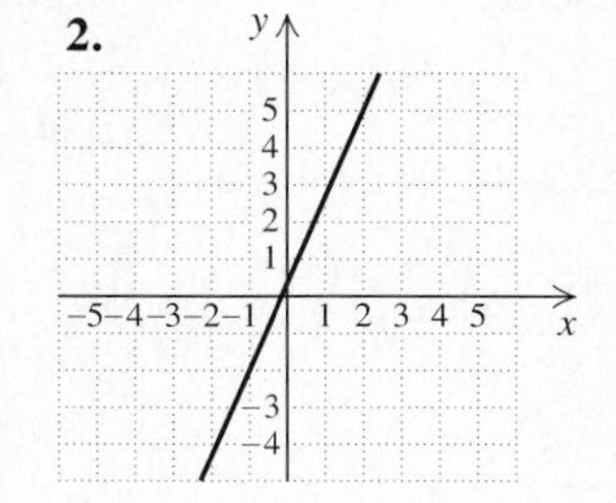

3.

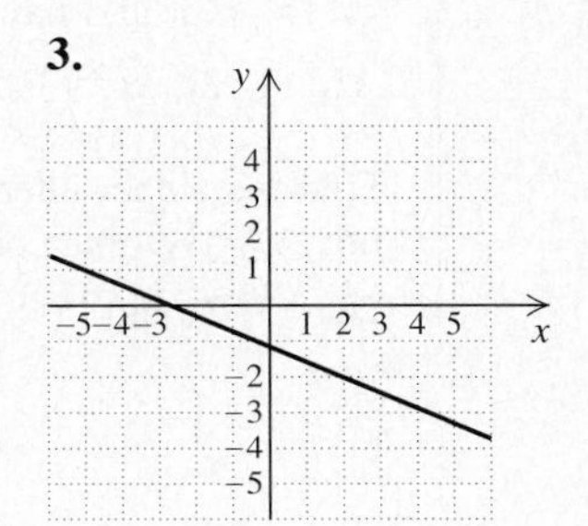

Find the slope of the line containing the given pair of points.

4. (6, 8) and (−2, −4)

5. (5, 1) and (−1, 1)

6. (−3, 0) and (−3, 5)

7. (−8.3, 4.6) and (−9.9, 1.4)

8. A road drops 369.6 ft vertically over a horizontal distance of 5280 ft. What is the grade of the road?

Find the slope of the line.

9. $y = -6$

10. $3x - 5y = 4$

11. $2x + y = 6$

12. $x = 90$

Find the slope and the y-intercept of the line.

13. $y = -9x + 46$

14. $x + y = 9$

15. $2x - 6y = 4$

Find the slope–intercept equation for the line with the indicated slope and y-intercept.

16. Slope −2; y-intercept (0, −4)

17. Slope 1.5; y-intercept (0, 1)

Graph.

18. $y = -\frac{3}{4}x - 2$

19. $y + \frac{1}{2}x = 2$

20. $y - 2 = 3(x - 6)$

Determine whether the pair of equations represents parallel lines.

21. $4x + y = 6,$
$4x = 8 - y$

22. $3x - y = 6,$
$3x + y = 8$

Find a point–slope equation for the line containing the given point and having the given slope.

23. (1, 2), $m = 3$

24. (−2, −5), $m = \frac{2}{3}$

Find the slope–intercept equation for the line containing the given pair of points.

25. (5, 7) and (−1, 1)

26. (2, 0) and (−4, −3)

Graph on a plane.

27. $x \leqslant y$

28. $x + 2y \geqslant 4$

29. $x > -2$

30. If y varies inversely as x and $y = 81$ when $x = 3$, find the equation of variation.

31. The number of sandwiches S that can be made at a buffet varies directly as the number of pounds of cold cuts C in the buffet. From 6 lb of cold cuts, 25 sandwiches can be made. How many pounds of cold cuts are needed for 40 sandwiches?

Skill Maintenance

Multiply.

32. $(x - 2)(3x^2 - 2x + 1)$

33. $\left(\frac{1}{2}y + \frac{1}{4}\right)^2$

34. Factor: $x^3 - x^2 + 2x - 2$.

35. Simplify: $\dfrac{a^2 - 4}{2a^2 - 3a - 2}$.

Synthesis

36. ◈ Describe a situation in which point–slope form would be more useful than slope–intercept form.

37. ◈ Why is the boundary line part of the graph when a linear inequality contains the symbol $\leqslant$ or $\geqslant$, but not when it contains the symbol $<$ or $>$?

38. Find an equation of the line having the same y-intercept as the line $2x - y = 3$ and the same slope as the line $2x + y = 3$.

39. Find an equation of the line that contains the point (−1, 2) and is perpendicular to the line $2x + 3y = 1$. (See Exercises 67–72 in Section 7.2.)

40. Find an equation of the line for which the second coordinate is the opposite of the first coordinate.

41. Find the slope and the intercepts of a line whose equation is

$$\frac{x}{a} + \frac{y}{b} = 1, \quad a \neq 0 \text{ and } b \neq 0.$$

CHAPTER TEST 7

Find the slope of the line containing the pair of points.

1. (4, 7) and (4, −1)
2. (9, 2) and (−3, −5)

Find the slope of the line.

3. $2x + y = \frac{1}{3}$
4. $y = -7$
5. $x = 6$

Find the slope and the y-intercept of the line.

6. $y = 2x - \frac{1}{4}$
7. $-4x + 3y = -6$

Find the slope–intercept equation for the line with the indicated slope and y-intercept.

8. Slope $\frac{1}{2}$; y-intercept (0, −7)
9. Slope −4; y-intercept (0, 3)

Find a point–slope equation for the line containing the given point and having the given slope.

10. (3, 5), $m = 1$
11. (−2, 0), $m = -3$

Find the slope–intercept equation for the line containing the given pair of points.

12. (1, 1) and (2, −2)
13. (4, −1) and (−4, −3)

Graph.

14. $2x + y = 5$
15. $y - 5 = \frac{2}{3}(x - 6)$
16. Determine whether the following pair of equations represents parallel lines:

$$2x + y = 8,$$
$$2x + y = 4.$$

Graph on a plane.

17. $y > x - 1$
18. $2x - y \leq 4$
19. If y varies directly as x and $y = 9$ when $x = 2$, find the equation of variation.
20. It takes 45 min for 2 people to shovel a driveway. How long would it take 5 people to shovel the same driveway?

Skill Maintenance

Multiply.

21. $(-3y^4)(y^3 - 3y + 7)$
22. $(x + 0.1)(x - 0.1)$
23. Factor: $6x^3 + 3x^2 - 3x$.
24. Simplify: $\dfrac{-2x + 8}{2x}$.

Synthesis

25. Find the slope–intercept equation of the line that contains the point (−4, 1) and has the same slope as the line $2x - 3y = -6$.
26. Graph on a plane: $|x| \leq 4$.

CHAPTER 8

Systems of Equations and Problem Solving

AN APPLICATION

The Java Joint wants to mix Kenyan beans that sell for $7.25 per pound with Venezuelan beans that sell for $8.50 per pound to form a 50-lb batch of Morning Blend that sells for $8.00 per pound. How many pounds of Kenyan beans and how many pounds of Venezuelan beans should go into the blend?

This problem appears as Example 3 in Section 8.4.

Curtis Hooper
ROASTMASTER

"Everyone here at Green Mountain Coffee Roasters, from the factory workers to the president, uses spreadsheets on a daily basis. Every division of the company relies on math to know how much coffee to roast and how much to charge the customer."

In fields such as sociology, psychology, business, engineering, and science, problems frequently occur that are most easily solved using a *system of equations.* In this chapter, we study three methods for solving systems of two equations in which two variables appear. We then use these methods in a variety of problem-solving situations.

In addition to material from this chapter, the review and test for Chapter 8 include material from Sections 4.2, 5.3, 5.4, and 6.5.

8.1 Systems of Equations and Graphing

Solutions of Systems • Graphing Systems of Equations

In Section 2.5, the following problem appeared as Example 5:

> A rectangular community garden is to be enclosed with 92 m of fencing. In order to allow for compost storage, the garden must be 4 m longer than it is wide. Determine the dimensions of the garden.

When solving this problem, we used the formula for the perimeter of a rectangle, $2l + 2w = P$. Since 92 m of fencing would outline the garden, we had $2l + 2w = 92$, an equation in two variables. Because we could solve only equations in one variable at the time, we made use of the fact that the length was 4 m greater than the width and wrote $2(w + 4) + 2w = 92$. Had we wished to write two equations in two variables, we could have formed the following *system* of two equations:

$$2l + 2w = 92,$$
$$l = w + 4. \quad \text{The length is 4 more than the width.}$$

Most people find that it is often easier to translate a problem into two equations in two variables than one equation in one variable. Thus it is important to know how to solve such a system when one appears.

Solutions of Systems

Solution of a System

A *solution* of a system of two equations is an ordered pair that makes both equations true.

EXAMPLE 1

Consider the system

$$2l + 2w = 92,$$
$$l = w + 4.$$

Determine if the given pair is a solution of the system: **(a)** (25, 21); **(b)** (16, 12).

Solution

a) We check by substituting (alphabetically) 25 for l and 21 for w:

$$\begin{array}{c|l} 2l + 2w = 92 & \\ \hline 2\cdot 25 + 2\cdot 21 & ?\ 92 \\ 50 + 42 & \\ 92 & 92 \quad \text{TRUE} \end{array} \qquad \begin{array}{c|l} l = w + 4 & \\ \hline 25 & ?\ 21 + 4 \\ 25 & 25 \quad \text{TRUE} \end{array}$$

Since (25, 21) checks in *both* equations, it is a solution of the system.

b) We substitute 16 for l and 12 for w:

$$\begin{array}{c|l} 2l + 2w = 92 & \\ \hline 2\cdot 16 + 2\cdot 12 & ?\ 92 \\ 32 + 24 & \\ 56 & 92 \quad \text{FALSE} \end{array} \qquad \begin{array}{c|l} l = w + 4 & \\ \hline 16 & ?\ 12 + 4 \\ 16 & 16 \quad \text{TRUE} \end{array}$$

Since (16, 12) is not a solution of $2l + 2w = 92$, it is *not* a solution of the system. ❑

Graphing Systems of Equations

Recall that a graph of an equation is a set of points representing its solution set. Each point on the graph corresponds to an ordered pair that is a solution of the equation. By graphing two equations on the same set of axes, we can identify a solution of both equations by looking for a point of intersection.

EXAMPLE 2

Solve this system of equations by graphing:

$$x + y = 6,$$
$$y = x - 2.$$

Solution We graph the equations using any method studied earlier. The equation $x + y = 6$ can be graphed easily using the intercepts, (0, 6) and (6, 0). The equation $y = x - 2$ is in slope–intercept form, so we graph the line by plotting its y-intercept, (0, −2), and "counting off" a slope of 1.

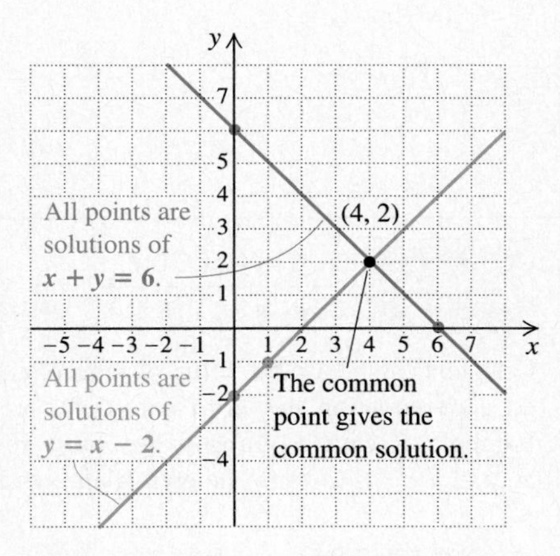

The "apparent" solution of the system, (4, 2), should be checked in both equations:

Check:

$$\begin{array}{c|l} x + y = 6 & \\ \hline 4 + 2 & ?\ 6 \\ 6 & 6 \quad \text{TRUE} \end{array} \qquad \begin{array}{c|l} y = x - 2 & \\ \hline 2 & ?\ 4 - 2 \\ 2 & 2 \quad \text{TRUE} \end{array}$$

Since it checks in both equations, (4, 2) is a solution of the system. ❑

A system of equations that has at least one solution, like the systems in Examples 1 and 2, is said to be **consistent.** A system for which there is no solution is said to be **inconsistent.**

EXAMPLE 3

Solve this system of equations by graphing:

$$y = 3x + 4,$$
$$y = 3x - 3.$$

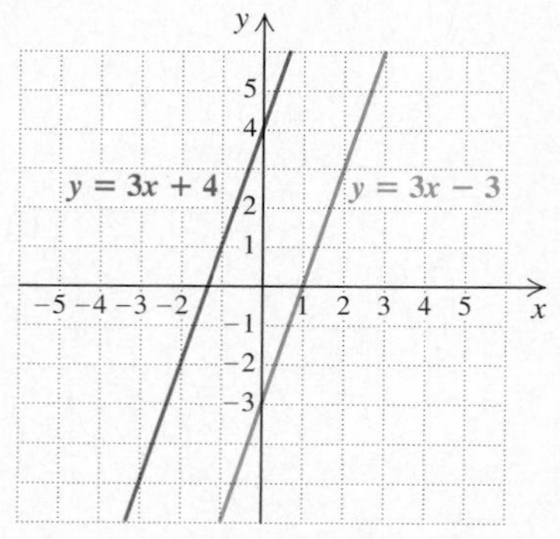

Solution Both equations are in slope–intercept form so it is easy to see that both lines have the same slope, 3. The y-intercepts differ so the lines are parallel, as shown in the figure.

Because the lines are parallel, there is no point of intersection. Thus the system is inconsistent and has no solution. ☐

Sometimes both equations in a system have the same graph.

EXAMPLE 4

Solve this system by graphing:

$$2x + 3y = 6,$$
$$-8x - 12y = -24.$$

Solution Graphing the equations, we see that they both represent the same line. This can also be seen by solving each equation for y, obtaining the equivalent slope–intercept form, $y = -\frac{2}{3}x + 2$. Since the equations are equivalent, any solution of one equation is a solution of the other equation as well. We show four such solutions on the graph at the top of the next page.

TECHNOLOGY CONNECTION

Graphers can be used to solve systems of equations, provided each equation has been solved for y. Thus, to solve Example 2 with a grapher, we must rewrite $x + y = 6$ as $y = -x + 6$ to form the equivalent system

$$y = -x + 6,$$
$$y = x - 2.$$

We then enter the two equations as y_1 and y_2. After the equations have been graphed, we use the Trace and Zoom features to magnify the portion of the graph in which the intersection occurs. Should your grapher lack a Zoom feature, adjust the Range accordingly.

TC1. Use a grapher to solve Example 2. You may need to magnify the area containing the intersection many times before the exact solution, $x = 4$, $y = 2$, appears.

TC2. Graph the system

$$y = 0.23x + 1.49,$$
$$y = 0.23x + 3.49.$$

Show that the vertical separation between the two lines is uniform by using the Trace feature to show that for any given x-value, the y-values on the two lines differ by exactly 2. Does this system have a solution? Why or why not?

We check one solution, (0, 2), the y-intercept of each line.

$2x + 3y = 6$		$-8x - 12y = -24$	
$2(0) + 3(2)$	? 6	$-8(0) - 12(2)$	? -24
$0 + 6$		$0 - 24$	
6	6 TRUE	-24	-24 TRUE

On your own, check that (3, 0) is also a solution of the system. If two points are solutions, then all points on the line containing them are solutions. The lines coincide, so there are infinitely many solutions. ❑

The equations in Example 4 are equivalent. When equivalent equations appear in a system, we call the equations **dependent.** Thus the equations in Example 4 are dependent, but those in Examples 2 and 3 are **independent.**

When a system of two linear equations is graphed, one of the following must occur:

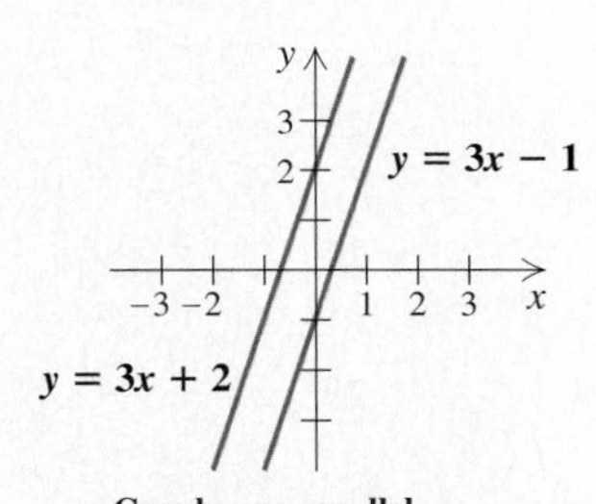

Graphs are parallel.
The system is *inconsistent* because there is no solution. The equations are *independent* since the graphs differ.

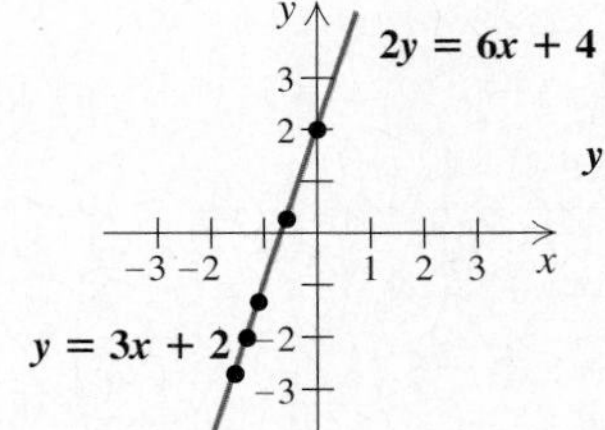

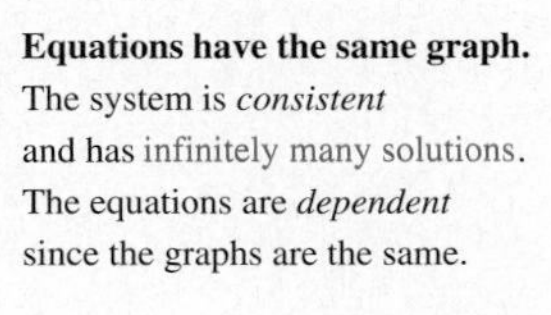

Equations have the same graph.
The system is *consistent* and has infinitely many solutions. The equations are *dependent* since the graphs are the same.

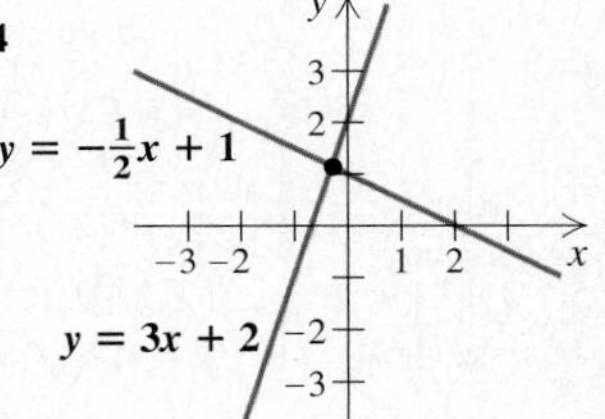

Graphs intersect.
The system is *consistent* and has one solution. Since the graphs differ, the equations are *independent*.

Although graphing lets us "see" the solution of a system easily, it is sometimes difficult to pinpoint the solution accurately. For example, the solution of the system

$$3x + 7y = 5,$$
$$6x - 7y = 1$$

is $\left(\frac{2}{3}, \frac{3}{7}\right)$, but finding that precise solution from a graph—*even with a computer or graphing calculator*—is quite difficult. Fortunately, systems like this can be solved with complete precision using methods discussed in Sections 8.2 and 8.3.

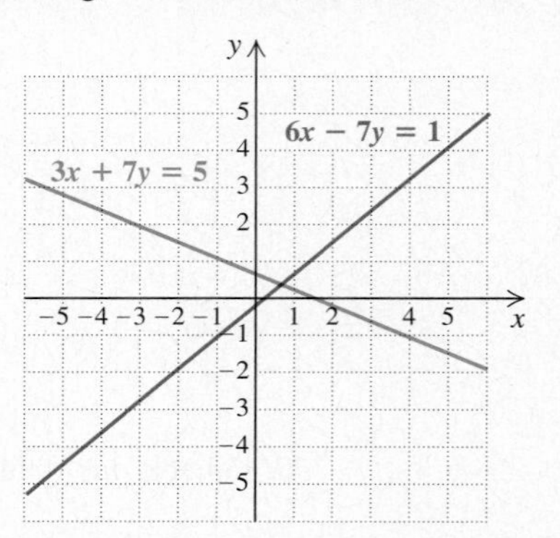

EXERCISE SET | 8.1

Determine whether the given ordered pair is a solution of the system of equations. Use alphabetical order of the variables.

1. (3, 2); $2x + 3y = 12,$ $x - 4y = -5$

2. (1, 5); $5x - 2y = -5,$ $3x - 7y = -32$

3. (3, 2); $3b - 2a = 0,$ $b + 2a = 15$

4. (2, −2); $b + 2a = 2,$ $b - a = -4$

5. (15, 20); $3x - 2y = 5,$ $6x - 5y = -10$

6. (−1, −3); $3r + s = -6,$ $2r = 1 + s$

7. (−1, 1); $x = -1,$ $x - y = -2$

8. (−3, 4); $2x = -y - 2,$ $y = -4$

9. (12, 3); $y = \frac{1}{4}x,$ $3x - y = 33$

10. (−3, 1); $y = -\frac{1}{3}x,$ $3y = -5x - 12$

11. (10, −3); $x + 2y = 4,$ $18 = (x - 7)y$

12. (−4, 5); $3y = 7 - 2x,$ $x(1 + y) = -24$

Solve the system of equations by graphing. If there is no solution or infinitely many solutions, state so.

13. $x + y = 3,$ $x - y = 1$

14. $x - y = 2,$ $x + y = 6$

15. $x + 2y = 10,$ $3x + 4y = 8$

16. $x - 2y = 6,$ $2x - 3y = 5$

17. $y = 2x - 5,$ $-x + 7 = y$

18. $y = -x - 2,$ $\frac{1}{3}x + 4 = y$

19. $x = y,$ $4x = 2y - 6$

20. $x = 3y,$ $3y - 6 = 2x$

21. $x = -y,$ $x + y = 4$

22. $-3x = 5 - y,$ $2y = 6x + 10$

23. $y = \frac{1}{2}x + 1,$ $y = 2x - 2$

24. $y = 3x - 6,$ $y = -2x - 1$

25. $y = 3,$ $x = 5$

26. $y = 3x,$ $y = -3x + 2$

27. $x + y = 9,$ $3x + 3y = 27$

28. $x + y = 4,$ $x + y = -4$

29. $x = 5,$ $x = -3$

30. $x = 5,$ $y = -3$

Skill Maintenance

Simplify.

31. $\dfrac{x + 2}{x - 4} - \dfrac{x + 1}{x + 4}$

32. $\dfrac{2x^2 - x - 15}{x^2 - 9}$

Classify the polynomial as a monomial, binomial, trinomial, or none of these.

33. $5x^2 - 3x + 7$

34. $4x^3 - 2x^2$

Synthesis

35. Explain in your own words the strengths and weaknesses of using graphing to solve a system.

36. Suppose that the equations in a linear system of two equations are dependent. Does it follow that the system is consistent? Why or why not?

37. Which of the systems in Exercises 13–30 contain dependent equations?

38. Which of the systems in Exercises 13–30 are consistent?

39. Which of the systems in Exercises 13–30 are inconsistent?

40. Which of the systems in Exercises 13–30 contain independent equations?

41. Find a system of equations with $(2, -4)$ as the solution. Answers may vary.

42. Find an equation to pair with $5x + 2y = 11$ so that the solution of the system is $(3, -2)$. Answers may vary.

43. The solution of the following system is $(2, -3)$. Find A and B.

$$Ax - 3y = 13,$$
$$x - By = 8$$

44. The pair $(-1, -5)$ is the solution of the system

$$4x - y = 1,$$
$$-x + y = -4,$$
$$2x + 3y = -17.$$

Without drawing a graph, describe what the graph of the system must look like.

45. Use a grapher to solve the system

$$y = 1.2x - 32.7,$$
$$y = -0.7x + 46.15.$$

You may need to Zoom out and use the Trace feature to find an approximate solution and then Zoom in repeatedly to find the precise answer.

8.2 Systems of Equations and Substitution

The Substitution Method • Solving for the Variable First • Problem Solving

Near the end of Section 8.1, we discussed the fact that graphing can be an imprecise method for solving systems. We now learn methods of finding exact solutions using algebra.

The Substitution Method

One nongraphical method for solving systems is known as the **substitution method.** It uses algebra and is thus considered an *algebraic* method.

EXAMPLE 1

Solve the system

$$x + y = 6, \quad (1)$$
$$y = x - 2. \quad (2)$$

Solution The second equation says that y and $x - 2$ represent the same value. Thus, in the first equation, we can substitute $x - 2$ for y:

$$x + y = 6 \qquad \text{Equation (1)}$$
$$x + x - 2 = 6. \qquad \text{Substituting } x - 2 \text{ for } y$$

The equation $x + x - 2 = 6$ has only one variable, for which we now solve:

$2x - 2 = 6$ Collecting like terms

$2x = 8$ Adding 2 on both sides

$x = 4.$ Dividing by 2

We have found the x-value of the solution. To find the y-value, we return to the original pair of equations. Substituting into either equation will give us the y-value. We choose Equation (1):

$x + y = 6$ Equation (1)

$4 + y = 6$ Substituting 4 for x

$y = 2.$ Subtracting 4

The ordered pair (4, 2) appears to be a solution. We check:

$x + y = 6$			$y = x - 2$		
$4 + 2$	?	6	2	?	$4 - 2$
6		6 TRUE	2		2 TRUE

Since (4, 2) checks, it is the solution. For this particular system, we can also check by examining Example 2 in Section 8.1. There we found the same solution by graphing. ❑

EXAMPLE 2

Solve:

$$x = 13 - 3y, \quad (1)$$
$$x + y = 5. \quad (2)$$

Solution We substitute $13 - 3y$ for x in the second equation:

$x + y = 5$ Equation (2)

$13 - 3y + y = 5.$ Substituting $13 - 3y$ for x

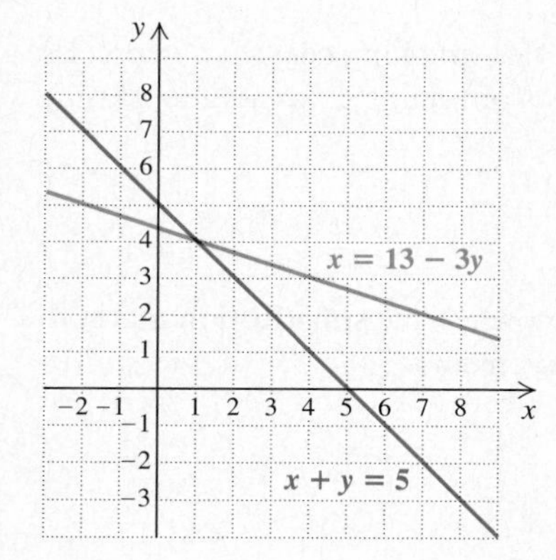

Now we solve for y:

$13 - 2y = 5$ Collecting like terms

$-2y = -8$ Subtracting 13

$y = 4.$ Dividing by -2

Next we substitute 4 for y in Equation (2) of the original system:

$x + y = 5$ Equation (2)

$x + 4 = 5$ Substituting 4 for y

$x = 1.$ Subtracting 4

The pair (1, 4) is the solution. A graph is shown in the margin as a check. ❑

Solving for the Variable First

Sometimes neither equation has a variable alone on one side. In that case, we solve one equation for one of the variables and proceed as before.

EXAMPLE 3

Solve:

$$x - 2y = 6, \quad (1)$$
$$3x + 2y = 4. \quad (2)$$

Solution We solve one equation for one variable. Since the coefficient of x is 1 in Equation (1), we can easily solve that equation for x:

$$x - 2y = 6 \qquad \text{Equation (1)}$$
$$x = 6 + 2y. \qquad \text{Adding } 2y \qquad (3)$$

We substitute $6 + 2y$ for x in Equation (2) of the original pair and solve for y:

$$3x + 2y = 4 \qquad \text{Equation (2)}$$
$$3(6 + 2y) + 2y = 4 \qquad \text{Substituting } 6 + 2y \text{ for } x$$

Remember to use parentheses when you substitute.

$$18 + 6y + 2y = 4 \qquad \text{Using the distributive law}$$
$$18 + 8y = 4 \qquad \text{Collecting like terms}$$
$$8y = -14 \qquad \text{Subtracting 18}$$
$$y = \frac{-14}{8} = -\frac{7}{4}. \qquad \text{Dividing by 8}$$

To find x, we can substitute $-\frac{7}{4}$ for y in Equation (1), (2), or (3). Because it is generally easier to use an equation in which we have solved for a specific variable, we elect to use Equation (3):

$$x = 6 + 2y = 6 + 2\left(-\tfrac{7}{4}\right) = 6 - \tfrac{7}{2}, \quad \text{or } \tfrac{5}{2}.$$

We check the ordered pair $\left(\frac{5}{2}, -\frac{7}{4}\right)$.

Check:

$x - 2y = 6$	
$\frac{5}{2} - 2\left(-\frac{7}{4}\right)$	? 6
$\frac{5}{2} + \frac{7}{2}$	
$\frac{12}{2}$	
6	6 TRUE

$3x + 2y = 4$	
$3 \cdot \frac{5}{2} + 2\left(-\frac{7}{4}\right)$	? 4
$\frac{15}{2} - \frac{7}{2}$	
$\frac{8}{2}$	
4	4 TRUE

Since $\left(\frac{5}{2}, -\frac{7}{4}\right)$ checks, it is the solution. ❑

Some systems have no solution and some have infinitely many solutions.

EXAMPLE 4

Solve the system.

a) $y = 3x + 4, \quad (1)$
$y = 3x - 3 \quad (2)$

b) $2y = 6x + 4, \quad (1)$
$y = 3x + 2 \quad (2)$

Solution

a) We solved this system graphically in Example 3 of Section 8.1. The lines are parallel and the system has no solution. Let's see what happens if we try to solve this system using substitution. We substitute $3x - 3$ for y in the first equation:

$$
\begin{aligned}
y &= 3x + 4 && \text{Equation (1)} \\
3x - 3 &= 3x + 4 && \text{Substituting } 3x - 3 \text{ for } y \\
-3 &= 4. && \text{Subtracting } 3x \text{ on both sides}
\end{aligned}
$$

When we subtract $3x$ on both sides, we obtain a *false* equation. In such a case, when solving algebraically leads to a false equation, we state that the system has no solution and thus is *inconsistent*.

b) The graph of this system is shown at the bottom of p. 355. The equations have the same graph so there are infinitely many solutions. Were we to use substitution, we would replace y in the first equation with $3x + 2$:

$$
\begin{aligned}
2y &= 6x + 4 && \text{Equation (1)} \\
2(3x + 2) &= 6x + 4 && \text{Substituting } 3x + 2 \text{ for } y \\
6x + 4 &= 6x + 4.
\end{aligned}
$$

Since this last equation is true for *any* choice of x, the system has infinitely many solutions. In this text, whenever solving a system algebraically leads to an equation that is always true, we state that the equations in the system are *dependent*. ❑

Problem Solving

Now let's use the substitution method in problem solving.

EXAMPLE 5

Two angles are supplementary. One angle is 30° more than two times the other. Find the angles.

Solution

1. FAMILIARIZE. Recall that two angles are supplementary if their sum is 180°. We could try to guess a solution, but instead we make a sketch and translate. Let x = one angle and y = the other angle.

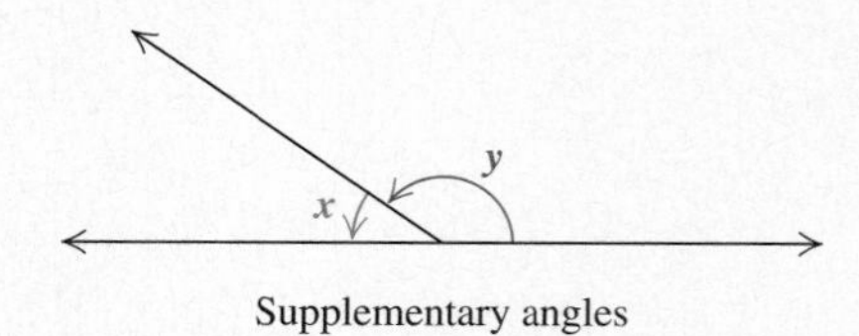

Supplementary angles

2. TRANSLATE. Since we are told that the angles must be supplementary, one equation is

$$x + y = 180. \qquad (1)$$

The second sentence can be translated as follows:

One angle is 30° more than two times the other.

$$y \quad = \quad 2x + 30 \qquad (2)$$

We have now translated the problem to a system of equations:

$$x + y = 180, \quad (1)$$
$$y = 2x + 30. \quad (2)$$

3. CARRY OUT. We solve the system using substitution:

$$x + (2x + 30) = 180 \qquad \text{Substituting } 2x + 30 \text{ for } y \text{ in Equation (1)}$$
$$3x + 30 = 180$$
$$3x = 150 \qquad \text{Subtracting 30}$$
$$x = 50. \qquad \text{Dividing by 3}$$

Substituting 50 for x in Equation (1) then gives us

$$x + y = 180 \qquad \text{Equation (1)}$$
$$50 + y = 180 \qquad \text{Substituting 50 for } x$$
$$y = 130.$$

4. CHECK. If one angle is 50° and the other is 130°, then the sum of the angles is 180°. Thus the angles are supplementary. If 30° is added to twice the smaller angle, we have $2 \cdot 50° + 30°$, or 130°, which is the other angle. The pair of angles checks.

5. STATE. One angle is 50° and the other is 130°. ❑

EXERCISE SET | 8.2

Solve the system using the substitution method. If a system has no solution or infinitely many solutions, state so.

1. $x + y = 4,$ $y = 2x + 1$

2. $x + y = 10,$ $y = x + 8$

3. $y = x + 1,$ $2x + y = 4$

4. $y = x - 6,$ $x + y = -2$

5. $y = 2x - 5,$ $3y - x = 5$

6. $y = 2x + 1,$ $x + y = -2$

7. $x = -2y,$ $x + 4y = 2$

8. $r = -3s,$ $r + 4s = 10$

9. $y = x - 6,$ $3x + 2y = 8$

10. $x = y - 8,$ $3x + 2y = 1$

11. $x = 2y + 1,$ $3x - 6y = 2$

12. $y = 3x - 1,$ $6x - 2y = 2$

13. $s + t = -4,$ $s - t = 2$

14. $x - y = 6,$ $x + y = -2$

15. $y - 2x = -6,$ $2y - x = 5$

16. $x - y = 5,$ $x + 2y = 7$

17. $x - 4y = 3,$ $2x - 6 = 8y$

18. $x - 2y = 7,$ $3x - 21 = 6y$

19. $y = 2x + 5,$ $y = 2x - 5$

20. $y = -2x + 3,$ $2y = -4x + 6$

21. $2x + 3y = -2,$ $2x - y = 9$

22. $x + 2y = 10,$ $3x + 4y = 8$

23. $x - y = -3,$ $2x + 3y = -6$

24. $3b + 2a = 2,$ $-2b + a = 8$

25. $r - 2s = 0,$ $4r - 3s = 15$

26. $y - 2x = 0,$ $3x + 7y = 17$

27. $x - 3y = 7,$ $-4x + 12y = 28$

28. $8x + 2y = 6,$ $4x = 3 - y$

29. $x - 2y = 5,$ $2y - 3x = 1$

30. $y - 3x = -1,$ $5x - 2y = 4$

31. $2x = y - 3,$ $2x = y + 5$

32. $5x - y = 0,$ $5x - y = -2$

Problem Solving

33. The sum of two numbers is 27. One number is 3 more than the other. Find the numbers.

34. The sum of two numbers is 36. One number is 2 more than the other. Find the numbers.

35. Find two numbers whose sum is 58 and whose difference is 16.

36. Find two numbers whose sum is 66 and whose difference is 8.

37. The difference between two numbers is 16. Three times the larger number is seven times the smaller. What are the numbers?

38. The difference between two numbers is 18. Twice the smaller number plus three times the larger is 74. What are the numbers?

39. Two angles are supplementary. One angle is 30° less than twice the other. Find the angles.

40. Two angles are supplementary. One angle is 8° less than three times the other. Find the angles.

41. Two angles are complementary. Their difference is 34°. Find the angles. (*Complementary angles* are angles whose sum is 90°.)

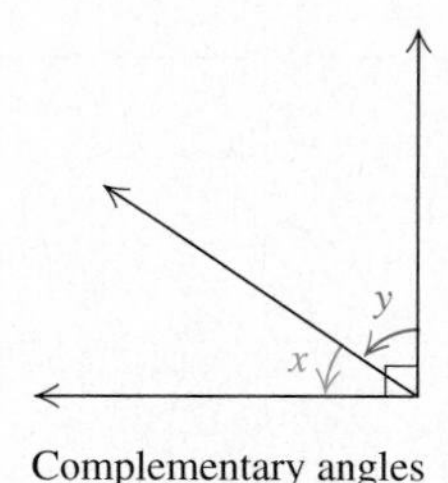

Complementary angles

42. Two angles are complementary. One angle is 42° more than one-half the other. Find the angles.

43. The state of Wyoming is rectangular with a perimeter of 1280 mi. The width is 90 mi less than the length. Find the length and the width.

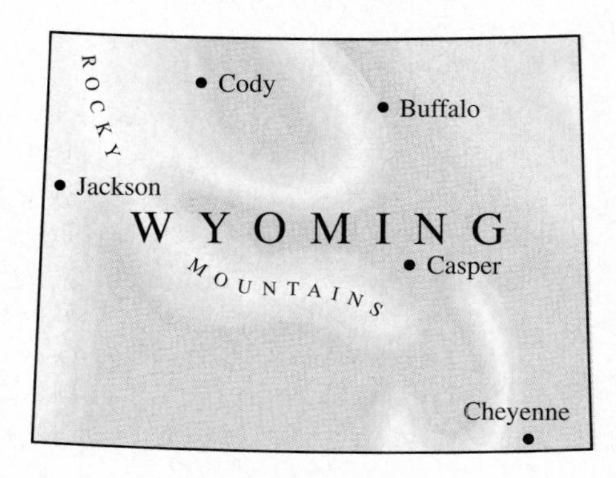

44. The state of Colorado is rectangular with a perimeter of 1300 mi. The length is 110 mi more than the width. Find the length and the width.

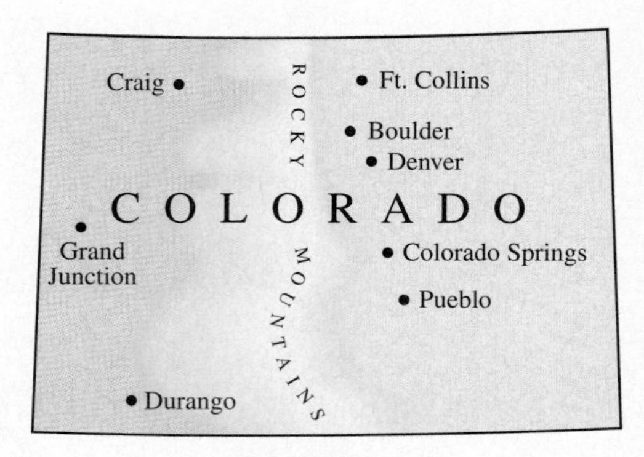

45. The perimeter of a rectangle is 400 m. The length is 3 m more than twice the width. Find the length and the width.

46. The perimeter of a rectangle is 876 cm. The length is 1 cm less than three times the width. Find the length and the width.

47. The perimeter of a soccer field is 340 yd. The length exceeds the width by 50 yd. Find the length and the width.

48. The perimeter of a football field (including the end zones) is $346\frac{2}{3}$ yd. The length is $66\frac{2}{3}$ yd longer than the width. Find the length and the width.

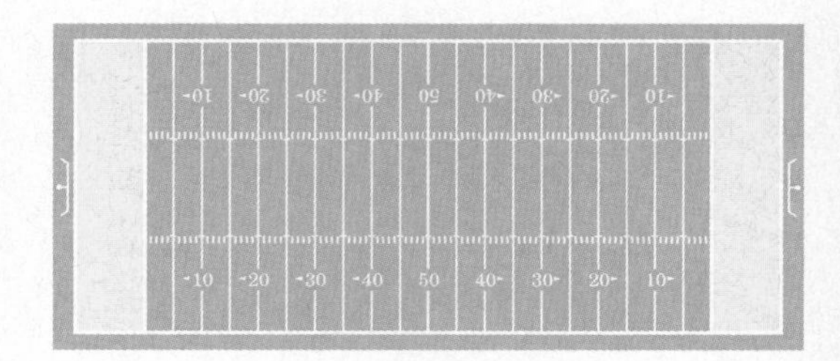

Skill Maintenance

Factor completely. If a polynomial is prime, state so.

49. $6x^2 - 13x + 6$

50. $4p^2 - p - 3$

51. $4x^2 + 3x + 2$

52. $9a^2 - 25$

Synthesis

53. Describe two advantages of the substitution method over the graphing method for solving systems of equations.

54. Under what circumstances can a system of equations be solved more easily by graphing than by substitution?

55. Which systems in Exercises 1–32 contain dependent equations?

56. Which systems in Exercises 1–32 are inconsistent?

Solve by the substitution method.

57. $y - 2.35x = -5.97,$
$2.14y - x = 4.88$

58. $\frac{1}{4}(a - b) = 2,$
$\frac{1}{6}(a + b) = 1$

59. $\frac{x}{2} + \frac{3y}{2} = 2,$
$\frac{x}{5} - \frac{y}{2} = 3$

60. $0.4x + 0.7y = 0.1,$
$0.5x - 0.1y = 1.1$

Exercises 61 and 62 contain systems of three equations in three variables. A solution is an ordered triple of the form (x, y, z). Use the substitution method to solve.

61. $x + y + z = 4,$
$x - 2y - z = 1,$
$y = -1$

62. $x + y + z = 180,$
$x = z - 70,$
$2y - z = 0$

63. Solve Example 3 by first solving for $2y$ in Equation (1) and then substituting for $2y$ in Equation (2). Is this method easier than the procedure used in Example 3? Why or why not?

64. Write a system of two linear equations that can be solved more quickly—but still precisely—by a grapher than by substitution. Time yourself using both methods to solve the system.

8.3 Systems of Equations and Elimination

Solving by the Elimination Method • Comparing the Three Methods • Problem Solving

We have seen that graphing is not always a precise method of solving a system of equations, especially when fractional solutions are involved. The substitution method, considered in Section 8.2, is precise but sometimes difficult to use. For example, to solve the system

$$2x + 3y = 13, \quad (1)$$
$$4x - 3y = 17 \quad (2)$$

by substitution, we would need to first solve for a variable in one of the equations. Were we to solve Equation (1) for y, we would find (after several steps) that $y = \frac{13}{3} - \frac{2}{3}x$. We could then use the expression $\frac{13}{3} - \frac{2}{3}x$ in Equation (2) as a replacement for y:

$$4x - 3\left(\tfrac{13}{3} - \tfrac{2}{3}x\right) = 17.$$

As you can see, although substitution *could* be used to solve this system, to do so would be fairly complicated. Fortunately, another method, *elimination,* can be used to solve systems and, on problems like this, is simpler to use.

Solving by the Elimination Method

The **elimination method** for solving systems of equations makes use of the addition principle. To see how it works, we use it to solve the system discussed above.

EXAMPLE 1

Solve the system

$$2x + 3y = 13, \quad (1)$$
$$4x - 3y = 17. \quad (2)$$

Solution According to Equation (2), $4x - 3y$ and 17 are the same number. Thus we can add $4x - 3y$ to the left side of Equation (1) and 17 to the right side:

$$\begin{array}{ll} 2x + 3y = 13 & (1) \\ \underline{4x - 3y = 17} & (2) \\ 6x + 0y = 30. & \text{Adding. Note that } y \text{ has been ``eliminated.''} \end{array}$$

The resulting equation has just one variable:

$$6x = 30.$$

We now solve for x and obtain $x = 5$.

Next we substitute 5 for x in either of the original equations:

$$\begin{array}{rl} 2x + 3y = 13 & \text{Equation (1)} \\ 2 \cdot 5 + 3y = 13 & \text{Substituting 5 for } x \\ 10 + 3y = 13 & \\ 3y = 3 & \\ y = 1. & \text{Solving for } y \end{array}$$

We check the ordered pair (5, 1).

Check:

$$\begin{array}{r|l} \multicolumn{2}{c}{2x + 3y = 13} \\ \hline 2(5) + 3(1) & ?\ 13 \\ 10 + 3 & \\ 13 & 13 \text{ TRUE} \end{array} \qquad \begin{array}{r|l} \multicolumn{2}{c}{4x - 3y = 17} \\ \hline 4(5) - 3(1) & ?\ 17 \\ 20 - 3 & \\ 17 & 17 \text{ TRUE} \end{array}$$

Since (5, 1) checks, it is the solution. ❑

The system in Example 1 is easier to solve by elimination than by substitution because the term $-3y$ in Equation (2) is the opposite of the term $3y$ in Equation (1). When a system has no pair of terms that are opposites, we need to multiply one or both of the equations by some number to make sure that elimination can occur.

EXAMPLE 2

Solve:

$$2x + 3y = 8, \quad (1)$$
$$x + 3y = 7. \quad (2)$$

Solution Note that in this case, if we add, we will not eliminate a variable. However, if the $3y$ were $-3y$ in one equation, we could eliminate y. We multiply on both sides of Equation (2) by -1 and then add:

$$\begin{array}{rl} 2x + 3y = 8 & \text{Equation (1)} \\ -x - 3y = -7 & \text{Multiplying Equation (2) by } -1 \\ \hline x \quad = 1. & \text{Adding} \end{array}$$

Now we substitute 1 for x in either of the original equations:

$$\begin{array}{rl} x + 3y = 7 & \text{Equation (2)} \\ 1 + 3y = 7 & \text{Substituting 1 for } x \\ 3y = 6 & \\ y = 2. & \text{Solving for } y \end{array}$$

We can check the ordered pair (1, 2).

Check:

$$\begin{array}{r|l} 2x + 3y & 8 \\ \hline 2 \cdot 1 + 3 \cdot 2 & ?\ 8 \\ 2 + 6 & \\ 8 & 8 \text{ TRUE} \end{array} \qquad \begin{array}{r|l} x + 3y & 7 \\ \hline 1 + 3 \cdot 2 & ?\ 7 \\ 1 + 6 & \\ 7 & 7 \text{ TRUE} \end{array}$$

Since (1, 2) checks, it is the solution. ❑

In Example 2, we used the multiplication principle, multiplying by -1. We often need to multiply by a number other than -1.

EXAMPLE 3

Solve:

$$\begin{array}{ll} 3x + 6y = -6, & (1) \\ 5x - 2y = 14. & (2) \end{array}$$

Solution Looking at the terms with variables, we see that if $-2y$ were $-6y$, we would have terms that are opposites. To get $-6y$, we multiply both sides of Equation (2) by 3. Then we add:

$$\begin{array}{rl} 3x + 6y = -6 & \text{Equation (1)} \\ 15x - 6y = 42 & \text{Multiplying Equation (2) by 3} \\ \hline 18x \quad = 36 & \text{Adding} \\ x = 2. & \text{Solving for } x \end{array}$$

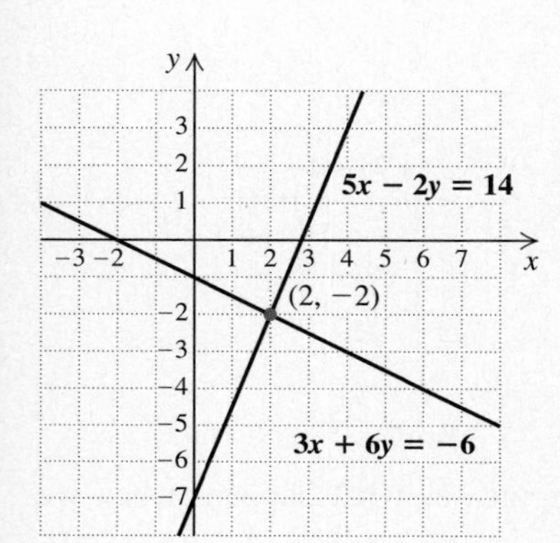

We then go back to Equation (1) and substitute 2 for x:

$$\begin{array}{rl} 3 \cdot 2 + 6y = -6 & \text{Substituting 2 for } x \text{ in Equation (1)} \\ 6 + 6y = -6 & \\ 6y = -12 & \\ y = -2. & \text{Solving for } y \end{array}$$

We leave it to the student to confirm that $(2, -2)$ checks and is the solution. The graph in the margin also serves as a check. ❑

CAUTION! A solution of a system of equations in two variables is a *pair* of numbers. Once you have solved for one variable, don't forget the other.

EXAMPLE 4

Solve:

$$3y + 1 + 2x = 0,$$
$$5x = 7 - 4y.$$

Solution It is often helpful to write equations in the form $Ax + By = C$ before attempting to eliminate a variable:

$2x + 3y = -1,$ Subtracting 1 on both sides and rearranging the terms of the first equation

$5x + 4y = 7.$ Adding $4y$ on both sides of the second equation

Since neither coefficient of x is a multiple of the other and neither coefficient of y is a multiple of the other, we use the multiplication principle with *both* equations:

$2x + 3y = -1,$ (1)

$5x + 4y = 7.$ (2)

Note that the LCM of 2 and 5 is 10.

We can eliminate the x-term by multiplying on both sides of Equation (1) by 5 and on both sides of Equation (2) by -2:

$10x + 15y = -5$ Multiplying on both sides of Equation (1) by 5

$-10x - 8y = -14$ Multiplying on both sides of Equation (2) by -2

$7y = -19$ Adding

$y = \frac{-19}{7} = -\frac{19}{7}.$ Dividing by 7

We substitute $-\frac{19}{7}$ for y in Equation (1):

$2x + 3y = -1$ Equation (1)

$2x + 3\left(-\frac{19}{7}\right) = -1$ Substituting $-\frac{19}{7}$ for y

$$2x - \frac{57}{7} = -1$$
$$2x = -1 + \frac{57}{7}$$
$$2x = -\frac{7}{7} + \frac{57}{7}$$
$$2x = \frac{50}{7}$$

$x = \frac{50}{7} \cdot \frac{1}{2} = \frac{25}{7}.$ Solving for x

We check the ordered pair $\left(\frac{25}{7}, -\frac{19}{7}\right)$.

Check:

$3y + 1 + 2x = 0$	
$3\left(-\frac{19}{7}\right) + 1 + 2 \cdot \frac{25}{7}$	? 0
$-\frac{57}{7} + \frac{7}{7} + \frac{50}{7}$	
0	0 TRUE

$5x = 7 - 4y$	
$5 \cdot \frac{25}{7}$?	$7 - 4\left(-\frac{19}{7}\right)$
$\frac{125}{7}$	$\frac{49}{7} + \frac{76}{7}$
$\frac{125}{7}$	$\frac{125}{7}$ TRUE

The solution is $\left(\frac{25}{7}, -\frac{19}{7}\right)$. ❑

Next we consider a system with no solution and see what happens when we apply the elimination method.

EXAMPLE 5

Solve:

$y - 3x = 2,$ (1)

$y - 3x = 1.$ (2)

Solution To eliminate y, we multiply on both sides of Equation (2) by -1. Then we add:

$$\begin{aligned} y - 3x &= 2 \\ -y + 3x &= -1 \quad \text{Multiplying on both sides of Equation (2) by } -1 \\ \hline 0 &= 1. \quad \text{Adding} \end{aligned}$$

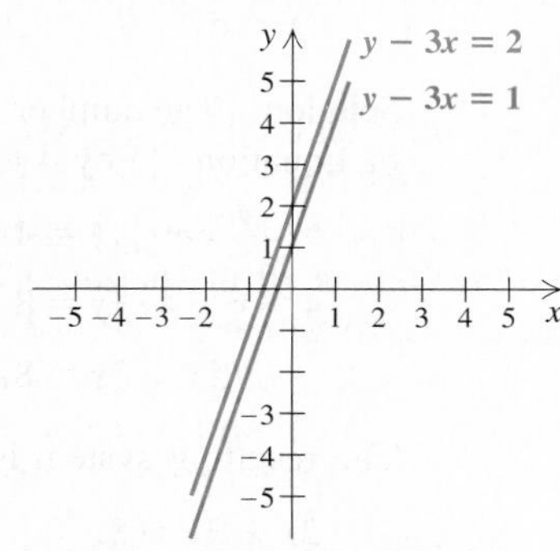

A visualization of Example 5

Note that in eliminating y, we eliminated x as well. The resulting equation, $0 = 1$, is false for any pair (x, y), so there is *no solution*. ❑

Sometimes there is an infinite number of solutions. Consider a system that we graphed in Example 4 of Section 8.1.

EXAMPLE 6

Solve:

$$\begin{aligned} 2x + 3y &= 6, \quad &(1) \\ -8x - 12y &= -24. \quad &(2) \end{aligned}$$

Solution To eliminate x, we multiply on both sides of Equation (1) by 4 and then add the two equations:

$$\begin{aligned} 8x + 12y &= 24 \quad \text{Multiplying on both sides of Equation (1) by 4} \\ -8x - 12y &= -24 \\ \hline 0 &= 0. \quad \text{Adding} \end{aligned}$$

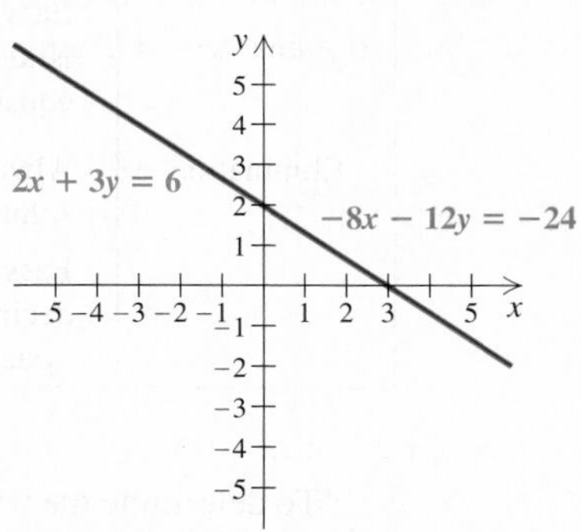

A visualization of Example 6

Again, we have eliminated *both* letters. The resulting equation, $0 = 0$, is always true, so there are infinitely many solutions. The equations are dependent. ❑

children. The total receipts were then \$2450 + \$228.75, or \$2678.75. The numbers check.

5. **STATE.** The movie was attended by 350 adults and 61 children. ❑

EXAMPLE 3

The Java Joint wants to mix Kenyan beans that sell for \$7.25 per pound with Venezuelan beans that sell for \$8.50 per pound to form a 50-lb batch of Morning Blend that sells for \$8.00 per pound. How many pounds of Kenyan beans and how many pounds of Venezuelan beans should go into the blend?

Solution

1. **FAMILIARIZE.** This problem seems similar to Example 2. Instead of adults and children, we have pounds of Kenyan coffee and pounds of Venezuelan coffee. Instead of two different prices of admission, we have two different prices per pound. Finally, instead of having the total receipts, we know the weight and price per pound of the batch of Morning Blend that is to be made. Note that we can easily find the value of the batch of Morning Blend by multiplying 50 lb times \$8.00 per pound. We let $k =$ the number of pounds of Kenyan coffee used and $v =$ the number of pounds of Venezuelan coffee used.

2. **TRANSLATE.** Since a 50-lb batch is being made, we must have

$$k + v = 50.$$

To find a second equation, we consider the total value of the 50-lb batch. That value must be the same as the value of the Kenyan beans and the value of the Venezuelan beans that go into the blend:

Reword:	The value of the Kenyan beans	plus	the value of the Venezuelan beans	is	the value of the Morning Blend.
	↓	↓	↓	↓	↓
Translate:	$k \cdot 7.25$	$+$	$v \cdot 8.50$	$=$	$50 \cdot 8.00$.

This information can be presented in a table.

	Kenyan	Venezuelan	Morning Blend	
Price per Pound	\$7.25	\$8.50	\$8.00	
Number of Pounds	k	v	50	⟶ $k + v = 50$
Value of Beans	$7.25k$	$8.50v$	400	⟶ $7.25k + 8.50v = 400$

We have translated to a system of equations:

$$k + v = 50, \quad (1)$$
$$7.25k + 8.50v = 400. \quad (2)$$

3. **CARRY OUT.** When Equation (1) is solved for k, we have $k = 50 - v$. We then substitute $50 - v$ for k in Equation (2):

$$7.25(50 - v) + 8.50v = 400. \quad \text{Solving by substitution}$$

Then

$$362.50 - 7.25v + 8.50v = 400 \quad \text{Using the distributive law}$$

$$1.25v = 37.50 \quad \text{Collecting like terms; subtracting 362.50 on both sides}$$

$$v = 30. \quad \text{Dividing by 1.25 on both sides}$$

If $v = 30$, we see from Equation (1) that $k = 20$.

4. CHECK. If 20 lb of Kenyan beans and 30 lb of Venezuelan beans are mixed, a 50-lb blend will result. The value of 20 lb of Kenyan beans is 20($7.25), or $145. The value of 30 lb of Venezuelan beans is 30($8.50), or $255, so the combined value for the blend is $145 + $255 = $400. A 50-lb blend priced at $8.00 a pound would also be worth $400, so our answer checks.

5. STATE. The Morning Blend should be made by combining 20 lb of Kenyan beans with 30 lb of Venezuelan beans. ❑

EXAMPLE 4

At a local ''paint swap,'' Gayle found large supplies of Skylite Pink (12.5% red pigment) and MacIntosh Red (20% red pigment). How many gallons of each color should Gayle take in order to mix a 10-gallon batch of Summer Rose (17% red pigment)?

Solution

1. FAMILIARIZE. We make a picture of the situation. Note that the pigment in the paint can be thought of as a solid that would eventually settle to the bottom of the can. We let p = the number of gallons of Skylite Pink needed and m = the number of gallons of MacIntosh Red needed.

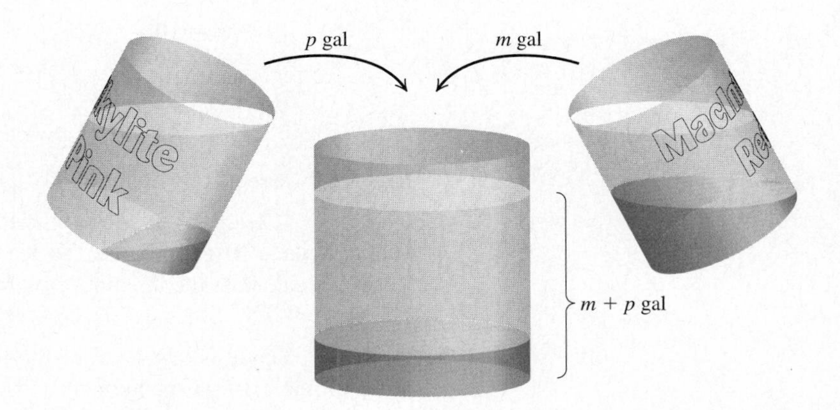

Suppose that 2 gal of Skylite Pink and 8 gal of MacIntosh Red are mixed. The Skylite Pink would contribute 12.5% of 2 gal, or 0.25 gal of pigment, while the MacIntosh red would contribute 20% of 8 gal, or 1.6 gal of pigment. Thus the 10-gal mixture would contain $0.25 + 1.6 = 1.85$ gal of pigment. Since Gayle wants the 10 gal of Summer Rose to be 17% pigment, and since 17% of 10 gal is 1.7 gal, our guess is incorrect. Note that it is not enough just to find two numbers that add to 10. We must also make sure that the amount of pigment in the Summer Rose mixture is 17% of 10 gal.

This information can be arranged in a table.

	Skylight Pink	**MacIntosh Red**	**Summer Rose**
Amount of Paint (in Gallons)	p	m	10
Percent of Pigment	12.5%	20%	17%
Amount of Pigment (in Gallons)	$0.125p$	$0.2m$	0.17×10, or 1.7

2. TRANSLATE. A system of two equations can be formed by reading across the first and third rows of the table. Since Gayle needs 10 gal of mixture, we must have

$$p + m = 10. \qquad \leftarrow \text{Total amount of paint}$$

Since the pigment in the Summer Rose paint comes from the pigment in both the Skylite Pink and the MacIntosh Red paint, we have

$$0.125p + 0.2m = 1.7. \qquad \leftarrow \text{Total amount of pigment}$$

We have translated to a system of equations:

$$\begin{aligned} p + \quad m &= 10, & (1) \\ 0.125p + 0.2m &= 1.7. & (2) \end{aligned}$$

3. CARRY OUT. We multiply on both sides of Equation (2) by -5 to eliminate m:

$$\begin{aligned} p + m &= 10 \\ -0.625p - m &= -8.5 \qquad \text{We observed that } (-5)(0.2m) = -m. \\ \hline 0.375p \quad &= 1.5 \\ p &= 4. \qquad \text{Dividing on both sides by } 0.375 \end{aligned}$$

If $p = 4$, we see from Equation (1) that $m = 6$.

4. CHECK. Clearly, 4 gal of Skylite Pink and 6 gal of MacIntosh Red do combine to make a 10-gal mixture. To see if the mixture is the right color, Summer Rose, we calculate the amount of pigment in the mixture: $0.125 \cdot 4 + 0.2 \cdot 6 = 0.5 + 1.2 = 1.7$. Since 1.7 is 17% of 10, the mixture is the correct color.

5. STATE. Gayle needs 4 gal of Skylite Pink and 6 gal of MacIntosh Red in order to make 10 gal of Summer Rose. ❑

Re-examine Examples 1–4, looking for similarities. Examples 3 and 4 are often called *mixture problems,* but they still share much in common with Examples 1 and 2.

Problem-Solving Tip

When solving a problem, see if it is patterned or modeled after a problem that you have already solved.

EXERCISE SET 8.4

Solve. Use the five steps for problem solving.

1. A firm sells cars and trucks. There is room on its lot for 510 vehicles. From experience, they know that profits will be greatest if there are 190 more cars than trucks on the lot. How many of each vehicle should the firm have on the lot for the greatest profit?
2. A family drove and hiked 45 km to reach a campsite. They drove 23 km more than they walked to get there. How far did they walk?
3. Mr. Frank's charges $1.99 for a slice of pizza and a soda and $5.48 for three slices of pizza and two sodas. Determine the cost of one soda and the cost of one slice of pizza.
4. A fast-food restaurant is running a promotion in which a burger and two pieces of chicken cost $2.39 and a burger and one piece of chicken cost $1.69. Determine the cost of one burger.
5. The Hendersons generate two and a half times as much trash as their neighbors, the Savickis. Together, the two households produce 14 bags of trash each month. How much trash does each household produce?
6. The Mazzas' attic required three and a half times as much insulation as did the Kranepools'. Together, the two attics required 36 rolls of insulation. How much insulation did each attic require?
7. In a recent basketball game, the Hot Shots scored 69 points on 41 shots. If only two-pointers and foul shots (each foul shot is worth one point) were taken, how many foul shots were made?
8. In a recent basketball game, the Slammers made 48 shots for a total of 107 points. If only two-point and three-point shots were attempted, how many of each type did the Slammers make?
9. A busload of campers stopped at a dairy stand for ice cream. They ordered 75 cones, some soft-serve at $1.25 and the rest hard-pack at $1.50. If the total bill was $104.25, how many of each type of cone were ordered?
10. The Dixville Cub Scout troop collected 436 returnable bottles and cans, some worth 5 cents each and the rest worth 10 cents each. If the total value of the cans and bottles was $26.60, how many 5-cent bottles or cans and how many 10-cent bottles or cans were collected?
11. There were 429 people at a play. Admission was $4 each for adults and $3 each for children. The receipts were $1490. How many adults and how many children attended?
12. The attendance at a school concert was 578. Admission was $2 each for adults and $1.50 each for children. The receipts were $985. How many adults and how many children attended?
13. There were 200 tickets sold for a women's basketball game. Tickets for students were $2 each and for adults were $3 each. The total amount of money collected was $530. How many of each type of ticket were sold?
14. There were 203 tickets sold for a volleyball game. For activity-card holders the price was $1.25, and for noncard holders the price was $2. The total amount of money collected was $310. How many of each type of ticket were sold?
15. The Nuthouse has 10 kg of mixed cashews and pecans worth $8.40 per kilogram. Cashews alone sell for $8.00 per kilogram, and pecans sell for $9.00 per kilogram. How many kilograms of each are in the mixture?
16. A coffee shop mixes Brazilian coffee worth $10 per kilogram with Turkish coffee worth $16 per kilogram. The mixture should be worth $14 per kilogram. How much of each type of coffee should be used to make a 300-kg mixture?
17. Sunflower seed is worth $1.00 per pound and rolled oats are worth $1.35 per pound. How much of each would you use to make 50 lb of a mixture worth $1.14 per pound?
18. A grocer wishes to mix some nuts worth $2.52 per pound and some worth $3.80 per pound to make 480 lb of a mixture worth $3.44 per pound. How much of each should be used?
19. Solution A is 50% acid and solution B is 80% acid. How much of each should be used to make 100 L of a solution that is 68% acid? (*Hint:* 68% of what

is acid?) Complete the following to aid in the familiarization.

Type of Solution	A	B	Mix
Amount of Solution	x	y	
Percent of Acid	50%		68%
Amount of Acid in Solution		$0.8y$	

20. Solution A is 30% alcohol and solution B is 75% alcohol. How much of each should be used to make 100 L of a solution that is 50% alcohol?

21. A solution containing 30% insecticide is to be mixed with a solution containing 50% insecticide to make 200 L of a solution containing 42% insecticide. How much of each solution should be used?

22. A solution containing 28% fungicide is to be mixed with a solution containing 40% fungicide to make 300 L of a solution containing 36% fungicide. How much of each solution should be used?

23. A chemist has one solution that is 80% base and another solution that is 30% base. What is needed is 200 L of a solution that is 62% base. The chemist will prepare it by mixing the two solutions on hand. How much of each should be used?

24. One solution is 50% alcohol and a second is 70% alcohol. How much of each should be mixed to make 30 L of a solution that is 55% alcohol?

25. You are taking a test in which items of type A are worth 10 points and items of type B are worth 15 points. You answer 16 questions and score 180 points. How many questions of each type did you answer correctly?

26. A goldsmith has two alloys that are different purities of gold. The first is three-fourths pure gold and the second is five-twelfths pure gold. How many ounces of each should be melted and mixed in order to obtain a 60-ounce mixture that is two-thirds pure gold?

27. A collection of dimes and quarters is worth \$15.25. There are 103 coins in all. How many of each are there?

28. A collection of quarters and nickels is worth \$1.25. There are 13 coins in all. How many of each are there?

29. A collection of nickels and dimes is worth \$25. There are three times as many nickels as dimes. How many of each are there?

30. A collection of nickels and dimes is worth \$2.90. There are 19 more nickels than dimes. How many of each are there?

31. Campus Painters has two kinds of paint. If 9 gal of the inexpensive paint is mixed with 7 gal of the expensive paint, the mixture will be worth \$19.70 per gallon. If 3 gal of the inexpensive paint is mixed with 5 gal of the expensive paint, the mixture will be worth \$19.825 per gallon. What is the price per gallon of each type of paint?

32. A printer knows that a page of print contains 1300 words if large type is used and 1850 words if small type is used. A document containing 18,526 words fills exactly 12 pages. How many pages are in the large type? How many in the small type?

Skill Maintenance

Factor.

33. $25x^2 - 81$

34. $36 - a^2$

Solve.

35. $\dfrac{x^2}{x+4} = \dfrac{16}{x+4}$

36. $x^2 - 10x + 25 = 0$

Synthesis

37. ◈ Write a problem for a classmate to solve by translating to a system of two equations in two unknowns.

38. ◈ What characteristics do Examples 1–4 share when they are being translated to systems of equations?

39. A total of \$27,000 is invested, part of it at 12% and the rest at 13%. The total yield after one year is \$3385. How much was invested at each rate?

40. A student earned \$288 on investments. If \$1100 was invested at one yearly rate and \$1800 at a rate that was 1.5% higher, find the two rates of interest.

41. A two-digit number is six times the sum of its digits. The ten's digit is 1 more than the one's digit. Find the number.

42. The sum of the digits of a two-digit number is 12. When the digits are reversed, the number is decreased by 18. Find the original number.

43. Farmer Jones has 100 L of milk that is 4.6% butterfat. How much skim milk (no butterfat) should be added to make milk that is 3.2% butterfat?

44. A tank contains 8000 L of a solution that is 40% acid. How much water should be added in order to make a solution that is 30% acid?

45. An automobile radiator contains 16 L of antifreeze and water. This mixture is 30% antifreeze. How much of this mixture should be drained and replaced with pure antifreeze so that the mixture will be 50% antifreeze?

46. Ace Engineering pays a total of $325 an hour when employing some workers at $20 an hour and others at $25 an hour. When the number of $20 workers is increased by 50% and the number of $25 workers is decreased by 20%, the cost per hour is $400. Find how many were originally employed at each rate.

47. In a two-digit number, the sum of the one's digit and the number is 43 more than five times the ten's digit. The sum of the digits is 11. Find the number.

48. The sum of the digits of a three-digit number is 9. If the digits are reversed, the number increases by 495. The sum of the ten's and hundred's digits is half the one's digit. Find the number.

49. Together, a bat, ball, and glove cost $99.00. The bat costs $9.95 more than the ball, and the glove costs $65.45 more than the bat. How much does each cost?

50. In Lewis Carroll's *Through the Looking Glass,* Tweedledum says to Tweedledee, "The sum of your weight and twice mine is 361 pounds." Then Tweedledee says to Tweedledum, "Contrariwise, the sum of your weight and twice mine is 362 pounds." Find the weights of Tweedledum and Tweedledee.

8.5 Systems of Linear Inequalities

Graphing Systems of Inequalities • Locating Solution Sets

Linear inequalities in two variables were first graphed in Section 7.4. We now consider *systems of linear inequalities* in two variables, such as

$$x + y \leq 3,$$
$$x - y < 3.$$

When systems of equations were solved graphically in Section 8.1, we searched for any points common to both lines. To solve a system of inequalities graphically, we again look for points common to both graphs. We do so by graphing each inequality and determining where the graphs overlap.

EXAMPLE 1

Graph the solutions of the system

$$x + y \leq 3,$$
$$x - y < 3.$$

Solution To graph the inequality $x + y \leq 3$, we draw the graph of $x + y = 3$ using a solid line (see the graph at the top of the following page). Next, since (0, 0) is a solution of $x + y \leq 3$, we shade (in red) all points on that side of the line. The arrows near the ends of the line also indicate the region that contains solutions.

Next we superimpose the graph of $x - y < 3$, using a dashed line for $x - y = 3$ and again using (0, 0) as a test point. Since (0, 0) is a solution, we shade (in blue) the region on the same side of the dashed line as (0, 0).

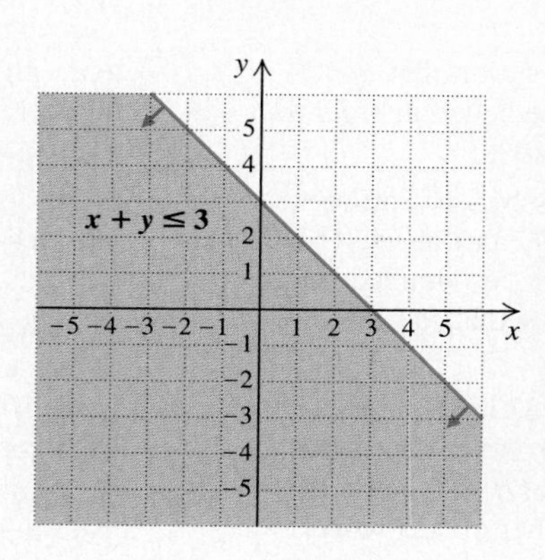

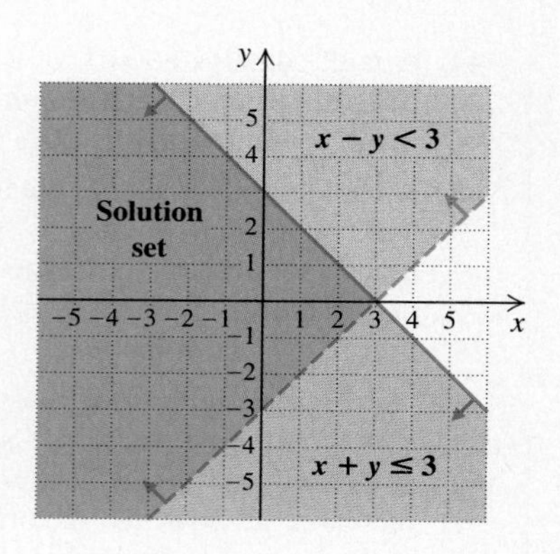

The solution set of the system is the region shaded purple along with the purple portion of the line $x + y = 3$. ❑

EXAMPLE 2

Graph the solutions of the system

$$x \geqslant 3,$$
$$x - 3y < 6.$$

Solution We draw the graphs of $x \geqslant 3$ and $x - 3y < 6$ using the same set of axes. The solution set is the purple region along with the purple portion of the solid line.

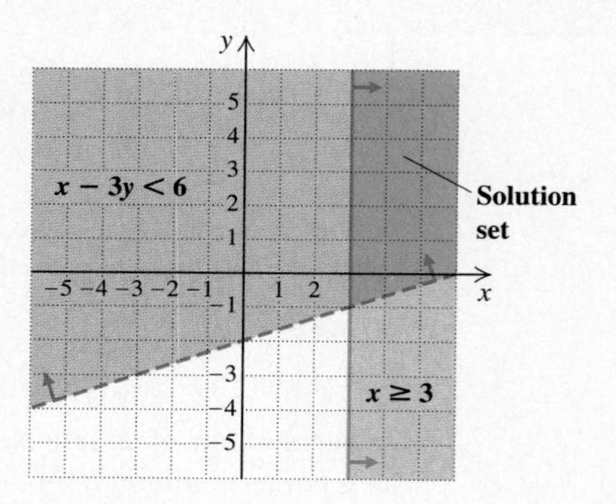

❑

EXAMPLE 3

Graph the solutions of the system

$$x - 2y < 0,$$
$$-2x + y > 2.$$

Solution We graph $x - 2y < 0$ using red and $-2x + y > 2$ using blue. The region that is purple is the solution set of the system since those points solve both inequalities.

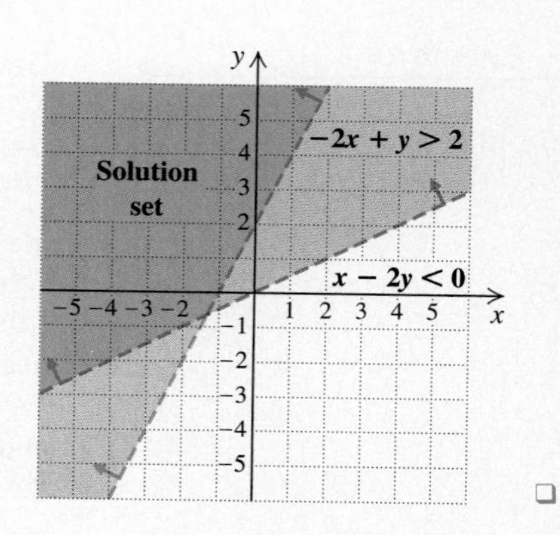

❑

EXERCISE SET 8.5

Graph the solutions of the system.

1. $x + y \leq 1,$
   $x - y \leq 5$
2. $x + y \leq 3,$
   $x - y \leq 4$
3. $y - 2x > 1,$
   $y - 2x < 3$
4. $x + y < 2,$
   $x + y > 0$
5. $y \geq 1,$
   $x > 2 + y$
6. $x > 3,$
   $x + y \leq 4$
7. $y > 4x - 1,$
   $y < -2x + 3$
8. $y \geq x,$
   $y \leq 1 - x$
9. $2x - 3y \geq 9,$
   $2y + x > 6$
10. $3x - 2y \leq 8,$
    $2x + y > 6$
11. $y > 5x + 2,$
    $y \leq 1 - x$
12. $y > 4,$
    $2y + x \leq 4$
13. $x \leq 3,$
    $y \leq 4$
14. $x \geq -2,$
    $y \geq -3$
15. $x \leq 0,$
    $y \leq 0$
16. $x \geq 0,$
    $y \geq 0$
17. $x + y \leq 6,$
    $x \geq 0,$
    $y \geq 0,$
    $y \leq 5$
18. $x + 2y \leq 8,$
    $x \leq 6,$
    $x \geq 0,$
    $y \geq 0$
19. $y - x \geq 1,$
    $y - x \leq 3,$
    $x \leq 5,$
    $x \geq 2$
20. $x - 2y \leq 0,$
    $y - 2x \leq 2,$
    $x \leq 2,$
    $y \leq 2$
21. $y \leq x,$
    $x \geq -2,$
    $x \leq -y$
22. $y > 0,$
    $2y + x \leq -6,$
    $x + 2 \leq 2y$

Skill Maintenance

Subtract.

23. $\dfrac{3}{x^2 - 4} - \dfrac{2}{3x + 6}$

24. $\dfrac{5}{4a^3 - 12a^2} - \dfrac{2}{a^2 - 4a + 3}$

Evaluate the polynomial when $x = -3$.

25. $x^2 - 5x + 2$

26. $4x^3 - 5x^2$

Synthesis

27. ◈ Does the solution set of Example 1 include any points that are on the solid line? If so, which points and why? If not, why not?

28. ◈ Under what condition(s) will a system of two linear inequalities have no solution?

Graph the solutions of the system.

29. $5a + 3b \geq 30,$
    $2a + 3b \geq 21,$
    $3a + 6b \geq 36,$
    $a \geq 0,$
    $b \geq 0$
30. $2u + v \geq 8,$
    $4u + 3v \geq 22,$
    $2u + 5v \geq 18,$
    $u \geq 0,$
    $v \geq 0$

SUMMARY AND REVIEW 8

KEY TERMS

System of equations, p. 352
Consistent, p. 354
Inconsistent, p. 354
Dependent, p. 355
Independent, p. 355
Substitution method, p. 357
Supplementary angles, p. 360
Complementary angles, p. 362
Elimination method, p. 364
System of linear inequalities, p. 379

IMPORTANT PROPERTIES AND FORMULAS

When graphing a system of two linear equations, one of the following must occur:

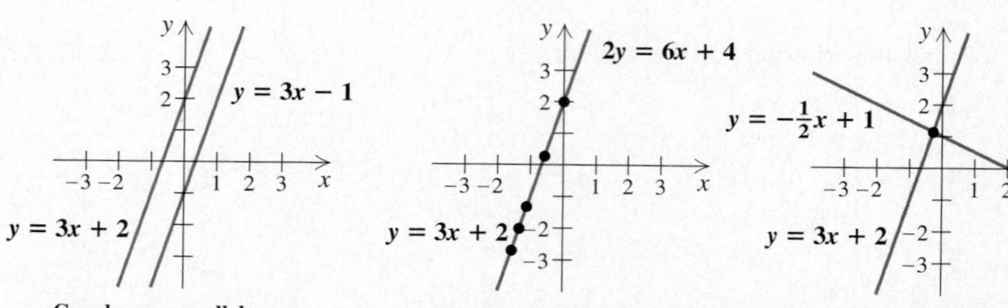

Graphs are parallel.
The system is *inconsistent* because there is no solution.
The equations are *independent* since the graphs differ.

Equations have the same graph.
The system is *consistent* and has infinitely many solutions.
The equations are *dependent* since the graphs are the same.

Graphs intersect.
The system is *consistent* and has one solution.
Since the graphs differ, the equations are *independent*.

A Comparison of Methods for Solving Systems of Linear Equations

Method	Strengths	Weaknesses
Graphical	Gives solutions that can be seen visually.	Inexact when solution involves numbers that are not integers or are very large and off the graph.
Substitution	Always yields exact solutions. Easy to use when a variable is alone on one side.	Introduces extensive computations with fractions when solving more complicated systems. Cannot see the solution.
Elimination	Always yields exact solutions. Easy to use when fractions or decimals appear in the system.	Cannot see the solution.

REVIEW EXERCISES

This chapter's review and test include Skill Maintenance exercises from Sections 4.2, 5.3, 5.4, and 6.5.

Determine whether the given ordered pair is a solution of the system of equations.

1. $(6, -1)$; $x - y = 3$,
$2x + 5y = 6$

2. $(2, -3)$; $2x + y = 1$,
$x - y = 5$

3. $(-2, 1)$; $x + 3y = 1$,
$2x - y = -5$

4. $(-4, -1)$; $x - y = 3$,
$x + y = -5$

Solve the system of equations by graphing.

5. $x + y = 4$,
$x - y = 8$

6. $x + 3y = 12$,
$2x - 4y = 4$

7. $y = 5 - x$,
$3x - 4y = -20$

8. $3x - 2y = -4$,
$2y - 3x = -2$

Solve using the substitution method.

9. $y = 5 - x$,
$4x + 5y = 22$

10. $x + 2y = 6$,
$2x + 3y = 8$

11. $3x + y = 1$,
$x - 2y = 5$

12. $x + y = 6$,
$y = 3 - 2x$

13. $s + t = 5$,
$s = 13 - 3t$

14. $x - y = 4$,
$y = 2 - x$

Solve using the elimination method.

15. $x + y = 4$,
$2x - y = 5$

16. $x + 2y = 9$,
$3x - 2y = -5$

17. $x - y = 8$,
$2x + y = 7$

18. $\frac{2}{3}x + y = -\frac{5}{3}$,
$x - \frac{1}{3}y = -\frac{13}{3}$

19. $2x + 3y = 8$,
$5x + 2y = -2$

20. $5x - 2y = 2$,
$3x - 7y = 36$

21. $-x - y = -5$,
$2x - y = 4$

22. $6x + 2y = 4$,
$10x + 7y = -8$

23. $-6x - 2y = 5$,
$12x + 4y = -10$

Problem Solving

24. The sum of two numbers is 8. Their difference is 12. Find the numbers.

25. The sum of two numbers is 27. One half of the first number plus one third of the second number is 11. Find the numbers.

26. The perimeter of a rectangle is 96 cm. The length is 27 cm more than the width. Find the length and the width.

27. In a recent basketball game, the Hoopsters made 48 shots for a total of 81 points. If only two-pointers and foul shots (each foul shot is worth one point) were taken, how many foul shots were made?

28. There were 508 people at an organ recital. Orchestra seats cost \$5.00 per person and balcony seats cost \$3.00 each. The total receipts were \$2118. Find the number of orchestra seats and the number of balcony seats sold.

29. Solution A is 30% alcohol and solution B is 60% alcohol. How much of each is needed to make 80 L of a solution that is 45% alcohol?

Graph the solutions of the system.

30. $x \geq 1$,
$y \leq -2$

31. $x - y > 2$,
$x + y < 1$

Skill Maintenance

32. Collect like terms and arrange in descending order:

$$2x^3 - x^4 + x^3 + 2x^4.$$

33. Add and simplify, if possible:

$$\frac{2}{t^2 - 4} + \frac{3}{t - 2}.$$

Factor completely.

34. $2t^3 - 5t^2 - 3t$

35. $9y^2 + 36y + 36$

Synthesis

36. ◈ Explain why the solution of a system of equations is the point of intersection of the graphs of the lines.

37. ◈ A student sketches the boundary lines of a system of two linear inequalities and notes that the lines are parallel. Since there is no point of intersection, the student concludes that the solution set is empty. What is wrong with this conclusion?

38. The solution of the following system is (6, 2). Find C and D.

$$2x - Dy = 6,$$
$$Cx + 4y = 14$$

39. Solve using the substitution method:

$$x - y + 2z = -3,$$
$$2x + y - 3z = 11,$$
$$z = -2.$$

40. Solve:

$$3(x - y) = 4 + x,$$
$$x = 5y + 2.$$

41. For a two-digit number, the sum of the one's digit and the ten's digit is 6. When the digits are reversed, the new number is 18 more than the original number. Find the original number.

42. A stable boy agreed to work for one year. At the end of that time, he was to receive \$240 and one horse. After 7 months, the boy quit the job, but still received the horse and \$100. What was the value of the boy's yearly salary?

CHAPTER TEST 8

1. Determine whether the given ordered pair is a solution of the system of equations.

$(-2, -1)$; $x = 4 + 2y,$
$2y - 3x = 4$

2. Solve by graphing:

$$x - y = 3,$$
$$x - 2y = 4.$$

Solve using the substitution method.

3. $y = 6 - x,$ $2x - 3y = 22$

4. $x + 2y = 5,$ $x + y = 2$

5. $y = 5x - 2,$ $y - 2 = 5x$

Solve using the elimination method.

6. $x - y = 6,$ $3x + y = -2$

7. $\frac{1}{2}x - \frac{1}{3}y = 8,$ $\frac{2}{3}x + \frac{1}{2}y = 5$

8. $4x + 5y = 5,$ $6x + 7y = 7$

9. $2x + 3y = 13,$ $3x - 5y = 10$

10. Two angles are complementary. One angle is 18° less than twice the other. Find the angles.

11. Solution A is 25% acid and solution B is 40% acid. How much of each is needed to make 60 L of a solution that is 30% acid?

12. Tyler bought 18 sheets of plywood for \$750. Oak plywood cost \$45 for a sheet and pine cost \$25. How many sheets of each kind of plywood did he buy?

Graph the solutions of the system.

13. $y \geq x,$ $y < x + 1$

14. $x + y \leq 6,$ $x \geq 0,$ $y \geq 0$

Skill Maintenance

15. Subtract: $\frac{1}{x^2 - 16} - \frac{x - 4}{x^2 - 3x - 4}$.

16. Evaluate $-3x^4 + 4x^3 - 5x + 2$ when $x = -1$.

Factor completely.

17. $18y^2 + 3y - 3$

18. $5x^2 + 30x + 45$

Synthesis

19. Find the numbers C and D such that $(-2, 3)$ is a solution of the system

$$Cx - 4y = 7,$$
$$3x + Dy = 8.$$

20. You are in line at a ticket window. There are two more people ahead of you in line than there are behind you. In the entire line, there are three times as many people as there are behind you. How many are in the line?

CHAPTER 9

Radical Expressions and Equations

AN APPLICATION

A new water pipe is being prepared so that it will run diagonally under a kitchen floor. If the kitchen is 8 ft wide and 12 ft long, how long should the pipe be?

This problem appears as Exercise 21 in Section 9.6.

Joseph Gossett
PLUMBER AND PIPE FITTER

"It might come as a surprise to some that plumbers use math every day. Calculations ranging from water pressure to rate of flow and cost estimates all involve math. I use the Pythagorean theorem at least five times a week to calculate pipe length."

Many of us already have some familiarity with the notion of *square roots.* For example, 3 is a square root of 9 because $3^2 = 9$. In this chapter, we learn how to manipulate square roots of polynomials and rational expressions. Later in this chapter, these *radical expressions* will appear in equations and in problem-solving situations.

In addition to the material from this chapter, the review and test for Chapter 9 include material from Sections 6.2, 6.3, 7.2, and 7.5.

9.1 Introduction to Square Roots and Radical Expressions

Square Roots • Radicands and Radical Expressions • Irrational Numbers • Approximating Square Roots • Square Roots and Absolute Value • Problem Solving

We begin our study of square roots by examining square roots of numbers, square roots of variable expressions, and an application involving a formula.

Square Roots

Often in this text we have found the result of squaring a number. When the process is reversed, we say that we are looking for a number's *square root*.

Square Root

The number c is a *square root* of a if $c^2 = a$.

Every positive number has two square roots. For example, the square roots of 25 are 5 and -5 because $5^2 = 25$ and $(-5)^2 = 25$.

EXAMPLE 1

Find the square roots of each number: **(a)** 81; **(b)** 100.

Solution

a) The square roots of 81 are 9 and -9. To check, note that $9^2 = 81$ and $(-9)^2 = (-9)(-9) = 81$.

b) The square roots of 100 are 10 and -10, since $10^2 = 100$ and $(-10)^2 = 100$. ❑

The positive square root of a number is also called the **principal square root.** A **radical sign,** $\sqrt{}$, is often used when finding square roots and refers to the principal root. Thus, $\sqrt{25} = 5$ and $\sqrt{25} \neq -5$.

EXAMPLE 2

Find each of the following: **(a)** $\sqrt{225}$; **(b)** $-\sqrt{64}$.

Solution

a) The principal square root of 225 is its positive square root, so $\sqrt{225} = 15$.

b) The symbol $-\sqrt{64}$ represents the opposite of $\sqrt{64}$. Since $\sqrt{64} = 8$, we have $-\sqrt{64} = -8$. ❑

Radicands and Radical Expressions

A **radical expression** is an algebraic expression that contains at least one radical sign. Here are some examples:

$$\sqrt{14}, \qquad 7 + \sqrt{2x}, \qquad \sqrt{x^2 + 4}, \qquad \sqrt{\frac{x^2 - 5}{2}}.$$

The expression under the radical is called the **radicand.**

EXAMPLE 3

Identify the radicand in each expression: **(a)** $\sqrt{x}$; **(b)** $\sqrt{y^2 - 5}$.

Solution

a) In $\sqrt{x}$, the radicand is x.
b) In $\sqrt{y^2 - 5}$, the radicand is $y^2 - 5$. ❑

The square of any nonzero number is always positive. For example, $8^2 = 64$ and $(-11)^2 = 121$. No real number, squared, is equal to a negative number. Thus the following expressions are not real numbers:

$$\sqrt{-100}, \qquad \sqrt{-49}, \qquad -\sqrt{-3}.$$

In Chapter 10, we learn that numbers like $\sqrt{-100}$, $\sqrt{-49}$, and $-\sqrt{-3}$ are imaginary numbers.

In Sections 9.1–9.6, we will assume that all radicands are nonnegative.

Irrational Numbers

In Section 1.4, we learned that numbers like $\sqrt{2}$ cannot be written as a ratio of two integers. These numbers are real but not rational. We call numbers like $\sqrt{2}$ *irrational*. The square root of any whole number that is not a perfect square is irrational.

> If a is a whole number that is a perfect square, then $\sqrt{a}$ is a rational number.
>
> If a is a whole number that is not a perfect square, then $\sqrt{a}$ is an irrational number.

EXAMPLE 4

Determine whether each of the following numbers is rational or irrational: **(a)** $\sqrt{3}$; **(b)** $\sqrt{25}$; **(c)** $\sqrt{35}$; **(d)** $-\sqrt{9}$.

Solution

a) $\sqrt{3}$ is irrational, since 3 is not a perfect square.
b) $\sqrt{25}$ is rational, since 25 is a perfect square: $\sqrt{25} = 5$.
c) $\sqrt{35}$ is irrational, since 35 is not a perfect square.
d) $-\sqrt{9}$ is rational, since 9 is a perfect square: $-\sqrt{9} = -3$. ❑

Approximating Square Roots

We often need to approximate square roots that are irrational. Such approximations can be found using a calculator with a square root key $\boxed{\sqrt{\ }}$. They can also be found using Table 1 at the back of the book.

EXAMPLE 5

Use your calculator or Table 1 to approximate $\sqrt{10}$. Round to three decimal places.

Solution Calculators vary in their methods of operation. In most cases, however, we simply enter the number and then press $\boxed{\sqrt{\ }}$:

$$\sqrt{10} \approx 3.162277660. \quad \text{Using a calculator with a 10-digit readout}$$

Recall that, as decimals, all irrational numbers are nonrepeating and nonending. Rounding to the third decimal place, we have $\sqrt{10} \approx 3.162$. ❑

Square Roots and Absolute Value

Note that $\sqrt{(-5)^2} = \sqrt{25} = 5$ and $\sqrt{5^2} = \sqrt{25} = 5$, so it appears that squaring a number and then taking the square root is the same as taking the absolute value of the number: $|-5| = 5$ and $|5| = 5$. This situation can be generalized as follows.

For any real number A,

$$\sqrt{A^2} = |A|.$$

(The principal square root of the square of A is the absolute value of A.)

EXAMPLE 6

Simplify $\sqrt{(3x)^2}$ given that x is any real number.

Solution If x is a negative number, then $3x$ is negative. Since the principal square root is always positive, to write $\sqrt{(3x)^2} = 3x$ would be incorrect. Instead, we write

$$\sqrt{(3x)^2} = |3x|.$$ ❑

Fortunately, in many uses of radicals, it can be assumed that radicands never represent the square of a negative number. When this assumption is made, the need for absolute-value symbols disappears.

For any nonnegative real number A,

$$\sqrt{A^2} = A.$$

(For $A \geqslant 0$, the principal square root of the square of A is A.)

EXAMPLE 7

Simplify each expression. Assume that all variables represent nonnegative numbers.

a) $\sqrt{(3x)^2}$ **b)** $\sqrt{a^2b^2}$

Solution

a) $\sqrt{(3x)^2} = 3x$ Since $3x$ is assumed to be nonnegative, $|3x| = 3x$.

b) $\sqrt{a^2b^2} = \sqrt{(ab)^2} = ab$ Since ab is assumed to be nonnegative, $|ab| = ab$. ❑

To reduce the need for absolute-value symbols, except for our study in Section 9.7, we will assume that all variables represent nonnegative numbers.

Problem Solving

Radical expressions often appear in applications.

EXAMPLE 8

Parking-lot arrival spaces. The attendants at a parking lot use spaces to leave cars before they are taken to long-term parking stalls. The number N of such spaces needed is approximated by the formula

$$N = 2.5\sqrt{A},$$

where A is the average number of arrivals in peak hours. Find the number of spaces needed when the average number of arrivals in peak hours is 77.

Solution We substitute 77 into the formula. We use a calculator or Table 1 to find an approximation:

$$N = 2.5\sqrt{77} \approx 2.5(8.775) \approx 21.938 \approx 22.$$

Note that we round up to 22 spaces because 21.938 spaces would give us part of a space, which we could not use. For an average of 77 arrivals, 22 spaces are needed. ❑

Calculator note. In most situations when using a calculator for a calculation like that in Example 8, we round at the end of our calculations. Thus, on a calculator, we might find

$$N = 2.5\sqrt{77} \approx 2.5(8.774964387) = 21.93741097.$$

Note that this gives a variance in the third decimal place. When using a calculator for approximation, be aware of possible variance in answers. You may get answers different than those given at the back of the text. Answers to the exercises have been found by rounding at the end of the calculations.

EXERCISE SET 9.1

Find the square roots.

1. 1 **2.** 4 **3.** 16
4. 9 **5.** 49 **6.** 121
7. 169 **8.** 144

Simplify.

9. $\sqrt{4}$ **10.** $\sqrt{1}$ **11.** $-\sqrt{9}$
12. $-\sqrt{25}$ **13.** $\sqrt{0}$ **14.** $-\sqrt{81}$
15. $-\sqrt{121}$ **16.** $\sqrt{400}$ **17.** $\sqrt{361}$
18. $\sqrt{441}$ **19.** $\sqrt{144}$ **20.** $\sqrt{169}$
21. $-\sqrt{625}$ **22.** $-\sqrt{900}$

Identify the radicand.

23. $\sqrt{a-4}$ **24.** $\sqrt{t+3}$
25. $5\sqrt{t^2+1}$ **26.** $8\sqrt{x^2+5}$
27. $x^2y\sqrt{\frac{3}{x+2}}$ **28.** $ab^2\sqrt{\frac{a}{a-b}}$

Identify the square root as either rational or irrational.

29. $\sqrt{7}$ **30.** $\sqrt{6}$ **31.** $\sqrt{8}$
32. $\sqrt{10}$ **33.** $\sqrt{49}$ **34.** $\sqrt{100}$
35. $\sqrt{98}$ **36.** $\sqrt{75}$ **37.** $-\sqrt{4}$
38. $-\sqrt{1}$ **39.** $-\sqrt{12}$ **40.** $-\sqrt{14}$

Use a calculator or Table 1 to approximate the square root. Round to three decimal places.

41. $\sqrt{5}$ **42.** $\sqrt{6}$ **43.** $\sqrt{17}$
44. $\sqrt{19}$ **45.** $\sqrt{93}$ **46.** $\sqrt{43}$

Simplify. Remember that we have assumed that all variables represent nonnegative numbers.

47. $\sqrt{t^2}$ **48.** $\sqrt{x^2}$ **49.** $\sqrt{9x^2}$
50. $\sqrt{4a^2}$ **51.** $\sqrt{(ab)^2}$ **52.** $\sqrt{(6y)^2}$
53. $\sqrt{(34d)^2}$ **54.** $\sqrt{(53b)^2}$ **55.** $\sqrt{(5ab)^2}$
56. $\sqrt{(7xy)^2}$

Problem Solving

Use the formula $N = 2.5\sqrt{A}$ of Example 8.

57. Find the number of spaces needed when the average number of arrivals is **(a)** 25; **(b)** 89.

58. Find the number of spaces needed when the average number of arrivals is **(a)** 62; **(b)** 100.

Skill Maintenance

59. Find the slope of the line containing the points $(-3, 4)$ and $(5, -6)$.

60. Find the slope of the line $-3x + 5y = 15$.

61. Find an equation of the line containing the point $(-3, 4)$ with slope 2.

62. Find an equation of the line containing the points $(-3, 4)$ and $(5, -6)$.

Synthesis

63. ◈ Explain in your own words why $\sqrt{A^2} \neq A$ when A is negative.

64. ◈ Which is the more exact way to write the square root of 12: 3.464101615 or $\sqrt{12}$? Why?

Simplify.

65. $\sqrt{\sqrt{16}}$ **66.** $\sqrt{3^2 + 4^2}$

67. Between what two consecutive integers is $-\sqrt{33}$?

68. Find a number that is the square of an integer and the cube of a different integer.

Solve.

69. $\sqrt{x^2} = 6$

70. $\sqrt{y^2} = -7$

71. $-\sqrt{x^2} = -3$

72. $t^2 = 49$

Simplify. Assume that all variables represent positive numbers.

73. $\sqrt{(5a^2b)^2}$

74. $(\sqrt{3a})^2$

75. $\sqrt{\dfrac{144x^8}{36y^6}}$

76. $\sqrt{\dfrac{y^{12}}{8100}}$

77. $\sqrt{\dfrac{169}{m^{16}}}$

78. $\sqrt{\dfrac{p^2}{3600}}$

79. Use the graph of $y = \sqrt{x}$, shown in the right-hand column, to find rational approximations for each of the following numbers: **(a)** $\sqrt{3}$; **(b)** $\sqrt{5}$; **(c)** $\sqrt{7}$. Answers may vary.

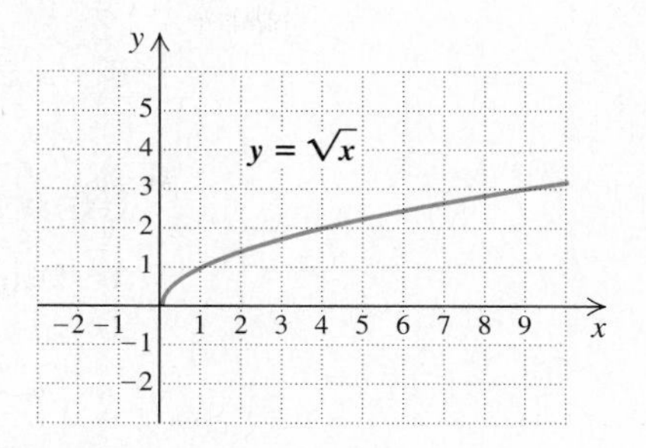

80. Use a grapher to draw the graphs of $y_1 = \sqrt{x-2}$, $y_2 = \sqrt{x+7}$, $y_3 = 5 + \sqrt{x}$, and $y_4 = -4 + \sqrt{x}$. If possible, graph all four equations using a $[-10, 10, -10, 10]$ window. Then determine which equation corresponds to each curve.

9.2 Multiplying and Simplifying Radical Expressions

Multiplying • Simplifying and Factoring • Simplifying Square Roots of Powers • Multiplying and Simplifying

We now learn to multiply and simplify radical expressions.

Multiplying

To see how to multiply with radical notation, consider the following:

$\sqrt{9} \cdot \sqrt{4} = 3 \cdot 2 = 6;$ This is a product of square roots.

$\sqrt{9 \cdot 4} = \sqrt{36} = 6.$ This is the square root of a product.

Note that $\sqrt{9} \cdot \sqrt{4} = \sqrt{9 \cdot 4}$.

The Product Rule for Radicals

For any real numbers $\sqrt{A}$ and $\sqrt{B}$,

$$\sqrt{A} \cdot \sqrt{B} = \sqrt{A \cdot B}.$$

(To multiply square roots, multiply the radicands and take the square root.)

EXAMPLE 1

Multiply: **(a)** $\sqrt{5}\,\sqrt{7}$; **(b)** $\sqrt{8}\,\sqrt{8}$; **(c)** $\sqrt{\frac{2}{3}}\,\sqrt{\frac{7}{5}}$; **(d)** $\sqrt{2x}\,\sqrt{3y}$.

Solution

a) $\sqrt{5}\,\sqrt{7} = \sqrt{5 \cdot 7} = \sqrt{35}$

b) $\sqrt{8}\sqrt{8} = \sqrt{8 \cdot 8} = \sqrt{64} = 8$ Try to do this directly: $\sqrt{8}\sqrt{8} = 8$.

c) $\sqrt{\frac{2}{3}}\sqrt{\frac{7}{5}} = \sqrt{\frac{2}{3} \cdot \frac{7}{5}} = \sqrt{\frac{14}{15}}$

d) $\sqrt{2x}\sqrt{3y} = \sqrt{6xy}$ ❑

Simplifying and Factoring

To factor radical expressions, we can use the product rule for radicals in reverse. That is,

$$\sqrt{AB} = \sqrt{A}\sqrt{B}.$$

In some cases, we can simplify after factoring.

> A radical expression is simplified when its radicand has no factors other than 1 that are perfect squares.

When a radicand is a perfect square, we simplify by writing the square root. For example, $\sqrt{36} = 6$. When the radicand is not a perfect square but has a perfect-square factor, we use the preceding rule. For example, $\sqrt{48} = \sqrt{16}\sqrt{3} = 4\sqrt{3}$.

Compare the following:

$$\sqrt{50} = \sqrt{10 \cdot 5} = \sqrt{10}\sqrt{5};$$

$$\sqrt{50} = \sqrt{25 \cdot 2} = \sqrt{25}\sqrt{2} = 5\sqrt{2}.$$

The second factorization is more useful because of the perfect-square factor 25. If you do not recognize any perfect-square factors, try factoring the radicand into prime factors. For example,

$\sqrt{50} = \sqrt{2 \cdot 5 \cdot 5}$ Factoring into prime factors

$= \sqrt{2} \cdot \sqrt{5 \cdot 5}$ $5 \cdot 5$ is a perfect square.

$= \sqrt{2} \cdot 5.$

To avoid any uncertainty as to what is under the radical sign, it is customary to write the radical factor last. Thus, $\sqrt{50} = 5\sqrt{2}$.

A radical expression, like $\sqrt{26}$, in which the radicand has no perfect-square factors, is considered to be in simplest form.

EXAMPLE 2

Simplify by factoring (remember that all variables are assumed to represent nonnegative numbers): **(a)** $\sqrt{18}$; **(b)** $\sqrt{48t}$; **(c)** $\sqrt{a^2b}$; **(d)** $\sqrt{20t^2}$.

Solution

a) $\sqrt{18} = \sqrt{9 \cdot 2}$ Identifying a perfect-square factor and factoring the radicand. The factor 9 is a perfect square.

$= \sqrt{9}\sqrt{2}$ Factoring into a product of radicals

$= 3\sqrt{2}$ The radicand has no factors that are perfect squares.

b) $\sqrt{48t} = \sqrt{16 \cdot 3t}$ Identifying a perfect-square factor and factoring the radicand. The factor 16 is a perfect square.

$= \sqrt{16}\sqrt{3t}$ Factoring into a product of radicals

$= 4\sqrt{3t}$

c) $\sqrt{a^2b} = \sqrt{a^2}\sqrt{b}$ Identifying a perfect-square factor and factoring into a product of radicals

$= a\sqrt{b}$ No absolute-value signs are necessary since a is assumed to be nonnegative.

d) $\sqrt{20t^2} = \sqrt{4 \cdot t^2 \cdot 5}$ Identifying perfect-square factors and factoring the radicand. The factors 4 and t^2 are perfect squares.

$= \sqrt{4}\sqrt{t^2}\sqrt{5}$ Factoring into a product of radicals

$= 2t\sqrt{5}$ Taking square roots. No absolute-value signs are necessary since t is assumed to be nonnegative. ❑

Simplifying Square Roots of Powers

To take the square root of an even power such as x^{10}, note that $x^{10} = (x^5)^2$. Then

$$\sqrt{x^{10}} = \sqrt{(x^5)^2} = x^5.$$

The square root can be found by taking half the exponent. That is,

$$\sqrt{x^{10}} = x^5. \longleftarrow \tfrac{1}{2}(10) = 5$$

EXAMPLE 3

Simplify: **(a)** $\sqrt{x^6}$; **(b)** $\sqrt{x^8}$; **(c)** $\sqrt{t^{22}}$.

Solution

a) $\sqrt{x^6} = \sqrt{(x^3)^2} = x^3$ Half of 6 is 3.

b) $\sqrt{x^8} = x^4$ *Check:* $(x^4)(x^4) = x^8$.

c) $\sqrt{t^{22}} = t^{11}$ ❑

If a radicand is an odd power, we can simplify by factoring, as in the following example.

EXAMPLE 4

Simplify: **(a)** $\sqrt{x^9}$; **(b)** $\sqrt{32x^{15}}$.

Solution

a) $\sqrt{x^9} = \sqrt{x^8 \cdot x}$ x^8 is the largest perfect-square factor of x^9.

$= \sqrt{x^8}\sqrt{x}$

$= x^4\sqrt{x}$

This illustrates that the square root of x^9 *is not* x^3.

b) $\sqrt{32x^{15}} = \sqrt{16x^{14} \cdot 2x}$ 16 is the largest perfect-square factor of 32; x^{14} is the largest perfect-square factor of x^{15}.

$= \sqrt{16}\sqrt{x^{14}}\sqrt{2x}$

$= 4x^7\sqrt{2x}$ Simplifying ❑

Multiplying and Simplifying

Sometimes we can simplify after multiplying. To do so, we again try to locate any perfect-square factors of the radicand.

EXAMPLE 5

Multiply and then simplify by factoring: **(a)** $\sqrt{2}\sqrt{14}$; **(b)** $\sqrt{3x^4}\sqrt{9x^3}$.

Solution

a) $\sqrt{2}\sqrt{14} = \sqrt{28}$ — Multiplying. Note that 4 is a perfect-square factor of 28.

$= \sqrt{4 \cdot 7}$ — Factoring

$= \sqrt{4}\sqrt{7}$

$= 2\sqrt{7}$ — Simplifying

b) $\sqrt{3x^4}\sqrt{9x^3} = \sqrt{27x^7}$ — The perfect-square factors are 9 and x^6.

$= \sqrt{9x^6 \cdot 3x}$ — Factoring

$= \sqrt{9}\sqrt{x^6}\sqrt{3x}$ — You might perform this step mentally.

$= 3x^3\sqrt{3x}$ — Simplifying ❑

EXERCISE SET 9.2

Multiply.

1. $\sqrt{2}\sqrt{3}$
2. $\sqrt{3}\sqrt{5}$
3. $\sqrt{4}\sqrt{3}$
4. $\sqrt{2}\sqrt{9}$
5. $\sqrt{\frac{2}{5}}\sqrt{\frac{3}{4}}$
6. $\sqrt{\frac{3}{8}}\sqrt{\frac{1}{5}}$
7. $\sqrt{17}\sqrt{17}$
8. $\sqrt{18}\sqrt{18}$
9. $\sqrt{25}\sqrt{3}$
10. $\sqrt{36}\sqrt{2}$
11. $\sqrt{2}\sqrt{x}$
12. $\sqrt{3}\sqrt{a}$
13. $\sqrt{3}\sqrt{2x}$
14. $\sqrt{5}\sqrt{4x}$
15. $\sqrt{x}\sqrt{7y}$
16. $\sqrt{5m}\sqrt{2n}$
17. $\sqrt{3a}\sqrt{2c}$
18. $\sqrt{3x}\sqrt{yz}$

Simplify by factoring.

19. $\sqrt{12}$
20. $\sqrt{8}$
21. $\sqrt{20}$
22. $\sqrt{45}$
23. $\sqrt{200}$
24. $\sqrt{300}$
25. $\sqrt{9x}$
26. $\sqrt{4y}$
27. $\sqrt{75a}$
28. $\sqrt{40m}$
29. $\sqrt{16a}$
30. $\sqrt{49b}$
31. $\sqrt{64y^2}$
32. $\sqrt{9x^2}$
33. $\sqrt{13x^2}$
34. $\sqrt{29t^2}$
35. $\sqrt{8t^2}$
36. $\sqrt{125a^2}$
37. $\sqrt{180}$
38. $\sqrt{98}$
39. $\sqrt{288y}$
40. $\sqrt{363p}$
41. $\sqrt{x^{20}}$
42. $\sqrt{x^{30}}$
43. $\sqrt{x^{12}}$
44. $\sqrt{x^{16}}$
45. $\sqrt{x^5}$
46. $\sqrt{x^3}$
47. $\sqrt{t^{19}}$
48. $\sqrt{p^{17}}$
49. $\sqrt{36m^3}$
50. $\sqrt{250y^3}$
51. $\sqrt{8a^5}$
52. $\sqrt{12b^7}$
53. $\sqrt{104p^{17}}$
54. $\sqrt{90m^{23}}$

Multiply and, if possible, simplify.

55. $\sqrt{3}\sqrt{6}$
56. $\sqrt{5}\sqrt{10}$
57. $\sqrt{15}\sqrt{6}$
58. $\sqrt{3}\sqrt{27}$
59. $\sqrt{3x}\sqrt{12y}$
60. $\sqrt{5x}\sqrt{20y}$
61. $\sqrt{10}\sqrt{10}$
62. $\sqrt{11}\sqrt{11x}$
63. $\sqrt{5b}\sqrt{15b}$
64. $\sqrt{6a}\sqrt{18a}$
65. $\sqrt{2t}\sqrt{2t}$
66. $\sqrt{3a}\sqrt{3a}$
67. $\sqrt{ab}\sqrt{ac}$
68. $\sqrt{xy}\sqrt{xz}$
69. $\sqrt{2x}\sqrt{4x^5}$
70. $\sqrt{15m^6}\sqrt{5m^2}$
71. $\sqrt{x^2y^3}\sqrt{xy^4}$
72. $\sqrt{x^3y^2}\sqrt{xy}$
73. $\sqrt{50ab}\sqrt{10a^2b^4}$
74. $\sqrt{10xy^2}\sqrt{5x^2y^3}$

Problem Solving

Speed of a skidding car. How do police determine the speed of a car after an accident? The formula

$$r = 2\sqrt{5L}$$

can be used to approximate the speed r, in miles per hour, of a car that has left a skid mark of length L, in feet.

75. What was the speed of a car that left skid marks of 20 ft? of 150 ft?

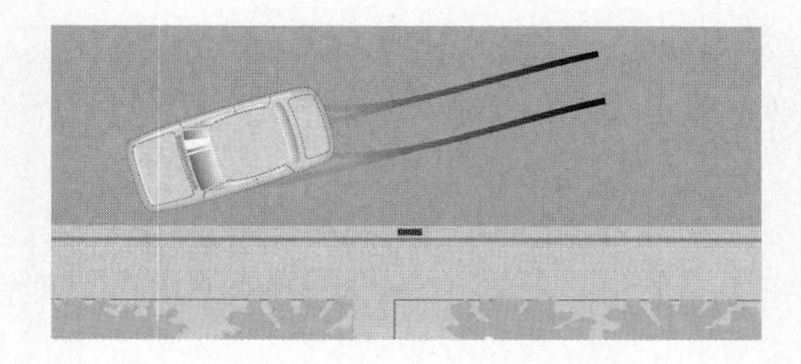

76. What was the speed of a car that left skid marks of 30 ft? of 70 ft?

Skill Maintenance

77. A car leaves Hereford traveling north at a speed of 56 km/h. Another car leaves Hereford one hour later, traveling north at 84 km/h. How far from Hereford will the second car overtake the first?

78. An airplane flew for 5 hr with a 25-km/h tail wind. The return flight against the same wind took 6 hr. Find the speed of the airplane in still air.

Synthesis

79. ◈ Explain why $\sqrt{16x^4} = 4x^2$, but $\sqrt{4x^{16}} \neq 2x^4$.

80. ◈ What is wrong with the following?
$\sqrt{x^2 - 25} = \sqrt{x^2} - \sqrt{25} = x - 5$

Simplify.

81. $\sqrt{0.01}$

82. $\sqrt{0.25}$

83. $\sqrt{0.0625}$

84. $\sqrt{0.000001}$

85. ◈ Simplify $\sqrt{49}$, $\sqrt{490}$, $\sqrt{4900}$, $\sqrt{49{,}000}$, and $\sqrt{490{,}000}$; describe the pattern you see.

Use the proper symbol (>, <, or =) between each pair of values to make a true sentence. Do not use a calculator.

86. $15 \quad 4\sqrt{14}$

87. $15\sqrt{2} \quad \sqrt{450}$

88. $16 \quad \sqrt{15}\sqrt{17}$

89. $3\sqrt{11} \quad 7\sqrt{2}$

90. $5\sqrt{7} \quad 4\sqrt{11}$

91. $8 \quad \sqrt{15} + \sqrt{17}$

Multiply and then simplify by factoring.

92. $\sqrt{27(x+1)}\sqrt{12y(x+1)^2}$

93. $\sqrt{18(x-2)}\sqrt{20(x-2)^3}$

94. $\sqrt{x}\sqrt{2x}\sqrt{10x^5}$

95. $\sqrt{2^{109}}\sqrt{x^{306}}\sqrt{x^{11}}$

Simplify.

96. $\sqrt{x^{8n}}$

97. $\sqrt{0.04x^{4n}}$

98. Simplify $\sqrt{y^n}$, where n is an odd whole number greater than or equal to 3.

9.3 Quotients Involving Square Roots

Dividing Radical Expressions • Rationalizing Denominators • Simplified Form

In this section, we divide radical expressions and simplify quotients containing radicals.

Dividing Radical Expressions

To see how to divide with radical notation, consider the following.

$$\frac{\sqrt{25}}{\sqrt{16}} = \frac{5}{4} \text{ since } \sqrt{25} = 5 \text{ and } \sqrt{16} = 4;$$

$$\sqrt{\frac{25}{16}} = \frac{5}{4} \text{ because } \frac{5}{4} \cdot \frac{5}{4} = \frac{25}{16}.$$

We see that $\dfrac{\sqrt{25}}{\sqrt{16}} = \sqrt{\dfrac{25}{16}}$.

The Quotient Rule for Radicals

For any real numbers $\sqrt{A}$ and $\sqrt{B}$, $B \neq 0$,

$$\frac{\sqrt{A}}{\sqrt{B}} = \sqrt{\frac{A}{B}}.$$

(To divide two square roots, divide the radicands and take the square root.)

EXAMPLE 1

Divide and simplify: **(a)** $\dfrac{\sqrt{27}}{\sqrt{3}}$; **(b)** $\dfrac{\sqrt{8a^7}}{\sqrt{2a}}$.

Solution

a) $\dfrac{\sqrt{27}}{\sqrt{3}} = \sqrt{\dfrac{27}{3}}$

$= \sqrt{9} = 3$

b) $\dfrac{\sqrt{8a^7}}{\sqrt{2a}} = \sqrt{\dfrac{8a^7}{2a}}$

$= \sqrt{4a^6} = 2a^3$ ❑

The quotient rule for radicals can also be read from right to left:

$$\sqrt{\frac{A}{B}} = \frac{\sqrt{A}}{\sqrt{B}}.$$

EXAMPLE 2

Simplify by taking the square roots of the numerator and the denominator separately.

a) $\sqrt{\dfrac{25}{9}}$ **b)** $\sqrt{\dfrac{1}{16}}$ **c)** $\sqrt{\dfrac{49}{t^2}}$

Solution

a) $\sqrt{\dfrac{25}{9}} = \dfrac{\sqrt{25}}{\sqrt{9}} = \dfrac{5}{3}$ Taking the square root of the numerator and the square root of the denominator

b) $\sqrt{\dfrac{1}{16}} = \dfrac{\sqrt{1}}{\sqrt{16}} = \dfrac{1}{4}$ Taking the square root of the numerator and the square root of the denominator

c) $\sqrt{\dfrac{49}{t^2}} = \dfrac{\sqrt{49}}{\sqrt{t^2}} = \dfrac{7}{t}$ ❑

Sometimes a rational expression can be simplified to one that has a perfect-square numerator and a perfect-square denominator.

EXAMPLE 3

Simplify: **(a)** $\sqrt{\dfrac{18}{50}}$; **(b)** $\sqrt{\dfrac{48x^3}{3x^7}}$.

Solution

a) $\sqrt{\frac{18}{50}} = \sqrt{\frac{9 \cdot 2}{25 \cdot 2}}$

$= \sqrt{\frac{9 \cdot \cancel{2}}{25 \cdot \cancel{2}}}$ Removing a factor of 1: $\frac{2}{2} = 1$

$= \frac{\sqrt{9}}{\sqrt{25}} = \frac{3}{5}$

b) $\sqrt{\frac{48x^3}{3x^7}} = \sqrt{\frac{16 \cdot \cancel{3x^3}}{x^4 \cdot \cancel{3x^3}}}$ Removing a factor of 1: $\frac{3x^3}{3x^3} = 1$

$= \frac{\sqrt{16}}{\sqrt{x^4}} = \frac{4}{x^2}$ ❑

Rationalizing Denominators

A procedure for finding an equivalent expression without a radical in the denominator is sometimes useful. This makes long division involving decimal approximations easier to perform. The procedure is called **rationalizing the denominator** and involves multiplying by 1.

EXAMPLE 4

Rationalize the denominator: **(a)** $\frac{8}{\sqrt{7}}$; **(b)** $\sqrt{\frac{2}{3}}$; **(c)** $\sqrt{\frac{5}{x}}$.

Solution

a) We multiply by 1, choosing $\sqrt{7}/\sqrt{7}$ for 1:

$\frac{8}{\sqrt{7}} = \frac{8}{\sqrt{7}} \cdot \frac{\sqrt{7}}{\sqrt{7}}$ Multiplying by 1

$= \frac{8\sqrt{7}}{7}$. $\sqrt{7} \cdot \sqrt{7} = 7$

b) $\sqrt{\frac{2}{3}} = \frac{\sqrt{2}}{\sqrt{3}}$ The square root of a quotient is the quotient of the square roots.

$= \frac{\sqrt{2}}{\sqrt{3}} \cdot \frac{\sqrt{3}}{\sqrt{3}}$ Multiplying by 1

$= \frac{\sqrt{6}}{3}$ $\sqrt{3} \cdot \sqrt{3} = 3$

c) $\sqrt{\frac{5}{x}} = \frac{\sqrt{5}}{\sqrt{x}}$ The square root of a quotient is the quotient of the square roots.

$= \frac{\sqrt{5}}{\sqrt{x}} \cdot \frac{\sqrt{x}}{\sqrt{x}}$ Multiplying by 1

$= \frac{\sqrt{5x}}{x}$ $\sqrt{x} \cdot \sqrt{x} = x$ ❑

It is usually easiest to rationalize a denominator when the expression has already been simplified.

EXAMPLE 5

Rationalize the denominator: **(a)** $\sqrt{\frac{5}{18}}$; **(b)** $\frac{\sqrt{7a}}{\sqrt{20}}$; **(c)** $\sqrt{\frac{x^2}{3}}$.

Solution

a) $\sqrt{\frac{5}{18}} = \frac{\sqrt{5}}{\sqrt{18}}$ — The square root of a quotient is the quotient of the square roots.

$= \frac{\sqrt{5}}{\sqrt{9}\sqrt{2}}$

$= \frac{\sqrt{5}}{3\sqrt{2}}$ — Simplifying the denominator. Note that 9 is a perfect square.

$= \frac{\sqrt{5}}{3\sqrt{2}} \cdot \frac{\sqrt{2}}{\sqrt{2}}$ — Multiplying by 1

$= \frac{\sqrt{10}}{3 \cdot 2} = \frac{\sqrt{10}}{6}$

b) $\frac{\sqrt{7a}}{\sqrt{20}} = \frac{\sqrt{7a}}{\sqrt{4}\sqrt{5}}$

$= \frac{\sqrt{7a}}{2\sqrt{5}}$ — Simplifying the denominator. Note that 4 is a perfect square.

$= \frac{\sqrt{7a}}{2\sqrt{5}} \cdot \frac{\sqrt{5}}{\sqrt{5}}$ — Multiplying by 1

$= \frac{\sqrt{35a}}{2 \cdot 5}$

$= \frac{\sqrt{35a}}{10}$

c) $\sqrt{\frac{x^2}{3}} = \frac{\sqrt{x^2}}{\sqrt{3}}$

$= \frac{x}{\sqrt{3}}$ — Simplifying the numerator. Remember that we assume $x \geqslant 0$.

$= \frac{x}{\sqrt{3}} \cdot \frac{\sqrt{3}}{\sqrt{3}}$ — Multiplying by 1

$= \frac{x\sqrt{3}}{3}$ ❑

CAUTION! Our solutions in Example 5 cannot be simplified any further. A common mistake is to remove a factor of 1 that does not exist. For example $\sqrt{10}/6$ *cannot* be simplified to $\sqrt{5}/3$ since $\sqrt{10}$ and 6 do not share a common factor.

Simplified Form

The following guidelines summarize various ways in which radical expressions can be simplified.

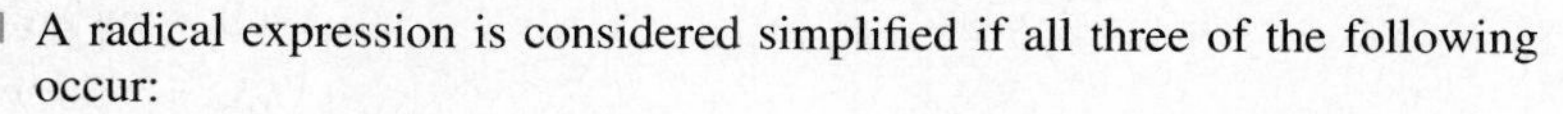

A radical expression is considered simplified if all three of the following occur:

1. No radicand contains any perfect-square factors (other than 1).
2. The radicand does not have a denominator.
3. No denominator contains a radical.

EXERCISE SET 9.3

Divide and simplify.

1. $\frac{\sqrt{18}}{\sqrt{2}}$
2. $\frac{\sqrt{20}}{\sqrt{5}}$
3. $\frac{\sqrt{60}}{\sqrt{15}}$
4. $\frac{\sqrt{108}}{\sqrt{3}}$
5. $\frac{\sqrt{75}}{\sqrt{15}}$
6. $\frac{\sqrt{18}}{\sqrt{3}}$
7. $\frac{\sqrt{3}}{\sqrt{75}}$
8. $\frac{\sqrt{3}}{\sqrt{48}}$
9. $\frac{\sqrt{12}}{\sqrt{75}}$
10. $\frac{\sqrt{18}}{\sqrt{32}}$
11. $\frac{\sqrt{8x}}{\sqrt{2x}}$
12. $\frac{\sqrt{18b}}{\sqrt{2b}}$
13. $\frac{\sqrt{63y^3}}{\sqrt{7y}}$
14. $\frac{\sqrt{48x^3}}{\sqrt{3x}}$
15. $\frac{\sqrt{27x^5}}{\sqrt{3x}}$
16. $\frac{\sqrt{20a^8}}{\sqrt{5a^2}}$
17. $\frac{\sqrt{75x}}{\sqrt{3x^7}}$
18. $\frac{\sqrt{6x^5}}{\sqrt{54x}}$
19. $\frac{\sqrt{20a^{10}}}{\sqrt{10a^2}}$
20. $\frac{\sqrt{35x^{11}}}{\sqrt{7x}}$

Simplify.

21. $\sqrt{\frac{9}{49}}$
22. $\sqrt{\frac{16}{25}}$
23. $\sqrt{\frac{1}{36}}$
24. $\sqrt{\frac{1}{4}}$
25. $-\sqrt{\frac{16}{81}}$
26. $-\sqrt{\frac{25}{49}}$
27. $\sqrt{\frac{64}{144}}$
28. $\sqrt{\frac{81}{121}}$
29. $\sqrt{\frac{1690}{1210}}$
30. $\sqrt{\frac{1440}{6250}}$
31. $\sqrt{\frac{36}{a^2}}$
32. $\sqrt{\frac{25}{x^2}}$
33. $\sqrt{\frac{9a^2}{625}}$
34. $\sqrt{\frac{x^2y^2}{144}}$

Rationalize the denominator.

35. $\sqrt{\frac{2}{5}}$
36. $\sqrt{\frac{2}{7}}$
37. $\sqrt{\frac{3}{8}}$
38. $\sqrt{\frac{7}{8}}$
39. $\sqrt{\frac{7}{20}}$
40. $\sqrt{\frac{1}{12}}$
41. $\sqrt{\frac{1}{18}}$
42. $\sqrt{\frac{7}{18}}$
43. $\frac{3}{\sqrt{5}}$
44. $\frac{4}{\sqrt{3}}$
45. $\sqrt{\frac{8}{3}}$
46. $\sqrt{\frac{12}{5}}$
47. $\sqrt{\frac{3}{x}}$
48. $\sqrt{\frac{2}{x}}$
49. $\sqrt{\frac{x}{y}}$
50. $\sqrt{\frac{a}{b}}$
51. $\frac{\sqrt{7}}{\sqrt{3}}$
52. $\frac{\sqrt{11}}{\sqrt{7}}$
53. $\frac{\sqrt{9}}{\sqrt{8}}$
54. $\frac{\sqrt{4}}{\sqrt{27}}$
55. $\frac{\sqrt{2}}{\sqrt{13}}$
56. $\frac{\sqrt{3}}{\sqrt{2}}$
57. $\frac{2}{\sqrt{2}}$
58. $\frac{3}{\sqrt{3}}$
59. $\frac{\sqrt{5}}{\sqrt{27}}$
60. $\frac{\sqrt{7}}{\sqrt{11}}$
61. $\frac{\sqrt{7}}{\sqrt{12}}$
62. $\frac{\sqrt{5}}{\sqrt{18}}$
63. $\frac{\sqrt{x}}{\sqrt{32}}$
64. $\frac{\sqrt{a}}{\sqrt{40}}$

65. $\frac{\sqrt{8}}{\sqrt{18}}$

66. $\frac{\sqrt{3}}{\sqrt{14}}$

67. $\frac{\sqrt{3}}{\sqrt{x}}$

68. $\frac{\sqrt{2}}{\sqrt{y}}$

69. $\frac{4y}{\sqrt{3}}$

70. $\frac{8x}{\sqrt{5}}$

71. $\frac{\sqrt{6a}}{\sqrt{8}}$

72. $\frac{\sqrt{3x}}{\sqrt{27}}$

73. $\frac{\sqrt{50}}{\sqrt{12x}}$

74. $\frac{\sqrt{45}}{\sqrt{8a}}$

75. $\frac{\sqrt{27c}}{\sqrt{32c^3}}$

76. $\frac{\sqrt{7x^3}}{\sqrt{12x}}$

Problem Solving

The period T of a pendulum is the time it takes to move from one side to the other and back. A formula for the period is

$$T = 2\pi\sqrt{\frac{L}{32}},$$

where T is in seconds and L is in feet. Use 3.14 for π.

77. Find the periods of pendulums of lengths 2 ft, 8 ft, 64 ft, and 100 ft.

78. Find the period of a pendulum of length $\frac{2}{3}$ in.

79. The pendulum of a grandfather clock is $32/\pi^2$ ft long. How long does it take to swing from one side to the other?

80. The pendulum of a grandfather clock is $45/\pi^2$ ft long. How long does it take to swing from one side to the other?

Skill Maintenance

Solve.

81. $x = y + 2,$
$x + y = 6$

82. $2x - 3y = 7,$
$2x + 3y = 9$

Find equations in which y varies directly as x and the following are true.

83. $y = 30$ when $x = 9$

84. $y = 3$ when $x = 12$

Synthesis

85. ◈ Explain why it is easier to approximate $\sqrt{2}/2$ by long division than it is to approximate $1/\sqrt{2}$.

86. ◈ Why is it important to know how to multiply radical expressions before learning how to divide them?

Rationalize the denominator.

87. $\sqrt{\frac{5}{1600}}$

88. $\sqrt{\frac{3}{1000}}$

89. $\sqrt{\frac{1}{5x^3}}$

90. $\sqrt{\frac{3x^2y}{a^2x^5}}$

91. $\sqrt{\frac{3a}{b}}$

92. $\sqrt{\frac{1}{5zw^2}}$

Simplify.

93. $\sqrt{\frac{1}{x^2} - \frac{2}{xy} + \frac{1}{y^2}}$

94. $\sqrt{2 - \frac{4}{z^2} + \frac{2}{z^4}}$

9.4 More Operations with Radicals

Adding and Subtracting Radical Expressions • More with Multiplication • More with Rationalizing Denominators

We now consider addition and subtraction of radical expressions as well as some new types of multiplication and simplification.

Adding and Subtracting Radical Expressions

The sum of a rational number and an irrational number, like $5 + \sqrt{2}$, *cannot* be simplified. However, the sum of **like radicals**—that is, radical expressions that have a common radical factor—*can* be simplified.

EXAMPLE 1

Add: $3\sqrt{5} + 4\sqrt{5}$.

Solution Recall that to simplify an expression like $3x + 4x$, we use the distributive law, as follows:

$$3x + 4x = (3 + 4)x$$
$$= 7x.$$

In this example, x is replaced by $\sqrt{5}$:

$$3\sqrt{5} + 4\sqrt{5} = (3 + 4)\sqrt{5}$$ Using the distributive law to factor out $\sqrt{5}$

$$= 7\sqrt{5}.$$ $3\sqrt{5}$ and $4\sqrt{5}$ are like radicals. ❑

To simplify in this manner, the radical factors must be the same.

EXAMPLE 2

Simplify.

a) $9\sqrt{17} - 3\sqrt{17}$

b) $7\sqrt{x} + \sqrt{x}$

c) $5\sqrt{2} - \sqrt{18}$

d) $3\sqrt{5} + \sqrt{7}$

Solution

a) $9\sqrt{17} - 3\sqrt{17} = (9 - 3)\sqrt{17}$ Using the distributive law. Try to do this mentally.

$$= 6\sqrt{17}$$

b) $7\sqrt{x} + \sqrt{x} = (7 + 1)\sqrt{x}$ Using the distributive law. Try to do this mentally.

$$= 8\sqrt{x}$$

c) $5\sqrt{2} - \sqrt{18} = 5\sqrt{2} - \sqrt{9 \cdot 2}$
$= 5\sqrt{2} - \sqrt{9}\sqrt{2}$ } Simplifying $\sqrt{18}$
$= 5\sqrt{2} - 3\sqrt{2}$ We now have like radicals.
$= 2\sqrt{2}$ Using the distributive law mentally: $5\sqrt{2} - 3\sqrt{2} = (5 - 3)\sqrt{2}$

d) $3\sqrt{5} + \sqrt{7}$ cannot be simplified. ❑

CAUTION! It is *not true* that the sum of two square roots is the square root of the sum: $\sqrt{A} + \sqrt{B} \neq \sqrt{A + B}$. For example, $\sqrt{9} + \sqrt{16} \neq \sqrt{9 + 16}$.

More with Multiplication

Radical expressions with more than one term are multiplied in much the same way that polynomials with more than one term are multiplied.

EXAMPLE 3

Multiply.

a) $\sqrt{2}(\sqrt{3} + \sqrt{5})$

b) $(4 + \sqrt{7})(2 + \sqrt{7})$

c) $(2 - \sqrt{5})(2 + \sqrt{5})$

d) $(2 + \sqrt{3})(5 - 4\sqrt{3})$

e) $(3 - \sqrt{p})^2$

Solution

a) $\sqrt{2}(\sqrt{3}+\sqrt{5}) = \sqrt{2}\sqrt{3}+\sqrt{2}\sqrt{5}$ — Using the distributive law

$= \sqrt{6}+\sqrt{10}$ — Using the product rule for radicals

b) $(4+\sqrt{7})(2+\sqrt{7}) = 4\cdot 2 + 4\cdot\sqrt{7} + \sqrt{7}\cdot 2 + \sqrt{7}\cdot\sqrt{7}$ — Multiplying two binomials

$= 8 + 4\sqrt{7} + 2\sqrt{7} + 7$

$= 15 + 6\sqrt{7}$ — Collecting like terms

c) $(2-\sqrt{5})(2+\sqrt{5}) = 2^2 - (\sqrt{5})^2$ — Using $(A-B)(A+B) = A^2 - B^2$

$= 4 - 5$

$= -1$

d) $(2+\sqrt{3})(5-4\sqrt{3}) = 2\cdot 5 - 2\cdot 4\sqrt{3} + \sqrt{3}\cdot 5 - \sqrt{3}\cdot 4\sqrt{3}$ — Using FOIL

$= 10 - 8\sqrt{3} + 5\sqrt{3} - 4\cdot 3$ — $2\cdot 5 = 10$, $2\cdot 4 = 8$, and $\sqrt{3}\cdot\sqrt{3} = 3$

$= 10 - 3\sqrt{3} - 12$ — Adding like radicals

$= -2 - 3\sqrt{3}$

e) $(3-\sqrt{p})^2 = 3^2 - 2\cdot 3\cdot\sqrt{p} + (\sqrt{p})^2$ — Using $(A-B)^2 = A^2 - 2AB + B^2$

$= 9 - 6\sqrt{p} + p$ ❑

More with Rationalizing Denominators

Note in Example 3(c) that the result has no radicals. This will happen whenever we multiply expressions such as $\sqrt{a}-\sqrt{b}$ and $\sqrt{a}+\sqrt{b}$, where a and b are rational numbers:

$$(\sqrt{a}+\sqrt{b})(\sqrt{a}-\sqrt{b}) = (\sqrt{a})^2 - (\sqrt{b})^2 = a - b.$$

Expressions such as $\sqrt{3}-\sqrt{5}$ and $\sqrt{3}+\sqrt{5}$ are said to be **conjugates** of each other. So too are expressions like $2+\sqrt{7}$ and $2-\sqrt{7}$. Conjugates are used to rationalize certain denominators.

EXAMPLE 4

Rationalize the denominator: **(a)** $\dfrac{3}{2+\sqrt{5}}$; **(b)** $\dfrac{\sqrt{3}}{\sqrt{3}-\sqrt{7}}$.

Solution

a) We multiply by 1 using the conjugate of $2+\sqrt{5}$, which is $2-\sqrt{5}$, as the numerator and the denominator:

$$\frac{3}{2+\sqrt{5}} = \frac{3}{2+\sqrt{5}}\cdot\frac{2-\sqrt{5}}{2-\sqrt{5}}$$ — Multiplying by 1

$$= \frac{3(2-\sqrt{5})}{(2+\sqrt{5})(2-\sqrt{5})}$$

$$= \frac{6-3\sqrt{5}}{2^2-(\sqrt{5})^2}.$$ — Using $(A+B)(A-B) = A^2 - B^2$

Then

$$\frac{3}{2+\sqrt{5}} = \frac{6-3\sqrt{5}}{4-5} \qquad \text{The denominator is free of radicals.}$$

$$= \frac{6-3\sqrt{5}}{-1}$$

$$= -6+3\sqrt{5}. \qquad \text{Dividing } \textit{both} \text{ terms in the numerator by } -1$$

b) $$\frac{\sqrt{3}}{\sqrt{3}-\sqrt{7}} = \frac{\sqrt{3}}{\sqrt{3}-\sqrt{7}} \cdot \frac{\sqrt{3}+\sqrt{7}}{\sqrt{3}+\sqrt{7}} \qquad \text{Multiplying by 1}$$

$$= \frac{\sqrt{3}(\sqrt{3}+\sqrt{7})}{(\sqrt{3}-\sqrt{7})(\sqrt{3}+\sqrt{7})}$$

$$= \frac{\sqrt{3}\,\sqrt{3}+\sqrt{3}\,\sqrt{7}}{(\sqrt{3})^2-(\sqrt{7})^2} \qquad \text{Using } (A-B)(A+B) = A^2-B^2$$

$$= \frac{3+\sqrt{21}}{3-7} \qquad \text{The denominator is free of radicals.}$$

$$= \frac{3+\sqrt{21}}{-4} = -\frac{3+\sqrt{21}}{4} \qquad \text{Rewriting with a positive denominator}$$

❑

When a result can be simplified by removing a factor of 1, we should do so.

EXAMPLE 5

Rationalize the denominator: $\dfrac{\sqrt{7}-\sqrt{5}}{\sqrt{7}+\sqrt{5}}$.

Solution We multiply by 1, using the conjugate of $\sqrt{7}+\sqrt{5}$, which is $\sqrt{7}-\sqrt{5}$, as the numerator and the denominator:

$$\frac{\sqrt{7}-\sqrt{5}}{\sqrt{7}+\sqrt{5}} = \frac{\sqrt{7}-\sqrt{5}}{\sqrt{7}+\sqrt{5}} \cdot \frac{\sqrt{7}-\sqrt{5}}{\sqrt{7}-\sqrt{5}} \qquad \text{Multiplying by 1}$$

$$= \frac{(\sqrt{7}-\sqrt{5})^2}{(\sqrt{7}+\sqrt{5})(\sqrt{7}-\sqrt{5})}$$

$$= \frac{(\sqrt{7})^2-2\sqrt{7}\,\sqrt{5}+(\sqrt{5})^2}{(\sqrt{7})^2-(\sqrt{5})^2} \qquad \begin{array}{l}\leftarrow \text{Using } (A-B)^2 = A^2-2AB+B^2 \\ \leftarrow \text{Using } (A+B)(A-B) = A^2-B^2\end{array}$$

$$= \frac{7-2\sqrt{35}+5}{7-5}$$

$$= \frac{12-2\sqrt{35}}{2} \qquad \text{This result can be simplified since 2 is a common factor of all terms.}$$

$$= \frac{\cancel{2}(6-\sqrt{35})}{\cancel{2}\cdot 1} \qquad \text{Factoring and removing a factor of 1: } \tfrac{2}{2}=1$$

$$= 6-\sqrt{35}.$$

❑

EXERCISE SET 9.4

Add or subtract. Simplify by collecting like radical terms, if possible.

1. $3\sqrt{2} + 4\sqrt{2}$
2. $8\sqrt{3} + 3\sqrt{3}$
3. $7\sqrt{5} - 3\sqrt{5}$
4. $8\sqrt{2} - 5\sqrt{2}$
5. $6\sqrt{x} + 7\sqrt{x}$
6. $9\sqrt{y} + 3\sqrt{y}$
7. $9\sqrt{x} - 11\sqrt{x}$
8. $6\sqrt{a} - 14\sqrt{a}$
9. $5\sqrt{2a} + 3\sqrt{2a}$
10. $7\sqrt{6x} + 2\sqrt{6x}$
11. $9\sqrt{10y} - \sqrt{10y}$
12. $12\sqrt{14y} - \sqrt{14y}$
13. $5\sqrt{7} + 2\sqrt{7} + 4\sqrt{7}$
14. $3\sqrt{5} + 7\sqrt{5} + 5\sqrt{5}$
15. $8\sqrt{2} - 11\sqrt{2} + 4\sqrt{2}$
16. $3\sqrt{10} - 7\sqrt{10} + 5\sqrt{10}$
17. $5\sqrt{3} + \sqrt{8}$
18. $2\sqrt{5} + \sqrt{45}$
19. $\sqrt{x} - \sqrt{9x}$
20. $\sqrt{25a} - \sqrt{a}$
21. $5\sqrt{8} + 15\sqrt{2}$
22. $3\sqrt{12} + 2\sqrt{3}$
23. $\sqrt{27} - 2\sqrt{3}$
24. $7\sqrt{50} - 3\sqrt{2}$
25. $\sqrt{45} - \sqrt{20}$
26. $\sqrt{27} - \sqrt{12}$
27. $\sqrt{72} + \sqrt{98}$
28. $\sqrt{45} + \sqrt{80}$
29. $2\sqrt{12} + \sqrt{27} - \sqrt{48}$
30. $9\sqrt{8} - \sqrt{72} + \sqrt{98}$
31. $3\sqrt{18} - 2\sqrt{32} - 5\sqrt{50}$
32. $\sqrt{18} - 3\sqrt{8} + \sqrt{50}$
33. $\sqrt{9x} + \sqrt{49x} - 9\sqrt{x}$
34. $\sqrt{16a} - 4\sqrt{a} + \sqrt{25a}$

Multiply.

35. $\sqrt{3}(\sqrt{5} + \sqrt{7})$
36. $\sqrt{5}(\sqrt{2} + \sqrt{11})$
37. $\sqrt{7}(\sqrt{6} - \sqrt{15})$
38. $\sqrt{6}(\sqrt{15} - \sqrt{7})$
39. $(3 + \sqrt{2})(5 + \sqrt{2})$
40. $(5 + \sqrt{11})(3 + \sqrt{11})$
41. $(\sqrt{6} - 2)(\sqrt{6} - 5)$
42. $(\sqrt{10} + 4)(\sqrt{10} - 7)$
43. $(\sqrt{5} + 7)(\sqrt{5} - 7)$
44. $(1 + \sqrt{5})(1 - \sqrt{5})$
45. $(\sqrt{6} - \sqrt{3})(\sqrt{6} + \sqrt{3})$
46. $(\sqrt{2} + \sqrt{6})(\sqrt{2} - \sqrt{6})$
47. $(5 + 3\sqrt{2})(1 - \sqrt{2})$
48. $(4 - \sqrt{7})(3 + 2\sqrt{7})$
49. $(6 + \sqrt{3})^2$
50. $(2 + \sqrt{5})^2$
51. $(5 - 2\sqrt{3})^2$
52. $(6 - 3\sqrt{5})^2$
53. $(\sqrt{x} - \sqrt{10})^2$
54. $(\sqrt{a} - \sqrt{6})^2$

Rationalize the denominator.

55. $\dfrac{5}{1 + \sqrt{2}}$
56. $\dfrac{2}{3 + \sqrt{5}}$
57. $\dfrac{4}{2 - \sqrt{5}}$
58. $\dfrac{3}{7 - \sqrt{2}}$
59. $\dfrac{2}{\sqrt{7} + 3}$
60. $\dfrac{6}{\sqrt{10} + 5}$
61. $\dfrac{\sqrt{6}}{\sqrt{6} - 5}$
62. $\dfrac{\sqrt{10}}{\sqrt{10} - 7}$
63. $\dfrac{\sqrt{5}}{\sqrt{5} - \sqrt{3}}$
64. $\dfrac{\sqrt{7}}{\sqrt{7} - \sqrt{5}}$
65. $\dfrac{\sqrt{14}}{\sqrt{10} + \sqrt{14}}$
66. $\dfrac{\sqrt{6}}{\sqrt{14} - \sqrt{6}}$
67. $\dfrac{\sqrt{3} - \sqrt{2}}{\sqrt{3} + \sqrt{2}}$
68. $\dfrac{\sqrt{5} + \sqrt{3}}{\sqrt{5} - \sqrt{3}}$
69. $\dfrac{\sqrt{6} + \sqrt{5}}{\sqrt{6} - \sqrt{5}}$
70. $\dfrac{\sqrt{10} - \sqrt{7}}{\sqrt{10} + \sqrt{7}}$
71. $\dfrac{1 - \sqrt{7}}{3 + \sqrt{7}}$
72. $\dfrac{2 + \sqrt{5}}{4 - \sqrt{5}}$

Skill Maintenance

73. The time t it takes a bus to travel a fixed distance varies inversely as its speed r. At a speed of 40 mph, it takes 2 hr to travel a fixed distance. How long will it take to travel the same distance at 60 mph? Describe the variation constant.
74. Solution A is 3% alcohol and solution B is 6% alcohol. A service station attendant wants to mix the two to get 80 gal of a solution that is 5.4% alcohol. How many gallons of each should the attendant use?

Synthesis

75. ◈ Explain why it is important for the signs within each pair of conjugates to differ.
76. ◈ Why must you know how to add and subtract radical expressions before you can rationalize denominators with two terms?
77. Three students were asked to simplify $\sqrt{10} + \sqrt{50}$. Their answers were $\sqrt{10}(1 + \sqrt{5})$, $\sqrt{10} + 5\sqrt{2}$, and $\sqrt{2}(5 + \sqrt{5})$. Which, if any, are correct?

78. Can you find any pairs of nonnegative numbers a and b for which $\sqrt{a}+\sqrt{b}=\sqrt{a+b}$? If so, name them.

Add or subtract.

79. $\sqrt{125}-\sqrt{45}+2\sqrt{5}$

80. $3\sqrt{\frac{1}{2}}+\frac{5}{2}\sqrt{18}+\sqrt{98}$

81. $\frac{3}{5}\sqrt{24}+\frac{2}{5}\sqrt{150}-\sqrt{96}$

82. $\frac{1}{3}\sqrt{27}+\sqrt{8}+\sqrt{300}-\sqrt{18}-\sqrt{162}$

83. $\sqrt{ab^6}+b\sqrt{a^3}+a\sqrt{a}$

84. $x\sqrt{2y}-\sqrt{8x^2y}+\frac{x}{3}\sqrt{18y}$

85. $7x\sqrt{12xy^2}-9y\sqrt{27x^3}+5\sqrt{300x^3y^2}$

86. $\sqrt{x}+\sqrt{\frac{1}{x}}$

Wind chill temperature. The **wind chill temperature** is what the temperature would have to be with no wind in order to give the same chilling effect as when there is wind. A formula for finding the wind chill temperature, T_w, is

$$T_w = 91.4 - \frac{(10.45 + 6.68\sqrt{v} - 0.447v)(457 - 5T)}{110},$$

where T is the actual temperature as given by a thermometer, in degrees Fahrenheit, and v is the wind speed, in miles per hour. Using a calculator, find the wind chill temperature in each case. Round to the nearest degree.

87. $T = 30°\text{F}$, $v = 25$ mph

88. $T = 20°\text{F}$, $v = 20$ mph

89. $T = 20°\text{F}$, $v = 40$ mph

90. $T = -10°\text{F}$, $v = 30$ mph

9.5 Radical Equations

Solving Radical Equations • Problem Solving and Applications

An equation in which the variable being solved for appears in a radicand is called a **radical equation.** The following are examples:

$$\sqrt{2x}-4=7, \qquad 2\sqrt{x+2}=\sqrt{x+10}, \quad \text{and} \quad 3+\sqrt{27-3x}=x.$$

In this section, we solve such equations and use them in problem solving.

Solving Radical Equations

An equation with a square root can be rewritten without the radical by using *the principle of squaring*.

The Principle of Squaring

If $a = b$, then $a^2 = b^2$.

If, before using the principle of squaring, we isolate a radical on one side of an equation, that radical will be eliminated when both sides are squared.

EXAMPLE 1

Solve: $\sqrt{x} + 3 = 7$.

Solution

$$\sqrt{x} + 3 = 7$$

$$\sqrt{x} = 4 \qquad \text{Subtracting 3 to get the radical alone on one side}$$

$$(\sqrt{x})^2 = 4^2 \qquad \text{Squaring both sides (using the principle of squaring)}$$

$$x = 16$$

Check:

$\sqrt{x} + 3$	$= 7$
$\sqrt{16} + 3$	? 7
$4 + 3$	
7	7 TRUE

The solution is 16. ❑

EXAMPLE 2

Solve: $3\sqrt{x} = \sqrt{x + 32}$.

Solution Each radical is already isolated on a side of the equation. We proceed with the principle of squaring:

$$3\sqrt{x} = \sqrt{x + 32}$$

$$(3\sqrt{x})^2 = (\sqrt{x + 32})^2 \qquad \text{Squaring both sides (using the principle of squaring)}$$

$$3^2(\sqrt{x})^2 = x + 32 \qquad \text{Squaring the product on the left; simplifying on the right}$$

$$9x = x + 32 \qquad \text{Simplifying on the left}$$

$$\left.\begin{aligned} 8x &= 32 \\ x &= 4. \end{aligned}\right\} \quad \text{Solving for } x$$

Check:

$3\sqrt{x}$	$= \sqrt{x + 32}$
$3\sqrt{4}$	? $\sqrt{4 + 32}$
$3 \cdot 2$	$\sqrt{36}$
6	6 TRUE

The number 4 checks. The solution is 4. ❑

The principle of squaring states that whenever $a = b$ is true, $a^2 = b^2$ is also true. It does *not* say that whenever $a^2 = b^2$ is true, $a = b$ is also true. For example, if a is replaced with -5 and b with 5, the equation $a^2 = b^2$ is true ($(-5)^2 = 5^2$), but the equation $a = b$ is false ($-5 \neq 5$). Thus, although the principle of squaring enables us to find any solutions that a radical equation might have, it can also lead us to numbers that are not solutions of the original equation.

When the principle of squaring is used to solve an equation, possible solutions *must* be checked in the original equation!

EXAMPLE 3

Solve: $\sqrt{2x} = -5$.

Solution

$$(\sqrt{2x})^2 = (-5)^2 \quad \text{Squaring both sides (using the principle of squaring)}$$

$$\left.\begin{aligned} 2x &= 25 \\ x &= \tfrac{25}{2} \end{aligned}\right\} \quad \text{Solving for } x$$

Check:

$$\begin{array}{c|c} \sqrt{2x} & -5 \\ \hline \sqrt{2 \cdot \frac{25}{2}} & ? \; -5 \\ \sqrt{25} & \\ 5 & -5 \end{array} \quad \text{FALSE}$$

There are no solutions. You might have suspected this from the start since a principal square root is never negative. ❑

Sometimes we may need to apply the principle of zero products (see Section 5.6) after squaring.

EXAMPLE 4

Solve: $x - 5 = \sqrt{x + 7}$.

Solution

$$x - 5 = \sqrt{x + 7}$$

$$(x - 5)^2 = (\sqrt{x + 7})^2 \quad \text{Using the principle of squaring}$$

$$x^2 - 10x + 25 = x + 7 \quad \text{Squaring a binomial on the left side}$$

$$x^2 - 11x + 18 = 0 \quad \text{Adding } -x - 7 \text{ on both sides}$$

$$(x - 9)(x - 2) = 0 \quad \text{Factoring}$$

$$x - 9 = 0 \quad or \quad x - 2 = 0 \quad \text{Using the principle of zero products}$$

$$x = 9 \quad or \quad x = 2$$

Check:

For 9:

$$\begin{array}{c|c} x - 5 & \sqrt{x + 7} \\ \hline 9 - 5 & ? \; \sqrt{9 + 7} \\ 4 & 4 \end{array} \quad \text{TRUE}$$

For 2:

$$\begin{array}{c|c} x - 5 & \sqrt{x + 7} \\ \hline 2 - 5 & ? \; \sqrt{2 + 7} \\ -3 & 3 \end{array}$$

The number 9 checks, but 2 does not. Thus the solution is 9. ❑

EXAMPLE 5

Solve: $3 + \sqrt{27 - 3x} = x$.

Solution

$$3 + \sqrt{27 - 3x} = x$$

$$\sqrt{27 - 3x} = x - 3 \quad \text{Subtracting 3 to isolate the radical}$$

$$(\sqrt{27 - 3x})^2 = (x - 3)^2 \quad \text{Using the principle of squaring}$$

$$27 - 3x = x^2 - 6x + 9$$

$$0 = x^2 - 3x - 18 \quad \text{Adding } 3x - 27 \text{ on both sides}$$

$$0 = (x - 6)(x + 3) \quad \text{Factoring}$$

$$x - 6 = 0 \quad or \quad x + 3 = 0 \quad \text{Using the principle of zero products}$$

$$x = 6 \quad or \quad x = -3$$

Check: For 6:

$$\begin{array}{r|l} 3 + \sqrt{27 - 3x} & = x \\ \hline 3 + \sqrt{27 - 3 \cdot 6} \;?\; & 6 \\ 3 + \sqrt{9} & \\ 3 + 3 & \\ 6 & 6 \quad \text{TRUE} \end{array}$$

For -3:

$$\begin{array}{r|l} 3 + \sqrt{27 - 3x} & = x \\ \hline 3 + \sqrt{27 - 3 \cdot (-3)} \;?\; & -3 \\ 3 + \sqrt{27 + 9} & \\ 3 + \sqrt{36} & \\ 3 + 6 & \\ 9 & -3 \quad \text{FALSE} \end{array}$$

The number 6 checks, but -3 does not. The solution is 6. ❑

Problem Solving and Applications

How far can you see from a given height? There is a formula to determine this. At a height of h meters, you can see V kilometers to the horizon. These numbers are related as follows:

$$V = 3.5\sqrt{h}.$$

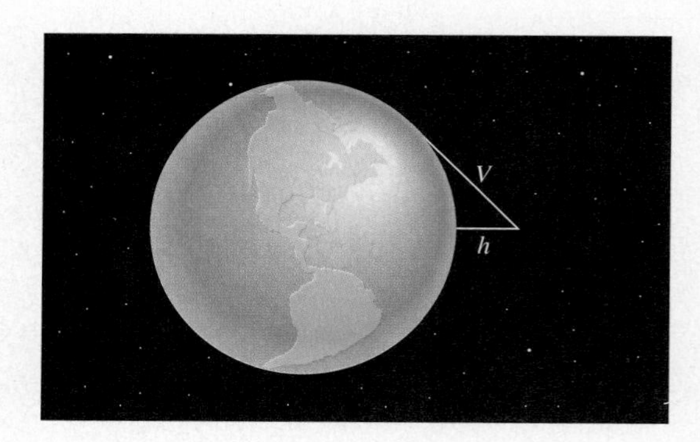

EXAMPLE 6

Elaine can see 50.4 km to the horizon from the top of a cliff. What is the altitude of Elaine's eyes?

Solution

1. **FAMILIARIZE.** A sketch can be helpful here. The altitude of Elaine's eyes, in meters, is labeled h.

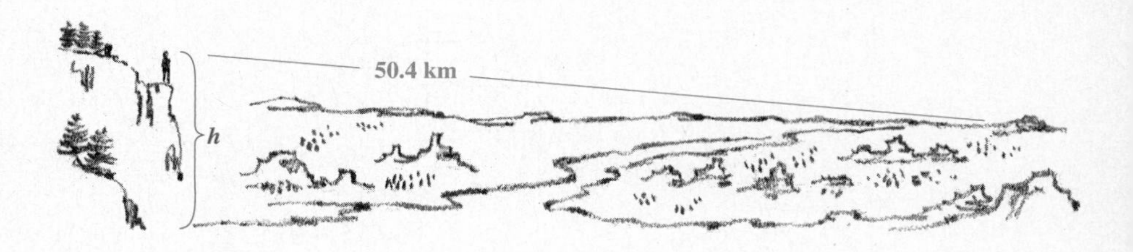

2. **TRANSLATE.** We substitute 50.4 for V in the equation $V = 3.5\sqrt{h}$:

$$50.4 = 3.5\sqrt{h}.$$

3. CARRY OUT. We solve the equation for h:

$$50.4 = 3.5\sqrt{h}$$
$$\frac{50.4}{3.5} = \sqrt{h}$$
$$14.4 = \sqrt{h}$$
$$(14.4)^2 = (\sqrt{h})^2 \qquad \text{Using the principle of squaring}$$
$$207.36 = h.$$

4. CHECK. We leave the check to the student.

5. STATE. The altitude of Elaine's eyes is about 207 m. ❑

EXERCISE SET | 9.5

Solve.

1. $\sqrt{x} = 5$
2. $\sqrt{x} = 7$
3. $\sqrt{x} = 1.5$
4. $\sqrt{x} = 4.3$
5. $\sqrt{x+3} = 8$
6. $\sqrt{x+4} = 11$
7. $\sqrt{2x+4} = 9$
8. $\sqrt{2x+1} = 13$
9. $3 + \sqrt{x-1} = 5$
10. $4 + \sqrt{y-3} = 11$
11. $6 - 2\sqrt{3n} = 0$
12. $8 - 4\sqrt{5n} = 0$
13. $\sqrt{5x-7} = \sqrt{x+10}$
14. $\sqrt{4x-5} = \sqrt{x+9}$
15. $\sqrt{x} = -7$
16. $\sqrt{x} = -5$
17. $\sqrt{2y+6} = \sqrt{2y-5}$
18. $2\sqrt{3x-2} = \sqrt{2x-3}$
19. $x - 7 = \sqrt{x-5}$
20. $\sqrt{x+7} = x - 5$
21. $\sqrt{x+18} = x - 2$
22. $x - 9 = \sqrt{x-3}$
23. $2\sqrt{x-1} = x - 1$
24. $x + 4 = 4\sqrt{x+1}$
25. $\sqrt{5x+21} = x + 3$
26. $\sqrt{27-3x} = x - 3$
27. $x = 1 + 6\sqrt{x-9}$
28. $\sqrt{2x-1} + 2 = x$
29. $\sqrt{x^2+6} - x + 3 = 0$
30. $\sqrt{x^2+5} - x + 2 = 0$
31. $\sqrt{(p+6)(p+1)} - 2 = p + 1$
32. $\sqrt{(4x+5)(x+4)} = 2x + 5$
33. $\sqrt{2-x} = \sqrt{3x-7}$
34. $\sqrt{4x-10} = \sqrt{2-x}$
35. $1 + \sqrt{1-x} = x$
36. $\sqrt{(x-2)(x+1)} = x - 1$

Problem Solving

Use $V = 3.5\sqrt{h}$ from Example 6 for Exercises 37–40.

37. A steeplejack can see 17.5 km to the horizon from the top of a building. What is the altitude of the steeplejack's eyes?

38. A scout can see 70 km to the horizon from the top of a hill. What is the altitude of the scout's eyes?

39. A person can see 371 km to the horizon from an airplane window. How high is the airplane?

40. A sailor can see 99.4 km to the horizon from the top of a mast. How high is the mast?

The formula $r = 2\sqrt{5L}$ can be used to approximate the speed r, in miles per hour, of a car that has left a skid mark of length L, in feet.

41. How far will a car skid at 50 mph? at 70 mph?

42. How far will a car skid at 60 mph? at 100 mph?

43. Find a number such that twice its square root is 14.

44. Find a number such that the additive inverse of three times its square root is -33.

45. Find a number such that the square root of 4 more than five times the number is 8.

46. Find a number such that 1 less than the square root of twice the number is 7.

The formula $T = 2\pi\sqrt{L/32}$ can be used to find the period T, in seconds, of a pendulum of length L, in feet.

47. What is the length of a pendulum that has a period of 1.6 sec? Use 3.14 for π.

48. What is the length of a pendulum that has a period of 3 sec? Use 3.14 for π.

Skill Maintenance

Multiply and simplify.

49. $\dfrac{7x^9}{27} \cdot \dfrac{9}{7x^3}$

50. $\dfrac{3}{x^2 - 9} \cdot \dfrac{x^2 - 6x + 9}{12}$

Add or subtract as indicated. Simplify, if possible.

51. $\dfrac{x}{x+5} + \dfrac{x^2 - 20}{x+5}$

52. $\dfrac{9}{a-1} - \dfrac{4}{1-a}$

Synthesis

53. ◈ Explain what would have happened in Example 1 if we had not isolated the radical before squaring. Could we still have solved the equation? Why or why not?

54. ◈ Explain in your own words why possible solutions of radical equations must be checked.

Sometimes the principle of squaring must be used more than once in order to solve an equation. Solve Exercises 55–62 by using the principle of squaring as often as necessary.

55. $\sqrt{x+9} = 1 + \sqrt{x}$

56. $\sqrt{x-5} = 5 - \sqrt{x}$

57. $\sqrt{3x+1} = 1 - \sqrt{x+4}$

58. $\sqrt{y+8} - \sqrt{y} = 2$

59. $4 + \sqrt{19 - x} = 6 + \sqrt{4 - x}$

60. $\sqrt{y+1} - \sqrt{2y-5} = \sqrt{y-2}$

61. $2\sqrt{x-1} - \sqrt{3x-5} = \sqrt{x-9}$

62. $x = (x-2)\sqrt{x}$

63. A mountain climber stops at some point in the climb and views the horizon. The mountain climber uses the formula $V = 3.5\sqrt{h}$ to determine the distance to the horizon. After climbing another 100 m, the climber again computes the distance to the horizon, and finds that it is 20 km farther than before. At what height was the climber when the first computation was made? (*Hint:* Use a system of equations.)

64. Solve $A = \sqrt{1 + \sqrt{a/b}}$ for b.

Graph. Use a calculator or Table 1 to find y-values.

65. $y = \sqrt{x}$

66. $y = \sqrt{x-5}$

67. $y = \sqrt{x-1}$

68. $y = \sqrt{x+1}$

69. Graph $y = x - 7$ and $y = \sqrt{x-5}$ using the same set of axes. Determine where the graphs intersect in order to estimate a solution of $x - 7 = \sqrt{x-5}$.

70. Graph $y = x - 5$ and $y = \sqrt{x+7}$ using the same set of axes. Determine where the graphs intersect in order to estimate a solution of $x - 5 = \sqrt{x+7}$.

Use a grapher and the procedure discussed in Exercises 69 and 70 to solve the following equations. Use the Trace and Zoom features to round answers to the nearest hundredth.

71. $\sqrt{x+3} = 2x - 1$

72. $-\sqrt{x+3} = 2x - 1$

9.6 Right Triangles and Problem Solving

Right Triangles • **Problem Solving**

Radicals frequently occur in problem-solving situations in which the Pythagorean theorem is used.

The Pythagorean Theorem*

In any right triangle, if a and b are the lengths of the legs and c is the length of the hypotenuse, then

$$a^2 + b^2 = c^2.$$

Right Triangles

When the Pythagorean theorem is used to find the length of a triangle's side, we need not concern ourselves with negative square roots, since length cannot be negative.

EXAMPLE 1

Find the length of the hypotenuse of this right triangle. Give an exact answer and an approximation to three decimal places.

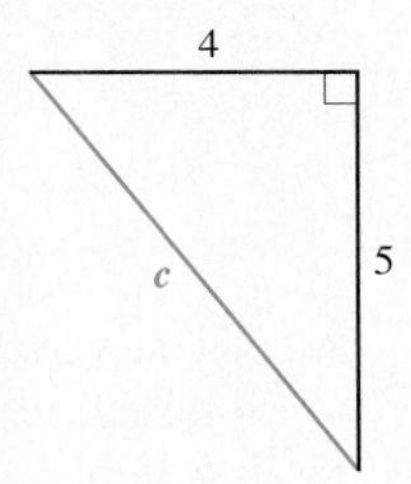

Solution We have

$$\begin{aligned} a^2 + b^2 &= c^2 \\ 4^2 + 5^2 &= c^2 \quad \text{Substituting} \\ 16 + 25 &= c^2 \\ 41 &= c^2. \end{aligned}$$

Since c is a length, it follows that c is the principal square root of 41:

$c = \sqrt{41}$ This is an exact answer.

$c \approx 6.403.$ Using a calculator or Table 1 for an approximation ❑

*The converse of the Pythagorean theorem also holds. That is, if a, b, and c are the lengths of the sides of a triangle and $a^2 + b^2 = c^2$, then the triangle is a right triangle.

EXAMPLE 2

Find the length of the indicated leg in the triangle. In each case, give an exact answer and an approximation to three decimal places.

a) (Right triangle with legs 10 and b, hypotenuse 12)

b) (Right triangle with legs $\sqrt{19}$ and a, hypotenuse 12)

Solution

a)

$$10^2 + b^2 = 12^2$$ Substituting in the Pythagorean equation

$$100 + b^2 = 144$$

$$b^2 = 44$$ Subtracting 100 on both sides

$$b = \sqrt{44}$$ The exact answer is $\sqrt{44}$. Since length is positive, only the principal square root is used.

$$b \approx 6.633$$ Approximating $\sqrt{44}$ with a calculator or Table 1

b)

$$a^2 + (\sqrt{19})^2 = 12^2$$ Substituting

$$a^2 + 19 = 144$$

$$a^2 = 125$$ Subtracting 19 on both sides

$$a = \sqrt{125}$$ The exact answer is $\sqrt{125}$.

$$a \approx 11.180$$ Using a calculator ❑

To approximate $\sqrt{125}$ using Table 1, you will need to simplify before finding an approximation:

$$\begin{aligned}\sqrt{125} &= \sqrt{25 \cdot 5}\\ &= 5\sqrt{5}\\ &\approx 5(2.236)\\ &= 11.180.\end{aligned}$$

A possible variance in answers can occur depending on the procedure used. Answers to exercises will be found using a calculator and rounding to three decimal places in the last step.

Problem Solving

Our five-step process and the Pythagorean theorem can be used for problem solving.

EXAMPLE 3

A 32-ft ladder is leaning against a house. The bottom of the ladder is 7 ft from the house. How high is the top of the ladder? Give an exact answer and an approximation to three decimal places.

Solution

1. FAMILIARIZE. First we make a sketch. In it, there is a right triangle. We label the unknown height h.

2. TRANSLATE. We use the Pythagorean theorem, substituting 7 for a, h for b, and 32 for c:

$$7^2 + h^2 = 32^2.$$

3. CARRY OUT. We solve the equation:

$$\begin{aligned} 7^2 + h^2 &= 32^2 \\ 49 + h^2 &= 1024 \\ h^2 &= 975 \\ h &= \sqrt{975} && \text{This answer is exact.} \\ h &\approx 31.225. && \text{Approximating with a calculator} \end{aligned}$$

4. CHECK. We check by substituting 7, $\sqrt{975}$, and 32 into the Pythagorean equation:

$$\begin{array}{r|l} \multicolumn{2}{c}{a^2 + b^2 = c^2} \\ \hline 7^2 + (\sqrt{975})^2 & ?\ 32^2 \\ 49 + 975 & 1024 \\ 1024 & 1024 \quad \text{TRUE} \end{array}$$

5. STATE. The top of the ladder is $\sqrt{975}$, or about 31.225 ft from the ground. ❑

EXAMPLE 4

A slow-pitch softball diamond is a square 65 ft on a side. How far is it from home plate to second base? Give an exact answer and an approximation to three decimal places.

Solution

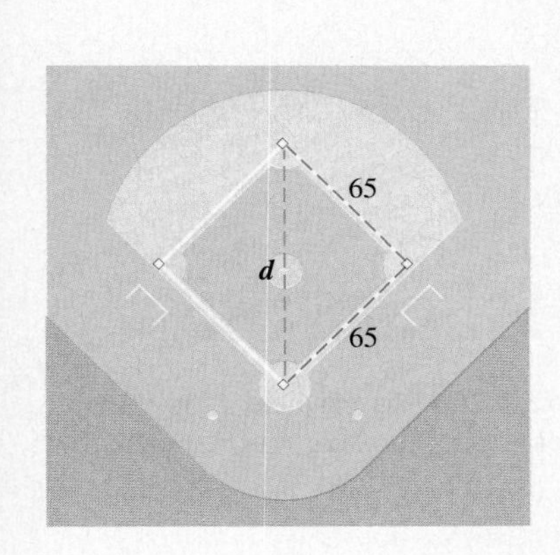

1. FAMILIARIZE. We first make a drawing. We note that the first- and second-base lines, together with a line from home to second, form a right triangle. We label the unknown distance d.

2. TRANSLATE. We substitute 65 for a, 65 for b, and d for c in the Pythagorean equation:

$$65^2 + 65^2 = d^2.$$

3. CARRY OUT. We solve the equation:

$$\begin{aligned} 65^2 + 65^2 &= d^2 \\ 4225 + 4225 &= d^2 \\ 8450 &= d^2 \\ \sqrt{8450} &= d && \text{This is exact.} \\ 91.924 &\approx d. && \text{This is approximate.} \end{aligned}$$

4. CHECK. We check by substituting 65, 65, and $\sqrt{8450}$ into the Pythagorean equation:

$$\begin{array}{r|l} \multicolumn{2}{c}{a^2 + b^2 = c^2} \\ \hline 65^2 + 65^2 & ?\ (\sqrt{8450})^2 \\ 4225 + 4225 & 8450 \\ 8450 & 8450 \quad \text{TRUE} \end{array}$$

5. STATE. The distance from home plate to second base is $\sqrt{8450}$, or about 91.924 ft. ❑

EXERCISE SET 9.6

Find the length of the third side of the triangle. Give an exact answer and, where appropriate, an approximation to three decimal places.

1.

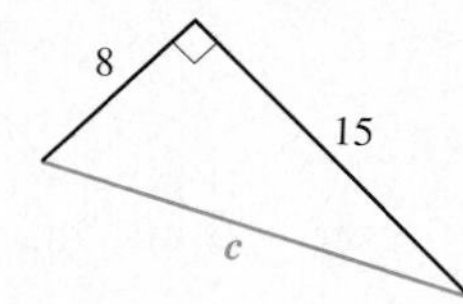

2.

3.

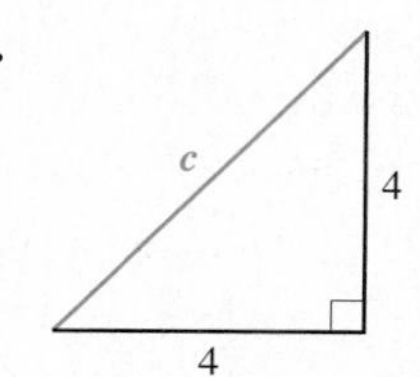

4.

5.

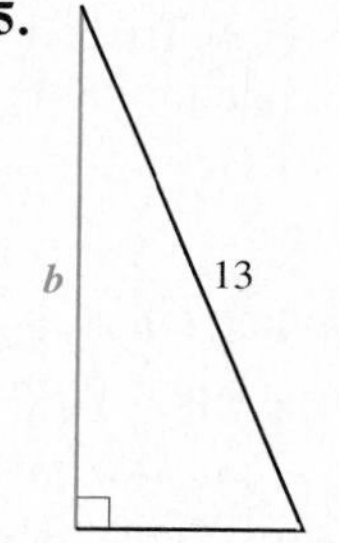

6.

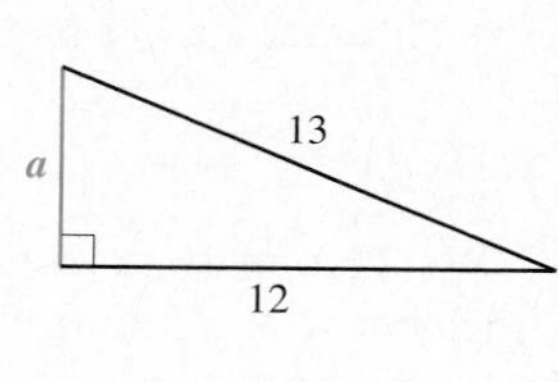

7.

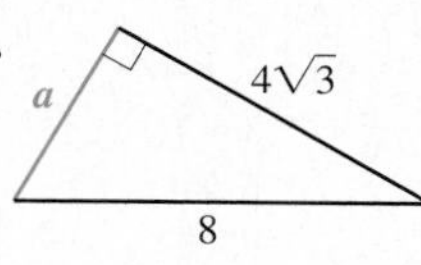

8.

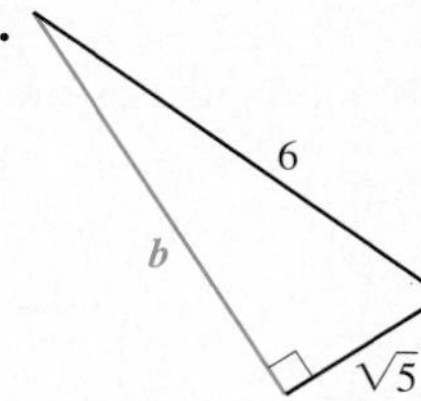

In a right triangle, find the length of the side not given. Give an exact answer and, where appropriate, an approximation to three decimal places. Keep in mind that a and b are the lengths of the legs and c is the length of the hypotenuse.

9. $a = 10$, $b = 24$

10. $a = 5$, $b = 12$

11. $a = 9$, $c = 15$

12. $a = 18$, $c = 30$

13. $b = 1$, $c = \sqrt{5}$

14. $b = 1$, $c = \sqrt{2}$

15. $a = 1$, $c = \sqrt{3}$

16. $a = \sqrt{3}$, $b = \sqrt{5}$

17. $c = 10$, $b = 5\sqrt{3}$

18. $a = 5$, $b = 5$

Problem Solving

Don't forget to make drawings. Give an exact answer and an approximation to three decimal places.

19. A 10-m ladder is leaning against a building. The bottom of the ladder is 5 m from the building. How high is the top of the ladder?

20. Find the length of a diagonal of a square whose sides are 3 cm long.

21. A new water pipe is being prepared so that it will run diagonally under a kitchen floor. If the kitchen is 8 ft wide and 12 ft long, how long should the pipe be?

22. How long must a wire be in order to reach from the top of a 13-m telephone pole to a point on the ground 9 m from the foot of the pole?

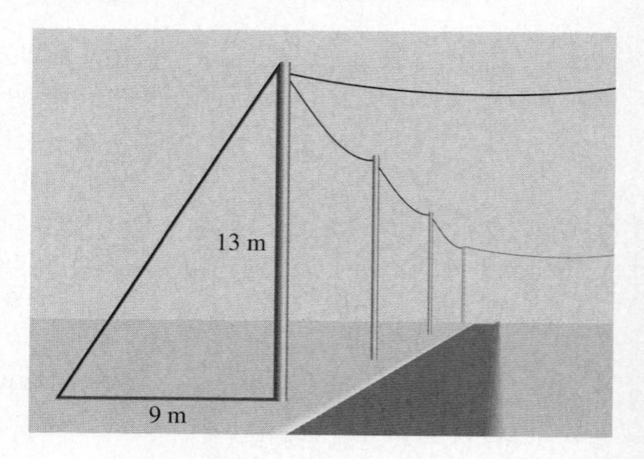

23. The smallest regulation soccer field is 50 yd wide and 100 yd long. Find the length of a diagonal of such a field.

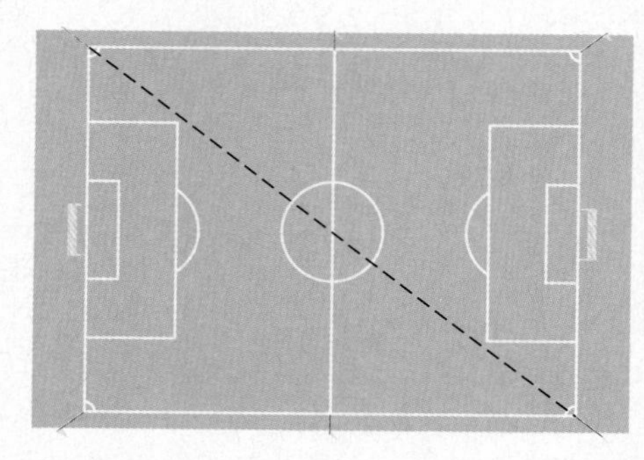

24. The largest regulation soccer field is 100 yd wide and 130 yd long. Find the length of a diagonal of such a field.

25. A Little-League baseball diamond is a square 60 ft on a side. How far is it from home plate to second base?

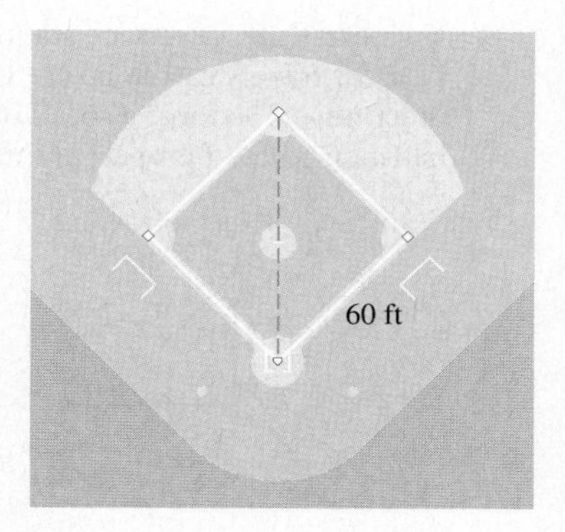

26. A baseball diamond is a square 90 ft on a side. How far is it from first base to third base?

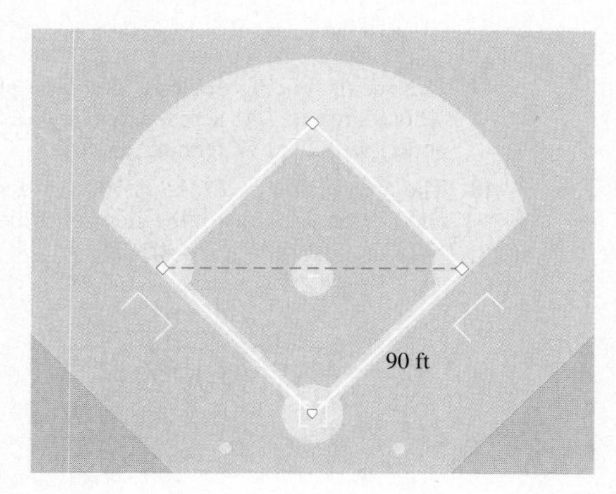

27. A surveyor had poles located at points P, Q, and R. The distances that the surveyor was able to measure are marked in the figure. What is the approximate length of the lake?

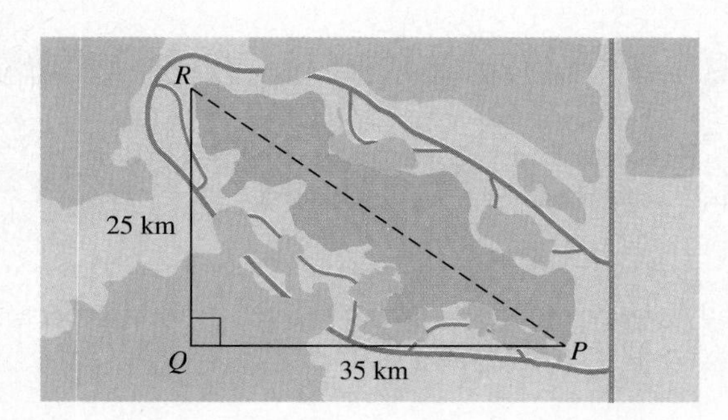

28. An airplane is flying at an altitude of 4100 ft. The slanted distance directly to the airport is 15,100 ft. How far is the airplane horizontally from the airport?

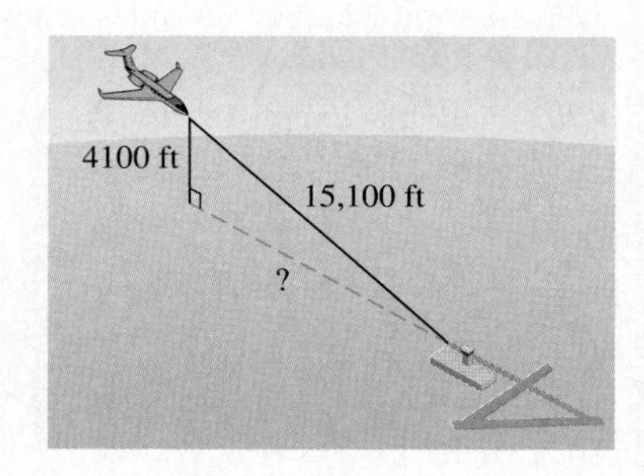

Skill Maintenance

29. Find the slope of the line $4 - x = 3y$.

30. Find the slope of the line containing the points $(8, -3)$ and $(0, -8)$.

Solve.

31. $-\frac{3}{5}x < 15$

32. $-2x + 6 \geqslant 7x - 3$

Synthesis

33. ◈ In an *equilateral triangle,* all sides have the same length. Can a right triangle be equilateral? Why or why not?

34. ◈ In an *isosceles triangle,* two sides have the same length. Can a right triangle be isosceles? Why or why not?

35. Suppose a fielder catches a ball on the third-base line on a baseball diamond about 40 ft behind third base. About how far would a throw to first base be?

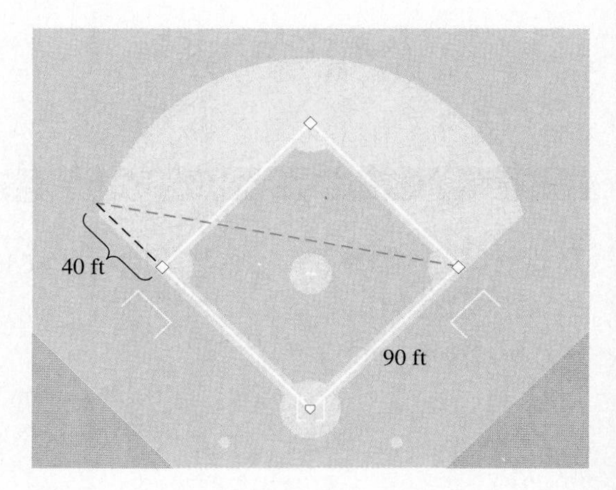

36. The diagonal of a square has a length of $8\sqrt{2}$ ft. Find the length of a side of the square.

37. Find the length of a side of a square that has an area of 7 m^2.

38. A right triangle has sides whose lengths are consecutive integers. Find the lengths of the sides.

39. Find the length of the diagonal of a cube with sides of length s.

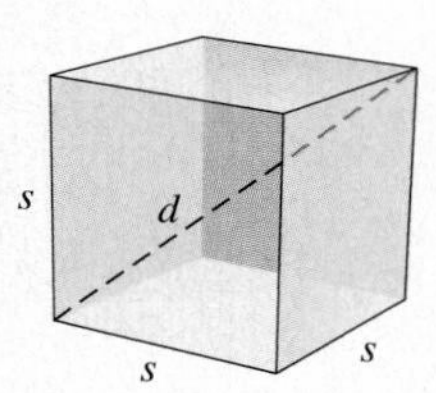

40. Figure $ABCD$ is a square. Find AC.

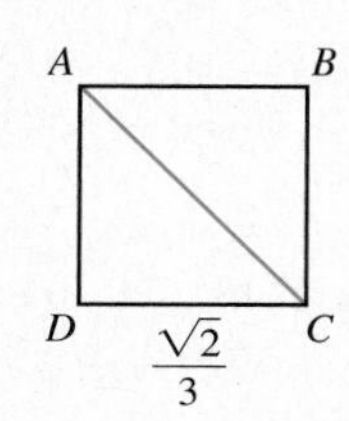

An equilateral triangle is shown below.

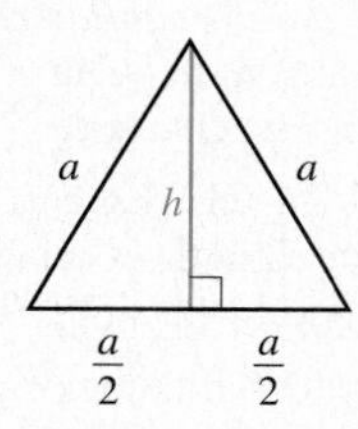

41. Find an expression for its height h in terms of a.

42. Find an expression for its area A in terms of a.

43. The length and the width of a rectangle are given by consecutive integers. The area of the rectangle is 90 cm^2. Find the length of the diagonal of the rectangle.

44. Two cars leave a service station at the same time. One car travels east at a speed of 50 mph, and the other travels south at a speed of 60 mph. After one half hour, how far apart are they?

Find x in each of the following.

45.

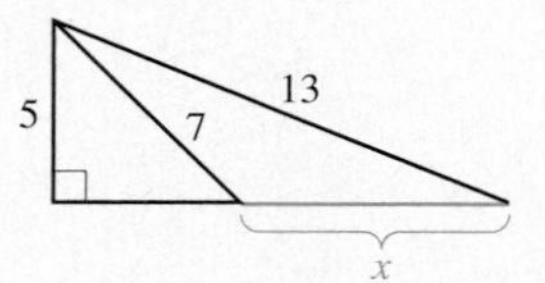

46.

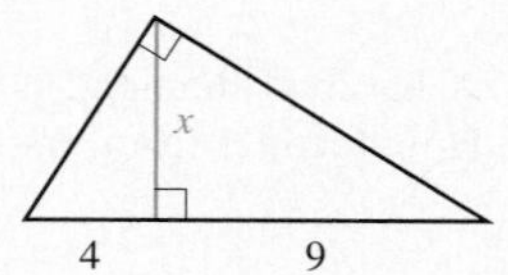

47. If 2 mi of fencing encloses a square plot of land whose area is 160 acres, how large a square, in acres, will 4 mi of fencing enclose?

48. The area of square $PQRS$ is 100 ft^2, and A, B, C, and D are midpoints of the sides on which they lie. Find the area of square $ABCD$.

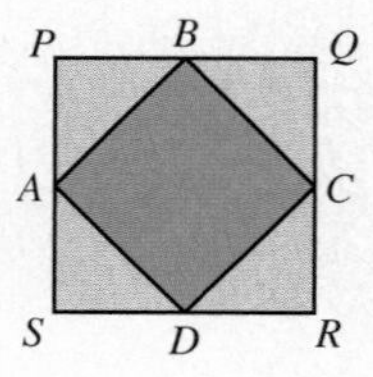

9.7 Higher Roots and Rational Exponents

Higher Roots • Products and Quotients Involving Higher Roots • Rational Exponents • Calculators

In this section, we study *higher* roots, such as cube roots or fourth roots, and exponents that are not integers.

Higher Roots

Recall that c is a square root of a if $c^2 = a$. A similar definition exists for *cube roots*.

Cube Root

> The number c is the *cube root* of a if $c^3 = a$.

The symbolism $\sqrt[3]{a}$ is used to represent the cube root of a. In the radical $\sqrt[3]{a}$, the number 3 is called the **index,** and a is called the **radicand.**

EXAMPLE 1

Find the cube root of each number: **(a)** 8; **(b)** -125.

Solution

a) The cube root of 8 is the number whose cube is 8. Since $2^3 = 2 \cdot 2 \cdot 2 = 8$, the cube root of 8 is 2: $\sqrt[3]{8} = 2$.

b) The cube root of -125 is the number whose cube is -125. Since $(-5)^3 = (-5)(-5)(-5) = -125$, the cube root of -125 is -5: $\sqrt[3]{-125} = -5$. ❑

Positive numbers always have *two* nth roots when n is even but when we refer to *the* nth root of a positive number a, denoted $\sqrt[n]{a}$, we mean the *positive* nth root. Thus, although -3 and 3 are both fourth roots of 81 (since $(-3)^4 = 3^4 = 81$), 3 is considered *the* fourth root of 81. In symbols,

$$\sqrt[4]{81} = 3.$$

nth Root

> The number c is the *nth root* of a if $c^n = a$.
>
> If n is odd, then there is exactly one nth root and the symbol $\sqrt[n]{a}$ represents that root.
>
> If n is even, then $\sqrt[n]{a}$ represents the nonnegative nth root.
>
> Even roots of negative numbers are not real numbers.

EXAMPLE 2

Find each root: **(a)** $\sqrt[4]{16}$; **(b)** $\sqrt[5]{-32}$; **(c)** $\sqrt[4]{-16}$; **(d)** $-\sqrt[3]{64}$.

Solution

a) $\sqrt[4]{16} = 2$ — Since $2^4 = 2 \cdot 2 \cdot 2 \cdot 2 = 16$

b) $\sqrt[5]{-32} = -2$ — Since $(-2)^5 = (-2)(-2)(-2)(-2)(-2) = -32$

c) $\sqrt[4]{-16}$ is not a real number, because it is an even root of a negative number.

d) $-\sqrt[3]{64} = -(\sqrt[3]{64})$ — This is the opposite of $\sqrt[3]{64}$.

$= -4$ — $4^3 = 4 \cdot 4 \cdot 4 = 64$ ❑

The following roots occur so frequently that you may want to memorize them.

Square Roots		Cube Roots	Fourth Roots	Fifth Roots
$\sqrt{1} = 1$	$\sqrt{4} = 2$	$\sqrt[3]{1} = 1$	$\sqrt[4]{1} = 1$	$\sqrt[5]{1} = 1$
$\sqrt{9} = 3$	$\sqrt{16} = 4$	$\sqrt[3]{8} = 2$	$\sqrt[4]{16} = 2$	$\sqrt[5]{32} = 2$
$\sqrt{25} = 5$	$\sqrt{36} = 6$	$\sqrt[3]{27} = 3$	$\sqrt[4]{81} = 3$	$\sqrt[5]{243} = 3$
$\sqrt{49} = 7$	$\sqrt{64} = 8$	$\sqrt[3]{64} = 4$	$\sqrt[4]{256} = 4$	
$\sqrt{81} = 9$	$\sqrt{100} = 10$	$\sqrt[3]{125} = 5$	$\sqrt[4]{625} = 5$	
$\sqrt{121} = 11$	$\sqrt{144} = 12$	$\sqrt[3]{216} = 6$		

Products and Quotients Involving Higher Roots

The rules for working with products and quotients of square roots can be extended to products and quotients of nth roots.

$$\sqrt[n]{AB} = \sqrt[n]{A}\,\sqrt[n]{B}, \qquad \sqrt[n]{\frac{A}{B}} = \frac{\sqrt[n]{A}}{\sqrt[n]{B}}$$

EXAMPLE 3

Simplify: **(a)** $\sqrt[3]{40}$; **(b)** $\sqrt[3]{\frac{125}{216}}$; **(c)** $\sqrt[4]{1250}$; **(d)** $\sqrt[5]{\frac{2}{243}}$.

Solution

a) $\sqrt[3]{40} = \sqrt[3]{8 \cdot 5}$ 8 is a perfect cube.
$= \sqrt[3]{8} \cdot \sqrt[3]{5}$
$= 2\sqrt[3]{5}$

b) $\sqrt[3]{\frac{125}{216}} = \frac{\sqrt[3]{125}}{\sqrt[3]{216}} = \frac{5}{6}$ 125 and 216 are perfect cubes.

c) $\sqrt[4]{1250} = \sqrt[4]{625 \cdot 2}$ 625 is a perfect fourth power.
$= \sqrt[4]{625} \cdot \sqrt[4]{2}$
$= 5\sqrt[4]{2}$

d) $\sqrt[5]{\frac{2}{243}} = \frac{\sqrt[5]{2}}{\sqrt[5]{243}} = \frac{\sqrt[5]{2}}{3}$ 243 is a perfect fifth power. ❑

Rational Exponents

Expressions containing rational exponents, like $8^{1/3}$, $4^{5/2}$, and $81^{-3/4}$, are defined in a manner that ensures that the laws of exponents still hold. For example, if the product rule, $a^m \cdot a^n = a^{m+n}$, is to hold, then

$$a^{1/2} \cdot a^{1/2} = a^{1/2+1/2} = a^1 = a.$$

This says that the square of $a^{1/2}$ is a. Thus we are led to the following:

$$a^{1/2} = \sqrt{a}.$$

This idea is generalized in the following:

For any nonnegative number a and any index n, $a^{1/n}$ means $\sqrt[n]{a}$. Note that $\sqrt[2]{a}$ is written $\sqrt{a}$. If a is negative, then n must be odd.

EXAMPLE 4

Simplify: **(a)** $8^{1/3}$; **(b)** $100^{1/2}$; **(c)** $81^{1/4}$; **(d)** $(-243)^{1/5}$.

Solution

a) $8^{1/3} = \sqrt[3]{8} = 2$

b) $100^{1/2} = \sqrt{100} = 10$

c) $81^{1/4} = \sqrt[4]{81} = 3$

d) $(-243)^{1/5} = \sqrt[5]{-243} = -3$ ❑

In assigning meaning to an expression like $8^{2/3}$, we wish to preserve the power rule: $(a^m)^n = a^{mn}$. To do this, we write $8^{2/3}$ as $(8^{1/3})^2$ or $(8^2)^{1/3}$ since $\frac{1}{3} \cdot 2 = 2 \cdot \frac{1}{3} = \frac{2}{3}$. Thus,

$$8^{2/3} = (\sqrt[3]{8})^2, \quad \text{or} \quad 8^{2/3} = \sqrt[3]{8^2}.$$

For any natural numbers m and n ($n \neq 1$) and any real number a for which $\sqrt[n]{a}$ exists,

$$a^{m/n} = (a^{1/n})^m = (\sqrt[n]{a})^m, \qquad (1)$$

or equivalently,

$$a^{m/n} = (a^m)^{1/n} = \sqrt[n]{a^m}. \qquad (2)$$

Usually simplifications are most easily carried out by using Equation (1).

EXAMPLE 5

Simplify: **(a)** $27^{2/3}$; **(b)** $8^{5/3}$; **(c)** $81^{3/4}$.

Solution

a) $27^{2/3} = (27^{1/3})^2 = (\sqrt[3]{27})^2 = 3^2 = 9$

b) $8^{5/3} = (8^{1/3})^5 = (\sqrt[3]{8})^5 = 2^5 = 32$

c) $81^{3/4} = (81^{1/4})^3 = (\sqrt[4]{81})^3 = 3^3 = 27$ ❑

Negative rational exponents are defined in much the same way that negative integer exponents are.

For any rational number m/n and any nonzero real number a for which $a^{m/n}$ exists,

$$a^{-m/n} = \frac{1}{a^{m/n}}.$$

EXAMPLE 6

Simplify: **(a)** $16^{-1/2}$; **(b)** $27^{-1/3}$; **(c)** $32^{-2/5}$; **(d)** $64^{-3/2}$.

Solution

a) $16^{-1/2} = \frac{1}{16^{1/2}} = \frac{1}{\sqrt{16}} = \frac{1}{4}$

b) $27^{-1/3} = \frac{1}{27^{1/3}} = \frac{1}{\sqrt[3]{27}} = \frac{1}{3}$

c) $32^{-2/5} = \frac{1}{32^{2/5}} = \frac{1}{(32^{1/5})^2} = \frac{1}{(\sqrt[5]{32})^2} = \frac{1}{2^2} = \frac{1}{4}$

d) $64^{-3/2} = \frac{1}{64^{3/2}} = \frac{1}{(\sqrt{64})^3} = \frac{1}{8^3} = \frac{1}{512}$ ❑

Calculators

A calculator with a key for finding powers can be used to approximate numbers like $\sqrt[5]{8}$. Generally such keys are labeled [xʸ], [aˣ], or [^]. We find approximations by entering the radicand, pressing the power key, entering the exponent, and then pressing [=] or [ENTER]. Thus, $\sqrt[5]{8}$ can be approximated by entering 8, pressing the power key, entering 0.2 (or (1 ÷ 5)), and pressing [=] to get $\sqrt[5]{8} \approx 1.515716567$. Consult an owner's manual or your instructor if your calculator works differently.

EXERCISE SET | 9.7

Simplify. If an expression does not represent a real number, state so.

1. $\sqrt[3]{-8}$
2. $\sqrt[3]{64}$
3. $\sqrt[3]{1000}$
4. $\sqrt[3]{-27}$
5. $-\sqrt[3]{125}$
6. $-\sqrt[3]{8}$
7. $\sqrt[3]{216}$
8. $\sqrt[3]{-343}$
9. $\sqrt[4]{625}$
10. $\sqrt[4]{81}$
11. $\sqrt[5]{0}$
12. $\sqrt[5]{1}$
13. $\sqrt[5]{-1}$
14. $\sqrt[5]{-243}$
15. $\sqrt[4]{-81}$
16. $\sqrt[4]{-1}$
17. $\sqrt[4]{10{,}000}$
18. $\sqrt[5]{100{,}000}$
19. $\sqrt[3]{5^3}$
20. $\sqrt[4]{7^4}$
21. $\sqrt[6]{64}$
22. $\sqrt[6]{1}$
23. $\sqrt[3]{x^3}$
24. $\sqrt[5]{t^5}$
25. $\sqrt[3]{32}$
26. $\sqrt[3]{54}$
27. $\sqrt[4]{48}$

28. $\sqrt[5]{160}$ 29. $\sqrt[3]{\frac{27}{64}}$ 30. $\sqrt[3]{\frac{125}{64}}$

31. $\sqrt[4]{\frac{256}{625}}$ 32. $\sqrt[5]{\frac{243}{32}}$ 33. $\sqrt[3]{\frac{17}{8}}$

34. $\sqrt[5]{\frac{11}{32}}$ 35. $\sqrt[4]{\frac{13}{81}}$ 36. $\sqrt[3]{\frac{10}{27}}$

Simplify.

37. $25^{1/2}$ 38. $9^{1/2}$ 39. $1000^{1/3}$

40. $125^{1/3}$ 41. $16^{1/4}$ 42. $32^{1/5}$

43. $16^{5/4}$ 44. $8^{4/3}$ 45. $4^{5/2}$

46. $9^{3/2}$ 47. $64^{2/3}$ 48. $32^{2/5}$

49. $8^{2/3}$ 50. $16^{3/4}$ 51. $25^{5/2}$

52. $4^{-1/2}$ 53. $36^{-1/2}$ 54. $32^{-1/5}$

55. $256^{-1/4}$ 56. $100^{-3/2}$ 57. $81^{-3/4}$

58. $121^{-1/2}$ 59. $144^{-1/2}$ 60. $16^{-3/4}$

61. $81^{-5/4}$ 62. $32^{-2/5}$ 63. $8^{-2/3}$

64. $625^{-3/4}$

Skill Maintenance

65. The amount F that a family spends on food varies directly as its income I. A family making $29,400 a year will spend $7644 on food. At this rate, how much would a family making $41,000 spend on food?

66. Find an equation of variation in which y varies directly as x and $y = 2.5$ when $x = 4$.

Perform the indicated operation and simplify, if possible.

67. $\frac{a+5}{a^2-25} - \frac{3a+15}{a^2-25}$ 68. $\frac{x}{x^2-9} \cdot \frac{x^2-4x+3}{3x^4}$

Synthesis

69. ◈ Explain in your own words why $\sqrt[n]{a}$ is negative when n is odd and a is negative.

70. ◈ Expressions of the form $a^{m/n}$ can be rewritten as $(\sqrt[n]{a})^m$ or $\sqrt[n]{a^m}$. Which radical expression would you use when simplifying $25^{3/2}$ and why?

Using a calculator, approximate each of the following to three decimal places.

71. $10^{3/5}$ 72. $24^{1/4}$

73. $10^{3/2}$ 74. $36^{5/8}$

Simplify.

75. $(x^{2/3})^{5/3}$ 76. $a^{1/4}\, a^{1/2}$

77. $\frac{p^{4/5}}{p^{2/3}}$ 78. $m^{-2/3}\, m^{3/4}\, m^{1/2}$

Graph.

79. $y = \sqrt[3]{x}$ 80. $y = \sqrt[4]{x}$

81. Use a grapher to draw the graphs of $y_1 = x^{2/3}$, $y_2 = x^1$, $y_3 = x^{5/4}$, and $y_4 = x^{3/2}$. Use the window [−1, 17, −1, 32] and the simultaneous mode. Then determine which curve corresponds to each equation.

SUMMARY AND REVIEW | 9

KEY TERMS

Square root, p. 386
Principal square root, p. 386
Radical sign, p. 386
Radical expression, p. 387
Radicand, p. 387
Irrational number, p. 387
Rationalize the denominator, p. 397
Like radicals, p. 400
Conjugates, p. 402
Radical equation, p. 405
Principle of squaring, p. 405
Pythagorean theorem, p. 411
Higher root, p. 416
Cube root, p. 417
Index, p. 417
nth root, p. 417
Rational exponent, p. 418

IMPORTANT PROPERTIES AND FORMULAS

Products: $\sqrt{A}\,\sqrt{B} = \sqrt{A \cdot B}$; $\sqrt[n]{AB} = \sqrt[n]{A}\,\sqrt[n]{B}$

Quotients: $\dfrac{\sqrt{A}}{\sqrt{B}} = \sqrt{\dfrac{A}{B}}$; $\sqrt[n]{\dfrac{A}{B}} = \dfrac{\sqrt[n]{A}}{\sqrt[n]{B}}$

Simplified Form

A radical expression for a square root is considered simplified if all three of the following occur:

1. No radicand contains any perfect-square factors (other than 1).
2. The radicand does not have a denominator.
3. No denominator contains a radical.

The principle of squaring: If $a = b$, then $a^2 = b^2$.

Pythagorean theorem: $a^2 + b^2 = c^2$, where a and b are the lengths of the legs of a right triangle and c is the length of the hypotenuse.

nth roots: $a^{1/n} = \sqrt[n]{a}$

Rational exponents: $a^{m/n} = (\sqrt[n]{a})^m = \sqrt[n]{a^m}$; $a^{-m/n} = \dfrac{1}{a^{m/n}}$

REVIEW EXERCISES

This chapter's review and test include Skill Maintenance exercises from Sections 6.2, 6.3, 7.2, and 7.5.

Find the square roots.

1. 64 **2.** 25

3. 196 **4.** 400

Simplify.

5. $\sqrt{36}$ **6.** $-\sqrt{81}$

7. $\sqrt{49}$ **8.** $-\sqrt{169}$

Identify the radicand.

9. $\sqrt{x^2 + 4}$ **10.** $\sqrt{5ab^3}$

Identify the square root as either rational or irrational.

11. $\sqrt{3}$ **12.** $\sqrt{36}$

13. $-\sqrt{12}$ **14.** $-\sqrt{4}$

Use the calculator or Table 1 to approximate the square root. Round to three decimal places.

15. $\sqrt{3}$ **16.** $\sqrt{99}$

17. $\sqrt{13}$ **18.** $\sqrt{57}$

Simplify. Remember for Exercises 19–40 that we assume all variables represent nonnegative numbers.

19. $\sqrt{m^2}$ **20.** $\sqrt{49t^2}$

21. $\sqrt{p^2}$ **22.** $\sqrt{(ac)^2}$

Multiply.

23. $\sqrt{3}\,\sqrt{7}$ **24.** $\sqrt{a}\,\sqrt{t}$ **25.** $\sqrt{2x}\,\sqrt{3y}$

Simplify by factoring.

26. $\sqrt{48}$ **27.** $\sqrt{32t^2}$ **28.** $\sqrt{98p}$

29. $\sqrt{x^8}$ **30.** $\sqrt{12a^{14}}$ **31.** $\sqrt{m^{15}}$

Multiply and simplify.

32. $\sqrt{6}\sqrt{10}$

33. $\sqrt{5x}\sqrt{8x}$

34. $\sqrt{5x}\sqrt{10xy^2}$

35. $\sqrt{20a^3b}\sqrt{5a^2b^2}$

Divide and simplify.

36. $\dfrac{\sqrt{35}}{\sqrt{45}}$

37. $\dfrac{\sqrt{30y^9}}{\sqrt{54y}}$

Simplify.

38. $\sqrt{\dfrac{25}{64}}$

39. $\sqrt{\dfrac{20}{45}}$

40. $\sqrt{\dfrac{49}{t^2}}$

Rationalize the denominator.

41. $\sqrt{\dfrac{1}{2}}$

42. $\sqrt{\dfrac{5}{8}}$

43. $\sqrt{\dfrac{5}{y}}$

44. $\dfrac{2}{\sqrt{3}}$

Simplify.

45. $10\sqrt{5} + 3\sqrt{5}$

46. $\sqrt{80} - \sqrt{45}$

47. $2\sqrt{x} - \sqrt{25x}$

48. $(2 + \sqrt{3})^2$

49. $(2 + \sqrt{3})(2 - \sqrt{3})$

50. $(1 + 2\sqrt{7})(3 - \sqrt{7})$

Rationalize the denominator.

51. $\dfrac{4}{2 + \sqrt{3}}$

52. $\dfrac{1 + \sqrt{5}}{2 - \sqrt{5}}$

Solve.

53. $\sqrt{x - 3} = 7$

54. $\sqrt{5x + 3} = \sqrt{2x - 1}$

55. $\sqrt{x + 5} = x - 1$

56. $1 + x = \sqrt{1 + 5x}$

57. The formula $r = 2\sqrt{5L}$ can be used to approximate the speed r, in miles per hour, of a car that has left a skid mark of length L, in feet. How far will a car skid at a speed of 90 mph?

In a right triangle, find the length of the side not given. Give an exact answer and, where appropriate, an approximation to three decimal places. Keep in mind that a and b are the lengths of the legs and c is the length of the hypotenuse.

58. $a = 15,\ c = 25$

59. $a = 1,\ b = \sqrt{2}$

60. Find the length of the diagonal of a square whose sides are 7 m long.

Simplify. If an expression does not represent a real number, state so.

61. $\sqrt[5]{32}$

62. $\sqrt[4]{-16}$

63. $\sqrt[3]{-27}$

64. $\sqrt[4]{32}$

Simplify.

65. $100^{1/2}$

66. $9^{-1/2}$

67. $16^{3/2}$

68. $81^{-3/4}$

Skill Maintenance

69. A person's paycheck P varies directly as the number of hours worked H. For working 15 hr, the pay is \$168.75 Find the pay for 40 hr of work.

70. Find the slope and the y-intercept of the graph of $2x - 5y = 14$.

71. Multiply and simplify, if possible:

$$\frac{x^2 + x}{x^2 - 4} \cdot \frac{x^2 + 4x + 4}{x^2}.$$

72. Subtract and simplify, if possible.

$$\frac{2a - 5}{a + 1} - \frac{a - 7}{a + 1}.$$

Synthesis

73. ◈ Explain why a square root of an even power can be found by taking half the exponent.

74. ◈ Why should you simplify each term in a radical expression before attempting to collect like radical terms?

75. Simplify: $\sqrt{\sqrt{\sqrt{256}}}$.

76. Solve: $\sqrt{x^2} = -10$.

77. Use square roots to factor $x^2 - 5$.

78. Solve $A = \sqrt{a^2 + b^2}$ for b.

CHAPTER TEST 9

1. Find the square roots of 81.

Simplify.

2. $\sqrt{64}$

3. $-\sqrt{25}$

4. Identify the radicand in $\sqrt{4 - y^3}$.

Identify the square root as either rational or irrational.

5. $\sqrt{10}$

6. $\sqrt{16}$

Approximate using a calculator or Table 1. Round to three decimal places.

7. $\sqrt{87}$

8. $\sqrt{7}$

Simplify.

9. $\sqrt{a^2}$

10. $\sqrt{36y^2}$

Multiply.

11. $\sqrt{5}\sqrt{6}$

12. $\sqrt{7a}\sqrt{2b}$

Simplify by factoring.

13. $\sqrt{27}$

14. $\sqrt{t^5}$

15. $\sqrt{18a^6}$

Multiply and simplify.

16. $\sqrt{5}\sqrt{10}$

17. $\sqrt{5t}\sqrt{5t}$

18. $\sqrt{3ab}\sqrt{6ab^3}$

Divide and simplify.

19. $\dfrac{\sqrt{18}}{\sqrt{32}}$

20. $\dfrac{\sqrt{35x}}{\sqrt{80xy^2}}$

Simplify.

21. $\sqrt{\dfrac{27}{12}}$

22. $\sqrt{\dfrac{144}{a^2}}$

Rationalize the denominator.

23. $\sqrt{\dfrac{2}{5}}$

24. $\dfrac{2x}{\sqrt{y}}$

Simplify.

25. $3\sqrt{18} - 5\sqrt{18}$

26. $\sqrt{27} + 2\sqrt{12}$

27. $(4 - \sqrt{5})^2$

28. $(4 - \sqrt{5})(4 + \sqrt{5})$

29. Rationalize the denominator: $\dfrac{10}{4 - \sqrt{5}}$.

30. The legs of a right triangle are 8 cm and 4 cm long. Find the length of the hypotenuse. Give an exact answer and an approximation to three decimal places.

Solve.

31. $\sqrt{3x} + 2 = 14$

32. $\sqrt{6x + 13} = x + 3$

33. A person can see 247.49 km to the horizon from an airplane window. How high is the airplane? Use the formula $V = 3.5\sqrt{h}$.

Simplify. If an expression does not represent a real number, state so.

34. $\sqrt[4]{16}$

35. $-\sqrt[6]{1}$

36. $\sqrt[3]{-64}$

37. $\sqrt[4]{-81}$

38. $9^{1/2}$

39. $27^{-1/3}$

40. $100^{3/2}$

41. $16^{-5/4}$

Skill Maintenance

42. The number N of switches that a production line can make varies directly as the time it operates. It can make 7224 switches in 6 hr. How many can it make in 13 hr?

43. Find the slope–intercept equation for a line with slope $-\frac{1}{2}$ and y-intercept $(0, -1)$.

44. Divide and simplify, if possible:

$$\frac{a^2 - 1}{a} \div \frac{a - 1}{a + 1}.$$

45. Add and simplify, if possible:

$$\frac{2x + 1}{3x - 5} + \frac{x - 3}{5 - 3x}.$$

Synthesis

46. Solve: $\sqrt{1 - x} + 1 = \sqrt{6 - x}$.

47. Simplify: $\sqrt{y^{16n}}$.

CHAPTER 10

Quadratic Equations

AN APPLICATION

The formula

$$A = P(1 + r)^t$$

is used to find the amount A in an account when P dollars is invested for t years at a yearly interest rate r.

If $2560 grows to $3610 in 2 years, determine the interest rate r.

This problem appears as Exercise 63 in Section 10.3.

Annette M. Hawley
BANK OPERATIONS OFFICER

"I encourage the tellers I train to gain as strong a background in math as possible. I myself use math regularly when calculating monthly and yearly interest accruals and when recalculating interest payments for customers. My department also puts together audit schedules for annual review by an independent accounting firm."

Quadratic equations first appeared in Section 5.6. At that time, we used the principle of zero products because all of the equations could be solved by factoring. In this chapter, we will learn methods for solving *any* quadratic equation. These methods are then used to solve applications and to assist us in graphing.

In addition to the material from this chapter, the review and test for Chapter 10 include material from Sections 7.3, 8.2, 9.4, and 9.6.

10.1 Solving Quadratic Equations: The Principle of Square Roots

The Principle of Square Roots • Solving Quadratic Equations of the Type $(x + k)^2 = p$

The following are examples of quadratic equations:

$$x^2 - 7x + 9 = 0, \qquad 5t^2 - 4t = 8, \qquad 6y^2 = -9y, \qquad m^2 = 49.$$

We saw in Chapter 5 that one way to solve an equation like $m^2 = 49$ is to subtract 49 on both sides, factor, and then use the principle of zero products:

$$m^2 - 49 = 0$$
$$(m + 7)(m - 7) = 0$$
$$m + 7 = 0 \quad or \quad m - 7 = 0$$
$$m = -7 \quad or \quad m = 7.$$

This approach relies on our ability to factor. By using the *principle of square roots,* we can develop a method for solving equations like $m^2 = 49$ that can be used to solve equations for which factoring is impractical.

The Principle of Square Roots

One way to solve an equation like $m^2 = 49$ is to search for a number whose square is 49. There are two such numbers, -7 and 7. Note that -7 and 7 are the square roots of 49.

The Principle of Square Roots

For any nonnegative real number k, if $x^2 = k$, then $x = \sqrt{k}$ or $x = -\sqrt{k}$.

EXAMPLE 1

Solve: $x^2 = 16$.

Solution We use the principle of square roots:

$$x^2 = 16$$
$$x = \sqrt{16} \quad or \quad x = -\sqrt{16} \qquad \text{Using the principle of square roots}$$
$$x = 4 \quad or \quad x = -4. \qquad \text{Simplifying}$$

We check mentally that $4^2 = 16$ and $(-4)^2 = 16$. The solutions are 4 and -4. ❑

Unlike the principle of zero products, the principle of square roots can be easily used to solve quadratic equations that have irrational solutions.

EXAMPLE 2

Solve: **(a)** $x^2 = 17$; **(b)** $5x^2 = 15$; **(c)** $-3x^2 + 7 = 0$.

Solution

a)

$$x^2 = 17$$

$$x = \sqrt{17} \quad or \quad x = -\sqrt{17} \qquad \text{Using the principle of square roots}$$

Check: For $\sqrt{17}$:

$$\begin{array}{c|l} x^2 & 17 \\ \hline (\sqrt{17})^2 & ?\ 17 \\ 17 & 17 \quad \text{TRUE} \end{array}$$

For $-\sqrt{17}$:

$$\begin{array}{c|l} x^2 & 17 \\ \hline (-\sqrt{17})^2 & ?\ 17 \\ 17 & 17 \quad \text{TRUE} \end{array}$$

The solutions are $\sqrt{17}$ and $-\sqrt{17}$.

b)

$$5x^2 = 15$$

$$x^2 = 3 \qquad \text{Solving for } x^2$$

$$x = \sqrt{3} \quad or \quad x = -\sqrt{3} \qquad \text{Using the principle of square roots}$$

We leave the check to the student. The solutions are $\sqrt{3}$ and $-\sqrt{3}$.

c)

$$-3x^2 + 7 = 0$$

$$-3x^2 = -7 \qquad \text{Subtracting 7}$$

$$x^2 = \frac{-7}{-3} \qquad \text{Dividing by } -3$$

$$x^2 = \frac{7}{3}$$

$$x = \sqrt{\frac{7}{3}} \quad or \quad x = -\sqrt{\frac{7}{3}} \qquad \text{Using the principle of square roots}$$

$$x = \frac{\sqrt{7}}{\sqrt{3}} \cdot \frac{\sqrt{3}}{\sqrt{3}} \quad or \quad x = -\frac{\sqrt{7}}{\sqrt{3}} \cdot \frac{\sqrt{3}}{\sqrt{3}} \qquad \text{Rationalizing the denominators}$$

$$x = \frac{\sqrt{21}}{3} \quad or \quad x = -\frac{\sqrt{21}}{3}$$

Check: For $\frac{\sqrt{21}}{3}$:

$$\begin{array}{c|l} -3x^2 + 7 & 0 \\ \hline -3\left(\frac{\sqrt{21}}{3}\right)^2 + 7 & ?\ 0 \\ -3 \cdot \frac{21}{9} + 7 & \\ -7 + 7 & \\ 0 & 0 \quad \text{TRUE} \end{array}$$

For $-\frac{\sqrt{21}}{3}$:

$$\begin{array}{c|l} -3x^2 + 7 & 0 \\ \hline -3\left(-\frac{\sqrt{21}}{3}\right)^2 + 7 & ?\ 0 \\ -3 \cdot \frac{21}{9} + 7 & \\ -7 + 7 & \\ 0 & 0 \quad \text{TRUE} \end{array}$$

The solutions are $\frac{\sqrt{21}}{3}$ and $-\frac{\sqrt{21}}{3}$. ❑

Solving Quadratic Equations of the Type $(x + k)^2 = p$

Equations like $(x - 5)^2 = 9$ or $(x + 2)^2 = 7$ are of the form $(x + k)^2 = p$. The principle of square roots can be used to solve such equations.

EXAMPLE 3

Solve: **(a)** $(x - 5)^2 = 9$; **(b)** $(x + 2)^2 = 7$.

Solution

a) $(x - 5)^2 = 9$

$x - 5 = 3$ *or* $x - 5 = -3$ Using the principle of square roots

$x = 8$ *or* $x = 2$

The solutions are 8 and 2.

b) $(x + 2)^2 = 7$

$x + 2 = \sqrt{7}$ *or* $x + 2 = -\sqrt{7}$ Using the principle of square roots

$x = -2 + \sqrt{7}$ *or* $x = -2 - \sqrt{7}$

The solutions are $-2 + \sqrt{7}$ and $-2 - \sqrt{7}$, or simply $-2 \pm \sqrt{7}$ (read "-2 plus or minus $\sqrt{7}$"). ❑

In Example 3, the left sides of the equations are squares of binomials. If we can express an equation in such a form, we can proceed as we did in that example.

EXAMPLE 4

Solve by factoring and using the principle of square roots: **(a)** $x^2 + 8x + 16 = 49$; **(b)** $x^2 + 6x + 9 = 10$.

Solution

a) $x^2 + 8x + 16 = 49$ The left side is a trinomial square.

$(x + 4)^2 = 49$ A trinomial square is a binomial squared.

$x + 4 = 7$ *or* $x + 4 = -7$ Using the principle of square roots

$x = 3$ *or* $x = -11$

The solutions are 3 and -11.

b) $x^2 + 6x + 9 = 10$ The left side is a trinomial square.

$(x + 3)^2 = 10$ A trinomial square is a binomial squared.

$x + 3 = \sqrt{10}$ *or* $x + 3 = -\sqrt{10}$ Using the principle of square roots

$x = -3 + \sqrt{10}$ *or* $x = -3 - \sqrt{10}$

The solutions are $-3 + \sqrt{10}$ and $-3 - \sqrt{10}$, or simply $-3 \pm \sqrt{10}$. ❑

EXERCISE SET 10.1

Solve. Use the principle of square roots.

1. $x^2 = 25$ **2.** $x^2 = 36$ **3.** $a^2 = 81$

4. $a^2 = 121$ **5.** $m^2 = 15$ **6.** $t^2 = 10$

7. $x^2 = 19$ **8.** $a^2 = 29$ **9.** $5a^2 = 35$

10. $3x^2 = 84$ **11.** $7x^2 = 140$ **12.** $9m^2 = 72$

13. $4t^2 - 25 = 0$ **14.** $9a^2 - 4 = 0$

15. $3x^2 - 49 = 0$
16. $5x^2 - 16 = 0$
17. $4y^2 - 3 = 9$
18. $49y^2 - 5 = 15$
19. $25x^2 - 35 = 0$
20. $5x^2 - 120 = 0$
21. $(x - 2)^2 = 49$
22. $(x + 1)^2 = 25$
23. $(x + 3)^2 = 36$
24. $(x - 4)^2 = 81$
25. $(m + 3)^2 = 21$
26. $(m - 3)^2 = 6$
27. $(a + 13)^2 = 8$
28. $(a - 13)^2 = 64$
29. $(x - 7)^2 = 12$
30. $(x + 1)^2 = 14$
31. $(x + 9)^2 = 34$
32. $(t + 2)^2 = 25$
33. $(x + \frac{3}{2})^2 = \frac{7}{2}$
34. $(y - \frac{3}{4})^2 = \frac{17}{16}$
35. $x^2 - 6x + 9 = 64$
36. $x^2 - 10x + 25 = 100$
37. $y^2 + 14y + 49 = 4$
38. $p^2 + 8p + 16 = 1$
39. $m^2 - 2m + 1 = 5$
40. $t^2 + 6t + 9 = 13$
41. $x^2 + 4x + 4 = 12$
42. $x^2 - 12x + 36 = 18$

Skill Maintenance

Determine whether the pair of equations represents parallel lines.

43. $y + 5 = 2x,$
$y - 2x = 7$

44. $y - 4x = 6,$
$x + 4y = 8$

45. $3x - 2y = 9,$
$4y - 6x = 8$

46. $2x - 3y = 12,$
$3y - 2x = 24$

Synthesis

47. Explain why 9 is not *the* solution of $x^2 = 81$.

48. Write a quadratic equation that is most easily solved using the principle of square roots. Explain why the principle of zero products would not work as easily on the equation you wrote.

Factor the left side of the equation. Then solve.

49. $x^2 + 5x + \frac{25}{4} = \frac{13}{4}$
50. $x^2 - \frac{7}{3}x + \frac{49}{36} = \frac{7}{36}$
51. $m^2 - \frac{3}{2}m + \frac{9}{16} = \frac{17}{16}$
52. $t^2 + 3t + \frac{9}{4} = \frac{49}{4}$
53. $x^2 + 0.5x + 0.0625 = 13.69$
54. $x^2 + 2.5x + 1.5625 = 9.61$
55. $a^2 - 3.8a + 3.61 = 27.04$
56. $a^2 - 5.2a + 6.76 = 53.29$

10.2

Solving Quadratic Equations: Completing the Square

Completing the Square • Solving by Completing the Square

In Section 10.1, we solved equations like $(x - 5)^2 = 7$ using the principle of square roots. Equations like $x^2 + 8x + 16 = 12$ were also solved using the principle of square roots because the expression on the left side is a trinomial square. We now learn to solve equations like $x^2 + 10x = 4$, in which the left side is not (yet) a trinomial square. The new procedure involves *completing the square* and enables us to solve any quadratic equation.

Completing the Square

Consider the following quadratic equation:

$$x^2 + 10x = 4.$$

We would like to add some constant to both sides of the equation that would make the left side a perfect-square trinomial. To determine what that constant should be, recall that after a binomial of the form $x + a$ is squared, the coefficient of the x-term is $2a$:

$$(x + a)^2 = x^2 + 2ax + a^2.$$

In the trinomial above, a^2 is the square of half the coefficient of x. Returning to the equation $x^2 + 10x = 4$, note that half the coefficient of x is 5, and $5^2 = 25$. This suggests that we should add 25 on both sides:

$$x^2 + 10x = 4$$

$$x^2 + 10x + 25 = 4 + 25 \qquad \text{Adding 25 on both sides}$$

$$(x + 5)^2 = 29. \qquad \text{Factoring the trinomial square}$$

By adding 25 to $x^2 + 10x$, we have *completed the square*. The resulting equation contains the square of a binomial on one side. Thus we can find the solutions by using the principle of square roots, as in Section 10.1:

$$(x + 5)^2 = 29$$

$$x + 5 = \sqrt{29} \quad or \quad x + 5 = -\sqrt{29} \qquad \text{Using the principle of square roots}$$

$$x = -5 + \sqrt{29} \quad or \quad x = -5 - \sqrt{29}.$$

The solutions are $-5 \pm \sqrt{29}$.

The key to solving $x^2 + 10x = 4$ was adding 25 on both sides. Once we did so, the left side became a trinomial square.

Completing the Square

To *complete the square* for an expression like $x^2 + bx$, take half the coefficient of x and square it. Then add that number, which is $(b/2)^2$.

A visual interpretation of completing the square is sometimes helpful. Consider the following figures.

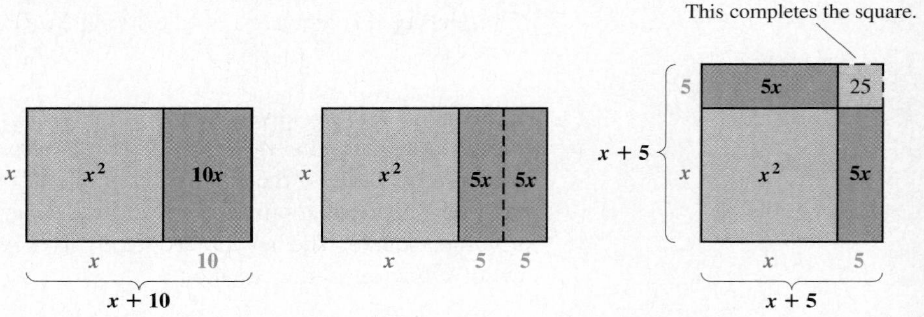

This completes the square.

In all figures, the sum of the red and purple areas is $x^2 + 10x$. However, by splitting the purple area in half, we can "complete" a square by adding the blue area. The blue area is $5 \cdot 5$, or 25 square units.

EXAMPLE 1

Complete the square: **(a)** $x^2 - 12x$; **(b)** $x^2 + 5x$.

Solution

a) To complete the square for $x^2 - 12x$, note that the coefficient of x is -12. Half of -12 is -6 and $(-6)^2$ is 36. Thus, $x^2 - 12x$ becomes a trinomial square when

36 is added:

$x^2 - 12x + 36$ is the square of $x - 6$.

That is,

$$x^2 - 12x + 36 = (x - 6)^2.$$

b) To complete the square for $x^2 + 5x$, we take half the coefficient of x and square it:

$$(\tfrac{5}{2})^2 = \tfrac{25}{4}.$$

The trinomial $x^2 + 5x + \frac{25}{4}$ is the square of $x + \frac{5}{2}$. That is,

$$x^2 + 5x + \tfrac{25}{4} = (x + \tfrac{5}{2})^2.$$ ❑

Solving by Completing the Square

The steps that we used to solve $x^2 + 10x = 4$ can be used in a wide variety of problems.

EXAMPLE 2

Solve by completing the square: **(a)** $x^2 + 6x = -8$; **(b)** $x^2 - 10x + 14 = 0$.

Solution

a) To solve $x^2 + 6x = -8$, we take half of 6 and square it, to get 9. Then we add 9 on both sides of the equation. This makes the left side the square of a binomial. Now we can solve:

$$x^2 + 6x + 9 = -8 + 9$$ Adding 9 on both sides to complete the square

$$(x + 3)^2 = 1$$ Writing the trinomial as a binomial squared

$$x + 3 = 1 \quad or \quad x + 3 = -1$$ Using the principle of square roots

$$x = -2 \quad or \quad x = -4.$$

The solutions are -2 and -4.

b) We have

$$x^2 - 10x + 14 = 0$$

$$x^2 - 10x = -14$$ Subtracting 14 on both sides

$$x^2 - 10x + 25 = -14 + 25$$ Adding 25 on both sides to complete the square: $(-10/2)^2 = 25$

$$(x - 5)^2 = 11$$ Factoring

$$x - 5 = \sqrt{11} \quad or \quad x - 5 = -\sqrt{11}$$ Using the principle of square roots

$$x = 5 + \sqrt{11} \quad or \quad x = 5 - \sqrt{11}.$$

The solutions are $5 + \sqrt{11}$ and $5 - \sqrt{11}$, or simply $5 \pm \sqrt{11}$. ❑

To complete the square, we must be sure that the coefficient of x^2 is 1. When the x^2-coefficient is not 1, we can multiply or divide on both sides to find an equivalent equation with an x^2-coefficient of 1.

EXAMPLE 3

Solve by completing the square: **(a)** $3x^2 + 24x = 3$; **(b)** $2x^2 - 3x - 1 = 0$.

Solution

a)

$$3x^2 + 24x = 3$$

$$\left.\begin{aligned}\tfrac{1}{3}(3x^2 + 24x) &= \tfrac{1}{3}\cdot 3\\ x^2 + 8x &= 1\end{aligned}\right\}$$ We multiply by $\frac{1}{3}$ to ensure an x^2-coefficient of 1.

$$x^2 + 8x + 16 = 1 + 16$$ Adding 16 on both sides to complete the square: $\left(\frac{8}{2}\right)^2 = 16$

$$(x + 4)^2 = 17$$ Factoring

$$x + 4 = \sqrt{17} \quad or \quad x + 4 = -\sqrt{17}$$

$$x = -4 + \sqrt{17} \quad or \quad x = -4 - \sqrt{17}$$

The solutions are $-4 \pm \sqrt{17}$.

b)

$$2x^2 - 3x - 1 = 0$$

$$\frac{1}{2}(2x^2 - 3x - 1) = \frac{1}{2}\cdot 0$$ Multiplying by $\frac{1}{2}$ to make the x^2-coefficient 1

$$x^2 - \frac{3}{2}x - \frac{1}{2} = 0$$

$$x^2 - \frac{3}{2}x = \frac{1}{2}$$ Adding $\frac{1}{2}$ on both sides

$$x^2 - \frac{3}{2}x + \frac{9}{16} = \frac{1}{2} + \frac{9}{16}$$ Adding $\frac{9}{16}$ on both sides: $\left[\frac{1}{2}\left(-\frac{3}{2}\right)\right]^2 = \left[-\frac{3}{4}\right]^2 = \frac{9}{16}$. This completes the square on the left side.

$$\left(x - \frac{3}{4}\right)^2 = \frac{8}{16} + \frac{9}{16}$$ Factoring and finding a common denominator

$$\left(x - \frac{3}{4}\right)^2 = \frac{17}{16}$$

$$x - \frac{3}{4} = \frac{\sqrt{17}}{4} \quad or \quad x - \frac{3}{4} = -\frac{\sqrt{17}}{4}$$ Using the principle of square roots

$$x = \frac{3}{4} + \frac{\sqrt{17}}{4} \quad or \quad x = \frac{3}{4} - \frac{\sqrt{17}}{4}$$

The solutions are $\dfrac{3 \pm \sqrt{17}}{4}$. ❑

The steps used in Example 3 can be used to solve any quadratic equation.

Solving by Completing the Square

To solve a quadratic equation $ax^2 + bx + c = 0$ by completing the square:

1. If $a \neq 1$, multiply by $1/a$ on both sides so that the x^2-coefficient is 1.
2. When the x^2-coefficient is 1, rewrite the equation in the form

$$x^2 + bx = -c, \quad \text{or} \quad x^2 + \frac{b}{a}x = -\frac{c}{a} \quad \text{if step (1) has been applied.}$$

3. Take half of the x-coefficient and square it. Add the result on both sides of the equation.
4. Express the side with the variables as the square of a binomial.
5. Use the principle of square roots and complete the solution.

EXERCISE SET 10.2

Complete the square.

1. $x^2 - 2x$
2. $x^2 - 4x$
3. $x^2 + 18x$
4. $x^2 + 22x$
5. $x^2 - x$
6. $x^2 + x$
7. $t^2 + 5t$
8. $y^2 - 9y$
9. $x^2 - \frac{3}{2}x$
10. $x^2 + \frac{4}{3}x$
11. $m^2 + \frac{9}{2}m$
12. $r^2 - \frac{2}{5}r$

Solve by completing the square.

13. $x^2 - 6x - 16 = 0$
14. $x^2 + 8x + 15 = 0$
15. $x^2 + 22x + 21 = 0$
16. $x^2 + 14x - 15 = 0$
17. $3x^2 - 6x - 15 = 0$
18. $3x^2 - 12x - 33 = 0$
19. $x^2 - 22x + 102 = 0$
20. $x^2 - 18x + 74 = 0$
21. $x^2 + 10x - 4 = 0$
22. $x^2 - 10x - 4 = 0$
23. $x^2 - 7x - 2 = 0$
24. $x^2 + 7x - 2 = 0$
25. $2x^2 + 6x - 56 = 0$
26. $2x^2 - 6x - 56 = 0$
27. $x^2 + \frac{3}{2}x - \frac{1}{2} = 0$
28. $x^2 - \frac{3}{2}x - 2 = 0$
29. $2x^2 + 3x - 17 = 0$
30. $2x^2 - 3x - 7 = 0$
31. $3x^2 + 4x - 1 = 0$
32. $3x^2 - 4x - 3 = 0$
33. $2x^2 = 9x + 5$
34. $2x^2 = 5x + 12$
35. $4x^2 + 12x = 7$
36. $6x^2 + 11x = 10$

Skill Maintenance

Solve the system using the substitution method.

37. $y - x = 5,$
$y + 2x = 7$
38. $2x + 3y = 8,$
$x = y - 6$

Graph.

39. $y = \frac{3}{5}x - 1$
40. $y - 4 = \frac{2}{3}(x + 1)$

Synthesis

41. ◈ Explain in your own words how completing the square enables us to solve equations we could not otherwise have solved.
42. ◈ A student states that "since solving a quadratic equation by completing the square relies on the principle of square roots, the solutions are always opposites of each other." Is the student correct? Why or why not?

Find b such that the trinomial is a square.

43. $x^2 + bx + 36$
44. $x^2 + bx + 55$
45. $x^2 + bx + 128$
46. $4x^2 + bx + 16$
47. $x^2 + bx + c$
48. $ax^2 + bx + c$

We can use a grapher to solve equations by letting y_1 represent the left side of an equation and y_2 represent the right side, and graphing y_1 and y_2 on the same set of axes. The Trace and Zoom features can then be used to determine the x-coordinate at any point of intersection. Use this approach to find solutions, accurate to two decimal places, of each of the following equations.

49. $(x - 5)^2 = 9$
50. $(x + 3)^2 = 25$
51. $(x + 4)^2 = 13$
52. $(x - 6)^2 = 2$
53. $x^2 - 7x - 2 = 0$ (Exercise 23)
54. $x^2 + 7x - 2 = 0$ (Exercise 24)
55. $2x^2 = 9x + 5$ (Exercise 33)
56. $2x^2 = 5x + 12$ (Exercise 34)

10.3 The Quadratic Formula and Problem Solving

The Quadratic Formula • Problem Solving

We now derive the *quadratic formula*. This formula enables us to solve quadratic equations more quickly than the method of completing the square.

The Quadratic Formula

When mathematicians use a procedure repeatedly on a wide variety of problems, they generally try to find a formula for the procedure. The quadratic formula condenses the many steps used to solve a quadratic equation by completing the square.

Consider a quadratic equation in *standard form*, $ax^2 + bx + c = 0$, with $a > 0$. Our plan is to solve this equation for x by completing the square. As the steps are performed, compare them with those in Example 3(b) on page 432:

$$\frac{1}{a}(ax^2 + bx + c) = \frac{1}{a} \cdot 0 \qquad \text{Multiplying by } \frac{1}{a} \text{ to make the } x^2\text{-coefficient 1}$$

$$x^2 + \frac{b}{a}x + \frac{c}{a} = 0$$

$$x^2 + \frac{b}{a}x = -\frac{c}{a}. \qquad \text{Adding } -\frac{c}{a} \text{ on both sides}$$

Half of b/a is $b/(2a)$, and the square of $b/(2a)$ is $b^2/(4a^2)$. We add $b^2/(4a^2)$ on both sides:

$$x^2 + \frac{b}{a}x + \frac{b^2}{4a^2} = -\frac{c}{a} + \frac{b^2}{4a^2} \qquad \text{Adding } \frac{b^2}{4a^2} \text{ on both sides. This completes the square on the left side.}$$

$$\left(x + \frac{b}{2a}\right)^2 = -\frac{4ac}{4a^2} + \frac{b^2}{4a^2} \qquad \text{Factoring and finding a common denominator}$$

$$\left(x + \frac{b}{2a}\right)^2 = \frac{b^2 - 4ac}{4a^2}$$

$$x + \frac{b}{2a} = \sqrt{\frac{b^2 - 4ac}{4a^2}} \quad \text{or} \quad x + \frac{b}{2a} = -\sqrt{\frac{b^2 - 4ac}{4a^2}}. \qquad \text{Using the principle of square roots}$$

Since $a > 0$, $\sqrt{4a^2} = 2a$, so we can simplify as follows:

$$x + \frac{b}{2a} = \frac{\sqrt{b^2 - 4ac}}{2a} \quad \text{or} \quad x + \frac{b}{2a} = -\frac{\sqrt{b^2 - 4ac}}{2a}.$$

Thus,

$$x = -\frac{b}{2a} + \frac{\sqrt{b^2 - 4ac}}{2a} \quad \text{or} \quad x = -\frac{b}{2a} - \frac{\sqrt{b^2 - 4ac}}{2a},$$

so

$$x = -\frac{b}{2a} \pm \frac{\sqrt{b^2 - 4ac}}{2a},$$

or

$$x = \frac{-b \pm \sqrt{b^2 - 4ac}}{2a}.$$

This last equation is the result we were after. It is so useful that it is worth memorizing.

The Quadratic Formula

The solutions of $ax^2 + bx + c = 0$ are given by

$$x = \frac{-b \pm \sqrt{b^2 - 4ac}}{2a}.$$

Note that the formula also holds when $a < 0$. A similar proof would show this, but we will not consider it here.

EXAMPLE 1

Solve using the quadratic formula: **(a)** $4x^2 + 5x - 6 = 0$; **(b)** $x^2 = 4x + 7$; **(c)** $x^2 + x = -1$.

Solution

a) We identify a, b, and c and substitute into the quadratic formula:

$$\underset{a}{\underset{\uparrow}{4}}x^2 + \underset{b}{\underset{\uparrow}{5}}x \underset{c}{\underset{\uparrow}{-\,6}} = 0$$

$$x = \frac{-b \pm \sqrt{b^2 - 4ac}}{2a}$$

$$x = \frac{-5 \pm \sqrt{5^2 - 4 \cdot 4(-6)}}{2 \cdot 4}$$

Be sure to write the fraction bar all the way across.

$$x = \frac{-5 \pm \sqrt{25 - (-96)}}{8}$$

$$x = \frac{-5 \pm \sqrt{121}}{8}$$

$$x = \frac{-5 \pm 11}{8}$$

$$x = \frac{-5 + 11}{8} \quad or \quad x = \frac{-5 - 11}{8}$$

$$x = \frac{6}{8} \quad or \quad x = \frac{-16}{8}$$

$$x = \frac{3}{4} \quad or \quad x = -2.$$

The solutions are $\frac{3}{4}$ and -2.

b) We rewrite $x^2 = 4x + 7$ in standard form, identify a, b, and c, and solve using the quadratic formula:

$$\underset{a}{\underset{\uparrow}{1}}x^2 \underset{b}{\underset{\uparrow}{-\,4}}x \underset{c}{\underset{\uparrow}{-\,7}} = 0$$

$$x = \frac{-(-4) \pm \sqrt{(-4)^2 - 4(1)(-7)}}{2 \cdot 1}$$ Substituting into the quadratic formula

$$x = \frac{4 \pm \sqrt{16 + 28}}{2}$$

$$x = \frac{4 \pm \sqrt{44}}{2}.$$

Since $\sqrt{44}$ can be simplified, we have

$$x = \frac{4 \pm \sqrt{4}\sqrt{11}}{2}$$

$$x = \frac{4 \pm 2\sqrt{11}}{2}.$$

Finally, since 2 is a common factor of 4 and $2\sqrt{11}$, we can simplify the fraction by removing a factor of 1:

$$x = \frac{2(2 \pm \sqrt{11})}{2 \cdot 1} \quad \text{Factoring}$$

$$x = \frac{\cancel{2}}{\cancel{2}} \cdot \frac{2 \pm \sqrt{11}}{1}. \quad \text{Removing a factor of 1: } \frac{2}{2} = 1$$

The solutions are $2 + \sqrt{11}$ and $2 - \sqrt{11}$, or $2 \pm \sqrt{11}$.

c) We rewrite $x^2 + x = -1$ in standard form and use the quadratic formula:

$$\underset{\substack{\uparrow \\ a}}{1}x^2 + \underset{\substack{\uparrow \\ b}}{1}x + \underset{\substack{\uparrow \\ c}}{1} = 0$$

$$x = \frac{-1 \pm \sqrt{1^2 - 4 \cdot 1 \cdot 1}}{2 \cdot 1} \quad \text{Substituting into the quadratic formula}$$

$$x = \frac{-1 \pm \sqrt{1 - 4}}{2}$$

$$x = \frac{-1 \pm \sqrt{-3}}{2}.$$

Since the radicand, -3, is negative, there are no real-number solutions and we state this as the answer. In Section 10.4, we will develop a number system in which solutions of this equation can be found. ❑

The following are general guidelines for solving a quadratic equation.

To solve a quadratic equation:

1. If it is in the form $ax^2 = p$ or $(x + k)^2 = p$, use the principle of square roots as in Section 10.1.
2. When it is not in the form of (1), write it in standard form,

   $$ax^2 + bx + c = 0.$$

3. Try to factor and use the principle of zero products.
4. If it is not possible to factor or if factoring seems difficult, use the quadratic formula.

The solutions of a quadratic equation can always be found using the quadratic formula. They cannot always be found by factoring. When the radicand, $b^2 - 4ac$, is nonnegative, the equation has real-number solutions. When $b^2 - 4ac$ is negative, the equation has no real-number solutions.

Problem Solving

EXAMPLE 2

The number of diagonals d of a polygon of n sides is given by the formula

$$d = \frac{n^2 - 3n}{2}.$$

If a polygon has 27 diagonals, how many sides does it have?

Solution

1. FAMILIARIZE. A sketch can help us to become familiar with the problem. We draw a hexagon (6 sides) and count the diagonals. As the formula predicts, for $n = 6$, there are

$$\frac{6^2 - 3 \cdot 6}{2} = \frac{36 - 18}{2} = \frac{18}{2} = 9 \text{ diagonals.}$$

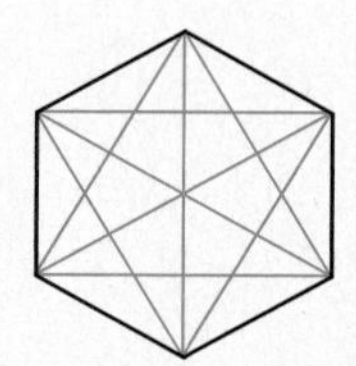

We might wonder if, when there are three times as many diagonals, there are three times as many sides. Using the above formula, you can confirm that this is *not* the case. Rather than continue guessing, we proceed to a translation.

2. TRANSLATE. Since the number of diagonals is 27, we substitute 27 for d:

$$27 = \frac{n^2 - 3n}{2}.$$

This gives us a translation.

3. CARRY OUT. We solve the equation for n, first reversing the equation for convenience:

$$\frac{n^2 - 3n}{2} = 27$$

$$n^2 - 3n = 54 \quad \text{Multiplying by 2 to clear fractions}$$

$$n^2 - 3n - 54 = 0 \quad \text{Subtracting 54 on both sides}$$

$$(n - 9)(n + 6) = 0 \quad \text{Factoring. There is no need for the quadratic formula here.}$$

$$n - 9 = 0 \quad or \quad n + 6 = 0$$

$$n = 9 \quad or \quad n = -6.$$

4. CHECK. Since the number of sides cannot be negative, -6 cannot be a solution. We leave it to the student to show by substitution that 9 checks.

5. STATE. The polygon has 9 sides (it is a nonagon). ❑

EXAMPLE 3

The World Trade Center in New York City is 1368 ft tall. How many seconds will it take an object to fall from the top? Round to the nearest hundredth.

Solution

1. FAMILIARIZE. If we did not know anything about this problem, we might consider looking up a formula in a mathematics or physics book. A formula that

fits this situation is

$$s = 16t^2,$$

where s is the distance, in feet, traveled by a body falling freely from rest in t seconds. This formula is actually an approximation in that it does not account for air resistance. In this problem, we know the distance s to be 1368. We want to determine the time t for the object to reach the ground. If we check a couple of guesses, we can see that the time t must be between 5 and 10 sec.

2. TRANSLATE. The distance is 1368 ft and we need to solve for t. We substitute 1368 for s in the formula above to get the following translation:

$$1368 = 16t^2.$$

3. CARRY OUT. Because there is no t-term, we can use the principle of square roots to solve:

$$1368 = 16t^2$$

$$\frac{1368}{16} = t^2 \quad \text{Solving for } t^2$$

$$\sqrt{\frac{1368}{16}} = t \quad or \quad -\sqrt{\frac{1368}{16}} = t \quad \text{Using the principle of square roots}$$

$$\frac{\sqrt{1368}}{4} = t \quad or \quad \frac{-\sqrt{1368}}{4} = t$$

$$9.25 \approx t \quad or \quad -9.25 \approx t. \quad \text{Using a calculator or Table 1 and rounding to the nearest hundredth}$$

4. CHECK. The number -9.25 cannot be a solution because time cannot be negative in this situation. We substitute 9.25 in the original equation:

$$s = 16(9.25)^2 = 16(85.5625) = 1369.$$

This is close. Remember that we approximated a solution. As we expected in step (1), the solution is between 5 and 10 sec.

5. STATE. It takes about 9.25 sec for the object to fall to the ground from the top of the World Trade Center.

EXAMPLE 4

The hypotenuse of a right triangle is 6 m long. One leg is 1 m longer than the other. Find the lengths of the legs. Round to the nearest hundredth.

Solution

1. FAMILIARIZE. We first make a drawing and label it. We let $s =$ the length of one leg. Then $s + 1 =$ the length of the other leg.

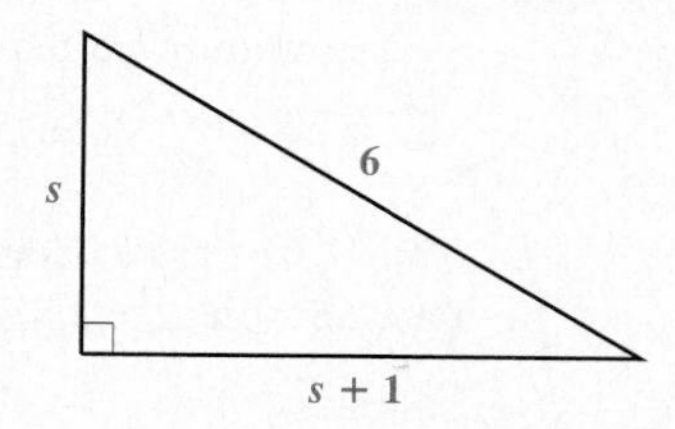

Note that if $s = 3$, then $s + 1 = 4$ and $3^2 + 4^2 = 25 \neq 6^2$. Thus, because of the Pythagorean theorem, we see that $s \neq 3$. Another guess, $s = 4$, is too big since

$4^2 + (4+1)^2 = 41 \neq 6^2$. Although we have not guessed the solution, we expect the value of s to be between 3 and 4.

2. TRANSLATE. To translate, we use the Pythagorean theorem:

$$s^2 + (s+1)^2 = 6^2.$$

3. CARRY OUT. We solve the equation:

$$s^2 + (s+1)^2 = 6^2$$

$$s^2 + s^2 + 2s + 1 = 36$$

$$\underset{a}{\underset{\uparrow}{2}}s^2 + \underset{b}{\underset{\uparrow}{2}}s \underset{c}{\underset{\uparrow}{-\,35}} = 0$$ This cannot be factored so we use the quadratic formula.

$$s = \frac{-2 \pm \sqrt{2^2 - 4\cdot 2(-35)}}{2\cdot 2}$$ Remember: $s = \dfrac{-b \pm \sqrt{b^2 - 4ac}}{2a}$.

$$s = \frac{-2 \pm \sqrt{4 + 280}}{4} = \frac{-2 \pm \sqrt{284}}{4}$$

$s \approx 3.71$ *or* $s \approx -4.71$. Using a calculator or Table 1 and rounding to the nearest hundredth

4. CHECK. Length cannot be negative, so -4.71 does not check. But 3.71 does check. If the smaller leg is 3.71, the other leg is 4.71. Then

$$(3.71)^2 + (4.71)^2 = 13.7641 + 22.1841 = 35.9482$$

and since $35.9482 \approx 6^2$, our approximation checks. Also, note that the value for s, 3.71, is between 3 and 4, as we expected from step (1).

5. STATE. One leg is about 3.71 m long; the other is about 4.71 m long. ❑

EXERCISE SET | 10.3

Solve. Try factoring first. If factoring is not posssible or is difficult, use the quadratic formula. If no real-number solutions exist, state so.

1. $x^2 - 4x = 21$
2. $x^2 + 7x = 18$
3. $x^2 = 6x - 9$
4. $x^2 = 8x - 16$
5. $3y^2 - 2y - 8 = 0$
6. $3y^2 - 7y + 4 = 0$
7. $4x^2 + 12x = 7$
8. $4x^2 + 4x = 15$
9. $x^2 - 9 = 0$
10. $x^2 - 4 = 0$
11. $x^2 - 2x - 2 = 0$
12. $x^2 - 4x - 7 = 0$
13. $y^2 - 10y + 22 = 0$
14. $y^2 + 6y - 1 = 0$
15. $x^2 + 4x + 4 = 7$
16. $x^2 - 2x + 1 = 5$
17. $3x^2 + 8x + 2 = 0$
18. $3x^2 - 4x - 2 = 0$
19. $2x^2 - 5x = 1$
20. $2x^2 + 2x = 3$
21. $4y^2 - 4y - 1 = 0$
22. $4y^2 + 4y - 1 = 0$
23. $2t^2 + 6t + 5 = 0$
24. $4y^2 + 3y + 2 = 0$
25. $3x^2 = 5x + 4$
26. $2x^2 + 3x = 1$
27. $2y^2 - 6y = 10$
28. $5m^2 = 3 + 11m$
29. $6x^2 - 9x = 0$
30. $7x^2 + 2 = 6x$
31. $5t^2 - 7t = -4$
32. $15t^2 + 10t = 0$
33. $4x^2 = 90$
34. $5t^2 = 100$

Solve using the quadratic formula. Use a calculator or Table 1 to approximate the solutions to the nearest thousandth.

35. $x^2 - 4x - 7 = 0$
36. $x^2 + 2x - 2 = 0$
37. $y^2 - 6y - 1 = 0$
38. $y^2 + 10y + 22 = 0$
39. $4x^2 + 4x = 1$
40. $4x^2 = 4x + 1$

Solve. Should irrational answers occur, round to the nearest hundredth.

41. An octagon is a figure with 8 sides. How many diagonals does an octagon have?

42. A decagon is a figure with 10 sides. How many diagonals does a decagon have?

43. A polygon has 14 diagonals. How many sides does it have?

44. A polygon has 5 diagonals. How many sides does it have?

45. The height of the Amoco building in Chicago is 1136 ft. How long would it take an object to fall from the top?

46. Library Square Tower, in Los Angeles, is 1012 ft tall. How long would it take an object to fall from the top?

47. The world record for free-fall to the ground without a parachute by a woman is 175 ft and is held by Kitty O'Neill. Approximately how long did the fall take?

48. The world record for free-fall to the ground without a parachute by a man is 311 ft and is held by Dar Robinson. Approximately how long did the fall take?

49. The hypotenuse of a right triangle is 25 ft long. One leg is 17 ft longer than the other. Find the lengths of the legs.

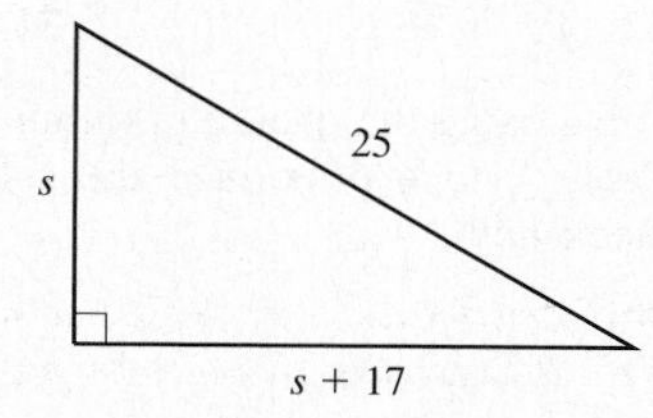

50. The hypotenuse of a right triangle is 26 yd long. One leg is 14 yd longer than the other. Find the lengths of the legs.

51. The length of a rectangle is 2 cm greater than the width. The area is 80 cm^2. Find the length and the width.

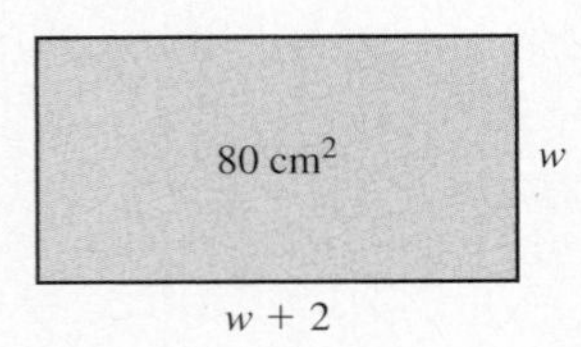

52. The length of a rectangle is 3 m greater than the width. The area is 70 m^2. Find the length and the width.

53. A water pipe runs diagonally under a rectangular yard that is 5 m longer than it is wide. If the pipe is 25 m long, determine the dimensions of the yard.

54. A 26-ft long guy wire is anchored 10 ft from the base of a telephone pole. How far up the pole does the wire reach?

55. The area of a right triangle is 13 m^2. One leg is 2.5 m longer than the other. Find the lengths of the legs.

56. The area of a right triangle is 15.5 cm^2. One leg is 1.2 cm longer than the other. Find the lengths of the legs.

57. The length of a rectangle is 2 in. greater than the width. The area is 20 in^2. Find the length and the width.

58. The length of a rectangle is 3 ft greater than the width. The area is 15 ft^2. Find the length and the width.

59. The length of a rectangle is twice the width. The area is 10 m^2. Find the length and the width.

60. The length of a rectangle is twice the width. The area is 20 cm^2. Find the length and the width.

The formula $A = P(1 + r)^t$ is used to find the value A to which P dollars grows when invested for t years at an annual interest rate r. In Exercises 61–66, find the interest rate for the information provided.

61. \$1000 grows to \$1210 in 2 years

62. \$1000 grows to \$1440 in 2 years

63. \$2560 grows to \$3610 in 2 years

64. \$4000 grows to \$4410 in 2 years

65. \$6250 grows to \$7290 in 2 years

66. \$6250 grows to \$6760 in 2 years

67. Laura has enough mulch to cover 250 ft^2 of garden space. How wide is the largest circular flower garden that Laura can cover with mulch?

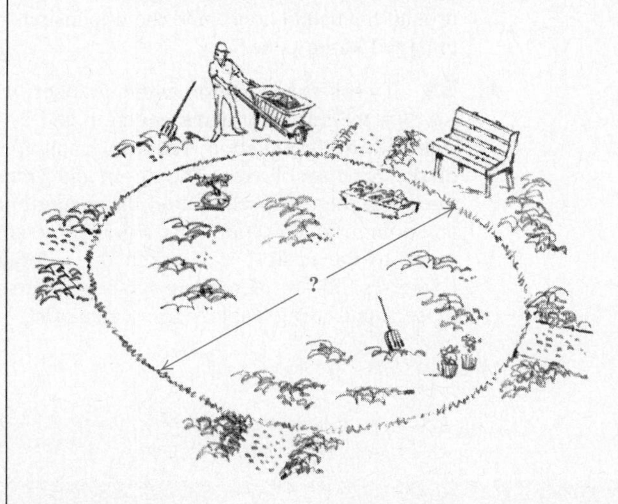

68. A circular oil slick is 20,000 m^2 in area. How wide is the oil slick?

Skill Maintenance

Solve.

69. $5(2x - 3) + 4x = 9 - 6x$

70. $\dfrac{3}{x-4} = \dfrac{2}{x+4}$

Simplify.

71. $\sqrt{40} - 2\sqrt{10} + \sqrt{90}$

72. $\sqrt{9000x^{10}}$

Synthesis

73. A student claims to be able to solve any quadratic equation by completing the square. The same student claims to be incapable of understanding why the quadratic formula works. Does this strike you as odd? Why or why not?

74. Write a problem for a classmate to solve. Devise the problem so that a quadratic equation is used to solve it and the solution is an irrational number.

Solve.

75. $5x + x(x - 7) = 0$

76. $x(3x + 7) - 3x = 0$

77. $3 - x(x - 3) = 4$

78. $x(5x - 7) = 1$

79. $(y + 4)(y + 3) = 15$

80. $x^2 + (x + 2)^2 = 7$

81. $(x + 2)^2 + (x + 1)^2 = 0$

82. $(x + 3)^2 + (x + 1)^2 = 0$

83. $\dfrac{x^2}{x-4} - \dfrac{7}{x-4} = 0$

84. $\dfrac{x^2}{x+3} - \dfrac{5}{x+3} = 0$

85. $\dfrac{1}{x} + \dfrac{1}{x+6} = \dfrac{1}{5}$

86. $\dfrac{1}{x} + \dfrac{1}{x+1} = \dfrac{1}{3}$

87. Find r in this figure. Round to the nearest hundredth.

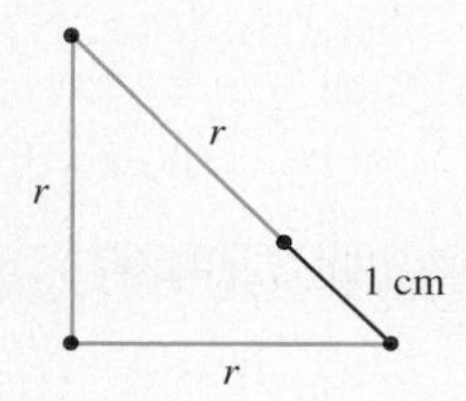

88. Two consecutive integers have squares that differ by 25. Find the integers.

89. Find the area of a square for which the diagonal is one unit longer than the length of the sides.

90. A 20-ft pole is struck by lightning and, while not completely broken, falls over and touches the ground 10 ft from the bottom of the pole. How high up did the pole break?

91. What should the diameter d of a pizza be so that it has the same area as two 10-in. pizzas? Do you get more to eat with a 13-in. pizza or with two 10-in. pizzas?

$d = ?$ = 10 in. + 10 in.

92. How long is the side of a square whose diagonal is 3 cm longer than a side?

93. \$4000 is invested at interest rate r. In 2 years, it grows to \$5267.03. What is the interest rate?

94. In 2 years, you want to have \$3000. How much do you need to invest now if you can get an interest rate of 15.75% compounded annually?

95. In baseball, a batter's strike zone is a rectangular area about 15 in. wide and 40 in. high. Many batters subconsciously enlarge this area by 40% when

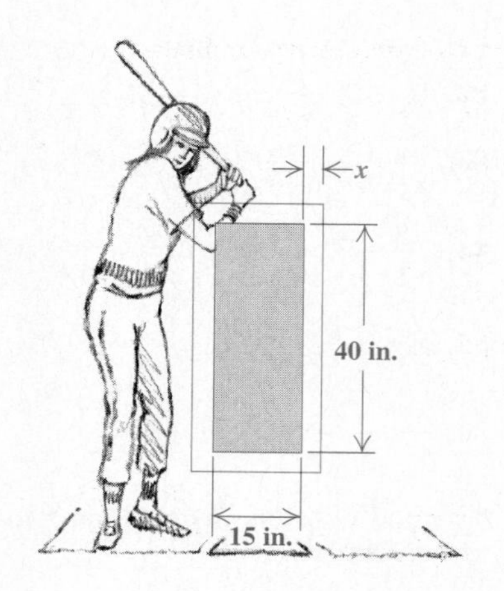

fearful that if they don't swing, the umpire will call the pitch a strike. Assuming that the strike zone is enlarged by an invisible band of uniform width around the actual zone, find the dimensions of the enlarged strike zone.

96. To solve an equation with a grapher, we can let y_1 = the left side of the equation and y_2 = the right side of the equation, and then graph y_1 and y_2 on the same set of axes. The Zoom and Trace features can then be used to find the x-coordinate at any point of intersection. Use a grapher to approximate to the nearest thousandth the solutions of Exercises 35–40. Compare your answers with those found using Table 1 or a calculator.

10.4 Complex Numbers as Solutions of Quadratic Equations

The Complex-Number System • Solutions of Equations

The Complex-Number System

Because negative numbers do not have square roots that are real numbers, mathematicians have devised a larger set of numbers known as *complex numbers*. In the complex-number system, the number i is used to represent the square root of -1.

The Number i

We define the number $i = \sqrt{-1}$. That is, $i = \sqrt{-1}$ and $i^2 = -1$.

EXAMPLE 1

Express in terms of i: **(a)** $\sqrt{-3}$; **(b)** $\sqrt{-25}$; **(c)** $-\sqrt{-10}$.

Solution

a) $\sqrt{-3} = \sqrt{-1 \cdot 3} = \sqrt{-1} \cdot \sqrt{3} = i\sqrt{3}$, or $\sqrt{3}i$ ← i is *not* under the radical.

b) $\sqrt{-25} = \sqrt{-1 \cdot 25} = \sqrt{-1} \cdot \sqrt{25} = i \cdot 5 = 5i$

c) $-\sqrt{-10} = -\sqrt{-1 \cdot 10} = -\sqrt{-1} \cdot \sqrt{10} = -i\sqrt{10}$, or $-\sqrt{10}i$ ❑

Imaginary Numbers

An *imaginary* number* is a number that can be written in the form $a + bi$, where a and b are real numbers and $b \neq 0$.

*The name "imaginary" should not lead you to believe that these numbers are not useful. Imaginary numbers have important applications in engineering and the physical sciences.

The following are examples of imaginary numbers:

$3 + 8i$, $\sqrt{7} - 2i$, $4 + \sqrt{6}i$, and $4i$ (here $a = 0$).

The imaginary numbers together with the real numbers form the set of **complex numbers.**

Complex Numbers

A *complex number* is any number that can be written as $a + bi$, where a and b are real numbers. (Note that a and b both can be 0.)

It may help to remember that every real number is a complex number ($a + bi$ with $b = 0$), but not every complex number is real ($a + bi$ with $b \neq 0$). For example, numbers like $2 + 3i$ and $-7i$ are complex but not real.

Solutions of Equations

As we saw in Example 1(c) of Section 10.3, not all quadratic equations have real-number solutions. All quadratic equations *do,* however, have complex-number solutions. These solutions are usually written in the form $a + bi$ unless a or b is zero.

EXAMPLE 2

Solve: **(a)** $x^2 + 3x + 4 = 0$; **(b)** $x^2 + 2 = 2x$.

Solution

a) We use the quadratic formula:

$$1x^2 + 3x + 4 = 0$$

$$x = \frac{-3 \pm \sqrt{3^2 - 4 \cdot 1 \cdot 4}}{2 \cdot 1}$$

Remember: $x = \dfrac{-b \pm \sqrt{b^2 - 4ac}}{2a}$

$$x = \frac{-3 \pm \sqrt{-7}}{2}$$

Simplifying

$$x = \frac{-3 \pm \sqrt{-1}\sqrt{7}}{2} = \frac{-3 \pm i\sqrt{7}}{2}$$

$$x = -\frac{3}{2} \pm \frac{\sqrt{7}}{2}i$$

Writing in the form $a + bi$

The solutions are $-\dfrac{3}{2} + \dfrac{\sqrt{7}}{2}i$ and $-\dfrac{3}{2} - \dfrac{\sqrt{7}}{2}i$.

b)

$$x^2 + 2 = 2x$$

$$1x^2 - 2x + 2 = 0$$

Rewriting in standard form

$$x = \frac{-(-2) \pm \sqrt{(-2)^2 - 4 \cdot 1 \cdot 2}}{2 \cdot 1}$$

Remember: $x = \dfrac{-b \pm \sqrt{b^2 - 4ac}}{2a}$.

$$x = \frac{2 \pm \sqrt{-4}}{2}$$

Simplifying

Then

$$x = \frac{2 \pm \sqrt{-1}\sqrt{4}}{2}$$

$$x = \frac{2 \pm i2}{2} \qquad \sqrt{-1} = i \text{ and } \sqrt{4} = 2$$

$$x = \frac{2}{2} \pm \frac{2i}{2} \qquad \text{Rewriting in the form } a + bi$$

$$x = 1 \pm i. \qquad \text{Simplifying}$$

The solutions are $1 + i$ and $1 - i$. ❑

EXERCISE SET 10.4

Express in terms of i.

1. $\sqrt{-1}$
2. $\sqrt{-9}$
3. $\sqrt{-49}$
4. $\sqrt{-16}$
5. $\sqrt{-8}$
6. $\sqrt{-20}$
7. $-\sqrt{-12}$
8. $-\sqrt{-45}$
9. $-\sqrt{-27}$
10. $-\sqrt{-4}$
11. $\sqrt{-50}$
12. $-\sqrt{-72}$
13. $-\sqrt{-300}$
14. $\sqrt{-162}$
15. $7 + \sqrt{-16}$
16. $-8 - \sqrt{-36}$
17. $3 - \sqrt{-98}$
18. $-2 + \sqrt{-125}$

Solve.

19. $x^2 + 4 = 0$
20. $x^2 + 9 = 0$
21. $x^2 = -12$
22. $x^2 = -48$
23. $x^2 - 4x + 6 = 0$
24. $x^2 + 4x + 5 = 0$
25. $(x - 2)^2 = -16$
26. $(x + 1)^2 = -25$
27. $x^2 + 2x + 2 = 0$
28. $x^2 + 2 = 2x$
29. $x^2 + 7 = 4x$
30. $x^2 + 7 + 4x = 0$
31. $2t^2 + 6t + 5 = 0$
32. $4y^2 + 3y + 2 = 0$
33. $1 + 2m + 3m^2 = 0$
34. $4p^2 + 3 = 6p$

Skill Maintenance

Rationalize the denominator.

35. $\dfrac{3}{1 - \sqrt{2}}$
36. $\dfrac{5}{4 + \sqrt{7}}$
37. $\dfrac{\sqrt{2} + \sqrt{5}}{\sqrt{2} - \sqrt{5}}$
38. $\dfrac{\sqrt{5} - \sqrt{7}}{\sqrt{5} + \sqrt{7}}$

Synthesis

39. ◈ When using the quadratic formula, why is it only necessary to examine $b^2 - 4ac$ to determine if the solutions are imaginary?

40. ◈ Is it possible for a quadratic equation to have one imaginary-number solution and one real-number solution? Why or why not?

Solve.

41. $(x + 2)^2 + (x + 1)^2 = 0$
42. $(p + 3)^2 + (p + 1)^2 = 0$
43. $\dfrac{2x - 1}{5} - \dfrac{2}{x} = \dfrac{x}{2}$
44. $\dfrac{1}{a - 1} - \dfrac{2}{a - 1} = 3a$
45. A grapher can be used to indicate when a quadratic equation has no real-number solutions. To do this, recall that real-number solutions of $ax^2 + bx + c = 0$ can be found by locating the x-intercepts of the graph of $y = ax^2 + bx + c$. When no x-intercepts exist, the equation $ax^2 + bx + c = 0$ has no real-number solution. Use a grapher to confirm that there are no real-number solutions of Examples 2(a) and (b).

10.5 Graphs of Quadratic Equations

Graphing Parabolas • Finding the x-intercepts of a Parabola • Graphing Equations of the Form $y = ax^2 + bx + c$

In this section, we will graph equations of the form

$$y = ax^2 + bx + c, \quad a \neq 0.$$

The polynomial on the right side of the equation is of second degree, or **quadratic.** Examples of the types of equations we will graph are

$$y = x^2, \qquad y = x^2 + 2x - 3, \quad \text{and} \quad y = -2x^2 + 3.$$

Graphing Parabolas

Graphs of quadratic equations of the type $y = ax^2 + bx + c$ (where $a \neq 0$) are always cupped upward or downward. These graphs are symmetric with respect to a **line of symmetry,** as shown in the figures below. If the graph of a quadratic equation is folded along its line of symmetry, the two halves match exactly.

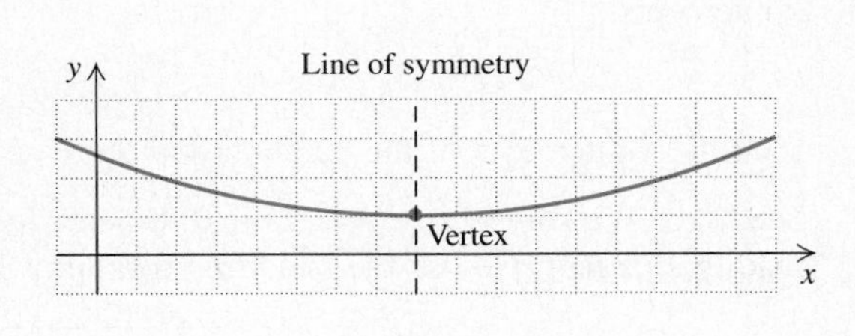

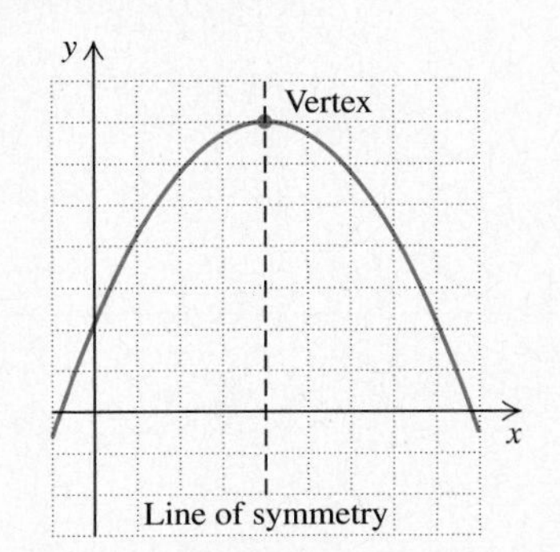

The point at which the graph of a quadratic equation crosses the line of symmetry is called the **vertex** (plural, vertices). The second coordinate of the vertex is either the graph's largest value of y (if the curve opens downward) or the smallest value of y (if the curve opens upward). Graphs of quadratic equations are called **parabolas.**

To graph a quadratic equation, we begin by choosing some numbers for x and computing the corresponding values of y.

EXAMPLE 1

Graph: $y = x^2$.

Solution We choose numbers for x and find the corresponding values for y:

If $x = -2$, then $y = (-2)^2 = 4$. We get the pair $(-2, 4)$.

If $x = -1$, then $y = (-1)^2 = 1$. We get the pair $(-1, 1)$.

If $x = 0$, then $y = 0^2 = 0$. We get the pair $(0, 0)$.

If $x = 1$, then $y = 1^2 = 1$. We get the pair $(1, 1)$.

If $x = 2$, then $y = 2^2 = 4$. We get the pair $(2, 4)$.

The following table lists these solutions of the equation $y = x^2$. After several ordered pairs are found, we plot them and connect them with a smooth curve.

x	y $y = x^2$	(x, y)
-2	4	$(-2, 4)$
-1	1	$(-1, 1)$
0	0	$(0, 0)$
1	1	$(1, 1)$
2	4	$(2, 4)$

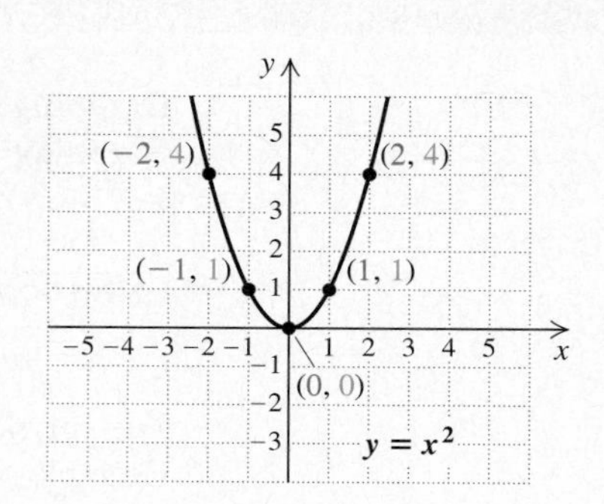

❑

In Example 1, the vertex is (0, 0) and the y-axis is the line of symmetry. The second coordinate of the vertex, 0, is the graph's smallest y-value. For any parabola with an equation of the form $y = ax^2$, the vertex is always (0, 0) and the y-axis is the line of symmetry.

Finding the x-intercepts of a Parabola

In Section 5.6, we found the x-intercepts of the parabola $y = ax^2 + bx + c$ by factoring to solve $ax^2 + bx + c = 0$. Now that we have studied the quadratic formula, it is possible to determine the x-intercepts of *any* parabola $y = ax^2 + bx + c$ that has x-intercepts.

EXAMPLE 2

Find the x-intercepts of the parabola $y = 2x^2 - x - 28$.

Solution We solve $2x^2 - x - 28 = 0$. Since a factorization of $2x^2 - x - 28$ is not readily apparent, we use the quadratic formula with $a = 2$, $b = -1$, and $c = -28$:

$$x = \frac{-(-1) \pm \sqrt{(-1)^2 - 4 \cdot 2(-28)}}{2 \cdot 2}$$

Remember: $x = \dfrac{-b \pm \sqrt{b^2 - 4ac}}{2a}$.

$$x = \frac{1 \pm \sqrt{1 + 224}}{4}$$

$$x = \frac{1 \pm \sqrt{225}}{4} = \frac{1 \pm 15}{4}$$

$$x = \frac{1 + 15}{4} \quad or \quad x = \frac{1 - 15}{4}$$

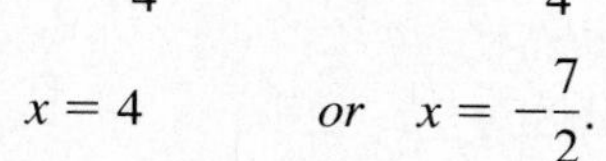

$$x = \frac{16}{4} \quad or \quad x = \frac{-14}{4}$$

$$x = 4 \quad or \quad x = -\frac{7}{2}.$$

The x-intercepts are $\left(-\frac{7}{2}, 0\right)$ and $(4, 0)$.

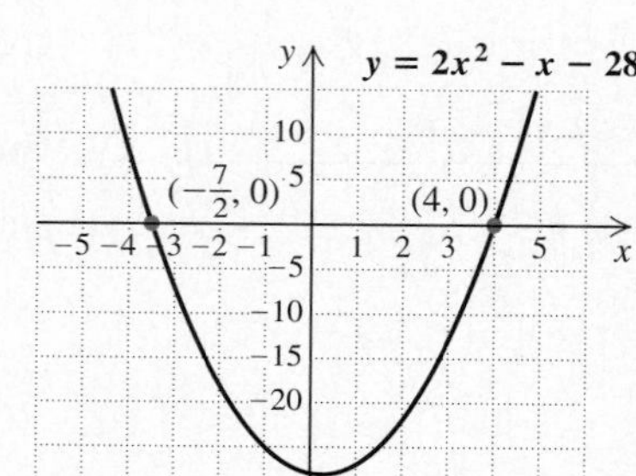

❑

The radicand $b^2 - 4ac$ is called the *discriminant*. Its sign determines how many real-number solutions the equation $0 = ax^2 + bx + c$ has, so it also tells how many x-intercepts there are.

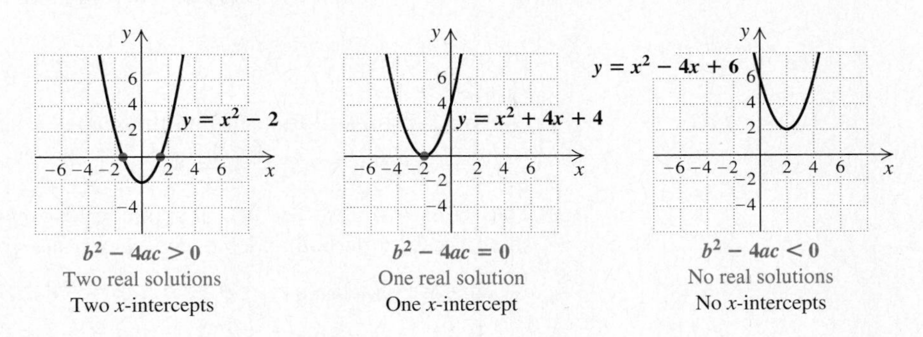

$b^2 - 4ac > 0$	$b^2 - 4ac = 0$	$b^2 - 4ac < 0$
Two real solutions	One real solution	No real solutions
Two x-intercepts	One x-intercept	No x-intercepts

Graphing Equations of the Form $y = ax^2 + bx + c$

Although we were not asked to graph the equation in Example 2, we did so for illustrative purposes. Note that the x-coordinate of the vertex is exactly midway between the x-intercepts. If one x-intercept is determined by $(-b - \sqrt{b^2 - 4ac})/(2a)$ and the other by $(-b + \sqrt{b^2 - 4ac})/(2a)$, then the average of these two values can be used to find the x-coordinate of the vertex:

$$x = \left(\frac{-b - \sqrt{b^2 - 4ac}}{2a} + \frac{-b + \sqrt{b^2 - 4ac}}{2a}\right) \div 2 \qquad \text{Finding the average by adding and dividing by 2}$$

$$x = \left(\frac{-b - \sqrt{b^2 - 4ac} - b + \sqrt{b^2 - 4ac}}{2a}\right) \div 2 \qquad \text{Adding fractions}$$

$$x = \frac{-2b}{2a} \div 2 \qquad -\sqrt{b^2 - 4ac} + \sqrt{b^2 - 4ac} = 0$$

$$x = -\frac{b}{a} \cdot \frac{1}{2} \qquad \text{Simplifying and multiplying by the reciprocal of 2}$$

$$x = -\frac{b}{2a}.$$

The formula $x = -b/(2a)$ can be used to find the x-coordinate of the vertex of *any* parabola given by an equation of the form $y = ax^2 + bx + c$.

Finding the Vertex

For a parabola given by the quadratic equation $y = ax^2 + bx + c$:

1. The x-coordinate of the vertex is $-\dfrac{b}{2a}$.

2. The y-coordinate of the vertex is found by substituting $-\dfrac{b}{2a}$ for x and computing y.

EXAMPLE 3

Graph: $y = -2x^2 + 3$.

Solution Note that $y = -2x^2 + 3$ can be rewritten as $y = -2x^2 + 0x + 3$. Thus, $a = -2$, $b = 0$, and $c = 3$. The x-coordinate of the vertex is

$$-\frac{b}{2a} = -\frac{0}{2(-2)} = 0.$$

We substitute 0 for x to find the second coordinate of the vertex:

$$y = -2x^2 + 3 = -2(0)^2 + 3 = 3.$$

The vertex is $(0, 3)$. The line of symmetry is $x = 0$, which is the y-axis. We choose some x-values on both sides of the vertex and graph the parabola.

If $x = -2$, then $y = -2(-2)^2 + 3 = -8 + 3 = -5$.
If $x = -1$, then $y = -2(-1)^2 + 3 = -2 + 3 = 1$.
If $x = 1$, then $y = -2 \cdot 1^2 + 3 = -2 + 3 = 1$.
If $x = 2$, then $y = -2 \cdot 2^2 + 3 = -8 + 3 = -5$.

x	y $y = -2x^2 + 3$	(x, y)	
-2	-5	$(-2, -5)$	
-1	1	$(-1, 1)$	
0	3	$(0, 3)$	← This is the vertex.
1	1	$(1, 1)$	
2	-5	$(2, -5)$	

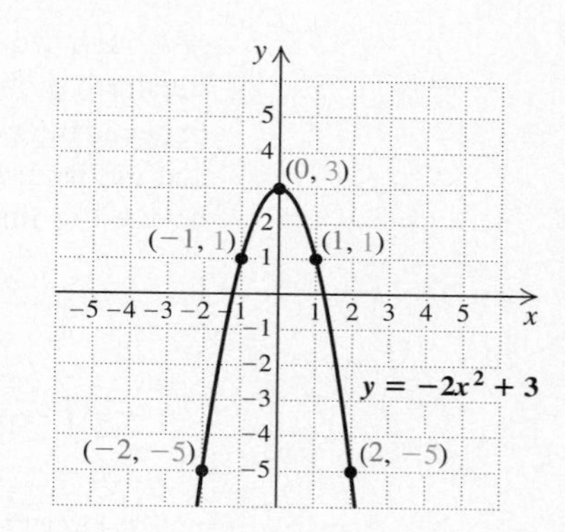

❑

We have found that we can graph equations of the type $y = ax^2 + bx + c$ by determining the coordinates of the vertex, a few points on either side of the vertex, and in some cases, the x-intercepts.

There are two tips you might use when graphing quadratic equations. The first involves the coefficient of x^2. Note that a in $y = ax^2 + bx + c$ tells us whether the graph opens upward or downward. When a is positive, as in Examples 1 and 2, the graph opens upward; when a is negative, as in Example 3, the graph opens downward. It is also helpful to plot the y-intercept. It occurs when $x = 0$.

EXAMPLE 4

Graph: $y = x^2 + 2x - 3$.

Solution We first find the vertex. The x-coordinate of the vertex is

$$-\frac{b}{2a} = -\frac{2}{2(1)} = -1.$$

Substituting -1 for x into the equation gives the second coordinate of the vertex:

$$y = x^2 + 2x - 3 = (-1)^2 + 2(-1) - 3 = 1 - 2 - 3 = -4.$$

The vertex is $(-1, -4)$. The line of symmetry is $x = -1$.

We choose some x-values on both sides of the vertex and graph the parabola. Since the coefficient of x^2 is 1, which is positive, we know that the graph opens upward.

x	y $y = x^2 + 2x - 3$	(x, y)	
-4	5	$(-4, 5)$	
-3	0	$(-3, 0)$	
-2	-3	$(-2, -3)$	
-1	-4	$(-1, -4)$	← Vertex
0	-3	$(0, -3)$	← y-intercept
1	0	$(1, 0)$	
2	5	$(2, 5)$	

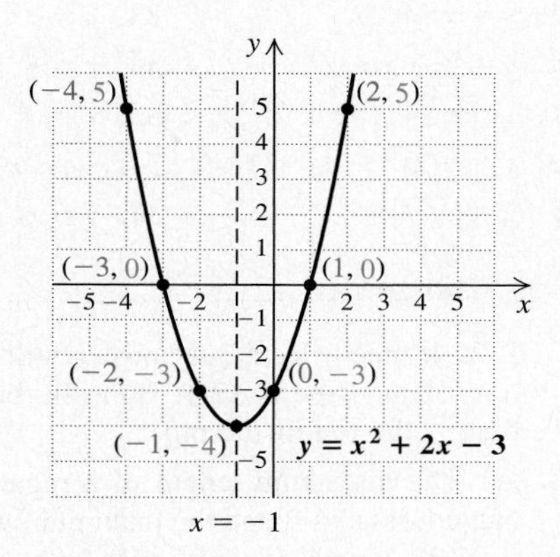

In Examples 1–4, you may have noticed that x-values to the left of the vertex were paired with the same y-values as x-values the same distance to the right of the vertex. Thus, since the vertex for Example 4 is $(-1, -4)$, we would *expect* the x-values -2 and 0 to both be paired with the same y-value. By being aware of symmetries, you can plot two points after calculating just one and also check your work as you proceed.

Guidelines for Graphing Quadratic Equations

1. Graphs of quadratic equations $y = ax^2 + bx + c$ $(a \neq 0)$ are parabolas. They are cupped upward if $a > 0$ and downward if $a < 0$.
2. Find the y-intercept. It occurs when $x = 0$.
3. Use the formula $x = -b/(2a)$ to find the vertex. Plot the vertex and some points on either side of it.
4. After a point has been graphed, a second point with the same y-coordinate can be found on the opposite side of the line of symmetry.

EXERCISE SET 10.5

Graph the quadratic equation. List the ordered pair for the vertex.

1. $y = x^2 - 1$
2. $y = 2x^2$
3. $y = -1 \cdot x^2$
4. $y = x^2 + 1$
5. $y = -x^2 + 2x$
6. $y = x^2 + x - 6$
7. $y = 8 - 6x - x^2$
8. $y = x^2 + 2x + 1$
9. $y = x^2 - 2x + 1$
10. $y = -\frac{1}{2}x^2$
11. $y = -x^2 + 2x + 3$
12. $y = -x^2 - 2x + 3$
13. $y = -2x^2 - 4x + 1$
14. $y = 2x^2 + 4x - 1$
15. $y = \frac{1}{4}x^2$
16. $y = -0.1x^2$
17. $y = 3 - x^2$
18. $y = x^2 + 3$
19. $y = -x^2 + x - 1$
20. $y = x^2 + 2x$
21. $y = -2x^2$
22. $y = -x^2 - 1$
23. $y = x^2 - x - 6$
24. $y = 8 + x - x^2$

For each of the equations $y = ax^2 + bx + c$ in Exercises 25–36, find the x-intercepts. If none exist, state so.

25. $y = x^2 - 5$

26. $y = x^2 - 3$

27. $y = x^2 + 2x$

28. $y = x^2 - 2x$

29. $y = 8 - x - x^2$

30. $y = 8 + x - x^2$

31. $y = x^2 + 10x + 25$

32. $y = x^2 - 8x + 16$

33. $y = -2x^2 - 4x + 1$

34. $y = 2x^2 + 4x - 1$

35. $y = x^2 + 5$

36. $y = x^2 + 3$

Skill Maintenance

37. A 24-ft pipe is leaning against a building. The bottom of the pipe is 12 ft from the building. How high is the top of the pipe?

38. The maximum length of a regulation basketball court is 94 ft and the maximum width is 50 ft. Find the length of a diagonal for such a court.

39. Write a slope–intercept equation for the line containing the points $(-2, 7)$ and $(4, -3)$.

40. Find the slope of the line given by the equation $8 - 6x = 2y$.

Synthesis

41. ◈ Suppose that both x-intercepts of a parabola are known. What is the easiest way to find the coordinates of the vertex?

42. ◈ Does the graph of every equation of the type $y = ax^2 + bx + c$ have a y-intercept? Why or why not? What if $a = 0$?

43. Using the same set of axes, graph $y = x^2$, $y = 2x^2$, $y = 3x^2$, $y = -x^2$, $y = -2x^2$, $y = -3x^2$, $y = \frac{1}{2}x^2$, and $y = -\frac{1}{2}x^2$. Describe the change in the graph of a quadratic equation $y = ax^2$ as $|a|$ increases.

44. Using the same set of axes, graph $y = x^2$, $y = (x - 3)^2$, and $y = (x + 1)^2$. Compare the three graphs. How can the graph of $y = (x - h)^2$ be obtained from the graph of $y = x^2$?

45. *Seller's supply.* As the price of a product increases, the seller is willing to sell, or *supply,* more of the product. Suppose that the supply for a certain product is given by

$$S = p^2 + p + 10,$$

where p is the price in dollars and S is the number sold, in thousands, at that price. Graph the equation for values of p such that $0 \leq p \leq 6$.

46. *Consumer's demand.* As the price of a product increases, consumers purchase, or *demand,* less of the product. Suppose that the demand for a certain product is given by

$$D = (p - 6)^2,$$

where p is the price in dollars and D is the number sold, in thousands, at that price. Graph the equation for values of p such that $0 \leq p \leq 6$.

47. *Equilibrium point.* The price p at which the consumer and the seller agree determines what is called the *equilibrium point.* Find p such that

$$D = S$$

for the demand and supply curves in Exercises 45 and 46. How many units of the product will be sold at that price?

48. *Stopping distance.* In how many feet can a car stop if it is traveling at a speed of r mph? One estimate, developed in Britain, is as follows. The distance d, in feet, is given by

$$d = \underbrace{\text{Thinking distance (in ft)}}_{r} + \underbrace{\text{Stopping distance (in ft)}}_{0.05r^2}$$

$$d = r + 0.05r^2.$$

a) How many feet would it take to stop the car at a speed of 25 mph? 40 mph? 55 mph? 65 mph? 75 mph? 100 mph?

b) Graph the equation, assuming $r \geq 0$.

49. Using the same set of axes, graph $y = x^2$, $y = x^2 - 5$, and $y = x^2 + 2$. Compare the graphs. How can the graph of $y = x^2 + k$ be obtained from the graph of $y = x^2$?

50. Use a grapher to draw the graph of $y = x^2 - 5$ and then, using the graph, estimate $\sqrt{5}$ to four decimal places.

51. *Height of a projectile.* The height H, in feet, of a projectile with an initial velocity of 96 ft/sec is given by the equation

$$H = -16t^2 + 96t,$$

where t is the number of seconds from launch. Use

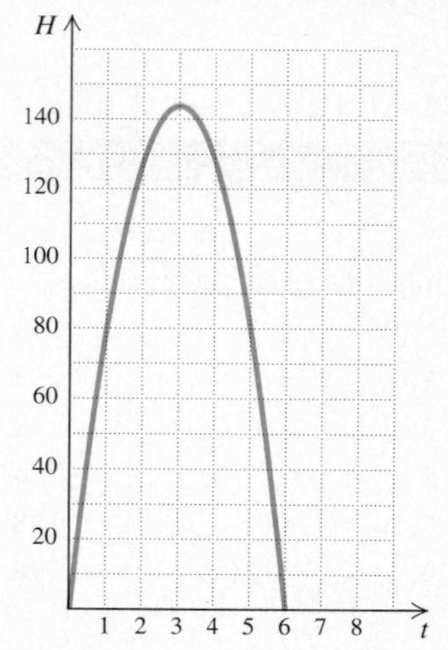

the graph of this function, shown in the figure on the preceding page, or any equation-solving technique to answer the following questions.

a) How many seconds after launch is the projectile 128 ft above ground?

b) When does the projectile reach its maximum height?

c) How many seconds after launch does the projectile return to the ground?

10.6 Functions

Identifying Functions • Functions Written as Formulas • Function Notation • Graphs of Functions • Recognizing Graphs of Functions

Functions are enormously important in modern mathematics. The more mathematics you study, the more you will use functions.

Identifying Functions

Functions appear regularly in magazines and newspapers although they are usually not referred to as such. Consider the following table.

Year	Global Average Temperature (in Degrees Celsius)
1983	12.30
1984	12.12
1985	12.12
1986	12.17
1987	12.33
1988	12.35
1989	12.25
1990	12.47
1991	12.41

Note that to each year there corresponds *exactly one* temperature. A correspondence of this sort is called a **function.**

Function

A *function* is a correspondence (or rule) that assigns to each member of some set (called the *domain*) exactly one member of a set (called the *range*).

Sometimes the members of the domain are also called **inputs,** and the members of the range are called **outputs.**

Domain (Set of Inputs)	**Range** (Set of Outputs)
1983	→ 12.30
1984	→ 12.12
1985	→ 12.12
1986	→ 12.17
1987	→ 12.33
1988	→ 12.35
1989	→ 12.25
1990	→ 12.47
1991	→ 12.41

Note that each input has exactly one output, even though one of the outputs, 12.12, is used twice.

EXAMPLE 1

Determine whether or not each of the following correspondences is a function:

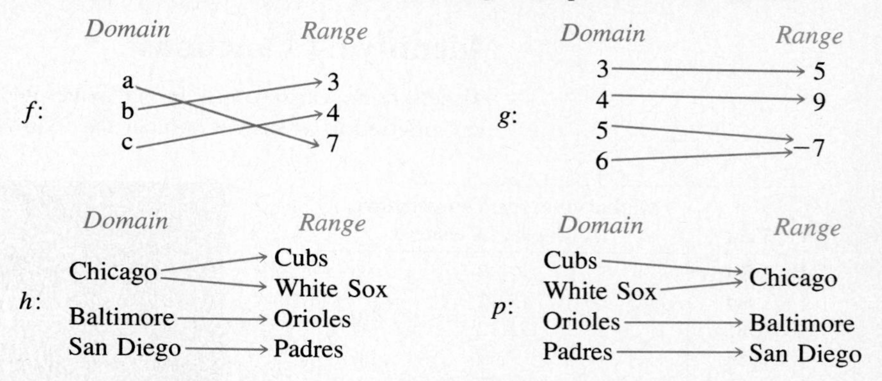

Solution The correspondence f is a function because each member of the domain is matched to only one member of the range.

The correspondence g is also a function because each member of the domain is matched to only one member of the range.

The correspondence h is *not* a function because one member of the domain, Chicago, is matched to more than one member of the range.

The correspondence p is a function because each member of the domain is paired with only one member of the range. ❑

Functions Written as Formulas

Many functions are described by formulas. Equations like $y = x + 3$ and $y = 4x^2$ are examples of such formulas. To find an output, we substitute a member of the domain for x.

EXAMPLE 2

During a thunderstorm, it is possible to calculate how far away lightning is by using the formula

$$M = \tfrac{1}{5}t.$$

Here M is the distance, in miles, that a storm is from an observer when the sound of thunder arrives t seconds after the lightning has been sighted.

Complete the following table for this function.

t (seconds)	0	1	2	3	4	5	6	10
M (miles)	0	$\frac{1}{5}$						

Solution To complete the table, we successively substitute values of t and compute M.

For $t = 2$, $M = \frac{1}{5} \cdot 2 = \frac{2}{5}$;
For $t = 3$, $M = \frac{1}{5} \cdot 3 = \frac{3}{5}$;
For $t = 4$, $M = \frac{1}{5} \cdot 4 = \frac{4}{5}$;
For $t = 5$, $M = \frac{1}{5} \cdot 5 = 1$;
For $t = 6$, $M = \frac{1}{5} \cdot 6 = \frac{6}{5}$, or $1\frac{1}{5}$;
For $t = 10$, $M = \frac{1}{5} \cdot 10 = 2$ ❑

Function Notation

In Example 2, it was somewhat time-consuming to repeatedly write "For $t = ■$, $M = \frac{1}{5} \cdot ■$." Function notation clearly and concisely presents inputs and outputs together. The notation $M(t)$, read "M of t," denotes the output that is paired with the input t by the function M. Thus, in Example 2, we could write

$$M(2) = \frac{1}{5} \cdot 2 = \frac{2}{5}, \quad M(3) = \frac{1}{5} \cdot 3 = \frac{3}{5}, \quad \text{and, in general,} \quad M(t) = \frac{1}{5} \cdot t.$$

The notation $M(4) = \frac{4}{5}$ makes clear that when 4 is the input, $\frac{4}{5}$ is the output. The notation $M(4)$ *does not* mean $M \cdot 4$.

Formulas for functions are commonly written in function notation. Thus, the formula $f(x) = x + 2$, read "f of x equals x plus 2," can be used instead of $y = x + 2$ when we are discussing functions, although both equations describe the same correspondence.

EXAMPLE 3

For the function given by $f(x) = x + 2$, find: **(a)** $f(8)$; **(b)** $f(-3)$; **(c)** $f(0)$.

Solution

a) $f(8) = 8 + 2$, or 10 — $f(8)$ is read "f of 8"; $f(8)$ does not mean "f times 8"! $f(8)$ is the output corresponding to the input 8.

b) $f(-3) = -3 + 2$, or -1

c) $f(0) = 0 + 2$, or 2 ❑

It is sometimes helpful to think of a function as a machine that gives an output for each input that enters the machine. The following diagram is one way in which the function given by $g(t) = 2t^2 + 5$ can be illustrated.

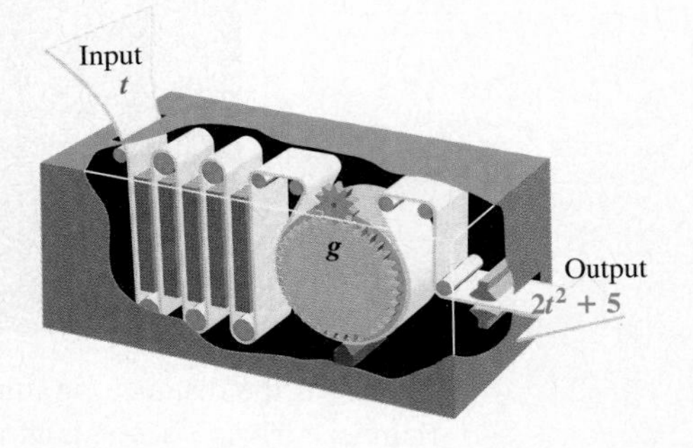

EXAMPLE 4

For the function $g(t) = 2t^2 + 5$, find: **(a)** $g(3)$; **(b)** $g(0)$; **(c)** $g(-2)$.

Solution

a) $g(3) = 2 \cdot 3^2 + 5 = 2 \cdot 9 + 5 = 23$

b) $g(0) = 2 \cdot 0^2 + 5 = 5$

c) $g(-2) = 2(-2)^2 + 5 = 2 \cdot 4 + 5 = 13$ ❑

Outputs are also called **function values.** In Example 4, $g(-2) = 13$. We could say that the "function value is 13 at -2," or "when x is -2, the value of the function is 13."

Graphs of Functions

To graph a function, we find ordered pairs (x, y) or $(x, f(x))$, plot them, and connect the points. Note that y and $f(x)$ are used interchangeably when working with functions and their graphs.

EXAMPLE 5

Graph: $f(x) = x + 2$.

Solution A list of some function values is shown in this table. We plot the points and connect them. The graph is a straight line.

x	$f(x)$
-4	-2
-3	-1
-2	0
-1	1
0	2
1	3
2	4
3	5
4	6

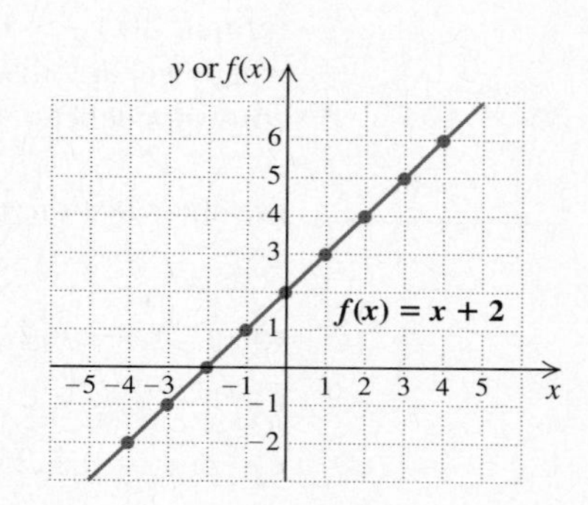

❑

EXAMPLE 6

Graph: $g(x) = 4 - x^2$.

Solution Recall from Section 10.5 that the graph is a parabola. We calculate some function values and draw the curve.

$$g(0) = 4 - 0^2 = 4 - 0 = 4,$$
$$g(-1) = 4 - (-1)^2 = 4 - 1 = 3,$$
$$g(2) = 4 - (2)^2 = 4 - 4 = 0,$$
$$g(-3) = 4 - (-3)^2 = 4 - 9 = -5$$

x	$g(x)$
−3	−5
−2	0
−1	3
0	4
1	3
2	0
3	−5

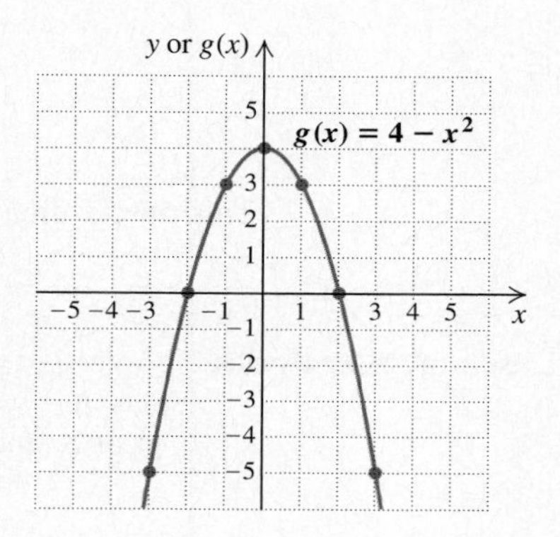

❑

EXAMPLE 7

Graph: $h(x) = |x|$.

Solution A list of some function values is shown in the following table. We plot the points and connect them. The graph is a V-shaped ''curve'' that rises on either side of the vertical axis.

x	$h(x)$
−3	3
−2	2
−1	1
0	0
1	1
2	2
3	3

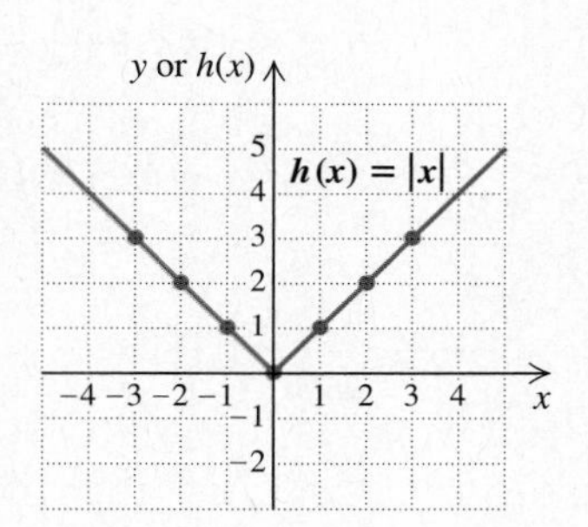

❑

Recognizing Graphs of Functions

Consider the function f described by $f(x) = x^2 - 5$. Its graph is shown at left. It is also the graph of the equation $y = x^2 - 5$.

To find a function value, like $f(3)$, from a graph, we locate the input on the horizontal axis, move vertically to the graph of the function, and then horizontally to find the output on the vertical axis, where members of the range can be found.

Recall that when one member of the domain is paired with two or more different members of the range, the correspondence is not a function. Thus, when a graph contains two or more different points with the same first coordinate, the graph cannot represent a function. Points sharing a common first coordinate are vertically above or below each other (see the graph on the following page).

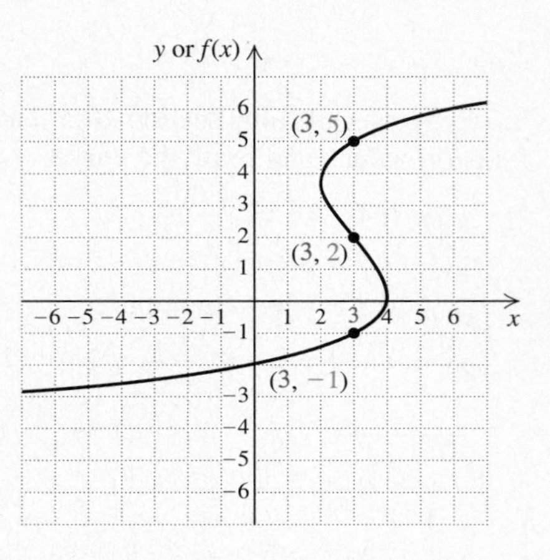

Since 3 is paired with more than one member of the range, the graph does not represent a function.

This observation leads to the *vertical-line test*.

The Vertical-Line Test

If it is possible for any vertical line to intersect the graph at more than one point, then the graph is not the graph of a function. Otherwise, the graph is that of a function.

EXAMPLE 8

Determine whether each of the following is the graph of a function.

a)

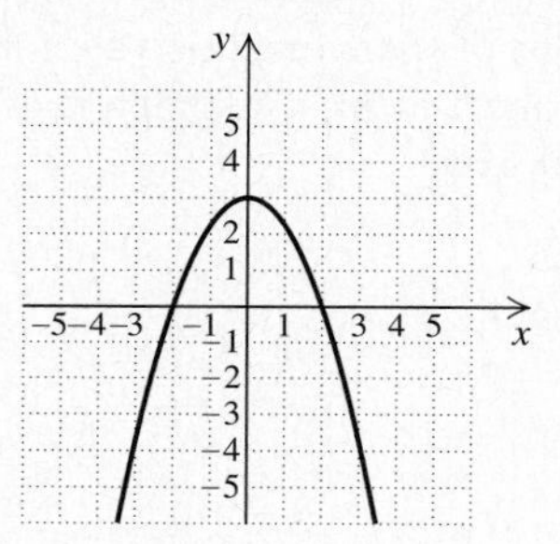

b)

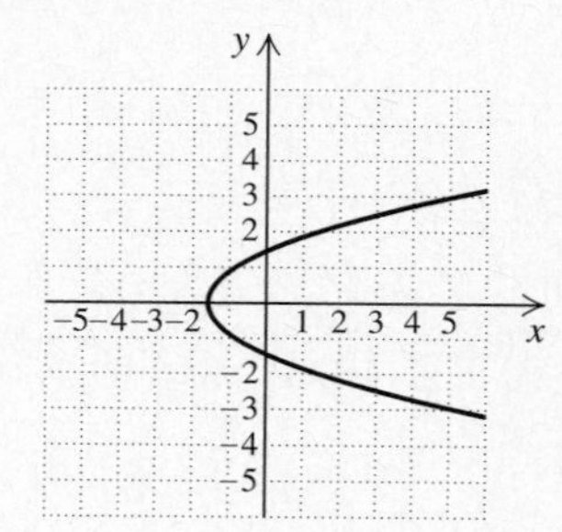

c)

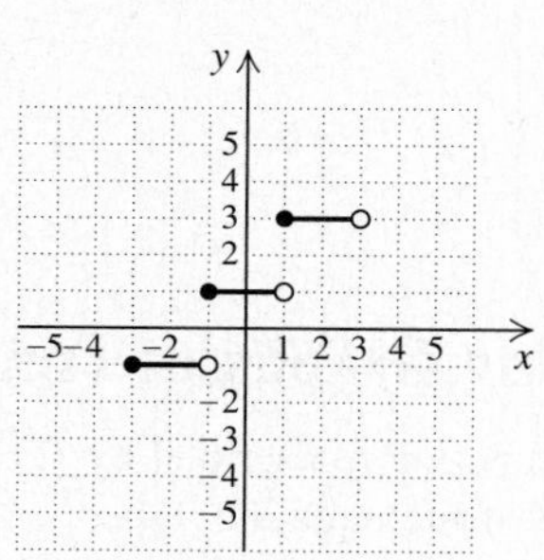

d)

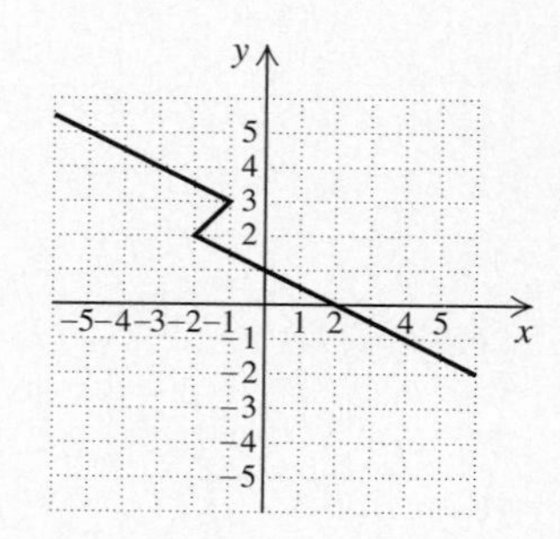

Solution

a) The graph *is* that of a function because no vertical line can cross the graph at more than one point. This can be confirmed with a ruler or straight edge.

b) The graph *is not* that of a function because a vertical line—say, $x = 1$—crosses the graph at more than one point (see the graph at the top of the following page).

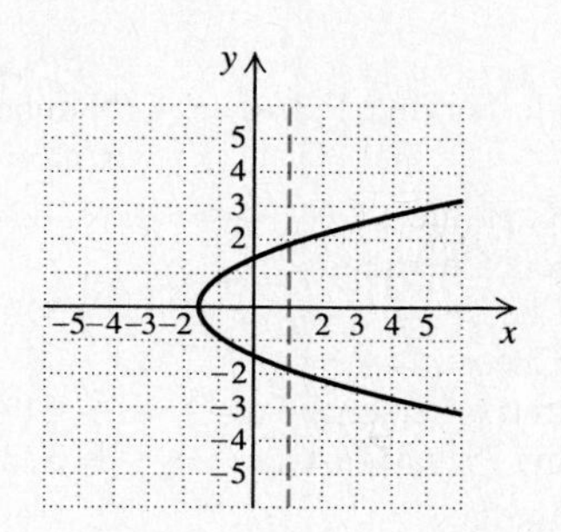

c) The graph *is* that of a function because no vertical line can cross the graph at more than one point. Note that the open dots indicate the absence of a point.

d) The graph *is not* that of a function because a vertical line—say, $x = -2$—crosses the graph at more than one point.

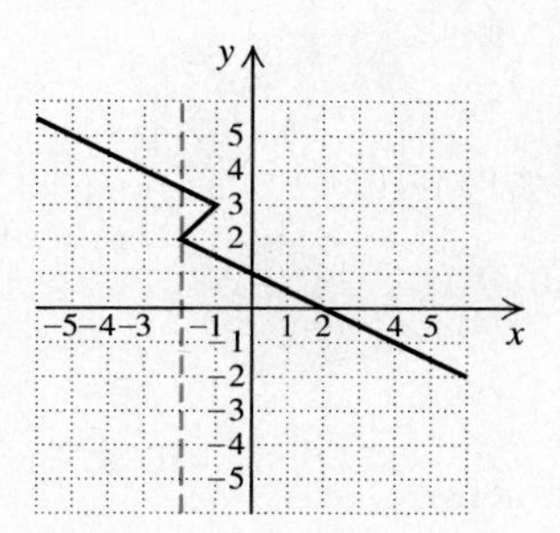

❑

EXERCISE SET 10.6

Determine whether the correspondence is a function.

1. *Domain* *Range*

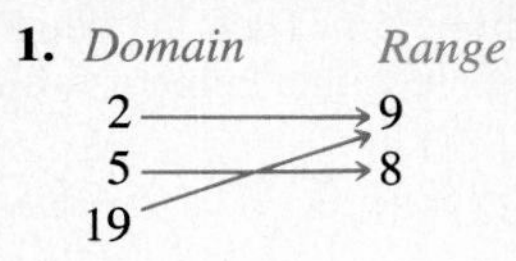

2. *Domain* *Range*

5 → 3
−3 → 7
7 → 3
−7 → 7

3. *Domain* *Range*

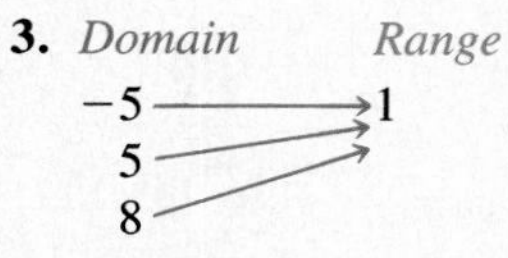

4. *Domain* *Range*

6 → −6
6 → −7
7 → −7
3 → −3

5. *Domain* *Range*

6. *Domain* *Range*

Dodgers → Los Angeles
Rams → Los Angeles
Lakers → Los Angeles
Yankees → New York
Giants → New York
Knickerbockers → New York

7. *Domain* (Year) *Range* (World Population Increase in Millions)

1985 → 81
1986 → 83
1987 → 88
1988 → 89
1989 → 90
1990 → 90
1991 → 92

8.

Domain (Brand of Single-Serving Pizza)	*Range* (Number of Calories)
Old Chicago Pizza-lite	→ 324
Weight Watchers Cheese	→ 310
Banquet Zap Cheese	→ 310
Lean Cuisine Cheese	→ 310
Pizza Hut Supreme Personal Pan	→ 647
Celeste Suprema Pizza-For-One	→ 678

Find the indicated outputs.

9. $f(3)$, $f(7)$, and $f(-9)$

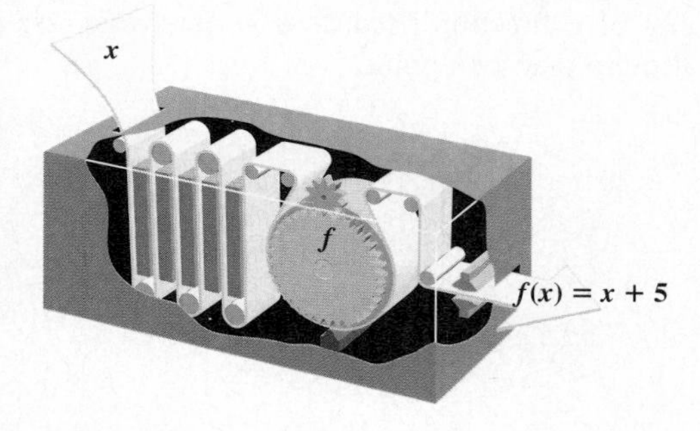

10. $g(0)$, $g(6)$, and $g(18)$

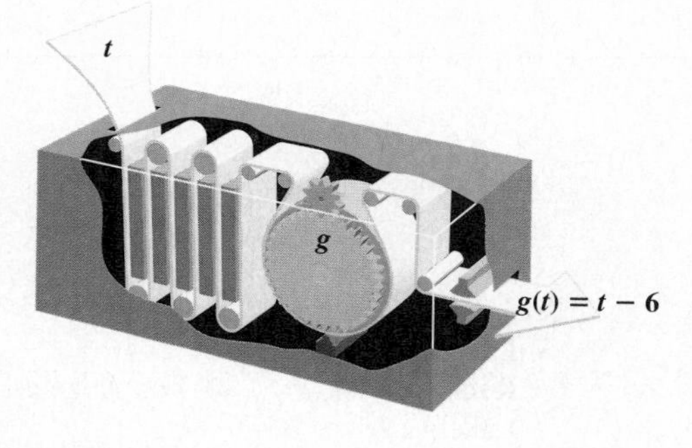

11. $h(-2)$, $h(5)$, and $h(24)$

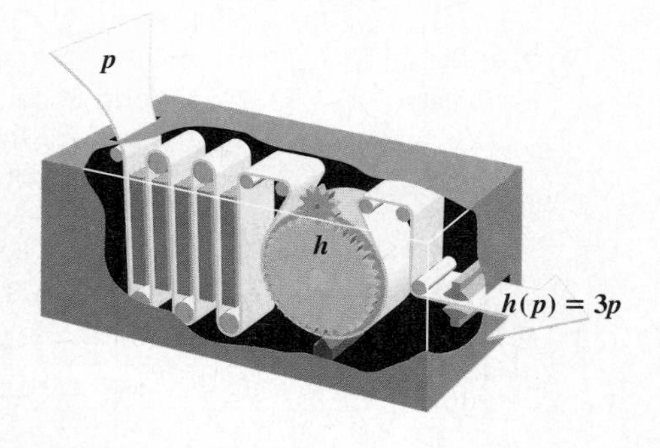

12. $f(6)$, $f\left(-\frac{1}{2}\right)$, and $f(20)$

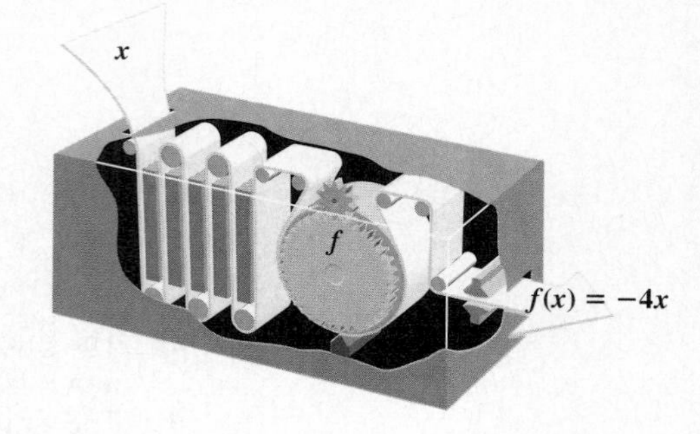

13. $g(s) = 2s + 4$; find $g(1)$, $g(-7)$, and $g(6.7)$.

14. $h(x) = 19$; find $h(4)$, $h(-6)$, and $h(12.5)$.

15. $F(x) = 2x^2 - 3x + 2$; find $F(0)$, $F(-1)$, and $F(2)$.

16. $P(x) = 3x^2 - 2x + 5$; find $P(0)$, $P(-2)$, and $P(3)$.

17. $h(x) = |x|$; find $h(-4)$, $h\left(\frac{2}{3}\right)$, and $h(-3.8)$.

18. $f(t) = |t| + 1$; find $f(-5)$, $f(0)$, and $f\left(-\frac{9}{4}\right)$.

19. $f(x) = |x| - 2$; find $f(-3)$, $f(93)$, and $f(-100)$.

20. $g(t) = t^3 + 3$; find $g(1)$, $g(-5)$, and $g(0)$.

21. $h(x) = x^4 - 3$; find $h(0)$, $h(-1)$, and $h(3)$.

22. $f(x) = 2/x$; find $f(-3)$, $f(-2)$, $f(-1)$, $f\left(\frac{1}{2}\right)$, $f(1)$, $f(2)$, $f(3)$, and $f(10)$.

23. *Predicting heights in anthropology.* An anthropologist can use certain functions to estimate the height of a male or a female, given the lengths of certain bones. A *humerus* is the bone from the elbow to the shoulder. Let x = the length of the humerus, in centimeters. Then the height, in centimeters, of a female with a humerus of length x is given by

$$F(x) = 2.75x + 71.48.$$

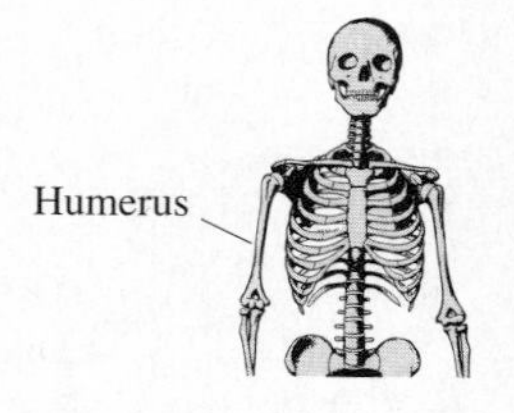

If a humerus is known to be from a female, how tall was the female if the bone is **(a)** 30 cm long? **(b)** 35 cm long?

24. When a humerus (see Exercise 23) is from a male, the function $M(x) = 2.89x + 70.64$ can be used to find the male's height, in centimeters. If a humerus is known to be from a male, how tall was the male if the bone is **(a)** 30 cm long? **b)** 35 cm long?

25. *Pressure at sea depth.* The function $P(d) = 1 + (d/33)$ gives the pressure, in *atmospheres* (atm), at a depth d, in feet, in the sea. Note that $P(0) = 1$ atm, $P(33) = 2$ atm, and so on. Find the pressure at 20 ft, 30 ft, and 100 ft.

26. *Record revolutions.* The function $R(t) = 33\frac{1}{3}t$ gives the number of revolutions of a $33\frac{1}{3}$ RPM record as a function of t, the number of minutes that it turns. Find the number of revolutions at 5 min, 20 min, and 25 min.

27. *Temperature as a function of depth.* The function $T(d) = 10d + 20$ gives the temperature, in degrees Celsius, inside the earth as a function of the depth d, in kilometers. Find the temperature at 5 km, 20 km, and 1000 km.

28. *Melting snow.* The function $W(d) = 0.112d$ approximates the amount, in centimeters, of water W that will melt from snow that is d cm deep. Find the amount of water that results from snow melting from depths of 16 cm, 25 cm, and 100 cm.

29. *Consumer's demand function.* As the price of a product increases, consumers purchase, or *demand,* less of the product. Suppose that the demand for sugar were related to price by the following demand function:

$$D(p) = -2.7p + 16.3,$$

where p is the price of a 5-lb bag of sugar and $D(p)$ is the quantity of 5-lb bags, in millions, purchased at price p. Find the quantity purchased when the price is \$1 per 5-lb bag, \$2 per 5-lb bag, \$3 per 5-lb bag, \$4 per 5-lb bag, and \$5 per 5-lb bag.

30. *Seller's supply function.* As the price of a product increases, the seller is willing to sell, or *supply*, more of the product. Suppose that the supply of sugar were related to price by the following supply function:

$$S(p) = 7p - 5.9,$$

where p is the price of a 5-lb bag of sugar and $S(p)$ is the quantity of 5-lb bags, in millions, that a seller allows to be purchased at price p. Find the quantity supplied for sale when the price is \$1 per 5-lb bag, \$2 per 5-lb bag, \$3 per 5-lb bag, \$4 per 5-lb bag, and \$5 per 5-lb bag.

Graph the function.

31. $f(x) = x + 4$

32. $g(x) = x + 3$

33. $h(x) = 2x - 3$

34. $f(x) = 3x - 1$

35. $g(x) = x - 6$

36. $h(x) = x - 5$

37. $f(x) = 2x - 7$

38. $g(x) = 4x - 13$

39. $f(x) = \frac{1}{2}x + 1$

40. $f(x) = -\frac{3}{4}x - 2$

41. $g(x) = 2|x|$

42. $h(x) = -|x|$

43. $g(x) = x^2$

44. $f(x) = x^2 - 1$

45. $f(x) = x^2 - x - 2$

46. $g(x) = x^2 + 6x + 5$

Determine whether the graph is that of a function.

47.

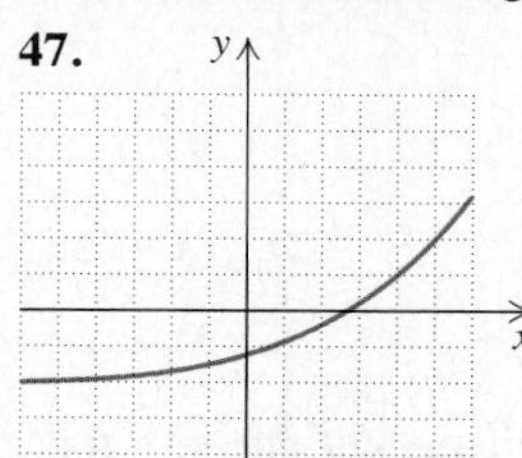

48.

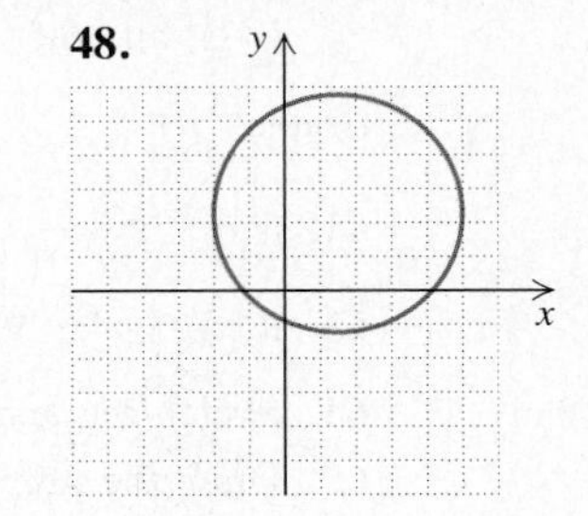

49.

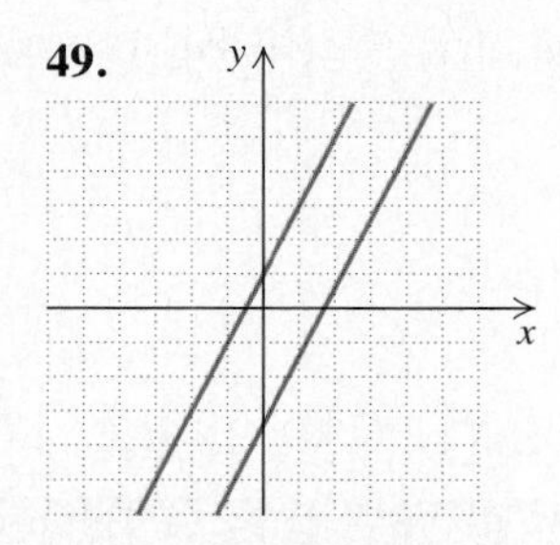

50.

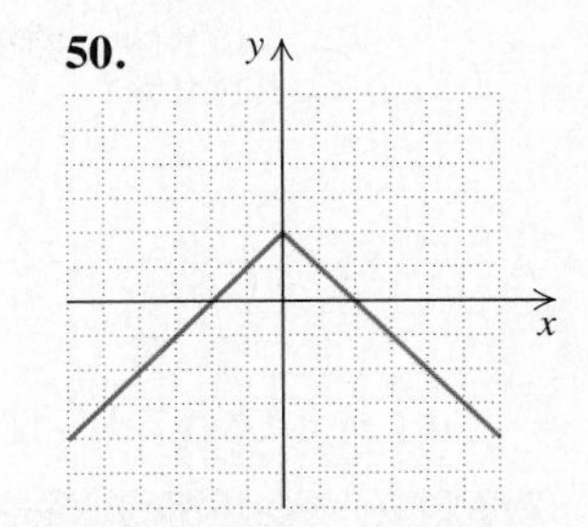

51.

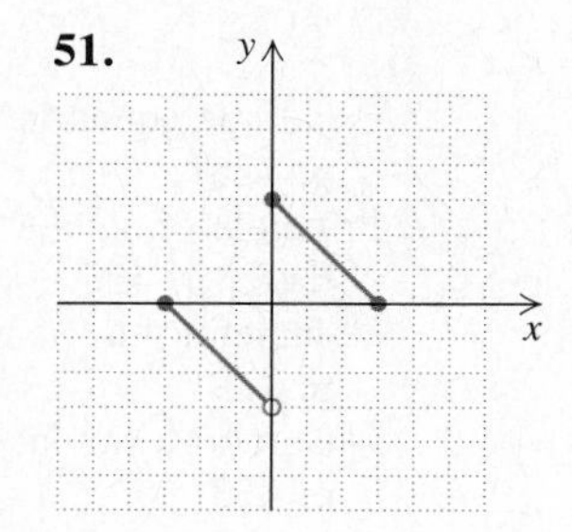

52.

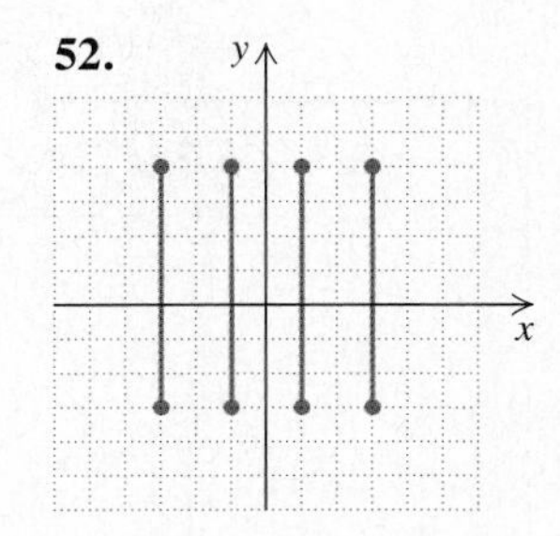

Skill Maintenance

Determine whether the pair of equations represents parallel lines.

53. $y = \frac{3}{4}x - 7,$
$3x + 4y = 7$

54. $y = \frac{3}{5},$
$y = -\frac{5}{3}$

Solve the system using the substitution method.

55. $2x - y = 6,$
$4x - 2y = 5$

56. $x - 3y = 2,$
$3x - 9y = 6$

Synthesis

57. Is it possible for a function to have a larger set of inputs than outputs? Why or why not?

58. Look up the word "function" in a dictionary. Explain how that definition might be related to the mathematical one given in this section.

59. For $g(x) = |x| + x$, find $g(-3)$, $g(3)$, $g(-2)$, $g(2)$, and $g(0)$.

60. For $h(x) = |x| - x$, find $h(-3)$, $h(3)$, $h(-2)$, $h(2)$, and $h(0)$.

Graph.

61. $f(x) = \dfrac{|x|}{x}$

62. $g(x) = |x| + x$

63. $h(x) = |x| - x$

64. Sketch a graph that is not that of a function.

65. Draw the graph of $|y| = x$. Is this the graph of a function?

66. Draw the graph of $y^2 = x$. Is this the graph of a function?

67. If $f(-1) = -7$ and $f(3) = 8$, find a linear equation for $f(x)$.

68. If $g(0) = -4$, $g(-2) = 0$, and $g(2) = 0$, find a quadratic equation for $g(x)$.

Find the range of the function for the given domain.

69. $f(x) = 3x + 5$, when the domain is the set of whole numbers less than 4

70. $g(t) = t^2 - 5$, when the domain is the set of integers between -4 and 2

71. $h(x) = |x| - x$, when the domain is the set of integers between -2 and 20

72. $f(m) = m^3 + 1$, when the domain is the set of integers between -3 and 3

73. Graphers are often useful when graphing functions that are given by a formula. Use a grapher and the window $[-5, 5, -25, 25]$ to draw the graph of $f(x) = x^4 - x^3 - 11x^2 + 9x + 18$. Then use the Zoom and Trace features to approximate the x-intercepts to the nearest hundredth.

SUMMARY AND REVIEW 10

KEY TERMS

Quadratic equation, p. 426
Principle of square roots, p. 426
Completing the square, p. 429
Quadratic formula, p. 433
Standard form, p. 434
Imaginary number, p. 442
Complex number, p. 443
Line of symmetry, p. 445
Vertex (plural, vertices), p. 445
Parabola, p. 445
Discriminant, p. 447
Function, p. 451
Domain, p. 451
Range, p. 451
Input, p. 452
Output, p. 452
Function value, p. 454
Vertical-line test, p. 456

IMPORTANT PROPERTIES AND FORMULAS

Principle of square roots: If $x^2 = k$, then $x = \sqrt{k}$ or $x = -\sqrt{k}$.

Quadratic formula: $x = \dfrac{-b \pm \sqrt{b^2 - 4ac}}{2a}$

To solve a quadratic equation:

1. If it is in the form $ax^2 = p$ or $(x + k)^2 = p$, use the principle of square roots as in Section 10.1.
2. When it is not in the form of (1), write it in standard form, $ax^2 + bx + c = 0$.
3. Try to factor and use the principle of zero products.
4. If it is not possible to factor or if factoring seems difficult, use the quadratic formula.

The solutions of a quadratic equation can always be found using the quadratic formula. They cannot always be found by factoring. When the radicand, $b^2 - 4ac$, is nonnegative, the equation has real-number solutions. When $b^2 - 4ac$ is negative, the equation has no real-number solutions.

Guidelines for Graphing Quadratic Equations

1. Graphs of quadratic equations $y = ax^2 + bx + c$ $(a \neq 0)$ are parabolas. They are cupped upward if $a > 0$ and downward if $a < 0$.
2. Find the y-intercept. It occurs when $x = 0$.
3. Use the formula $x = -b/(2a)$ to find the vertex. Plot the vertex and some points on either side of it.
4. After a point has been graphed, a second point with the same y-coordinate can be found on the opposite side of the line of symmetry.

The Vertical-Line Test

If it is possible for any vertical line to intersect the graph at more than one point, then the graph is not the graph of a function. Otherwise, the graph is that of a function.

REVIEW EXERCISES

This chapter's review and test include Skill Maintenance exercises from Sections 7.3, 8.2, 9.4, and 9.6.

Solve.

1. $8x^2 = 24$

2. $5x^2 - 8x + 3 = 0$

3. $x^2 - 2x - 10 = 0$

4. $3y^2 + 5y = 2$

5. $(x + 8)^2 = 13$

6. $9x^2 = 0$

7. $9x^2 - 6x - 9 = 0$

8. $x^2 + 6x = 9$

9. $1 + 4x^2 = 8x$

10. $6 + 3y = y^2$

11. $3m = 4 + 5m^2$

12. $40 = 5y^2$

13. $x^2 + 1 = 2x$

14. $x^2 + 2 = 2x$

Solve by completing the square.

15. $3x^2 - 2x - 5 = 0$

16. $x^2 - 5x + 2 = 0$

Approximate the solutions to the nearest thousandth.

17. $x^2 - 5x + 2 = 0$

18. $4y^2 + 8y + 1 = 0$

19. The hypotenuse of a right triangle is 5 m long. One leg is 3 m longer than the other. Find the lengths of the legs. Round to the nearest tenth.

20. \$1000 is invested at interest rate r, compounded annually. In 2 years, it grows to \$1060.90. What is the interest rate?

21. The length of a rectangle is 3 m greater than the width. The area is 70 m^2. Find the length and the width.

22. The height of the Lake Point Towers in Chicago is 645 ft. How long would it take an object to fall from the top?

Express in terms of i.

23. $\sqrt{-64}$

24. $-\sqrt{-24}$

Solve. Express solutions in the form $a + bi$.

25. $x^2 + 64 = 0$

26. $x^2 = -54$

27. $(x - 3)^2 = -4$

28. $x^2 - 10x + 26 = 0$

29. $x^2 + x + 1 = 0$

Graph.

30. $y = 2 - x^2$

31. $y = x^2 - 4x - 2$

32. Find the x-intercepts of $y = x^2 - 4x - 2$.

33. If $g(x) = 2x - 5$, find $g(2)$, $g(-1)$, and $g(3.5)$.

34. If $h(x) = |x| - 1$, find $h(1)$, $h(-1)$, and $h(-20)$.

35. If you are moderately active, you need each day about 15 calories per pound of body weight. The function $C(p) = 15p$ approximates the number of calories C that are needed to maintain body weight p, in pounds. How many calories are needed to maintain a body weight of 180 lb?

Graph.

36. $g(x) = x + 7$

37. $f(x) = x^2 - 3$

38. $h(x) = 3|x|$

Determine if the graph is that of a function.

39.

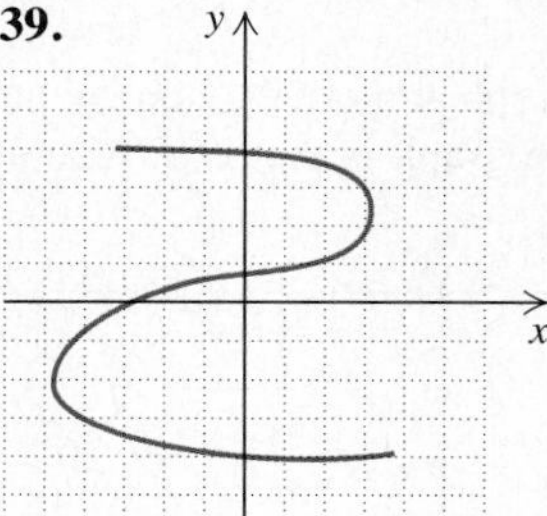

40.

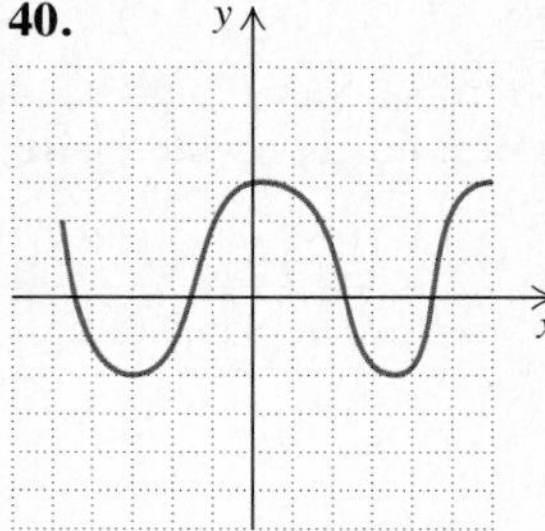

Skill Maintenance

41. Write a slope–intercept equation for the line containing the points $(-2, 5)$ and $(4, -3)$.

42. The difference between two numbers is 8. Twice the smaller one is 1 less than the larger one. What are the numbers?

43. Simplify: $(3 + \sqrt{2})(2 - \sqrt{2})$.

44. The sides of a rectangle are 1 and $\sqrt{2}$. Find the length of a diagonal.

Synthesis

45. ◈ Suppose you know one imaginary-number solution of a quadratic equation with real coefficients. Explain how you can find the other solution.

46. ◈ Explain how the radicand in the quadratic formula, $b^2 - 4ac$, can indicate how many x-intercepts the graph of a quadratic equation has.

47. Two consecutive integers have squares that differ by 63. Find the integers.

48. Find b such that the trinomial $x^2 + bx + 49$ is a square.

49. Solve: $x - 4\sqrt{x} - 5 = 0$.

50. A square with sides of length s has the same area as a circle with radius 5 in. Find s.

CHAPTER TEST 10

Solve.

1. $7x^2 = 35$

2. $7x^2 + 8x = 0$

3. $48 = t^2 + 2t$

4. $3y^2 + 4y = 15$

5. $(x - 2)^2 = 5$

6. $x^2 = x + 3$

7. $m^2 - 3m = 7$

8. $10 = 4x + x^2$

9. $3x^2 - 7x + 1 = 0$

10. $x^2 - 4x = -4$

11. $x^2 - 4x = -8$

12. Solve by completing the square:

$$x^2 - 4x - 10 = 0.$$

13. Approximate the solutions to the nearest thousandth:

$$x^2 - 4x - 10 = 0.$$

14. The width of a rectangle is 4 m less than the length. The area is 16.25 m^2. Find the length and the width.

15. A polygon has 20 diagonals. How many sides does it have? (*Hint:* $d = (n^2 - 3n)/2$.)

Express in terms of i.

16. $\sqrt{-49}$

17. $-\sqrt{-32}$

Solve. Express solutions in the form $a + bi$.

18. $x^2 + 25 = 0$

19. $x^2 = -49$

20. $x^2 + 8x + 17 = 0$

21. $x^2 - x + 2 = 0$

22. Graph: $y = -x^2 + x + 5$.

23. Find the x-intercepts of $y = -x^2 + x + 5$.

24. If $f(x) = \frac{1}{2}x + 1$, find $f(0)$, $f(1)$, and $f(2)$.

25. If $g(t) = -2|t| + 3$, find $g(-1)$, $g(0)$, and $g(3)$.

26. The world record for the 10,000-m run has been decreasing steadily since 1940. The record is approximately 30.18 min minus 0.06 times the number of years since 1940. The function $R(t) = 30.18 - 0.06t$ estimates the record R, in minutes, as a function of t, the time in years since 1940. Predict what the record will be in 1998.

Graph.

27. $h(x) = x - 4$

28. $g(x) = x^2 - 4$

Determine if the graph is that of a function.

29.

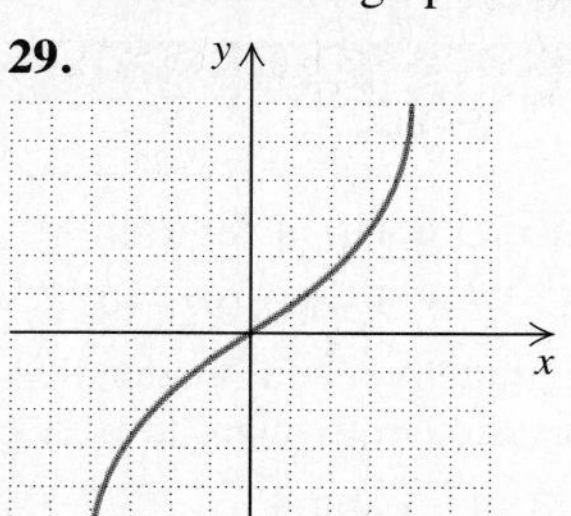

30.

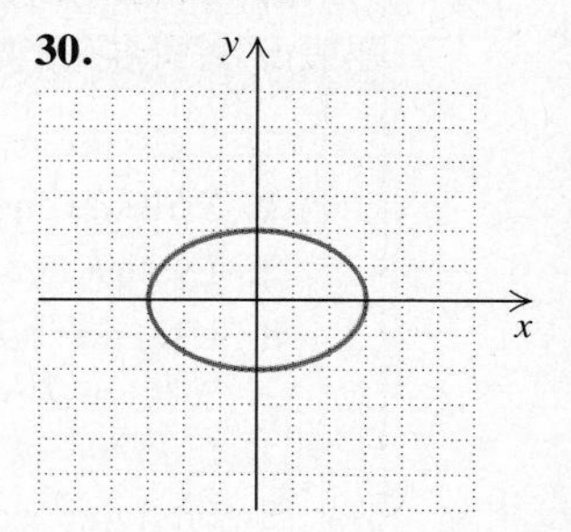

Skill Maintenance

31. Write a slope–intercept equation for the line of slope -3 passing through the point $(-5, 4)$.

32. Solve:

$$x + 2y = 10,$$
$$3x - y = 2.$$

33. Simplify: $2\sqrt{3} - \sqrt{27}$.

34. The sides of a rectangle are $\sqrt{2}$ and $\sqrt{3}$. Find the length of a diagonal.

Synthesis

35. Find the side of a square whose diagonal is 5 ft longer than a side.

36. Solve this system for x. Use the substitution method.

$$x - y = 2,$$
$$xy = 4$$

CUMULATIVE REVIEW 1–10

1. Write exponential notation for $x \cdot x \cdot x$.
2. Evaluate $(x - 3)^2 + 5$ when $x = 10$.
3. Use the commutative and associative laws to write an expression equivalent to $6(xy)$.
4. Find the LCM of 15 and 48.
5. Find the absolute value: $|-7|$.

Compute and simplify.

6. $-6 + 12 + (-4) + 7$
7. $2.8 - (-12.2)$
8. $-\frac{3}{8} \div \frac{5}{2}$
9. $13 \cdot 6 \div 3 \cdot 2 \div 13$
10. Remove parentheses and simplify:
    $4m + 9 - (6m + 13)$.

Solve.

11. $3x = -24$
12. $3x + 7 = 2x - 5$
13. $3(y - 1) - 2(y + 2) = 0$
14. $x^2 - 8x + 15 = 0$
15. $y - x = 1,$
    $y = 3 - x$
16. $x + y = 17,$
    $x - y = 17$
17. $4x - 3y = 3,$
    $3x - 2y = 4$
18. $x^2 - x - 6 = 0$
19. $x^2 + 3x = 5$
20. $3 - x = \sqrt{x^2 - 3}$
21. $5 - 9x \leq 19 + 5x$
22. $-\frac{7}{8}x + 7 = \frac{3}{8}x - 3$
23. $0.6x - 1.8 = 1.2x$
24. $-3x > 24$
25. $x^2 + 2x + 5 = 0$
26. $3y^2 = 30$
27. $(x - 3)^2 = 6$
28. $\frac{6x - 2}{2x - 1} = \frac{9x}{3x + 1}$
29. $\frac{2x}{x + 1} = 2 - \frac{5}{2x}$
30. $\frac{2x}{x + 3} + \frac{6}{x} + 7 = \frac{18}{x^2 + 3x}$
31. $\sqrt{x + 9} = \sqrt{2x - 3}$

Solve the formula for the given letter.

32. $A = \frac{4s + 3}{t}$, for t
33. $\frac{1}{t} = \frac{1}{m} - \frac{1}{n}$, for m

Simplify. Write scientific notation for the result.

34. $(2.1 \times 10^7)(1.3 \times 10^{-12})$
35. $\frac{5.2 \times 10^{-1}}{2.6 \times 10^{-15}}$

Simplify.

36. $x^{-6} \cdot x^2$
37. $\frac{y^3}{y^{-4}}$
38. $(2y^6)^2$
39. Collect like terms and arrange in descending order:
    $2x - 3 + 5x^3 - 2x^3 + 7x^3 + x$.

Perform the indicated operation and simplify.

40. $(4x^3 + 3x^2 - 5) + (3x^3 - 5x^2 + 4x - 12)$
41. $(6x^2 - 4x + 1) - (-2x^2 + 7)$
42. $-2y^2(4y^2 - 3y + 1)$
43. $(2t - 3)(3t^2 - 4t + 2)$
44. $\left(t - \frac{1}{4}\right)\left(t + \frac{1}{4}\right)$
45. $(3m - 2)^2$
46. $(15x^2y^3 + 10xy^2 + 5) - (5xy^2 - x^2y^2 - 2)$
47. $(x^2 - 0.2y)(x^2 + 0.2y)$
48. $(3p + 4q^2)^2$
49. $\frac{4}{2x - 6} \cdot \frac{x - 3}{x + 3}$
50. $\frac{3a^4}{a^2 - 1} \div \frac{2a^3}{a^2 - 2a + 1}$
51. $\frac{3}{3x - 1} + \frac{4}{5x}$
52. $\frac{2}{x^2 - 16} - \frac{x - 3}{x^2 - 9x + 20}$
53. $(x^3 + 7x^2 - 2x + 3) \div (x - 2)$

Factor.

54. $8x^2 - 4x$

55. $25x^2 - 4$

56. $6y^2 - 5y - 6$

57. $m^2 - 8m + 16$

58. $x^3 - 8x^2 - 5x + 40$

59. $3a^4 + 6a^2 - 72$

60. $16x^3 - x$

61. $49a^2b^2 - 4$

62. $9x^2 + 30xy + 25y^2$

63. $2ac - 6ab - 3db + dc$

64. $15x^2 + 14xy - 8y^2$

Simplify.

65. $\dfrac{\frac{3}{x} + \frac{1}{2x}}{\frac{1}{3x} - \frac{3}{4x}}$

66. $\sqrt{49}$

67. $-\sqrt[4]{625}$

68. $\sqrt{64x^2}$

69. Multiply: $\sqrt{a+b}\,\sqrt{a-b}$.

70. Multiply and simplify: $\sqrt{32ab}\,\sqrt{6a^4b^2}$.

Simplify.

71. $8^{-2/3}$

72. $\sqrt{243x^3y^2}$

73. $\sqrt{\dfrac{100}{81}}$

74. $(2 + \sqrt{5})(2 - \sqrt{5})$

75. $4\sqrt{12} + 2\sqrt{48}$

76. Divide and simplify: $\dfrac{\sqrt{72}}{\sqrt{45}}$.

77. The hypotenuse of a right triangle is 41 cm long and the length of a leg is 9 cm. Find the length of the other leg.

Graph in the plane.

78. $y = \frac{1}{3}x - 2$

79. $2x + 3y = -6$

80. $y = -3$

81. $4x - 3y > 12$

82. $y = x^2 + 2x + 1$

83. $x \geq -3$

84. Solve $9x^2 - 12x - 2 = 0$ by completing the square.

85. Approximate the solutions of $4x^2 = 4x + 1$ to the nearest thousandth.

86. Find the x-intercepts of $y = 4x^2 - 4x - 1$ exactly.

Problem Solving

87. What percent of 52 is 13?

88. Find all numbers such that the sum of the number and 10 is less than three times the number.

89. The product of two consecutive even integers is 224. Find the integers.

90. The length of a rectangle is 7 m more than the width. The length of a diagonal is 13 m. Find the length.

91. Three-fifths of the automobiles entering the city each morning will be parked in city parking lots. There are 3654 such parking spaces. How many cars enter the city each morning?

92. A candy shop mixes nuts worth \$1.10 per pound with another variety worth \$0.80 per pound to make 42 lb of a mixture worth \$0.90 per pound. How many pounds of each kind of nut should be used?

93. In checking records, a contractor finds that crew A can pave a certain length of highway in 8 hr. Crew B can do the same job in 10 hr. How long would they take if they worked together?

94. A student's paycheck varies directly as the number of hours worked. The pay was \$242.52 for 43 hr of work. What would the pay be for 80 hr of work? Explain the meaning of the variation constant.

95. For the function f described by

$$f(x) = 2x^2 + 7x - 4,$$

find $f(0)$, $f(-4)$, and $f(\frac{1}{2})$.

96. Determine whether the graphs of the following equations are parallel:

$$y - x = 4,$$
$$3y = 5 + 3x.$$

97. Graph the following system of inequalities:

$$x + y \leq 6,$$
$$x + y \geq 2,$$
$$3 \geq x \geq 1.$$

98. Find the slope and the y-intercept:

$$-6x + 3y = -24.$$

99. Find the slope of the line containing the points $(-5, -6)$ and $(-4, 9)$.

100. Find a point–slope equation for the line containing $(1, -3)$ and having slope $m = -\frac{1}{2}$.

101. Simplify: $-\sqrt{-25}$.

Synthesis

102. Find b such that the trinomial $x^2 - bx + 225$ is a square.

103. Find x.

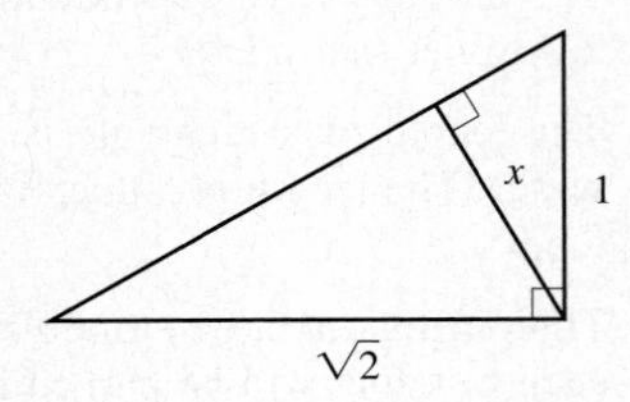

Determine whether the pair of expressions is equivalent.

104. $x^2 - 9, \quad (x - 3)(x + 3)$

105. $\dfrac{x + 3}{3}, \quad x$

106. $(x + 5)^2, \quad x^2 + 25$

107. $\sqrt{x^2 + 16}, \quad x + 4$

108. $\sqrt{x^2}, \quad |x|$

APPENDIXES

A

Sets

Naming Sets • Membership • Subsets • Intersections • Unions

The concept of set is used frequently in more advanced mathematics. We provide a basic introduction to sets in this appendix.

Naming Sets

To name the set of whole numbers less than 6, we can use the *roster method,* as follows:

$\{0, 1, 2, 3, 4, 5\}$.

The set of real numbers x such that x is less than 6 cannot be named by listing all its members because there are infinitely many. We name such a set using *set-builder notation,* as follows:

$\{x| \, x < 6\}$.

This is read

"The set of all x such that x is less than 6."

See Section 2.6 for more on this notation.

Membership

The symbol $\in$ means *is a member of* or *belongs to,* or *is an element of*. Thus,

$x \in A$

means

x is a member of A

or

x belongs to A

or

x is an element of A.

EXAMPLE 1

Classify each of the following as true or false.

a) $1 \in \{1, 2, 3\}$
b) $1 \in \{2, 3\}$
c) $4 \in \{x | x \text{ is an even whole number}\}$
d) $5 \in \{x | x \text{ is an even whole number}\}$

Solution

a) Since 1 *is* listed as a member of the set, $1 \in \{1, 2, 3\}$ is true.
b) Since 1 *is not* a member of $\{2, 3\}$, the statement $1 \in \{2, 3\}$ is false.
c) Since 4 *is* an even whole number, $4 \in \{x | x \text{ is an even whole number}\}$ is a true statement.
d) Since 5 is not even, $5 \in \{x | x \text{ is an even whole number}\}$ is false. ❑

Set membership can be illustrated with a diagram, as shown below.

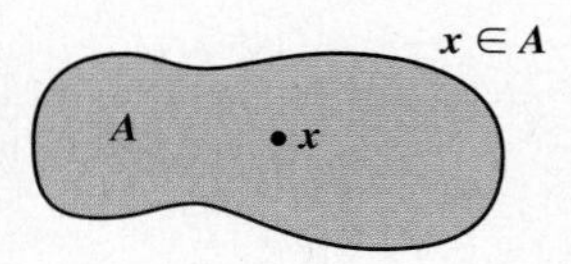

Subsets

If every element of A is an element of B, then A is a *subset* of B. This is denoted $A \subseteq B$.

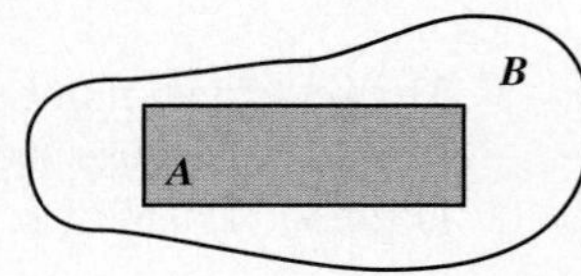

The set of whole numbers is a subset of the set of integers. The set of rational numbers is a subset of the set of real numbers.

EXAMPLE 2

Classify each of the following as true or false.

a) $\{1, 2\} \subseteq \{1, 2, 3, 4\}$
b) $\{p, q, r, w\} \subseteq \{a, p, r, z\}$
c) $\{x | x < 6\} \subseteq \{x | x \leq 11\}$

Solution

a) Since every element of $\{1, 2\}$ is in the set $\{1, 2, 3, 4\}$, the statement $\{1, 2\} \subseteq \{1, 2, 3, 4\}$ is true.
b) Since $q \in \{p, q, r, w\}$, but $q \notin \{a, p, r, z\}$, the statement $\{p, q, r, w\} \subseteq \{a, p, r, z\}$ is false.
c) Since every number that is less than 6 is also less than 11, the statement $\{x | x < 6\} \subseteq \{x | x \leq 11\}$ is true. ❑

Intersections

The *intersection* of sets A and B, denoted $A \cap B$, is the set of members common to both sets.

EXAMPLE 3

Find the intersection.

a) $\{0, 1, 3, 5, 25\} \cap \{2, 3, 4, 5, 6, 7, 9\}$
b) $\{a, p, q, w\} \cap \{p, q, t\}$

Solution

a) $\{0, 1, 3, 5, 25\} \cap \{2, 3, 4, 5, 6, 7, 9\} = \{3, 5\}$
b) $\{a, p, q, w\} \cap \{p, q, t\} = \{p, q\}$ ❑

Set intersection can be illustrated with a diagram, as shown below.

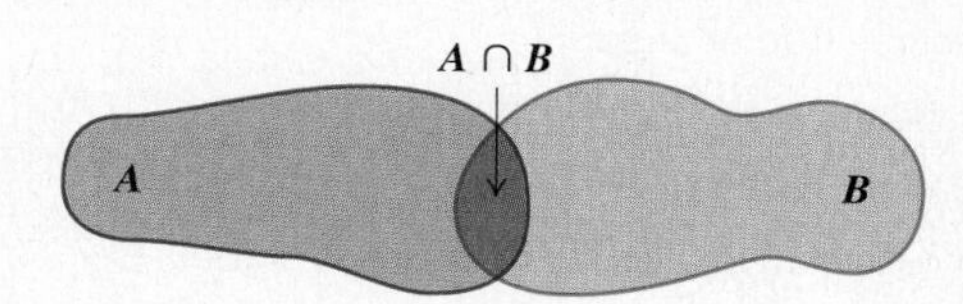

The set without members is known as the *empty set*, and is often named $\varnothing$, and sometimes { }. Each of the following is a description of the empty set:

The set of all six-eyed algebra teachers;
$\{2, 3\} \cap \{5, 6, 7\}$;
$\{x|\ x$ is an even natural number$\} \cap \{x|\ x$ is an odd natural number$\}$.

Unions

Two sets A and B can be combined to form a set that contains the members of A as well as those of B. The new set is called the *union* of A and B, denoted $A \cup B$.

EXAMPLE 4

Find the union.

a) $\{0, 5, 7, 13, 27\} \cup \{0, 2, 3, 4, 5\}$ **b)** $\{a, c, e, g\} \cup \{b, d, f\}$

Solution

a) $\{0, 5, 7, 13, 27\} \cup \{0, 2, 3, 4, 5\} = \{0, 2, 3, 4, 5, 7, 13, 27\}$

Note that the 0 and the 5 are *not* listed twice in the solution.

b) $\{a, c, e, g\} \cup \{b, d, f\} = \{a, b, c, d, e, f, g\}$ ❑

Set union can be illustrated with a diagram, as shown below.

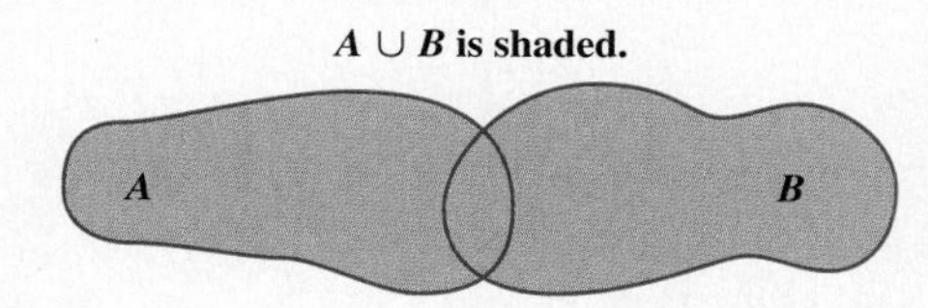

The solution set of the equation $(x - 3)(x + 2) = 0$ is $\{3, -2\}$. This set is the union of the solution sets of $x - 3 = 0$ and $x + 2 = 0$, which are $\{3\}$ and $\{-2\}$.

EXERCISE SET A

Name the set using the roster method.

1. The set of whole numbers 3 through 8

2. The set of whole numbers 101 through 107

3. The set of odd numbers between 40 and 50

4. The set of multiples of 5 between 10 and 40

5. $\{x \mid \text{the square of } x \text{ is } 9\}$

6. $\{x \mid x \text{ is the cube of } 0.2\}$

Determine whether true or false.

7. $2 \in \{x \mid x \text{ is an odd number}\}$

8. $7 \in \{x \mid x \text{ is an odd number}\}$

9. Bruce Springsteen $\in$ The set of all rock stars

10. Apple $\in$ The set of all fruit

11. $-3 \in \{-4, -3, 0, 1\}$

12. $0 \in \{-4, -3, 0, 1\}$

13. $\frac{2}{3} \in \{x \mid x \text{ is a rational number}\}$

14. Heads $\in$ The set of outcomes of flipping a penny

15. $\{4, 5, 8\} \subseteq \{1, 3, 4, 5, 6, 7, 8, 9\}$

16. The set of vowels $\subseteq$ The set of consonants

17. $\{-1, -2, -3, -4, -5\} \subseteq \{-1, 2, 3, 4, 5\}$

18. The set of integers $\subseteq$ The set of rational numbers

Find the intersection.

19. $\{a, b, c, d, e\} \cap \{c, d, e, f, g\}$

20. $\{a, e, i, o, u\} \cap \{q, u, i, c, k\}$

21. $\{1, 2, 5, 10\} \cap \{0, 1, 7, 10\}$

22. $\{0, 1, 7, 10\} \cap \{0, 1, 2, 5\}$

23. $\{1, 2, 5, 10\} \cap \{3, 4, 7, 8\}$

24. $\{a, e, i, o, u\} \cap \{m, n, f, g, h\}$

Find the solution of the system and interpret it in terms of set intersection.

25. $x + 2y = 7,$
$2x + 4y = 2$

26. $y = 2x + 7,$
$y = 2x - 9$

Find the union.

27. $\{a, e, i, o, u\} \cup \{q, u, i, c, k\}$

28. $\{a, b, c, d, e\} \cup \{c, d, e, f, g\}$

29. $\{0, 1, 7, 10\} \cup \{0, 1, 2, 5\}$

30. $\{1, 2, 5, 10\} \cup \{0, 1, 7, 10\}$

31. $\{a, e, i, o, u\} \cup \{m, n, f, g, h\}$

32. $\{1, 2, 5, 10\} \cup \{a, b\}$

Solve the equation and interpret it in terms of set union.

33. $(x - 3)(x + 5) = 0$

34. $x^2 + 2x - 3 = 0$

Synthesis

35. Find the union of the set of integers and the set of whole numbers.

36. Find the intersection of the set of odd integers and the set of even integers.

37. Find the union of the set of rational numbers and the set of irrational numbers.

38. Find the intersection of the set of even integers and the set of positive rational numbers.

39. Find the intersection of the set of rational numbers and the set of irrational numbers.

40. Find the union of the set of negative integers, the set of positive integers, and the set containing 0.

41. For a set A, find each of the following.

a) $A \cup \varnothing$
b) $A \cup A$
c) $A \cap A$
d) $A \cap \varnothing$

42. A set is *closed* under an operation if, when the operation is performed on its members, the result is in the set. For example, the set of real numbers is closed under the operation of addition since the sum of any two real numbers is a real number.

a) Is the set of even numbers closed under addition?

b) Is the set of odd numbers closed under addition?

c) Is the set $\{0, 1\}$ closed under addition?

d) Is the set $\{0, 1\}$ closed under multiplication?

e) Is the set of real numbers closed under multiplication?

f) Is the set of integers closed under division?

43. Experiment with sets of various types and determine whether the following distributive law for sets is true:

$$A \cap (B \cup C) = (A \cap B) \cup (A \cap C).$$

B

Factoring Sums or Differences of Cubes

Factoring a Sum of Two Cubes • Factoring a Difference of Two Cubes

Although a sum of two squares cannot be factored unless a common factor exists, a sum of two cubes can always be factored. A difference of two cubes can also be factored.

Consider the following products:

$$\begin{aligned}(A+B)(A^2-AB+B^2) &= A(A^2-AB+B^2)+B(A^2-AB+B^2)\\ &= A^3-A^2B+AB^2+A^2B-AB^2+B^3\\ &= A^3+B^3\end{aligned}$$

and

$$\begin{aligned}(A-B)(A^2+AB+B^2) &= A(A^2+AB+B^2)-B(A^2+AB+B^2)\\ &= A^3+A^2B+AB^2-A^2B-AB^2-B^3\\ &= A^3-B^3.\end{aligned}$$

These equations show how we can factor a sum or a difference of two cubes.

$$A^3+B^3=(A+B)(A^2-AB+B^2),$$
$$A^3-B^3=(A-B)(A^2+AB+B^2)$$

This table of cubes will help in the following examples.

N	0.2	0.1	0	1	2	3	4	5	6	7	8
N^3	0.008	0.001	0	1	8	27	64	125	216	343	512

EXAMPLE 1

Factor: x^3-8.

Solution We have

$$\begin{array}{c} x^3 - 8 = x^3 - 2^3 = (x-2)(x^2 + x\cdot 2 + 2^2).\\ \uparrow \quad \uparrow \quad \uparrow \quad \uparrow \quad \uparrow \quad \uparrow \quad \uparrow \quad \uparrow \\ \mathbf{A^3 - B^3 = (A - B)(A^2 + A\ \ B + B^2)}\end{array}$$

Thus, $x^3-8=(x-2)(x^2+2x+4)$. Note that we cannot factor x^2+2x+4. (It is not a trinomial square nor can it be factored by trial and error.) ❑

EXAMPLE 2

Factor: x^3+125.

Solution We have

$$x^3 + 125 = x^3 + 5^3 = (x + 5)(x^2 - x \cdot 5 + 5^2).$$

$$A^3 + B^3 = (A + B)(A^2 - A\ B + B^2)$$

Thus, $x^3 + 125 = (x + 5)(x^2 - 5x + 25)$. □

EXAMPLE 3

Factor: $16a^7b + 54ab^7$.

Solution We first look for a common factor:

$$16a^7b + 54ab^7 = 2ab[8a^6 + 27b^6]$$

$$= 2ab[(2a^2)^3 + (3b^2)^3]$$ This is of the form $A^3 + B^3$, where $A = 2a^2$ and $B = 3b^2$.

$$= 2ab[(2a^2 + 3b^2)(4a^4 - 6a^2b^2 + 9b^4)].$$

EXAMPLE 4

Factor: $y^3 - 0.001$.

Solution Since $0.001 = (0.1)^3$, we have a difference of cubes:

$$y^3 - 0.001 = (y - 0.1)(y^2 + 0.1y + 0.01).$$ □

Remember the following about factoring sums or differences of squares and cubes:

Difference of cubes: $A^3 - B^3 = (A - B)(A^2 + AB + B^2)$,
Sum of cubes: $A^3 + B^3 = (A + B)(A^2 - AB + B^2)$,
Difference of squares: $A^2 - B^2 = (A + B)(A - B)$,
Sum of squares: $A^2 + B^2$ **cannot be factored unless a common factor exists.**

EXERCISE SET B

Factor completely.

1. $t^3 + 27$
2. $p^3 + 8$
3. $a^3 - 1$
4. $w^3 - 64$
5. $z^3 + 125$
6. $x^3 + 1$
7. $8a^3 - 1$
8. $27x^3 - 1$
9. $y^3 - 27$
10. $p^3 - 8$
11. $64 + 125x^3$
12. $8 + 27b^3$
13. $125p^3 - 1$
14. $64w^3 - 1$
15. $27m^3 + 64$
16. $8t^3 + 27$
17. $p^3 - q^3$
18. $a^3 + b^3$
19. $x^3 + \frac{1}{8}$
20. $y^3 + \frac{1}{27}$
21. $2y^3 - 128$
22. $3z^3 - 3$
23. $24a^3 + 3$
24. $54x^3 + 2$
25. $rs^3 + 64r$
26. $ab^3 + 125a$
27. $5x^3 - 40z^3$
28. $2y^3 - 54z^3$
29. $x^3 + 0.001$
30. $y^3 + 0.125$
31. $64x^6 - 8t^6$
32. $125c^6 - 8d^6$
33. $x^3 - 0.027$
34. $y^3 + 0.008$

Synthesis

Factor. Assume that variables in exponents represent natural numbers.

35. $x^{6a} + y^{3b}$
36. $a^3x^3 - b^3y^3$
37. $3x^{3a} + 24y^{3b}$
38. $\frac{8}{27}x^3 + \frac{1}{64}y^3$
39. $\frac{1}{24}x^3y^3 + \frac{1}{3}z^3$
40. $\frac{1}{16}x^{3a} + \frac{1}{2}y^{6a}z^{9b}$

TABLES

Fractional and Decimal Equivalents

Fractional Notation	$\frac{1}{10}$	$\frac{1}{8}$	$\frac{1}{6}$	$\frac{1}{5}$	$\frac{1}{4}$	$\frac{3}{10}$	$\frac{1}{3}$	$\frac{3}{8}$	$\frac{2}{5}$	$\frac{1}{2}$	$\frac{3}{5}$	$\frac{5}{8}$	$\frac{2}{3}$	$\frac{7}{10}$	$\frac{3}{4}$	$\frac{4}{5}$	$\frac{5}{6}$	$\frac{7}{8}$	$\frac{9}{10}$	$\frac{1}{1}$
Decimal Notation	0.1	0.125	$0.16\overline{6}$	0.2	0.25	0.3	$0.333\overline{3}$	0.375	0.4	0.5	0.6	0.625	$0.666\overline{6}$	0.7	0.75	0.8	$0.83\overline{3}$	0.875	0.9	1
Percent Notation	10%	12.5% or $12\frac{1}{2}$%	$16.6\overline{6}$% or $16\frac{2}{3}$%	20%	25%	30%	$33.3\overline{3}$% or $33\frac{1}{3}$%	37.5% or $37\frac{1}{2}$%	40%	50%	60%	62.5% or $62\frac{1}{2}$%	$66.6\overline{6}$% or $66\frac{2}{3}$%	70%	75%	80%	$83.3\overline{3}$% or $83\frac{1}{3}$%	87.5% or $87\frac{1}{2}$%	90%	100%

TABLE 1 Squares and Square Roots

N	$\sqrt{N}$	N^2	N	$\sqrt{N}$	N^2	N	$\sqrt{N}$	N^2	N	$\sqrt{N}$	N^2
2	1.414	4	27	5.196	729	52	7.211	2704	77	8.775	5929
3	1.732	9	28	5.292	784	53	7.280	2809	78	8.832	6084
4	2	16	29	5.385	841	54	7.348	2916	79	8.888	6241
5	2.236	25	30	5.477	900	55	7.416	3025	80	8.944	6400
6	2.449	36	31	5.568	961	56	7.483	3136	81	9	6561
7	2.646	49	32	5.657	1024	57	7.550	3249	82	9.055	6724
8	2.828	64	33	5.745	1089	58	7.616	3364	83	9.110	6889
9	3	81	34	5.831	1156	59	7.681	3481	84	9.165	7056
10	3.162	100	35	5.916	1225	60	7.746	3600	85	9.220	7225
11	3.317	121	36	6	1296	61	7.810	3721	86	9.274	7396
12	3.464	144	37	6.083	1369	62	7.874	3844	87	9.327	7569
13	3.606	169	38	6.164	1444	63	7.937	3969	88	9.381	7744
14	3.742	196	39	6.245	1521	64	8	4096	89	9.434	7921
15	3.873	225	40	6.325	1600	65	8.062	4225	90	9.487	8100
16	4	256	41	6.403	1681	66	8.124	4356	91	9.539	8281
17	4.123	289	42	6.481	1764	67	8.185	4489	92	9.592	8464
18	4.243	324	43	6.557	1849	68	8.246	4624	93	9.644	8649
19	4.359	361	44	6.633	1936	69	8.307	4761	94	9.695	8836
20	4.472	400	45	6.708	2025	70	8.367	4900	95	9.747	9025
21	4.583	441	46	6.782	2116	71	8.426	5041	96	9.798	9216
22	4.690	484	47	6.856	2209	72	8.485	5184	97	9.849	9409
23	4.796	529	48	6.928	2304	73	8.544	5329	98	9.899	9604
24	4.899	576	49	7	2401	74	8.602	5476	99	9.950	9801
25	5	625	50	7.071	2500	75	8.660	5625	100	10	10,000
26	5.099	676	51	7.141	2601	76	8.718	5776			

TABLE 2 Geometric Formulas

Plane Geometry:

Rectangle
Area: $A = lw$
Perimeter: $P = 2l + 2w$

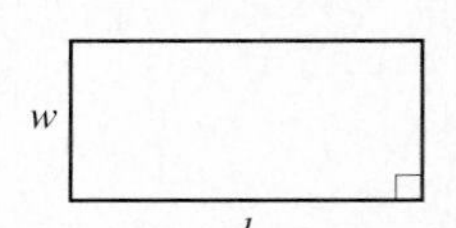

Square
Area: $A = s^2$
Perimeter: $P = 4s$

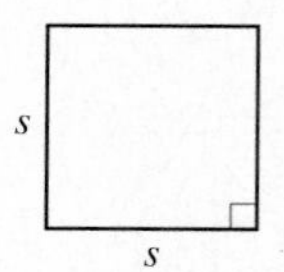

Triangle
Area: $A = \frac{1}{2}bh$

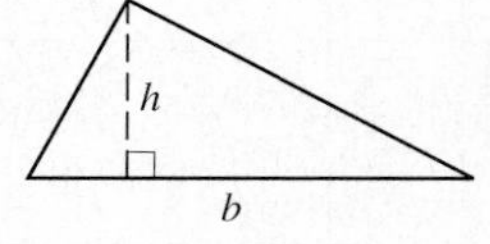

Triangle
Sum of Angle Measures:
$A + B + C = 180°$

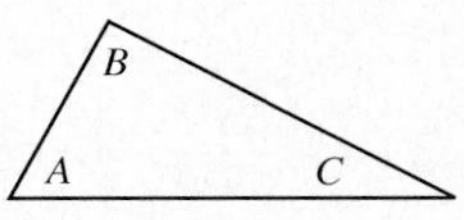

Right Triangle
Pythagorean Theorem
(Equation):
$a^2 + b^2 = c^2$

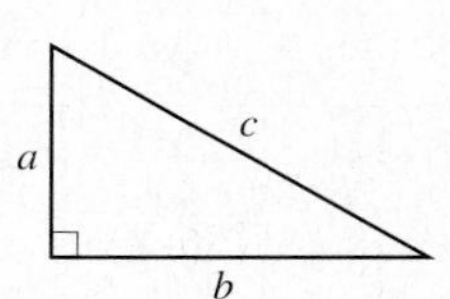

Parallelogram
Area: $A = bh$

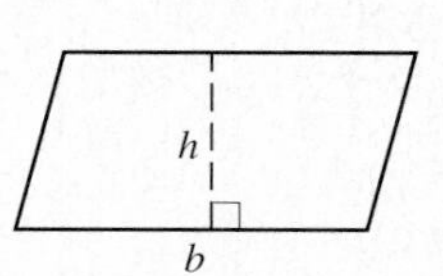

Trapezoid
Area: $A = \frac{1}{2}h(b_1 + b_2)$

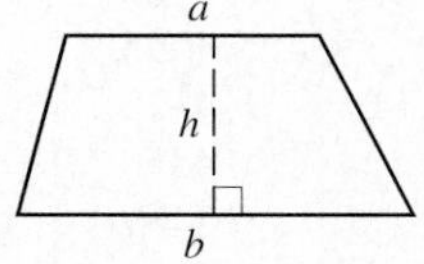

Circle
Area: $A = \pi r^2$
Circumference:
$C = \pi D = 2\pi r$
($\frac{22}{7}$ and 3.14 are different approximations for π)

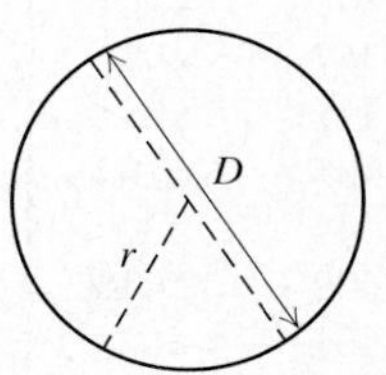

Solid Geometry:

Rectangular Solid
Volume: $V = lwh$

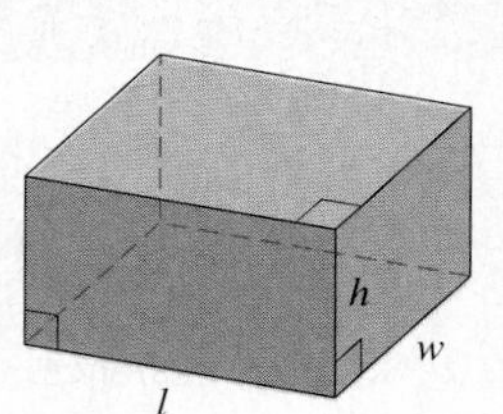

Cube
Volume: $V = s^3$

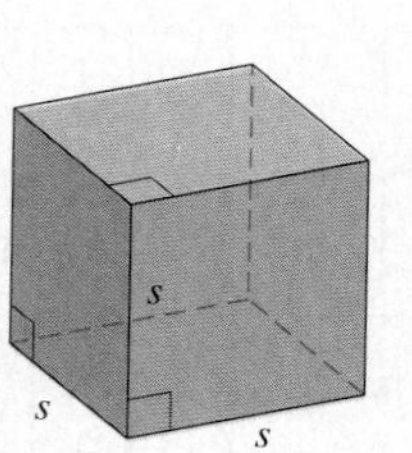

Right Circular Cylinder
Volume: $V = \pi r^2 h$
Total Surface Area:
$S = 2\pi rh + 2\pi r^2$

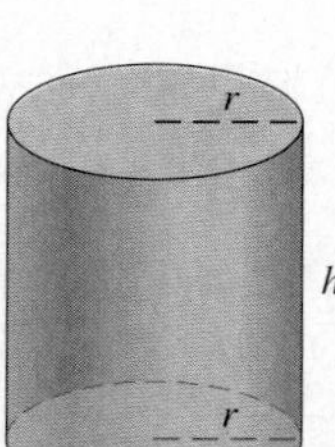
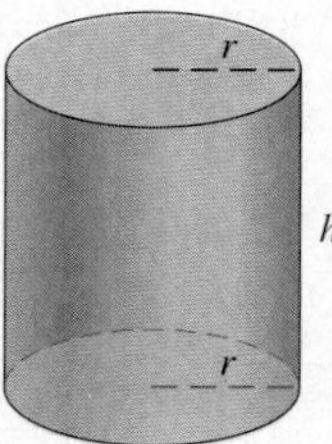

Right Circular Cone
Volume: $V = \frac{1}{3}\pi r^2 h$
Total Surface Area:
$S = \pi r^2 + \pi rs$
Slant Height:
$s = \sqrt{r^2 + h^2}$

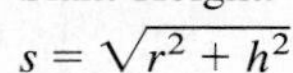

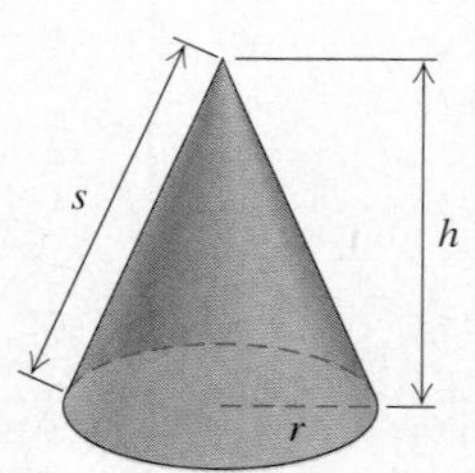

Sphere
Volume: $V = \frac{4}{3}\pi r^3$
Surface Area: $S = 4\pi r^2$

ANSWERS

CHAPTER 1

Exercise Set 1.1, pp. 6–8

1. 42 **3.** 16 **5.** 1 **7.** 6 **9.** 2 **11.** $\frac{1}{2}$ **13.** 20
15. 220 mi **17.** 100.1 sq cm
19. 150 sec, 450 sec, 10 min **21.** $b + 6$, or $6 + b$
23. $c - 9$ **25.** $6 + q$, or $q + 6$ **27.** $b + a$, or $a + b$
29. $y - x$ **31.** $x \div w$, or $\frac{x}{w}$ **33.** $n - m$
35. $r + s$, or $s + r$ **37.** $2x$ **39.** $\frac{1}{3}t$, or $\frac{t}{3}$
41. $97\%n$, or $0.97n$ **43.** $\$d - \$29.95$ **45.** Yes
47. No **49.** No **51.** Yes **53.** No
55. $x + 60 = 112$ **57.** $42y = 2352$ **59.** $s + 35 = 64$
61. $4c = \$7.96$ **63.** ◈ **65.** $y + 2x$ **67.** $2x - 3$
69. $b - 2$ **71.** $s + s + s + s$, or $4s$ **73.** 6 **75.** 6
77. $w + 4$ **79.** $t - 3$, $t + 3$

Exercise Set 1.2, pp. 13–14

1. $5 + y$ **3.** $ab + 5$ **5.** $3y + 9x$ **7.** $2(3 + a)$
9. tr **11.** $a5$ **13.** $5 + ba$ **15.** $(a + 3)2$
17. $(x + y) + 2$ **19.** $9 + (m + 2)$ **21.** $ab + (c + d)$
23. $(5a)b$ **25.** $6(mn)$ **27.** $(3 \cdot 2)(a + b)$
29. $2 + (b + a)$, $(2 + a) + b$; answers may vary
31. $(ba)7$, $a(7b)$; answers may vary

33.

$(3a)4 = 4(3a)$	Commutative law
$= (4 \cdot 3)a$	Associative law
$= 12a$	Simplifying

35.

$5 + (2 + x) = (5 + 2) + x$	Associative law
$= x + (5 + 2)$	Commutative law
$= x + 7$	Simplifying

37. $2b + 10$ **39.** $7 + 7t$ **41.** $3x + 3$ **43.** $4 + 4y$
45. $30x + 12$ **47.** $7x + 28 + 42y$ **49.** $2a + 2b$
51. $5x + 5y + 10$ **53.** $2(x + y)$ **55.** $5(1 + y)$
57. $3(x + 4y)$ **59.** $5(x + 2 + 3y)$ **61.** $9(x + 1)$
63. $3(3x + y)$ **65.** $2(a + 8b + 32)$
67. $11(x + 4y + 11)$ **69.** $t - 9$ **71.** ◈
73. Yes; commutative law of addition
75. Yes; distributive law and commutative laws of addition and multiplication **77.** ◈ **79.** ◈

Exercise Set 1.3, pp. 20–22

1. $4 \cdot 14$, $7 \cdot 8$ **3.** $1 \cdot 93$, $3 \cdot 31$ **5.** $2 \cdot 7$ **7.** $3 \cdot 11$
9. $3 \cdot 3$ **11.** $7 \cdot 7$ **13.** $2 \cdot 3 \cdot 3$ **15.** $2 \cdot 2 \cdot 2 \cdot 5$
17. $2 \cdot 3 \cdot 3 \cdot 5$ **19.** $2 \cdot 3 \cdot 5 \cdot 7$ **21.** Prime **23.** $7 \cdot 17$
25. $\frac{2}{5}$ **27.** $\frac{7}{2}$ **29.** $\frac{1}{7}$ **31.** 8 **33.** $\frac{1}{4}$ **35.** 5 **37.** $\frac{17}{21}$
39. $\frac{13}{7}$ **41.** $\frac{4}{3}$ **43.** $\frac{1}{8}$ **45.** $\frac{51}{8}$ **47.** 1 **49.** $\frac{7}{6}$ **51.** $\frac{3b}{7a}$
53. $\frac{5}{x}$ **55.** $\frac{5}{6}$ **57.** $\frac{1}{2}$ **59.** $\frac{5}{18}$ **61.** $\frac{31}{60}$ **63.** $\frac{35}{18}$ **65.** $\frac{10}{3}$
67. $\frac{1}{2}$ **69.** $\frac{5}{36}$ **71.** 500 **73.** $\frac{3}{40}$ **75.** $\frac{5b}{3a}$ **77.** $\frac{x - 2}{6}$
79. $5(3 + x)$; answers may vary **81.** ◈ **83.** $\frac{2}{3}$
85. $\frac{3sb}{2}$ **87.** $\frac{r}{g}$ **89.** 24 in. **91.** $\frac{28}{45}$ m^2 **93.** $\frac{20}{9}$ m
95. ◈

Exercise Set 1.4, pp. 29–30

1. 5, -12 **3.** -170, 950 **5.** -1286, 29,028
7. 750, -125 **9.** 20, -150, 300
11. **13.**

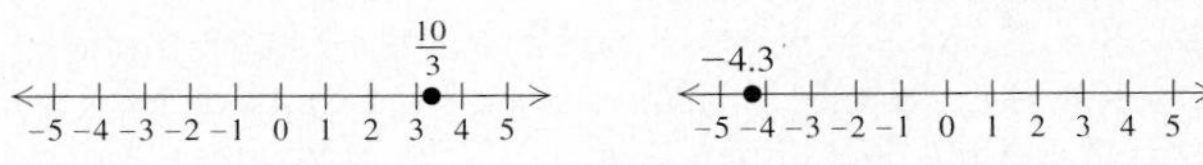

15.

17. -0.375 **19.** $1.\overline{6}$
21. $1.1\overline{6}$ **23.** $0.\overline{6}$
25. -0.5 **27.** 0.1
29. $>$ **31.** $<$ **33.** $<$
35. $<$ **37.** $>$ **39.** $<$ **41.** $>$ **43.** $<$
45. $x < -6$ **47.** $y \geq -10$ **49.** True **51.** False
53. True **55.** $-5 > x$ **57.** $120 > -20$

59. $-500{,}000 < 1{,}000{,}000$ **61.** $s \leq 95$
63. $p \leq 15{,}000$ **65.** 3 **67.** 10 **69.** 0 **71.** 24
73. $\frac{2}{3}$ **75.** 43.9 **77.** 5
79. Answers may vary. $-\frac{9}{7}$, 0, $4\frac{1}{2}$, -1.97, -491, 128, $\frac{3}{11}$, $-\frac{1}{7}$, 0.000011, $-26\frac{1}{3}$
81. Answers may vary. $-\pi$, $\sqrt{42}$, 8.4262262226 . . .
83. $\frac{3}{5}$ **85.** $5 + ab$, $ba + 5$, or $5 + ba$ **87.** ◈
89. -17, -12, 5, 13 **91.** $-\frac{4}{3}, \frac{4}{9}, \frac{4}{8}, \frac{4}{6}, \frac{4}{5}, \frac{4}{3}, \frac{4}{2}$ **93.** $>$
95. $=$ **97.** $<$ **99.** 7, -7
101. **(a)** $\frac{5}{9}$; **(b)** $\frac{1}{9}$; **(c)** $\frac{2}{9}$; **(d)** 1

Exercise Set 1.5, pp. 34–35

1. -7 **3.** -4 **5.** 0 **7.** -8 **9.** -7 **11.** -27
13. 0 **15.** -42 **17.** 0 **19.** 3 **21.** -9 **23.** 7
25. 2 **27.** -26 **29.** -22 **31.** 32 **33.** 0
35. 45 **37.** -1.8 **39.** -8.1 **41.** $-\frac{1}{5}$ **43.** $-\frac{8}{7}$
45. $-\frac{3}{8}$ **47.** $-\frac{29}{35}$ **49.** 39 **51.** 50 **53.** -1093
55. 8-yd gain **57.** 13-mb drop **59.** She owes $85.
61. $11a$ **63.** $13x$ **65.** $11x$ **67.** $-2m$ **69.** $4a$
71. $1 - 2x$ **73.** $16 + 11x$ **75.** $16 + 9m$
77. $21z + 7y + 14$ **79.** ◈ **81.** $$65\frac{1}{4}$ **83.** $-5y$
85. $-7m$ **87.** $3x$

Exercise Set 1.6, pp. 40–42

1. -24 **3.** 9 **5.** 26.9 **7.** -9 **9.** $\frac{14}{3}$
11. -0.101 **13.** -65 **15.** $\frac{5}{3}$ **17.** 1 **19.** -7
21. -4 **23.** -7 **25.** -6 **27.** 0 **29.** -4
31. -7 **33.** -6 **35.** 0 **37.** 0 **39.** 14 **41.** 11
43. -14 **45.** 5 **47.** -1 **49.** 18 **51.** -5
53. -3 **55.** -21 **57.** 5 **59.** -8 **61.** 12
63. -23 **65.** -68 **67.** -73 **69.** 116 **71.** 0
73. $-\frac{1}{4}$ **75.** $\frac{1}{12}$ **77.** $-\frac{17}{12}$ **79.** -2.8 **81.** -0.01
83. $-\frac{1}{2}$ **85.** $\frac{6}{7}$ **87.** $1.5 - (-3.5)$; 5
89. $-79 - (114)$; -193 **91.** -54 **93.** 34
95. Negative three point two minus five point eight; -9
97. Negative two hundred thirty minus negative five hundred; 270 **99.** 37 **101.** -62 **103.** -139
105. 6 **107.** $3x$, $-2y$ **109.** -5, $3m$, $-6mn$
111. 5, $-a$, $-6b$, 2 **113.** $-5a$ **115.** $-2m - 5$
117. $-6x + 5$ **119.** $-7 - 8t$ **121.** $10x + 7$
123. $15x + 66$ **125.** $-$330.54$ **127.** 50°C
129. 116 m **131.** 432 ft^2 **133.** ◈ **135.** True. For example, for $m = 5$ and $n = 3$, $5 > 3$ and $5 - 3 = 2 > 0$; for $m = -4$ and $n = -9$, $-4 > -9$ and $-4 - (-9) = 5 > 0$. **137.** False. For example, let $m = 2$ and $n = -2$. Then 2 and -2 are opposites, but $2 - (-2) = 4 \neq 0$.
139. ◈

Exercise Set 1.7, pp. 47–48

1. -16 **3.** -42 **5.** -24 **7.** -72 **9.** 16
11. 42 **13.** -120 **15.** -238 **17.** 1200 **19.** 98
21. -72 **23.** 21.7 **25.** $-\frac{2}{5}$ **27.** $\frac{1}{12}$ **29.** -17.01
31. $-\frac{5}{12}$ **33.** 420 **35.** $\frac{2}{7}$ **37.** -60 **39.** 150
41. 0 **43.** -720 **45.** $-30{,}240$ **47.** -6
49. -13 **51.** -2 **53.** 4 **55.** -8 **57.** 2
59. -12 **61.** -8 **63.** Undefined **65.** $-\frac{88}{9}$ **67.** 0
69. Indeterminate **71.** $\dfrac{-9}{5}, -\dfrac{9}{5}$ **73.** $\dfrac{36}{-11}, -\dfrac{36}{11}$
75. $\dfrac{-7}{3}, \dfrac{7}{-3}$ **77.** $\dfrac{x}{-2}, -\dfrac{x}{2}$ **79.** $-\dfrac{7}{3}$ **81.** $-\dfrac{13}{47}$
83. $-\dfrac{1}{10}$ **85.** $\dfrac{1}{4.3}$ **87.** $-\frac{3}{5}$ **89.** $-\frac{1}{1}$, or -1 **91.** $\frac{6}{35}$
93. $\frac{35}{12}$ **95.** $-\frac{11}{5}$ **97.** $\frac{5}{28}$ **99.** $-\frac{13}{7}$ **101.** $-\frac{9}{8}$
103. $-\frac{7}{36}$ **105.** -3 **107.** $\frac{5}{3}$ **109.** -2 **111.** $-\frac{5}{7}$
113. $-\frac{11}{9}$ **115.** $-\frac{3}{2}$ **117.** $\frac{5}{9}$ **119.** $-\frac{1}{2}$ **121.** $-\frac{1}{2}$
123. $\frac{22}{39}$ **125.** ◈ **127.** There are none.
129. Negative **131.** Positive **133.** Negative
135. Distributive law; law of opposites; multiplicative property of 0; law of opposites

Exercise Set 1.8, pp. 54–56

1. 10^3 **3.** x^7 **5.** $(3y)^4$ **7.** 16 **9.** 9 **11.** 1
13. 64 **15.** -64 **17.** 7 **19.** $16a^2$ **21.** $-343x^3$
23. 19 **25.** 86 **27.** 7 **29.** 5 **31.** 12 **33.** -8
35. -7 **37.** -7 **39.** -4 **41.** -334 **43.** 14
45. 1880 **47.** 16 **49.** 1 **51.** -26 **53.** 37
55. -6 **57.** 4 **59.** 144 **61.** 2 **63.** $-\frac{21}{38}$
65. $-\frac{4}{3}$ **67.** -8 **69.** 4 **71.** $\frac{55}{4}$ **73.** 6
75. $-2x - 7$ **77.** $-5x + 8$ **79.** $-4a + 3b - 7c$
81. $-3x^2 - 5x + 1$ **83.** $5x - 3$ **85.** $-3a + 9$
87. $5x - 6$ **89.** $-19x + 2y$ **91.** $9y - 25z$
93. $x^2 + 2$ **95.** $7x^3 - 5x$ **97.** $37a^2 - 23ab + 35b^2$
99. $-22t^3 - t^2 + 9t$ **101.** $12x + 30$ **103.** $3x^2 + 30$
105. $9x - 18$ **107.** $-4x^3 - 64$ **109.** $2x + 9$
111. ◈ **113.** $-4z$ **115.** $x - 3$ **117.** ◈
119. False **121.** False **123.** False **125.** True

Review Exercises: Chapter 1, pp. 57–58

1. [1.1] 15 **2.** [1.1] 6 **3.** [1.1] 5 **4.** [1.1] 4
5. [1.8] -15 **6.** [1.8] -5 **7.** [1.1] $z - 8$
8. [1.1] $3x$ **9.** [1.1] $\frac{1}{3}y$ **10.** [1.1] No
11. [1.1] $6x = 6768$ **12.** [1.2] $y + 2x$
13. [1.2] $x \cdot 2 + y$ **14.** [1.2] $2x + (y + z)$
15. [1.2] $4(yx)$, $(4y)x$, $y(4x)$ **16.** [1.2] $18x + 30y$
17. [1.2] $40x + 24y + 16$ **18.** [1.2] $7(3x + y)$
19. [1.2] $7(5x + 2 + y)$ **20.** [1.3] $\frac{5}{12}$ **21.** [1.3] 10
22. [1.3] $\frac{31}{36}$ **23.** [1.3] $\frac{1}{4}$ **24.** [1.3] $\frac{3}{5}$ **25.** [1.3] $\frac{72}{25}$
26. [1.4] -45, 72 **27.** [1.4] $x \leq 300$
28. [1.4]

$\frac{-1}{3}$
−5 −4 −3 −2 −1 0 1 2 3 4 5

29. [1.4] $x > -3$ **30.** [1.4] False **31.** [1.4] -0.875
32. [1.4] 1 **33.** [1.6] -5 **34.** [1.5] -3

35. [1.5] $-\frac{7}{12}$ **36.** [1.5] -4 **37.** [1.5] -5
38. [1.6] 4 **39.** [1.6] $-\frac{7}{5}$ **40.** [1.6] -7.9
41. [1.7] 54 **42.** [1.7] -9.18 **43.** [1.7] $-\frac{2}{7}$
44. [1.7] -210 **45.** [1.7] -7 **46.** [1.7] -3
47. [1.7] $\frac{3}{4}$ **48.** [1.8] 92 **49.** [1.8] 62 **50.** [1.8] 48
51. [1.8] 168 **52.** [1.8] $\frac{21}{8}$ **53.** [1.8] $\frac{103}{17}$
54. [1.5] $7a - 3b$ **55.** [1.6] $-2x + 5y$ **56.** [1.6] 7
57. [1.7] $-\frac{1}{7}$ **58.** [1.8] $(2x)^4$ **59.** [1.8] $-27y^3$
60. [1.8] $-3a + 9$ **61.** [1.8] $-2b + 21$
62. [1.8] $-3x + 9$ **63.** [1.8] $12y - 34$
64. [1.8] $5x + 24$ **65.** [1.8] $-15x + 25$
66. [1.2], [1.5], [1.8] ◈ The distributive law is used in factoring algebraic expressions, multiplying algebraic expressions, collecting like terms, finding the opposite of a sum, and subtracting algebraic expressions.
67. [1.8] ◈ A negative quantity raised to an even power is positive; a negative quantity raised to an odd power is negative.
68. [1.8] 25,281 **69.** [1.4] **(a)** $\frac{3}{11}$; **(b)** $\frac{10}{11}$
70. [1.8] $-\frac{5}{8}$ **71.** [1.8] -2.1

Test: Chapter 1, pp. 58–59

1. [1.1] 6 **2.** [1.1] $x - 9$ **3.** [1.1] 240 ft^2
4. [1.2] $q + 3p$ **5.** [1.2] $(x \cdot 4) \cdot y$ **6.** [1.2] $18 - 3x$
7. [1.2] $-5y + 5$ **8.** [1.2] $11(1 - 4x)$
9. [1.2] $7(x + 3 + 2y)$ **10.** [1.3] $2 \cdot 2 \cdot 3 \cdot 5 \cdot 5$
11. [1.4] $<$ **12.** [1.4] $>$ **13.** [1.4] $>$ **14.** [1.4] $<$
15. [1.4] 7 **16.** [1.4] $\frac{9}{4}$ **17.** [1.4] 2.7 **18.** [1.6] $-\frac{2}{3}$
19. [1.7] $-\frac{7}{4}$ **20.** [1.6] 8 **21.** [1.4] $-2 \geqslant x$
22. [1.6] 7.8 **23.** [1.5] -8 **24.** [1.5] $\frac{7}{40}$
25. [1.6] 10 **26.** [1.6] -2.5 **27.** [1.6] $\frac{7}{8}$
28. [1.7] -48 **29.** [1.7] $\frac{3}{16}$ **30.** [1.7] -9
31. [1.7] $\frac{3}{4}$ **32.** [1.7] -9.728 **33.** [1.8] -173
34. [1.6] 12 **35.** [1.8] -4 **36.** [1.8] 448
37. [1.6] $22y + 21a$ **38.** [1.8] $16x^4$ **39.** [1.8] $2x + 7$
40. [1.8] $9a - 12b - 7$ **41.** [1.8] $68y - 8$
42. [1.1] 15 **43.** [1.3] $\frac{23}{70}$ **44.** [1.8] 15 **45.** [1.8] $4a$

CHAPTER 2

Exercise Set 2.1, pp. 67–68

1. 4 **3.** -20 **5.** -14 **7.** -18 **9.** 15 **11.** -14
13. 2 **15.** 20 **17.** -6 **19.** $\frac{7}{3}$ **21.** $-\frac{7}{4}$ **23.** $\frac{41}{24}$
25. $-\frac{1}{20}$ **27.** 5.1 **29.** -5 **31.** 6 **33.** 9 **35.** 12
37. -40 **39.** 1 **41.** -7 **43.** -6 **45.** 6
47. -63 **49.** 36 **51.** -21 **53.** $-\frac{3}{5}$ **55.** $\frac{3}{2}$
57. $\frac{9}{2}$ **59.** 7 **61.** 4.5 **63.** -27 **65.** -12 **67.** $\frac{1}{2}$
69. -15 **71.** -2.5 **73.** $7x$ **75.** $x - 4$ **77.** ◈
79. 342.246 **81.** No solution **83.** 12, -12
85. All real numbers **87.** No solution **89.** 5
91. $\dfrac{b}{3a}$ **93.** $1 - c - a$ **95.** 11,074 **97.** ◈

Exercise Set 2.2, pp. 73–75

1. 5 **3.** 8 **5.** 10 **7.** 14 **9.** -8 **11.** -8
13. -7 **15.** 15 **17.** 6 **19.** 4 **21.** 6 **23.** -3
25. 1 **27.** -20 **29.** 6 **31.** 7 **33.** 2 **35.** 5
37. 2 **39.** 10 **41.** 4 **43.** 0 **45.** -1 **47.** $\frac{64}{3}$
49. $\frac{2}{5}$ **51.** -2 **53.** -4 **55.** 0.8 **57.** $-\frac{28}{27}$ **59.** 6
61. 2 **63.** 6 **65.** 8 **67.** 1 **69.** 17 **71.** $-\frac{5}{3}$
73. -3 **75.** 2 **77.** 6 **79.** $\frac{4}{7}$ **81.** 8 **83.** $\frac{11}{18}$
85. $-\frac{51}{31}$ **87.** 2 **89.** -6.5 **91.** $<$ **93.** ◈
95. 4.4233464 **97.** $-\frac{7}{2}$ **99.** $\dfrac{837{,}353}{1929}$ **101.** -2
103. 0 **105.** -2

Exercise Set 2.3, pp. 78–79

1. $b = \dfrac{A}{h}$ **3.** $r = \dfrac{d}{t}$ **5.** $P = \dfrac{I}{rt}$ **7.** $a = \dfrac{F}{m}$
9. $w = \dfrac{P - 2l}{2}$ **11.** $r^2 = \dfrac{A}{\pi}$ **13.** $b = \dfrac{2A}{h}$ **15.** $m = \dfrac{E}{c^2}$
17. $d = 2Q - c$ **19.** $b = 3A - a - c$ **21.** $t = \dfrac{3k}{v}$
23. $y = \dfrac{C - Ax}{B}$ **25.** $b = \dfrac{2A - ah}{h}$ **27.** $a = \dfrac{Q}{3 + 5c}$
29. $P = \dfrac{A}{1 + rt}$ **31.** $S = \dfrac{360A}{\pi r^2}$ **33.** $t = \dfrac{R - 3.85}{-0.0075}$
35. -42 **37.** -29 **39.** ◈ **41.** $y = \dfrac{z^2}{t}$
43. $t = \dfrac{q - rs}{r}$, or $t = \dfrac{q}{r} - s$ **45.** $x = \dfrac{a - cy}{c + b}$ **47.** ◈

Exercise Set 2.4, pp. 83–84

1. 0.76 **3.** 0.547 **5.** 1 **7.** 0.0061 **9.** 2.4
11. 0.0325 **13.** 454% **15.** 99.8% **17.** 200%
19. 7.2% **21.** 920% **23.** 0.68% **25.** 12.5%
27. 68% **29.** 75% **31.** 70% **33.** 60% **35.** $66\frac{2}{3}$%
37. 25% **39.** 24% **41.** 150 **43.** 2.5 **45.** 546
47. 125% **49.** 0.8 **51.** 5% **53.** 3.75% **55.** 27
57. 86.4% **59.** $36 **61.** 0.92 **63.** -13.2 **65.** ◈
67. Rollie's Music: $12.83; Sound Warp: $12.97
69. 62.5% **71.** $c = \dfrac{T}{1 + r}$

Exercise Set 2.5, pp. 92–95

1. $2x - 3$ **3.** $\frac{1}{2} \cdot 7x$ **5.** $5(a + 3)$ **7.** $L + 2$
9. $b - 30\%b$, or $b - 0.3b$, or $0.7b$
11. $x + (x + 2) + (x + 4)$ **13.** $34.95 + 0.27m$
15. **(a)** $2w$; **(b)** $\frac{1}{2}l$ **17.** Second: $3x$; third: $x + 30$
19. 14 **21.** 11 **23.** 19 **25.** -10 **27.** 40
29. 20 m, 40 m, 120 m **31.** 136 and 137
33. 56, 58 **35.** 35, 36, 37 **37.** 61, 63, 65

39. $l = 160$ ft, $w = 100$ ft; 16,000 ft^2
41. $l = 27.9$ cm; $w = 21.6$ cm **43.** 22.5° **45.** \$16
47. \$4400 **49.** 450.5 mi **51.** 28°, 84°, 68°
53. 2020 **55.** $3(x - 4y + 20)$ **57.** ◈ **59.** 20
61. 19¢ **63.** 12 cm, 9 cm **65.** 30
67. 0.726175 in. **69.** ◈

Exercise Set 2.6, pp. 101–102

1. **(a)** Yes; **(b)** yes; **(c)** no; **(d)** yes; **(e)** yes
3. **(a)** No; **(b)** no; **(c)** yes; **(d)** yes; **(e)** no
5.

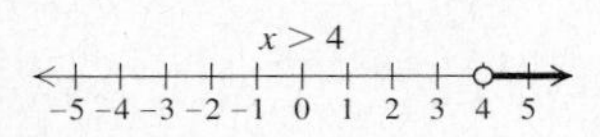

7.

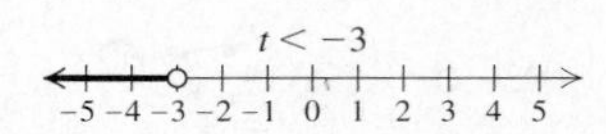

9.

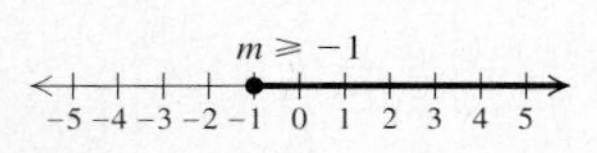

11.

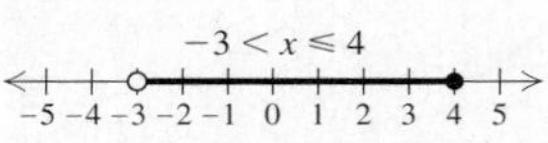

13.

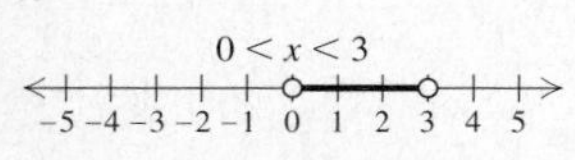

15. $\{x|x > -1\}$
17. $\{x|x \leqslant 2\}$
19. $\{x|x < -2\}$
21. $\{x|x \geqslant 0\}$

23. $\{y|y > 3\}$,

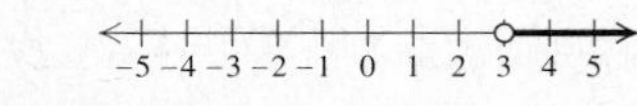

25. $\{x|x \leqslant -18\}$,

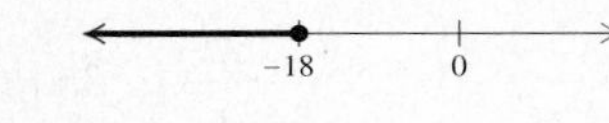

27. $\{x|x < 16\}$,

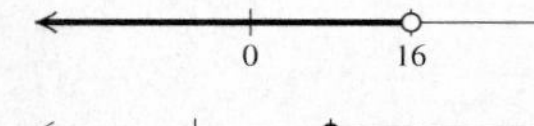

29. $\{x|x \geqslant 8\}$,

0 8

31. $\{y|y > -5\}$,

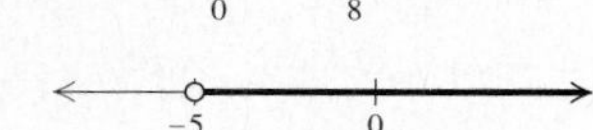

33. $\{x|x \leqslant 2\}$,

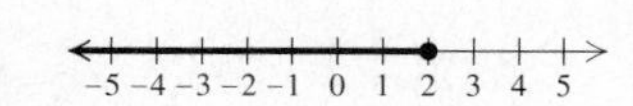

35. $\{x|x \geqslant 13\}$ **37.** $\{x|x < 4\}$ **39.** $\{c|c > 0\}$
41. $\{y|y \leqslant \frac{1}{4}\}$ **43.** $\{x|x > \frac{7}{12}\}$ **45.** $\{x|x > 0\}$

47. $\{x|x < 7\}$,

0 7

49. $\{y|y \leqslant 9\}$,

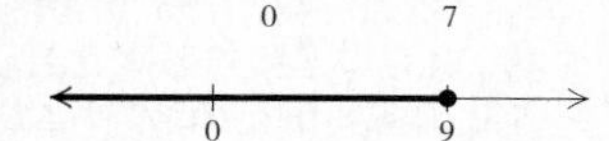

51. $\{x|x < \frac{13}{7}\}$,

13/7
−5 −4 −3 −2 −1 0 1 2 3 4 5

53. $\{x|x > -3\}$,

−5 −4 −3 −2 −1 0 1 2 3 4 5

55. $\{y|y \geqslant -\frac{2}{5}\}$ **57.** $\{x|x \geqslant -6\}$ **59.** $\{y|y \leqslant 4\}$
61. $\{x|x > \frac{17}{3}\}$ **63.** $\{y|y < -\frac{1}{14}\}$ **65.** $\{x|\frac{3}{10} \geqslant x\}$
67. $\{x|x < 8\}$ **69.** $\{y|y \geqslant 6\}$ **71.** $\{x|x \leqslant 6\}$
73. $\{x|x < -3\}$ **75.** $\{x|x \geqslant -2\}$ **77.** $\{y|y < -3\}$
79. $\{x|x > -3\}$ **81.** $\{y|y < -\frac{10}{3}\}$ **83.** $\{x|x > -10\}$

85. $\{y|y < 2\}$ **87.** $\{y|y \geqslant 3\}$ **89.** $\{y|y > -2\}$
91. $\{x|x < \frac{9}{5}\}$ **93.** $\{x|x > -4\}$ **95.** $\{n|n \geqslant 70\}$
97. $\{x|x \leqslant 9\}$ **99.** $\{y|y \leqslant -3\}$ **101.** $\{y|y < 6\}$
103. $\{d|d \geqslant 17\}$ **105.** $\{t|t < -\frac{5}{3}\}$ **107.** $\{r|r > -3\}$
109. $\{x|x \geqslant 8\}$ **111.** $\{x|x < \frac{11}{18}\}$ **113.** 32 **115.** ◈
117. $\{x|x \leqslant \frac{4}{7}\}$ **119.** $\{x|x \leqslant -4a\}$ **121.** $\left\{x\middle|x > \frac{y-b}{a}\right\}$
123. **(a)** Yes; **(b)** yes; **(c)** no; **(d)** no; **(e)** no; **(f)** yes; **(g)** yes

Exercise Set 2.7, pp. 105–107

1. $x > 4$ **3.** $x \leqslant -6$ **5.** $t \leqslant 80$ **7.** $75 < a < 100$
9. $p \geqslant 1200$ **11.** $y \leqslant 500$ **13.** $3x + 2 < 13$
15. $\{s|s \geqslant 97\}$ **17.** $\{m|m \leqslant 341.4 \text{ mi}\}$
19. $\{l|l < 21.5 \text{ cm}\}$ **21.** $\{t|t \geqslant 3.5 \text{ hr}\}$
23. $\{t|t > 1934\}$ **25.** $\{w|w > 18 \text{ wk}\}$
27. $\{C|C < 31.1°\}$ **29.** $\{n|n > 5\}$ **31.** $\{w|w > 14 \text{ cm}\}$
33. $\{b|b > 6 \text{ cm}\}$ **35.** $\{c|c \geqslant 21\}$
37. $\{s|s \leqslant \$49.02\}$. The most Angelo can spend for each sweater is \$49.02. **39.** $\{l\,|\,l \geqslant 64 \text{ km}\}$
41. $\{Y|Y \geqslant 2001\}$ **43.** -160 **45.** $4x^2 - 4x - 7$
47. ◈ **49.** $\{s|s \leqslant 8$ cm and s is positive$\}$
51. More than 6 hr **53.** Between 5 and 8 hr **55.** ◈

Review Exercises: Chapter 2, pp. 109–110

1. [2.1] -22 **2.** [2.1] 7 **3.** [2.1] -192 **4.** [2.1] 1
5. [2.1] $-\frac{7}{3}$ **6.** [2.1] 25 **7.** [2.1] $\frac{1}{2}$ **8.** [2.1] $-\frac{15}{64}$
9. [2.1] 9.99 **10.** [2.1] -8 **11.** [2.2] -5
12. [2.2] $-\frac{1}{3}$ **13.** [2.2] 4 **14.** [2.2] 3 **15.** [2.2] 4
16. [2.2] 16 **17.** [2.2] 6 **18.** [2.2] -3
19. [2.2] 12 **20.** [2.2] 4 **21.** [2.3] $d = \dfrac{C}{\pi}$
22. [2.3] $B = \dfrac{3V}{h}$ **23.** [2.3] $a = 2A - b$ **24.** [2.4] 0.007
25. [2.4] 44% **26.** [2.4] 20% **27.** [2.4] 360
28. [2.6] Yes **29.** [2.6] No **30.** [2.6] Yes
31. [2.6]

4x − 6 < x + 3
−5 −4 −3 −2 −1 0 1 2 3 4 5

32. [2.6]

−2 < x ≤ 5
−5 −4 −3 −2 −1 0 1 2 3 4 5

33. [2.6]

y > 0
−5 −4 −3 −2 −1 0 1 2 3 4 5

34. [2.6] $\{y|y \geqslant -\frac{1}{2}\}$
35. [2.6] $\{x|x \geqslant 7\}$
36. [2.6] $\{y|y > 2\}$
37. [2.6] $\{y|y \leqslant -4\}$ **38.** [2.6] $\{x|x < -11\}$
39. [2.6] $\{y|y > -7\}$ **40.** [2.6] $\{x|x > -6\}$
41. [2.6] $\{x|x > -\frac{9}{11}\}$ **42.** [2.6] $\{y|y \leqslant 7\}$
43. [2.6] $\{x|x \geqslant -\frac{1}{12}\}$ **44.** [2.5] \$591 **45.** [2.5] 27

46. [2.5] 3 m, 5 m **47.** [2.5] 9 **48.** [2.5] 57, 59
49. [2.5] Width: 11 cm; length: 17 cm **50.** [2.5] \$220
51. [2.5] \$26,087 **52.** [2.5] 35°, 85°, 60°
53. [2.7] 86 **54.** [2.7] $\{w \mid w > 17 \text{ cm}\}$ **55.** [1.1] 5
56. [1.2] $12t + 8 + 4s$ **57.** [1.7] -2.3
58. [1.8] $-43x + 8y$
59. [2.1], [2.6] ◈ Multiplying on both sides of an equation by *any* number results in an equivalent equation. When multiplying on both sides of an inequality, the sign of the number being multiplied by must be considered. If the number is 0, there is no equivalent inequality; if the number is positive, the direction of the inequality symbol remains unchanged; if the number is negative, the direction of the inequality symbol must be reversed to produce an equivalent inequality.
60. [2.1], [2,6] ◈ The solutions to an equation can each be checked. The solutions to an inequality are too numerous to check. Checking a few numbers from the solution set found cannot guarantee the answer is correct, although if any number does not check, the answer found is incorrect.
61. [2.5] Amazon: 6437 km; Nile: 6671 km
62. [2.5] \$14,150 **63.** [1.4], [2.2] 23, -23
64. [1.4], [2.1] 20, -20 **65.** [2.3] $a = \dfrac{y-3}{2-b}$

Test: Chapter 2, pp. 110–111

1. [2.1] 8 **2.** [2.1] 26 **3.** [2.1] -6 **4.** [2.1] 49
5. [2.2] -12 **6.** [2.2] 2 **7.** [2.1] -8 **8.** [2.1] $-\frac{7}{20}$
9. [2.2] 7 **10.** [2.2] -5 **11.** [2.2] 2.5
12. [2.6] $\{x \mid x \leqslant -4\}$ **13.** [2.6] $\{x \mid x > -13\}$
14. [2.6] $\{x \mid x \leqslant 5\}$ **15.** [2.6] $\{y \mid y \leqslant -13\}$
16. [2.6] $\{y \mid y \geqslant 8\}$ **17.** [2.6] $\left\{x \mid x \leqslant -\frac{1}{20}\right\}$
18. [2.6] $\{x \mid x < -6\}$ **19.** [2.6] $\{x \mid x \leqslant -1\}$
20. [2.3] $r = \dfrac{A}{2\pi h}$ **21.** [2.3] $l = \dfrac{2w - P}{-2}$ **22.** [2.4] 2
23. [2.4] 5.4% **24.** [2.4] 21 **25.** [2.4] 44%

26. [2.6]
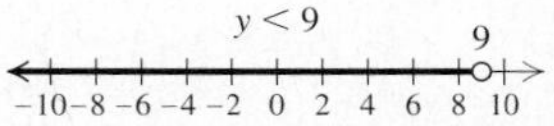

27. [2.6]
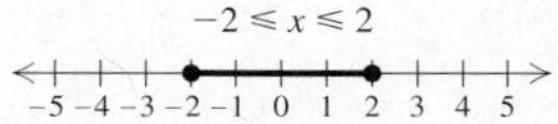

28. [2.5] Width: 7 cm; length: 11 cm **29.** [2.5] 6
30. [2.5] 81, 83, 85 **31.** [2.5] \$2500 **32.** [2.7] $\{x \mid x > 6\}$ **33.** [2.7] $\{l \mid l \geqslant 174 \text{ yd}\}$ **34.** [1.1] $x - 10$
35. [1.2] $3(a + 8b + 4)$ **36.** [1.7] $\frac{3}{10}$ **37.** [1.8] 5
38. [2.3] $d = \dfrac{ca - 1}{c}$ **39.** [1.4], [2.2] 15, -15
40. [2.5] 60

CHAPTER 3

Exercise Set 3.1, pp. 119–121

1. 3 **3.** 120 lb **5.** Boredom **7.** Approximately 5%
9. 1 **11.** The second month **13.** 12% **15.** \$672
17. 4% **19.** 1982 **21.** 1953, 1963, 1970 **23.** 1955
25. Approximately 5% **27.** 3.7% **29.** 70.1%
31. 1743; 360; 270

33.
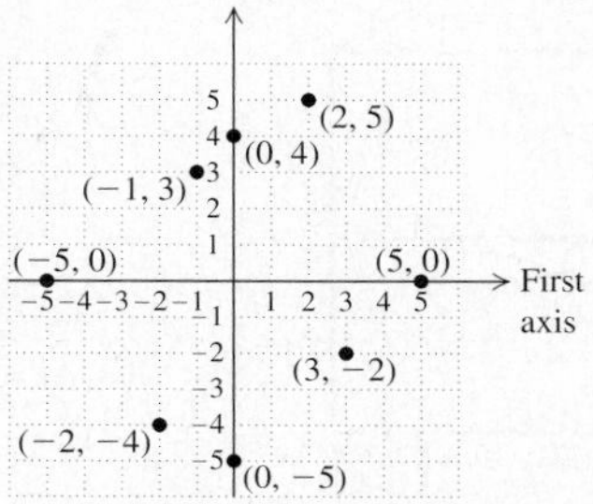

35. II **37.** IV **39.** III **41.** I
43. Negative, negative
45. *A*: (3, 3), *B*: (0, −4), *C*: (−5, 0), *D*: (−1, −1), *E*: (2, 0) **47.** $\frac{2}{3}$ **49.** $-\frac{13}{35}$ **51.** ◈ **53.** I or IV
55. I or III **57.** (−1, −5)

59.
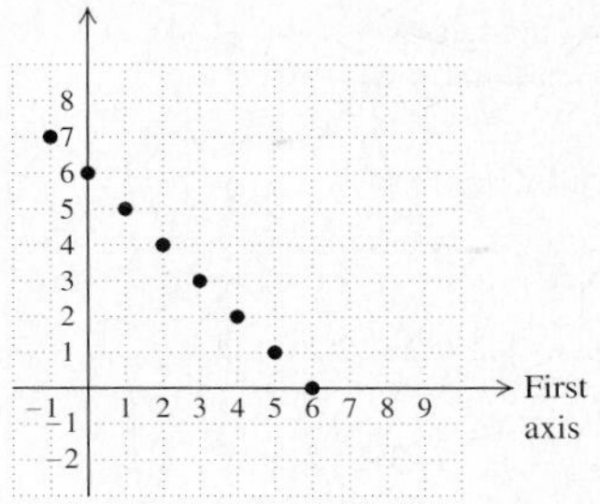

Answers may vary

61. 26
63. Latitude 32.5° North, longitude 64.5° West

Technology Connection, Section 3.2

TC1. $y = -5x + 6.5$
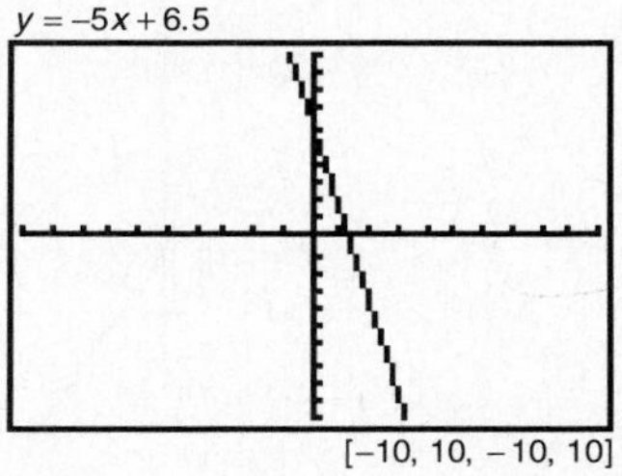
[−10, 10, −10, 10]

TC2. $y = 3x - 4.5$

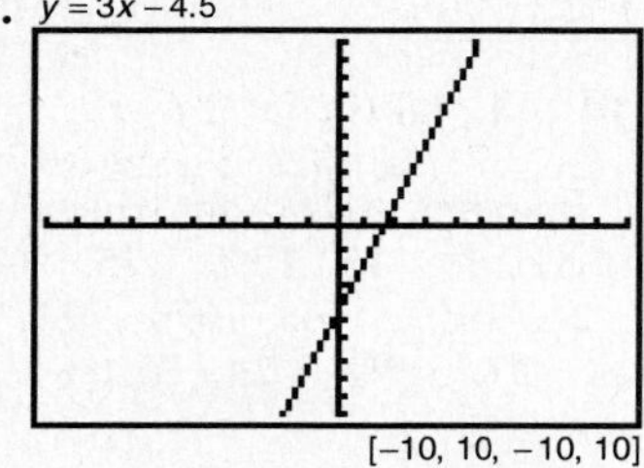

TC3. $y = \frac{4}{7}x - \frac{22}{7}$

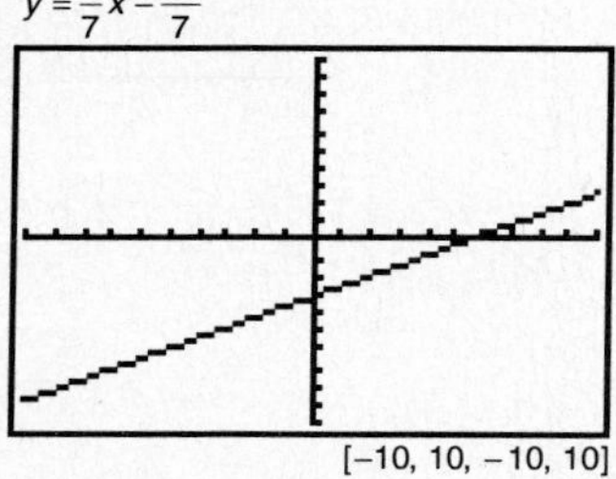

TC4. $y = -\frac{11}{5}x - 4$

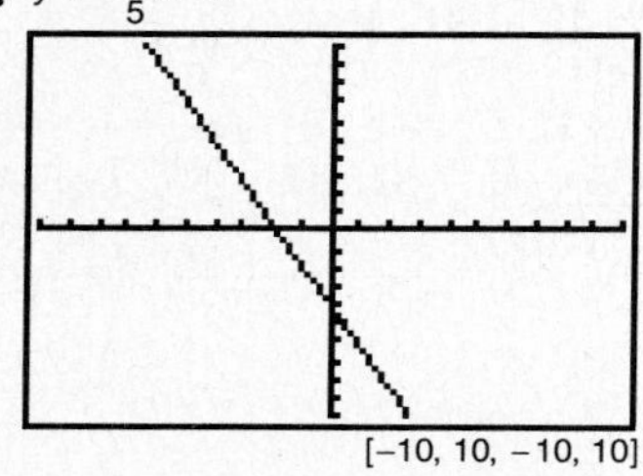

15.

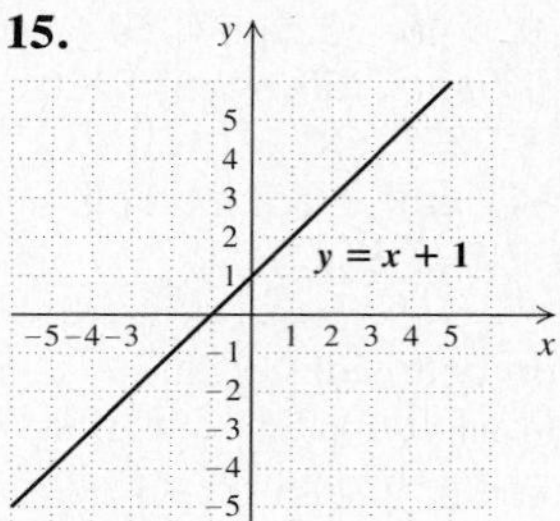

17.

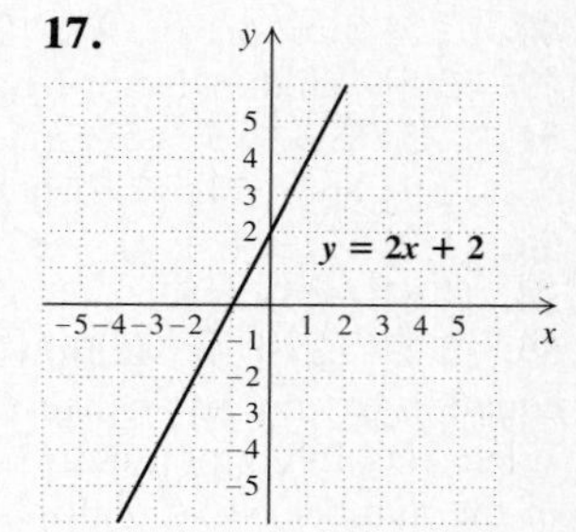

19.

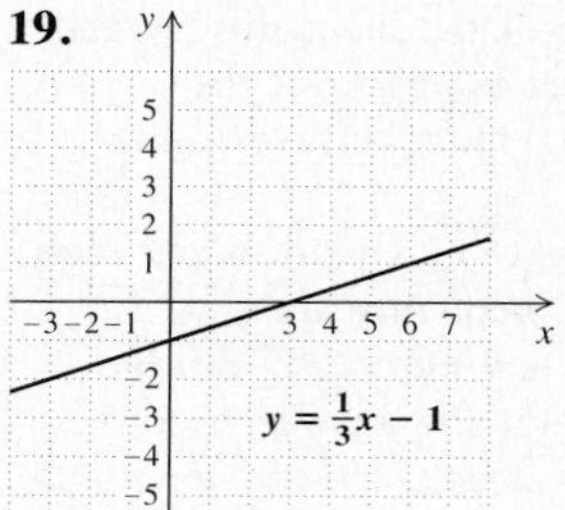

21.

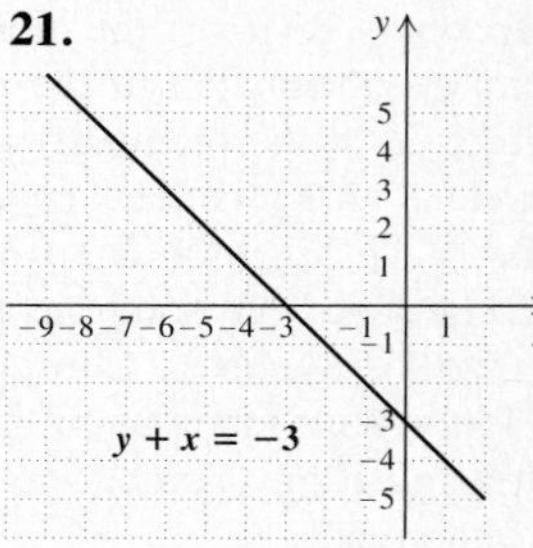

23.

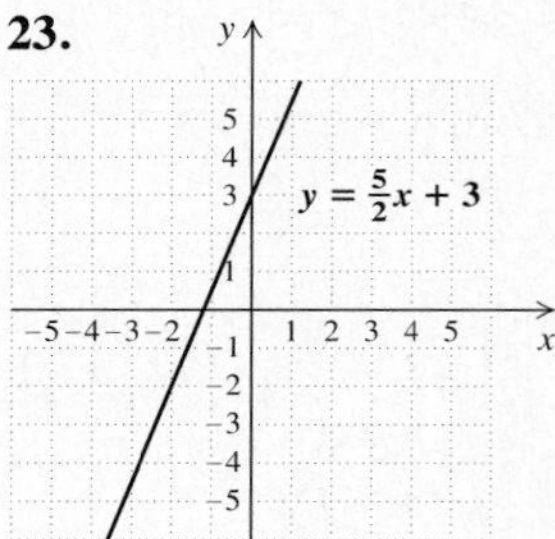

25.

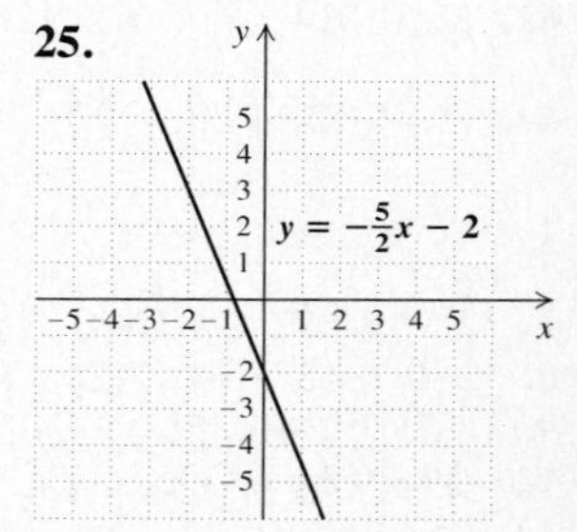

Exercise Set 3.2, p. 128–129

1. Yes **3.** No **5.** No

7.

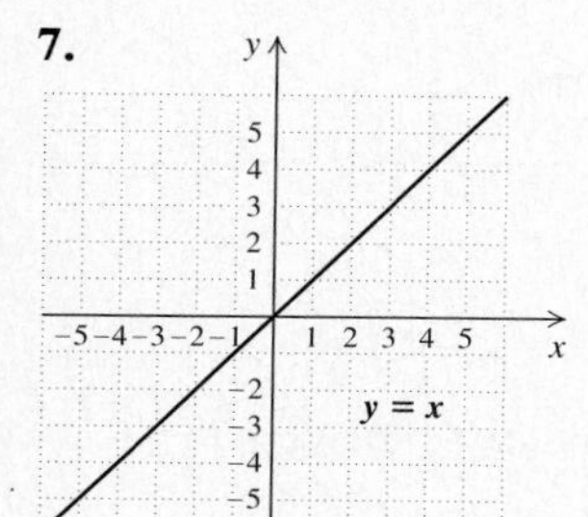

9.

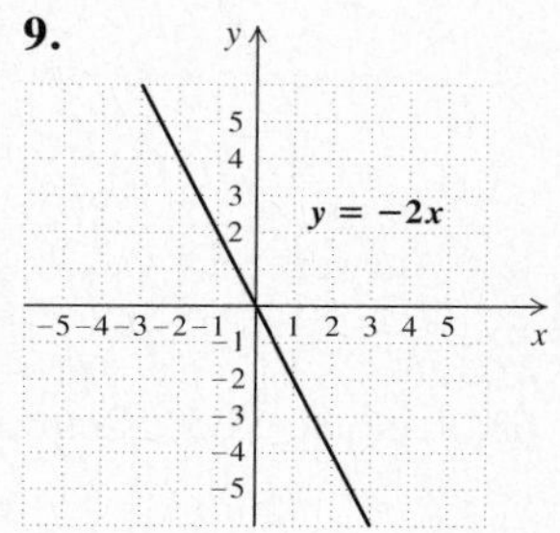

27.

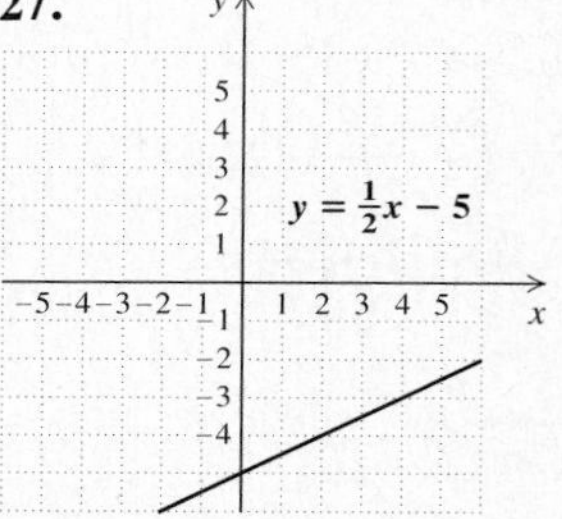

29.

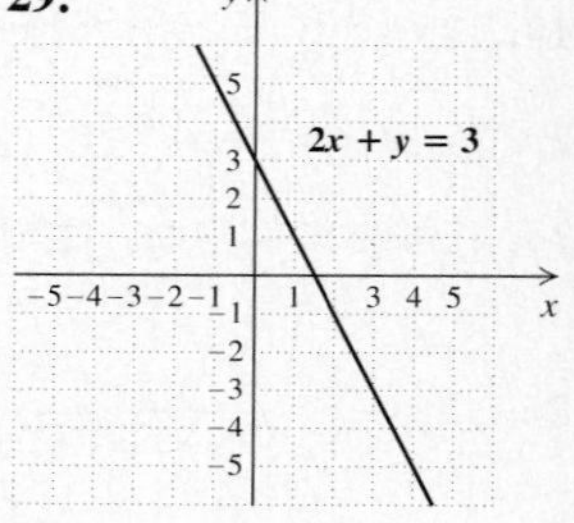

11.

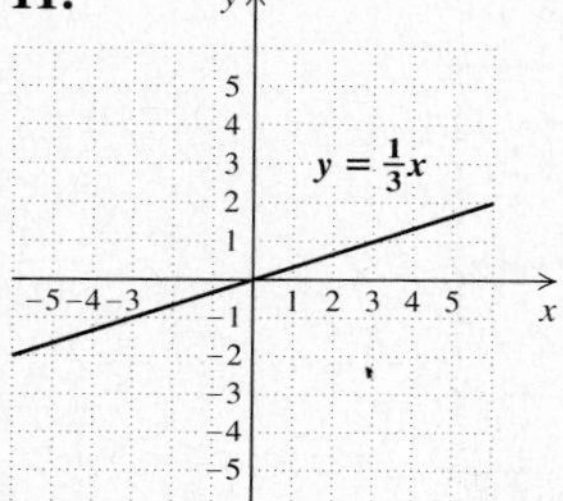

13.

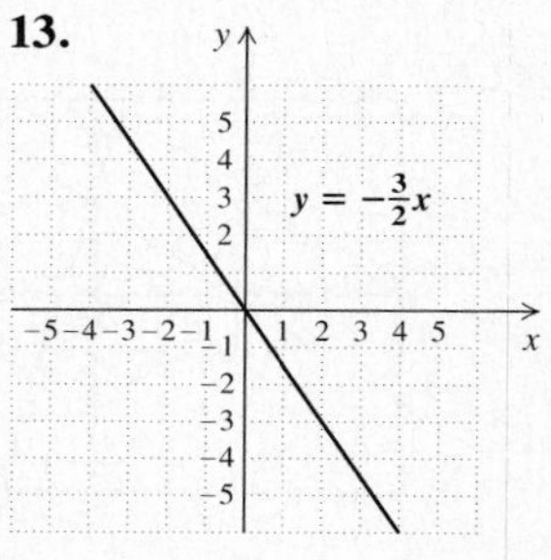

31.

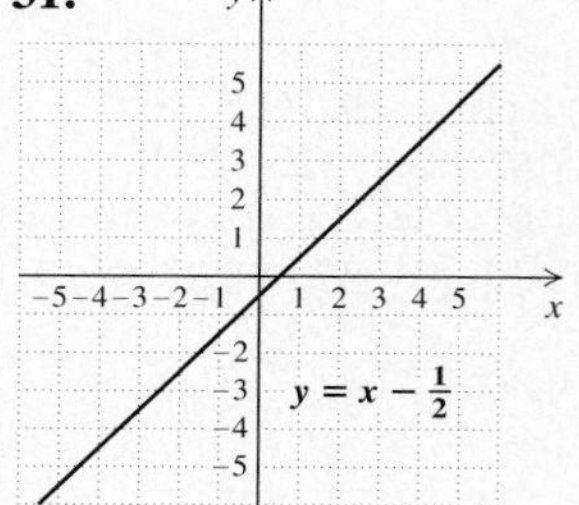

33.

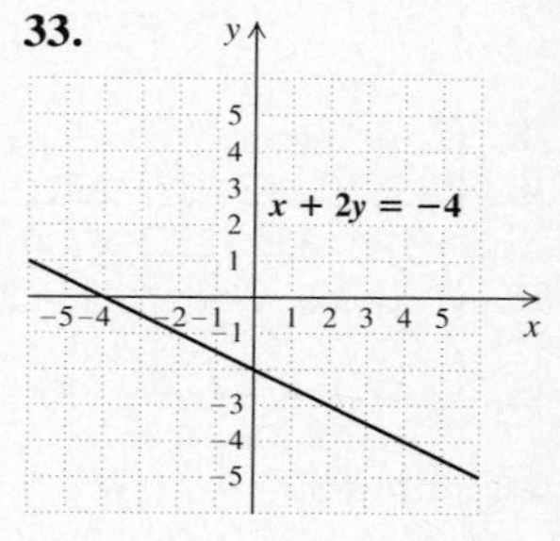

35.

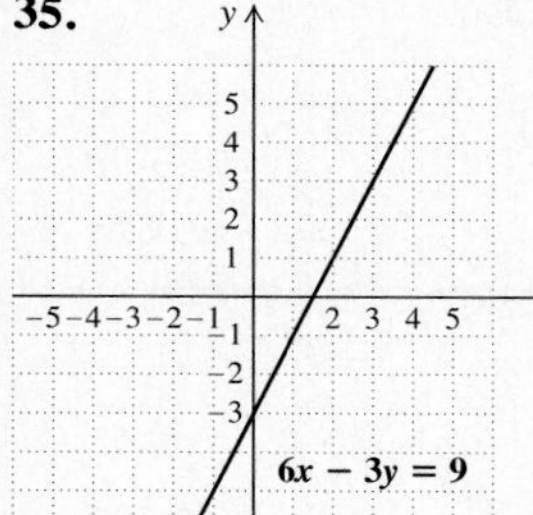

37.

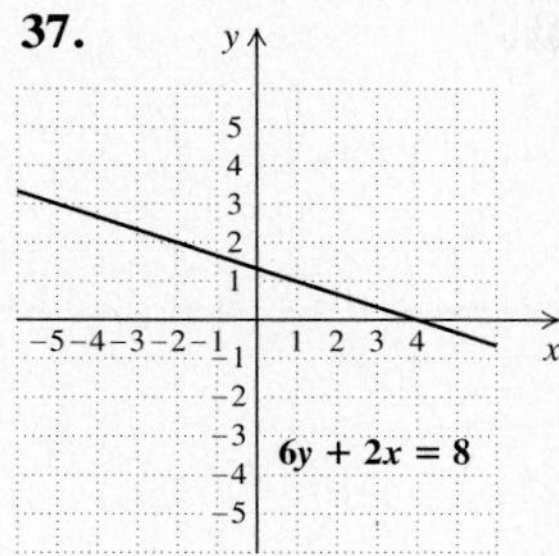

39. -9 **41.** $y = \dfrac{C - Ax}{B}$ **43.** ◈

45.

x	0	−1	1	−2	2	−3	3
y	1	2	2	5	5	10	10

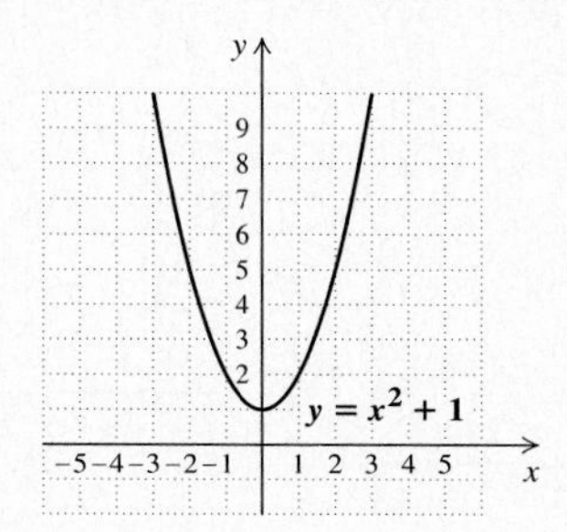

47. (15, 0), (12, 1), (9, 2), (6, 3), (3, 4), (0, 5)
49. $5n + 25q = 235$; (27, 4), (7, 8), (47, 0), answers may vary

51. $y = -2.8x + 3.5$

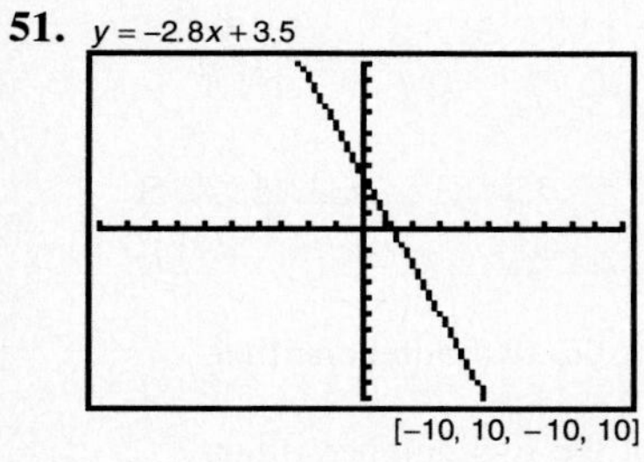

[−10, 10, −10, 10]

53. $y = \frac{2}{7}x - \frac{24}{5}$

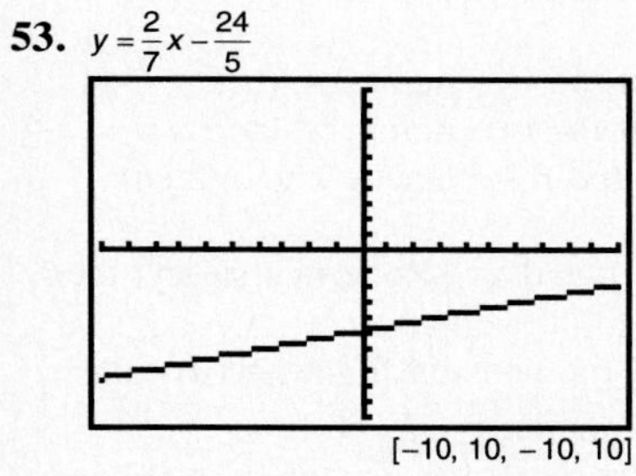

[−10, 10, −10, 10]

55. ◈

Exercise Set 3.3, p. 134–135

1. **(a)** (0, 3); **(b)** (4, 0) **3.** **(a)** (0, 5); **(b)** (−3, 0)
5. **(a)** (0, 4); **(b)** (10, 0) **7.** **(a)** (0, −8); **(b)** (6, 0)
9. **(a)** (0, 8); **(b)** $(-\frac{4}{3}, 0)$ **11.** **(a)** (0, 2); **(b)** $(-\frac{2}{3}, 0)$

13.

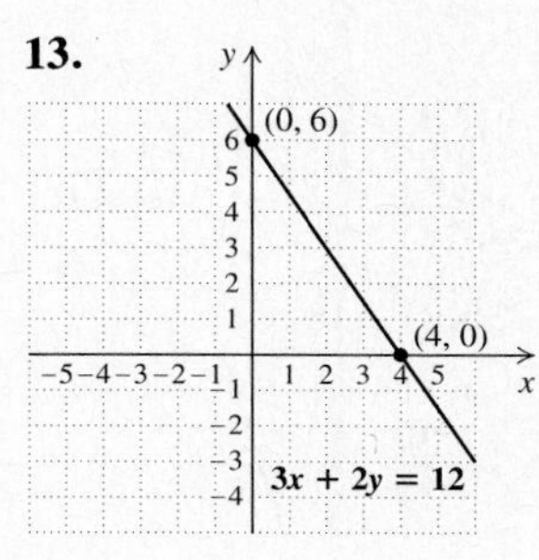

15.

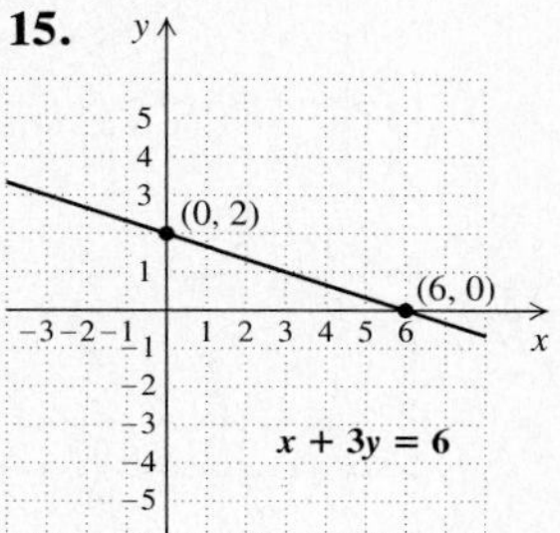

17.

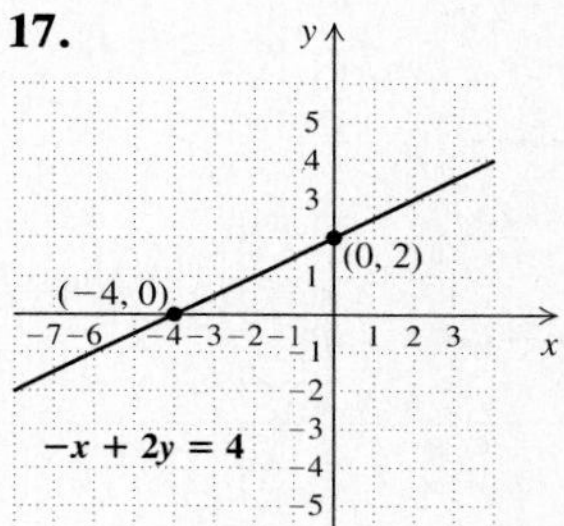

19.

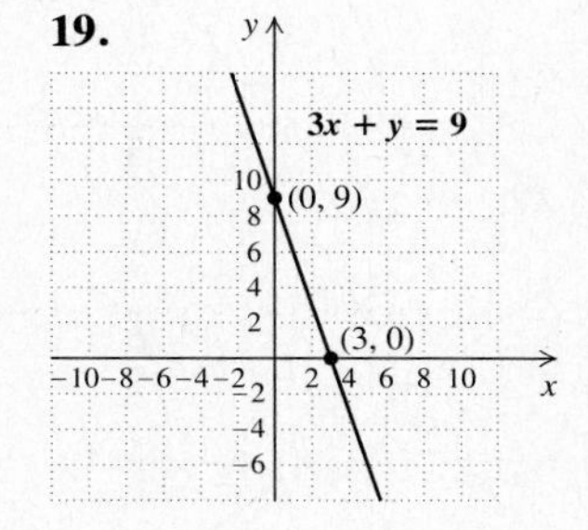

21.

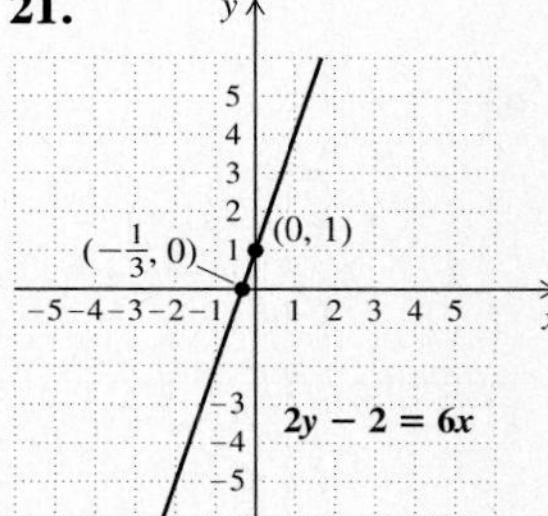

23.

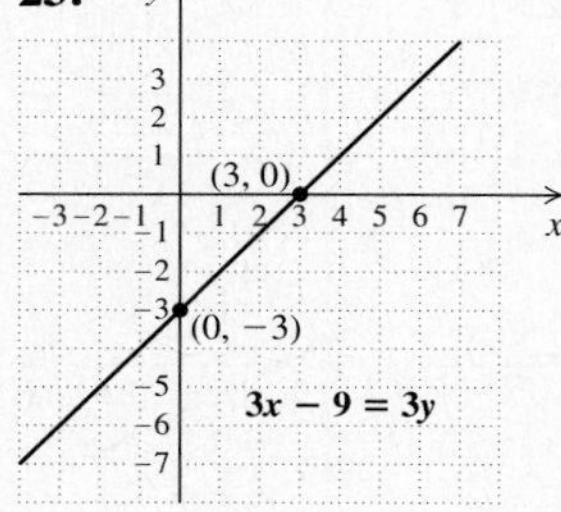

25.

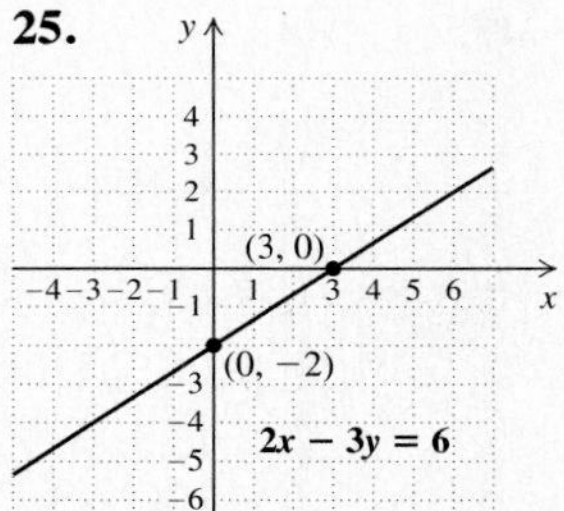

27.

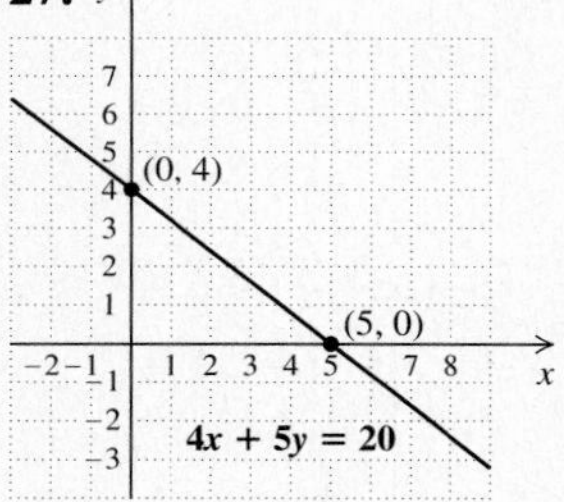

29.

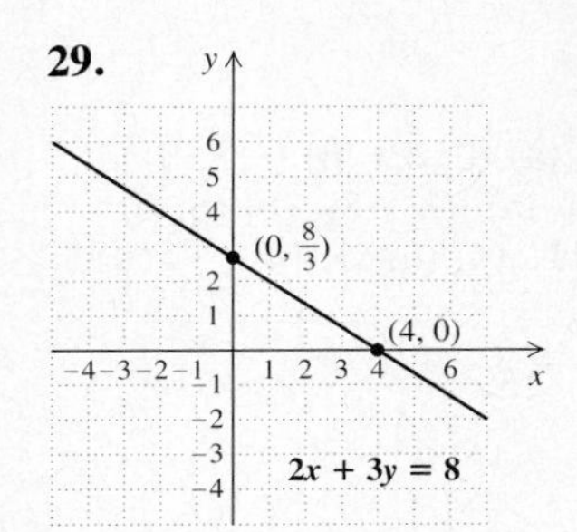

31.

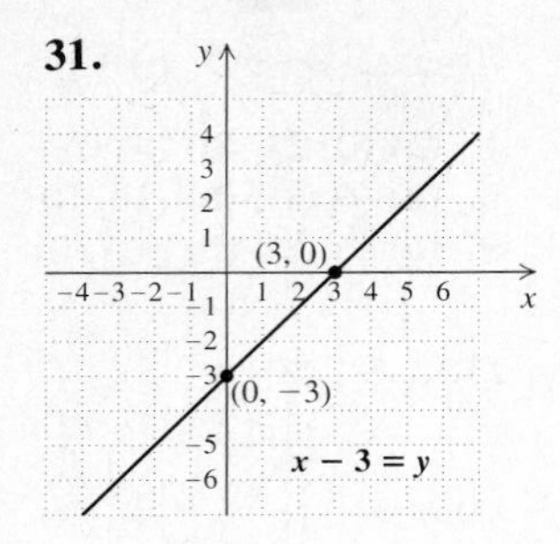

33.

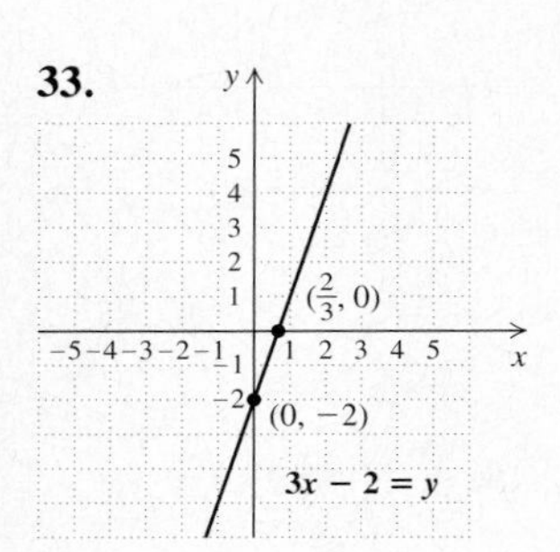

35.

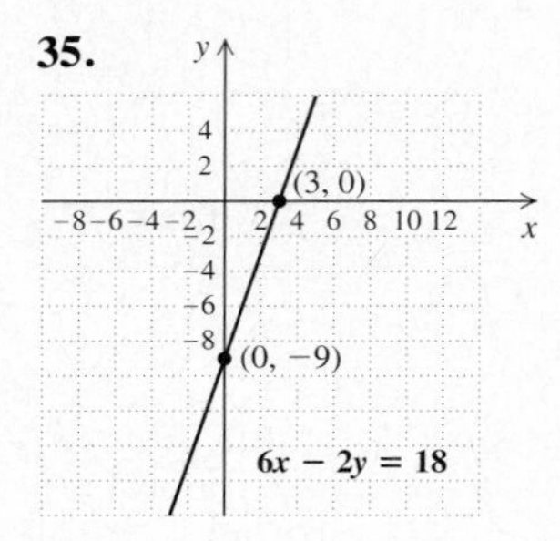

37.

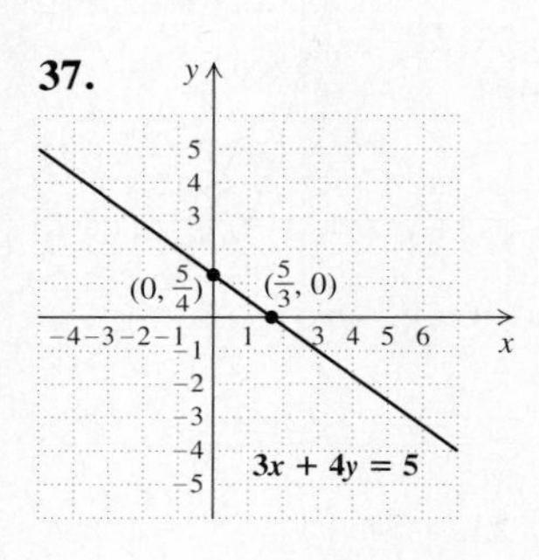

39.

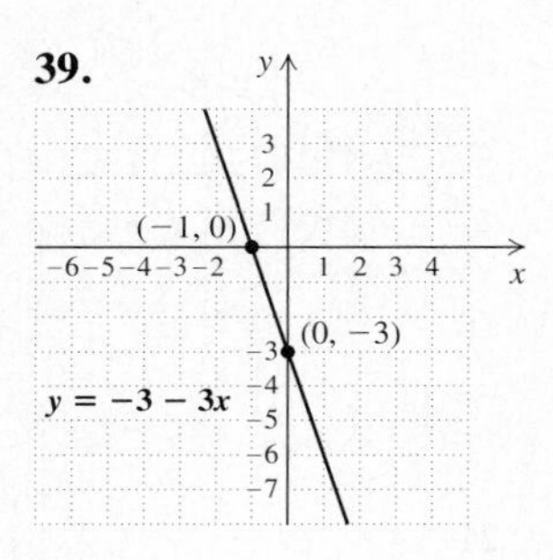

41.

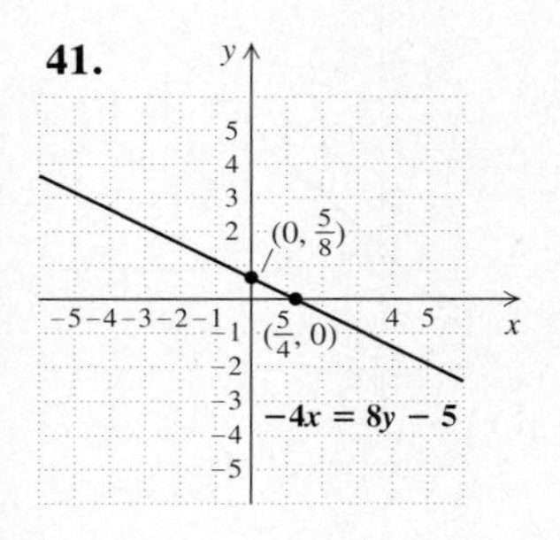

43.

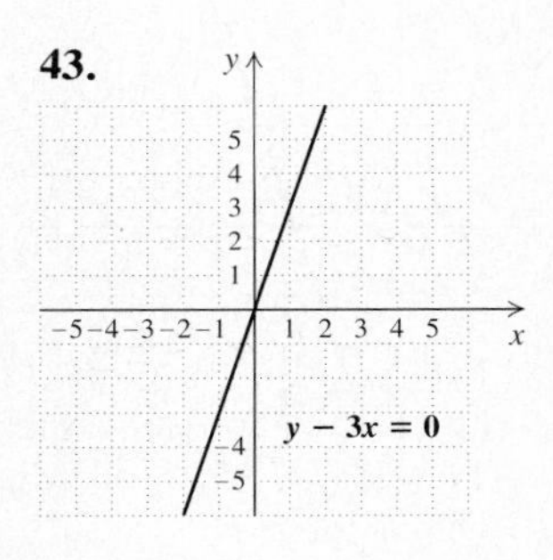

45.

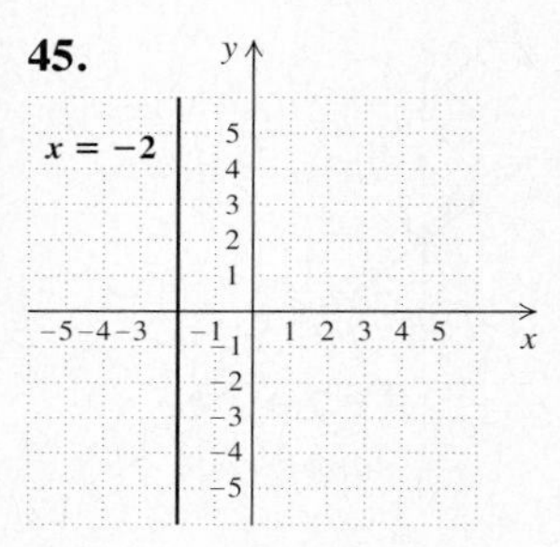

47.

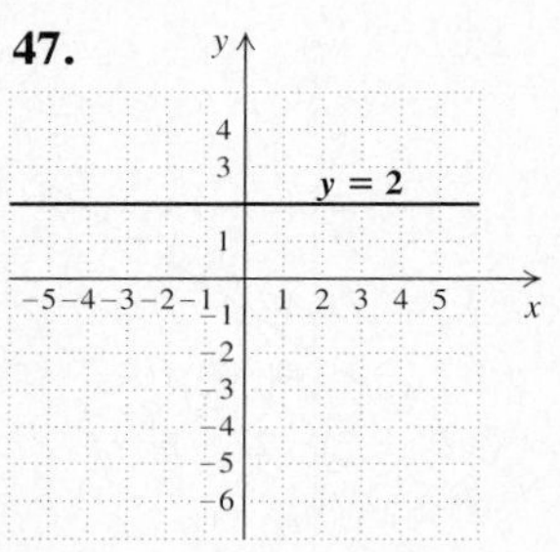

49.

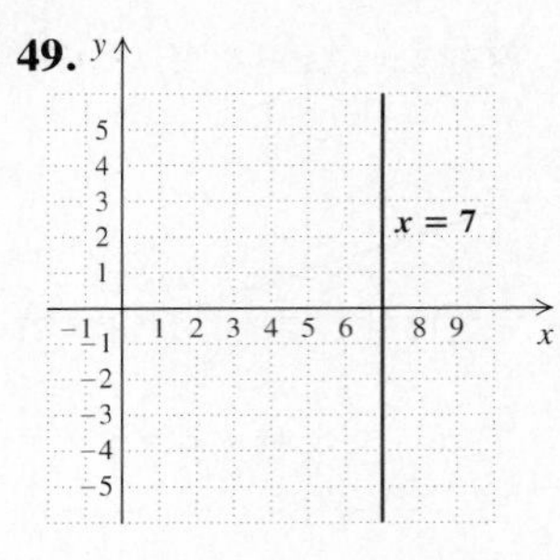

51.

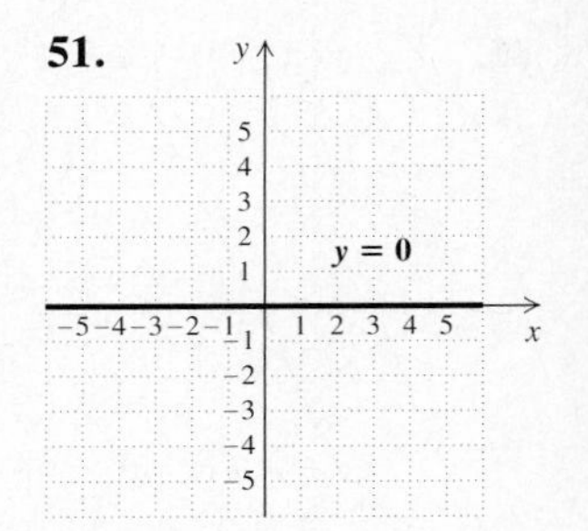

53.

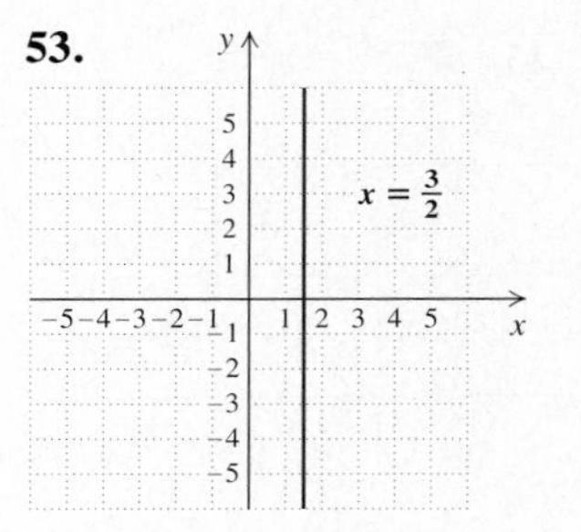

55.

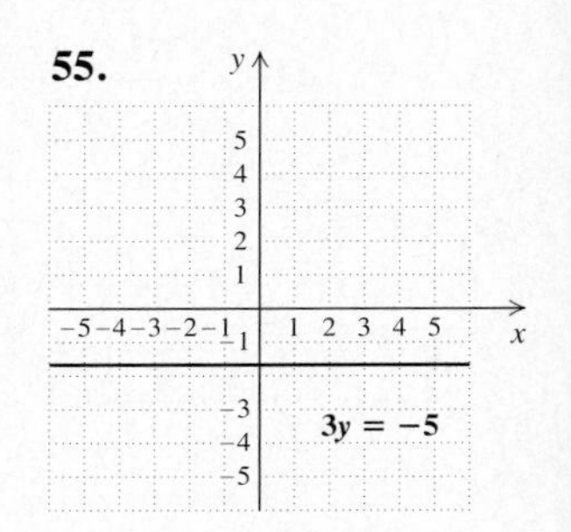

57.

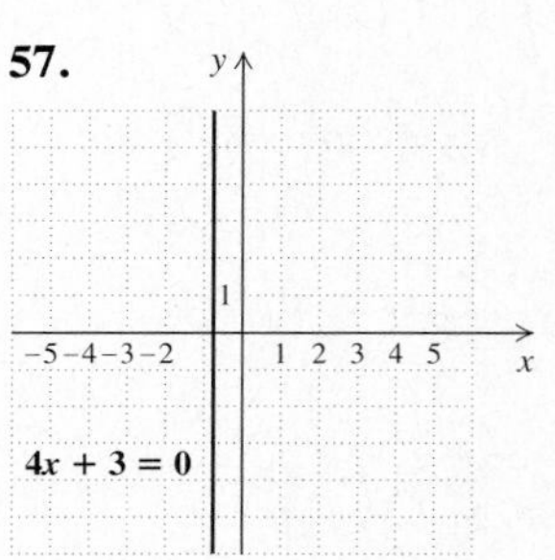

59.

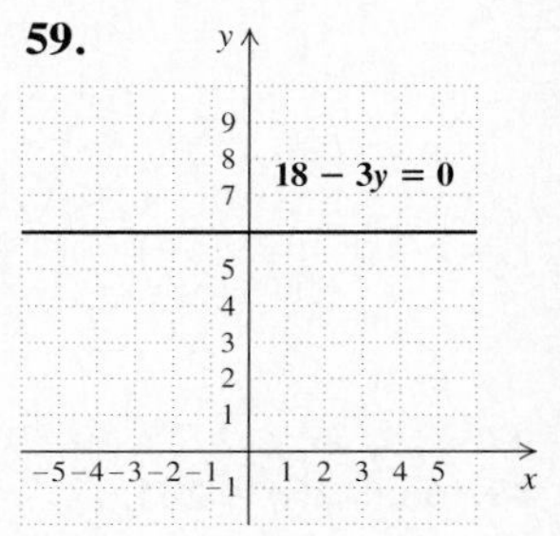

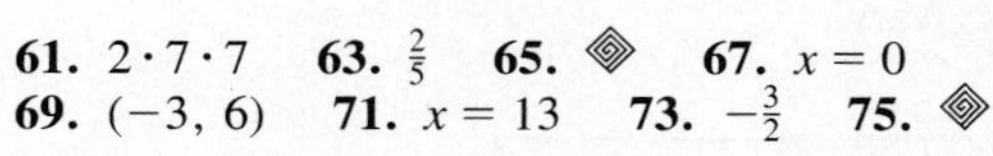

61. $2 \cdot 7 \cdot 7$ **63.** $\frac{2}{5}$ **65.** ◈ **67.** $x = 0$
69. $(-3, 6)$ **71.** $x = 13$ **73.** $-\frac{3}{2}$ **75.** ◈

Technology Connection, Section 3.4

TC1. \$116.45 **TC2.** 17,750 ft **TC3.** \$327.50

Exercise Set 3.4, pp. 139–142

1. Let x and y represent the two numbers; then $x + y = 27$.
3. Let x and y represent the two numbers; then $x + 2y = 65$.
5. Let x and y represent the two numbers; then $x = 3y$.
7. Let x and y represent the two numbers; then $x = y + 5$.
9. Let h = Hank's age and n = Nanette's age; then $h + 7 = 2n$.
11. Let x = Lois's salary and y = Roberta's salary; then $x = 3y + 170$.
13. Let n = the time of the nonstop flight and d = the time of the direct flight; then $n = \frac{1}{2}d + \frac{3}{4}$.
15. Let p = the cost of a pizza and s = the cost of a sandwich; then $3p + 2s = 37$.

17. **(a)** 55 mi, 110 mi, 275 mi, 550 mi; **(b)**

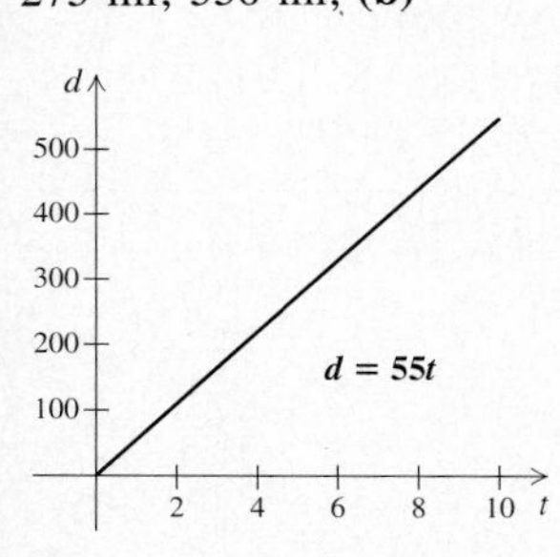

19. **(a)** 2, 3, 4, 5, 6; **(b)**

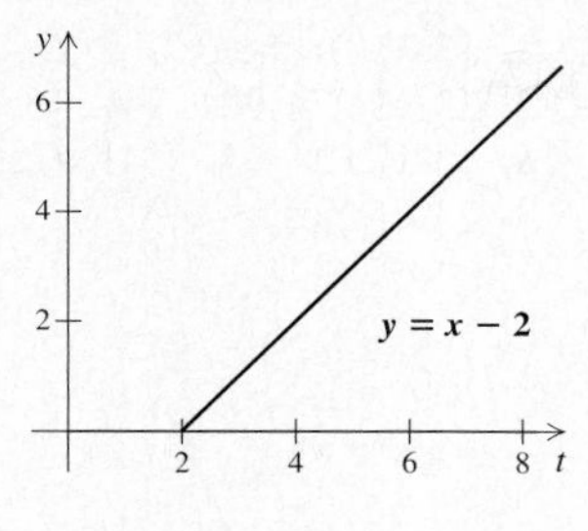

21. **(a)** $12\frac{1}{2}$ times; **(b)**

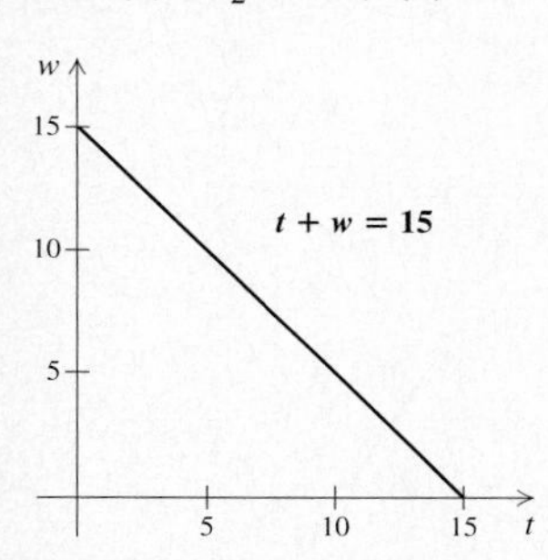

23.

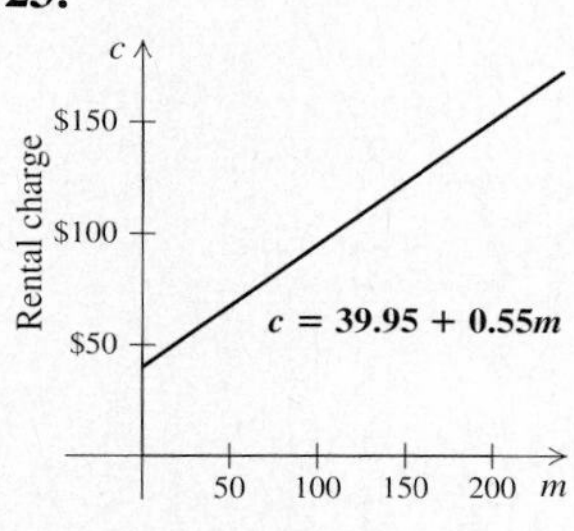

Approximately $140

25.

Wages

$w = 150 + 0.04s$

Sales

$330

27.

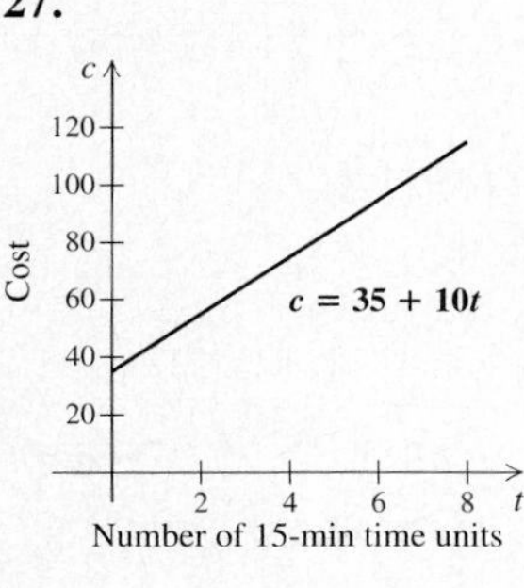

$95

29.

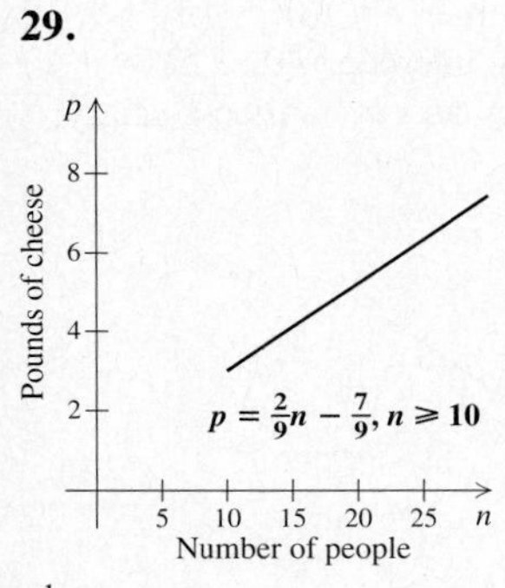

$5\frac{1}{2}$ lb

31.

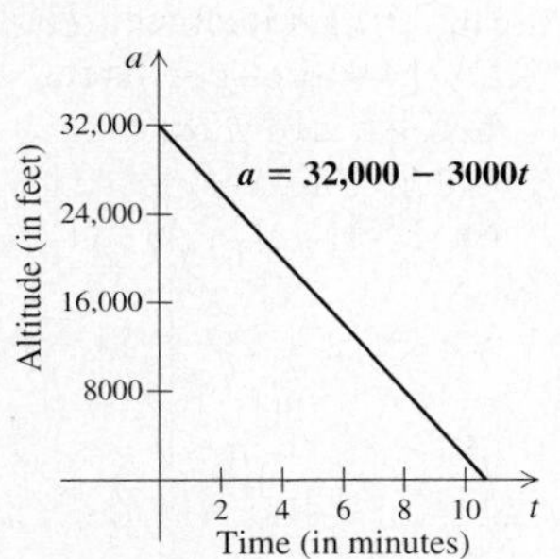

8000 ft

33. $t = \dfrac{s - d}{v}$ **35.** ◈

37.

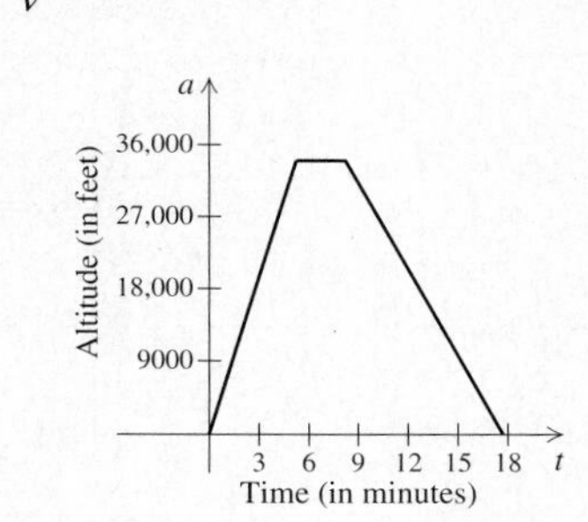

39.

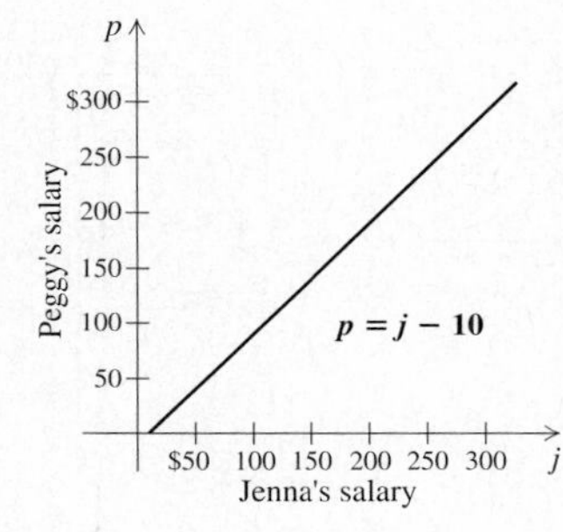

41. $280.13

Review Exercises: Chapter 3, p. 143

1. [3.1] 11% **2.** [3.1] Years 3 and 4

3.–5. [3.1]

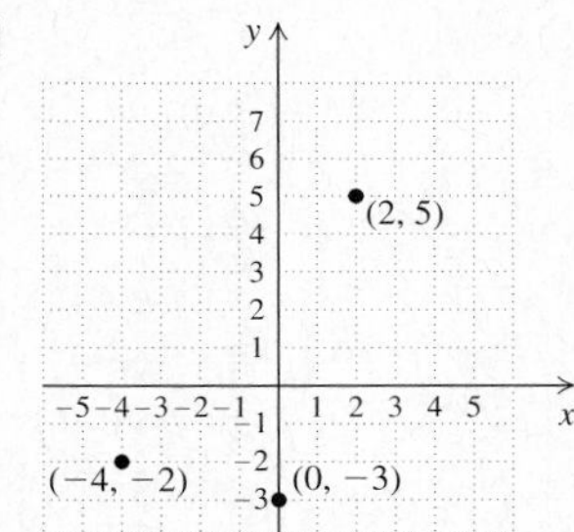

6. [3.1] IV **7.** [3.1] III **8.** [3.1] I
9. [3.1] (−5, −1) **10.** [3.1] (−2, 5)
11. [3.1] (3, 0) **12.** [3.2] No **13.** [3.2] Yes
14. [3.2]

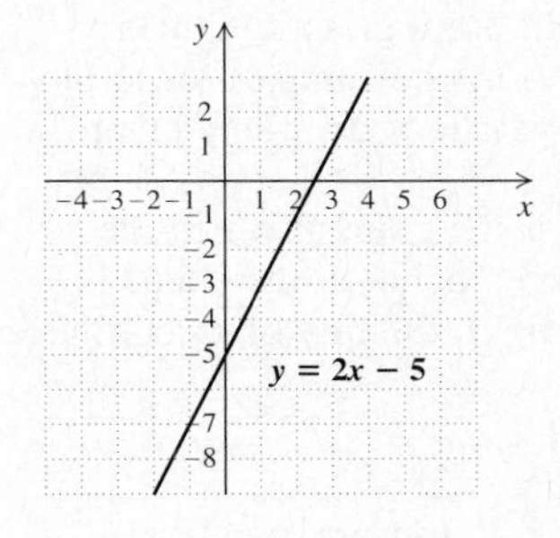

15. [3.2]

$y = -\frac{3}{4}x$

16. [3.2]

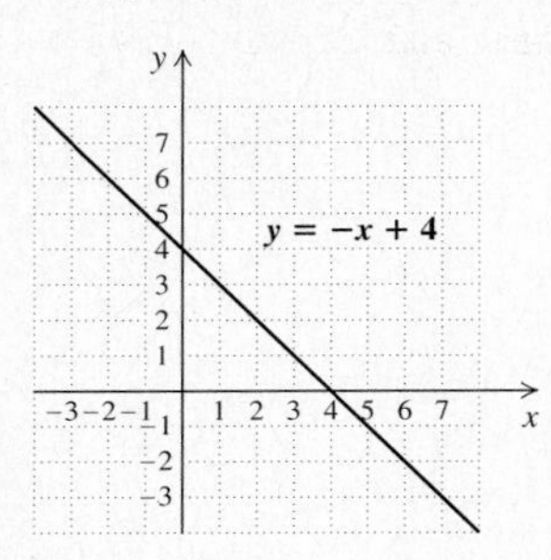

17. [3.2]

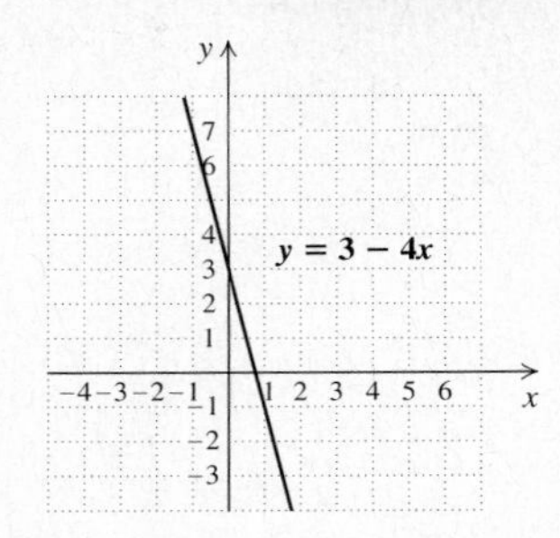

18. [3.3]

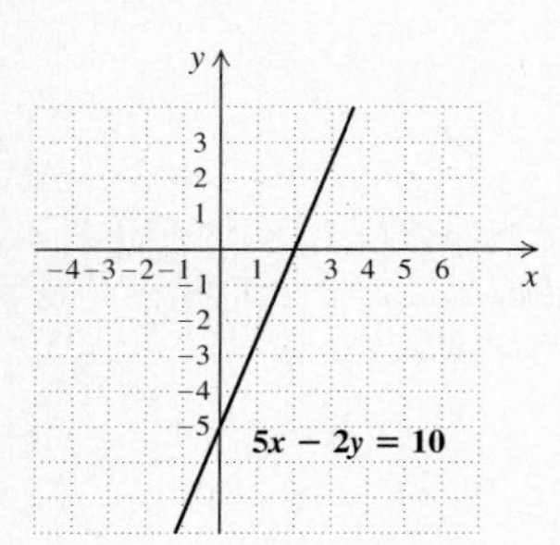

19. [3.3]

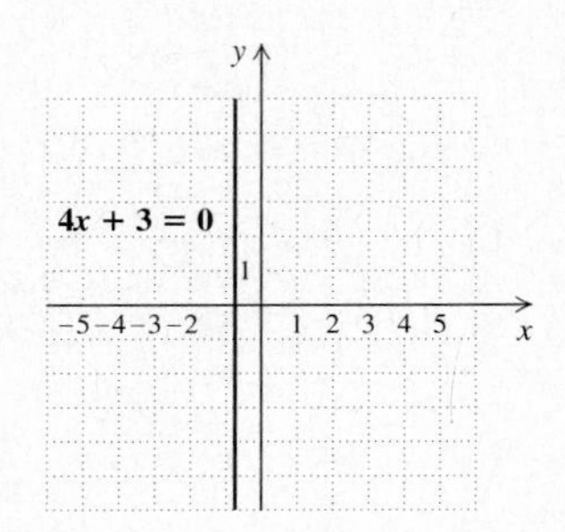
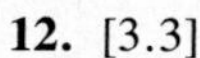

20. [3.3] x-intercept: (6, 0); y-intercept: (0, −3)
21. [3.3] x-intercept: (13.5, 0); y-intercept: (0, 9)
22. [3.4] Let x = the first number and y = the second; then $x = 2y - 3$.
23. [3.4] **(a)** 2.93 ft, 117.2 ft, 293 ft; **(b)**

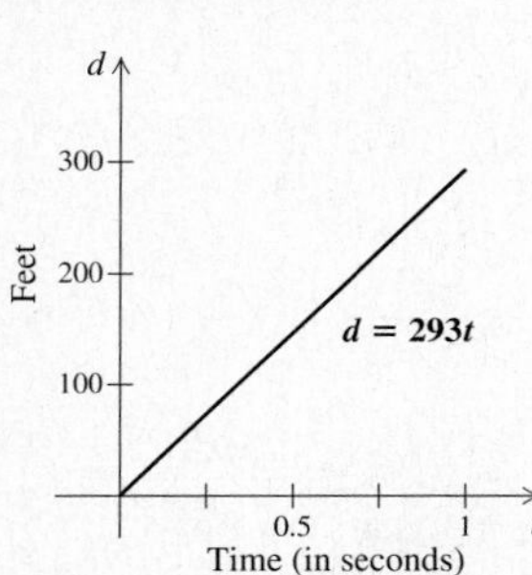

24. [3.4]

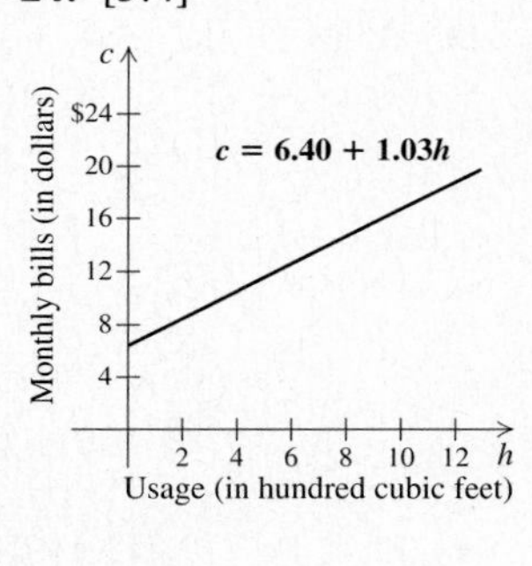

25. [1.3] $\frac{19}{24}$ **26.** [1.4] −0.875 **27.** [2.2] 8
28. [2.3] $m = 2A - n$
29. [3.4] ◈ A business might use a graph to quickly look up prices (as in the rental truck example), or to plot how total sales change from year to year. Many other applications exist.
30. [3.2] ◈ The y-intercept is the point at which the graph crosses the y-axis. Since a point on the y-axis is neither left nor right of the origin, the first or x-coordinate of the point is 0.
31. [3.2] −1 **32.** [3.2] 19
33. [3.1] Area = 45, perimeter = 28
34. [3.2] (0, 4), (1, 3), (−1, 3); answers may vary

Test: Chapter 3, p. 144

1. [3.1] Bachelor's **2.** [3.1] 581,991
3. [3.1] 274,901 **4.** [3.1] 417,197 **5.** [3.1] II
6. [3.1] III **7.** [3.1] (3, 4) **8.** [3.1] (0, −4)
9. [3.2] Yes
10. [3.2]

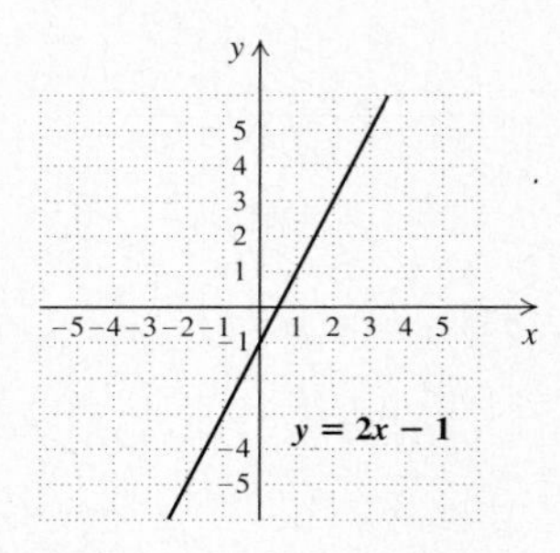

11. [3.3]

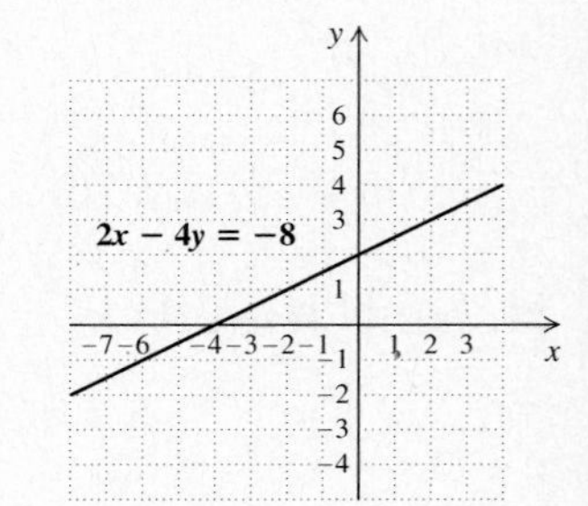

12. [3.3]

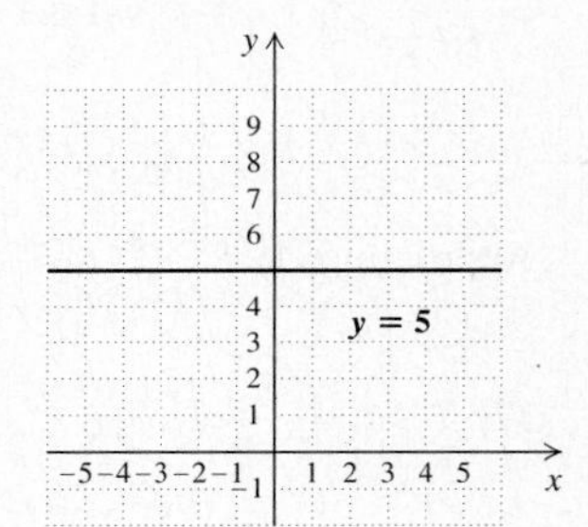

13. [3.2]

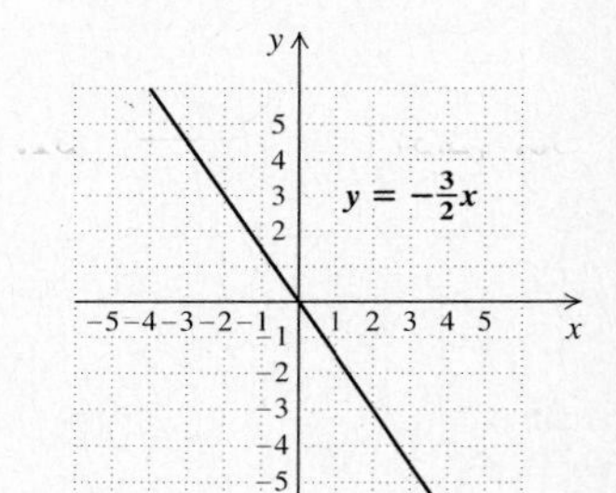

14. [3.3]

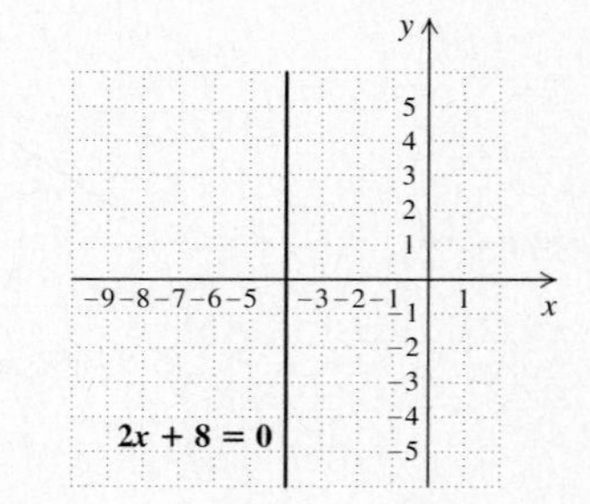

15. [3.3] x-intercept: (9, 0); y-intercept: (0, −15)
16. [3.3] x-intercept: (10, 0); y-intercept: (0, 2.5)
17. [3.4] Let g = Greta's salary and a = Alice's salary; then $g = 50 + 2a$.

18. [3.4]

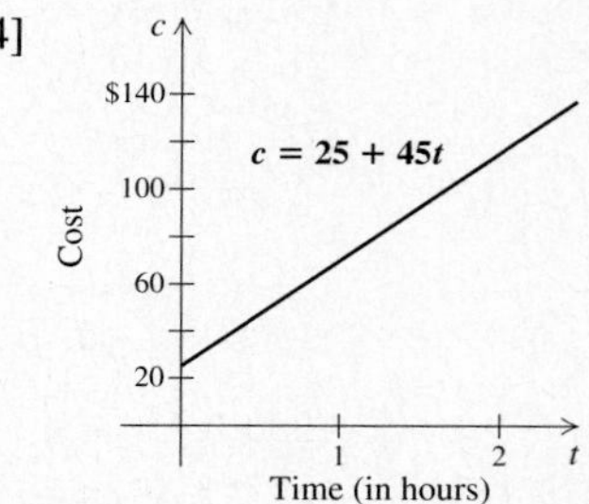

Approximately $50

19. [1.3] $\frac{11}{5}$ **20.** [1.4] $>$ **21.** [2.2] $\frac{4}{15}$

22. [2.3] $x = \dfrac{b}{m + n}$

23. [3.1] Area: 25, perimeter: 20 **24.** [3.3] $y = 3$

CUMULATIVE REVIEW: 1–3

1. [1.1] $\frac{5}{2}$ **2.** [1.2] $12x - 15y + 21$
3. [1.2] $3(x + 3y + 5)$ **4.** [1.3] $2 \cdot 3 \cdot 7$ **5.** [1.4] 0.45
6. [1.4] 4 **7.** [1.6] -5 **8.** [1.7] $\frac{1}{5}$ **9.** [1.6] $-x - y$
10. [2.4] 0.785 **11.** [1.3] $\frac{11}{60}$ **12.** [1.5] 2.6
13. [1.7] 7.28 **14.** [1.7] $-\frac{5}{12}$ **15.** [1.8] -2
16. [1.8] 27 **17.** [1.8] $-2y - 7$ **18.** [1.8] $5x + 11$
19. [2.1] -1.2 **20.** [2.1] -21 **21.** [2.2] 9
22. [2.1] $-\frac{20}{3}$ **23.** [2.2] 2 **24.** [2.1] $\frac{13}{8}$
25. [2.2] $-\frac{17}{21}$ **26.** [2.2] -17 **27.** [2.2] 2
28. [2.6] $\{x|x < 16\}$ **29.** [2.6] $\{x|x \leq -\frac{11}{8}\}$

30. [2.3] $h = \dfrac{A - \pi r^2}{2\pi r}$ **31.** [3.1] IV

32. [2.6] $-1 < x \leq 2$

33. [3.3]

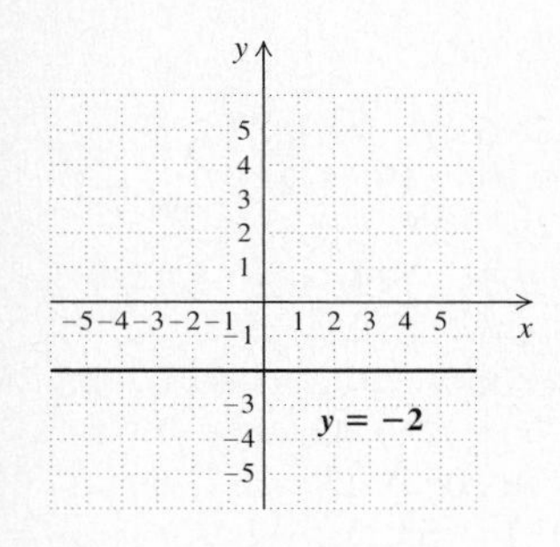

34. [3.3]

35. [3.2]

36. [3.2]

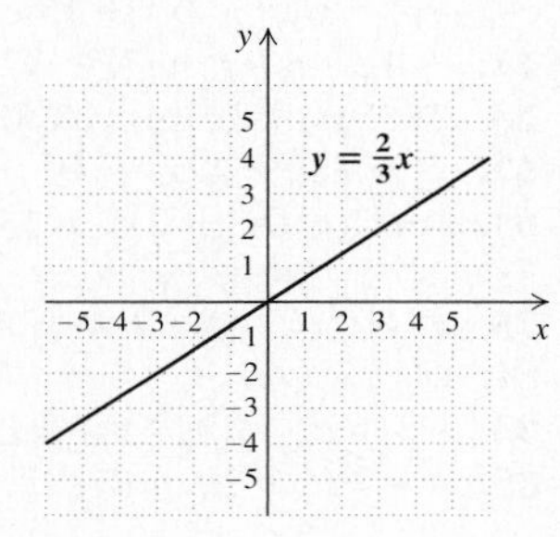

37. [3.3] x-intercept: (10.5, 0); y-intercept: (0, -3)
38. [3.3] x-intercept: $(-\frac{5}{4}, 0)$; y-intercept: (0, 5)
39. [2.4] \$13.60 **40.** [2.5] 154 **41.** [2.5] \$45
42. [2.5] 50 m, 53 m, 40 m **43.** [2.5] \$1500

44. [2.7] $x \geq 78$
45. [3.4] Let s = the cost of steak and c = the cost of chicken; then $s = 5 + 3c$.

46. [3.4]

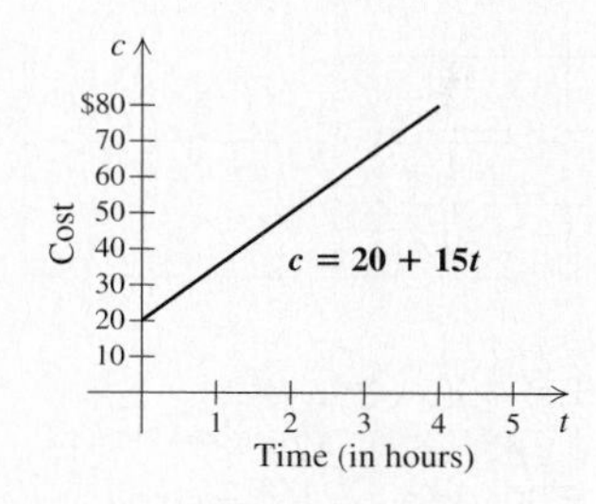

About \$55 for a $2\frac{1}{2}$ hr job

47. [2.4], [2.5] \$25,000 **48.** [1.4], [2.2] 4, -4
49. [2.2] All real numbers **50.** [2.2] No solution
51. [2.2] 3 **52.** [2.2] All real numbers

53. [2.3] $Q = \dfrac{2 - pm}{p}$

CHAPTER 4

Exercise Set 4.1, pp. 153–154

1. 2^7 **3.** 8^{14} **5.** x^7 **7.** 9^{38} **9.** $(3y)^{12}$
11. $(7y)^{17}$ **13.** a^5b^9 **15.** x^4y^{14} **17.** r^{12} **19.** x^5y^5
21. 7^3 **23.** 8^6 **25.** y^4 **27.** $5a$ **29.** 6^3x^5
31. $3m^3$ **33.** a^7b^6 **35.** m^9n^4 **37.** 1 **39.** 5
41. 1 **43.** 1 **45.** 4 **47.** x^{12} **49.** 2^{24} **51.** m^{35}
53. a^{75} **55.** $9x^2$ **57.** $-8a^3$ **59.** $16m^6$
61. $27a^6b^3$ **63.** $a^{15}b^{10}$ **65.** $25x^8y^{10}$

67. $\dfrac{a^3}{64}$ **69.** $\dfrac{49}{25a^2}$ **71.** $\dfrac{a^8}{b^{12}}$ **73.** $\dfrac{y^6}{4}$

75. $\dfrac{125x^6}{y^9}$ **77.** $\dfrac{a^{12}}{16b^{20}}$ **79.** $\dfrac{8a^6}{27b^{12}}$ **81.** $\dfrac{16x^6y^{10}}{9z^{14}}$

83. $3(s + t + 8)$ **85.** $5x$ **87.** ◈ **89.** 25 **91.** a^{2k}
93. 2 **95.** $>$ **97.** $<$
99. Let $x = 2$; then $3x^2 = 12$, and $(3x)^2 = 36$.

101. Let $x = 1$; then $\dfrac{x + 2}{2} = \dfrac{3}{2}$, and $x = 1$.

103. \$15,638.03 **105.** $25x^2$

Exercise Set 4.2, pp. 159–161

1. Trinomial **3.** None of these **5.** Binomial
7. Monomial **9.** $2, -3x, x^2$ **11.** $6x^2$ and $-3x^2$
13. $2x^4$ and $-3x^4$; $5x$ and $-7x$ **15.** -3, 6
17. 5, 3, 3 **19.** -7, 6, 3, 7 **21.** -5, 6, -3, 8, -2
23. 1, 0; 1 **25.** 2, 1, 0; 2 **27.** 3, 2, 1, 0; 3
29. 2, 1, 6, 4; 6

31.

Term	Coefficient	Degree of Term	Degree of Polynomial
$-7x^4$	-7	4	4
$6x^3$	6	3	
$-3x^2$	-3	2	
$8x$	8	1	
-2	-2	0	

33. $-3x$ **35.** $-8x$ **37.** $11x^3+4$ **39.** x^3-x
41. $4b^5$ **43.** $\frac{3}{4}x^5-2x-42$ **45.** x^4 **47.** $\frac{15}{16}x^3-\frac{7}{6}x^2$
49. x^6+x^4 **51.** $13x^3-9x+8$ **53.** $-5x^2+9x$
55. $12x^4-2x+\frac{1}{4}$ **57.** -18 **59.** 19 **61.** -12
63. 2 **65.** 4 **67.** 11 **69.** Approximately 449
71. 1112 ft **73.** \$18,750 **75.** \$155,000
77. 62.8 cm **79.** 78.5 m^2

81.

t	$-t^2+6t-4$
1	1
2	4
3	5
4	4
5	1

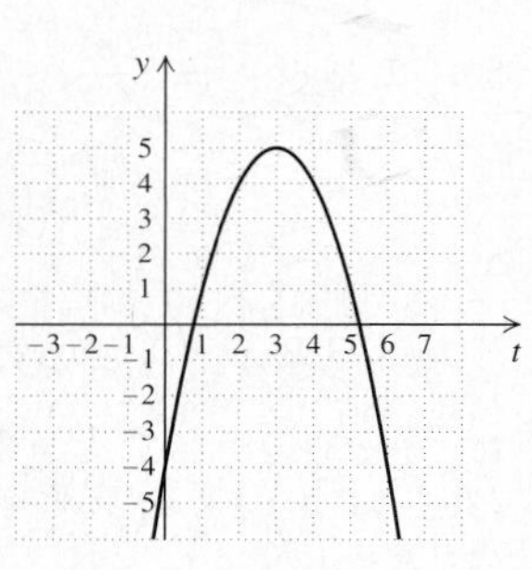

83. 274 and 275 **85.** ◈ **87.** $5x^9+4x^8+x^2+5x$
89. 99, -99; 50, -50; 99, -99
91. $-6x^5+14x^4-x^2+11$; answers may vary **93.** 10

95.

d	$-0.0064d^2+0.8d+2$
0	2
30	20.24
60	26.96
90	22.16
120	5.84

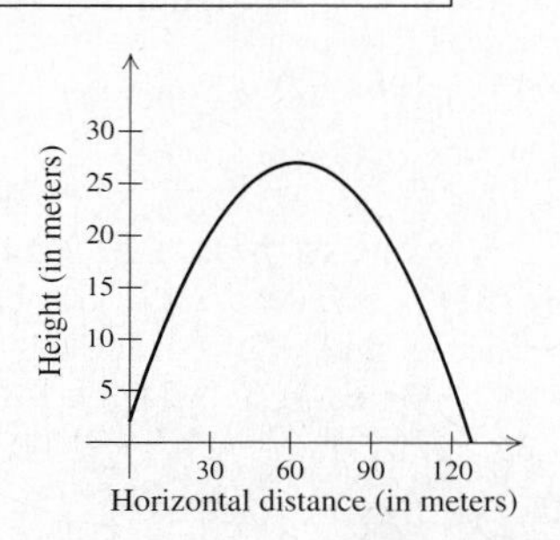

Exercise Set 4.3, pp. 166–169

1. $-x+5$ **3.** x^2-5x-1 **5.** $2x^2$
7. $5x^2+3x-30$ **9.** $-2.2x^3-0.2x^2-3.8x+23$
11. $12x^2+6$ **13.** $9x^8+8x^7-3x^4+2x^2-2x+5$
15. $-\frac{1}{2}x^4+\frac{2}{3}x^3+x^2$
17. $0.01x^5+x^4-0.2x^3+0.2x+0.06$
19. $-3x^4+3x^2+4x$
21. $1.05x^4+0.36x^3+14.22x^2+x+0.97$
23. $-(-5x)$, $5x$ **25.** $-(-x^2+10x-2)$, $x^2-10x+2$
27. $-(12x^4-3x^3+3)$, $-12x^4+3x^3-3$
29. $-3x+7$ **31.** $-4x^2+3x-2$
33. $4x^4-6x^2-\frac{3}{4}x+8$ **35.** $7x-1$
37. $-x^2-7x+5$ **39.** -18
41. $6x^4+3x^3-4x^2+3x-4$
43. $4.6x^3+9.2x^2-3.8x-23$ **45.** $2x^2+14$
47. $-2x^5-6x^4+x+2$ **49.** $9x^2+9x-8$
51. $\frac{3}{4}x^3-\frac{1}{2}x$ **53.** $0.06x^3-0.05x^2+0.01x+1$
55. $3x+6$ **57.** $11x^4+12x^3-9x^2-8x-9$
59. $-4x^5+9x^4+6x^2+16x+6$ **61.** $x^4-x^3+x^2-x$
63. (a) $5x^2+4x$; (b) 57, 352 **65.** $14y+17$
67. $(r+9)(r+11)$; $9r+99+r^2+11r$ **69.** $\pi r^2-9\pi$
71. $z^2-27z+72$ **73.** $144-4x^2$ **75.** $\frac{115}{22}$ **77.** 4
79. ◈ **81.** $11a^2-18a-4$ **83.** $-10y^2-2y-10$
85. $-3y^4-y^3+5y-2$ **87.** $569.607x^3-15.168x$
89. $28a+90$ **91.** (a) $P=-x^2+280x-5000$; (b) \$10,375; (c) \$13,000

Exercise Set 4.4, pp. 173–175

1. $42x^2$ **3.** x^4 **5.** $-x^8$ **7.** $6x^6$ **9.** $28t^8$
11. $-0.02x^{10}$ **13.** $\frac{1}{15}x^4$ **15.** 0 **17.** $-24x^{11}$
19. $-3x^2+15x$ **21.** $4x^2+4x$ **23.** $5x^2+35x$
25. x^5+x^2 **27.** $6x^3-18x^2+3x$ **29.** $12x^3+24x^2$
31. $-6x^4-6x^3$ **33.** $18y^6+24y^5$
35. $42x^{54}+60x^{15}+18x^{61}+180x^{19}$ **37.** $x^2+9x+18$
39. $x^2+3x-10$ **41.** $x^2-7x+12$ **43.** x^2-9
45. $25-15x+2x^2$ **47.** $4x^2+20x+25$
49. $9y^2-16$ **51.** $x^2-\frac{21}{10}x-1$ **53.** x^3-1
55. $4x^3+14x^2+8x+1$
57. $3y^4-6y^3-7y^2+18y-6$ **59.** $x^6+2x^5-x^3$
61. $-10x^5-9x^4+7x^3+2x^2-x$
63. x^4-x^2-2x-1 **65.** $4x^4+8x^3-9x^2-10x+8$
67. $x^4+8x^3+12x^2+9x+4$
69. $-4x^4+6x^3-2x^2-13x+10$
71. $2x^4-7x^3+2x^2-x-2$ **73.** x^4-1
75. $x^4-2x^3-4x^2+9$ **77.** $-\frac{3}{4}$ **79.** ◈
81. $84y^2-30y$
83. $V=4x^3-48x^2+144x$; $S=-4x^2+144$
85. $A=\frac{1}{2}b^2+2b$ **87.** $2x^2+18x+36$ **89.** $16x+16$

Exercise Set 4.5, pp. 181–183

1. x^3+x^2+3x+3 **3.** x^4+x^3+2x+2
5. y^2-y-6 **7.** $9x^2+15x+6$ **9.** $5x^2+4x-12$
11. $9t^2-1$ **13.** $4x^2-6x+2$ **15.** $p^2-\frac{1}{16}$

17. $x^2 - 0.01$ **19.** $2x^3 + 2x^2 + 6x + 6$
21. $-2x^2 - 11x + 6$ **23.** $a^2 + 14a + 49$
25. $1 - x - 6x^2$ **27.** $x^5 + 3x^3 - x^2 - 3$
29. $3x^6 - 2x^4 - 6x^2 + 4$ **31.** $6x^7 + 18x^5 + 4x^2 + 12$
33. $8x^6 + 65x^3 + 8$ **35.** $4x^3 - 12x^2 + 3x - 9$
37. $4y^6 + 4y^5 + y^4 + y^3$ **39.** $x^2 - 16$ **41.** $4x^2 - 1$
43. $25m^2 - 4$ **45.** $4x^4 - 9$ **47.** $9x^8 - 16$
49. $x^{12} - x^4$ **51.** $x^8 - 9x^2$ **53.** $x^{24} - 9$
55. $4y^{16} - 9$ **57.** $x^2 + 4x + 4$ **59.** $9x^4 + 6x^2 + 1$
61. $a^2 - a + \frac{1}{4}$ **63.** $9 + 6x + x^2$ **65.** $x^4 + 2x^2 + 1$
67. $4 - 12x^4 + 9x^8$ **69.** $25 + 60t^2 + 36t^4$
71. $49x^2 - 4.2x + 0.09$ **73.** $10a^5 - 5a^3$
75. $x^4 + x^3 - 6x^2 - 5x + 5$ **77.** $9 - 12x^3 + 4x^6$
79. $4x^3 + 24x^2 - 12x$ **81.** $4x^4 - 2x^2 + \frac{1}{4}$
83. $-1 + 9p^2$ **85.** $15t^5 - 3t^4 + 3t^3$
87. $36x^8 + 48x^4 + 16$ **89.** $12x^3 + 8x^2 + 15x + 10$
91. $64 - 96x^4 + 36x^8$ **93.** $t^3 - 1$ **95.** 25; 49
97. 56; 16 **99.** $x^2 + 6x + 9$ **101.** $t^2 + 7t + 12$
103. $a^2 + 8a + 7$
105. Television: 50 watts; lamps: 500 watts; air conditioner: 2000 watts
107. ◈ **109.** $8y^3 + 72y^2 + 160y$ **111.** $81x^4 - 16$
113. $625t^{12} - 450t^6 + 81$
115. $4567.0564x^2 + 435.891x + 10.400625$
117. $400 - 4 = 396$ **119.** -7
121. $V = w^3 + 3w^2 + 2w$ **123.** $V = h^3 - 3h^2 + 2h$
125. $F^2 - (F - 17)(F - 7)$; $24F - 119$
127. **(a)** $A^2 + AB$; **(b)** $AB + B^2$; **(c)** $A^2 - B^2$;
(d) $(A + B)(A - B) = A^2 - B^2$
129. $100x^2 + 100x + 25$, or $100(x^2 + x) + 25$. Add the first digit to its square, multiply by 100, and add 25.

Exercise Set 4.6, pp. 187–190

1. -1 **3.** -7 **5.** \$11,664 **7.** \$12,597.12
9. 20.60625 in^2
11. Coefficients: 1, −2, 3, −5; degrees: 4, 2, 2, 0; 4
13. Coefficients: 17, −3, −7; degrees: 5, 5, 0; 5
15. $-a - 2b$ **17.** $3x^2y - 2xy^2 + x^2$ **19.** $8u^2v - 5uv^2$
21. $20au + 10av$ **23.** $x^2 - 4xy + 3y^2$
25. $-4r^3 + 2rs - 9s^2$ **27.** $3r + s - 4$
29. $-x^2 - 8xy - y^2$ **31.** $2ab$
33. $-2a + 4b + 3c - 8d$ **35.** $-8x + 8y$
37. $6z^2 + 7zu - 3u^2$ **39.** $a^4b^2 - 7a^2b + 10$
41. $a^6 - b^2c^2$ **43.** $y^6x + y^4x + y^4 + 2y^2 + 1$
45. $12x^2y^2 + 2xy - 2$ **47.** $12 - c^2d^2 - c^4d^4$
49. $m^3 + m^2n - mn^2 - n^3$
51. $x^9y^9 - x^6y^6 + x^5y^5 - x^2y^2$ **53.** $x^2 + 2xh + h^2$
55. $r^6t^4 - 8r^3t^2 + 16$ **57.** $p^8 + 2m^2n^2p^4 + m^4n^4$
59. $4a^2 - b^2$ **61.** $c^4 - d^2$ **63.** $a^2b^2 - c^2d^4$
65. $x^2 + 2xy + y^2 - 9$ **67.** $x^2 - y^2 - 2yz - z^2$
69. $a^2 - b^2 - 2bc - c^2$ **71.** $x^2 + 2xy + y^2$
73. $x^2 - z^2$ **75.** \$60 per ton **77.** December 1987
79. ◈ **81.** $4xy - 4y^2$ **83.** $2xy + \pi x^2$
85. 3.535 L **87.** ◈

Exercise Set 4.7, pp. 194–195

1. $3x^4 - \frac{1}{2}x^3$ **3.** $1 - 2u - u^4$ **5.** $5t^2 + 8t - 2$
7. $-4x^4 + 4x^2 + 1$ **9.** $6x^2 - 10x + \frac{3}{2}$
11. $4x^2 - \frac{3}{2}x + \frac{1}{2}$ **13.** $x^2 + 3x + 2$ **15.** $-3rs - r + 2s$
17. $x + 2$ **19.** $x - 5 - \dfrac{50}{x - 5}$ **21.** $x - 2 - \dfrac{2}{x + 6}$
23. $x - 3$ **25.** $x^4 - x^3 + x^2 - x + 1$
27. $2x^2 - 7x + 4$ **29.** $x^3 - 6$ **31.** $x^3 + 2x^2 + 4x + 8$
33. $t^2 + 1$ **35.** 25,543.75 ft^2

37.

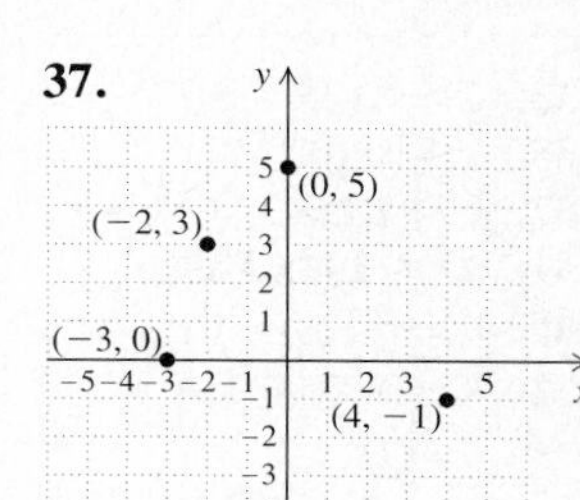

39. ◈ **41.** $x^2 + 5$
43. $a + 3 + \dfrac{5}{5a^2 - 7a - 2}$
45. $2x^2 + x - 3$
47. $3a^{2h} + 2a^h - 5$
49. 2

Exercise Set 4.8, pp. 201–203

1. $\dfrac{1}{3^2} = \dfrac{1}{9}$ **3.** $\dfrac{1}{10^4} = \dfrac{1}{10{,}000}$ **5.** $\dfrac{1}{7^3} = \dfrac{1}{343}$ **7.** $\dfrac{1}{a^3}$
9. y^4 **11.** z^n **13.** $\dfrac{1}{2}$ **15.** 16 **17.** 4^{-3} **19.** x^{-3}
21. a^{-4} **23.** p^{-n} **25.** 5^{-1} **27.** t^{-1} **29.** 3^3
31. x^{-1}, or $\dfrac{1}{x}$ **33.** x^{-13}, or $\dfrac{1}{x^{13}}$ **35.** m^{-6}, or $\dfrac{1}{m^6}$
37. $(8x)^{-4}$, or $\dfrac{1}{(8x)^4}$ **39.** 1 **41.** $a^{-7}b^{-11}$, or $\dfrac{1}{a^7b^{11}}$
43. x^9 **45.** z^{-4}, or $\dfrac{1}{z^4}$ **47.** x^3 **49.** x^2
51. a^{-15}, or $\dfrac{1}{a^{15}}$ **53.** 5^{-6}, or $\dfrac{1}{5^6}$ **55.** x^{12}
57. m^{-21}, or $\dfrac{1}{m^{21}}$ **59.** $a^{-3}b^{-3}$, or $\dfrac{1}{a^3b^3}$
61. $5^{-2}a^{-2}b^{-2}$, or $\dfrac{1}{25a^2b^2}$ **63.** $36x^{-10}$, or $\dfrac{36}{x^{10}}$
65. $x^{-12}y^{-15}$, or $\dfrac{1}{x^{12}y^{15}}$ **67.** $x^{24}y^8$
69. $9x^6y^{-16}z^{-6}$, or $\dfrac{9x^6}{y^{16}z^6}$ **71.** $x^{-1}y^{-6}z^4$, or $\dfrac{z^4}{xy^6}$
73. $m^5n^5p^{-7}$, or $\dfrac{m^5n^5}{p^7}$ **75.** $\dfrac{y^{-6}}{2^{-3}}$, or $\dfrac{8}{y^6}$ **77.** $\dfrac{27}{a^6}$
79. $\dfrac{x^6y^3}{z^3}$ **81.** $\dfrac{a^{-4}b^{-2}}{c^{-2}d^{-6}}$, or $\dfrac{c^2d^6}{a^4b^2}$ **83.** 2140

85. 0.00692 **87.** 784,000,000 **89.** 0.0000000008764
91. 100,000,000 **93.** 0.0001 **95.** 2.5×10^4
97. 3.71×10^{-3} **99.** 7.8×10^{10} **101.** 9.07×10^{17}
103. 3.74×10^{-6} **105.** 1.8×10^{-8} **107.** 10^7
109. 10^{-9} **111.** 6×10^9 **113.** 3.38×10^4
115. 8.1477×10^{-13} **117.** 2.5×10^{13}
119. 5×10^{-4} **121.** 3×10^{-21}
123. Approximately 1.231×10^{-5}
125. 2.3725×10^9 gal
127. 1.512×10^{10} cu ft, 1.324512×10^{14} cu ft
129. $8a$

131.

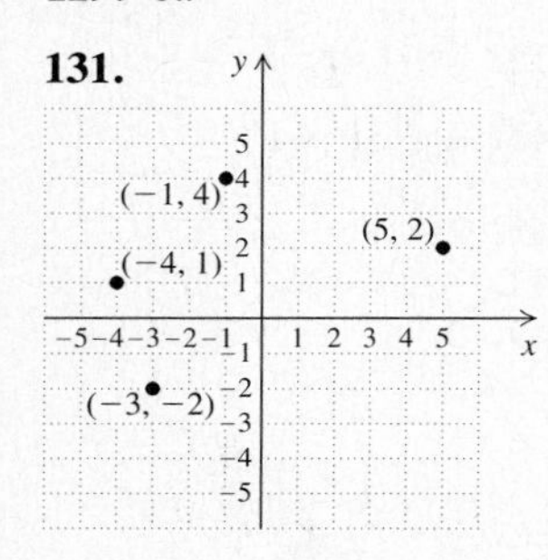

133. 2.478125×10^{-1}
135. 3.5×10^{-10}
137. 2 **139.** 5
141. a^n **143.** False
145. False

Review Exercises: Chapter 4, pp. 204–206

1. [4.1] y^{11} **2.** [4.1] $(3x)^{14}$ **3.** [4.1] t^8 **4.** [4.1] 4^3
5. [4.1] 1 **6.** [4.1] $\frac{9t^8}{4s^6}$ **7.** [4.1] $-8x^3y^6$
8. [4.1] $36x^8$ **9.** [4.1] $18x^8$ **10.** [4.2] $3x^2$, $6x$, $\frac{1}{2}$
11. [4.2] $-4y^5$, $7y^2$, $-3y$, -2 **12.** [4.2] 6, 17
13. [4.2] 4, 6, -5, $\frac{5}{3}$ **14.** [4.2] 3, 1, 0; 3
15. [4.2] 0, 4, 9, 6, 3; 9 **16.** [4.2] Binomial
17. [4.2] None of these **18.** [4.2] Monomial
19. [4.2] $-x^2 + 9x$ **20.** [4.2] $-\frac{1}{4}x^3 + 4x^2 + 7$
21. [4.2] $-3x^5 + 25$ **22.** [4.2] $-2x^2 - 3x + 2$
23. [4.2] $10x^4 - 7x^2 - x - \frac{1}{2}$ **24.** [4.2] -17
25. [4.2] 10 **26.** [4.3] $x^5 - 2x^4 + 6x^3 + 3x^2 - 9$
27. [4.3] $2x^5 - 6x^4 + 2x^3 - 2x^2 + 2$
28. [4.3] $2x^2 - 4x - 6$ **29.** [4.3] $x^5 - 3x^3 - 2x^2 + 8$
30. [4.3] $\frac{3}{4}x^4 + \frac{1}{4}x^3 - \frac{1}{3}x^2 - \frac{7}{4}x + \frac{3}{8}$
31. [4.3] $-x^5 + x^4 - 5x^3 - 2x^2 + 2x$
32. **(a)** [4.3] $4w + 8$; **(b)** [4.4] $w^2 + 4w$
33. [4.4] $-12x^3$ **34.** [4.5] $49x^2 + 14x + 1$
35. [4.5] $x^2 + \frac{7}{6}x + \frac{1}{3}$ **36.** [4.5] $0.3x^2 + 0.65x - 8.45$
37. [4.4] $12x^3 - 23x^2 + 13x - 2$
38. [4.5] $x^2 - 18x + 81$
39. [4.4] $15x^7 - 40x^6 + 50x^5 + 10x^4$
40. [4.5] $x^2 - 3x - 28$
41. [4.5] $x^2 - 1.05x + 0.225$
42. [4.4] $x^7 + x^5 - 3x^4 + 3x^3 - 2x^2 + 5x - 3$
43. [4.5] $9y^4 - 12y^3 + 4y^2$ **44.** [4.5] $2t^4 - 11t^2 - 21$
45. [4.4] $4x^5 - 5x^4 - 8x^3 + 22x^2 - 15x$
46. [4.5] $9x^4 - 16$ **47.** [4.5] $4 - x^2$
48. [4.5] $13x^2 - 172x + 39$ **49.** [4.6] 49
50. [4.6] Coefficients: 1, -7, 9, -8; degrees: 6, 2, 2, 0; 6
51. [4.6] Coefficients: 1, -1, 1; degrees: 16, 40, 23; 40
52. [4.6] $9w - y - 5$
53. [4.6] $m^6 - 2m^2n + 2m^2n^2 + 8n^2m - 6m^3$
54. [4.6] $-9xy - 2y^2$
55. [4.6] $11x^3y^2 - 8x^2y - 6x^2 - 6x + 6$ **56.** [4.6] $p^3 - q^3$
57. [4.6] $9a^8 - 2a^4b^3 + \frac{1}{9}b^6$ **58.** [4.7] $5x^2 - \frac{1}{2}x + 3$
59. [4.7] $3x^2 - 7x + 4 + \frac{1}{2x + 3}$ **60.** [4.7] $t^3 + 2t - 3$
61. [4.7] $2x^2 + 1 + \frac{x^2 + 2}{x^3 - 1}$ **62.** [4.8] $\frac{1}{y^4}$
63. [4.8] t^{-5} **64.** [4.8] $\frac{1}{7^2}$ **65.** [4.8] $\frac{1}{a^{13}b^7}$
66. [4.8] $\frac{1}{x^{12}}$ **67.** [4.8] $\frac{x^6}{4y^2}$ **68.** [4.8] $\frac{y^3}{8x^3}$
69. [4.8] 8,300,000 **70.** [4.8] 3.28×10^{-5}
71. [4.8] 2.09×10^4 **72.** [4.8] 5.12×10^{-5}
73. [4.8] 6.205×10^{10} **74.** [1.5] $13x + \frac{22}{15}$
75. [2.5] $w = 125.5$ m, $l = 144.5$ m
76. [2.6] $\{x|x \geq -2\}$ **77.** [3.1] IV
78. [4.1] ◈ In the expression $5x^3$, the exponent refers only to the x. In the expression $(5x)^3$, the entire expression within the parentheses is cubed.
79. [4.3] ◈ The sum of two polynomials of degree n will also have degree n, since only the coefficients are added and the variables remain unchanged. An exception to this occurs when the leading terms of the two polynomials are opposites. The sum of those terms is then zero and the sum of the polynomials will have a degree less than n.
80. [4.2], [4.5] **(a)** 3; **(b)** 2 **81.** [4.1], [4.2] $-28x^8$
82. [4.2] $8x^4 + 4x^3 + 5x - 2$
83. [4.5] $-4x^6 + 3x^4 - 20x^3 + x^2 - 16$
84. [2.2], [4.5] $\frac{94}{13}$

Test: Chapter 4, pp. 206–207

1. [4.1] x^9 **2.** [4.1] $(4a)^{11}$ **3.** [4.1] 3^3 **4.** [4.1] 1
5. [4.1] x^6 **6.** [4.1] $-27y^6$ **7.** [4.1] $-216x^{21}$
8. [4.1] $-24x^{21}$ **9.** [4.2] Binomial
10. [4.2] $\frac{1}{3}$, -1, 7 **11.** [4.2] 3, 0, 1, 6; 6

12. [4.2] -7 **13.** [4.2] $5a^2 - 6$ **14.** [4.2] $\frac{7}{4}y^2 - 4y$
15. [4.2] $x^5 + 2x^3 + 4x^2 - 8x + 3$
16. [4.3] $4x^5 + x^4 + 2x^3 - 8x^2 + 2x - 7$
17. [4.3] $5x^4 + 5x^2 + x + 5$
18. [4.3] $-4x^4 + x^3 - 8x - 3$
19. [4.3] $-x^5 + 0.7x^3 - 0.8x^2 - 21$
20. [4.4] $-12x^4 + 9x^3 + 15x^2$ **21.** [4.5] $x^2 - \frac{2}{3}x + \frac{1}{9}$
22. [4.5] $9x^2 - 100$ **23.** [4.5] $3b^2 - 4b - 15$
24. [4.5] $x^{14} - 4x^8 + 4x^6 - 16$
25. [4.5] $48 + 34y - 5y^2$
26. [4.4] $6x^3 - 7x^2 - 11x - 3$
27. [4.5] $25t^2 + 20t + 4$
28. [4.6] $-5x^3y - x^2y^2 + xy^3 - y^3 + 19$
29. [4.6] $8a^2b^2 + 6ab - 4b^3 + 6ab^2 + ab^3$
30. [4.6] $9x^{10} - 16y^{10}$ **31.** [4.7] $4x^2 + 3x - 5$
32. [4.7] $2x^2 - 4x - 2 + \dfrac{17}{3x + 2}$ **33.** [4.8] $\dfrac{1}{5^3}$
34. [4.8] y^{-8} **35.** [4.8] $\dfrac{1}{6^5}$ **36.** [4.8] $\dfrac{y^5}{x^5}$
37. [4.8] $\dfrac{b^4}{16a^{12}}$ **38.** [4.8] $\dfrac{c^3}{a^3b^3}$ **39.** [4.8] 3.9×10^9
40. [4.8] 0.00000005 **41.** [4.8] 1.75×10^{17}
42. [4.8] 1.296×10^{22}
43. [4.8] Approximately 2.49×10^2
44. [2.6] $\{x|x > 13\}$
45. [3.1]

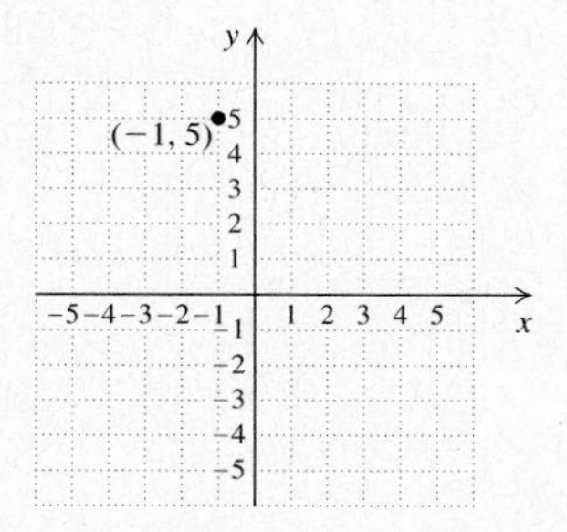

46. [1.5] $-\frac{7}{20}$ **47.** [2.5] 100°, 25°, 55°
48. [4.5] $V = l(l - 2)(l - 1) = l^3 - 3l^2 + 2l$
49. [2.2], [4.5] $\frac{100}{21}$

CHAPTER 5

Exercise Set 5.1, pp. 213–214

1. Answers may vary. $(6x)(x^2)$, $(3x^2)(2x)$, $(2x^2)(3x)$
3. Answers may vary. $(-3x^2)(3x^3)$, $(-x)(9x^4)$, $(3x^2)(-3x^3)$
5. Answers may vary. $(6x)(4x^3)$, $(-3x^2)(-8x^2)$, $(2x^3)(12x)$
7. $x(x - 4)$ **9.** $2x(x + 3)$ **11.** $x^2(x + 6)$
13. $8x^2(x^2 - 3)$ **15.** $2(x^2 + x - 4)$
17. $17xy(x^4y^2 + 2x^2y + 3)$ **19.** $x^2(6x^2 - 10x + 3)$
21. $x^2y^2(x^3y^3 + x^2y + xy - 1)$
23. $2x^3(x^4 - x^3 - 32x^2 + 2)$
25. $0.8x(2x^3 - 3x^2 + 4x + 8)$
27. $\frac{1}{3}x^3(5x^3 + 4x^2 + x + 1)$ **29.** $(y + 3)(y + 4)$
31. $(x + 3)(x^2 + 2)$ **33.** $(y + 8)(y^2 + 1)$
35. $(x + 3)(x^2 + 2)$ **37.** $(x + 3)(2x^2 + 1)$
39. $(2x - 3)(4x^2 + 3)$ **41.** $(3x - 4)(4x^2 + 1)$
43. $(x + 8)(x^2 - 3)$ **45.** $(w - 7)(w^2 + 4)$
47. Not factorable by grouping **49.** $(x - 4)(2x^2 - 9)$
51. $y = x - 6$
53. 12
55. $y^2 + 12y + 35$
57. $y^2 - 49$ **59.** ◈
61. $(2x^2 + 3)(2x^3 + 3)$
63. $(x^5 + 1)(x^7 + 1)$
65. Not factorable
67. ◈

Exercise Set 5.2, pp. 219–220

1. $(x + 5)(x + 3)$ **3.** $(x + 4)(x + 3)$ **5.** $(x - 3)^2$
7. $(x + 7)(x + 2)$ **9.** $(b + 4)(b + 1)$ **11.** $\left(x + \frac{1}{3}\right)^2$
13. $(d - 5)(d - 2)$ **15.** $(y - 10)(y - 1)$
17. $(x + 7)(x - 6)$ **19.** $2(x + 2)(x - 9)$
21. $x(x - 8)(x + 2)$ **23.** $(y + 5)(y - 9)$
25. $(x + 9)(x - 11)$ **27.** $c^2(c + 8)(c - 7)$
29. $2(a + 7)(a - 5)$ **31.** Not factorable
33. Not factorable **35.** $(x + 10)^2$
37. $3x(x - 25)(x + 4)$ **39.** $(x - 24)(x + 3)$
41. $(x - 16)(x - 9)$ **43.** $a^2(a + 12)(a - 11)$
45. $(x - 15)(x - 8)$ **47.** $(12 + x)(9 - x)$
49. $(y - 0.4)(y + 0.2)$ **51.** $(p + 5q)(p - 2q)$
53. Not factorable **55.** $(s - 5t)(s + 3t)$
57. $2x(x - 3)(x - 2)$ **59.** $7a^7(a + 1)(a - 5)$
61. $3x^2 + 22x + 24$ **63.** 29,443 **65.** ◈
67. 15, −15, 27, −27, 51, −51 **69.** $\left(x + \frac{1}{4}\right)\left(x - \frac{3}{4}\right)$
71. $(x + 5)\left(x - \frac{5}{7}\right)$ **73.** $(b^n + 5)(b^n + 2)$
75. $(x + 1)(a + 2)(a + 1)$ **77.** $2x^2(4 - \pi)$

Exercise Set 5.3, pp. 226–227

1. $(2x + 1)(x - 4)$ **3.** $(5x - 9)(x + 2)$
5. $(2x + 7)(3x + 1)$ **7.** $(3x + 1)(x + 1)$
9. $(2x + 5)(2x - 3)$ **11.** $(2x + 1)(x - 1)$
13. $(3x + 8)(3x - 2)$ **15.** $(3x + 1)(x - 2)$
17. $(3x + 4)(4x + 5)$ **19.** $(7x - 1)(2x + 3)$
21. $(3x + 4)(3x + 2)$ **23.** $(7 - 3x)^2$
25. $(x + 2)(24x - 1)$ **27.** $(7x + 4)(5x - 11)$
29. Prime **31.** $4(3x - 2)(x + 3)$
33. $6(5x - 9)(x + 1)$ **35.** $2(3x + 5)(x - 1)$
37. $(3x - 1)(x - 1)$ **39.** $4(3x + 2)(x - 3)$
41. $(2x + 1)(x - 1)$ **43.** $(3x - 8)(3x + 2)$
45. $5(3x + 1)(x - 2)$ **47.** $x(3x + 4)(4x + 5)$
49. $x^2(2x + 3)(7x - 1)$ **51.** $3x(8x - 1)(7x - 1)$
53. $(5x - 3)(3x - 2)$ **55.** $(5t + 8)^2$

57. $2x(3x + 5)(x - 1)$ **59.** $(25x + 64)(x + 1)$
61. Prime **63.** $(4m - 5n)(3m + 4n)$
65. $(3a - 5b)(2a + 3b)$ **67.** $(3a + 2b)(3a + 4b)$
69. $(5p + 2q)(7p + 4q)$ **71.** $6(3x - 4y)(x + y)$
73. $(y + 4)(y + 1)$ **75.** $(x - 4)(x - 1)$
77. $(3x + 2)(2x + 3)$ **79.** $(3x - 4)(x - 4)$
81. $(7x - 8)(5x + 3)$ **83.** $(2x + 3)(2x - 3)$
85. $(2x + 1)(x - 4)$ **87.** $(5x - 9)(x + 2)$
89. $(3x + 1)(2x + 7)$ **91.** $(3x + 1)(x + 1)$
93. $(2x + 5)(2x - 3)$ **95.** $(2x + 1)(x - 1)$
97. $(3x + 8)(3x - 2)$ **99.** $(3x + 1)(x - 2)$
101. $(4x + 5)(3x + 4)$ **103.** $(7x - 1)(2x + 3)$
105. $(3x + 2)(3x + 4)$ **107.** $(3x - 7)^2$
109. 6369 km, 3949 mi
111. 

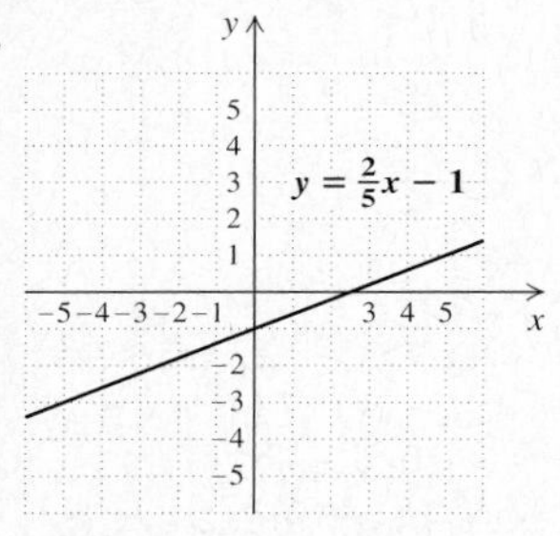

113. ◈ **115.** $(3x^5 - 2)^2$ **117.** $(10x^n + 3)(2x^n + 1)$
119. $(x^{3a} - 1)(3x^{3a} + 1)$
121. $-2(a + 1)^n(a + 3)^2(a + 6)$

Exercise Set 5.4, pp. 233–234

1. Yes **3.** No **5.** No **7.** No **9.** $(x - 7)^2$
11. $(x + 8)^2$ **13.** $(x - 1)^2$ **15.** $(x + 2)^2$
17. $(3x + 1)^2$ **19.** $(4y - 7)^2$, or $(7 - 4y)^2$
21. $2(x - 1)^2$ **23.** $x(x - 9)^2$ **25.** $5(2x + 5)^2$
27. $(7 - 3x)^2$ **29.** $5(y + 1)^2$ **31.** $2(1 + 5x)^2$
33. $(2p + 3q)^2$ **35.** $(a - 7b)^2$ **37.** $(8m + n)^2$
39. $(4s - 5t)^2$ **41.** Yes **43.** No **45.** No
47. Yes **49.** $(y + 2)(y - 2)$ **51.** $(p + 3)(p - 3)$
53. $(t + 7)(t - 7)$ **55.** $(a + b)(a - b)$
57. $(5t + m)(5t - m)$ **59.** $(10 + k)(10 - k)$
61. $(4a + 3)(4a - 3)$ **63.** $(2x + 5y)(2x - 5y)$
65. $2(2x + 7)(2x - 7)$ **67.** $x(6 + 7x)(6 - 7x)$
69. $(7a^2 + 9)(7a^2 - 9)$ **71.** $(x^2 + 1)(x + 1)(x - 1)$
73. $4(x^2 + 4)(x + 2)(x - 2)$
75. $(y^4 + 1)(y^2 + 1)(1 + y)(1 - y)$ **77.** $3x(x - 4)^2$
79. $(x^6 + 4)(x^3 + 2)(x^3 - 2)$ **81.** $\left(y + \frac{1}{4}\right)\left(y - \frac{1}{4}\right)$
83. $a^6(a - 1)^2$ **85.** $\left(5 + \frac{1}{7}x\right)\left(5 - \frac{1}{7}x\right)$
87. $(4m^2 + t^2)(2m + t)(2m - t)$ **89.** $s \geqslant 77$
91. $x^{12}y^{12}$ **93.** ◈ **95.** Prime **97.** $2x(3x + 1)^2$
99. $(x^4 + 2^4)(x^2 + 2^2)(x + 2)(x - 2)$
101. $3x^3(x + 2)(x - 2)$ **103.** $2x\left(3x + \frac{2}{5}\right)\left(3x - \frac{2}{5}\right)$
105. $p(0.7 + p)(0.7 - p)$ **107.** $x(x + 6)$
109. $\left(x + \dfrac{1}{x}\right)\left(x - \dfrac{1}{x}\right)$

111. $(9 + b^{2k})(3 + b^k)(3 - b^k)$ **113.** $(3b^n + 2)^2$
115. $(y + 4)^2$ **117.** $(3x + 7)(3x - 7)^2$
119. $(a + 4)(a - 2)$ **121.** 9
123. $(x + 1)^2 - x^2 = ((x + 1) + x)((x + 1) - x)$
$= ((x + 1) + x)(1) = (x + 1) + x$

Exercise Set 5.5, pp. 238–239

1. $2(x + 8)(x - 8)$ **3.** $(a - 5)^2$ **5.** $(2x - 3)(x - 4)$
7. $x(x + 12)^2$ **9.** $(x + 3)(x - 2)(x + 2)$
11. $6(2x + 3)(2x - 3)$ **13.** $4x(x - 2)(5x + 9)$
15. Prime **17.** $x(x^2 + 7)(x - 3)$ **19.** $x^3(x - 7)^2$
21. $-2(x - 2)(x + 5)$ **23.** Prime
25. $4(x^2 + 4)(x + 2)(x - 2)$
27. $(t^4 + 1)(t^2 + 1)(t + 1)(t - 1)$ **29.** $x^3(x - 3)(x - 1)$
31. $(x + y)(x - y)$ **33.** $12n^2(1 + 2n)$
35. $9xy(xy - 4)$ **37.** $2\pi r(h + r)$ **39.** $(a + b)(2x + 1)$
41. $(x + 1)(x - 1 - y)$ **43.** $(n + 2)(n + p)$
45. $(x - 2)(2x + z)$ **47.** $(x - y)^2$ **49.** $(3c + d)^2$
51. $7(p^2 + q^2)(p + q)(p - q)$ **53.** $(5z + y)^2$
55. $a^3(a - b)(a + 5b)$ **57.** $(a + b)(a - 2b)$
59. $(m + 20n)(m - 18n)$ **61.** $(mn - 8)(mn + 4)$
63. $a^3(ab + 5)(ab - 2)$ **65.** $(7m - 8n)^2$
67. $x^4(x + 2y)(x - y)$ **69.** $\left(6a - \frac{5}{4}\right)^2$ **71.** $\left(\frac{1}{2}a + \frac{1}{3}b\right)^2$
73. $(9a^2 + b^2)(3a - b)(3a + b)$
75. $(w - 7)(w + 2)(w - 2)$

77.

$y = -4x + 7$	
$11 \;?\; -4(-1) + 7$	
$4 + 7$	
$11 \;\big|\; 11$	TRUE

$y = -4x + 7$	
$7 \;?\; -4 \cdot 0 + 7$	
$0 + 7$	
$7 \;\big|\; 7$	TRUE

$y = -4x + 7$	
$-5 \;?\; -4 \cdot 3 + 7$	
$-12 + 7$	
$-5 \;\big|\; -5$	TRUE

79. $X = \dfrac{A + 7}{a + b}$
81. ◈ **83.** $-10 = -10$; probably correct
85. $(y - 2)(y + 3)(y - 3)$ **87.** $(a + 4)(a^2 + 1)$
89. $(x + 3)(x - 3)(x^2 + 2)$ **91.** $(x - 1)(x + 2)(x - 2)$
93. $(y - 1)^3$ **95.** $[(y + 4) + x]^2$
97. $(x^4 + 16)(x^2 + 4)(x + 2)(x - 2)$

Technology Connection, Section 5.6

TC1. $-4.65, 0.65$ **TC2.** $-0.37, 5.37$
TC3. $-4.56, -8.98$ **TC4.** No solution

Exercise Set 5.6, pp. 244–245

1. $-8, -6$ **3.** $3, -5$ **5.** $-12, 11$ **7.** $0, -5$
9. $0, -10$ **11.** $-\frac{5}{2}, -4$ **13.** $-\frac{1}{5}, 3$ **15.** $4, \frac{1}{4}$
17. $0, \frac{2}{3}$ **19.** $0, 18$ **21.** $-\frac{1}{10}, \frac{1}{27}$ **23.** $\frac{1}{3}, 20$
25. $0, \frac{2}{3}, \frac{1}{2}$ **27.** $-1, -5$ **29.** $-9, 2$ **31.** $3, 5$
33. $0, 8$ **35.** $0, -19$ **37.** $4, -4$ **39.** $\frac{2}{3}, -\frac{2}{3}$

41. -3 **43.** 4 **45.** $0, \frac{6}{5}$ **47.** $\frac{5}{3}, -1$ **49.** $\frac{2}{3}, -\frac{1}{4}$
51. $7, -2$ **53.** $\frac{9}{8}, -\frac{9}{8}$ **55.** $-3, 1$ **57.** $-\frac{2}{3}, -4$
59. $(3, 0), (-2, 0)$ **61.** $(2, 0), (-4, 0)$
63. $\left(\frac{3}{2}, 0\right), (-3, 0)$ **65.** $(a + b)^2$ **67.** $2x + 5 < 19$
69. ◈ **71.** ◈
73. **(a)** $x^2 - x - 12 = 0$; **(b)** $x^2 + 7x + 12 = 0$;
(c) $4x^2 - 4x + 1 = 0$; **(d)** $x^2 - 25 = 0$;
(e) $40x^3 - 14x^2 + x = 0$
75. 4 **77.** $\frac{1}{8}, -\frac{1}{8}$ **79.** $4, -4$
81. **(a)** $9x^2 - 12x + 24 = 0$; **(b)** $x^2 - 3x - 18 = 0$;
(c) $4x^2 + 8x + 36 = 0$; **(d)** $(2x + 8)(2x - 5) = 0$;
(e) $(x + 1)(5x - 5) = 0$; **(f)** $2x^2 + 20x - 4 = 0$
83. ◈ **85.** $-2.33, -6.77$ **87.** $-4.59, -9.15$
89. $-3.25, -6.75$

Exercise Set 5.7, pp. 250–253

1. $-\frac{3}{4}, 1$ **3.** 2, 4 **5.** 14 and 15
7. 12 and 14, -14 and -12
9. 15 and 17, -17 and -15
11. Length: 12 m; width: 8 m **13.** 5
15. Height: 4 cm; base: 14 cm **17.** 6 m **19.** 5 and 7
21. 506 **23.** 12 **25.** 780 **27.** 20
29. Hypotenuse: 17 ft; leg: 15 ft
31. Dining room: 12 ft by 12 ft; kitchen: 10 ft by 12 ft
33.

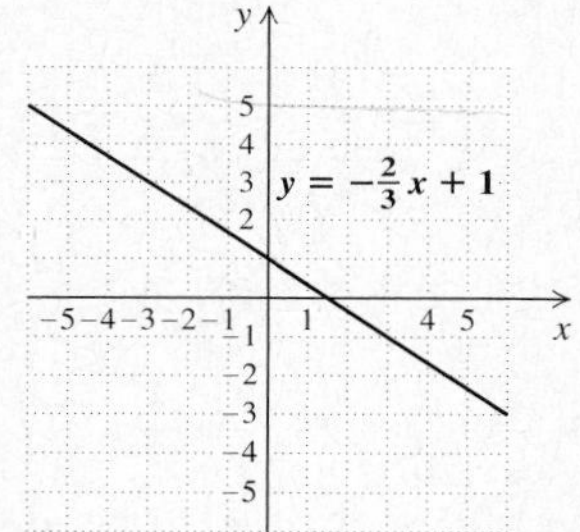

35. 7 **37.** ◈ **39.** 5 ft **41.** 37
43. 30 cm by 15 cm **45.** 100 cm^2; 225 cm^2

Review Exercises: Chapter 5, pp. 254–255

1. [5.1] Answers may vary. $(-5x)(2x)$, $(-x)(10x)$, $(-2x)(5x)$
2. [5.1] Answers may vary. $(4x^2)(9x^3)$, $(18x)(2x^4)$, $(-6x^3)(-6x^2)$
3. [5.4] $5(1 + 2x^3)(1 - 2x^3)$ **4.** [5.1] $x(x - 3)$
5. [5.4] $(3x + 2)(3x - 2)$ **6.** [5.2] $(x + 6)(x - 2)$
7. [5.4] $(x + 7)^2$ **8.** [5.1] $3x(2x^2 + 4x + 1)$
9. [5.1] $(x^2 + 3)(x + 1)$ **10.** [5.3] $(3x - 1)(2x - 1)$
11. [5.4] $(x^2 + 9)(x + 3)(x - 3)$
12. [5.3] $3x(3x - 5)(x + 3)$ **13.** [5.4] $2(x + 5)(x - 5)$
14. [5.1] $(x^3 - 2)(x + 4)$
15. [5.4] $(4x^2 + 1)(2x + 1)(2x - 1)$
16. [5.1] $4x^4(2x^2 - 8x + 1)$ **17.** [5.4] $3(2x + 5)^2$
18. [5.4] Prime **19.** [5.2] $x(x - 6)(x + 5)$
20. [5.4] $(2x + 5)(2x - 5)$ **21.** [5.4] $(3x - 5)^2$
22. [5.3] $2(3x + 4)(x - 6)$ **23.** [5.4] $(x - 3)^2$
24. [5.3] $(2x + 1)(x - 4)$ **25.** [5.4] $2(3x - 1)^2$
26. [5.4] $3(x + 3)(x - 3)$ **27.** [5.2] $(x - 5)(x - 3)$
28. [5.4] $(5x - 2)^2$ **29.** [5.2] $(xy + 4)(xy - 3)$
30. [5.4] $3(2a + 7b)^2$ **31.** [5.1] $(m + t)(m + 5)$
32. [5.4] $32(x^2 - 2y^2z^2)(x^2 + 2y^2z^2)$ **33.** [5.6] $1, -3$
34. [5.6] $-7, 5$ **35.** [5.6] $-4, 3$ **36.** [5.6] $\frac{2}{3}, 1$
37. [5.6] $\frac{3}{2}, -4$ **38.** [5.6] $8, -2$ **39.** [5.7] $3, -2$
40. [5.6] $(-1, 0), \left(\frac{5}{2}, 0\right)$
41. [5.7] -19 and -17, 17 and 19 **42.** [5.7] $\frac{5}{2}, -2$
43. [1.6] $\frac{1}{2}$
44. [3.2]

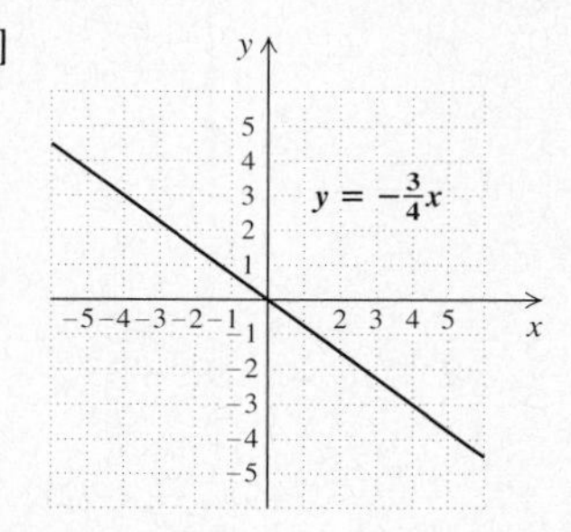

45. [4.1] m^2n^5 **46.** [2.7] $\frac{1}{2}x - 2 \geq 10$
47. [5.1], [5.5] ◈ Factorizations can be checked by multiplying or by evaluating. Evaluating a factorization for one number and comparing the result with that of evaluating the original polynomial for that number may be quicker than multiplying the factors, but it is only a partial check.
48. [5.6] ◈ The equations solved in this chapter have an x^2-term (are quadratic), whereas those solved previously have no x^2-term (are linear). The principle of zero products is used to solve quadratic equations and is not used to solve linear equations.
49. [5.7] $2\frac{1}{2}$ cm **50.** [5.7] 0, 2
51. [5.7] $l = 12$, $w = 6$ **52.** [5.6] No real solution
53. [5.6] $2, -3, \frac{5}{2}$

Test: Chapter 5, p. 255

1. [5.1] Answers may vary. $4x \cdot x^2$, $2x^2 \cdot 2x$, $2 \cdot 2x^3$
2. [5.2] $(x - 5)(x - 2)$ **3.** [5.4] $(x - 5)^2$
4. [5.1] $2y^2(3 - 4y + 2y^2)$ **5.** [5.1] $(x^2 + 2)(x + 1)$
6. [5.1] $x(x - 5)$ **7.** [5.2] $x(x + 3)(x - 1)$
8. [5.3] $2(5x - 6)(x + 4)$ **9.** [5.4] $(2x + 3)(2x - 3)$
10. [5.2] $(x - 4)(x + 3)$ **11.** [5.3] $3m(2m + 1)(m + 1)$
12. [5.4] $3(w + 5)(w - 5)$ **13.** [5.4] $5(3x + 2)^2$
14. [5.4] $3(x^2 + 4)(x + 2)(x - 2)$ **15.** [5.4] $(7x - 6)^2$
16. [5.3] $(5x - 1)(x - 5)$ **17.** [5.1] $(x^3 - 3)(x + 2)$
18. [5.4] $5(4 + x^2)(2 + x)(2 - x)$
19. [5.3] $(2x - 5)(2x + 3)$ **20.** [5.3] $3t(2t + 5)(t - 1)$
21. [5.2] $3(m + 2n)(m - 5n)$ **22.** [5.6] $5, -4$
23. [5.6] $-5, \frac{3}{2}$ **24.** [5.6] $-4, 7$ **25.** [5.7] $8, -3$
26. [5.7] $l = 10$ m, $w = 4$ m **27.** [1.6] -0.4
28. [2.7] $\{l \mid l < 13\}$

29. [3.2]

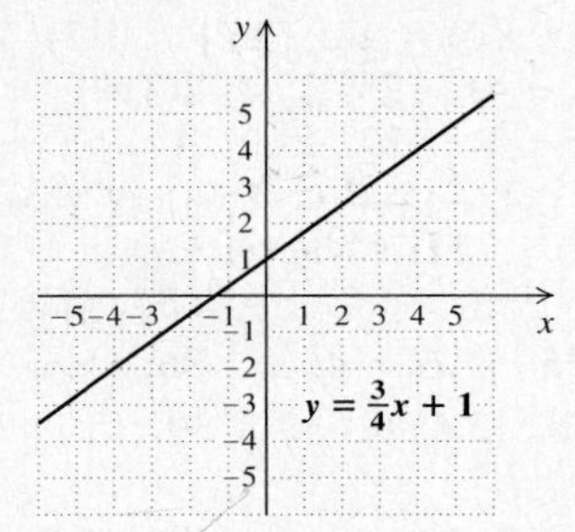

30. [4.1] $49a^6b^{10}$ **31.** [5.7] $l = 15$, $w = 3$
32. [5.2] $(a - 4)(a + 8)$

CHAPTER 6

Exercise Set 6.1, pp. 262–263

1. 0 **3.** 8 **5.** $-\frac{5}{2}$ **7.** 7, −4 **9.** −5, 5 **11.** $\frac{a^2}{3b}$

13. $\frac{5}{2xy^4}$ **15.** $\frac{3}{2}$ **17.** $\frac{a-5}{a+1}$ **19.** $\frac{8}{3x^2}$ **21.** $\frac{x-3}{x}$

23. $\frac{m+1}{2m+3}$ **25.** $\frac{a-3}{a+2}$ **27.** $\frac{t+2}{2(t-4)}$ **29.** $\frac{x+5}{x-5}$

31. $a+1$ **33.** $\frac{x^2+1}{x+1}$ **35.** $\frac{3}{2}$ **37.** $\frac{6}{t-3}$ **39.** $\frac{a-3}{a-4}$

41. $\frac{t-2}{t+2}$ **43.** −1 **45.** −1 **47.** −6 **49.** $-a-1$

51. $(x+7)(x+1)$

53.

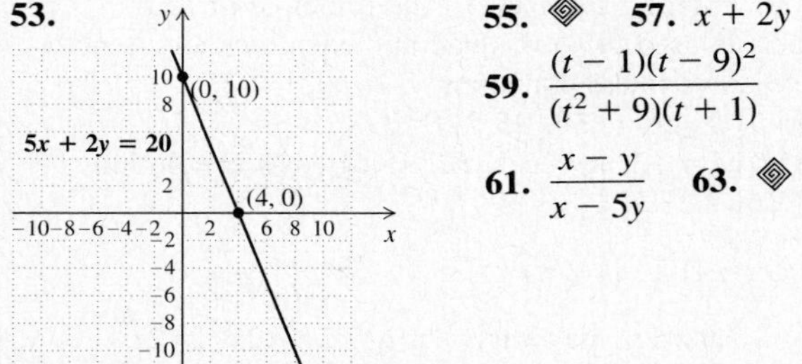

55. ◈ **57.** $x + 2y$ **59.** $\frac{(t-1)(t-9)^2}{(t^2+9)(t+1)}$

61. $\frac{x-y}{x-5y}$ **63.** ◈

Exercise Set 6.2, pp. 266–268

1. $\frac{3x(x+4)}{2(x-1)}$ **3.** $\frac{(x-1)(x+1)}{(x+2)(x+2)}$ **5.** $\frac{(2x+3)(x+1)}{4(x-5)}$

7. $\frac{(a-5)(a+2)}{(a^2+1)(a^2-1)}$ **9.** $\frac{(x+1)(x-1)}{(2+x)(x+1)}$ **11.** $\frac{56x}{3}$

13. $\frac{2}{dc^2}$ **15.** $\frac{x+2}{x-2}$ **17.** $\frac{(a+5)(a-5)(2a-5)}{(a-3)(a-1)(2a+5)}$

19. $\frac{(a+3)(a-3)}{a(a+4)}$ **21.** $\frac{2a}{a-2}$ **23.** $\frac{t-5}{t+5}$

25. $\frac{5(a+6)}{a-1}$ **27.** $\frac{(x-1)(x-3)^3}{(x+3)(x+1)}$

29. $\frac{(a+2)(a-2)(a-1)}{(a+1)^2(a^4+1)}$ **31.** $\frac{t-2}{t-1}$ **33.** $\frac{x}{4}$

35. $\frac{1}{x^2-y^2}$ **37.** $\frac{x^2-4x+7}{x^2+2x-5}$ **39.** $\frac{3}{10}$ **41.** $\frac{1}{4}$ **43.** $\frac{y^2}{x}$

45. $\frac{(a+2)(a+3)}{(a-3)(a-1)}$ **47.** $\frac{(x-1)^2}{x}$ **49.** $\frac{1}{2}$

51. $\frac{(y+3)(y^2+1)}{y+1}$ **53.** $\frac{15}{8}$ **55.** $\frac{15}{4}$ **57.** $\frac{a-5}{3(a-1)}$

59. $2x+1$ **61.** $\frac{(x+2)^2}{x}$ **63.** $\frac{3}{2}$ **65.** $\frac{c+1}{c-1}$

67. $\frac{y-3}{2y-1}$ **69.** $\frac{1}{(c-5)^2}$ **71.** $\frac{t+5}{t-5}$ **73.** 4

75. $8x^3 - 11x^2 - 3x + 12$ **77.** ◈

79. $\frac{a}{(c-3d)(2a+5b)}$ **81.** $-\frac{1}{b^2}$ **83.** x **85.** $\frac{4}{x+7}$

87. $\frac{(t-1)(t-9)(t-9)}{(t^2+9)(t+1)}$ **89.** $\frac{3(y+2)^3}{y(y-1)}$

Exercise Set 6.3, pp. 272–273

1. $\frac{8}{x}$ **3.** $\frac{3x+1}{15}$ **5.** $\frac{6}{a+3}$ **7.** $\frac{4}{a+6}$ **9.** $\frac{y+4}{y}$

11. $\frac{11x+8}{x+1}$ **13.** $\frac{7x+2}{x+1}$ **15.** $a+5$ **17.** $x-4$

19. $t+7$ **21.** $\frac{1}{x+2}$ **23.** $\frac{a+1}{a+6}$ **25.** $\frac{t-4}{t+3}$

27. $\frac{x+6}{x-5}$ **29.** $\frac{x}{4}$ **31.** $-\frac{1}{t}$ **33.** $\frac{-x+7}{x-6}$ **35.** $\frac{4a}{3}$

37. $\frac{13}{a}$ **39.** $\frac{4x-5}{4}$ **41.** $y+3$ **43.** $\frac{2b-14}{b^2-16}$

45. $\frac{-5}{t-4}$ **47.** $\frac{x-2}{x-7}$ **49.** $\frac{2x-16}{x^2-16}$ **51.** $\frac{2x-4}{x-9}$

53. $\frac{-4}{x-1}$

55.

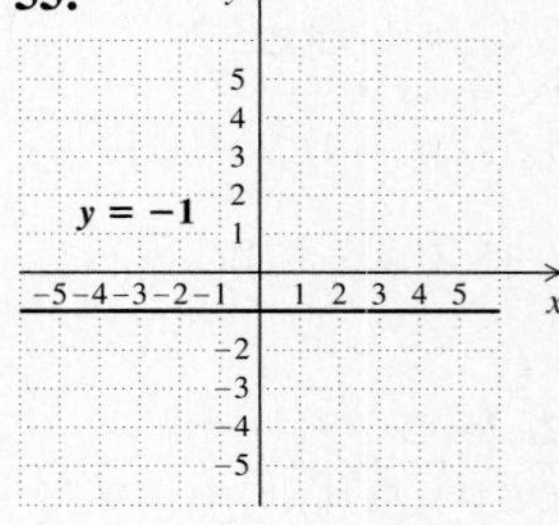

57.

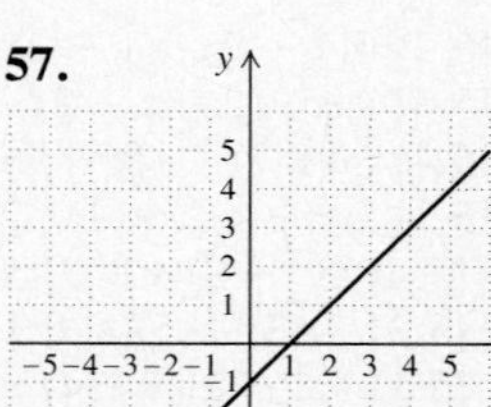

59. ◈ **61.** $\frac{18x+5}{x-1}$ **63.** 0 **65.** $\frac{20}{2y-1}$ **67.** 0
69. $\frac{x}{3x+1}$ **71.** ◈

Exercise Set 6.4, pp. 277–278

1. 108 **3.** 72 **5.** 126 **7.** 360 **9.** 420 **11.** $\frac{65}{72}$
13. $\frac{161}{600}$ **15.** $\frac{151}{180}$ **17.** $12x^3$ **19.** $18x^2y^2$
21. $6(y-3)$ **23.** $t(t+2)(t-2)$
25. $(x+2)(x-2)(x+3)$ **27.** $t(t+2)^2(t-4)$
29. $18a^5b^6$ **31.** $30x^2y^2z^3$ **33.** $(a-1)^2(a+1)$
35. $(m-2)^2(m-3)$ **37.** $(2+3x)(2-3x)$
39. $10v(v+3)(v+4)$ **41.** $18x^3(x-2)^2(x+1)$
43. $6x^3(x+2)^2(x-2)$ **45.** $\frac{14}{12x^5}, \frac{x^2y}{12x^5}$
47. $\frac{12b}{8a^2b^2}, \frac{5a}{8a^2b^2}$
49. $\frac{(x+3)(x+1)}{(x+3)(x+2)(x-2)}, \frac{(x-2)^2}{(x+3)(x+2)(x-2)}$
51. $\frac{3(t+2)(t-2)}{t(t+2)(t-2)}, \frac{4t(t-2)}{t(t+2)(t-2)}, \frac{t^2(t+2)}{t(t+2)(t-2)}$
53. $\frac{(x+1)(2x+3)}{(2x-3)(2x+3)}, \frac{x-2}{(2x+3)(2x-3)}, \frac{(x+1)(2x-3)}{(2x+3)(2x-3)}$
55. $(x-4)(x-15)$ **57.** $x^2-9x+18$ **59.** ◈
61. 1440 **63.** 24 min

Exercise Set 6.5, pp. 283–285

1. $\frac{2x+5}{x^2}$ **3.** $\frac{-1}{24r}$ **5.** $\frac{4x+6y}{x^2y^2}$ **7.** $\frac{4-3t}{18t^3}$
9. $\frac{5x+9}{24}$ **11.** $\frac{-x-4}{6}$ **13.** $\frac{a^2+16a+16}{16a^2}$
15. $\frac{7z-12}{12z}$ **17.** $\frac{x^2+4xy+y^2}{x^2y^2}$ **19.** $\frac{4x^2-13xt+9t^2}{3x^2t^2}$
21. $\frac{6x}{(x-2)(x+2)}$ **23.** $\frac{2x-40}{(x+5)(x-5)}$ **25.** $\frac{11x+2}{3x(x+1)}$
27. $\frac{3-5t}{2t(t-1)}$ **29.** $\frac{x^2+6x}{(x-4)(x+4)}$ **31.** $\frac{16}{3(z+4)}$
33. $\frac{3x-1}{(x-1)^2}$ **35.** $\frac{-t-9}{(t+3)(t-3)}$ **37.** $\frac{11a}{10(a-2)}$
39. $\frac{-2a^2}{(x+a)(x-a)}$ **41.** $\frac{2x^2+8x+16}{x(x+4)}$
43. $\frac{x-3}{(x+1)(x+3)}$ **45.** $\frac{x^2+5x+1}{(x+1)^2(x+4)}$
47. $\frac{x^2-48}{(x+7)(x+8)(x+6)}$ **49.** $\frac{3x^2+19x-20}{(x+3)(x-2)^2}$
51. $\frac{y^2+10y+11}{(y+7)(y-7)}$ **53.** $\frac{13x+20}{(4-x)(4+x)}$
55. $\frac{-a-2}{(a+1)(a-1)}$ **57.** $\frac{10x+6y}{(x+y)(x-y)}$ **59.** $\frac{2}{y(y-1)}$
61. $\frac{z-3}{2z-1}$ **63.** $\frac{-3x+1}{(2x-3)(x+1)}$ **65.** $\frac{1}{2c-1}$
67. $\frac{2}{x+y}$
69.

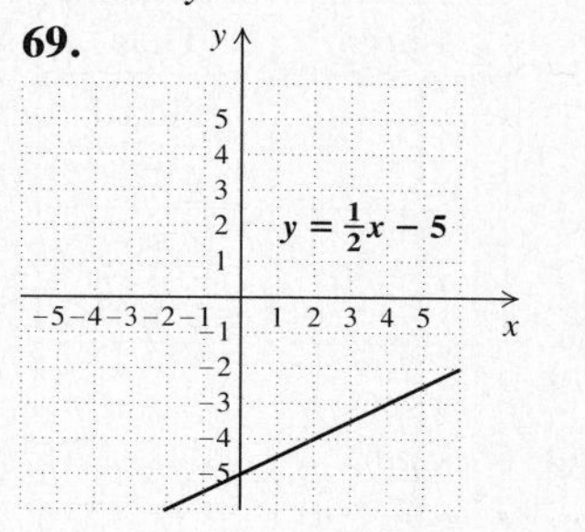

71.

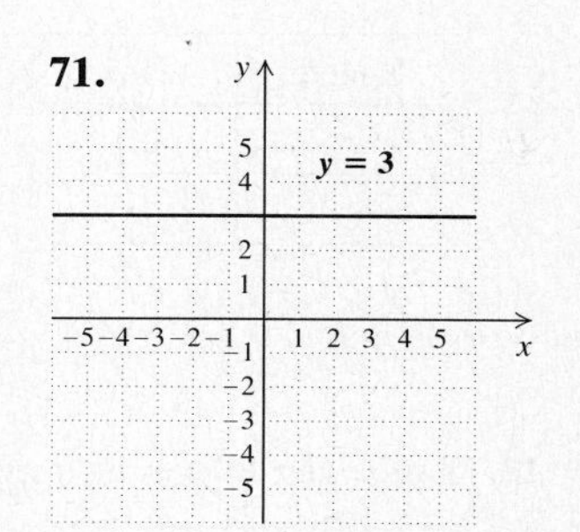

73. ◈ **75.** Perimeter: $\frac{16y+28}{15}$; area: $\frac{y^2+2y-8}{15}$
77. $\frac{(z+6)(2z-3)}{(z+2)(z-2)}$ **79.** $\frac{-3xy-3a+6x}{(a+2x)(a-2x)(y-3)^2}$
81. $\frac{a}{a-b}+\frac{3b}{b-a}$; answers may vary

Exercise Set 6.6, pp. 289–290

1. $\frac{25}{4}$ **3.** $\frac{1}{3}$ **5.** $\frac{1+3x}{1-5x}$ **7.** -6 **9.** $\frac{5}{3y^2}$ **11.** 8
13. $x-8$ **15.** $\frac{y}{y-1}$ **17.** $-\frac{1}{a}$ **19.** $\frac{x+y}{x}$
21. $\frac{3m^2+2}{4m-3m^2}$ **23.** $\frac{10-3x^2}{12x^2+6}$ **25.** $\frac{3a+8b}{15b^2-2}$
27. $\frac{2y^2+3xy}{2x+y^2}$ **29.** $\frac{3a^4-2}{2a^4+3a}$ **31.** $\frac{x^2+3}{x^2-2}$
33. $\frac{5x^3y+3x}{3+x}$ **35.** $\frac{x+5}{2x-3}$ **37.** $\frac{x-2}{x-3}$
39. $\frac{a^2+5a-3}{a^2-3a+5}$ **41.** $23x^4+50x^3+23x^2-163x+41$
43. ◈ **45.** $\frac{(x-1)(3x-2)}{5x-3}$ **47.** $-\frac{ac}{bd}$ **49.** $\frac{3x+2}{2x+1}$

Exercise Set 6.7, p. 295

1. $\frac{47}{2}$ **3.** -6 **5.** $\frac{24}{7}$ **7.** $-4, -1$ **9.** $4, -4$
11. 3 **13.** $\frac{14}{3}$ **15.** 10 **17.** 5 **19.** $\frac{5}{2}$ **21.** -1
23. $\frac{17}{2}$ **25.** No solution **27.** -5 **29.** $\frac{5}{3}$ **31.** $\frac{1}{2}$
33. No solution **35.** -13

37. $a^{-6}b^{-15}$, or $\frac{1}{a^6b^{15}}$ **39.** $\frac{16x^4}{t^8}$ **41.** ◈ **43.** 7
45. 3 **47.** 2, −2 **49.** 4 **51.** ▩

Exercise Set 6.8, pp. 303–306

1. −1, 2 **3.** 1 **5.** $2\frac{2}{9}$ hr **7.** $25\frac{5}{7}$ min **9.** $5\frac{1}{7}$ hr
11. $3\frac{3}{7}$ hr
13. 30 km/h, 70 km/h

Speed	Time
r	t
$r + 40$	t

Speed	Time
r	$\frac{150}{r}$
$r + 40$	$\frac{350}{r + 40}$

15. Passenger: 80 km/h; freight: 66 km/h

Speed	Time
$r - 14$	t
r	t

Speed	Time
$r - 14$	$\frac{330}{r - 14}$
r	$\frac{400}{r}$

17. $1\frac{7}{17}$ hr **19.** 3 hr **21.** 9 **23.** 2.3 km/h
25. 582 **27.** 702 km **29.** 1.92 g **31.** 10.5 **33.** $\frac{8}{3}$
35. 6.25 **37.** 287 **39.** 2074 **41.** 20
43. **(a)** 1.92 tons; **(b)** 28.8 lb **45.** $\frac{36}{68}$ **47.** 1
49. $13y^3 - 14y^2 + 12y - 73$ **51.** ◈ **53.** $\frac{3}{4}$
55. 2 mph **57.** $\frac{A}{C} = \frac{B}{D}$; $\frac{D}{B} = \frac{C}{A}$; $\frac{D}{C} = \frac{B}{A}$ **59.** $9\frac{3}{13}$ days
61. 45 mph **63.** $66\frac{2}{3}$ ft

Exercise Set 6.9, pp. 309–310

1. $r = \frac{S}{2\pi h}$ **3.** $b = \frac{2A}{h}$ **5.** $n = \frac{s}{180} + 2$, or $n = \frac{s + 360}{180}$
7. $b = \frac{3V - kB - 4kM}{k}$ **9.** $r = \frac{L}{l - S}$ **11.** $h = \frac{2A}{b_1 + b_2}$
13. $a = \frac{d}{b - c}$ **15.** $p = \frac{r}{q}$ **17.** $d = \frac{c}{a + b}$
19. $z = \frac{x - y}{p + q}$ **21.** $f = \frac{pq}{q + p}$ **23.** $p = \frac{ar}{v^2L}$
25. $n = \frac{a}{c(1 + b)}$ **27.** $b = \frac{a}{3S - 2}$ **29.** $F = \frac{9C + 160}{5}$
31. $g = \frac{mf + t}{m}$ **33.** $m = \frac{-t}{f - g}$, or $m = \frac{t}{g - f}$

35. $a = \frac{Kb}{C - 1}$ **37.** $-\frac{3}{4}$ **39.** $(x + 2)(x - 15)$ **41.** ◈
43. $T = \frac{FP}{u + EF}$ **45.** −40°

Review Exercises: Chapter 6, pp. 312–313

1. [6.1] 0 **2.** [6.1] 6 **3.** [6.1] −6, 6 **4.** [6.1] −6, 5
5. [6.1] −2 **6.** [6.1] 0, 3, 5 **7.** [6.1] $\frac{x - 2}{x + 1}$
8. [6.1] $\frac{7x + 3}{x - 3}$ **9.** [6.1] $\frac{y - 5}{y + 5}$ **10.** [6.2] $\frac{a - 6}{5}$
11. [6.2] $\frac{6}{2t - 1}$ **12.** [6.2] $-20t$ **13.** [6.2] $\frac{2x^2 - 2x}{x + 1}$
14. [6.2] $\frac{(x^2 + 1)(2x + 1)}{(x - 2)(x + 1)}$ **15.** [6.2] $\frac{(t + 4)^2}{t + 1}$
16. [6.4] $30x^2y^2$ **17.** [6.4] $x^4(x + 1)(x - 1)$
18. [6.4] $(y - 2)(y + 2)(y + 1)$ **19.** [6.3] $\frac{-3x + 18}{x + 7}$
20. [6.5] −1 **21.** [6.3] $\frac{4}{x - 4}$ **22.** [6.5] $\frac{x + 5}{2x}$
23. [6.3] $\frac{2x + 3}{x - 2}$ **24.** [6.5] $\frac{2a}{a - 1}$ **25.** [6.3] $d + c$
26. [6.5] $\frac{-x^2 + x + 26}{(x - 5)(x + 5)(x + 1)}$ **27.** [6.5] $\frac{2(x - 2)}{x + 2}$
28. [6.5] $\frac{8x + 3}{2x(2x + 1)}$ **29.** [6.6] $\frac{z}{1 - z}$
30. [6.6] $\frac{2x^4y^2 + x^3}{y + xy}$ **31.** [6.6] $c - d$ **32.** [6.7] 8
33. [6.7] $-\frac{1}{2}$ **34.** [6.7] 3, −5 **35.** [6.8] $5\frac{1}{7}$ hr
36. [6.8] 240 km/h, 280 km/h **37.** [6.8] −2
38. [6.8] 160 **39.** [6.8] $x = 6$ **40.** [6.9] $s = \frac{rt}{r - t}$
41. [6.9] $C = \frac{5}{9}(F - 32)$, or $C = \frac{5}{9}F - \frac{160}{9}$ **42.** [2.1] 11
43. [3.3] (3, 0), (0, −6)

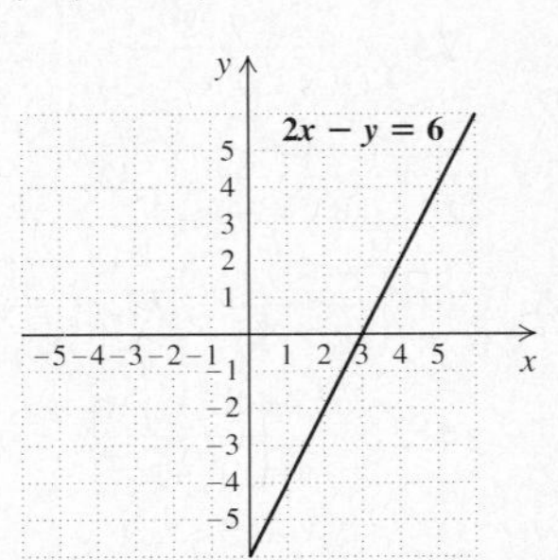

44. [5.2] $(x + 12)(x - 4)$
45. [4.3] $-2x^3 + 3x^2 + 12x - 18$

46. [6.5], [6.6], [6.7] ◈ A student should master factoring before beginning a study of rational equations because it is necessary to factor when finding the LCD of the rational expressions. It may also be necessary to factor to use the principle of zero products after fractions have been cleared.
47. [6.5] ◈ Although multiplying the denominators of the expressions being added results in a common denominator, it is often not the *least* common denominator. Using a common denominator other than the LCD makes the expressions more complicated, requires additional simplifying after the addition has been performed, and leaves more room for error.
48. [6.2] $\frac{5(a+3)^2}{a}$ **49.** [6.3] $\frac{10a}{(a-b)(b-c)}$

Test: Chapter 6, p. 314

1. [6.1] 0 **2.** [6.1] -8 **3.** [6.1] -7, 7
4. [6.1] 1, 2 **5.** [6.1] 1 **6.** [6.1] 0, -3, -5
7. [6.1] $\frac{3x+7}{x+3}$ **8.** [6.2] $\frac{a+5}{2}$ **9.** [6.2] $\frac{(5x+1)(x+1)}{3x(x+2)}$
10. [6.4] $(y-3)(y+3)(y+7)$ **11.** [6.3] $\frac{23-3x}{x^3}$
12. [6.3] $\frac{8-2t}{t^2+1}$ **13.** [6.3] $\frac{-3}{x-3}$ **14.** [6.3] $\frac{2x-5}{x-3}$
15. [6.5] $\frac{8t-3}{t(t-1)}$ **16.** [6.5] $\frac{-x^2-7x-15}{(x+4)(x-4)(x+1)}$
17. [6.5] $\frac{x^2+2x-7}{(x-1)^2(x+1)}$ **18.** [6.6] $\frac{3y+1}{y}$
19. [6.6] $\frac{3a^2b^2-2a^3}{a^3b^2+2b^2}$ **20.** [6.7] 12 **21.** [6.7] -3, 5
22. [6.8] 4 **23.** [6.8] 16 **24.** [6.8] 45 km/h, 65 km/h
25. [6.9] $t=\frac{d}{r+w}$ **26.** [2.1] $-\frac{3}{7}$
27. [3.3] (10, 0), (0, 4)

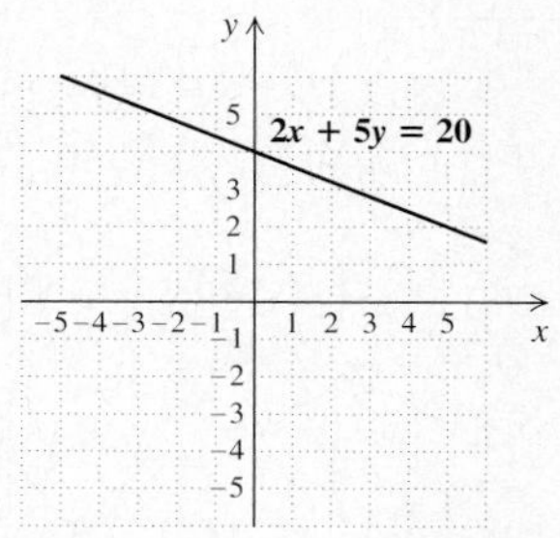

28. [5.2] $(x-9)(x+5)$ **29.** [4.3] $13x^2-29x+76$
30. [6.8] Reggie: 10 hr; Rema: 4 hr **31.** [6.6] $\frac{3a+2}{2a+1}$

CUMULATIVE REVIEW: 1–6

1. [1.2] $2b+a$ **2.** [1.4] $-3.1 > -3.15$ **3.** [1.8] 49
4. [1.8] $-8x+28$ **5.** [1.5] $-\frac{43}{8}$ **6.** [1.7] 1
7. [1.7] -6.2 **8.** [1.8] 8 **9.** [2.2] 10
10. [2.2] -3 **11.** [2.1] $\frac{9}{2}$ **12.** [2.2] -2
13. [2.2] $\frac{8}{3}$ **14.** [2.2] $-\frac{1}{4}$ **15.** [2.2] -8
16. [2.1] $-\frac{1}{2}$ **17.** [2.6] $\{y \mid y \leq -\frac{2}{3}\}$ **18.** [5.6] $\frac{4}{3}$, $-\frac{5}{2}$
19. [5.6] $\frac{1}{2}$, -4 **20.** [5.6] 4, -4 **21.** [6.7] 2
22. [6.7] 1 **23.** [6.7] -13 **24.** [6.9] $t=\frac{4b}{A}$
25. [6.9] $n=\frac{tm}{t-m}$ **26.** [6.9] $c=\frac{a-b}{r}$
27. [1.6] $\frac{3}{2}x+2y-3z$ **28.** [4.2] $-4x^3-\frac{1}{7}x^2-2$
29. [3.2]

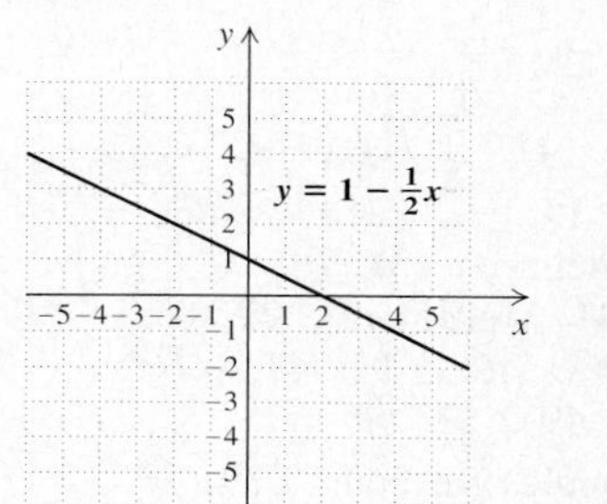

30. [3.3]

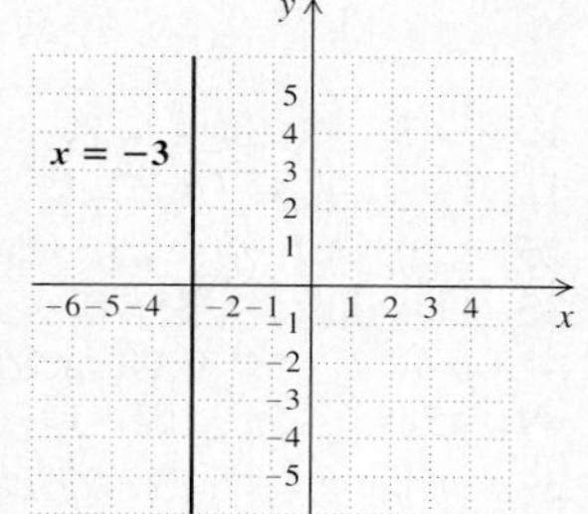

31. [3.3]

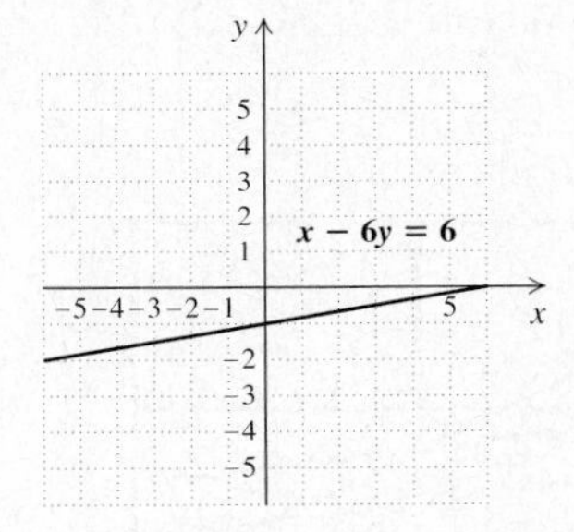

32. [4.1] x^{10} **33.** [4.8] z^{11} **34.** [4.1] $-27x^6y^3$
35. [4.3] $-y^3-2y^2-2y+7$
36. [1.2] $12x+16y+4z$ **37.** [4.5] a^2-9
38. [4.4] $2x^5+x^3-6x^2-x+3$
39. [4.5] $36x^2-60xy+25y^2$
40. [4.5] $6x^7-12x^5+9x^2-18$ **41.** [4.5] $4x^6-1$
42. [5.1] $2x(3-x-12x^3)$ **43.** [5.4] $(4x+9)(4x-9)$
44. [5.2] $(x-6)(x-4)$ **45.** [5.3] $(2x+1)(4x+3)$
46. [5.3] $2(3x-2)(x-4)$ **47.** [5.4] $2(x+3)(x-3)$
48. [5.4] $(4x+5)^2$ **49.** [5.3] $(3x-2)(x+4)$
50. [5.1] $(x^3-3)(x+2)$ **51.** [6.2] $\frac{y-6}{2}$
52. [6.2] 1 **53.** [6.5] $\frac{a^2+7ab+b^2}{a^2-b^2}$ **54.** [6.3] $\frac{-2x-5}{x-4}$

55. [6.6] $\dfrac{x}{x-2}$ **56.** [6.6] $\dfrac{t+2t^3}{t^3-2}$

57. [4.7] $5x^2 - 4x + 2 + \dfrac{2}{3x} + \dfrac{6}{x^2}$

58. [4.7] $15x^3 - 57x^2 + 177x - 529 + \dfrac{1605}{x+3}$

59. [2.5] -278 and -276 **60.** [2.4] 544.32
61. [2.5] -15 **62.** [2.7] $\{s \mid s \leq 225\}$ **63.** [5.7] 14 ft
64. [6.8] 40 km/h, 50 km/h **65.** [6.8] $3\frac{3}{7}$ hr
66. [4.3], [4.5] 12 **67.** [1.4], [2.2] -144, 144
68. [4.5] $16y^6 - y^4 + 6y^2 - 9$
69. [5.4] $2(a^{16} + 81b^{20})(a^8 + 9b^{10})(a^4 + 3b^5)(a^4 - 3b^5)$
70. [5.6] 4, -7, 12 **71.** [1.4], [1.6] -7

CHAPTER 7

Exercise Set 7.1, pp. 324–326

1. $\frac{2}{3}$ **3.** 1 **5.** $\frac{1}{3}$ **7.** 3 **9.** $-\frac{1}{2}$ **11.** $-\frac{3}{2}$ **13.** -2
15. Undefined **17.** $-\frac{3}{4}$ **19.** $-\frac{4}{5}$ **21.** 7 **23.** $-\frac{2}{3}$
25. -2 **27.** 0 **29.** Undefined **31.** 0
33. Undefined **35.** 0 **37.** Undefined **39.** 0
41. 6.7% **43.** 0.08 or 8% **45.** 3% **47.** 0.6 ft
49. $45x^2 - 15x$ **51.** $x^2 - 49$ **53.** ◈
55. $\{m \mid m \geq \frac{4}{3}\}$ **57.** 27 candles per hour **59.** ◈

Technology Connection, Section 7.2

TC1. $y_1 = -\frac{3}{4}x - 2$ $y_2 = -\frac{1}{5}x - 2$
$y_3 = -\frac{3}{4}x - 5$ $y_4 = -\frac{1}{5}x - 5$

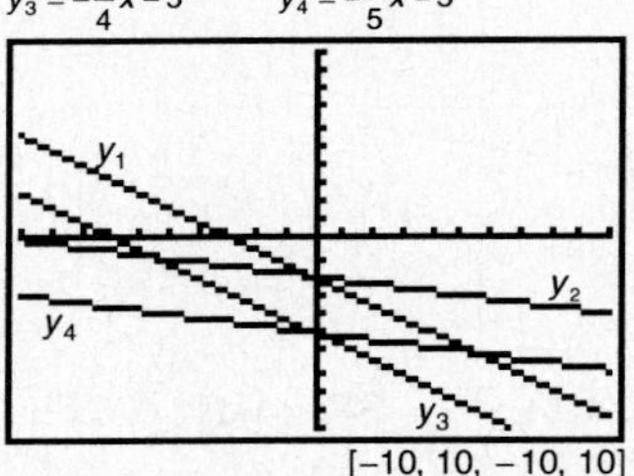

TC2. $y = 2x - 1$, $y = 2x + 4$, $y = -x + 3$, $y = 4x + 3$; Answers may vary.

Exercise Set 7.2, pp. 331–332

1.

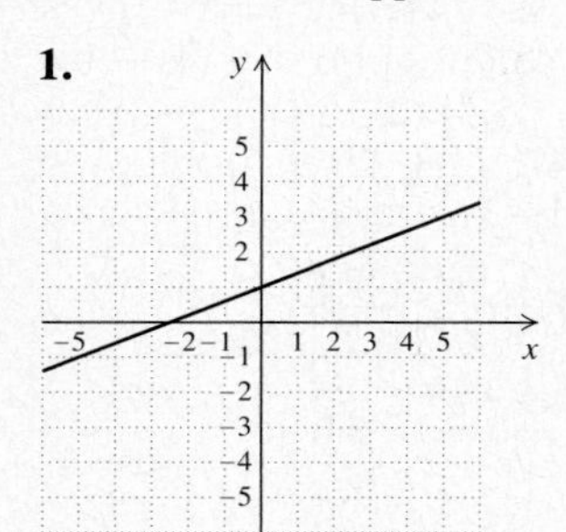

3.

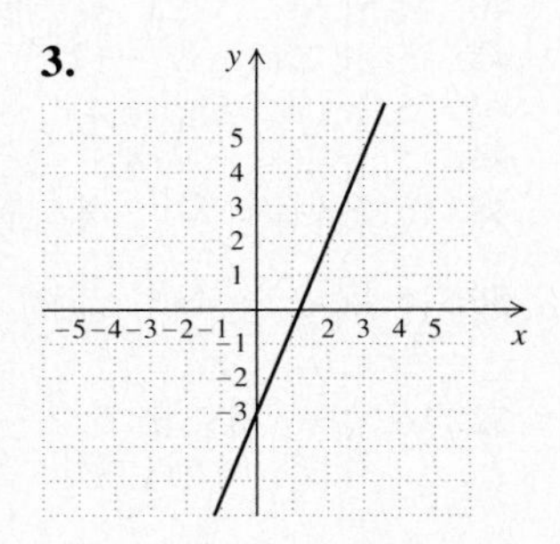

5.

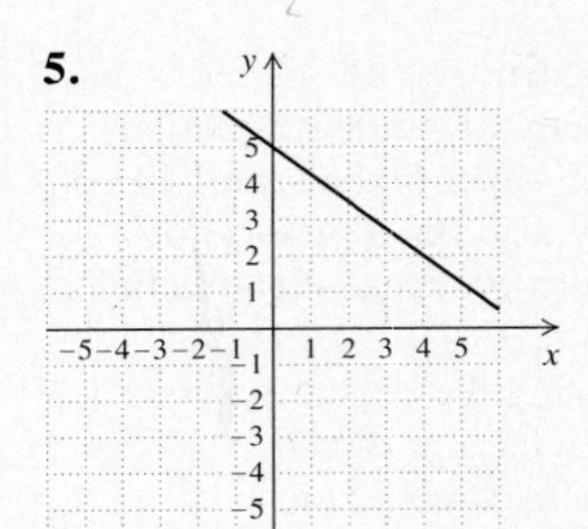

7.

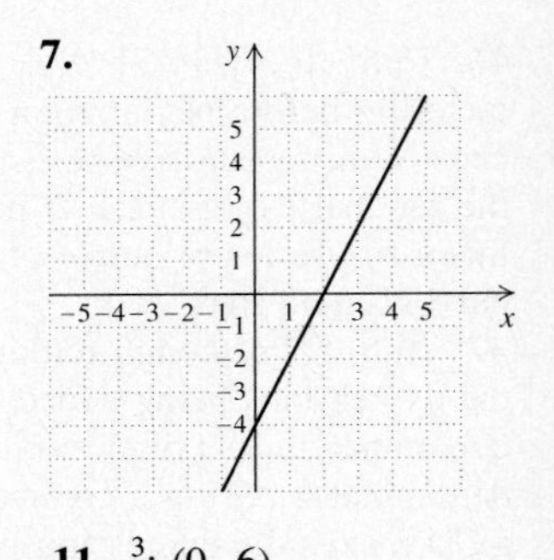

9.

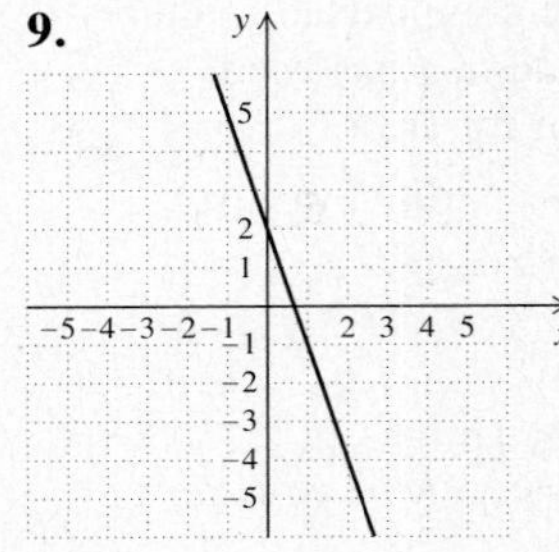

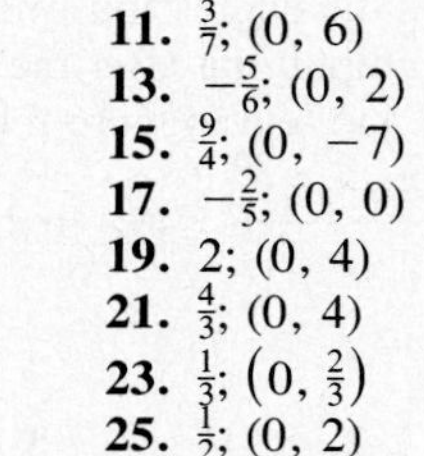

11. $\frac{3}{7}$; (0, 6)
13. $-\frac{5}{6}$; (0, 2)
15. $\frac{9}{4}$; (0, -7)
17. $-\frac{2}{5}$; (0, 0)
19. 2; (0, 4)
21. $\frac{4}{3}$; (0, 4)
23. $\frac{1}{3}$; $\left(0, \frac{2}{3}\right)$
25. $\frac{1}{2}$; (0, 2)
27. 0; (0, 5)

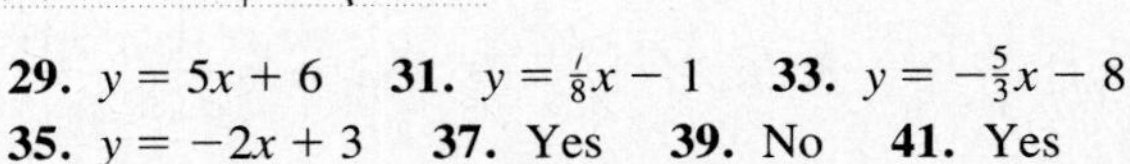

29. $y = 5x + 6$ **31.** $y = \frac{7}{8}x - 1$ **33.** $y = -\frac{5}{3}x - 8$
35. $y = -2x + 3$ **37.** Yes **39.** No **41.** Yes

43.

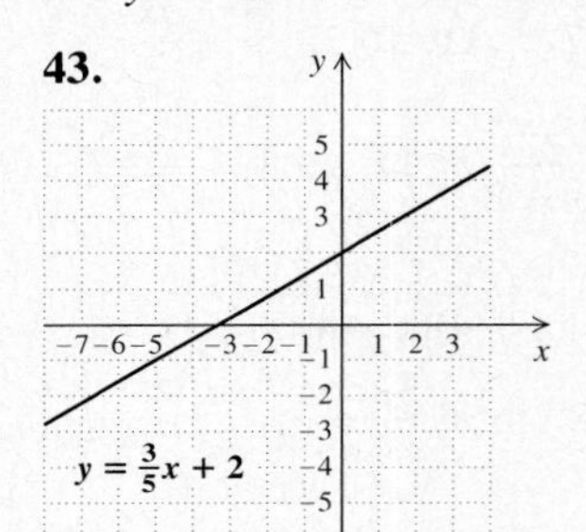

45.

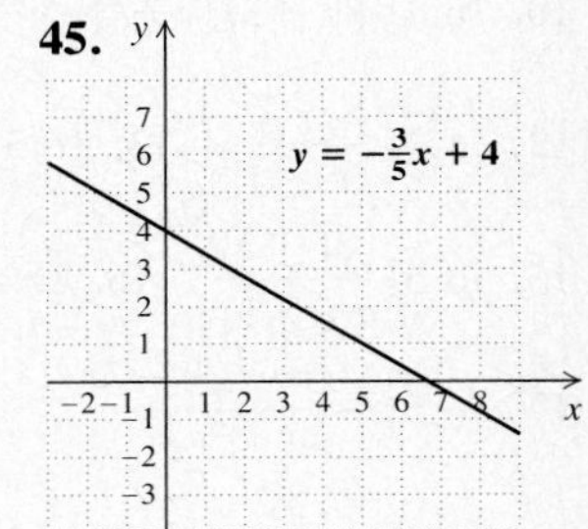

47.

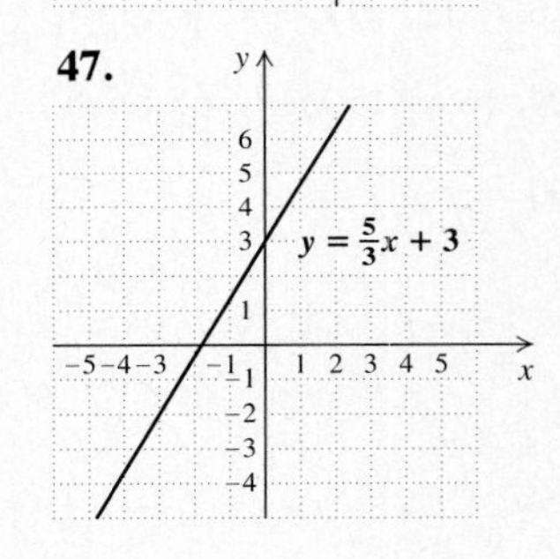

49.

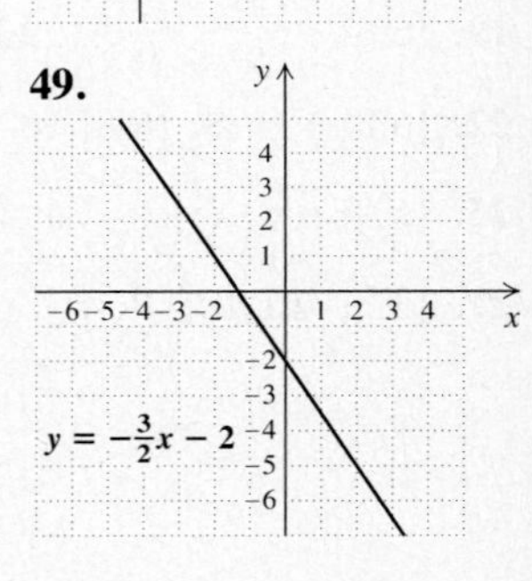

51.

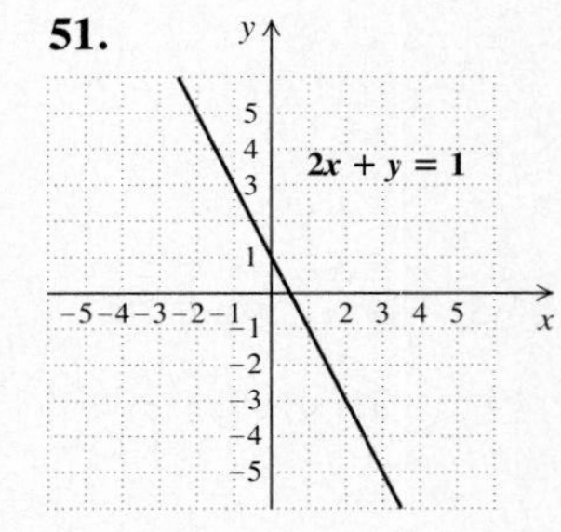

53.

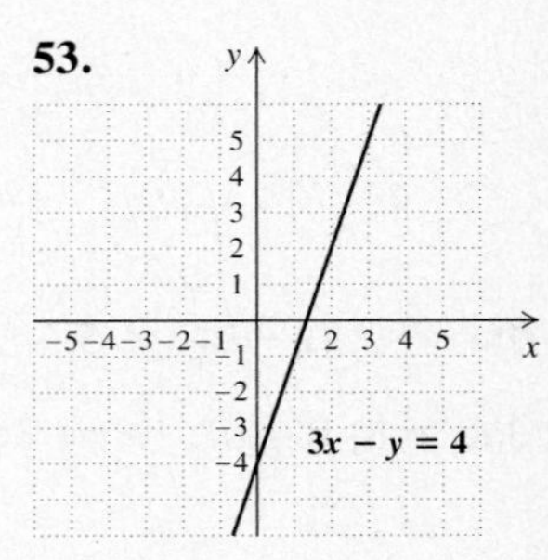

55.

$2x + 3y = 9$

57.

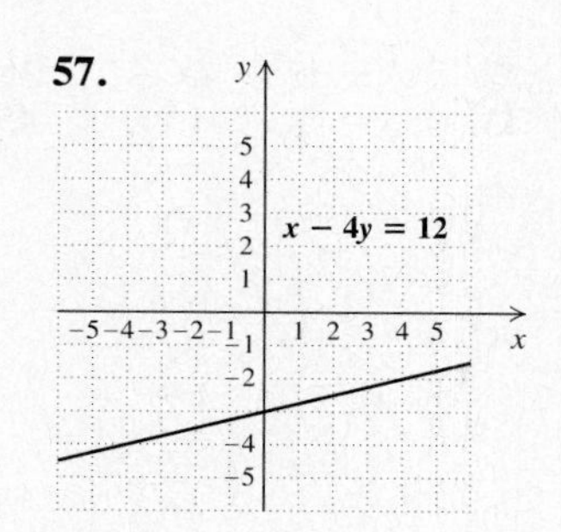

59.

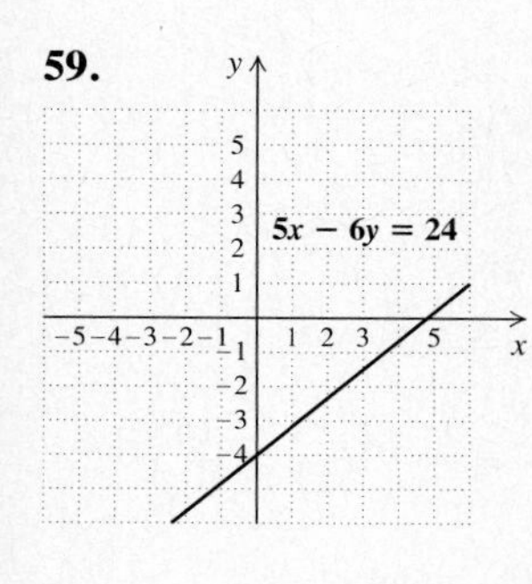

61. 0, −3
63. 13 and 15, −15 and −13 **65.** ◈
67. Yes **69.** No
71. Yes
73. When $x = 0$, $y = b$, so $(0, b)$ is on the line. When $x = 1$, $y = m + b$, so $(1, m + b)$ is on the line. Then

$$\text{slope} = \frac{(m + b) - b}{1 - 0} = m.$$

75. $y = \frac{3}{2}x - 2$

Exercise Set 7.3, pp. 335–336

1. $y - 5 = 5(x - 2)$ **3.** $y - 4 = \frac{3}{4}(x - 2)$
5. $y - (-6) = 1 \cdot (x - 2)$ **7.** $y - 0 = -3(x - (-3))$
9. $y - 6 = \frac{2}{3}(x - 5)$ **11.** $y = 2x + 1$ **13.** $y = -x + 9$
15. $y = \frac{1}{2}x + 4$ **17.** $y = -\frac{1}{3}x - 7$ **19.** $y = \frac{5}{4}x - 8$
21. $y = \frac{1}{4}x + \frac{5}{2}$ **23.** $y = -\frac{1}{2}x + 4$ **25.** $y = -\frac{3}{2}x + \frac{13}{2}$
27. $y = \frac{2}{5}x - 2$ **29.** $y = \frac{3}{4}x - \frac{5}{2}$

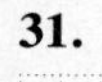
31.

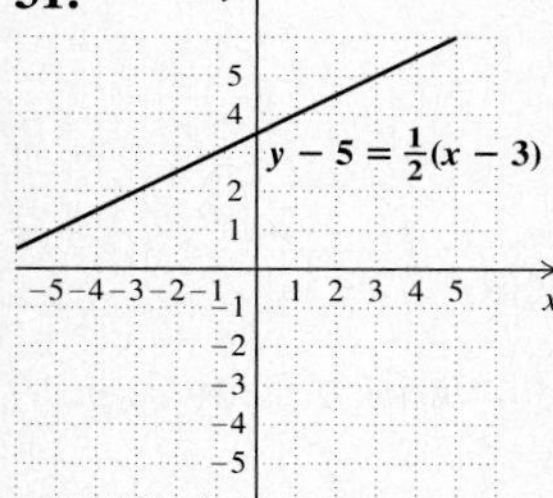

33.

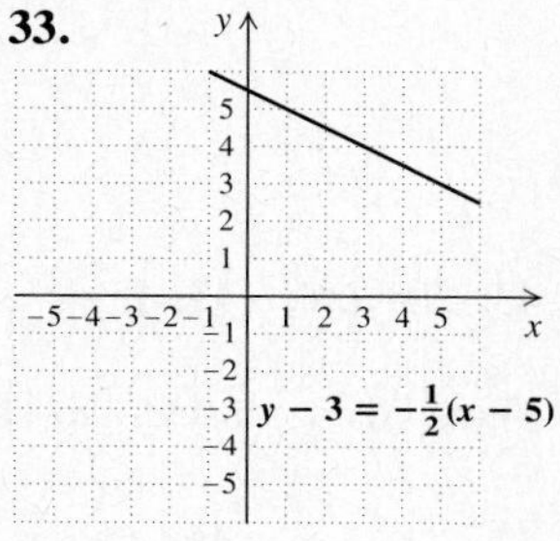

35.

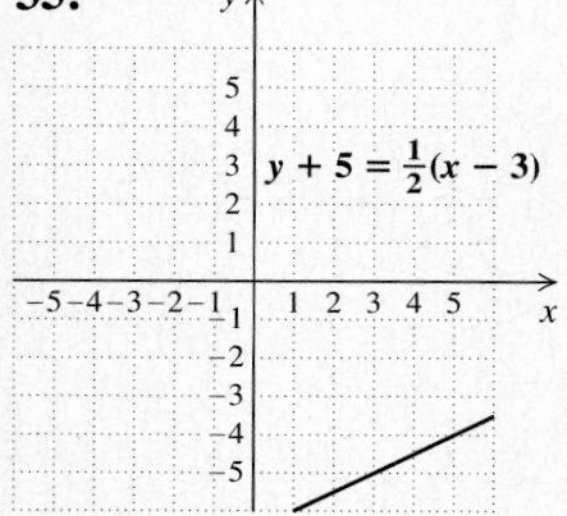

37.

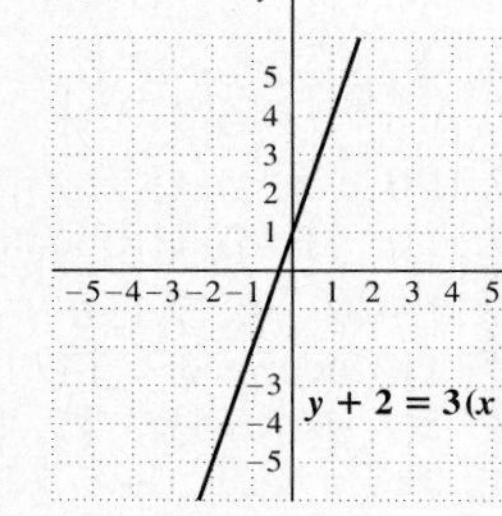
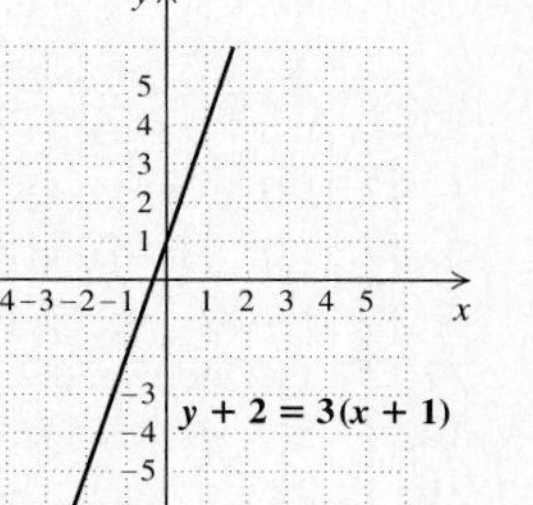

39.

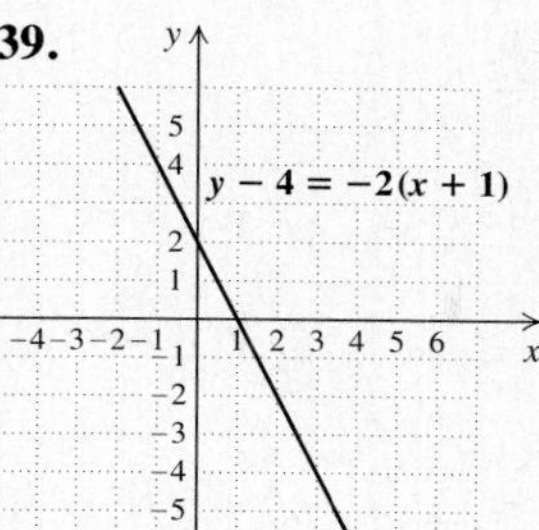

41.

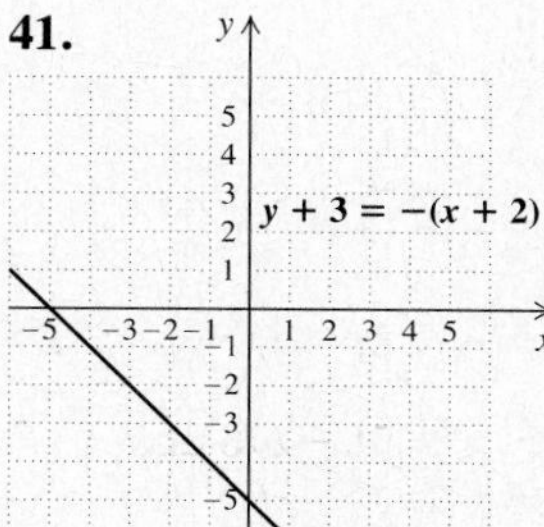

43. $7x^2y^2(x + 5y^4)$ **45.** $\frac{x + 1}{2x(x - 1)}$ **47.** ◈
49. $y = 3x - 9$ **51.** $y = \frac{1}{2}x + 1$

Exercise Set 7.4, p. 339

1. No **3.** No

5.

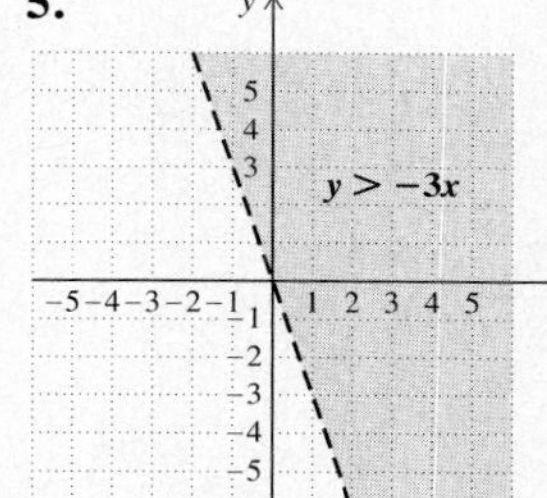

7.

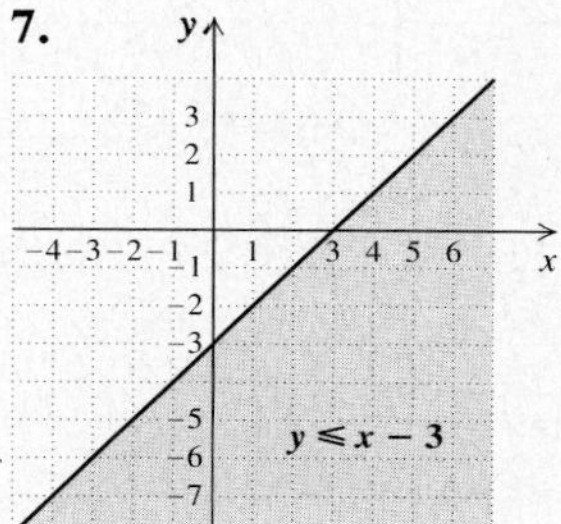

9.

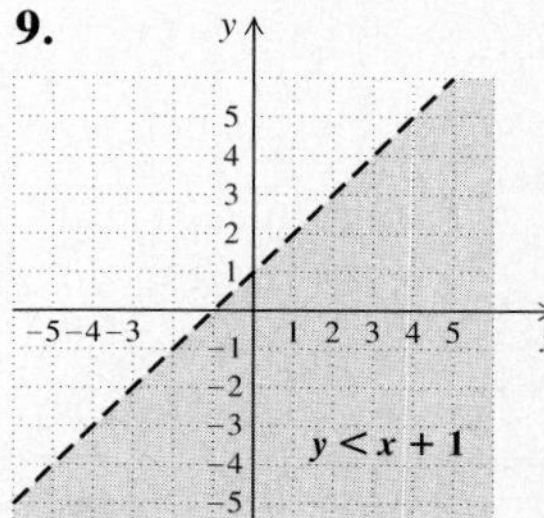

11.

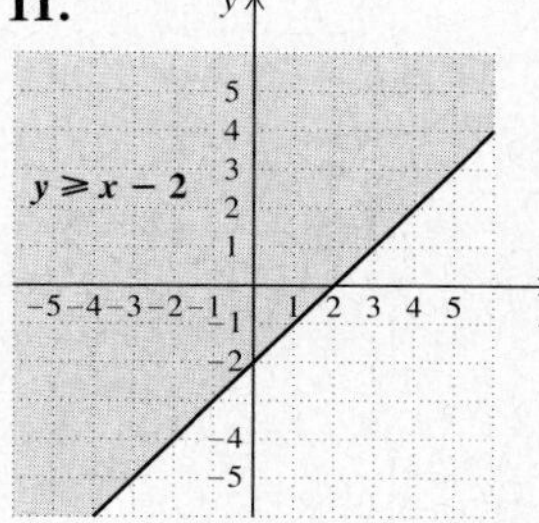

13.

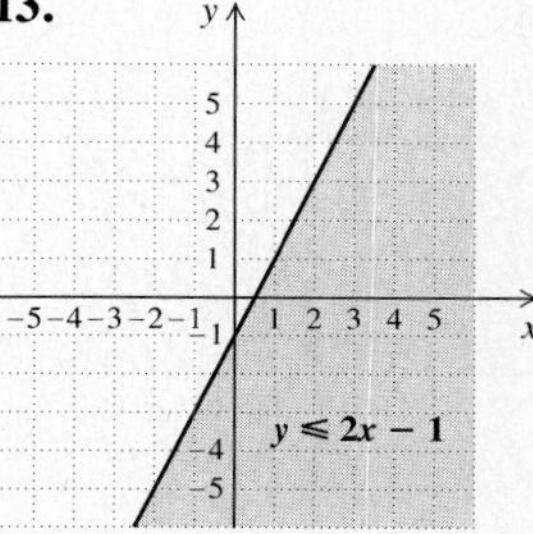

15.

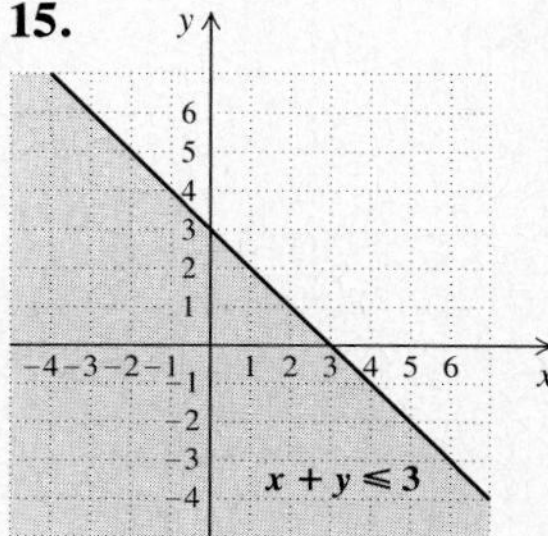

17.

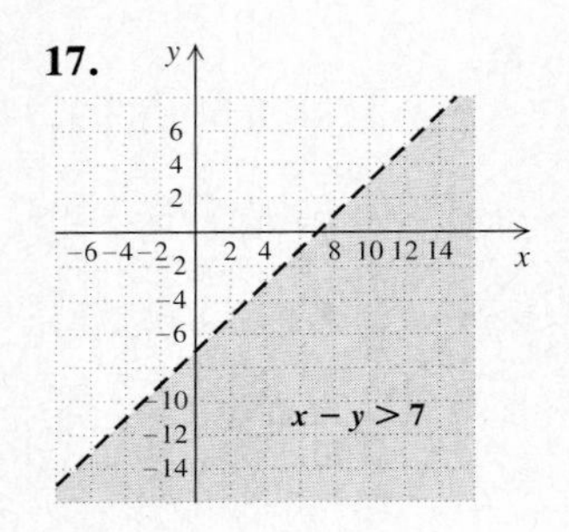

19.

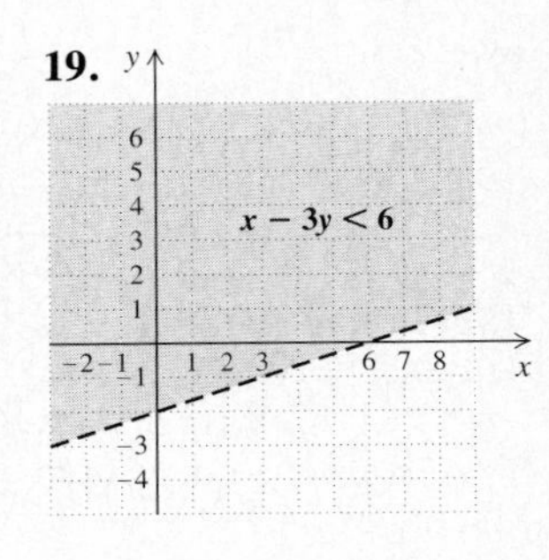

21.

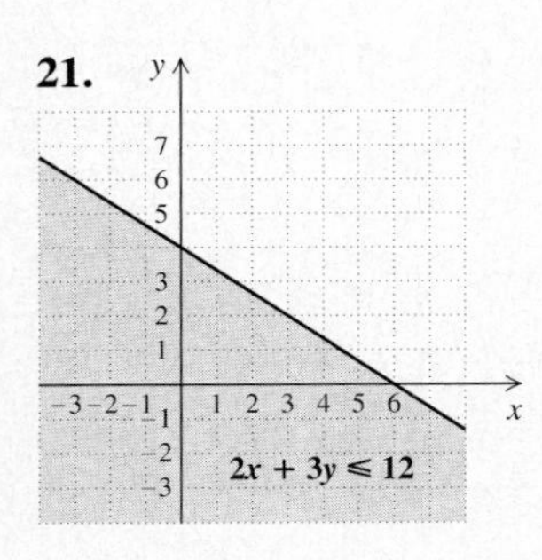

23.

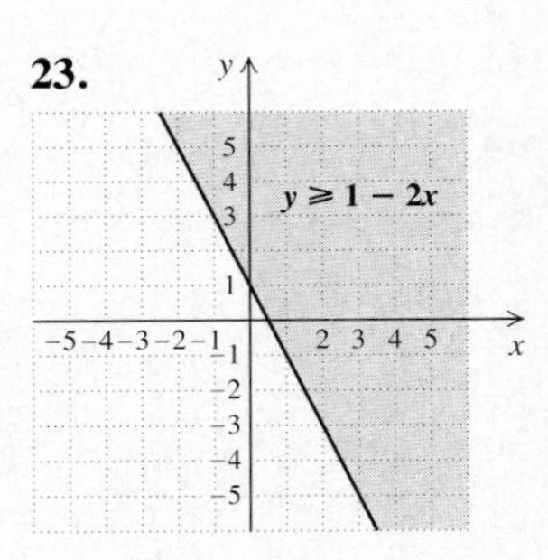

25.

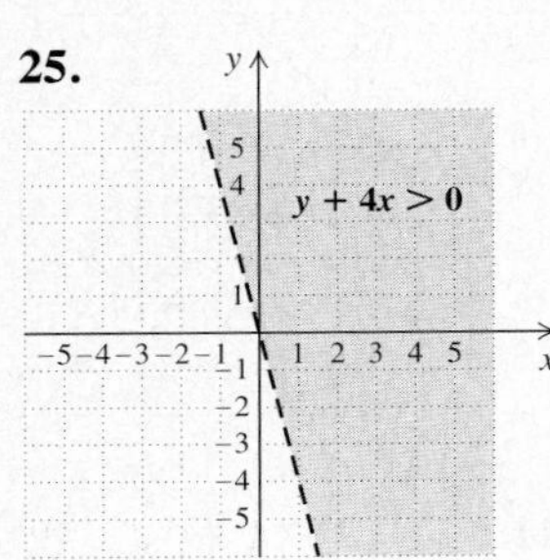

27.

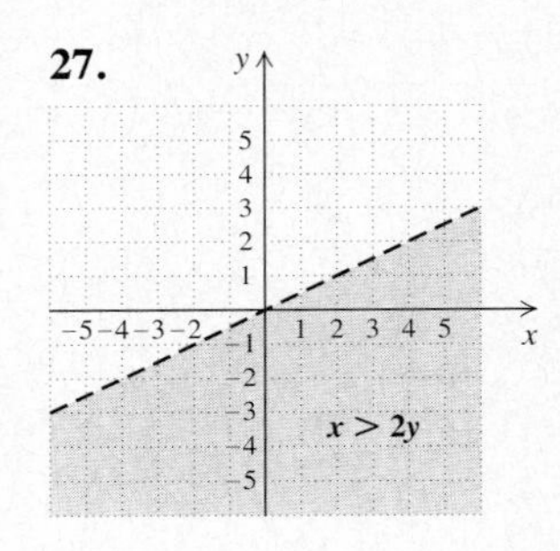

29.

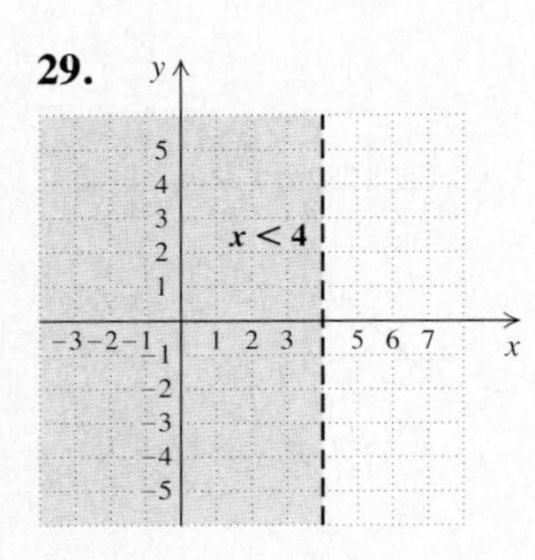

31.

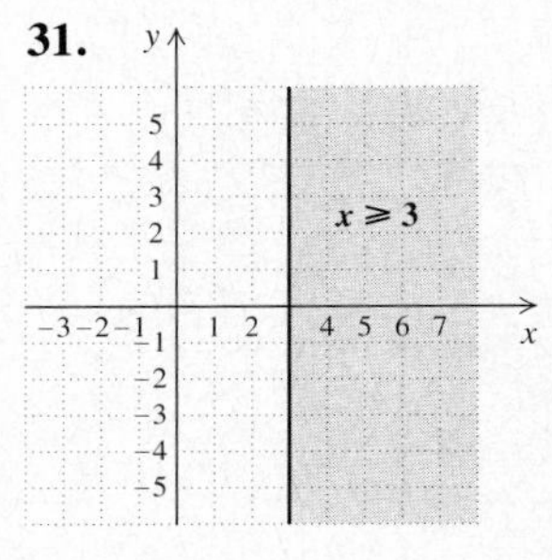

33.

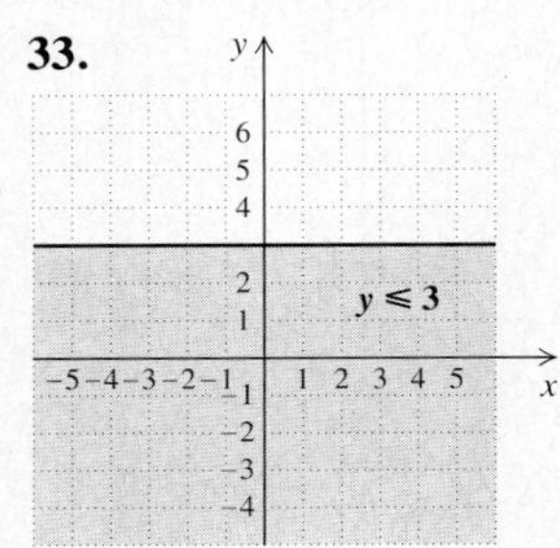

35.

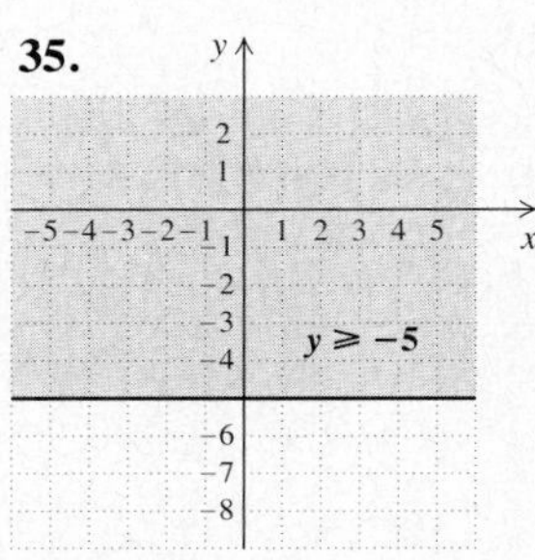

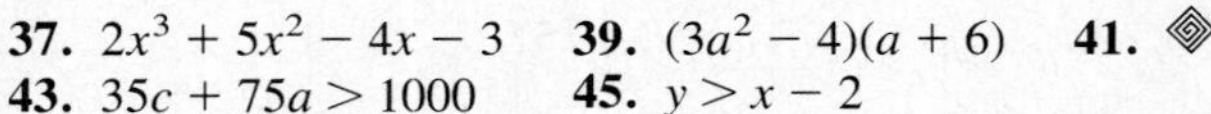

37. $2x^3 + 5x^2 - 4x - 3$ **39.** $(3a^2 - 4)(a + 6)$ **41.** ◈
43. $35c + 75a > 1000$ **45.** $y > x - 2$

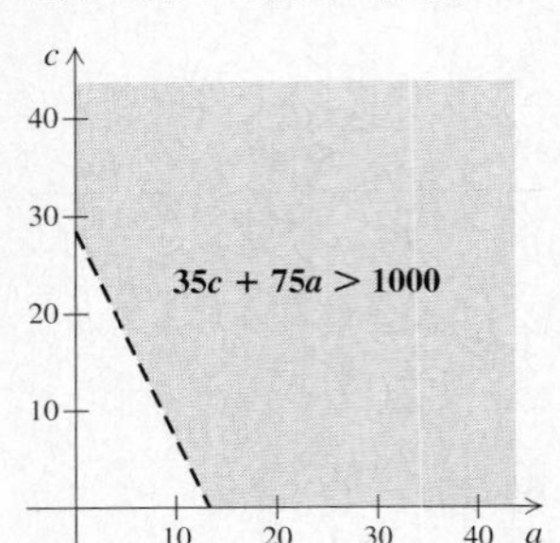

47.

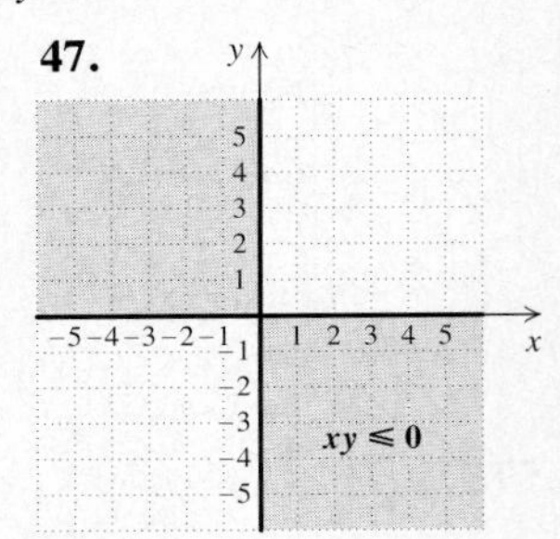

Exercise Set 7.5, pp. 344–346

1. $y = 4x$ **3.** $y = 1.75x$ **5.** $y = 3.2x$ **7.** $y = \frac{2}{3}x$
9. $y = \frac{75}{x}$ **11.** $y = \frac{80}{x}$ **13.** $y = \frac{1}{x}$ **15.** $y = \frac{1050}{x}$
17. $y = \frac{0.06}{x}$ **19.** **(a)** Direct; **(b)** about 69
21. **(a)** Inverse; **(b)** $5\frac{1}{3}$ hr **23.** \$183.75
25. $22\frac{6}{7}$ **27.** 320 cm³ **29.** $18.\overline{3}$ lb **31.** 2.4 ft
33. 46.7¢, 1.9¢ **35.** 10 **37.** ◈ **39.** ◈
41. $P = kS$, $k = 8$

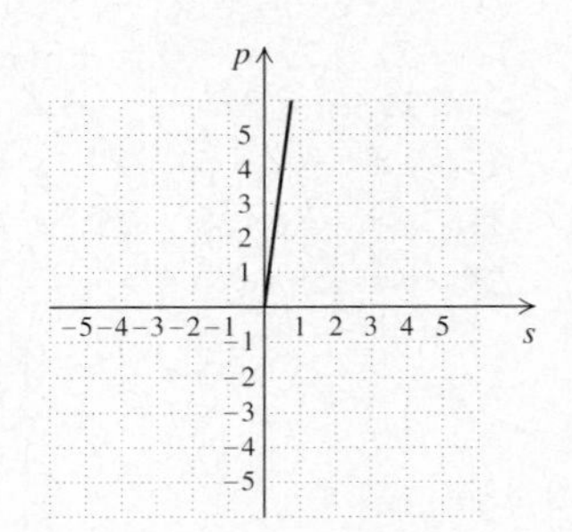

43. $B = kN$ **45.** If $p = kq$, then $q = \frac{1}{k}p$.
Since k is a constant, so is $\frac{1}{k}$, and q varies directly as p.
47. $S = kv^6$ **49.** $I = \frac{k}{d^2}$ **51.** $V = kr^3$ **53.** ◈

Review Exercises: Chapter 7, pp. 347–348

1. [7.1] 0 **2.** [7.1] $\frac{7}{3}$ **3.** [7.1] $-\frac{3}{7}$ **4.** [7.1] $\frac{3}{2}$
5. [7.1] 0 **6.** [7.1] Undefined **7.** [7.1] 2
8. [7.1] 7% **9.** [7.1] 0 **10.** [7.2] $\frac{3}{5}$ **11.** [7.2] −2
12. [7.1] Undefined **13.** [7.2] −9, (0, 46)
14. [7.2] −1, (0, 9) **15.** [7.2] $\frac{1}{3}$, $(0, -\frac{2}{3})$
16. [7.2] $y = -2x - 4$ **17.** [7.2] $y = 1.5x + 1$

18. [7.2]

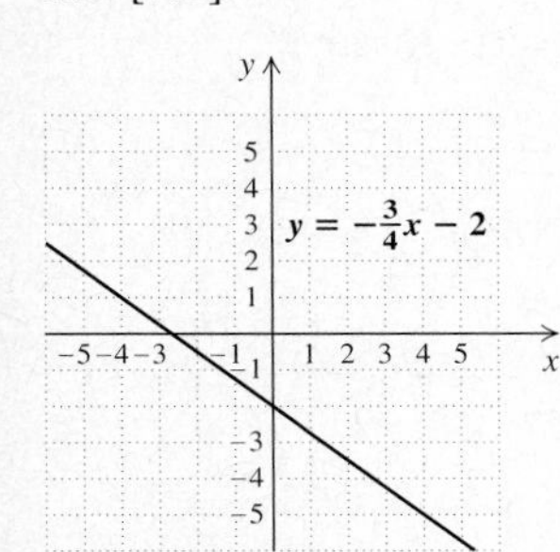

19. [7.2]

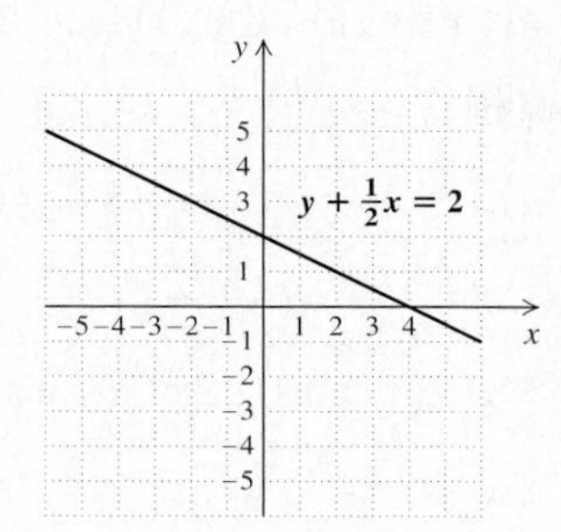

20. [7.3]

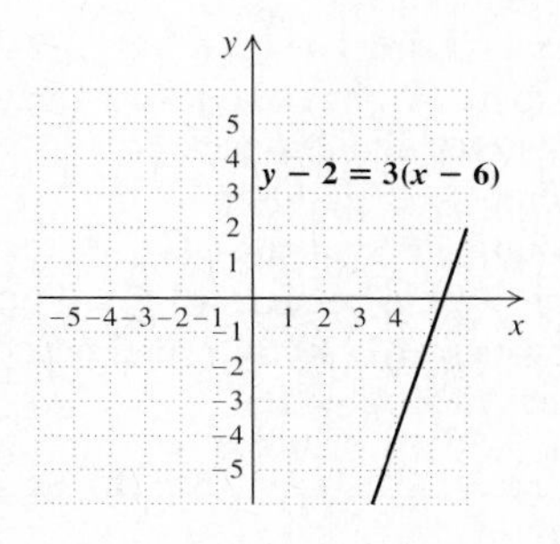

21. [7.2] Parallel **22.** [7.2] Not parallel
23. [7.3] $y - 2 = 3(x - 1)$ **24.** [7.3] $y - (-5) = \frac{2}{3}(x - (-2))$ **25.** [7.3] $y = x + 2$ **26.** [7.3] $y = \frac{1}{2}x - 1$
27. [7.4]

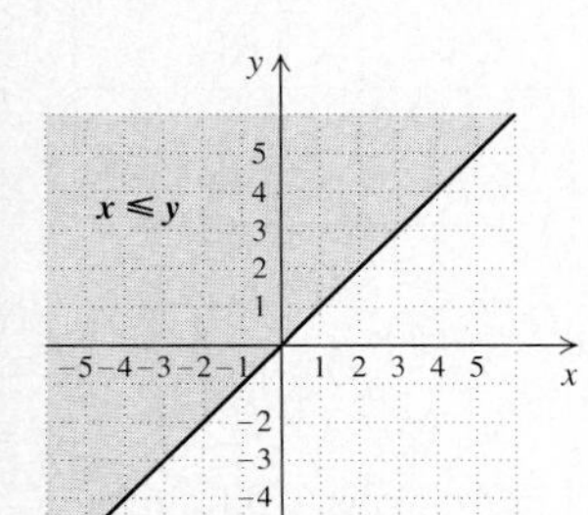

28. [7.4]

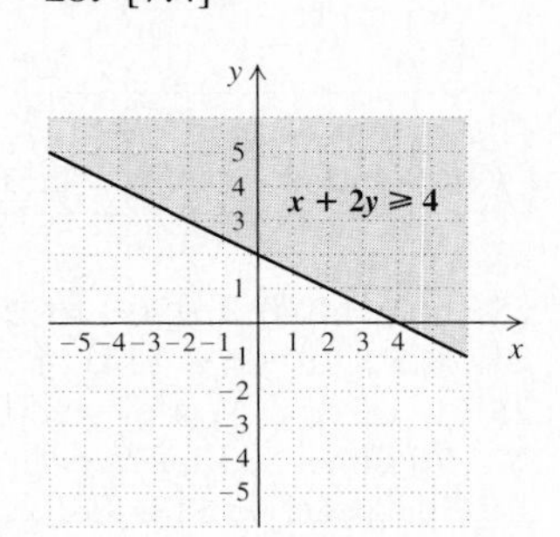

29. [7.4]

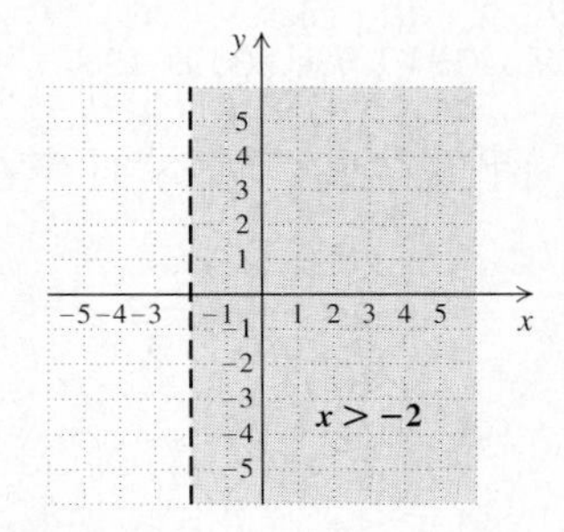

30. [7.5] $y = \dfrac{243}{x}$ **31.** [7.5] 9.6 lb
32. [4.4] $3x^3 - 8x^2 + 5x - 2$ **33.** [4.5] $\frac{1}{4}y^2 + \frac{1}{4}y + \frac{1}{16}$
34. [5.1] $(x^2 + 2)(x - 1)$ **35.** [6.1] $\dfrac{a + 2}{2a + 1}$

36. [7.3] ◈ Point–slope form would be more useful than slope–intercept form if we were asked to find an equation for a line with a specified slope that passes through a specified point that is not the y-intercept.
37. [7.4] ◈ The boundary line is part of the graph of a linear inequality $ax + by \leqslant c$ because the $\leqslant$ sign indicates that the graph of $ax + by < c$, as well as the graph of $ax + by = c$, form the solution set. The graph of $ax + by < c$ does not contain the graph of $ax + by = c$.
38. [7.2] $y = -2x - 3$ **39.** [7.2], [7.3] $y = \frac{3}{2}x + \frac{7}{2}$
40. [7.2] $y = -x$
41. [7.2] $-\dfrac{b}{a}$; $(0, b)$, $(a, 0)$

Test: Chapter 7, p. 349

1. [7.1] Undefined **2.** [7.1] $\frac{7}{12}$ **3.** [7.2] -2
4. [7.1] 0 **5.** [7.1] Undefined **6.** [7.2] 2, $\left(0, -\frac{1}{4}\right)$
7. [7.2] $\frac{4}{3}$, $(0, -2)$ **8.** [7.2] $y = \frac{1}{2}x - 7$
9. [7.2] $y = -4x + 3$ **10.** [7.3] $y - 5 = 1(x - 3)$
11. [7.3] $y = -3(x - (-2))$ **12.** [7.3] $y = -3x + 4$
13. [7.3] $y = \frac{1}{4}x - 2$
14. [7.2]

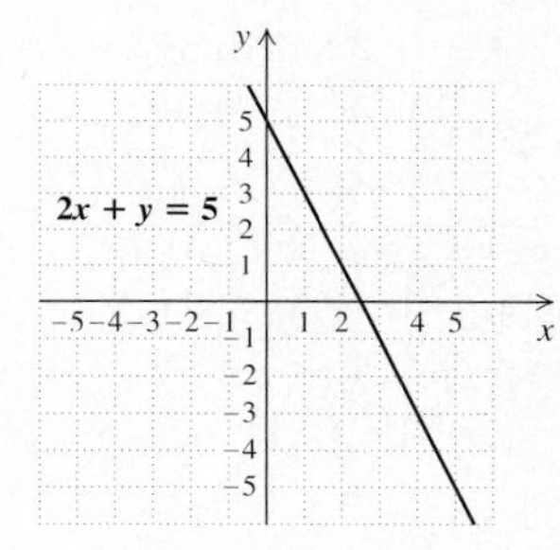

15. [7.3]

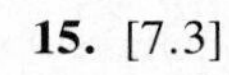

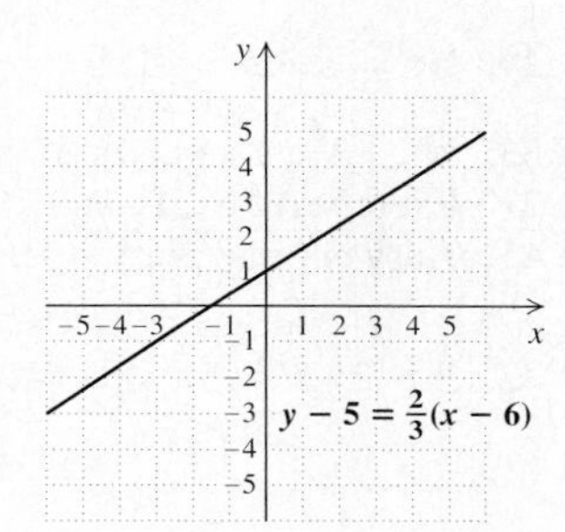

16. [7.2] Parallel
17. [7.4]

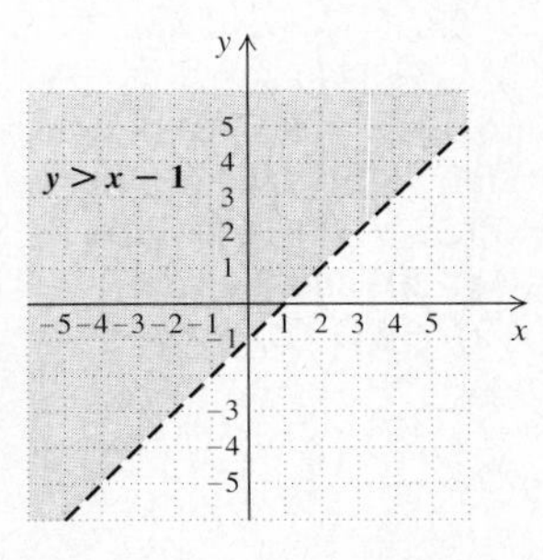

18. [7.4]

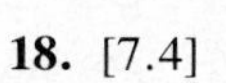

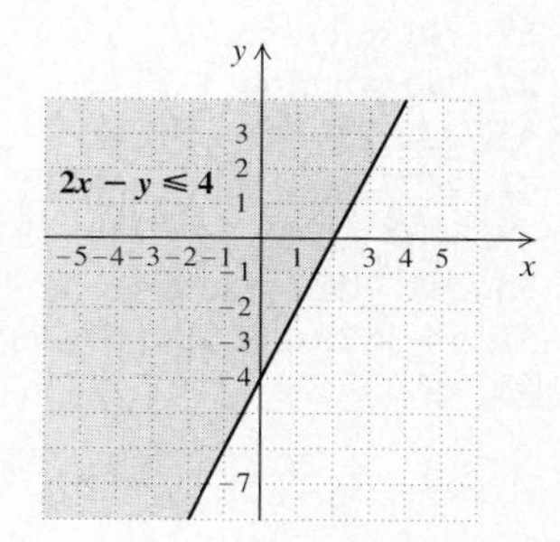

19. [7.5] $y = 4.5x$ **20.** [7.5] 18 min
21. [4.4] $-3y^7 + 9y^5 - 21y^4$ **22.** [4.5] $x^2 - 0.01$
23. [5.1] $3x(2x^2 + x - 1)$ **24.** [6.1] $\dfrac{-x + 4}{x}$
25. [7.3] $y = \frac{2}{3}x + \frac{11}{3}$

26. [7.4]

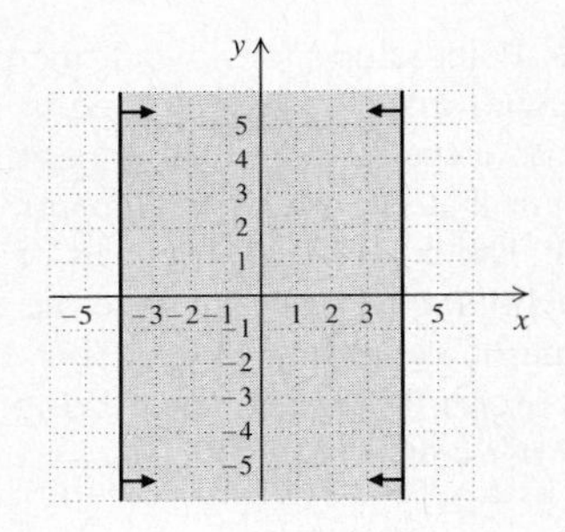

39. 58°, 32°
41. 480 acres Chardonnay, 340 acres Riesling
43. $1\frac{2}{3}$ ft by $3\frac{1}{3}$ ft **45.** a^3b^{-6}, or $\frac{a^3}{b^6}$
47. $(2x+5)^2$ **49.** ◈ **51.** (5, 2) **53.** (0, −1)
55. $\left(\frac{b-c}{1-a}, \frac{b-ac}{1-a}\right)$
57. 12 rabbits, 23 pheasants **59.** 45, 10

CHAPTER 8

Technology Connection, Section 8.1

TC1. [graphing calculator icon] **TC2.** No. The lines are parallel.

Exercise Set 8.1, pp. 356–357

1. Yes **3.** No **5.** Yes **7.** Yes **9.** Yes **11.** No
13. (2, 1) **15.** (−12, 11) **17.** (4, 3)
19. (−3, −3) **21.** No solution **23.** (2, 2)
25. (5, 3) **27.** Infinitely many solutions
29. No solution **31.** $\frac{9x+12}{(x+4)(x-4)}$ **33.** Trinomial
35. ◈ **37.** Exercises 22 and 27
39. Exercises 21, 28, and 29
41. Answers may vary. $2x - y = 8$, $x + 3y = -10$
43. $A = 2$, $B = 2$ **45.** (41.5, 17.1)

Exercise Set 8.2, pp. 361–363

1. (1, 3) **3.** (1, 2) **5.** (4, 3) **7.** (−2, 1)
9. (4, −2) **11.** No solution **13.** (−1, −3)
15. $\left(\frac{17}{3}, \frac{16}{3}\right)$ **17.** Infinitely many solutions
19. No solution **21.** $\left(\frac{25}{8}, -\frac{11}{4}\right)$ **23.** (−3, 0) **25.** (6, 3)
27. No solution **29.** (−3, −4) **31.** No solution
33. 15, 12 **35.** 37, 21 **37.** 28, 12 **39.** 70°, 110°
41. 62°, 28° **43.** 365 mi, 275 mi **45.** $134\frac{1}{3}$ m, $65\frac{2}{3}$ m
47. 110 yd, 60 yd **49.** $(3x-2)(2x-3)$ **51.** Prime
53. ◈ **55.** Exercises 12, 17, 18, 20, and 28
57. (4.382, 4.328) **59.** (10, −2) **61.** (2, −1, 3)
63. ◈

Exercise Set 8.3, pp. 370–371

1. (9, 1) **3.** (3, 5) **5.** (3, 0) **7.** $\left(-\frac{1}{2}, 3\right)$
9. $\left(-1, \frac{1}{5}\right)$ **11.** No solution **13.** (−3, −5)
15. (4, 5) **17.** (4, 1) **19.** (4, 3) **21.** (1, −1)
23. (−3, −1) **25.** (2, −2) **27.** $\left(5, \frac{1}{2}\right)$
29. Infinitely many solutions **31.** (2, −1)
33. $\left(\frac{231}{202}, \frac{117}{202}\right)$ **35.** 10 mi **37.** 75°, 105°

Exercise Set 8.4, pp. 377–379

1. 350 cars, 160 trucks **3.** Soda: \$0.49; pizza: \$1.50
5. Hendersons: 10 bags; Savickis: 4 bags **7.** 13
9. 33 soft-serve, 42 hard-pack
11. 203 adults, 226 children
13. 130 adults, 70 students
15. Cashews: 6 kg; pecans: 4 kg
17. Sunflower seeds: 30 lb; rolled oats: 20 lb
19. 40 L of A, 60 L of B;

Type of Solution	A	B	Mixture
Amount of Solution	x	y	100
Percent of Acid	50%	80%	68%
Amount of Acid in Solution	$0.5x$	$0.8y$	68

21. 80 L of 30%, 120 L of 50%
23. 128 L of 80%, 72 L of 30%
25. 12 of type A, 4 of type B
27. 70 dimes, 33 quarters **29.** 300 nickels, 100 dimes
31. Inexpensive: \$19.408; expensive: \$20.075
33. $(5x+9)(5x-9)$ **35.** 4 **37.** ◈
39. \$12,500 at 12%, \$14,500 at 13% **41.** 54
43. 43.75 L **45.** $4\frac{4}{7}$ L **47.** 74
49. Glove: \$79.95; bat: \$14.50; ball: \$4.55

Exercise Set 8.5, p. 381

1.

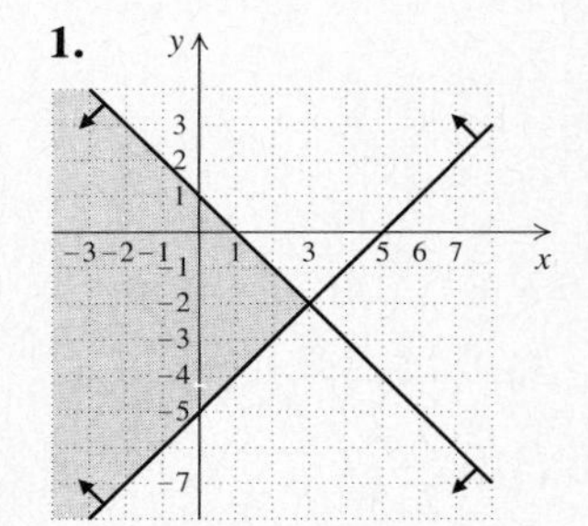

3.

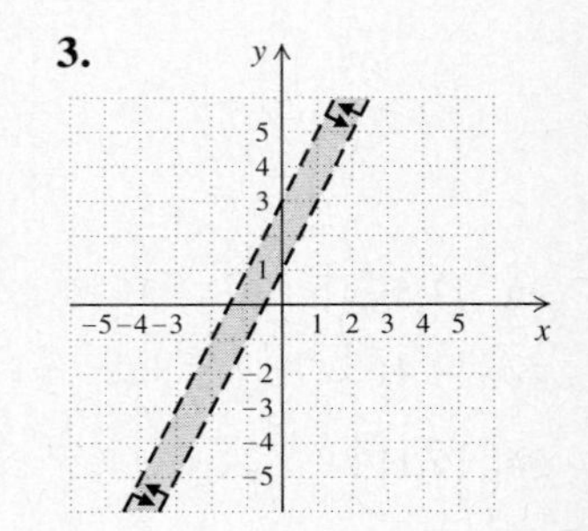

5.

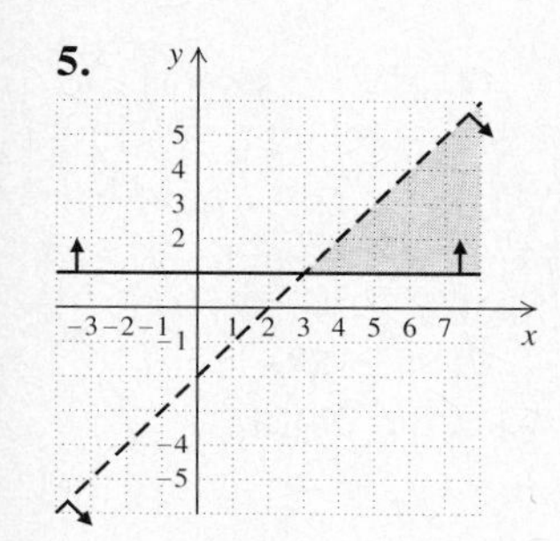

7.

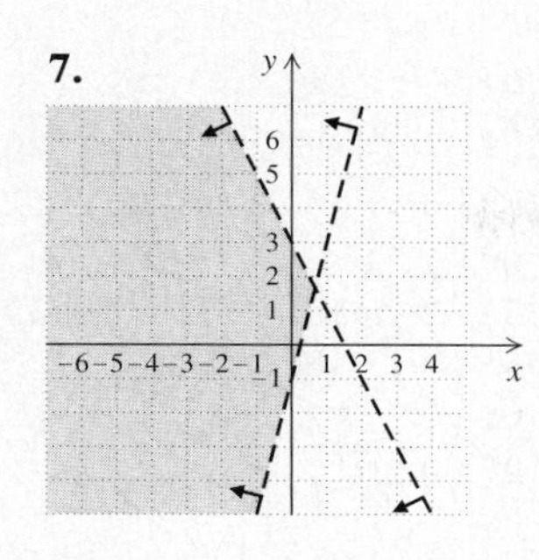

9.

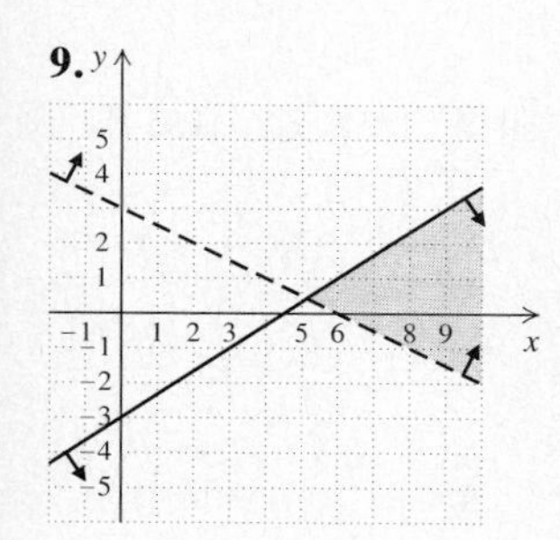

11.

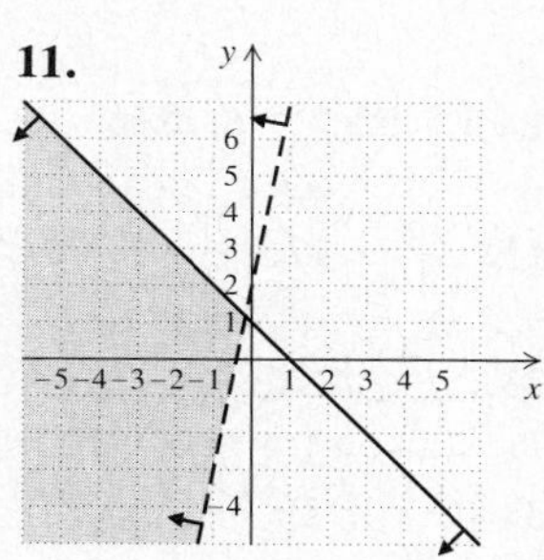

13.

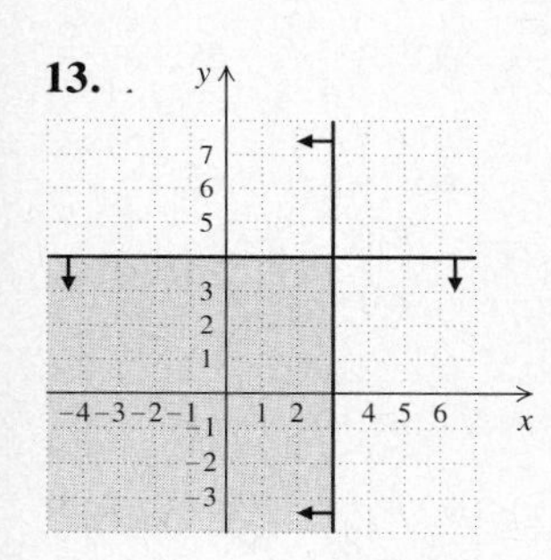

15.

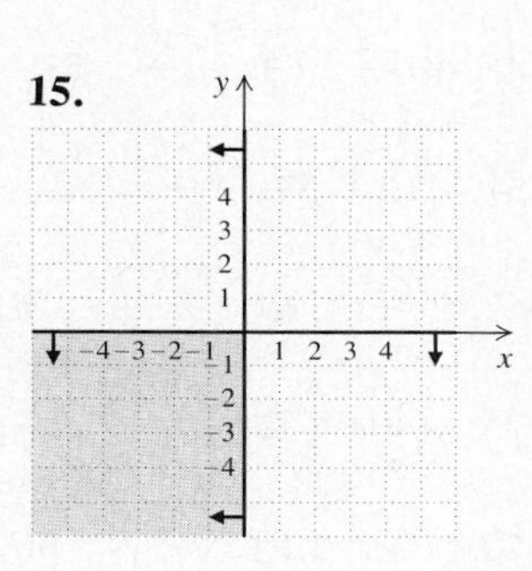

17.

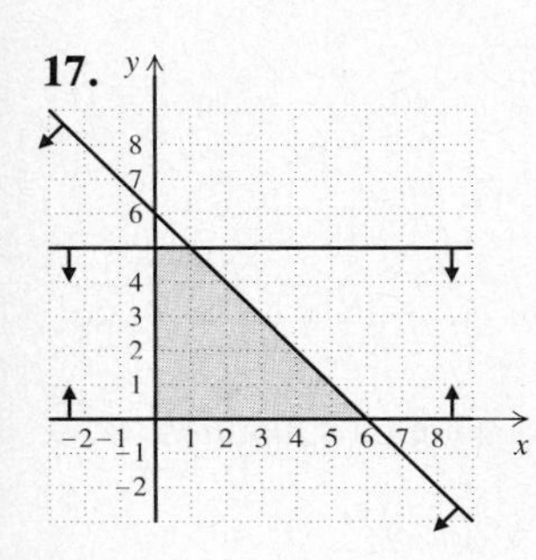

19.

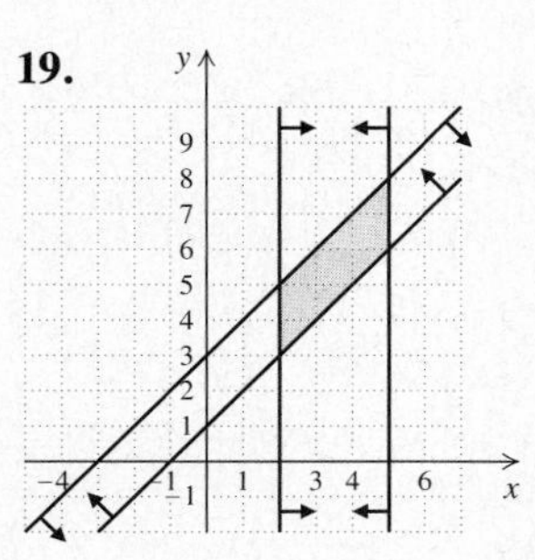

21.

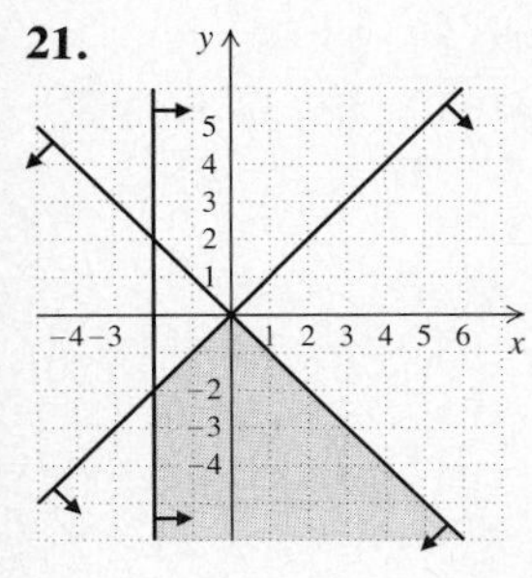

23. $\dfrac{13 - 2x}{3(x + 2)(x - 2)}$ **25.** 26 **27.** ◈

29.

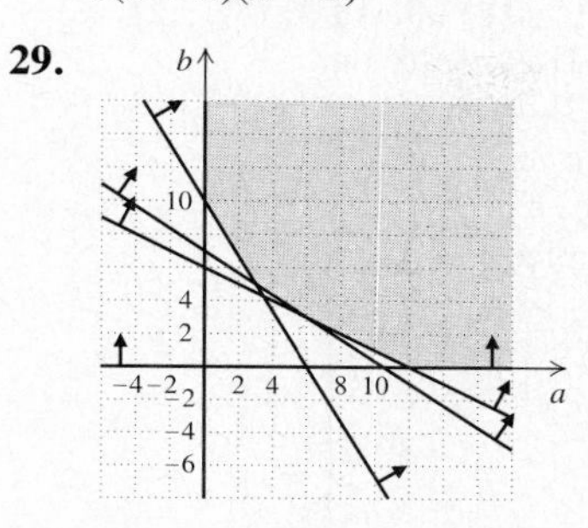

Review Exercises, Chapter 8, pp. 383–384

1. [8.1] No **2.** [8.1] Yes **3.** [8.1] Yes
4. [8.1] No **5.** [8.1] $(6, -2)$ **6.** [8.1] $(6, 2)$
7. [8.1] $(0, 5)$ **8.** [8.1] No solution; lines are parallel
9. [8.2] $(3, 2)$ **10.** [8.2] $(-2, 4)$ **11.** [8.2] $(1, -2)$
12. [8.2] $(-3, 9)$ **13.** [8.2] $(1, 4)$ **14.** [8.2] $(3, -1)$
15. [8.3] $(3, 1)$ **16.** [8.3] $(1, 4)$ **17.** [8.3] $(5, -3)$
18. [8.3] $(-4, 1)$ **19.** [8.3] $(-2, 4)$
20. [8.3] $(-2, -6)$ **21.** [8.3] $(3, 2)$
22. [8.3] $(2, -4)$ **23.** [8.3] Infinitely many solutions
24. [8.2] $10, -2$ **25.** [8.2] 12, 15
26. [8.2] $l = 37\frac{1}{2}$ cm, $w = 10\frac{1}{2}$ cm **27.** [8.4] 15
28. [8.4] 297 orchestra seats, 211 balcony seats
29. [8.4] 40 L of each
30. [8.5]

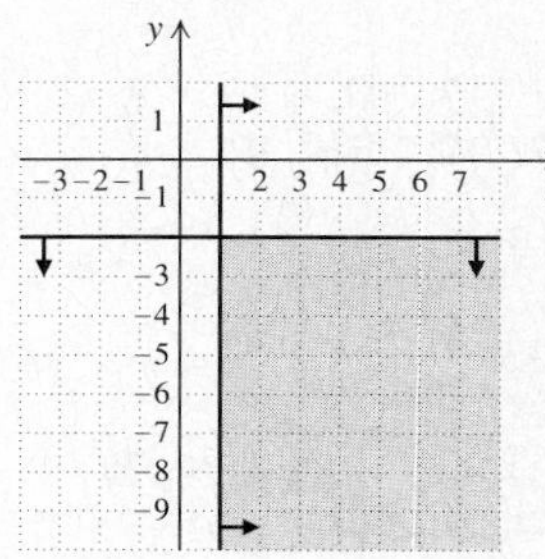

31. [8.5]

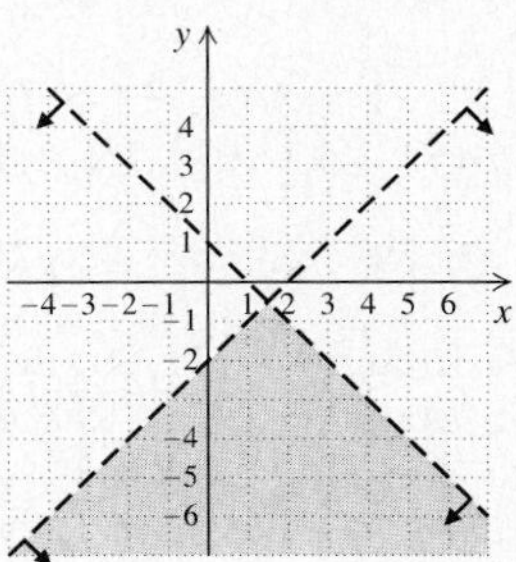

32. [4.2] $x^4 + 3x^3$ **33.** [6.5] $\dfrac{3t + 8}{(t + 2)(t - 2)}$
34. [5.3] $t(2t + 1)(t - 3)$ **35.** [5.4] $9(y + 2)^2$
36. [8.1] ◈ A solution of a system of two equations is an ordered pair that makes both equations true. The graph of an equation represents all ordered pairs that make that equation true. So for an ordered pair to make *both* equations true, it must be on both graphs.
37. [8.5] ◈ The solution sets of linear inequalities are regions, not lines. Thus the solution sets can intersect even if the boundary lines do not.
38. [8.1] $C = 1$, $D = 3$ **39.** [8.2] $(2, 1, -2)$
40. [8.2] $(2, 0)$ **41.** [8.4] 24 **42.** [8.4] \$336

Test: Chapter 8, p. 384

1. [8.1] No **2.** [8.1] $(2, -1)$ **3.** [8.2] $(8, -2)$
4. [8.2] $(-1, 3)$ **5.** [8.2] No solution
6. [8.3] $(1, -5)$ **7.** [8.3] $(12, -6)$ **8.** [8.3] $(0, 1)$
9. [8.3] $(5, 1)$ **10.** [8.3] 36°, 54°
11. [8.4] 40 L of A, 20 L of B
12. [8.4] Oak: 15 sheets; pine: 3 sheets
13. [8.5] **14.** [8.5]

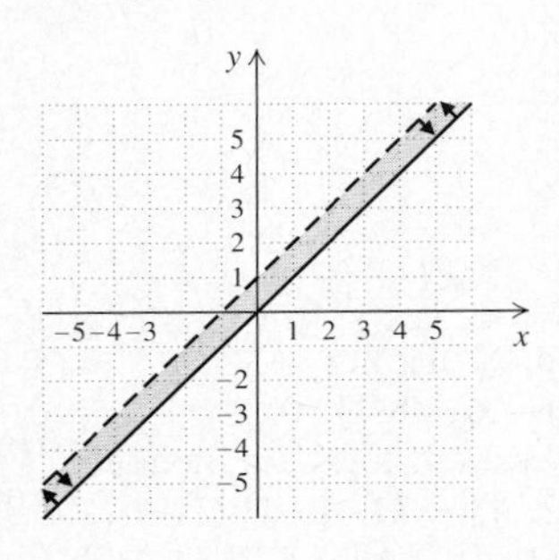

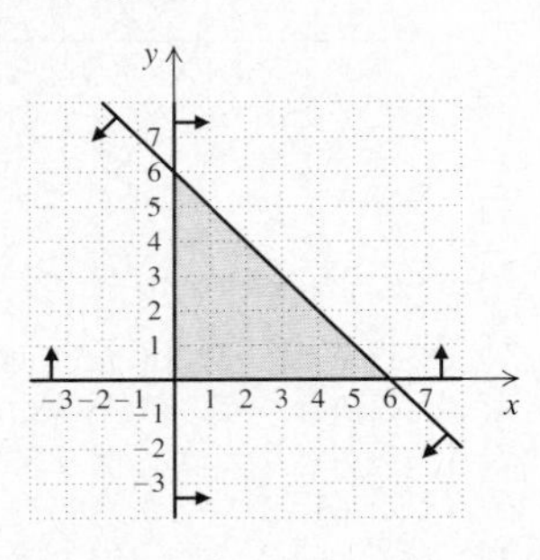

15. [6.5] $\dfrac{-x^2 + x + 17}{(x-4)(x+4)(x+1)}$ **16.** [4.2] 0
17. [5.3] $3(3y - 1)(2y + 1)$
18. [5.4] $5(x + 3)^2$ **19.** [8.1] $C = -\frac{19}{2}$, $D = \frac{14}{3}$
20. [8.4] 9

CHAPTER 9

Exercise Set 9.1, pp. 390–391

1. 1, −1 **3.** 4, −4 **5.** 7, −7 **7.** 13, −13 **9.** 2
11. −3 **13.** 0 **15.** −11 **17.** 19 **19.** 12
21. −25 **23.** $a - 4$ **25.** $t^2 + 1$ **27.** $\dfrac{3}{x+2}$
29. Irrational **31.** Irrational **33.** Rational
35. Irrational **37.** Rational **39.** Irrational
41. 2.236 **43.** 4.123 **45.** 9.644 **47.** t **49.** $3x$
51. ab **53.** $34d$ **55.** $5ab$ **57.** (a) 13; (b) 24
59. $-\frac{5}{4}$ **61.** $y = 2x + 10$ **63.** ◈ **65.** 2
67. −6 and −5 **69.** −6, 6 **71.** −3, 3 **73.** $5a^2b$
75. $\dfrac{2x^4}{y^3}$ **77.** $\dfrac{13}{m^8}$
79. (a) 1.7; (b) 2.2; (c) 2.6. Answers may vary.

Exercise Set 9.2, pp. 394–395

1. $\sqrt{6}$ **3.** $\sqrt{12}$, or $2\sqrt{3}$ **5.** $\sqrt{\dfrac{3}{10}}$ **7.** 17
9. $\sqrt{75}$, or $5\sqrt{3}$ **11.** $\sqrt{2x}$ **13.** $\sqrt{6x}$ **15.** $\sqrt{7xy}$
17. $\sqrt{6ac}$ **19.** $2\sqrt{3}$ **21.** $2\sqrt{5}$ **23.** $10\sqrt{2}$
25. $3\sqrt{x}$ **27.** $5\sqrt{3a}$ **29.** $4\sqrt{a}$ **31.** $8y$
33. $x\sqrt{13}$ **35.** $2t\sqrt{2}$ **37.** $6\sqrt{5}$ **39.** $12\sqrt{2y}$
41. x^{10} **43.** x^6 **45.** $x^2\sqrt{x}$ **47.** $t^9\sqrt{t}$ **49.** $6m\sqrt{m}$
51. $2a^2\sqrt{2a}$ **53.** $2p^8\sqrt{26p}$ **55.** $3\sqrt{2}$ **57.** $3\sqrt{10}$
59. $6\sqrt{xy}$ **61.** 10 **63.** $5b\sqrt{3}$ **65.** $2t$ **67.** $a\sqrt{bc}$
69. $2x^3\sqrt{2}$ **71.** $xy^3\sqrt{xy}$ **73.** $10ab^2\sqrt{5ab}$
75. 20 mph, 54.8 mph **77.** 168 km **79.** ◈
81. 0.1 **83.** 0.25 **85.** ◈ **87.** = **89.** >
91. > **93.** $6(x-2)^2\sqrt{10}$ **95.** $2^{54}x^{158}\sqrt{2x}$
97. $0.2x^{2n}$

Exercise Set 9.3, pp. 399–400

1. 3 **3.** 2 **5.** $\sqrt{5}$ **7.** $\dfrac{1}{5}$ **9.** $\dfrac{2}{5}$ **11.** 2 **13.** $3y$
15. $3x^2$ **17.** $\dfrac{5}{x^3}$ **19.** $a^4\sqrt{2}$ **21.** $\dfrac{3}{7}$ **23.** $\dfrac{1}{6}$
25. $-\dfrac{4}{9}$ **27.** $\dfrac{2}{3}$ **29.** $\dfrac{13}{11}$ **31.** $\dfrac{6}{a}$ **33.** $\dfrac{3a}{25}$ **35.** $\dfrac{\sqrt{10}}{5}$
37. $\dfrac{\sqrt{6}}{4}$ **39.** $\dfrac{\sqrt{35}}{10}$ **41.** $\dfrac{\sqrt{2}}{6}$ **43.** $\dfrac{3\sqrt{5}}{5}$ **45.** $\dfrac{2\sqrt{6}}{3}$
47. $\dfrac{\sqrt{3x}}{x}$ **49.** $\dfrac{\sqrt{xy}}{y}$ **51.** $\dfrac{\sqrt{21}}{3}$ **53.** $\dfrac{3\sqrt{2}}{4}$ **55.** $\dfrac{\sqrt{26}}{13}$
57. $\sqrt{2}$ **59.** $\dfrac{\sqrt{15}}{9}$ **61.** $\dfrac{\sqrt{21}}{6}$ **63.** $\dfrac{\sqrt{2x}}{8}$ **65.** $\dfrac{2}{3}$
67. $\dfrac{\sqrt{3x}}{x}$ **69.** $\dfrac{4y\sqrt{3}}{3}$ **71.** $\dfrac{\sqrt{3a}}{2}$ **73.** $\dfrac{5\sqrt{6x}}{6x}$
75. $\dfrac{3\sqrt{6}}{8c}$ **77.** 1.57 sec, 3.14 sec, 8.88 sec, 11.10 sec
79. 1 sec **81.** $(4, 2)$ **83.** $y = \dfrac{10}{3}x$ **85.** ◈
87. $\dfrac{\sqrt{5}}{40}$ **89.** $\dfrac{\sqrt{5x}}{5x^2}$ **91.** $\dfrac{\sqrt{3ab}}{b}$ **93.** $\dfrac{y-x}{xy}$, or $\dfrac{1}{x} - \dfrac{1}{y}$

Exercise Set 9.4, pp. 404–405

1. $7\sqrt{2}$ **3.** $4\sqrt{5}$ **5.** $13\sqrt{x}$ **7.** $-2\sqrt{x}$ **9.** $8\sqrt{2a}$
11. $8\sqrt{10y}$ **13.** $11\sqrt{7}$ **15.** $\sqrt{2}$
17. $5\sqrt{3} + \sqrt{8} = 5\sqrt{3} + 2\sqrt{2}$ cannot be simplified further.
19. $-2\sqrt{x}$ **21.** $25\sqrt{2}$ **23.** $\sqrt{3}$ **25.** $\sqrt{5}$
27. $13\sqrt{2}$ **29.** $3\sqrt{3}$ **31.** $-24\sqrt{2}$ **33.** $\sqrt{x}$
35. $\sqrt{15} + \sqrt{21}$ **37.** $\sqrt{42} - \sqrt{105}$ **39.** $17 + 8\sqrt{2}$
41. $16 - 7\sqrt{6}$ **43.** −44 **45.** 3 **47.** $-1 - 2\sqrt{2}$
49. $39 + 12\sqrt{3}$ **51.** $37 - 20\sqrt{3}$
53. $x - 2\sqrt{10x} + 10$ **55.** $-5 + 5\sqrt{2}$
57. $-8 - 4\sqrt{5}$ **59.** $-\sqrt{7} + 3$ **61.** $-\dfrac{6 + 5\sqrt{6}}{19}$
63. $\dfrac{5 + \sqrt{15}}{2}$ **65.** $-\dfrac{\sqrt{35} - 7}{2}$ **67.** $5 - 2\sqrt{6}$

69. $11 + 2\sqrt{30}$ **71.** $5 - 2\sqrt{7}$
73. $\frac{4}{3}$ hr; the variation constant is the fixed distance, 80 mi.
75. ◈ **77.** All of them **79.** $4\sqrt{5}$ **81.** $-\frac{4\sqrt{6}}{5}$
83. $(b^3 + ab + a)\sqrt{a}$ **85.** $37xy\sqrt{3x}$ **87.** 0°
89. −22°

Exercise Set 9.5, pp. 409–410

1. 25 **3.** 2.25 **5.** 61 **7.** $\frac{77}{2}$ **9.** 5 **11.** 3 **13.** $\frac{17}{4}$
15. No solution **17.** No solution **19.** 9 **21.** 7
23. 1, 5 **25.** 3 **27.** 13, 25 **29.** No solution
31. 3 **33.** No solution **35.** 1 **37.** 25 m
39. 11,236 m **41.** 125 ft, 245 ft **43.** 49 **45.** 12
47. About 2.08 ft **49.** $\frac{x^6}{3}$ **51.** $x - 4$ **53.** ◈
55. 16 **57.** No solution **59.** $-\frac{57}{16}$ **61.** 10
63. 34.726 m
65.

$y = \sqrt{x}$

67.

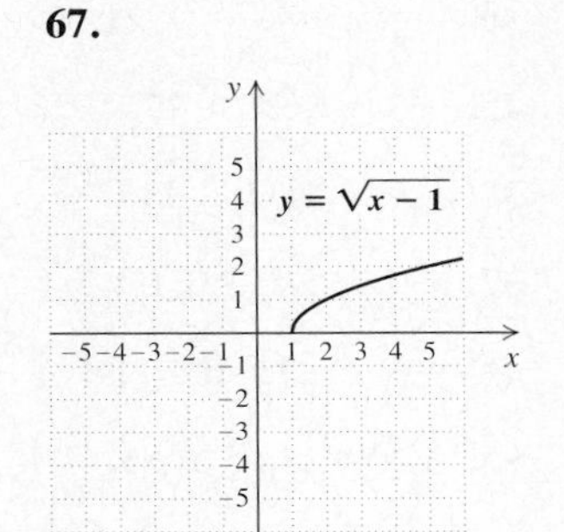

69.

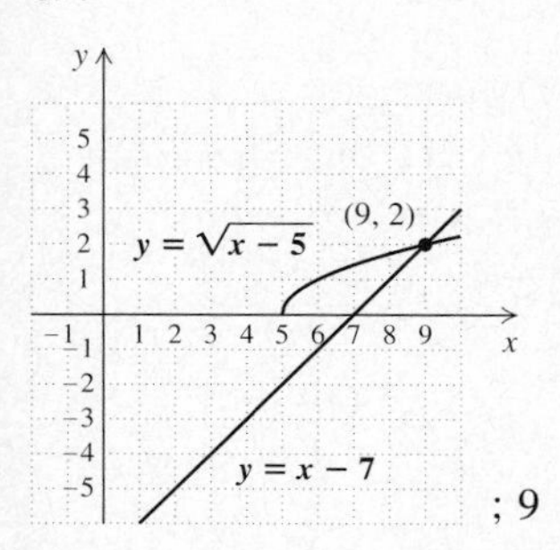

; 9

71. 1.57

Exercise Set 9.6, pp. 414–416

1. $c = 17$ **3.** $c = \sqrt{32} \approx 5.657$ **5.** $b = 12$
7. $a = 4$ **9.** $c = 26$ **11.** $b = 12$ **13.** $a = 2$
15. $b = \sqrt{2} \approx 1.414$ **17.** $a = 5$
19. $\sqrt{75} \approx 8.660$ m **21.** $\sqrt{208} = 4\sqrt{13} \approx 14.422$ ft
23. $\sqrt{12{,}500} \approx 111.803$ yd **25.** $60\sqrt{2} \approx 84.853$ ft
27. 43 km **29.** $-\frac{1}{3}$ **31.** $\{x \mid x > -25\}$ **33.** ◈
35. $50\sqrt{10} \approx 158$ ft **37.** $\sqrt{7} \approx 2.646$ m **39.** $s\sqrt{3}$
41. $h = \frac{a}{2}\sqrt{3}$ **43.** $\sqrt{181} \approx 13.454$ cm
45. $12 - 2\sqrt{6}$ **47.** 640 acres

Exercise Set 9.7, pp. 420–421

1. −2 **3.** 10 **5.** −5 **7.** 6 **9.** 5 **11.** 0
13. −1 **15.** Not a real number **17.** 10 **19.** 5
21. 2 **23.** x **25.** $2\sqrt[3]{4}$ **27.** $2\sqrt[4]{3}$ **29.** $\frac{3}{4}$
31. $\frac{4}{5}$ **33.** $\frac{\sqrt[3]{17}}{2}$ **35.** $\frac{\sqrt[4]{13}}{3}$ **37.** 5 **39.** 10
41. 2 **43.** 32 **45.** 32 **47.** 16 **49.** 4 **51.** 3125
53. $\frac{1}{6}$ **55.** $\frac{1}{4}$ **57.** $\frac{1}{27}$ **59.** $\frac{1}{12}$ **61.** $\frac{1}{243}$ **63.** $\frac{1}{4}$
65. $10,660 **67.** $\frac{-2}{a - 5}$ **69.** ◈ **71.** 3.981
73. 31.623 **75.** $x^{10/9}$ **77.** $p^{2/15}$
79.

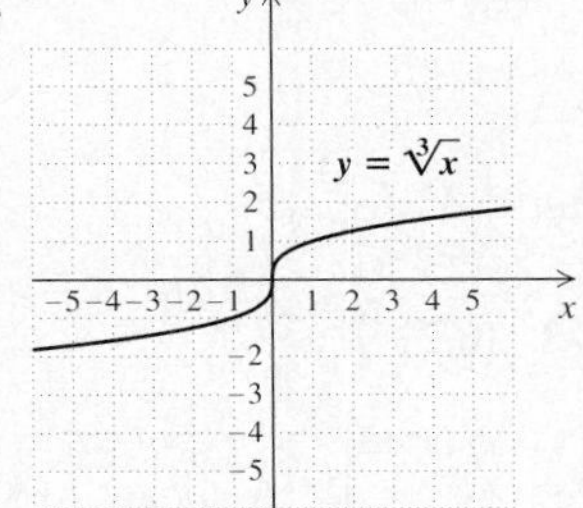

81. $y_1 = x^{2/3}$ $y_2 = x^1$ $y_3 = x^{5/4}$ $y_4 = x^{3/2}$

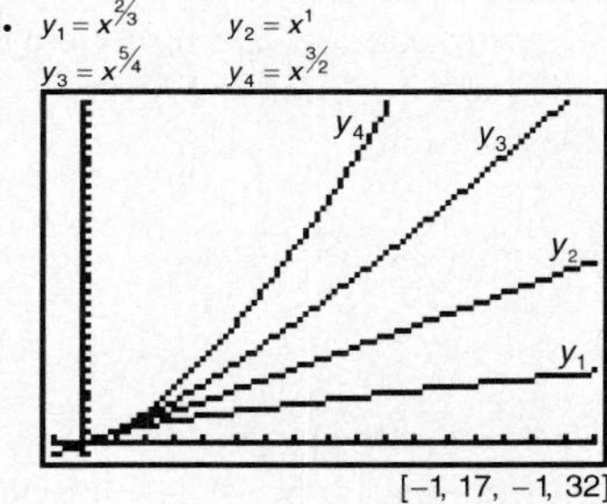

Review Exercises: Chapter 9, pp. 422–423

1. [9.1] −8, 8 **2.** [9.1] −5, 5 **3.** [9.1] −14, 14
4. [9.1] −20, 20 **5.** [9.1] 6 **6.** [9.1] −9
7. [9.1] 7 **8.** [9.1] −13 **9.** [9.1] $x^2 + 4$
10. [9.1] $5ab^3$ **11.** [9.1] Irrational
12. [9.1] Rational **13.** [9.1] Irrational
14. [9.1] Rational **15.** [9.1] 1.732 **16.** [9.1] 9.950
17. [9.1] 3.606 **18.** [9.1] 7.550 **19.** [9.1] m
20. [9.1] $7t$ **21.** [9.1] p **22.** [9.1] ac
23. [9.2] $\sqrt{21}$ **24.** [9.2] $\sqrt{at}$ **25.** [9.2] $\sqrt{6xy}$
26. [9.2] $4\sqrt{3}$ **27.** [9.2] $4t\sqrt{2}$ **28.** [9.2] $7\sqrt{2p}$

29. [9.2] x^4 **30.** [9.2] $2a^7\sqrt{3}$ **31.** [9.2] $m^7\sqrt{m}$
32. [9.2] $2\sqrt{15}$ **33.** [9.2] $2x\sqrt{10}$ **34.** [9.2] $5xy\sqrt{2}$
35. [9.2] $10a^2b\sqrt{ab}$ **36.** [9.3] $\frac{\sqrt{7}}{3}$ **37.** [9.3] $\frac{y^4\sqrt{5}}{3}$
38. [9.3] $\frac{5}{8}$ **39.** [9.3] $\frac{2}{3}$ **40.** [9.3] $\frac{7}{t}$ **41.** [9.3] $\frac{\sqrt{2}}{2}$
42. [9.3] $\frac{\sqrt{10}}{4}$ **43.** [9.3] $\frac{\sqrt{5y}}{y}$ **44.** [9.3] $\frac{2\sqrt{3}}{3}$
45. [9.4] $13\sqrt{5}$ **46.** [9.4] $\sqrt{5}$ **47.** [9.4] $-3\sqrt{x}$
48. [9.4] $7 + 4\sqrt{3}$ **49.** [9.4] 1
50. [9.4] $-11 + 5\sqrt{7}$ **51.** [9.4] $8 - 4\sqrt{3}$
52. [9.4] $-7 - 3\sqrt{5}$ **53.** [9.5] 52
54. [9.5] No solution **55.** [9.5] 4 **56.** [9.5] 0, 3
57. [9.5] 405 ft **58.** [9.6] $b = 20$
59. [9.6] $c = \sqrt{3} \approx 1.732$ **60.** [9.6] $7\sqrt{2} \approx 9.990$ m
61. [9.7] 2 **62.** [9.7] Not a real number
63. [9.7] -3 **64.** [9.7] $2\sqrt[4]{2}$ **65.** [9.7] 10
66. [9.7] $\frac{1}{3}$ **67.** [9.7] 64 **68.** [9.7] $\frac{1}{27}$
69. [7.5] \$450
70. [7.2] Slope: $\frac{2}{5}$; y-intercept: $\left(0, -\frac{14}{5}\right)$
71. [6.2] $\frac{(x+1)(x+2)}{x(x-2)}$ **72.** [6.3] $\frac{a+2}{a+1}$
73. [9.2] ◈ When a power is squared, the exponent is multiplied by 2. For example, $(x^3)^2 = x^6$. A square root of an even power can thus be found by using an exponent that is half the exponent of the radicand.
74. [9.4] ◈ Some radical terms that are like terms may not appear to be so until they are in simplified form.
75. [9.1] 2 **76.** [9.1] No solution
77. [9.4] $(x + \sqrt{5})(x - \sqrt{5})$
78. [9.5] $b = \sqrt{A^2 - a^2}$ or $b = -\sqrt{A^2 - a^2}$

Test: Chapter 9, p. 424

1. [9.1] $-9, 9$ **2.** [9.1] 8 **3.** [9.1] -5
4. [9.1] $4 - y^3$ **5.** [9.1] Irrational **6.** [9.1] Rational
7. [9.1] 9.327 **8.** [9.1] 2.646 **9.** [9.1] a
10. [9.1] $6y$ **11.** [9.2] $\sqrt{30}$ **12.** [9.2] $\sqrt{14ab}$
13. [9.2] $3\sqrt{3}$ **14.** [9.2] $t^2\sqrt{t}$ **15.** [9.2] $3a^3\sqrt{2}$
16. [9.2] $5\sqrt{2}$ **17.** [9.2] $5t$ **18.** [9.2] $3ab^2\sqrt{2}$
19. [9.3] $\frac{3}{4}$ **20.** [9.3] $\frac{\sqrt{7}}{4y}$ **21.** [9.3] $\frac{3}{2}$ **22.** [9.3] $\frac{12}{a}$
23. [9.3] $\frac{\sqrt{10}}{5}$ **24.** [9.3] $\frac{2x\sqrt{y}}{y}$ **25.** [9.4] $-6\sqrt{2}$
26. [9.4] $7\sqrt{3}$ **27.** [9.4] $21 - 8\sqrt{5}$ **28.** [9.4] 11
29. [9.4] $\frac{40 + 10\sqrt{5}}{11}$ **30.** [9.6] $c = \sqrt{80} \approx 8.944$ cm

31. [9.5] 48 **32.** [9.5] $-2, 2$ **33.** [9.5] About 5000 m
34. [9.7] 2 **35.** [9.7] -1 **36.** [9.7] -4
37. [9.7] Not a real number **38.** [9.7] 3 **39.** [9.7] $\frac{1}{3}$
40. [9.7] 1000 **41.** [9.7] $\frac{1}{32}$ **42.** [7.5] 15,652
43. [7.2] $y = -\frac{1}{2}x - 1$ **44.** [6.2] $\frac{(a+1)^2}{a}$
45. [6.3] $\frac{x+4}{3x-5}$ **46.** [9.5] -3 **47.** [9.2] y^{8n}

CHAPTER 10

Exercise Set 10.1, pp. 428–429

1. $5, -5$ **3.** $9, -9$ **5.** $\sqrt{15}, -\sqrt{15}$
7. $\sqrt{19}, -\sqrt{19}$ **9.** $\sqrt{7}, -\sqrt{7}$ **11.** $2\sqrt{5}, -2\sqrt{5}$
13. $\frac{5}{2}, -\frac{5}{2}$ **15.** $\frac{7\sqrt{3}}{3}, -\frac{7\sqrt{3}}{3}$ **17.** $\sqrt{3}, -\sqrt{3}$
19. $\frac{\sqrt{35}}{5}, -\frac{\sqrt{35}}{5}$ **21.** $9, -5$ **23.** $3, -9$
25. $-3 \pm \sqrt{21}$ **27.** $-13 \pm 2\sqrt{2}$ **29.** $7 \pm 2\sqrt{3}$
31. $-9 \pm \sqrt{34}$ **33.** $\frac{-3 \pm \sqrt{14}}{2}$ **35.** $11, -5$
37. $-9, -5$ **39.** $1 \pm \sqrt{5}$ **41.** $-2 \pm 2\sqrt{3}$ **43.** Yes
45. Yes **47.** ◈ **49.** $\frac{-5 \pm \sqrt{13}}{2}$ **51.** $\frac{3 \pm \sqrt{17}}{4}$
53. 3.45, -3.95 **55.** 7.1, -3.3

Exercise Set 10.2, p. 433

1. $x^2 - 2x + 1$ **3.** $x^2 + 18x + 81$ **5.** $x^2 - x + \frac{1}{4}$
7. $t^2 + 5t + \frac{25}{4}$ **9.** $x^2 - \frac{3}{2}x + \frac{9}{16}$ **11.** $m^2 + \frac{9}{2}m + \frac{81}{16}$
13. $-2, 8$ **15.** $-21, -1$ **17.** $1 \pm \sqrt{6}$
19. $11 \pm \sqrt{19}$ **21.** $-5 \pm \sqrt{29}$ **23.** $\frac{7 \pm \sqrt{57}}{2}$
25. $-7, 4$ **27.** $\frac{-3 \pm \sqrt{17}}{4}$ **29.** $\frac{-3 \pm \sqrt{145}}{4}$
31. $\frac{-2 \pm \sqrt{7}}{3}$ **33.** $-\frac{1}{2}, 5$ **35.** $-\frac{7}{2}, \frac{1}{2}$ **37.** $\left(\frac{2}{3}, \frac{17}{3}\right)$
39. $y = \frac{3}{5}x - 1$
41. ◈
43. $12, -12$
45. $\pm 16\sqrt{2}$
47. $\pm 2\sqrt{c}$
49. 8.00, 2.00
51. $-0.39, -7.61$
53. 7.27, -0.27
55. -0.50, 5.00

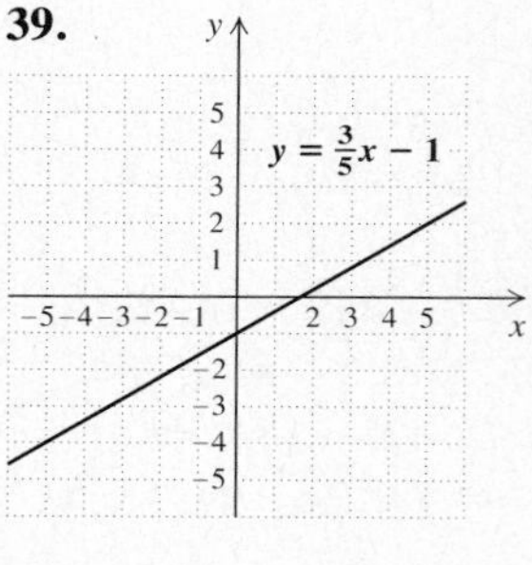

Exercise Set 10.3, pp. 439–442

1. 7, −3 **3.** 3 **5.** $-\frac{4}{3}$, 2 **7.** $\frac{1}{2}$, $-\frac{7}{2}$ **9.** −3, 3
11. $1 \pm \sqrt{3}$ **13.** $5 \pm \sqrt{3}$ **15.** $-2 \pm \sqrt{7}$
17. $\frac{-4 \pm \sqrt{10}}{3}$ **19.** $\frac{5 \pm \sqrt{33}}{4}$ **21.** $\frac{1 \pm \sqrt{2}}{2}$
23. No real-number solutions
25. $\frac{5 \pm \sqrt{73}}{6}$ **27.** $\frac{3 \pm \sqrt{29}}{2}$ **29.** 0, $\frac{3}{2}$
31. No real-number solutions **33.** $\pm \frac{3\sqrt{10}}{2}$
35. 5.317, −1.317 **37.** 6.162, −0.162
39. 0.207, −1.207 **41.** 20 **43.** 7
45. About 8.43 sec **47.** About 3.31 sec
49. 7 ft, 24 ft **51.** Width: 8 cm; length: 10 cm
53. 15 m by 20 m **55.** 4 m, 6.5 m
57. Width: 3.58 in.; length: 5.58 in.
59. Width: 2.24 m; length: 4.47 m **61.** 10%
63. 18.75% **65.** 8% **67.** 17.84 ft **69.** $\frac{6}{5}$
71. $3\sqrt{10}$ **73.** ◈ **75.** 0, 2 **77.** $\frac{3 \pm \sqrt{5}}{2}$
79. $\frac{-7 \pm \sqrt{61}}{2}$ **81.** No real-number solutions
83. $\pm \sqrt{7}$ **85.** $2 \pm \sqrt{34}$ **87.** $1 + \sqrt{2} \approx 2.41$ cm
89. $3 + 2\sqrt{2} \approx 5.828$
91. $10\sqrt{2} \approx 14.14$ in.; two 10-in. pizzas **93.** 14.75%
95. About 19 in. by 44 in.

Exercise Set 10.4, p. 444

1. i **3.** $7i$ **5.** $2i\sqrt{2}$ **7.** $-2i\sqrt{3}$ **9.** $-3i\sqrt{3}$
11. $5i\sqrt{2}$ **13.** $-10i\sqrt{3}$ **15.** $7 + 4i$ **17.** $3 - 7i\sqrt{2}$
19. $\pm 2i$ **21.** $\pm 2i\sqrt{3}$ **23.** $2 \pm \sqrt{2}i$ **25.** $2 \pm 4i$
27. $-1 \pm i$ **29.** $2 \pm i\sqrt{3}$ **31.** $-\frac{3}{2} \pm \frac{1}{2}i$
33. $-\frac{1}{3} \pm \frac{1}{3}\sqrt{2}i$ **35.** $-3 - 3\sqrt{2}$
37. $-\frac{7 + 2\sqrt{10}}{3}$ **39.** ◈ **41.** $-\frac{3}{2} \pm \frac{1}{2}i$
43. $-1 \pm \sqrt{19}i$

45.

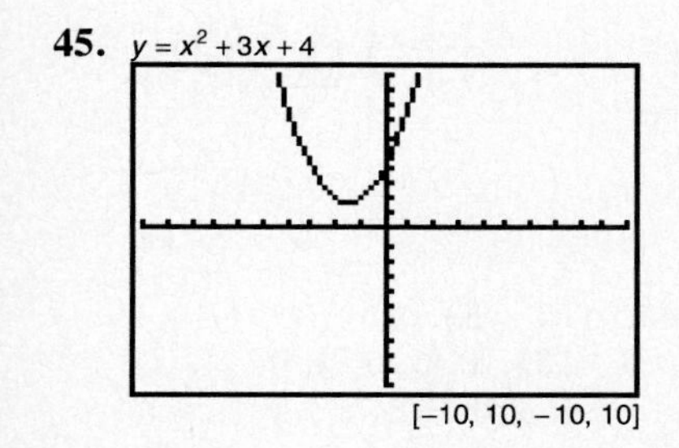

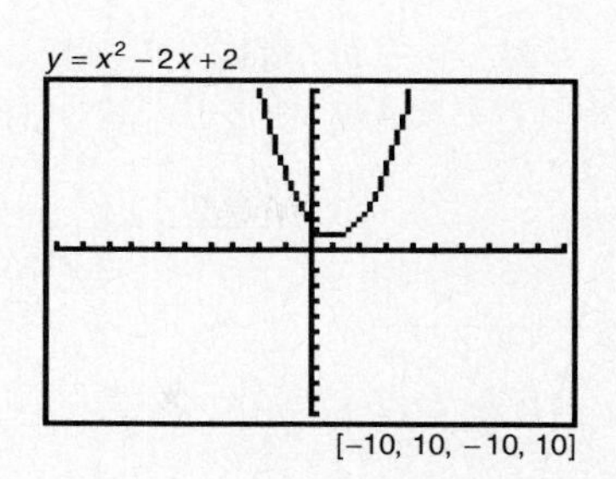

Exercise Set 10.5, pp. 449–451

1.

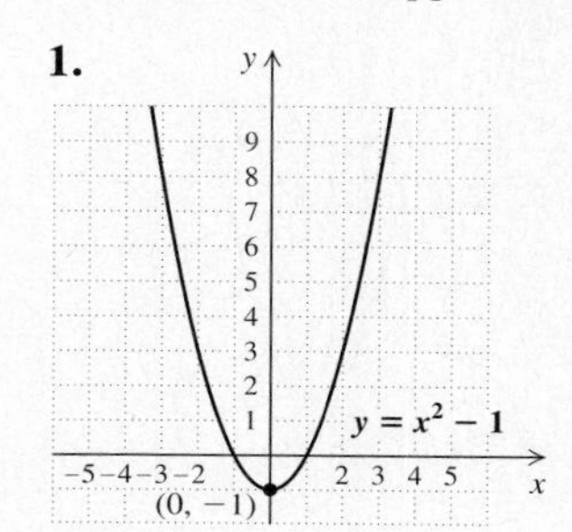

3.

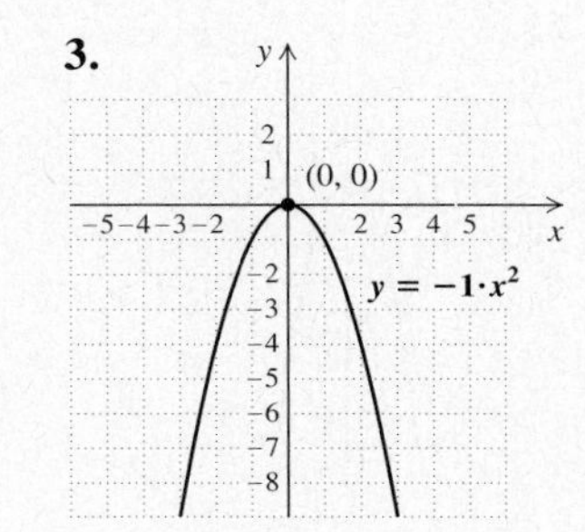

5.

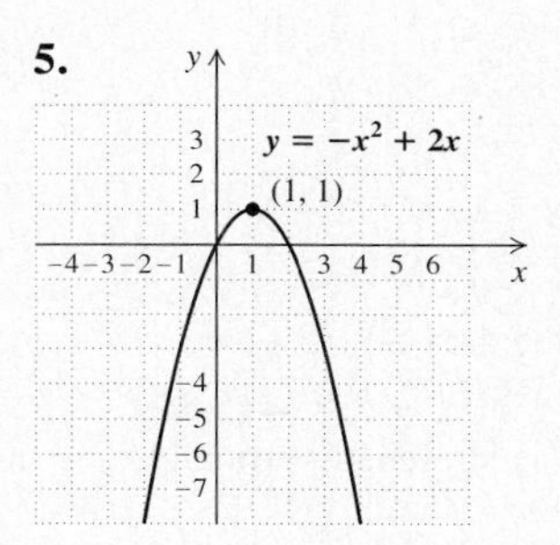

7.

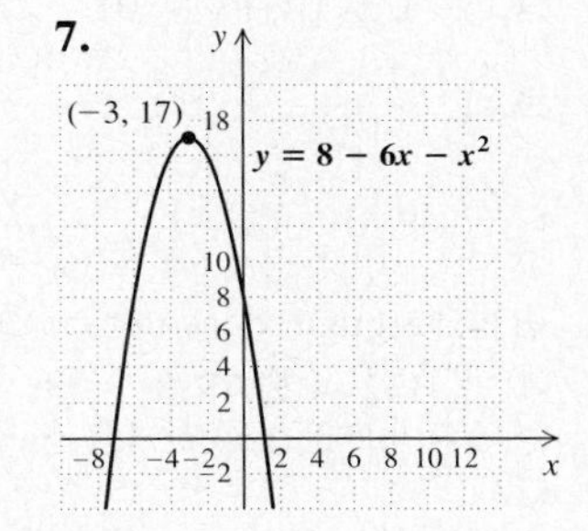

9.

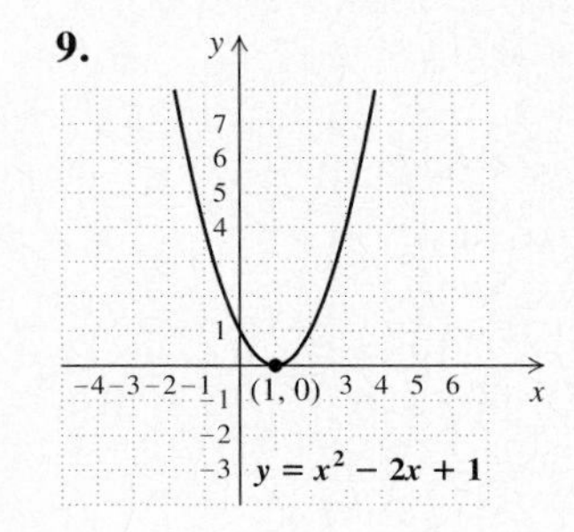

11.

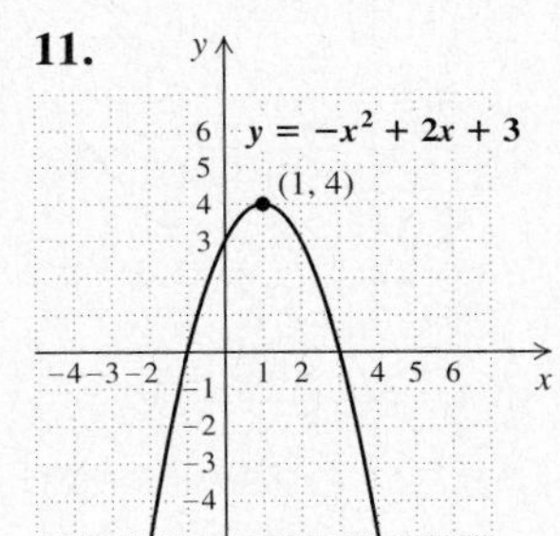

13.

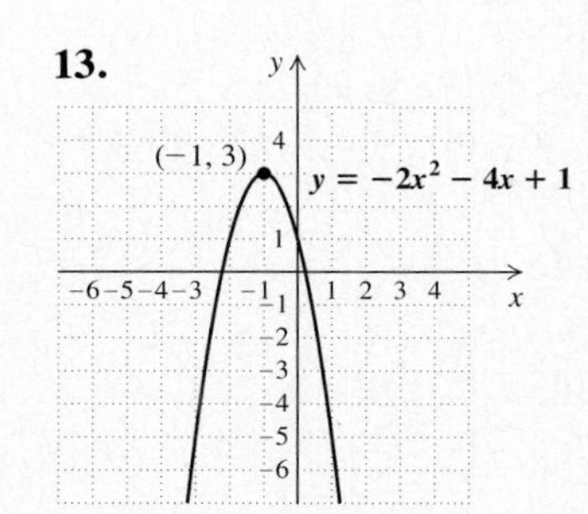

15.

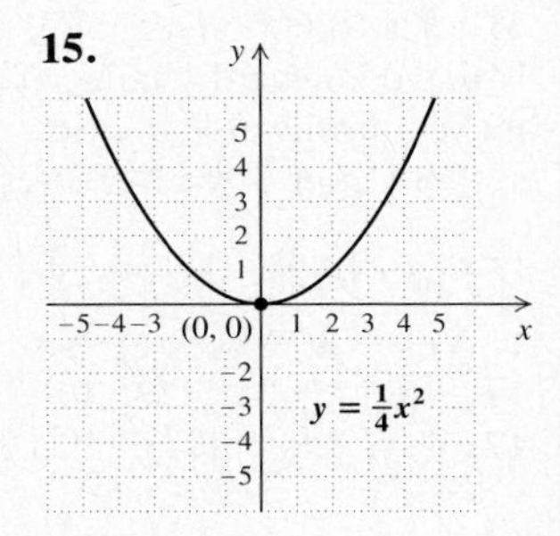

17.

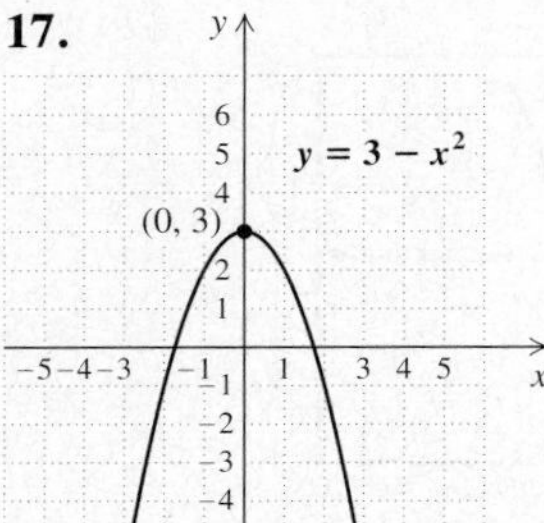

19.

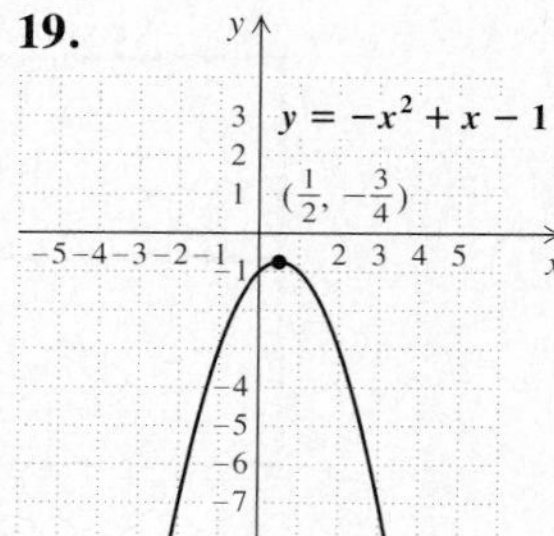

21.

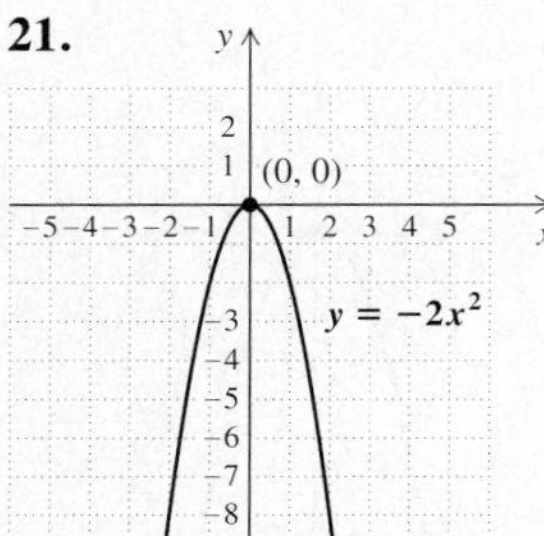

23.

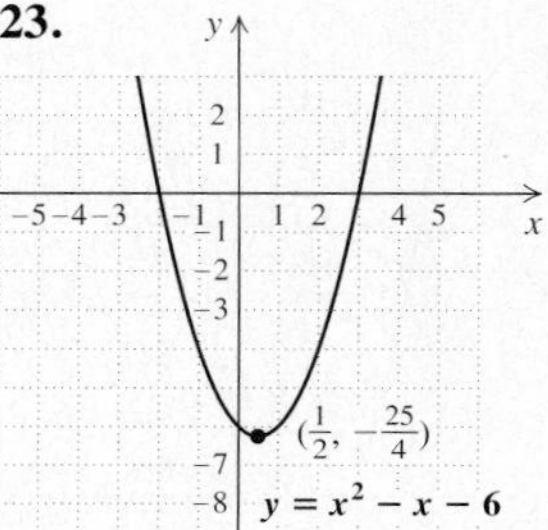

25. $(-\sqrt{5}, 0), (\sqrt{5}, 0)$ **27.** $(0, 0), (-2, 0)$

29. $\left(\frac{-1-\sqrt{33}}{2}, 0\right), \left(\frac{-1+\sqrt{33}}{2}, 0\right)$

31. $(-5, 0)$ **33.** $\left(\frac{-2-\sqrt{6}}{2}, 0\right), \left(\frac{-2+\sqrt{6}}{2}, 0\right)$

35. The graph does not cross the x-axis.

37. $\sqrt{432} \approx 20.78$ ft **39.** $y = -\frac{5}{3}x + \frac{11}{3}$ **41.** ◈

43. As $|a|$ increases, the graph is stretched vertically.

45.

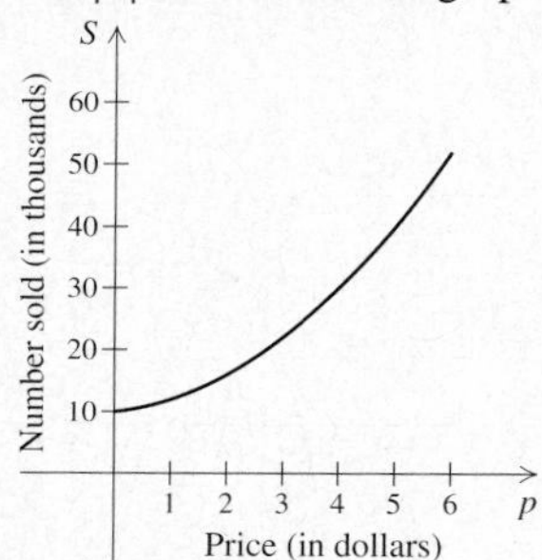

47. \$2; 16,000 units **49.** If $k \geq 0$, the graph of $y = x^2$ is moved upward k units. If $k < 0$, the graph of $y = x^2$ is moved downward $|k|$ units. **51.** **(a)** After 2 sec, after 4 sec; **(b)** after 3 sec; **(c)** after 6 sec

Exercise Set 10.6, pp. 457–460

1. Yes **3.** Yes **5.** No **7.** Yes **9.** 8, 12, −4
11. −6, 15, 72 **13.** 6, −10, 17.4 **15.** 2, 7, 4
17. 4, $\frac{2}{3}$, 3.8 **19.** 1, 91, 98 **21.** −3, −2, 78

23. **(a)** 153.98 cm; **(b)** 167.73 cm **25.** 1.606, 1.909, 4.03 **27.** 70°, 220°, 10,020° **29.** 13.6 million, 10.9 million, 8.2 million, 5.5 million, 2.8 million

31.

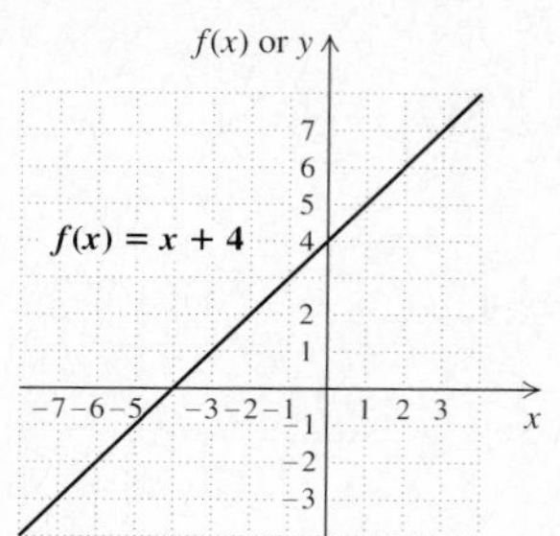

33.

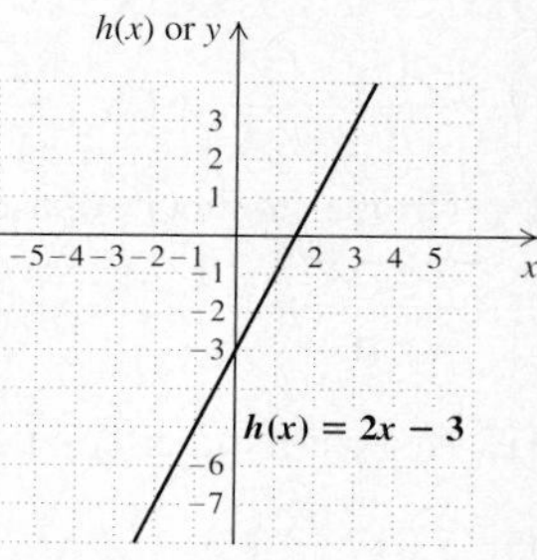

35.

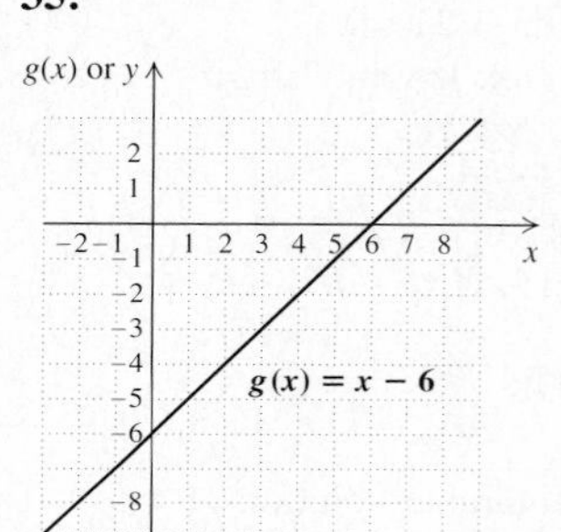

37.

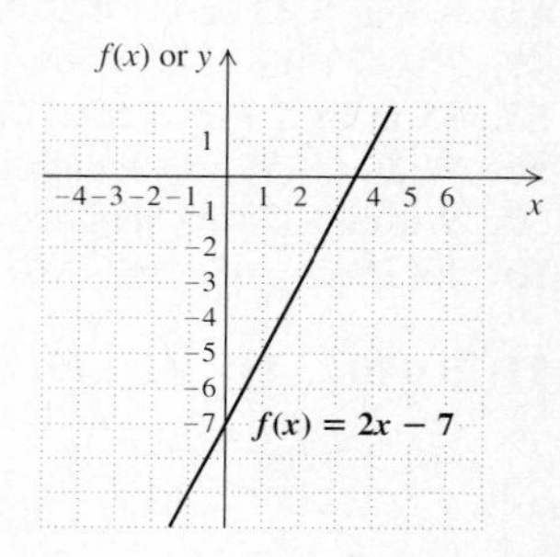

39.

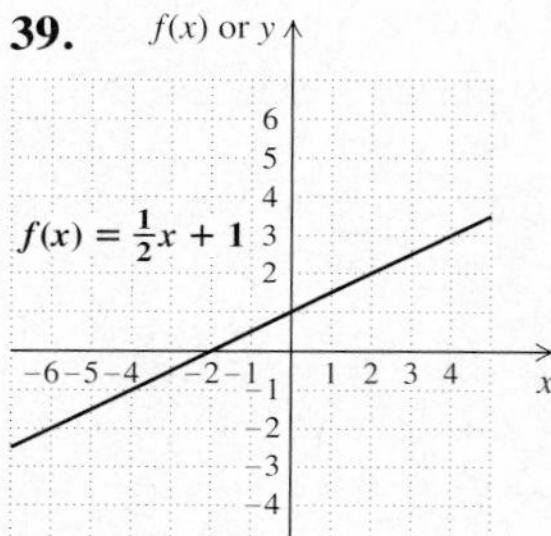

41.

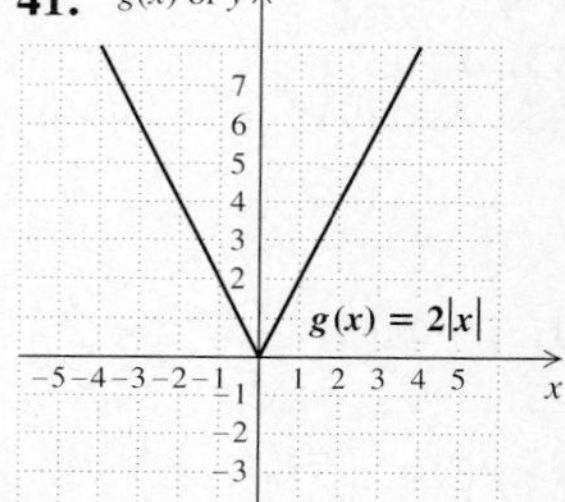

43.

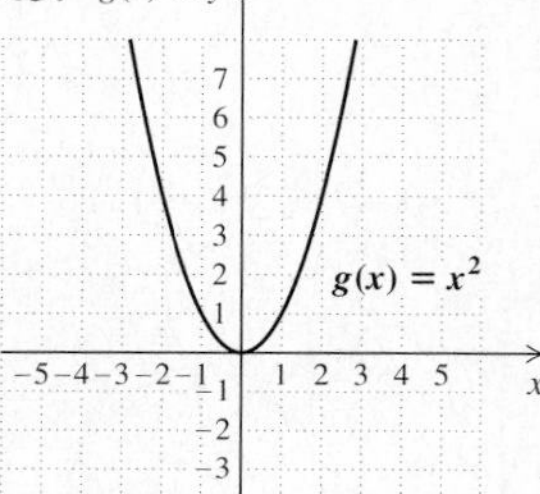

45.

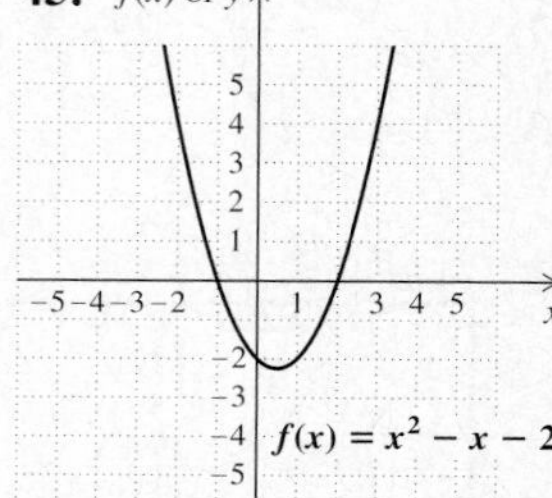

47. Yes **49.** No **51.** Yes **53.** No
55. No solution **57.** ◈ **59.** 0, 6, 0, 4, 0

61.

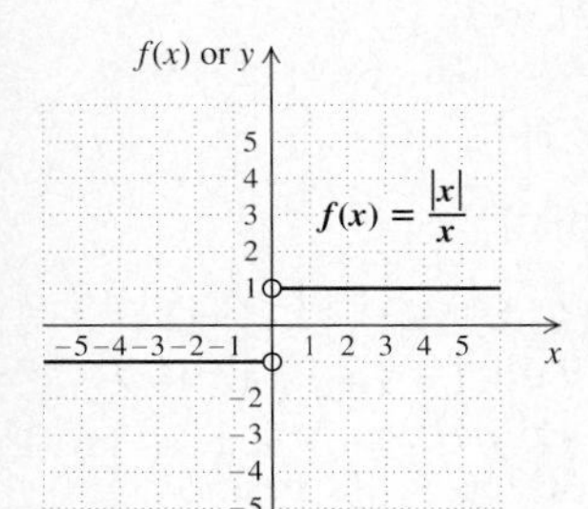

63.

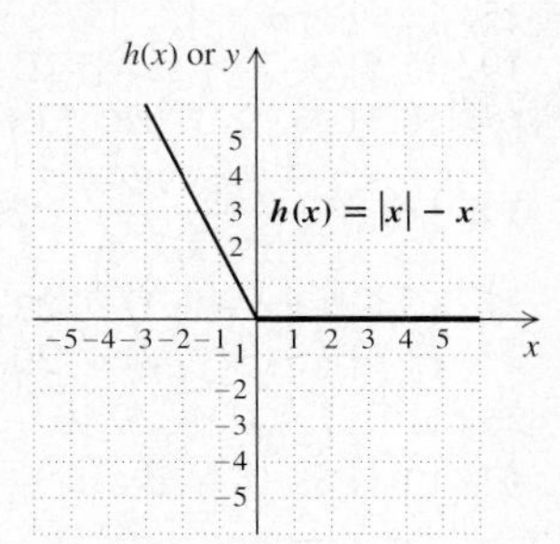

65.

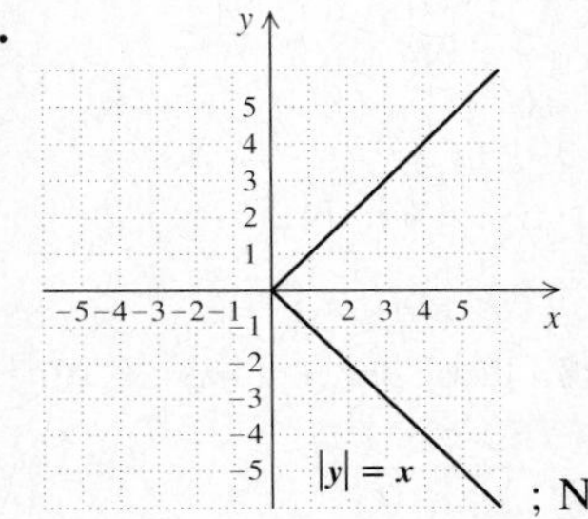

; No

67. $f(x) = \frac{15}{4}x - \frac{13}{4}$ **69.** $\{5, 8, 11, 14\}$ **71.** $\{0, 2\}$
73. $(-3.00, 0), (-1.00, 0), (2.00, 0), (3.00, 0)$

Review Exercises: Chapter 10, pp. 461–462

1. [10.1] $-\sqrt{3}, \sqrt{3}$ **2.** [10.3] $\frac{3}{5}, 1$
3. [10.3] $1 \pm \sqrt{11}$ **4.** [10.3] $\frac{1}{3}, -2$
5. [10.1] $-8 \pm \sqrt{13}$ **6.** [10.1] 0 **7.** [10.3] $\frac{1 \pm \sqrt{10}}{3}$
8. [10.3] $-3 \pm 3\sqrt{2}$ **9.** [10.3] $\frac{2 \pm \sqrt{3}}{2}$
10. [10.3] $\frac{3 \pm \sqrt{33}}{2}$ **11.** [10.4] $\frac{3 \pm i\sqrt{71}}{10}$
12. [10.1] $-2\sqrt{2}, 2\sqrt{2}$ **13.** [10.1] 1
14. [10.4] $1 \pm i$ **15.** [10.2] $\frac{5}{3}, -1$
16. [10.2] $\frac{5 \pm \sqrt{17}}{2}$ **17.** [10.3] 4.562, 0.438
18. [10.3] −0.134, −1.866 **19.** [10.3] 1.7 m, 4.7 m
20. [10.3] 3% **21.** [10.3] Width: 7 m; length: 10 m
22. [10.3] 6.3 sec **23.** [10.4] $8i$ **24.** [10.4] $-2i\sqrt{6}$
25. [10.4] $\pm 8i$ **26.** [10.4] $\pm 3i\sqrt{6}$
27. [10.4] $3 \pm 2i$ **28.** [10.4] $5 \pm i$
29. [10.4] $-\frac{1}{2} \pm \frac{\sqrt{3}}{2}i$

30. [10.5]

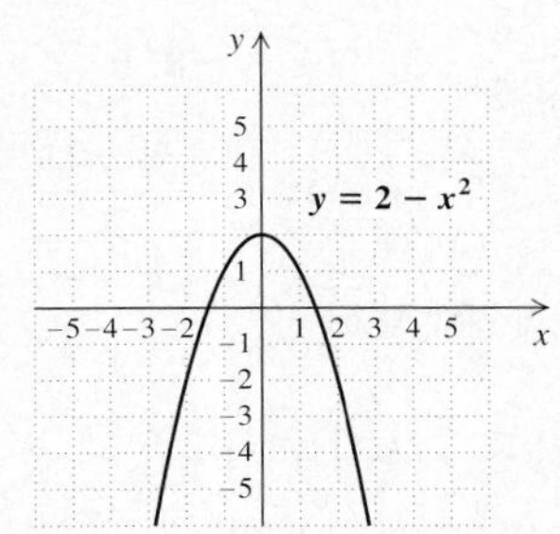

31. [10.5]

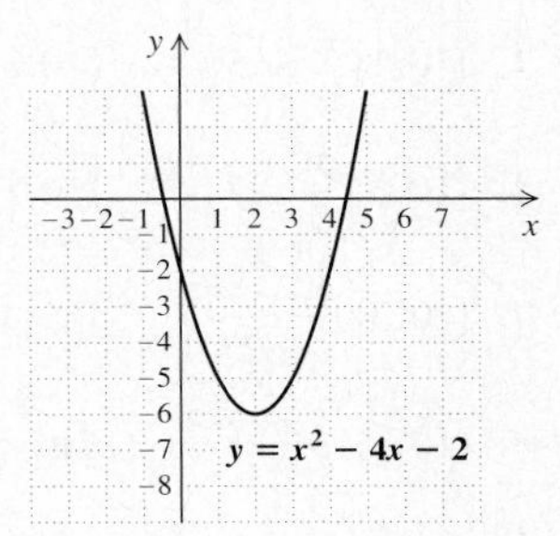

32. [10.5] $(2 - \sqrt{6}, 0), (2 + \sqrt{6}, 0)$
33. [10.6] −1, −7, 2 **34.** [10.6] 0, 0, 19
35. [10.6] 2700
36. [10.6]

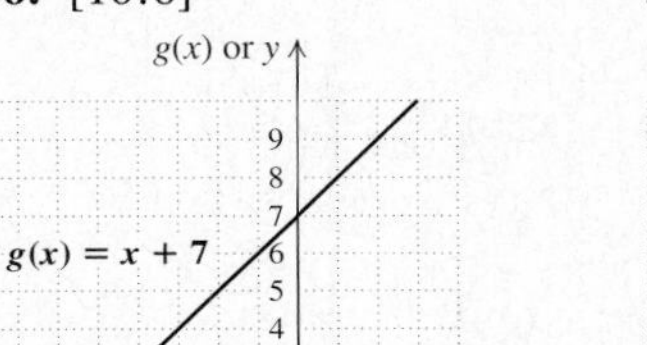

37. [10.6]

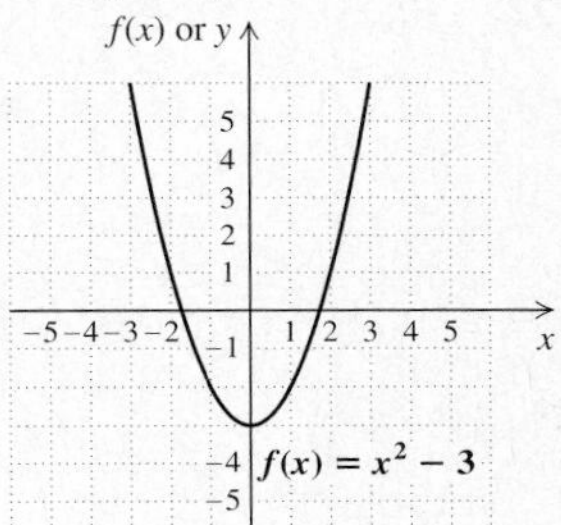

38. [10.6]

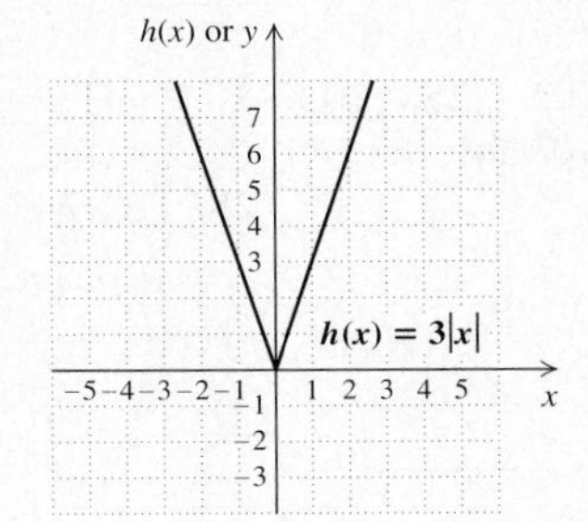

39. [10.6] No **40.** [10.6] Yes **41.** [7.3] $y = -\frac{4}{3}x + \frac{7}{3}$
42. [8.2] 7, 15 **43.** [9.4] $4 - \sqrt{2}$ **44.** [9.6] $\sqrt{3}$
45. [10.4] ◈ The other solution can be found by changing the sign of the i-term in the given imaginary-number solution. You can see this by examining the quadratic formula. **46.** [10.5] ◈ If the radicand is 0, the formula becomes $x = -b/(2a)$; thus there is only one x-intercept. If the radicand is negative, there are no real-number solutions and thus no x-intercepts. If the radicand is positive, there must be two x-intercepts. **47.** [10.3] 31 and 32; −32 and −31
48. [10.2] $b = 14$ or -14 **49.** [10.3] 25
50. [10.3] $s = 5\sqrt{\pi}$

Test: Chapter 10, pp. 462–463

1. [10.1] $\pm\sqrt{5}$ **2.** [10.3] $0, -\frac{8}{7}$ **3.** [10.3] $-8, 6$

4. [10.3] $\frac{5}{3}, -3$ **5.** [10.1] $2 \pm \sqrt{5}$ **6.** [10.3] $\frac{1 \pm \sqrt{13}}{2}$

7. [10.3] $\frac{3 \pm \sqrt{37}}{2}$ **8.** [10.3] $-2 \pm \sqrt{14}$

9. [10.3] $\frac{7 \pm \sqrt{37}}{6}$ **10.** [10.1] 2 **11.** [10.4] $2 \pm 2i$

12. [10.2] $2 \pm \sqrt{14}$ **13.** [10.3] 5.742, −1.742
14. [10.3] Width: 2.5 m; length: 6.5 m **15.** [10.3] 8
16. [10.4] $7i$ **17.** [10.4] $-4i\sqrt{2}$ **18.** [10.4] $\pm 5i$

19. [10.2] $\pm 7i$ **20.** [10.4] $-4 \pm i$ **21.** [10.4] $\frac{1}{2} \pm \frac{\sqrt{7}}{2}i$

22. [10.5]

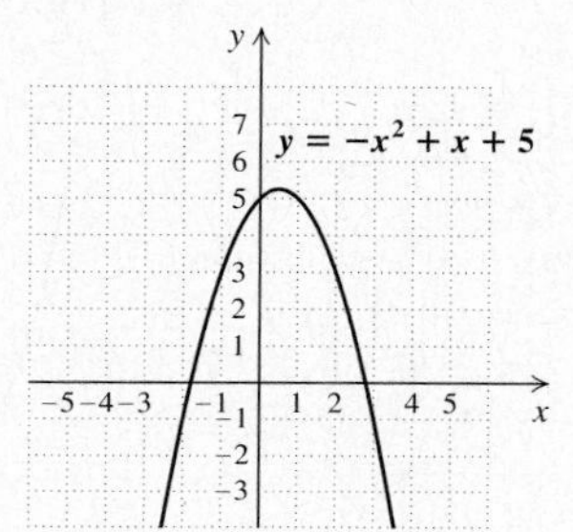

23. [10.5] $\left(\frac{1 - \sqrt{21}}{2}, 0\right), \left(\frac{1 + \sqrt{21}}{2}, 0\right)$

24. [10.6] $1, \frac{3}{2}, 2$ **25.** [10.6] $1, 3, -3$
26. [10.6] 26.70 min
27. [10.6]

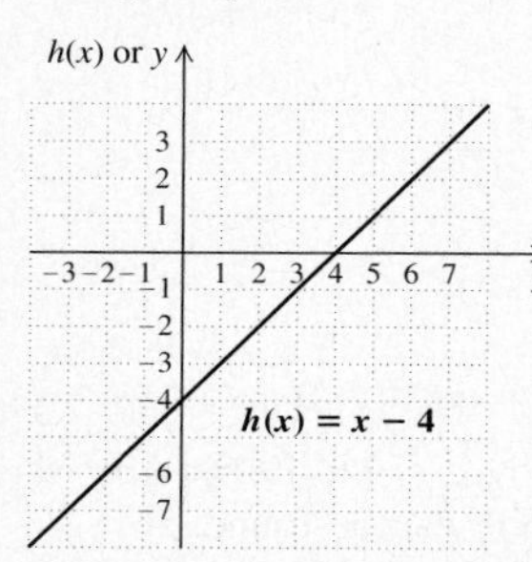

28. [10.6]

29. [10.6] Yes **30.** [10.6] No
31. [7.3] $y = -3x - 11$ **32.** [8.2] (2, 4)
33. [9.4] $-\sqrt{3}$ **34.** [9.6] $\sqrt{5}$
35. [10.3] $5 + 5\sqrt{2}$ ft **36.** [10.3] $1 \pm \sqrt{5}$

CUMULATIVE REVIEW: 1–10

1. [1.8] x^3 **2.** [1.8] 54
3. [1.2] $(6x)y$, $x(6y)$; there are other answers.
4. [6.4] 240 **5.** [1.4] 7 **6.** [1.5] 9 **7.** [1.6] 15
8. [1.7] $-\frac{3}{20}$ **9.** [1.8] 4 **10.** [1.8] $-2m - 4$
11. [2.1] −8 **12.** [2.2] −12 **13.** [2.2] 7
14. [5.6] 3, 5 **15.** [8.2] (1, 2) **16.** [8.3] (17, 0)
17. [8.3] (6, 7) **18.** [5.6] 3, −2

19. [10.3] $\frac{-3 \pm \sqrt{29}}{2}$ **20.** [9.5] 2

21. [2.6] $\{x | x \geq -1\}$ **22.** [2.2] 8 **23.** [2.2] −3
24. [2.6] $\{x | x < -8\}$ **25.** [10.4] $-1 \pm 2i$
26. [10.1] $-\sqrt{10}, \sqrt{10}$ **27.** [10.1] $3 \pm \sqrt{6}$
28. [6.7] $\frac{2}{9}$ **29.** [6.7] −5 **30.** [6.7] No solution

31. [9.5] 12 **32.** [6.9] $t = \frac{4s + 3}{A}$ **33.** [6.9] $m = \frac{tn}{t + n}$

34. [4.8] 2.73×10^{-5} **35.** [4.8] 2.0×10^{14}
36. [4.8] x^{-4} **37.** [4.8] y^7 **38.** [4.1] $4y^{12}$
39. [4.2] $10x^3 + 3x - 3$ **40.** [4.3] $7x^3 - 2x^2 + 4x - 17$
41. [4.3] $8x^2 - 4x - 6$ **42.** [4.4] $-8y^4 + 6y^3 - 2y^2$
43. [4.4] $6t^3 - 17t^2 + 16t - 6$ **44.** [4.5] $t^2 - \frac{1}{16}$
45. [4.5] $9m^2 - 12m + 4$
46. [4.6] $15x^2y^3 + x^2y^2 + 5xy^2 + 7$
47. [4.6] $x^4 - 0.04y^2$ **48.** [4.6] $9p^2 + 24pq^2 + 16q^4$

49. [6.2] $\frac{2}{x + 3}$ **50.** [6.2] $\frac{3a(a - 1)}{2(a + 1)}$

51. [6.5] $\frac{27x - 4}{5x(3x - 1)}$ **52.** [6.5] $\frac{-x^2 + x + 2}{(x + 4)(x - 4)(x - 5)}$

53. [4.7] $x^2 + 9x + 16 + \frac{35}{x - 2}$ **54.** [5.1] $4x(2x - 1)$

55. [5.4] $(5x - 2)(5x + 2)$ **56.** [5.3] $(3y + 2)(2y - 3)$
57. [5.4] $(m - 4)^2$ **58.** [5.1] $(x^2 - 5)(x - 8)$
59. [5.5] $3(a^2 + 6)(a + 2)(a - 2)$
60. [5.4] $x(4x + 1)(4x - 1)$
61. [5.4] $(7ab - 2)(7ab + 2)$ **62.** [5.4] $(3x + 5y)^2$
63. [5.1] $(2a + d)(c - 3b)$
64. [5.3] $(5x - 2y)(3x + 4y)$ **65.** [6.6] $-\frac{42}{5}$
66. [9.1] 7 **67.** [9.7] −5 **68.** [9.2] $8x$
69. [9.2] $\sqrt{a^2 - b^2}$ **70.** [9.2] $8a^2b\sqrt{3ab}$ **71.** [9.7] $\frac{1}{4}$
72. [9.2] $9xy\sqrt{3x}$ **73.** [9.3] $\frac{10}{9}$ **74.** [9.4] −1
75. [9.4] $16\sqrt{3}$

76. [9.3] $\frac{2\sqrt{10}}{5}$ **77.** [9.6] 40

78. [7.2]

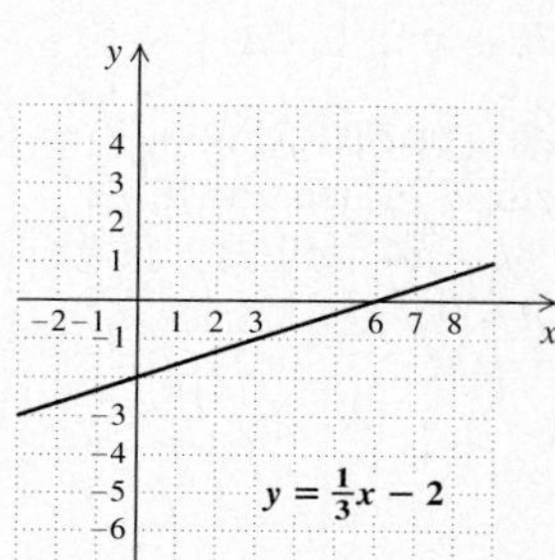

79. [3.3]

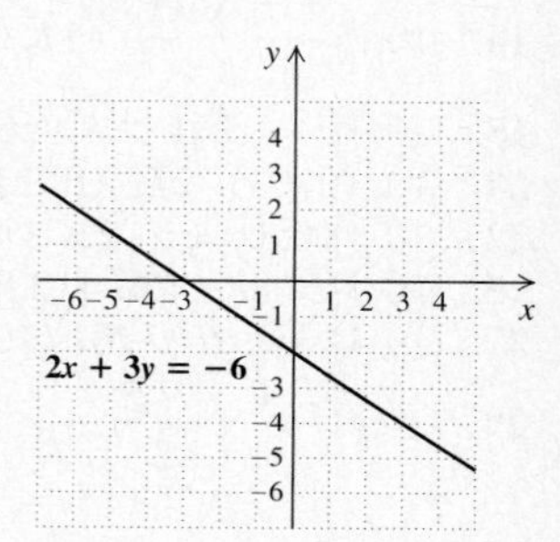

80. [3.3]

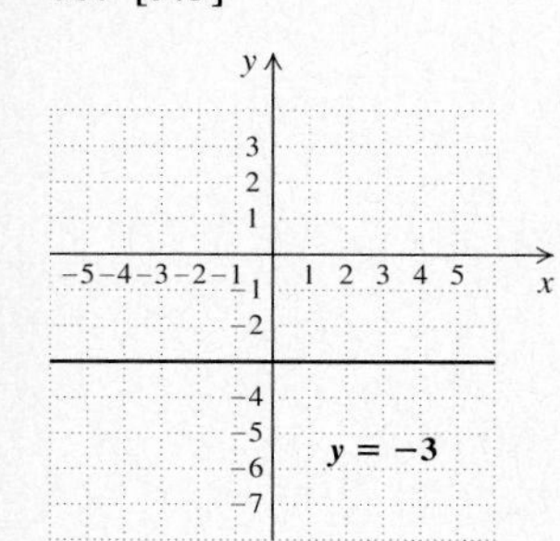

81. [7.4]

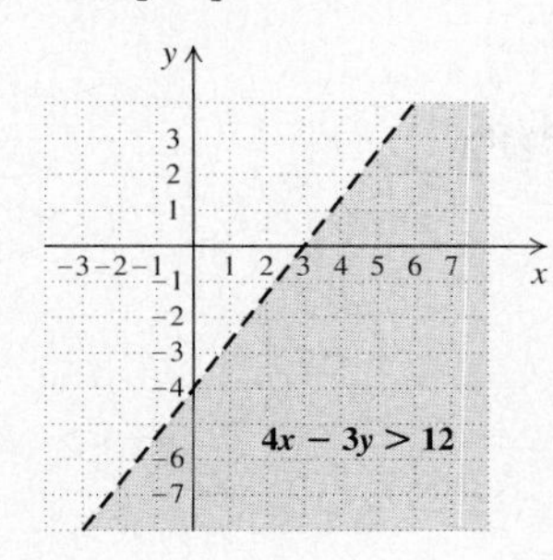

82. [10.5]

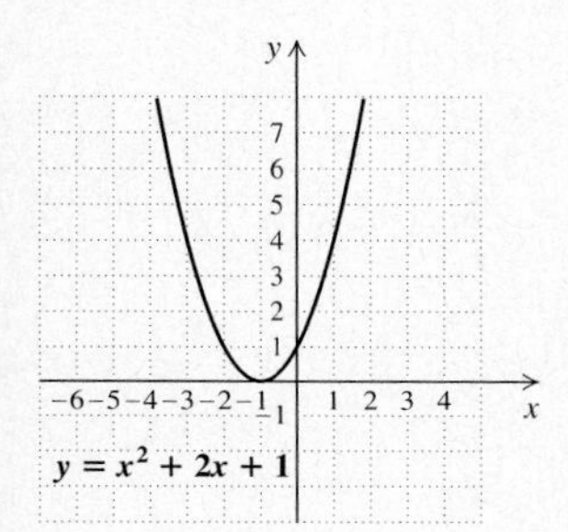

83. [7.4]

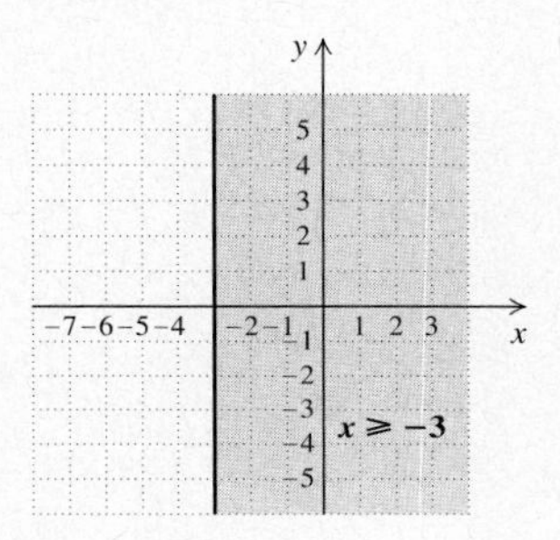

84. [10.2] $\dfrac{2 \pm \sqrt{6}}{3}$ **85.** [10.3] 1.207, −0.207

86. [10.5] $\left(\dfrac{1-\sqrt{2}}{2}, 0\right), \left(\dfrac{1+\sqrt{2}}{2}, 0\right)$

87. [2.4] 25% **88.** [2.7] $\{x \mid x > 5\}$

89. [5.7] 14, 16; −16, −14 **90.** [10.3] 12 m

91. [2.5] 6090

92. [8.4] \$1.10 per lb: 14 lb; \$0.80 per lb: 28 lb

93. [6.8] $4\frac{4}{9}$ hr

94. [7.5] \$451.20, variation constant is the amount earned per hour **95.** [10.6] −4, 0, 0

96. [7.2] Parallel

97. [8.5]

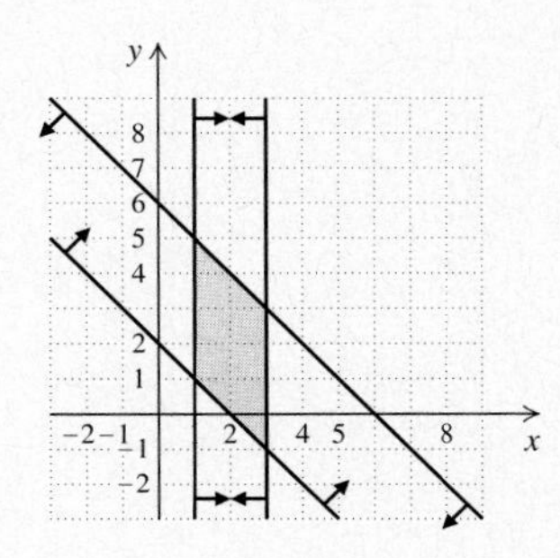

98. [7.2] 2, (0, −8) **99.** [7.1] 15

100. [7.3] $y + 3 = -\frac{1}{2}(x - 1)$ **101.** [10.4] $-5i$

102. [10.2] 30, −30

103. [9.6] $\dfrac{\sqrt{6}}{3}$ **104.** [4.5] Yes **105.** [6.1] No

106. [4.5] No **107.** [9.1] No **108.** [9.1] Yes

APPENDIXES

Exercise Set A, p. 470

1. {3, 4, 5, 6, 7, 8} **3.** {41, 43, 45, 47, 49}
5. {−3, 3} **7.** False **9.** True **11.** True **13.** True
15. True **17.** False **19.** $\{c, d, e\}$ **21.** {1, 10}
23. ∅
25. The system has no solution. The lines are parallel. Their intersection is empty.
27. $\{a, e, i, o, u, q, c, k\}$ **29.** {0, 1, 2, 5, 7, 10}
31. $\{a, e, i, o, u, m, n, f, g, h\}$
33. The solution set is {3, −5}. This set is the union of the solution sets of the equations $x - 3 = 0$ and $x + 5 = 0$, which are {3} and {−5}.
35. The set of integers **37.** The set of real numbers
39. ∅ **41.** **(a)** A; **(b)** A; **(c)** A; **(d)** ∅ **43.** True

Exercise Set B, p. 472

1. $(t + 3)(t^2 - 3t + 9)$ **3.** $(a - 1)(a^2 + a + 1)$
5. $(z + 5)(z^2 - 5z + 25)$ **7.** $(2a - 1)(4a^2 + 2a + 1)$
9. $(y - 3)(y^2 + 3y + 9)$ **11.** $(4 + 5x)(16 - 20x + 25x^2)$
13. $(5p - 1)(25p^2 + 5p + 1)$
15. $(3m + 4)(9m^2 - 12m + 16)$
17. $(p - q)(p^2 + pq + q^2)$ **19.** $\left(x + \frac{1}{2}\right)\left(x^2 - \frac{1}{2}x + \frac{1}{4}\right)$
21. $2(y - 4)(y^2 + 4y + 16)$
23. $3(2a + 1)(4a^2 - 2a + 1)$
25. $r(s + 4)(s^2 - 4s + 16)$
27. $5(x - 2z)(x^2 + 2xz + 4z^2)$
29. $(x + 0.1)(x^2 - 0.1x + 0.01)$
31. $8(2x^2 - t^2)(4x^4 + 2x^2t^2 + t^4)$
33. $(x - 0.3)(x^2 + 0.3x + 0.09)$
35. $(x^{2a} + y^b)(x^{4a} - x^{2a}y^b + y^{2b})$
37. $3(x^a + 2y^b)(x^{2a} - 2x^ay^b + 4y^{2b})$
39. $\frac{1}{3}\left(\frac{1}{2}xy + z\right)\left(\frac{1}{4}x^2y^2 - \frac{1}{2}xyz + z^2\right)$

INDEX

P

Q

R

T

U

V

W

X

Y

Z

CALCULATOR KEYS
RELEVANT TO ELEMENTARY ALGEBRA

SELECTED KEYS OF THE GRAPHING CALCULATOR

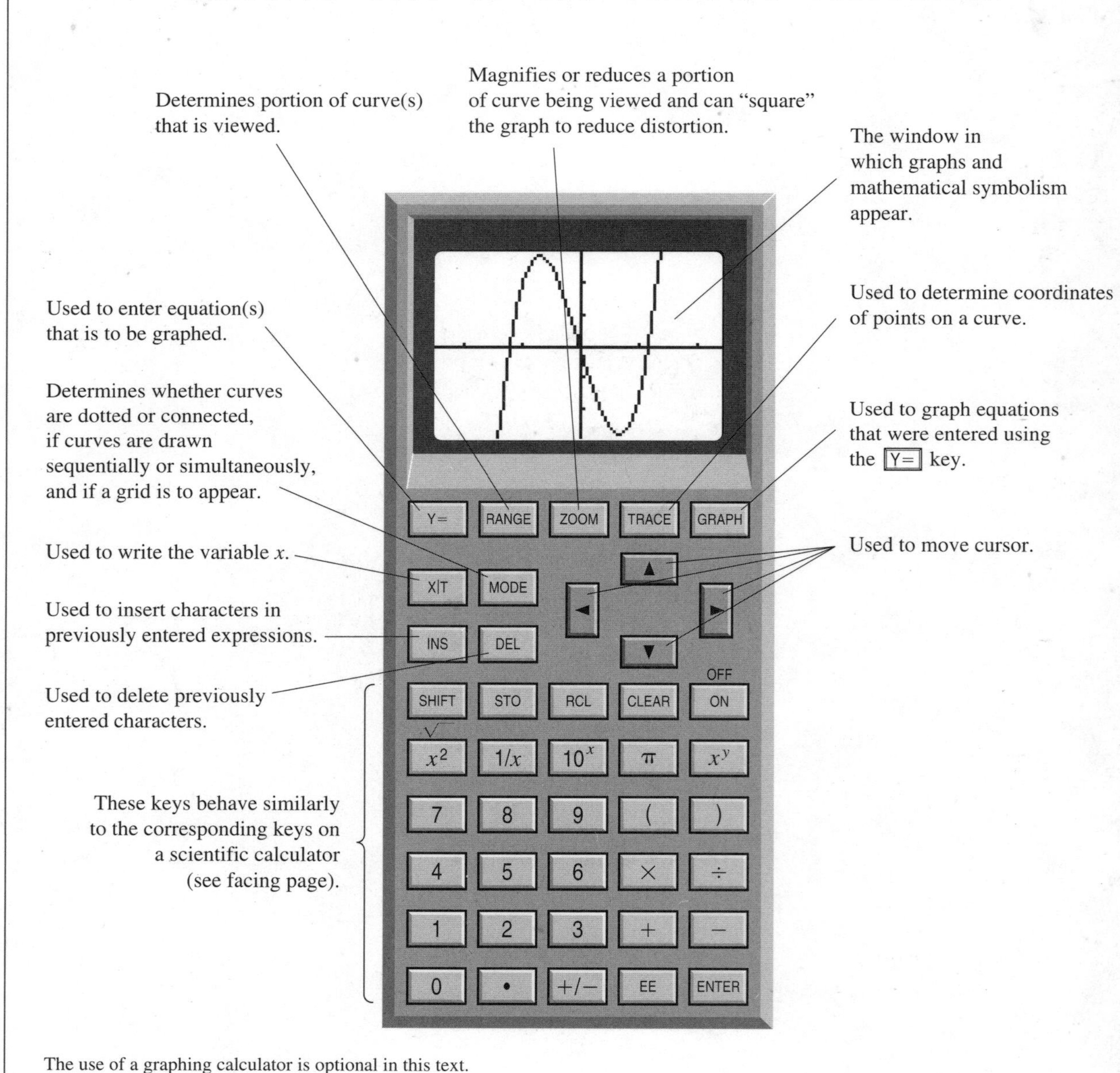

The use of a graphing calculator is optional in this text.